IMPORTANT NAMES, DATES AND ADDRESSES

This book belongs

____

____

____

phone ______________________

tape your cancelled check here

tape your proof of mailing here

examination date: ____________ hours: ________

examination location: ______________________________

____

____

____

phone number of your registration board: ______________________

address of your registration board: ______________________

____

____

names of contacts at your registration board: ______________________

____

____

date you sent your application: ______________

registered/certified mail receipt number: ______________

date confirmation was received: ______________

names of examination proctors: ______________________

____

____

booklet number: ____________ (A.M.) ____________ (P.M.)

problems you disagreed with on the examination

problem no.	reason

ELECTRICAL ENGINEERING REVIEW MANUAL

Fourth Edition

A complete review course for the P.E. examination for Electrical Engineers

Raymond B. Yarbrough, P. E.
Professor, Electrical Engineering
University of Santa Clara

PROFESSIONAL PUBLICATIONS, INC.
San Carlos, CA 94070

In the *ENGINEERING REVIEW MANUAL SERIES:*

Engineer-In-Training Review Manual
Quick Reference Cards for the E-I-T Exam
Mini-Exams for the E-I-T Exam
Civil Engineering Review Manual
Seismic Design for the California Civil P.E. Exam
Timber Design for the Civil P.E. Exam
Mechanical Engineering Review Manual
Electrical Engineering Review Manual
Chemical Engineering Review Manual
Expanded Interest Tables
Engineering Law, Design Liability, and Professional Ethics

Distributed by: Professional Publications, Inc.
Post Office Box 199
Department 77
San Carlos, CA 94070
(415) 593-9119

ELECTRICAL ENGINEERING REVIEW MANUAL

Printed in the United States of America

ISBN 0-932276-36-9

Professional Engineering Registration Program
Post Office Box 991, San Carlos, CA 94070

Current printing of this edition (last number) 5 4 3 2

TABLE OF CONTENTS

PREFACE xi

INTRODUCTION xiii

Purpose of Registration xiii
The Registration Procedure xiii
The National Council of Engineering Examiners xiv
Reciprocity Among States xiv
Examination Dates xiv
The Examination Format xiv
Grading the Examination xv
Preparing for the Exam xv
Steps to Take Before the Exam xvi
What to Do During the Examination xvi

ACKNOWLEDGEMENTS xix

1 MATHEMATICS

1 Real Numbers 1-1
2 Complex Numbers 1-1
3 Algebra 1-3
Quadratic Equations 1-3
Exponentiation 1-3
Logarithms 1-3
Partial Fraction Expansion 1-3
4 Trigonometric Identities 1-4
5 Analytic Geometry 1-5
6 Differential Calculus 1-5
7 Integral Calculus 1-6
8 Differential Equations 1-6
First Order Differential Equations 1-6
Second Order Differential Equations . . . 1-7
9 Fourier Analysis 1-8
10 Transient Analysis By Laplace Transforms 1-9
Initial Value Theorem 1-11
Final Value Theorem 1-11

2 LINEAR CIRCUIT ANALYSIS

1 Linear Circuit Elements 2-1
Resistance 2-1
Capacitance 2-2
Inductance 2-3
2 Linear Source Models 2-5
3 Two Terminal Networks 2-11
4 Two-Port Networks 2-13
5 Network Analysis 2-15
6 Impedance Analysis 2-18
Resistance 2-19
Inductance 2-19
Capacitance 2-19

3 MEASUREMENTS OF SIGNAL WAVEFORMS

Part 1: Periodic Signals 3-1

1 Average Values 3-1
2 RMS Values 3-4
3 Parameters of Fourier Series 3-6

Part 2: Measurement of DC and Periodic Signals

4 Introduction 3-8
5 Permanent Magnet Moving Coil Instruments 3-8
DC Voltmeters 3-8
DC Ammeters 3-9
6 The Electrodynamometer Movement . . . 3-9
7 Moving-Iron Instruments 3-10
8 Thermocouple Instruments 3-10
9 Rectifier Meters 3-10
10 Electronic Analog Meters 3-11
11 Digital Meters 3-11

Part 3: Power with Sinusoidal Circuits

12 Review of Phasors 3-12
13 Phasor Power 3-13
14 Power Factor Correction 3-16

4 TIME AND FREQUENCY RESPONSE

1 Introduction 4-1
2 Circuit Reduction Techniques 4-1
Source Conversion 4-1
Series and Parallel Combination Rules . . 4-2
Partial Reductions 4-2
Wye-Delta and Delta-Wye Conversion . . 4-4
Redundant Impedances 4-5
Voltage Dividers 4-6
3 First-Order Transient Analysis 4-6
Introduction 4-6
First-Order Solutions 4-7
First-Order Switching Transients 4-7
First-Order Pulse Transients 4-9
4 Second-Order Transient Analysis 4-10
Introduction 4-10
Step Response of Second-Order Systems . 4-13
Initial Conditions In Second-Order Systems 4-14
5 LaPlace Transform Analysis 4-16
Introduction 4-16
Capacitance 4-16
Inductance 4-17
First Order Systems 4-17
Second-Order Systems 4-19
Higher-Order Systems 4-20
6 Sinusoidal Analysis 4-20
Introduction 4-20
Transfer Functions 4-20
First-Order Response 4-20
Second-Order Response 4-20
Higher-Order Response 4-23

5 POWER SYSTEMS

1 Three-Phase Circuits 5-1
2 Power Transformers 5-5
Open Circuit Test 5-5
Short-Circuit Test 5-6
ABCD Parameters 5-7
3 The Per-Unit System 5-8
4 One-Line Diagrams 5-10
5 Distribution Transformers 5-11
Typical Per-Unit Values 5-11
Voltage Regulation 5-12
Power Factor Correction 5-13
Induction Motor Starting 5-14
6 Three-Phase Transformer Connections . . 5-14
7 Voltage Control 5-15
8 Representative Values of Power Device Parameters 5-16
9 Fault Calculations 5-17

6 TRANSMISSION LINES

1 Introduction 6-1
2 Skin Effect 6-2
3 External Inductance 6-3
4 Capacitance 6-4
5 Power Transmission Lines 6-4
6 Three-Phase Transmission Lines 6-6
Inductance 6-6
Capacitance 6-7
7 Power Transmission Line Representation . 6-8
Introduction 6-8
Delta-Wye Transformation 6-8
Short Transmission Lines 6-8
Medium-Length Transmission Lines . . . 6-9
Long Transmission Lines 6-10
8 High Frequency Transmission Lines 6-12
9 The Smith Chart 6-13

7 ROTATING MACHINES

1 Review of the Fundamentals of D.C. Machines 7-2
2 D.C. Generators 7-2
Introduction 7-2
Separately-Excited D.C. Generators . . . 7-3
Self-Excited Generators 7-3
Compound Generators 7-5
3 D.C. Motors 7-7
Introduction 7-7
Separately-Excited Motors 7-7
D.C. Shunt Motors 7-7
Compound D.C. Motors 7-10
D.C. Series Motors 7-11
4 Induction Motors 7-12
Introduction 7-12
Induction Motor Tests 7-14
Starting Induction Motors 7-15
5 Synchronous Machines 7-17
Introduction 7-17
Synchronous Motors 7-20
Power Factor Correction 7-21
Synchronous Capacitors 7-21
Synchronous Generators 7-21
Power Factor Relations for Synchronous Generators 7-23

8 LINEAR AMPLIFIERS

1 Ideal Operational Amplifier 8-2
Introduction 8-2
Ideal Operational Amplifier with Positive Feedback 8-2
Inverting Operational Amplifier with Negative Feedback 8-3
Non-Inverting Operational Amplifier with Negative Feedback 8-3

Voltage Follower 8-4
Operational Amplifier with Multiple Inputs 8-4
2 Active Filters with Inverting Operational Amplifiers 8-4
Introduction 8-4
Operational Amplifier Integrator 8-5
Single-Pole Operational Amplifier Low-Pass Filter 8-5
Single-Pole Operational Amplifier High-Pass Filter 8-5
Lead-Lag (or Lag-Lead) Operational Amplifier Filter 8-5
Operational Amplifier Band-Pass Filter . 8-6
A Two-Pole Operational Amplifier Low-Pass Filter 8-6
Two-Pole Operational Amplifier High-Pass Filter 8-6
3 Active Filters with Non-Inverting Operational Amplifiers 8-6
Introduction 8-6
Low-Pass Sallen and Key Circuit 8-6
High-Pass Non-Inverting Operator Amplifier Filter 8-6
Sallen and Key Bandpass Circuit 8-7
4 P-N Junction Characteristics 8-7
Introduction 8-7
Avalanche Breakdown 8-8
Very Low Frequency Characteristics of P-N Junctions 8-8
Junction Capacitances 8-9
5 Field Effect Transistors 8-9
Introduction 8-9
Junction FET's (JFET's) 8-9
Depletion Mode Pinchoff 8-10
Depletion Mode Turn-Off 8-10
JFET Drain-Source Characteristics . . . 8-11
JFET Circuit Symbols 8-11
6 Depletion Mode MOSFET's 8-11
Introduction 8-11
Characteristics of Depletion Mode MOSFET's 8-12
7 Depletion Mode FET Biasing 8-13
Introduction 8-13
A Simple Self-Biasing Circuit 8-13
Biasing For Device Characteristic Variations 8-14
8 Enhancement Mode MOSFET's 8-14
Introduction 8-14
Enhancement Mode MOSFET Circuit Symbols 8-15
Voltage Feedback Biasing 8-15
Current Feedback Biasing 8-16
Vertical Enhancement Mode MOSFET (VMOS) 8-16
9 Small-Signal FET Parameters 8-17
Introduction 8-17
Low-Frequency Small-Signal FET Equivalent Circuit 8-17
High-Frequency Small-Signal FET Equivalent Circuit 8-17
10 Bipolar Junction Transistors 8-17
Introduction 8-17
Bipolar Junction Transistor Biasing . . . 8-18
Stability 8-19
Parameter Variation 8-19
Small-Signal Parameters 8-20
11 Small-Signal Amplifiers 8-21
Introduction 8-21
Common Emitter Bipolar Junction Transistor Amplifier 8-22
The Common Base Amplifier 8-23
Common Source FET Amplifiers 8-23
Common Drain FET Amplifiers 8-24
12 Multiple Transistor Circuits 8-24
Introduction 8-24
Differential Amplifiers 8-24
The Darlington Circuit 8-26
Cascode Amplifiers 8-26
13 Frequency Effects 8-27
Low Frequency Behavior 8-27
High Frequency Bipolar Junction Transistor Behavior 8-28
High Frequency Effects on FET's 8-28
14 Practical Operational Amplifiers 8-29
15 Amplifier Noise 8-30
Shot Noise 8-30
Johnson or Thermal Noise 8-30
Low Frequency Noise 8-30
Interference Noise 8-30
Bandwidth Limiting of Noise 8-30
Noise Source Equivalent Circuits 8-31
Optimum Source Resistance 8-32
Noise Power to an Input Resistance . . . 8-33
Noise Figures for Cascaded Amplifiers . . 8-33
Noise Temperature and Noise Resistance 8-34

9 NONLINEAR WAVESHAPING AND SWITCHING

1 Diode Circuits 9-1
Introduction 9-1
The Precision Diode 9-1
The Clipper-Limiter Circuit 9-3
2 Voltage Regulators 9-5
Introduction 9-5
Ideal Zener Voltage Regulator 9-5
Practical Zener Diode Regulator 9-6
An Operational Amplifier Voltage Regulator 9-7
3 Logic Circuits 9-8

Introduction . . . 9-8
The Logical Not Inverter Circuit . . . 9-9
Gate Delays . . . 9-11
Logical Functions . . . 9-11
4 Types of Logic . . . 9-12
Resistor-Transistor Logic (RTL) . . . 9-12
Diode-Transistor Logic (DTL) . . . 9-13
Transistor-Transistor Logic (TTL) . . . 9-13
TTL Output Circuits . . . 9-14
Emitter-Coupled (ECL) or Current Mode (CML) Logic . . . 9-15
PMOS Logic Gates . . . 9-15
CMOS Logic Gates . . . 9-16
5 Analog Switches . . . 9-17
Introduction . . . 9-17
CMOS Transmission Switch . . . 9-17
JFET Analog Switches . . . 9-18
VMOS Analog Switches . . . 9-18
6 Power Switching . . . 9-18
Introduction . . . 9-18
Thyristor Control . . . 9-18
UJT Trigger for SCR's . . . 9-19
DIAC Trigger For TRIAC Circuits . . . 9-20
7 Non-Sinusoidal Waveform Generation . . . 9-21
Introduction . . . 9-21
UJT Relaxation Oscillator . . . 9-22
Neon Bulb Relaxation Oscillator . . . 9-22
Negative Resistance Oscillator . . . 9-23
Negative Conductance Oscillator . . . 9-25
Operational Amplifier Square-Wave Generator . . . 9-25
Astable Multivibrator . . . 9-26
A Monostable Multivibrator . . . 9-28

10 DIGITAL LOGIC

1 Fundamental Logical Operations . . . 10-1
Symbols, Truth Tables, and Expressions . . . 10-1
Logical Variables and Constants . . . 10-2
Logical Identities . . . 10-3
Logical Theorems . . . 10-3
Logical Functions . . . 10-3
Canonical SOP Realization, NAND-NAND . . . 10-4
Canonical POS Realization, NOR-NOR . . . 10-5
2 Logical Minimization Techniques . . . 10-6
3 Minimization of Functions . . . 10-8
4 Multiple Adjacencies and Veitch Diagrams . . . 10-9
5 Exclusive OR and Exclusive NOR Logic . 10-13
Introduction . . . 10-13
XOR Maps . . . 10-13
XOR with Minterms . . . 10-15
XOR with Maxterms . . . 10-16
6 Data Selectors . . . 10-17
Introduction . . . 10-17
2-Input MUX Logic . . . 10-18
4-Input MUX Logic . . . 10-18
8-Input and Higher MUX Logic . . . 10-20
7 Decoders/Demultiplexers . . . 10-20
8 Read-Only-Memories and Programmable Logic Arrays . . . 10-20
9 Flip-Flops . . . 10-20
The Set Reset (S-R) Flip-Flop . . . 10-20
The Toggle (T) Flip-Flop . . . 10-21
The Data (D) Flip-Flop . . . 10-22
J-K Flip-Flops . . . 10-22
Clocked Flip-Flops . . . 10-22
10 Flip-Flop Transition Tables . . . 10-23
11 Counters . . . 10-23
Synchronous Binary Counters . . . 10-24
An Adjacent-State Binary Counter . . 10-25
Counter Analysis . . . 10-26
12 Sequence Detection . . . 10-27

11 CONTROL SYSTEMS

1 General Feedback Theory . . . 11-1
2 Block Diagram Algebra . . . 11-4
3 Bode Diagrams . . . 11-6
4 The Nyquist Diagram . . . 11-9
5 Steady-State Errors . . . 11-10
Introduction . . . 11-10
Position Error . . . 11-10
Velocity Error . . . 11-10
Acceleration Error . . . 11-11
6 The Routh-Hurwitz Stability Criterion . 11-11
7 Root Locus . . . 11-12
Introduction . . . 11-12
Root Locus on the Real Axis . . . 11-13
Adjacent Poles with a Zero . . . 11-14
Higher Order Systems . . . 11-14
8 State Variable Representation . . . 11-15
Introduction . . . 11-15
Transfer Functions with Zeros . . . 11-16

12 ILLUMINATION

Definitions . . . 12-1
1 Lamps . . . 12-2
Relative Costs . . . 12-2
Lamp Classification . . . 12-2
2 Ballast . . . 12-3
3 Luminaires . . . 12-3
Coefficient of Utilization (CU) . . . 12-3
Luminaire Dirt Depreciation Factors (LDD) . . . 12-3
Lamp Lumen Depreciation Factor (LLD) . . . 12-3
Luminaire Determination . . . 12-6

4 Minimum Maintained Illumination 12-6
5 Lumen Method of Illumination Design . . 12-7
Zonal Cavity Ratios 12-7
Effective Ceiling Reflectance 12-8
Determining the Coefficient of Utilization 12-8
Determining the Number of Luminaires 12-9
Placement of Luminaires 12-10
6 Point Method of Interior Illumination Design 12-11
The Reflected Component 12-11
Reflected Radiation Coefficient (RRC) . 12-12
Reflected Illumination on Vertical Surfaces 12-12
Line Source Calculation of Direct Illumination Component 12-14
7 Street Lighting 12-15
Recommended Levels for Roadway Lighting 12-16
Average Illumination Design 12-16
Luminaire Classifications 12-18
Spacing and Mounting Height 12-18
Point Calculations 12-20
8 Floodlighting 12-20
9 Wiring Design and the National Electric Code 12-21
Service 12-21
Feeders and Branch Circuits 12-21
Conductors for General Wiring 12-21
Cables and Raceways 12-22
Overcurrent Protection [240] 12-22
Branch Circuit Design - Electric Discharge Lighting 12-23

13 ENGINEERING ECONOMIC ANALYSIS

1 Equivalence 13-1
2 Cash Flow Diagrams 13-1
3 Typical Problem Format 13-2
4 Calculating Equivalence 13-2
5 The Meaning of 'Present Worth' and 'i' . . 13-4
6 Choice Between Alternatives 13-4
7 Treatment of Salvage Value in Replacement Studies 13-6
8 Basic Income Tax Considerations 13-6
9 Depreciation 13-7
3-Year Personal Property 13-8
5-Year Personal Property 13-8
10-Year Personal and Real Property . . . 13-8
15-Year Real Property 13-9
10 Advanced Income Tax Considerations . 13-11
Immediate Write-off 13-11
Tax Credits 13-11
Gain on the Sale of a Depreciated Asset 13-11
Capital Gains and Losses 13-11
11 Rate and Period Changes 13-11
12 Probabilistic Problems 13-12
13 Estimating Economic Life 13-14
14 Basic Cost Accounting 13-14
15 Break-Even Analysis 13-15
16 Handling Inflation 13-16
17 Learning Curves 13-16
18 Economic Order Quantity 13-17
19 Consumer Loans 13-17
Simple Interest 13-17
Loans with Constant Amount Paid Towards Principal 13-18
Direct Reduction Loans 13-18
Direct Reduction Loan with Balloon Payment 13-18

14 SAMPLE EXAMINATION 14-1

17 SUBJECT INDEX 17-1

PREFACE TO THE FOURTH EDITION

This edition of the *Electrical Engineering Review Manual* will enable you to prepare effectively and efficiently for the National Council of Engineering Examiners (NCEE) license examination for Electrical Engineers.

This manual will be highly valuable as a reference during your preparation for the examination. The contents have been extensively tested in actual classroom examination review courses. Explanations, examples, and problems are completely up-to-date and are representative of the areas covered in recent examinations.

This manual will help you review the broad field of electrical engineering by emphasizing both fundamental and detailed subjects. Most solution techniques will be recognized by electrical engineers holding degrees earned in the past 20 years. However, the explanations are sufficiently detailed for learning solution techniques new to you.

In reviewing electrical engineering fundamentals, this book covers a majority of the problem types appearing on current NCEE examinations. However, some examinations contain a few problems requiring either specialized experience or advanced knowledge. Such problems are too diverse to be included.

This edition differs considerably from the third edition, both in appearance and in contents. More attention has been given to reviewing basic principles for engineers who have been out of school for a long time. All subjects have been expanded, but particular emphasis has been placed on solid state components and digital logic. Furthermore, the end-of-chapter practice problems have been increased in number and organized in order of increasing difficulty.

Raymond B. Yarbrough, P.E.

June, 1983

INTRODUCTION

Purpose of Registration

As an engineer, you may have to obtain your Professional Engineering license through procedures which have been established by the state in which you reside. These procedures are designed to protect the public by preventing unqualified individuals from legally practicing as engineers.

There are many reasons for wanting to become a Professional Engineer. Among them are the following:

- You may wish to become an independent consultant. By law, consulting engineers must be registered.

- Your company may require a professional engineering license as a requirement for employment or advancement.

- Your state may require registration as a Professional Engineer if you use the title *engineer*.

The Registration Procedure

The registration procedure is similar in most states. You probably will take two 8-hour written examinations. The first examination is the *Engineering-In-Training examination*, also known as the *Intern Engineer exam* and the *Fundamentals of Engineering exam*. The initals E-I-T, I.E., and F.E. also are used. The second examination is the *Professional Engineering (P.E.) exam*, also known as the *Principles and Practice exam*, which is different from the E-I-T exam in format and content.

If you have significant experience in engineering, you may be allowed skip the E-I-T examination. However, actual details of registration, experience requirement, minimum education levels, fees, and examination schedules vary from state to state. You should contact your state's Board of Registration for Professional Engineers.

Alabama	(205) 832-6100
Alaska	(907) 465-2540
Arizona	(602) 255-4053
Arkansas	(501) 371-2517
California	(916) 445-5544
Colorado	(303) 866-2396
Connecticut	(203) 566-3386
Delaware	(302) 656-7311
District of Columbia	(202) 727-3673
Florida	(904) 488-9912
Georgia	(404) 656-3926
Guam	(671) 646-8643
Hawaii	(808) 548-3086
Idaho	(208) 334-3860
Illinois	(217) 785-0800
Indiana	(317) 232-1840
Iowa	(515) 281-6566
Kansas	(913) 296-3053
Kentucky	(502) 564-2680
Louisiana	(504) 581-7938
Maine	(207) 289-3236
Maryland	(301) 659-6322
Massachusetts	(617) 727-3088
Michigan	(517) 373-3880
Minnesota	(612) 296-2388
Mississippi	(601) 354-7241
Missouri	(314) 751-2334
Montana	(406) 449-3737
Nebraska	(402) 471-2021
Nevada	(702) 329-1955
New Hampshire	(603) 271-2219
New Jersey	(201) 648-2660
New Mexico	(505) 827-2241
New York	(518) 474-3846
North Carolina	(919) 781-9499
North Dakota	(701) 852-1220
Ohio	(614) 466-3650
Oklahoma	(405) 521-2874
Oregon	(503) 378-4180
Pennsylvania	(717) 783-7049
Puerto Rico	(809) 725-7060
Rhode Island	(401) 277-2565

South Carolina	(803) 758-2855
South Dakota	(605) 394-2510
Tennessee	(615) 741-3221
Texas	(512) 475-3141
Utah	(801) 530-6628
Vermont	(802) 828-2363
Virgin Islands	(809) 774-1301
Virginia	(804) 786-8818
Washington	(206) 753-6966
West Virginia	(304) 822-4090
Wisconsin	(608) 266-1397
Wyoming	(307) 777-6156

The National Council of Engineering Examiners

Few states continue to write their own examination questions. Most use the exams prepared by the National Council of Engineering Examiners (NCEE).[1] These exams are given simultaneously in all of the participating states. This use of identical examinations has led to the term *Uniform Examination.*

NCEE, thus far, has not made current examinations available for review purposes. The formats of its examinations continually are changing and usually are considerably different from examinations previously developed by individual state boards. Therefore, exam review materials not based on the current NCEE exam format probably are out-of-date.

Reciprocity Among States

All states except Illinois use the NCEE P.E. examination. If you take and pass the P.E. examination in one state, your certificate probably will be honored by the other states using the NCEE examination. It will not be necessary to retake the examination.

Examination Dates

The national P.E. examination is administered on the same weekend in all states. Each state decides independently whether to offer examinations on Thursday, Friday, or Saturday of the examination period. The upcoming examination dates are given in the accompanying table.

National P.E. Examination Dates

year	Spring exam	Fall exam
1983	April 14-16	October 27-29
1984	April 12-14	October 25-27
1985	April 18-20	October 24-26
1986	April 10-12	October 23-25
1987	April 9-11	October 29-31
1988	April 14-16	October 27-29
1989	April 13-15	October 26-28

The Examination Format

The NCEE examination consists of 20 problems which are divided equally between morning and afternoon sessions of four hours each. During each session, you will be permitted to submit solutions to four of the available ten problems. Thus, you will have an average of one hour per problem.

Since the examination structure is not rigid, it is not possible to give the exact number of questions that will appear in each subject area. However, the NCEE Uniform Examination Committee has recommended the following distribution of problems for examinations starting in 1983.[2]

Subject	Approximate Number of Questions	
Power	8	
Electronics	6	
Control Systems	3	
Computers	2	
Engineering Economics	1	
	20	total

It is often difficult to categorize an examination problem. Different subjects may be mixed to various degrees within the same problem. Therefore, the NCEE recommendations should not be taken as rigid categories.

Each examinee is given an examination booklet which contains problems for mechanical, civil, chemical, and electrical engineers. Some states, such as California, will allow electrical examinees to work problems only from the electrical part of the examination booklet. Other states will allow examinees to work problems from the entire booklet. Read the examination instructions carefully on this point.

In past years, the examination always had contained two problems from the subject of engineering economics. This now has been reduced to one problem. This reduction is expected to be permanent, and you should expect only one problem in Engineering Economics.

[1] The National Council of Engineering Examiners (NCEE) in Seneca, South Carolina, produces, distributes, and grades the national P.E. examination used by all states except Illinois. It does not distribute applications to take the examinations.

[2] The breakdown of questions by type reflects NCEE's Task Analysis evaluation completed in 1982. The breakdown differs from previous years' categorizations which included such subjects as communication, instrumentation, and machines.

Recent examinations have had few illumination problems (one or none). The material in this book covers approximately 80% of the illumination problems that have occurred in recent examinations. However, many illumination problems are based to some degree on the National Electric Code (NEC). If you are planning to work illumination problems, you should also become familiar with the NEC.

Grading the Examination

Most states do not have any required problems on the examination. Therefore, you will be free to choose from among the twenty available problems. Full credit is achieved by correctly working four problems in the morning and four problems in the afternoon. You may not claim credit for more than eight worked problems or for more than four per session. All solutions are recorded in official solution booklets.

Each problem is worth ten points. Solutions are graded subjectively by hand. Scoring of each problem is on two scales, one recognizing method, the other accounting for mathematical errors. The *methods* scale runs from 0 to 10; the *errors* scale runs from 0 to minus 5. The sum of the two scales is your score for the problem.

The total number of possible points for both sessions is 80. A score of 56 (70%) generally is considered passing.

The examination is open book.[3] Usually, all forms of solutions aids are allowed in the examination, including nomographs, specialty slide-rules, and pre-programmed and programmable calculators. Since their use says little about the depth of your knowledge, such aids should be used only to check your work.

Grading and reporting of your score takes approximately four months. You probably will not be told your score if you pass the exam. If you fail, you will be told your score and it will be possible to review the grading. If you feel that you are entitled to more points than were awarded, you will be able to request that your solutions be re-evaluated.

Preparing for the Exam

You should develop an examination strategy early in the preparation process. This strategy will depend on your background. One of the following two general strategies is recommended.

(1) A broad approach has been successful for examinees who have recently completed academic studies. Their strategy has been to review the fundamentals of a broad range of undergraduate electrical engineering subjects. The examination includes enough fundamental problems to give merit to this strategy.

(2) Working engineers who have been away from classroom work for a long time have found it better to concentrate on the subjects in which they have had extensive professional experience. By studying the list of subjects, they have been able to choose those which will

[3]Check with your state board to see if examination review books and collections of solved problems are allowed in the exam.

Outline of Subjects for Self-Study

(Subjects do not have to be studied in the order listed.)

subject	chapter	recommended number of weeks	date to be started	date to be completed	check when complete
mathematics	1	1	______	______	[ ]
linear circuit analysis	2	1	______	______	[ ]
measurements of signal waveforms	3	1	______	______	[ ]
time and frequency response	4	1	______	______	[ ]
power systems	5	1	______	______	[ ]
transmission lines	6	1	______	______	[ ]
electric machines	7	1	______	______	[ ]
linear amplifiers	8	1	______	______	[ ]
nonlinear waveshaping and switching	9	1	______	______	[ ]
digital logic	10	1	______	______	[ ]
control systems	11	1	______	______	[ ]
illumination	12	1	______	______	[ ]
engineering economic analysis	13	1	______	______	[ ]
final exam	14	1	______	______	[ ]

give them a good probability of finding enough problems that they can solve.

Although this manual will provide an excellent review of examination subjects, you might want to enroll in a classroom review course to help pace your studies. Such courses also will enable you to obtain answers to your questions about difficult subjects.

Whether or not you enroll in a classroom review course, you should work sample problems (such as those contained in this book) in order to become familiar with the problem types and solution methods. You will usually not have sufficient time during the examination to derive solution methods; you must know them instinctively.

Steps to Take Before the Exam

The engineers who have taken the P.E. exam in previous years have developed the suggestions listed below. These suggestions will make your examination experience as successful as possible.

• Keep a copy of your examination application. Send the original by certified mail and request a receipt of delivery.

• Keep the application copy, receipt of delivery, and your cancelled check with the materials you will take to the examination.

• Visit the exam site the day before the examination. Find the examination room, the parking area, and the rest rooms.

• If you live a long distance from the examination site, consider obtaining a hotel room in which to spend the night before.

• Get a good night's sleep the night before. Don't cram the last night.

• Make arrangements for babysitters or transportation. Allow for a delayed completion.

• Organize the materials you are taking the examination. You may want to start with the following checklist.

[ ] The documents suggested above
[ ] The letter admitting you to the examination
[ ] Photographic identification
[ ] This book
[ ] Other reference books
[ ] Course notes
[ ] Calculators (primary and back-up)
[ ] Extra batteries or battery packs
[ ] Battery charger and extension cord
[ ] Chair cushions
[ ] Earplugs
[ ] A cardboard box cut to fit your books
[ ] Twist-to-advance pencils
[ ] Extra leads
[ ] Large eraser
[ ] Thermos filled with hot chocolate
[ ] A light lunch
[ ] Interim snacks (i.e., raisins, nuts, trail mix)
[ ] A collection of all required types of graph paper
[ ] Scotch and masking tape
[ ] Scissors, stapler, and staple remover
[ ] Extra prescription glasses
[ ] Sunglasses
[ ] Aspirin
[ ] English dictionary
[ ] Dictionary of scientific terms
[ ] $2 in change
[ ] A light sweater
[ ] Comfortable shoes or slippers
[ ] Raincoat, boots, gloves, hat, umbrella
[ ] Street maps of the examination site area
[ ] Note to the parking patrol for your windshield
[ ] Pad of scratch paper with holes in a 3-ring binder
[ ] Ruler, compass, protractor, French curves
[ ] Battery-power desk lamp
[ ] Watch
[ ] Travel pack of Kleenex
[ ] Wire coat hanger

Some states prohibit programmed or programmable calculators, extension cords, and electrical machines. Most states prohibit scratch paper.

What to Do During the Examination

Previous examinees have reported that the following strategies and techniques helped them considerably.

- Read through all of the problems before starting your first solution. In order to save you from re-reading and re-evaluating each problem later in the day, you should classify each problem at the beginning of the four hour session. The following categories are suggested:

  · problems you can easily do
  · problems you can do with effort
  · problems for which you might be able to get partial credit
  · problems you cannot do

- Do all of the problems in order of increasing difficulty. All problems on the examination are worth 10 points. There is nothing to be gained by attempting the difficult or long problems if easier or shorter problems are available.

- Follow these guidelines when solving a problem:

    - Do not rewrite the problem statement.
    - Do not redraw any figures unnecessarily
    - Use pencil only
    - Be neat. (Print all text. Use a straight-edge or template where possible.)
    - Draw a box around each answer.
    - Label each answer with a symbol.
    - Give the units.
    - List your sources whenever you use obscure solution methods or data.
    - Write on one side of the page only.
    - Start each problem on a new page, no matter how short the solution is.
    - Go through all calculations a second time and check for mathematical errors.

ACKNOWLEDGEMENTS

A textbook of this scope and character is a concentrated team effort. The road from manuscript to bound volume is littered along the way with thousands of pages of galley and page proofs. The editing, typesetting, and illustrating work in a technical book of this type is enormous. Therefore, I want to acknowledge the assistance of a few of the many people who helped me with the preparation of this book.

I was greatly influenced in the preparation of this book by the hundreds of students who reviewed for the P.E. examination in my classes throughout the years. Although space does not permit mentioning their names, I am grateful for the insight these individuals have given me into the needs of engineers who have been out of school for a long time.

These engineers also provided feedback about the examination contents. Over the years, they have helped me understand the pattern of examination questions... their complexity as well as the subject material covered.

For her skillful conversion of my manuscript into typeset book form, I would like to thank Joanne Bergeson. Her ability to recognize engineering symbols and reformat equations into acceptable typographical form is exceptional. For her generous assistance with editing, I am grateful to Gay Liesse. And, for never giving up on the seemingly endless number of figures in this book, I am indebted to Ken Liesse who turned my sketches into excellent illustrations.

For his constant encouragement, I should like to thank Michael R. Lindeburg, who also supplied the engineering economics chapter to this book.

For her interest, cooperation, and continued support through the entire project, I am especially indebted to my wife, Judith Mary Sherrill Yarbrough.

1 MATHEMATICS

1 REAL NUMBERS

Positional number systems are used to designate the value of a real number according to the positions of its digits. In the *decimal (base 10) system*, the digits are *zero* through *nine*. The position of a particular digit within a string determines the contribution of that digit to the total value of the number. For example *three* in the decimal number 347 has a place value of $3 \cdot 10^2$, *four* has a value of $4 \cdot 10^1$, and *seven* has a value of $7 \cdot 10^0$. Thus, the integer number 347 can be written as

$$(347)_{10} = (3 \cdot 10^2) + (4 \cdot 10^1) + (7 \cdot 10^0) = 300 + 40 + 7$$

This method can be applied to any number system. Digits of an integer are weighted by the base raised to powers determined by their positions in the string. At the right-most digit, the weighting factor is the base raised to the zero power. The digit second from the right is multiplied by the base raised to the first power. The third digit is multiplied by the base squared. Then, for the base b integer string, $N_n N_{n-1} N_{n-2} \ldots N_i \ldots N_2$-$N_1 N_0$, the value given to a digit n_i, is

$$\text{value of } i^{th} \text{ member} = N_i \cdot b^i \qquad 1.1$$

Fractional numbers can be similarly represented. The exponent i is negative, with a value equal to the digit's position from the right-most integer digit. For example, in the number 6.25, *five* is two places to the right of *six*, the right-most integer digit. So, the weighting for 5 is 10^{-2}. The proper name for the decimal point is the *radix point*. The radix point is used to separate the integer part of a number from its fractional part.

Example 1.1

Determine the decimal value of the *octal* (base 8) number 347.16.

$$\begin{aligned}(347.16)_8 &= (3 \cdot 8^2) + (4 \cdot 8^1) + (7 \cdot 8^0) + (1 \cdot 8^{-1}) + (6 \cdot 8^{-2}) \\ &= (3 \cdot 64) + (4 \cdot 8) + (7 \cdot 1) + (1 \cdot .125) + (6 \cdot .015625) \\ &= 192 + 32 + 7 + .125 + .09375 \\ &= (231.21875)_{10}\end{aligned}$$

2 COMPLEX NUMBERS

A *complex number* has a real part and an imaginary part. The *imaginary part* is identified by a coefficient j, where $j = \sqrt{-1}$. Thus a complex number can be written as $a + jb$. A complex number can be represented on a *complex plane* by graphing coordinates which correspond to the real and imaginary coefficients. Figure 1.1 shows a complex plane in which complex number $a + jb$ is represented as a point.

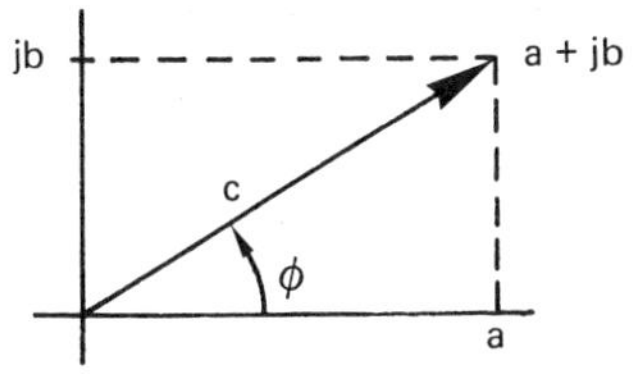

Figure 1.1 Complex Plane

A complex number can also be represented as a two-dimensional vector in the complex plane. Thus the complex number can be written in *polar* or *phasor form* as

$$a + jb = c\angle\phi$$

$$c = \sqrt{a^2 + b^2} \qquad 1.2a$$

$$\phi = \arctan\left(\frac{b}{a}\right) \qquad 1.2b$$

The orthogonal sides of the triangle can be found from trigonometry.

$$a = c\cos\phi \qquad 1.3$$

$$b = c\sin\phi \qquad 1.4$$

Combining equations 1.2, 1.3, and 1.4, the complex number can be written as equation 1.5.

$$\begin{aligned} a + jb &= c\cos\phi + jc\sin\phi \\ &= c(\cos\phi + j\sin\phi) \end{aligned} \qquad 1.5$$

Euler's equation ($e^{j\phi} = \cos\phi + j\sin\phi$) can be used to write this as an exponential.

$$a + jb = ce^{j\phi} \qquad 1.6$$

The notations $\angle\phi$ and $e^{j\phi}$ can generally be used interchangeably, although the exponential form requires that ϕ be in radians for computational purposes. The $\angle$ notation is not so constrained.

The complex conjugate of a complex number can be plotted on the complex plane as a mirror image with respect to the real axis. The conjugate of a complex number A is written as A^*. If $A = a + jb$, then $A^* = a - jb$. In the polar and exponential forms, the angle is negated to obtain the conjugate.

$$\begin{aligned} A &= a + jb = c\angle\phi \\ &= ce^{j\phi} = c(\cos\phi + j\sin\phi) \end{aligned} \qquad 1.7a$$

$$\begin{aligned} A^* &= a - jb = c\angle-\phi \\ &= ce^{-j\phi} = c(\cos\phi - j\sin\phi) \end{aligned} \qquad 1.7b$$

Addition and subtraction of complex numbers is most conveniently done in rectangular form. If the numbers are in polar form, they should be converted to rectangular form before being added.

Multiplication and division of complex numbers can be done in either polar or rectangular form, as long as both numbers are in the same form. Examples of each operation in rectangular and polar forms follow.

Example 1.2

If $A = 4 + j5$ and $B = 2 - j3$, what is $A \cdot B$?

$$\begin{aligned} &(4 + j5)\cdot(2 - j3) \\ &= (4)(2) + (4)(-j3) + (j5)(2) + (j5)(-j3) \\ &= 8 - j12 + j10 - j^2 15 \end{aligned}$$

However, $j^2 = -1$, so that $-j^2 15 = +15$.

$$\begin{aligned} A\cdot B &= (8 + 15) + j(-12 + 10) \\ &= 23 - j2 \end{aligned}$$

Example 1.3

Convert A and B in example 1.2 to phasor and exponential forms before multiplying.

$$\begin{aligned} A &= \sqrt{4^2 + 5^2}\angle\arctan\left(\frac{5}{4}\right) = 6.4\angle 51.34^o \\ &= 6.4e^{j.896} \\ B &= \sqrt{2^2 + 3^2}\angle\arctan\left(\frac{-3}{2}\right) = 3.6\angle - 56.31^o \\ &= 3.6e^{-j.983} \end{aligned}$$

Note that the angle in the phasor form is in degrees, and in the exponential form it is in radians.

$$\begin{aligned} A\cdot B &= (6.4\cdot 3.6)\angle(51.32 - 56.31) \\ &= 23.04\angle - 4.97 \\ &= 23.04(e^{-j.087}) \end{aligned}$$

Division of complex numbers can be considered as the multiplication of the numerator by the reciprocal of the denominator. Division of complex numbers in rectangular form is accomplished by multiplying numerator and denominator by the complex conjugate of the denominator.

Example 1.4

Find the reciprocal of $A = 4 + j5$.

$$\begin{aligned} A^{-1} &= \frac{1}{4 + j5}\cdot\frac{4 - j5}{4 - j5} = \frac{4 - j5}{4^2 + 5^2} \\ &= \frac{4 - j5}{41} = .0967 - j.1220 \end{aligned}$$

Example 1.5

Divide $B = 2 - j3$ by $A = 4 + j5$.

$$\begin{aligned} \frac{B}{A} &= \frac{2 - j3}{4 + j5}\cdot\frac{4 - j5}{4 - j5} = \frac{8 - j10 - j12 + j^2 15}{41} \\ &= \frac{-7 - j22}{41} \\ &= -.1707 - j.5366 \end{aligned}$$

Division of complex numbers in the phasor and exponential forms is much simpler, as illustrated by example 1.6.

Example 1.6

Divide $B = 3.6e^{j(-.983)}$ by $A = 6.4e^{j.896}$

$$\frac{B}{A} = \frac{3.6e^{j(-.983)}}{6.4e^{j.896}} = \frac{3.6}{6.4}e^{-j.983}e^{-j.896}$$
$$= .5625e^{-j1.879}$$

3 ALGEBRA

A. QUADRATIC EQUATIONS

Quadratic equations frequently arise in electrical problems. The usual form of a quadratic equation is

$$ax^2 + bx + c = 0 \qquad 1.8$$

The values of x which satisfy equation 1.8 are called the *roots* or *zeros* of the equation. These roots are found from equation 1.9.

$$r_1, r_2 = \frac{-b \pm \sqrt{b^2 - 4ac}}{2a} \qquad 1.9$$

Equation 1.8 can then be rewritten as equation 1.10.

$$a(x - r_1)(x - r_2) = 0 \qquad 1.10$$

The variables r_1 and r_2 can be real or imaginary depending on the coefficients a, b, and c.

- If $b^2 - 4ac > 0$ then there are two different real roots.
- If $b^2 - 4ac = 0$ then there are two identical real roots equal to $\frac{-b}{2a}$.
- If $b^2 - 4ac < 0$ then there are two complex conjugate roots with real part $\frac{-b}{2a}$ and imaginary parts of $\pm\sqrt{\left(\frac{c}{a}\right) - \left(\frac{b}{2a}\right)^2}$

B. EXPONENTIATION

$$x^a x^b = x^{a+b} \qquad 1.11$$

$$x^{-a} = \frac{1}{x^a} \qquad 1.12$$

$$x^{(1/a)} = \sqrt[a]{x} \qquad 1.13$$

$$(x^a)^b = x^{ab} \qquad 1.14$$

$$x^{(b/a)} = \sqrt[a]{x^b} = \left(\sqrt[a]{x}\right)^b \qquad 1.15$$

C. LOGARITHMS

If $b^x = y$, then $x = \log_b y$. For example, if $y = 10^x$, then $x = \log_{10} y$, ... or if, $e^x = y$, then $x = \log_e y = ln\ y$. Other identities that may be of value are listed here.

$$\log_b y^a = a \log_b y \qquad 1.16$$

$$a^x = b^{(x \log_b a)} \qquad 1.17$$

$$\log_a y = (\log_b y)(\log_a b) \qquad 1.18$$

$$\log_b xy = \log_b x + \log_b y \qquad 1.19$$

$$\begin{aligned}\log(x + jy) &= \log\left[\left(\sqrt{x^2 + y^2}\right)\left(e^{j[\arctan(\frac{y}{x})]}\right)\right] \\ &= \log\left(\sqrt{x^2 + y^2}\right) + \log\left(e^{j[\arctan(\frac{y}{x})]}\right) \\ &= \log\left(\sqrt{x^2 + y^2}\right) + j(\log e)\cdot\arctan\left(\frac{y}{x}\right)\end{aligned} \qquad 1.20$$

D. PARTIAL FRACTION EXPANSION

A ratio of two polynomials can be rewritten as a series of terms which are the partial fractions, each fraction having terms such as $(x - r)$ in the denominator, where the constants r are the roots of the original denominator.[1] For example, $x^3 + 4x^2 + 3x$ has roots at $x = 0$, $x = -1$, and $x = -3$.

$$\frac{6x^2 + 15x + 3}{x^3 + 4x^2 + 3x} = \frac{6x^2 + 15x + 3}{x(x+1)(x+3)}$$

$$\frac{6x^2 + 15x + 3}{x(x+1)(x+3)} = \frac{A}{x} + \frac{B}{x+1} + \frac{C}{x+3}$$

The partial fraction expansion is found by the method *of undetermined coefficients*. The terms in the right-hand side are cross multiplied and are then placed over a common denominator.

[1]Before a partial fraction expansion can be carried out, the degree of the numerator must be lower than the degree of the denominator. Otherwise, it will be necessary to reduce the numerator's degree through long division.

$$\frac{A}{x}+\frac{B}{x+1}+\frac{C}{x+3}=$$

$$=\frac{A(x^2+4x+3)+B(x^2+3x)+C(x^2+x)}{x(x+1)(x+3)}$$

$$=\frac{(A+B+C)x^2+(4A+3B+C)x+3A}{x(x+1)(x+3)}$$

Matching the coefficients of x^n in the numerator terms results in simultaneous equations in three unknowns.

$$A+B+C=6$$
$$4A+3B+C=15$$
$$3A=3$$

The solution to these simultaneous equations is $A=1$, $B=3$ and $C=2$. Therefore,

$$\frac{6x^2+15x+3}{x^3+4x^2+3x}=\frac{1}{x}+\frac{3}{x+1}+\frac{2}{x+3}$$

This partial fraction expansion procedure is adequate as explained when the roots of the denominator are real and distinctly different from one another. When the roots repeat, the denominator terms will be $(x-r)^n$, $(x-r)^{n-1}$, and so on, to and including $(x-r)$. For example,

$$\frac{x^2+4x+2}{x(x+1)^2}=\frac{A}{x}+\frac{B}{x+1}+\frac{C}{(x+1)^2}$$

$$=\frac{(A+B)x^2+(2A+B+C)x+A}{x(x+1)^2}$$

(The solution to this illustration is $A=2$, $B=-1$, and $C=1$. Try it.) When complex conjugate roots occur (i.e., the denominator will not factor into real terms), the form of the resulting partial fraction will be

$$\frac{Ax+B}{x^2+ax+b} \qquad 1.21$$

Partial fraction expansions are particularly useful in the solution of differential equations by the use of Laplace transforms.

Long division is required prior to partial fraction expansion whenever the degree of the numerator polynomial is greater than or equal to the degree of the denominator polynomial. This is illustrated by example 1.7.

Example 1.7

Use long division to reduce the degree of the numerator polynomial for

$$\frac{s^2+3s+2}{s^2+2s+1}$$

$$\begin{array}{r} 1 \\ s^2+2s+1 \overline{)\, s^2+3s+2} \\ \underline{s^2+2s+1} \\ s+1 \end{array}$$

$s+1$ is the remainder term. Since the degree of the remainder term is less than the degree of the numerator, the original expression can be written as

$$\frac{s^2+3s+2}{s^2+2s+1}=1+\frac{s+1}{s^2+2s+1}$$

4 TRIGONOMETRIC IDENTITIES

The common trigonometric functions can be approximated by infinite series when the angle x is expressed in radians. (Remember that $180^\circ=\pi$ radius.)

$$\sin x = x-\frac{x^3}{3!}+\frac{x^5}{5!}-\frac{x^5}{7!}+\cdots \qquad 1.22$$

$$\cos x = 1-\frac{x^2}{2!}+\frac{x^4}{4!}-\frac{x^6}{6!}+\cdots \qquad 1.23$$

$$\tan x = x+\frac{x^3}{3}+\frac{2x^5}{15}+\frac{17x^7}{315}+\cdots \qquad 1.24$$

$$\text{for}|x|<\frac{\pi}{2}$$

Additional identities are presented here for convenience.

$$\tan x=\frac{\sin x}{\cos x} \qquad 1.25$$

$$\cot x=\frac{1}{\tan x} \qquad 1.26$$

$$\csc x=\frac{1}{\sin x} \qquad 1.27$$

$$\sec x=\frac{1}{\cos x} \qquad 1.28$$

$$\sin^2 x+\cos^2 x=1 \qquad 1.29$$

$$1+\tan^2 x=\sec^2 x \qquad 1.30$$

$$1+\cot^2 x=\csc^2 x \qquad 1.31$$

$$\sin 2x=2\sin x\cos x \qquad 1.32$$

$$\tan 2x=\frac{2\tan x}{1-\tan^2 x} \qquad 1.33$$

$$\begin{aligned}\cos 2x&=2\cos^2 x-1\\&=\cos^2 x-\sin^2 x\\&=1-2\sin^2 x\end{aligned} \qquad 1.34$$

$$\sin(x+y)=\sin x\cos y+\cos x\sin y \qquad 1.35$$

$$\sin(x-y)=\sin x\cos y-\cos x\sin y \qquad 1.36$$

$$\cos(x+y)=\cos x\cos y-\sin x\sin y \qquad 1.37$$

$$\cos(x-y)=\cos x\cos y+\sin x\sin y \qquad 1.38$$

$$2[\sin x \sin y] = \cos(x-y) - \cos(x+y) \qquad 1.39$$
$$2\cos x \cos y = \cos(x-y) + \cos(x+y) \qquad 1.40$$
$$2\sin x \cos y = \sin(x+y) + \sin(x-y) \qquad 1.41$$

5 ANALYTIC GEOMETRY

A major use of analytic geometry in electrical engineering problems is mathematical representation of graphical information. Much of this involves straight-line approximations to device characteristics. Equations for a *straight line* in $x-y$ space can be in several forms. Refer to figure 1.2 which slows a straight line with y-intercept b, x-intercept a, and slope m. (A negative slope is shown.)

general form:

$$Ax + By + C = 0 \qquad 1.42$$

slope-intercept form:

$$y = mx + b \qquad 1.43$$

two-point form:

$$\frac{y-y_1}{x-x_1} = \frac{y_2-y_1}{x_2-x_1} = m \qquad 1.44$$

intercept form:

$$\frac{x}{a} + \frac{y}{b} = 1 \qquad 1.45$$

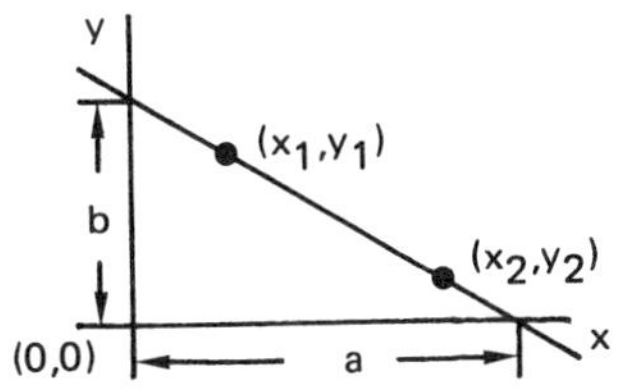

Figure 1.2 A straight line in $x-y$ space

6 DIFFERENTIAL CALCULUS

Some of the frequently encountered derivatives are listed here. The terms u and v are functions of x. The terms a and b are numerical constants. The term n is an integer constant.

$$\frac{d}{dx}a = 0 \qquad 1.46$$
$$\frac{d}{dx}x^n = n \cdot x^{n-1} \qquad 1.47$$
$$\frac{d}{dx}a \cdot u = a \cdot \frac{du}{dx} \qquad 1.48$$
$$\frac{d}{dx}u^n = n \cdot u^{n-1} \cdot \frac{du}{dx} \qquad 1.49$$
$$\frac{d}{dx}uv = u \cdot \frac{dv}{dx} + v \cdot \frac{du}{dx} \qquad 1.50$$
$$\frac{d}{dx}\frac{u}{v} = \frac{1}{v} \cdot \frac{du}{dx} - \frac{u}{v^2} \cdot \frac{dv}{dx} = \frac{v \cdot \frac{du}{dx} - u \cdot \frac{dv}{dx}}{v^2} \qquad 1.51$$
$$\frac{d}{dx}\sin u = \frac{du}{dx}\cos u \qquad 1.52$$
$$\frac{d}{dx}\cos u = -\frac{du}{dx} \cdot \sin u \qquad 1.53$$
$$\frac{d}{dx}\log_b u = \log_b(e) \cdot \frac{1}{u} \cdot \frac{du}{dx} \qquad 1.54$$
$$\frac{d}{dx}\ln u = \frac{1}{u} \cdot \frac{du}{dx} \qquad 1.55$$
$$\frac{d}{dx}b^u = b^u \cdot \ln b \cdot \frac{du}{dx} \qquad 1.56$$
$$\frac{d}{dx}e^u = e^u \cdot \frac{du}{dx} \qquad 1.57$$

Differentiation is used to find local *maxima* and *minima* of functions. At a maximum or minimum, a function will have a zero slope. Therefore, the first derivative of the function must be zero at a maximum or minimum. Another point which will have zero slope is a *point of inflection*. Here, the slope is the same sign on either side but zero is at the point of inflection.

To distinguish between maxima, minima, and points of inflection, it is necessary to evaluate the second derivative of the function. If the second derivative is negative, the point is a *local maximum*. If the second derivative is positive, the point is a *local minimum*. If the second derivative is zero, the point is a point of inflection.

Example 1.8

Find a maximum or minimum of $u = x^2 + 2x + 3$.

$$\frac{du}{dx} = \frac{d}{dx}(x^2 + 2x + 3) = 2x + 2$$

The point sought is where

$$\frac{du}{dx} = 0$$

or at $x = -1$.

To determine whether this point is a maximum or a minimum, the second derivative is taken.

$$\frac{d^2u}{dx^2} = \frac{d}{dx}(2x+2) = 2.$$

This is positive at $x = -1$, so u is minimum at $x = -1$.

7 INTEGRAL CALCULUS

Some integrals which occur frequently in electrical engineering problems are listed here. The letters u and v are functions of x, while the letters a, b, and c are constants. n is an integer constant.

$$\int x^n dx = \frac{x^{n+1}}{n+1} + c, \quad \text{for } n \neq -1 \qquad 1.58$$

$$\int \frac{1}{x} dx = ln\, x + c \qquad 1.59$$

$$\int u dv = uv - \int v du \qquad 1.60$$

$$\int e^{ax} dx = \frac{1}{a} e^{ax} + c \qquad 1.61$$

$$\int \sin(bx) dx = -\frac{1}{b}\cos(bx) + c \qquad 1.62$$

$$\int \cos(bx) dx = \frac{1}{b}\sin(bx) + c \qquad 1.63$$

$$\int e^{ax}\cos(bx) dx = \frac{e^{ax}}{a^2+b^2}[a\cos(bx) + b\sin(bx)] + c \qquad 1.64$$

$$\int e^{ax}\sin(bx) dx = \frac{e^{ax}}{a^2+b^2}[a\sin(bx) - b\cos(bx)] + c \qquad 1.65$$

A common application of integration is to find the *average value* of a function. For a function u, the average value from $x = a$ to $x = b$ is

$$\overline{u} = \frac{1}{b-a}\int_a^b u dx \qquad 1.66$$

Example 1.9

Find the average value of $u = 2x + 5\sin^2 x$ over the interval $x = -1$ to $x = 2$. Use the trigonometric identity $\cos 2x = 1 - 2\sin^2 x$ and solve for $\sin^2 x$. (x must be expressed in radians.)

$$\sin^2 x = \frac{1}{2}(1 - \cos 2x)$$

$$u = 2x + \frac{5}{2}[1 - \cos 2x] = 2x + 2.5 - 2.5\cos 2x$$

From equation 1.66 with $a = -1$ and $b = 2$, we obtain

$$\overline{u} = \frac{1}{2-(-1)}\left[2\left(\frac{x^2}{2}\right) + 2.5x - 2.5\left(\frac{1}{2}\sin 2x\right)\right]_{x=-1}^{x=2}$$

$$= \frac{1}{3}[(2^2 - (-1)^2) + 2.5(2-(-1)) - 1.25[\sin 4 - \sin(-2)]]$$

$$= \frac{1}{3}[3 + 7.5 - 1.25[-.757 - (-.909)]]$$

$$= \frac{1}{3}[10.5 - 1.25(.152)] = 3.436$$

8 DIFFERENTIAL EQUATIONS

First and second order linear differential equations can be solved by classical methods or with the aid of Laplace transforms. Classical solutions to linear differential equations of the first and second order are presented here. Such differential equations often occur in d.c. switching problems. In these problems, the equations will have the form of equation 1.67, where a, b, c, and d are constants.

$$a\frac{d^2x}{dt^2} + b\frac{dx}{dt} + cx = d \qquad 1.67$$

A. FIRST ORDER DIFFERENTIAL EQUATIONS

The first order differential equation contains only the first derivative of the independent variable (i.e., a in equation 1.67 is zero). The solution is

$$x = \frac{d}{c} + Ae^{-ct/b} \qquad 1.68$$

A is an unknown coefficient. In order to determine the value of A, it is necessary to know the value of x or one its derivatives for a known value of t. The initial value of x is frequently known. In this case, $x = x_0$ at $t = 0$.

$$x_0 = \frac{d}{c} + Ae^0 = \frac{d}{c} + A \qquad 1.69$$

So, the coefficient is

$$A = x_0 - \frac{d}{c} \qquad 1.70$$

The solution to the differential equation is

$$x = \frac{d}{c} + (x_0 - \frac{d}{c})e^{-ct/b} \qquad 1.71$$

Example 1.10

Given a current described by $2\frac{di}{dt} + 5i = 10$, determine i for $t > 0$, if $i = -1$ at $t = 0$.

Here $b = 2$, $c = 5$ and $d = 10$.

Then $c/b = 2.5$, $d/c = 2$, and $i_0 - (d/c) = -3$. From equation 1.71,

$$i = 2 - 3e^{-2.5t}$$

B. SECOND ORDER DIFFERENTIAL EQUATIONS

There are three distinct cases to be considered when the coefficient a in equation 1.67 is non-zero.

case 1: $b^2 - 4ac$ is positive (the overdamped case)
case 2: $b^2 - 4ac$ is zero (the critically damped case)
case 3: $b^2 - 4ac$ is negative (the underdamped case)

•Case 1: Overdamped: $b^2 - 4ac > 0$

If $b^2 - 4ac > 0$, then the solution to equation 1.67 is

$$x = \frac{d}{c} + Ae^{-t/\tau_1} + Be^{-t/\tau_2} \qquad 1.72$$

$$\frac{1}{\tau_1} = \frac{b + \sqrt{b^2 - 4ac}}{2a} \qquad 1.73$$

$$\frac{1}{\tau_2} = \frac{b - \sqrt{b^2 - 4ac}}{2a} \qquad 1.74$$

A and B are undetermined coefficients which must be evaluated from the initial conditions. The usual situation is to know x and its derivative at $t = 0$. From this information it is possible to obtain two equations in two unknowns and solve for the undetermined coefficients A and B. The derivative of equation 1.72 is

$$\frac{dx}{dt} = \frac{-A}{\tau_1}e^{-t/\tau_1} - \frac{B}{\tau_2}e^{-t/\tau_2} \qquad 1.75$$

Example 1.11

A current is described by

$$\frac{d^2i}{dt^2} + 5\frac{di}{dt} + 4i = 2$$

At $t = 0$, $i = 0$ and $di/dt = .8$. Determine the current for $t > 0$.

Referring to equation 1.67, $a = 1$, $b = 5$, $c = 4$, and $d = 2$. $b^2 - 4ac = 25 - 16 = 9$, so this is the overdamped case. From equations 1.73 and 1.74,

$$\frac{1}{\tau_1} = \frac{5+3}{2} = 4$$

$$\frac{1}{\tau_2} = \frac{5-3}{2} = 1$$

$$\text{and } \frac{d}{c} = .5$$

i and $\frac{di}{dt}$ can be found from equations 1.72 and 1.75 respectively.

$$i = .5 + Ae^{-4t} + Be^{-t}$$
$$\frac{di}{dt} = -4Ae^{-4t} - Be^{-t}$$

Substituting the initial conditions,

$$0 = .5 + Ae^0 + Be^0 = .5 + A + B$$
$$.8 = -4Ae^0 - Be^0 = -4A - B$$

The simultaneous solution to these two equations is $A = -.1$ and $B = -.4$. Therefore, the current as a function of time is

$$i = .5 + .1e^{-4t} - .4e^{-t}$$

•Case 2: Critically damped: $b^2 - 4ac = 0$

In the critically damped case, the solution to equation 1.67 is

$$x = \frac{d}{c} + Ae^{-bt/2a} + Bte^{-bt/2a} \qquad 1.76$$

A and B are undetermined coefficients which must be evaluated from the initial conditions. The derivative of equation 1.76 is

$$\frac{dx}{dt} = -\frac{bA}{2a}e^{-bt/2a} + Be^{-bt/2a} - \frac{bB}{2a}te^{-bt/2a} \qquad 1.77$$

If the conditions are given at $t = 0$, then

$$x_0 = \frac{d}{c} + A \qquad 1.78$$

$$\left.\frac{dx}{dt}\right|_{t=0} = -\frac{bA}{2a} + B \qquad 1.79$$

Example 1.12

A current is described by the differential equation

$$\frac{d^2i}{dt^2} + 2\frac{di}{dt} + i = .25$$

Determine i as a function of time if $i = 0$ and $\frac{di}{dt} = .8$ at $t = 0$.

This is the critically damped case since $a = 1$, $b = 2$, $c = 1$ and $d = .25$, so that $b^2 - 4ac = 0$. From equation 1.76, $b/2a = 1$, and $d/c = .25$, so that the skeleton solution is

$$i = .25 + Ae^{-t} + Bte^{-t}$$

At $t = 0$, $i = 0$. From equation 1.78,

$$i_0 = 0 = .25 + A$$
$$A = -.25$$

At $t = 0$, $\frac{di}{dt} = .8$. From equation 1.79,

$$\left.\frac{di}{dt}\right|_{t=0} = .8 = -A + B$$
$$B = .55$$

The current as a function of time is

$$i = .25 - .25e^{-t} + .55te^{-t}$$

•Case 3: Underdamped: $b^2 - 4ac < 0$

In this case the solution of equation 1.67 is

$$x = \frac{d}{c} + Ae^{-\alpha t}\cos\beta t + Be^{-\alpha t}\sin\beta t \qquad 1.80$$

$$\alpha = \frac{b}{2a} \qquad 1.81$$

$$\beta = \frac{\sqrt{4ac - b^2}}{2a} \qquad 1.82$$

The derivative of x is

$$\frac{dx}{dt} = (-\alpha A + \beta B)e^{-\alpha t}\cos\beta t - (\beta A + \alpha B)e^{-\alpha t}\sin\beta t \qquad 1.83$$

At $t = 0$, equations 1.80 and 1.83 become

$$x_0 = \frac{d}{c} + A \qquad 1.84$$

$$\left.\frac{dx}{dt}\right|_{t=0} = -\alpha A + \beta B \qquad 1.85$$

Example 1.13

A voltage is described by $d^2v/dt^2 + dv/dt + 2.5v = 50$. $v = 0$ and $dv/dt = 2.5$ at $t = 0$. Determine v for $t > 0$.

Referring to equation 1.67, $a = 1$, $b = 1$, $c = 2.5$ and $d = 50$. Then $b/2a = .5$, $b^2 - 4ac = -9$ (the underdamped case), and $d/c = 20$. From equation 1.80,

$$v = 20 + Ae^{-.5t}\cos 1.5t + Be^{-.5t}\sin 1.5t$$

Substituting the initial conditions in equations 1.84 and 1.85 respectively,

$$30 = 20 + A$$
$$2.5 = -.5 \times 10 + 1.5B$$

Therefore $A = 10$ and $B = 5$. The voltage as a function of time is

$$v = 20 + 10e^{-.5t}\cos 1.5t + 5e^{-.5t}\sin 1.5t$$

9 FOURIER ANALYSIS

Periodic functions are conveniently analyzed by means of *Fourier series*. *Phasor analysis* is performed for each component, and superposition is used to combine the components.

A periodic function with a period T can be represented by the Fourier series given in equation 1.86.

$$f(t) = \frac{a_0}{2} + \sum_{n=1}^{\infty} a_n \cos\frac{2\pi n}{T}t + \sum_{n=1}^{\infty} b_n \sin\frac{2\pi n}{T}t \qquad 1.86$$

The Fourier coefficients a_0, a_n, and b_n can be found from equations 1.87, 1.88, and 1.89.

$$a_0 = \frac{2}{T}\int_{t_1}^{t_1+T} f(t)\,dt \qquad 1.87$$

$$a_n = \frac{2}{T}\int_{t_1}^{t_1+T} f(t)\cos\frac{2\pi n}{T}t\,dt \qquad 1.88$$

$$b_n = \frac{2}{T}\int_{t_1}^{t_1+T} f(t)\sin\frac{2\pi n}{T}t\,dt \qquad 1.89$$

Example 1.14

A rectangular wave has voltage levels of 0 and 5 volts, equal times at both levels, and a period of 10 seconds (i.e., 5 seconds at 5 volts and 5 seconds at 0 volts). The $t = 0$ line bisects the 5 volt level. Obtain a Fourier series for this wave. One cycle of the wave can be represented by the function

$$f(t) = 0\,,\ -5 < t < -2.5$$
$$= 5\,,\ -2.5 < t < 2.5$$

Substituting $t_1 = -5$ and $T = 10$ into equations 1.87, 1.88 and 1.89 gives the coefficients a_0, a_n, and b_n.

$$a_0 = \frac{2}{10}\int_{-5}^{5} f(t)\,dt = \frac{1}{5}\int_{-2.5}^{2.5} 5dt = 5$$

$$a_n = \frac{1}{5}\int_{-2.5}^{2.5} 5\cos\frac{2\pi n}{10}t\,dt = \frac{10}{\pi n}\sin\frac{\pi n}{2}$$

$$b_n = \frac{1}{5}\int_{-2.5}^{2.5} 5\sin\frac{2\pi n}{10}t\,dt = \frac{-5}{n}\left(\cos\frac{\pi n}{2} - \cos\left(\frac{-\pi n}{2}\right)\right) = 0$$

Evaluating the first few a coefficients,

$$a_1 = \frac{10}{1\pi}\sin\frac{\pi}{2} = \frac{10}{\pi}$$
$$a_2 = \frac{10}{2\pi}\sin\pi = 0$$
$$a_3 = \frac{10}{3\pi}\sin\frac{3\pi}{2} = -\frac{10}{3\pi}$$
$$a_4 = \frac{10}{4\pi}\sin 2\pi = 0$$

In general, the a coefficients are

$$a_{2n} = 0$$
$$a_{2n-1} = \frac{10(-1)^{n-1}}{(2n-1)\pi}$$

The infinite Fourier series which approximates the original square wave is

$$f(t) = \frac{5}{2} + \sum_{n=1}^{\infty}\frac{10(-1)^{n-1}}{(2n-1)\pi}\cos\frac{(2n-1)2\pi}{10}t$$

10 TRANSIENT ANALYSIS BY LAPLACE TRANSFORMS

Laplace transform analysis is similar to operational calculus if the operator s is interpreted as the derivative operator $\frac{d}{dt}$ or **D**. The advantage of the Laplace method is that the initial conditions are automatically included in the analysis. The Laplace transform of a function $f(t)$ is written $F(s)$.

$$F(s) = \mathcal{L}\{f(t)\} = \int_0^{\infty} f(t)e^{-st}\,dt \qquad 1.90$$

The inverse Laplace transform of $F(s)$ is $f(t)$.

$$f(t) = \mathcal{L}^{-1}\{f(s)\} = \frac{1}{2\pi j}\int_{c-j\infty}^{c+j\infty} F(s)e^{st}\,ds \qquad 1.91$$

Equations 1.92, 1.93 and 1.94 are useful in writing network relations.

$$\mathcal{L}\left\{\frac{d}{dt}f(t)\right\} = sF(s) - f(0^+) \qquad 1.92$$

$$\mathcal{L}\left\{\int_0^t f(t)dt\right\} = \frac{1}{s}F(s) \qquad 1.93$$

The Laplace transform of the nth derivative is

$$\mathcal{L}\left\{\frac{d^n}{dt^n}f(t)\right\} = s^nF(s) - s^{n-1}f(0^+) - s^{n-2}\frac{df(0^+)}{dt} - s^{n-3}\frac{d^2f(0^+)}{dt^2} - \cdots - s\frac{d^{n-2}f(0^+)}{dt^{n-2}} - \frac{d^{n-1}f(0^+)}{dt^n - 1} \qquad 1.94$$

Table 1.1 provides a useful set of transform pairs.

Example 1.15

Obtain the Laplace transform of the differential equation

$$10 = (10^{-5})\frac{d^2V}{dt^2} + (5\cdot 10^{-3})\frac{dV}{dt} + V$$

The initial conditions are $v = 5$, $\frac{dv}{dt} = 10^3$ at $t = 0$.

The terms are transformed using table 1.1 to obtain

$$\frac{10}{s} = 10^{-5}[s^2V(s) - 5s - 1000] + 5\cdot 10^{-3}[sV(s) - 5] + V(s)$$

Once a differential equation has been transformed, it is necessary to solve for the transformed variable in terms of s.

Example 1.16

From example 1.15 obtain $V(s)$ as a function s.

First, multiply the answer of example 1.15 by 10^5.

$$\frac{10^6}{s} = s^2V(s) - 5s - 1000 + 500sV(s)2500 + 10^5V(s)$$

Next gather terms and solve for $V(s)$.

$$\frac{10^6}{s} + 5s + 3500 = (s^2 + 500s + 10^5)V(s)$$
$$V(s) = \frac{5s^2 + 3500s + 10^6}{s(s^2 + 500s + 10^5)}$$

The next step is to rearrange the function of s into terms that are shown in the right-hand column of table 1.1. This will typically require that function of s be factored into first and second order terms using the *partial-fraction expansion* method. Partial fractions have been covered earlier in this chapter.

Example 1.17

Separate the function $V(s)$ from example 1.16 into its partial fractions. From table 1.1, the useful forms are

$$\frac{1}{s};\ \frac{s+a}{(s+a)^2+b^2};\ \text{and}\ \frac{b}{(s+a)^2+b^2}$$

The function can be written as

$$V(s)=\frac{5s^2+3500s+10^6}{s[(s+250)^2+(10^5-250^2)]}$$

Therefore $a=250$ and $b^2=10^5-250^2$, or $b=194$. The partial fraction expansion is

$$V(s)=\frac{a}{s}+\frac{b(s+250)+c(194)}{(s+250)^2+194^2}$$

The two terms are cross multiplied to obtain the original denominator.

$$V(s)=\frac{as^2+500as+10^5a+bs^2+250bs+194cs}{s(s^2+500as+10^5)}$$

Next, equate the coefficients of the various powers of s in the numerator of this last expression with the coefficients from the original expressions.

$$\begin{aligned} s^2 &: \quad a+b=5 \\ s^1 &: \quad 500a+250b+194c=3500 \\ s^0 &: \quad 10^5a=10^6 \end{aligned}$$

a is easily found to be 10. Since $a+b=5$, then $b=-5$. c is subsequently found to equal -1.29. Finally $V(s)$ can be written as

$$V(s)=\frac{10}{s}-5\left[\frac{s+250}{(s+250)^2+194^2}\right]-1.29\left[\frac{194}{(s+250)^2+194^2}\right]$$

Table 1.1
Laplace Transform Pairs

$f(t)$	$F(s)$
$df(t)/dt$	$sF(s)-f(0^+)$
$d^2f(t)/dt^2$	$s^2F(s)-sf(0^+)-\frac{df(0^+)}{dt}$
$\int_0^t f(t)dt$	$(1/s)F(s)$
1 or $u(t)$ (unit step)	$1/s$
$u(t-\tau)$	$e^{-s\tau}/s$
$\frac{d}{dt}u(t)=\delta(t)$	1
$\delta(t-\tau)$	$e^{-s\tau}$
t	$1/s^2$
e^{-at}	$1/(s+a)$
$t\,e^{-at}$	$1/(s+a)^2$
$\sin bt$	$b/(s^2+b^2)$
$\cos bt$	$s/(s^2+b^2)$
$e^{-at}\sin bt$	$b/(s+a)^2+b^2$
$e^{-at}\cos bt$	$(s+a)/(s+a)^2+b^2$

The function $u(t)$ is zero before $t=0$ and one thereafter; the function $u(t-\tau)$ is defined as zero before $t=\tau$ and one thereafter. The function $\delta(t)$ is the time derivative of $u(t)$ and is infinite at $t=0$, but has an area of one.

The final step is to obtain the inverse transform from the transform pairs in table 1.1.

Example 1.18

Obtain $V(t)$ from $V(s)$ in example 1.17.

$$V(t) = 10 - 5e^{-250t}\cos 194t - 1.29e^{-250t}\sin 294t$$

A. INITIAL VALUE THEOREM

When a Laplace transform, $F(s)$, is known, a corresponding time function can be evaluated at $t = 0$ by the *initial value theorem*, equation 1.95.

$$f(0^+) = \lim_{s \to \infty} sF(s) \qquad 1.95$$

Example 1.19

Evaluate the function

$$V(s) = \frac{5s^2 + 3500s + 10^6}{s(s^2 + 500s + 10^5}$$

of example 1.16 at $t = 0^+$. Use the initial value theorem to obtain $v(t)$ at $t = 0^+$

$$v(0^+) = \lim_{s \to \infty} \frac{5s^3 + 3500s^2 + 10^6 s}{s^3 + 500s^2 + 10^5 s}$$

$$= \lim_{s \to \infty} \frac{5s^3}{s^3} = 5$$

B. FINAL VALUE THEOREM

When a Laplace transform $F(s)$, is known, the corresponding time function can be evaluated at $t = \infty$ by the *final value theorem*, equation 1.96.

$$f(\infty) = \lim_{s \to 0} sF(s) \qquad 1.96$$

Example 1.20

Determine $v(\infty)$ for the function $V(s)$ from example 1.16.

$$v(\infty) = \lim_{s \to 0} \frac{5s^3 + 3500s^2 + 10^6 s}{s^3 + 500s^2 + 10^5 s}$$

$$= \lim_{s \to 0} \frac{5s^2 + 3500s + 10^6}{s^2 + 500s + 10^5}$$

$$= \lim_{s \to 0} \frac{10^6}{10^5} = 10$$

Appendix A

Using Determinants To Solve Simultaneous Equations

1 DETERMINANTS

Matrixes are rectangular arrays of numbers which are associated with simultaneous linear equations. (In this appendix, symbols and subscripted letters are used to indicate the actual numbers.) The arrays are two dimensional, laid out as rows and columns. The subscripts identify the position of a particular member of the array. The first subscript gives the row position and the second gives the column position. Thus a member with the subscript 23 will belong in the second row and the third column.

Each position in the array is associated with a sign. Those positions whose subscripts add to an even number are associated with a positive sign, and those whose subscripts add to an odd number are associated with a negative sign. The arrays for which *determinants* are calculated must have an equal number of rows and columns.

As an example, consider a square 4 × 4 matrix called **A**.

$$\mathbf{A} = \begin{bmatrix} a_{11} & a_{12} & a_{13} & a_{14} \\ a_{21} & a_{22} & a_{23} & a_{24} \\ a_{31} & a_{32} & a_{33} & a_{34} \\ a_{41} & a_{42} & a_{43} & a_{44} \end{bmatrix}$$

The signs associated with the positions of **A** are

$$\begin{bmatrix} + & - & + & - \\ - & + & - & + \\ + & - & + & - \\ - & + & - & + \end{bmatrix}$$

The fundamental determinant operation can be carried out on a 2 × 2 matrix such as the matrix **B**. The determinant of **B** is indicated by writing the original matrix between straight vertical lines.

$$\mathbf{B} = \begin{bmatrix} b_{11} & b_{12} \\ b_{21} & b_{22} \end{bmatrix}$$

$$\det \mathbf{B} = \begin{vmatrix} b_{11} & b_{12} \\ b_{21} & b_{22} \end{vmatrix} = b_{11}b_{22} - b_{21}b_{12}$$

For a 3 × 3 determinant, an algorithm can be used to simplify the determinant calculation. (This algorithm is valid only for 3 × 3 determinants. It will produce erroneous results for any other sized matrix.)

For a 3 × 3 matrix, first copy columns 1 and 2 and place them in the positions of columns 4 and 5, respectively. Then, the determinant can be taken by adding all of the diagonals downward to the right and subtracting the diagonals upward to the right.

$$\mathbf{C} = \begin{bmatrix} c_{11} & c_{12} & c_{13} \\ c_{21} & c_{22} & c_{23} \\ c_{31} & c_{32} & c_{33} \end{bmatrix}$$

$$\begin{aligned} \det \mathbf{C} &= \begin{bmatrix} c_{11} c_{12} c_{13} & c_{11} & c_{12} \\ c_{21} c_{22} c_{23} & c_{21} & c_{22} \\ c_{31} c_{32} c_{33} & c_{31} & c_{32} \end{bmatrix} \\ &= c_{11}c_{22}c_{33} + c_{12}c_{23}c_{31} + c_{13}c_{21}c_{32} \\ &\quad - c_{31}c_{22}c_{13} - c_{32}c_{23}c_{11} - c_{33}c_{21}c_{12} \end{aligned}$$

In general, higher order determinants are found by the use of the so-called *minor*. Each position in a matrix has a minor. The minor of a position is a matrix of one less dimension created by eliminating the row and column of the position. Thus, the minor of position 11 in the matrix C above is obtained by eliminating row 1 and column 1 from C. This 2 × 2 minor is indicated by the symbol m with the subscripts of the position. For example, m_{32} is the minor of position row-3 column-2.

$$m_{11} = \begin{bmatrix} c_{22} & c_{23} \\ c_{32} & c_{33} \end{bmatrix}$$

$$m_{32} = -\begin{bmatrix} c_{11} & c_{13} \\ c_{21} & c_{23} \end{bmatrix}$$

The rigorous method of calculating a determinant involves expanding the matrix along any row or column. The determinant is found by summing the products of elements in any row or column and each element's minor, thus, assigning the sign according to the position. Det **C** can be obtained by either of the following:

$$\begin{aligned} \det \mathbf{C} &= c_{11}\det m_{11} + c_{12}\det m_{12} + c_{13}\det m_{13} \\ &= c_{12}\det m_{12} + c_{22}\det m_{22} + c_{32}\det m_{32} \end{aligned}$$

Any other expansion along other rows or columns will also work. The usual method is to expand along the first row or first column. The sign associated with each expansion term comes from the position–positive when the subscripts add to an even number, and negative when they add to an odd number.

Determinants of larger matrixes require multiple levels of expansion until the minors are reduced to two or three dimensions. Then, the methods given for calculating the determinants can be used. For example, the determinant of matrix **A** given previously can be found by expanding along the first column.

$$\det \mathbf{A} = a_{11}\det m_{11} + a_{21}\det m_{21} + a_{31}\det m_{31} + a_{41}\det m_{41}$$

$$\det \mathbf{A} = +a_{11}\begin{vmatrix} a_{22} & a_{23} & a_{24} \\ a_{32} & a_{33} & a_{34} \\ a_{42} & a_{43} & a_{44} \end{vmatrix} - a_{21}\begin{vmatrix} a_{12} & a_{13} & a_{14} \\ a_{32} & a_{33} & a_{34} \\ a_{42} & a_{43} & a_{44} \end{vmatrix} + a_{31}\begin{vmatrix} a_{12} & a_{13} & a_{14} \\ a_{22} & a_{23} & a_{24} \\ a_{42} & a_{43} & a_{44} \end{vmatrix} - a_{41}\begin{vmatrix} a_{12} & a_{13} & a_{14} \\ a_{22} & a_{23} & a_{24} \\ a_{32} & a_{33} & a_{34} \end{vmatrix}$$

The resulting 3×3 determinants can be obtained by the methods given above. For larger matrixes, the process becomes sufficiently involved as to warrant the use of a computer.

2 LINEAR SIMULTANEOUS EQUATIONS

System analysis involving linear or piecewise linear elements only will result in simultaneous linear equations with constant coefficients. For example, a set of four equations in four unknowns might be

$$f_1 = a_{11}x_1 + a_{12}x_2 + a_{13}x_3 + a_{14}x_4$$
$$f_2 = a_{21}x_1 + a_{22}x_2 + a_{23}x_3 + a_{24}x_4$$
$$f_3 = a_{31}x_1 + a_{32}x_2 + a_{33}x_3 + a_{34}x_4$$
$$f_4 = a_{41}x_1 + a_{42}x_2 + a_{43}x_3 + a_{44}x_4$$

The fs are known functions, and the x's are variables. This is set into a matrix equation as follows: The fs are formed into a column matrix on the left side of the equation. The a's are extracted into a 4×4 matrix which multiplies a column matrix of the x's.

$$\begin{bmatrix} f_1 \\ f_2 \\ f_3 \\ f_4 \end{bmatrix} = \begin{bmatrix} a_{11} & a_{12} & a_{13} & a_{14} \\ a_{21} & a_{22} & a_{23} & a_{24} \\ a_{31} & a_{32} & a_{33} & a_{34} \\ a_{41} & a_{42} & a_{43} & a_{44} \end{bmatrix} \begin{bmatrix} x_1 \\ x_2 \\ x_3 \\ x_4 \end{bmatrix}$$

The solution for any of the variables (x's) is obtained by *Kramer's rule*. This rule uses determinants as follows:

$$x_j = \frac{f_1\det m_{1j} + f_2\det m_{2j} + f_3\det m_{3j} + f_4\det m_{4j}}{\det A}$$

The determinants of the minors take on the positional signs, and A is the 4×4 matrix of the a terms. This is illustrated with a numerical example involving a 3×3 system of simultaneous equations.

Example 1.21

A system is described by the set of equations given. Determine the values of the variables x, y, and z which simultaneously satisfy the three equations.

$$10 = 3x + y + 2z$$
$$5t = x - 2y + z$$
$$\sin t = -x + y - z$$

These are placed in matrix form.

$$\begin{bmatrix} 10 \\ 5t \\ \sin t \end{bmatrix} = \begin{bmatrix} 3 & 1 & 2 \\ 1 & -2 & 1 \\ -1 & 1 & -1 \end{bmatrix} \begin{bmatrix} x \\ y \\ z \end{bmatrix}$$

Kramer's rule is used to find the simultaneous solution values.

$$x = \frac{+10\begin{vmatrix} -2 & 1 \\ 1 & -1 \end{vmatrix} - 5t\begin{vmatrix} 1 & 2 \\ 1 & -1 \end{vmatrix} + \sin t\begin{vmatrix} 1 & 2 \\ -2 & 1 \end{vmatrix}}{+3\begin{vmatrix} -2 & 1 \\ 1 & -1 \end{vmatrix} - 1\begin{vmatrix} 1 & 2 \\ 1 & -1 \end{vmatrix} + (-1)\begin{vmatrix} 1 & 2 \\ -2 & 1 \end{vmatrix}}$$

$$= \frac{10(2-1) - 5t(-1-2) + \sin t(1+5)}{3(2-1) - 1(-1-2) + (-1)(1+4)}$$

$$= \frac{10 + 15t + 6\sin t}{3 + 3 - 5} = 10 + 15t + 6\sin t$$

$$y = \frac{-10\begin{vmatrix} 1 & 1 \\ -1 & -1 \end{vmatrix} + 5t\begin{vmatrix} 3 & 2 \\ -1 & -1 \end{vmatrix} - \sin t\begin{vmatrix} 3 & 2 \\ 1 & 1 \end{vmatrix}}{\det A}$$

$$= \frac{-10(-1+1) + 5t(-3+2) - \sin t(3-2)}{1}$$

$$= 0 - 5t - \sin t$$

$$z = \frac{+10\begin{vmatrix} 1 & -2 \\ -1 & 1 \end{vmatrix} - 5t\begin{vmatrix} 3 & 1 \\ -1 & 1 \end{vmatrix} + \sin t\begin{vmatrix} 3 & 1 \\ 1 & -2 \end{vmatrix}}{1}$$

$$= \frac{+10(1-2) - 5t(3+1) + \sin t(-6-1)}{1}$$

$$= -10 - 20t - 7\sin t$$

Appendix B

prefix		symbol
tera	10^{12}	T
giga	10^{9}	G
mega	10^{6}	M
kilo	10^{3}	k
hecto	10^{2}	h
deka	10^{1}	da
deci	10^{-1}	d
centi	10^{-2}	c
milli	10^{-3}	m
micro	10^{-6}	μ
nano	10^{-9}	n
pico	10^{-12}	p
femto	10^{-15}	f
atto	10^{-18}	a

Practice Problems: MATHEMATICS

1. Determine the decimal value of the binary number 1101.011.

2. Given $A = 4 + j5$ and $B = 2 - j3$, obtain the product $A \cdot B^*$.

3. Given $A = 4 + j5$, obtain the product $A \cdot A^*$ using (a) polar and (b) rectangular operations.

4. Given $B = 2 - j3$, find the reciprocal B^{-1}.

5. Given $A = 4 + j5$ and $B = 2 - j3$, find the ratio $\frac{A}{B}$.

6. Given $A = 6.4e^{j0.896}$ and $B = 3.6e^{-j0.983}$ find the ratio $\frac{A}{B}$ and compare your answer to problem 5.

7. Obtain the partial fraction expansion of

$$\frac{1}{x(x+1)}$$

8. Obtain the partial fraction expansion of

$$\frac{4x^2 + 5x + 2}{x^2(x+1)}$$

9. Obtain the partial fraction expansion of

$$\frac{5x^2 + 11x + 20}{x(x^2 + 2x + 5)}$$

10. Use long division to reduce the degree of the numerator order for

$$\frac{s^3 + 4s^2 + 6s + 2}{(s+1)^2}$$

11. Use trigonometric identities to obtain expressions for (a) $\cos\frac{\theta}{2}$, and (b) $\sin\frac{\theta}{2}$.

12. Three curves are shown. Obtain straight-line approximations for them over the specified intervals.

- Curve (a) over the interval of $1<I<10$
- Curve (b) over the intervals $0<I<4.5$ and $5.5<I<10$.
- Curve (c) over the intervals $0<I<2$ and $3<I<4.5$.

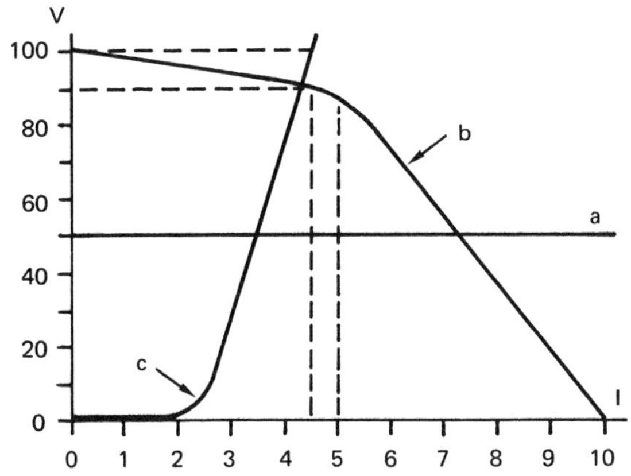

13. Find the local maximum or minimum of

$$u = 2\sin x + 3\cos x$$

in the region of $-90^o < x < 90^o$. Determine whether it is a minimum or a maximum.

14. Determine the average value of the function

$$e^{-2x}\cos\pi x$$

over the interval $x = 0$ to $x = 1$.

15. A voltage is described by $.01\, dv/dt + .0025v = 0$. $v = 100$ at $t = 0$. Determine v for $t > 0$.

16. Given a current described by

$$\frac{d^2i}{dt^2} + 5\frac{di}{dt} + 4i = 2$$

with $i = 1.7$ and $di/dt = 0$ at $t = 0$, determine the current for $t > 0$.

17. Given

$$\frac{d^2i}{dt^2} + 2\frac{di}{dt} + i = .25$$

and conditions at $t = 0$ of $i = .8$ and $di/dt = 0$, determine i for $t > 0$.

18. Given

$$\frac{d^2v}{dt^2} + 3\frac{dv}{dt} + 6.25v + 12.5 = 0$$

and conditions at $t = 0$ of $v = 10$ and $dv/dt = 0$, determine v for $t > 0$.

19. A rectangular wave has voltage levels of 0 and 5 volts, equal times at both levels, and a period of 10 seconds (i.e., 5 seconds at 5 volts and 5 seconds at 0 volts). Obtain a Fourier series for this wave, taking $t = 0$ as the instant that the voltage rises from the zero level to the 5 volt level.

20. Obtain the Laplace transformed equation for the differential equation

$$0.02 = 10^{-8}\frac{d^2 i}{dt^2} + 2 \times 10^{-4}\frac{di}{dt} + i$$

$$i(0) = 0.5$$

$$\frac{di}{dt}(0) = 50$$

21. From the answer to problem 20, obtain $I(s)$ as a function of s.

22. Using partial fraction expansion, separate the function $I(s)$ of problem 21 into its component parts.

23. Using table 1.1, obtain $i(t)$ from your solution of problem 22.

RESERVED FOR FUTURE USE

RESERVED FOR FUTURE USE

2 LINEAR CIRCUIT ANALYSIS

Nomenclature

A	area or chain parameter	meters^2,-
B	chain parameter	-
$\mathbf{B}$	magnetic flux density	teslas
C	capacitance or chain parameter	farads,-
C_{ab}	Seebeck coefficient	-
D	chain parameter	-
$\mathbf{E}$	electric field intensity	volts/meter
$\mathbf{F}$	force	newtons
g	hybrid parameter	-
h	hybrid parameter	-
i	instantaneous current	amps
I	electric current magnitude	amps
k	magnetic coupling coefficient, or constant multiplier	-
l	length	meters
L	inductance	henrys
M	mutual inductance	henrys
n	charge carrier density	meters^{-3}
N	number of turns	-
q	charge	coulombs
r	small signal resistance parameter	ohms
R	resistance	ohms
s	Laplace transform variable	-
t	time	seconds
T	temperature	degrees
v	small signal voltage	volts
$\mathbf{v}$	velocity	meters/second
V	voltage magnitude	volts
w	width	meters
W	energy	joules
y	small signal admittance parameter	mho's
Y	admittance	mho's
z	small signal impedance parameter	ohms
Z	impedance	ohms

Symbols

α	common base transistor current gain	
α_{20}	temperature coefficient of resistivity	1/°C
β	common emitter transistor current gain	-
ϵ	permittivity	farads/meter
μ	permeability	henrys/meter
ω	radian frequency	radians/second
ρ	resistivity	ohm-meters
ϕ	flux	webers

Subscripts

b	transistor base or common base
c	transistor collector, common collector, or cold
C	capacitance
e	transistor emitter or common emitter
f	forward
h	hot
i	input
L	inductance, load
m	maximum
o	output
r	reverse
R	resistance
s	source
Th	Thevenin
x	x-component
y	y-component
z	z-component

1 LINEAR CIRCUIT ELEMENTS

The basic circuit elements are resistance, capacitance, inductance, mutual inductance, and ideal transformers.

A. RESISTANCE

Resistance is the opposition of current flow in a conductor. It is the ratio of voltage to current for constant voltage and current. Resistance of a conductor depends on the *resistivity* ρ of the conductor material. However, the resistivity is a function of temperature and has been found to vary approximately linearly with temperature. Resistivity is usually referenced to 20°C and is written as ρ_{20}. ρ_{20} has values of 1.8×10^{-8} ohm-meters for copper and 1.08×10^{-6} ohm-meters for nichrome. The *temperature coefficient of resistivity* is also measured at 20°C and is given the name α_{20}. It has values of 3.9×10^{-3} per degree for copper and 17×10^{-3} per degree for nichrome. The expression for the resistivity at a temperature T other than 20°C is

$$\rho_T = \rho_{20}[1 + \alpha_{20}(T - 20)] \qquad 2.1$$

The resistance of a conductor of length l and cross-sectional area A is

$$R = \frac{\rho_T \times l}{A} \qquad 2.2$$

Ohm's law relates the voltage and current in a circuit to the circuit resistance.

$$V = IR \qquad 2.3$$

The *power* dissipated by a resistor is

$$P = I^2R = \frac{V^2}{R} \qquad 2.4$$

Resistances in series add linearly, and resistances in parallel add inversely. This second property gives rise to the use of the parameter *conductance* which is the reciprocal of resistance. Conductances in parallel add linearly.

$$R_{series} = R_1 + R_2 + R_3 + \cdots \qquad 2.5$$

$$\frac{1}{R_{parallel}} = \frac{1}{R_1} + \frac{1}{R_2} + \frac{1}{R_3} + \cdots \qquad 2.6$$

$$G = \frac{1}{R} \qquad 2.7$$

$$G_{parallel} = G_1 + G_2 + G_3 + \cdots \qquad 2.8$$

Example 2.1

A resistor constructed from AWG (American Wire Gage) #30 copper wire is to have a resistance of one ohm at 100°C. Determine the length of wire necessary to obtain this resistance. What will be the resistance of the wire at 20°?

From Appendix 1, AWG #30 wire has a cross-sectional area of 100 circular mils. *Circular measure* is the diameter squared, so the diameter is 10 mils (.01 inches). In SI units, the diameter is 2.54×10^{-4} meters and the cross-sectional area is $5.07 \times 10^{-8} m^2$.

$$\rho_{100} = 1.8 \times 10^{-8}[1 + 3.9 \times 10^{-3}(100 - 20)]$$
$$= 2.36 \times 10^{-8} \text{ohm-m}$$
$$l = \frac{R \times A}{\rho_{100}} = \frac{1.0 \times 5.07 \times 10^{-8}}{2.36 \times 10^{-8}}$$
$$= 2.15 \text{ m}$$
$$R_{20} = R_{100} \times \frac{\rho_{20}}{\rho_{100}} = 1 \times \frac{1.8}{2.36}$$
$$= .763 \text{ ohms}$$

B. CAPACITANCE

Capacitance is the ability to store electrical charge. It is calculated for parallel plates as the *permittivity* (ϵ) multiplied by the surface area of the plates and divided by the plate separation. The permittivity is the product of the *permittivity of free space* (ϵ_0) and the dielectric constant (ϵ_r) of the material separating the parallel plates. The permittivity of free space has a value of $\frac{1}{36\pi \times 10^9}$. The dielectric constant for linear materials varies between 1 and 10.

$$\epsilon = \epsilon_0 \epsilon_r \qquad 2.9$$

The current into the positive terminal of a capacitor is the time rate of increase in stored charge. If the current direction is chosen as positive when flowing into the positive terminal of the capacitor as shown in figure 2.1, the voltage-current relationship for a capacitor is

$$i = C\frac{dv_C}{dt} \qquad 2.10$$

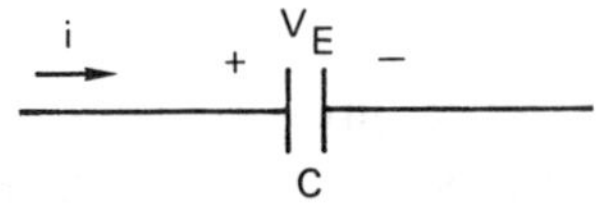

Figure 2.1. Voltage polarity and current directions for capacitors

The energy stored in a capacitor is

$$W = \frac{1}{2}Cv_C^2 \qquad 2.11$$

As energy cannot change instantaneously, the voltage cannot change instantaneously either. This fact is of major importance in determining the transient behavior of circuits containing capacitors.

Example 2.2

A voltage source is connected to a parallel combination of a resistance and a capacitance as shown. Assume that the switch has been open for a long time so that the capacitor has no charge on it prior to $t = 0$ when the switch is closed. Determine (a) the current through the capacitor at the instant the switch is closed, (b) the time derivative of the capacitor voltage at the instant the switch is closed, and (c) the voltage that will exist across the capacitor after a long time.

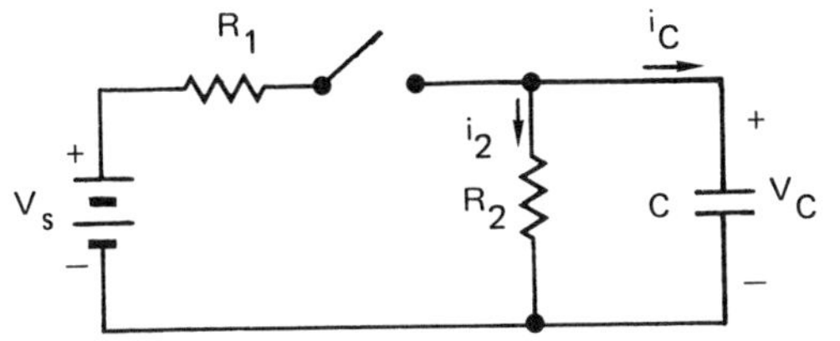

As the capacitor is uncharged at the instant before the switch closes ($t = 0^-$), it must also be uncharged at the instant after the switch closes ($t = 0^+$). Thus the voltage across the capacitance remains zero. This causes the capacitor to act as a short circuit across R_2. All of the current flows initially through the capacitor instantaneously. This current is

$$i_C(0^+) = \frac{V_S}{R_1}$$

Since $i = C\frac{dv}{dt}$,

$$\frac{dv(0^+)}{dt} = \frac{V_S}{R_1 C}$$

The capacitance will become fully charged after a long time, and the voltage across it will no longer be changing. Thus $\frac{dv}{dt}$ and i_C are zero. All the current flows through R_2, and the capacitor acts as an open circuit to the current. The current through R_2 is

$$\frac{V_S}{R_1 + R_2}$$

The voltage across the capacitor is the same as across R_2.

$$V_C(\infty) = i_2 R_2 = V_S \frac{R_2}{R_1 + R_2}$$

C. INDUCTANCE

Inductance is a property which resists *changes* in current. Inductance is measured in units of *henrys*. For a closed magnetic circuit such as the toroid shown in figure 2.2, the inductance is

$$L = \mu \frac{N^2 A}{l} \qquad 2.12$$

μ is the *permeability* of the medium, N is the number of turns on the winding, A is the core cross-section area, and l is the average flux path length.

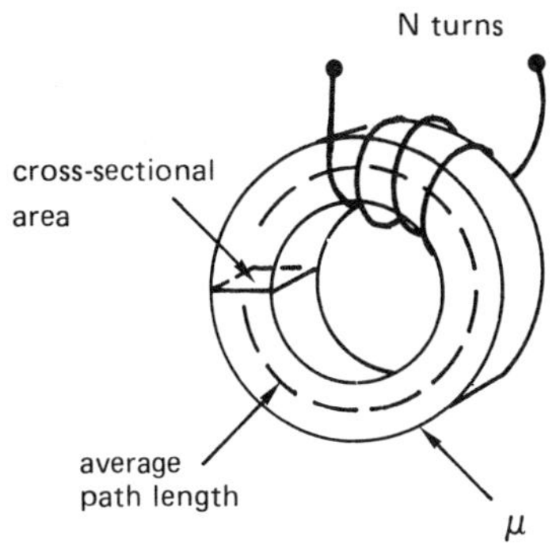

Figure 2.2 Inductance of a toroid with N turns

The general voltage-current relationship is

$$V = L\frac{di}{dt} \qquad 2.13$$

The energy stored in an inductance is

$$W = \frac{Li^2}{2} \qquad 2.14$$

Energy is unable to change instantaneously and, consequently, current is also unable to change instantaneously. This concept is used to determine the transient behavior of circuits containing inductors.

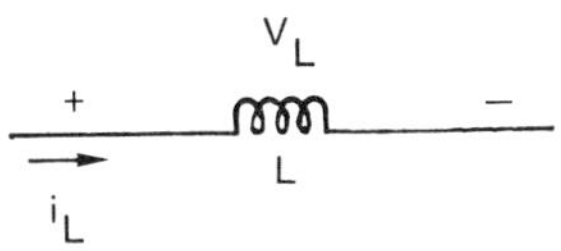

Figure 2.3 Defined voltage and current directions for an inductance

Example 2.3

A voltage source is connected to a parallel combination of resistance and inductance by a switch which closes at $t = 0$. Assume that the switch has been open for a long time so no energy is stored in the inductance just prior to the closing of the switch. Determine (a) the inductance current, (b) the inductance voltage, and (c), the inductance time rate of change of current immediately after the switch closing. Determine (d) the current in the inductance, and (e) the voltage across the inductance after a long period of time.

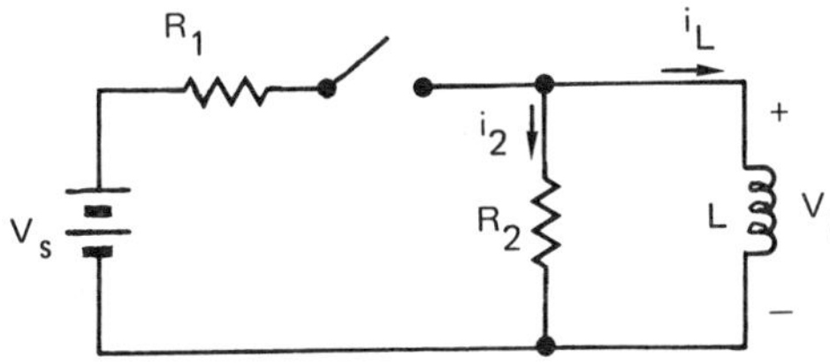

As there is no stored energy just prior to the switch closing, there is no current flowing at that time, nor is any current flowing at the instant the switch is closed. Therefore,

$$i_L(0^+) = 0$$

The inductance acts as an open circuit at the instant the switch closes, so the current is all shunted through R_2. The current is then

$$\frac{V_S}{R_1 + R_2}$$

The voltage across R_2 is the current times its resistance. The same voltage appears across the inductance

$$v_L(0^+) = V_S \frac{R_2}{R_1 + R_2}$$

As $v_L = L\frac{di}{dt}$, the rate of current change in the inductor is

$$\frac{di_L(0^+)}{dt} = \left(\frac{V_S}{L}\right)\left(\frac{R_2}{R_1 + R_2}\right)$$

After a long time the current will settle down to a steady value, so $\frac{di_L}{dt}$ is zero. There will be no voltage drop across the inductance, and the inductance will behave as a short circuit across R_2. All the current will then flow through the inductance.

$$i_L(\infty) = \frac{V_S}{R_1}$$
$$v_L(\infty) = 0$$

Mutual inductance occurs when two or more coils share a magnetic flux. This is the case when two or more wires are wound on the same magnetic core. The reference direction of the flux is arbitrarily chosen, and each winding is *phased* by determining the current direction which will cause flux in the reference direction. Then a symbol (e.g., a dot) is placed on all terminals which will cause positive flux when a current flows into that terminal.

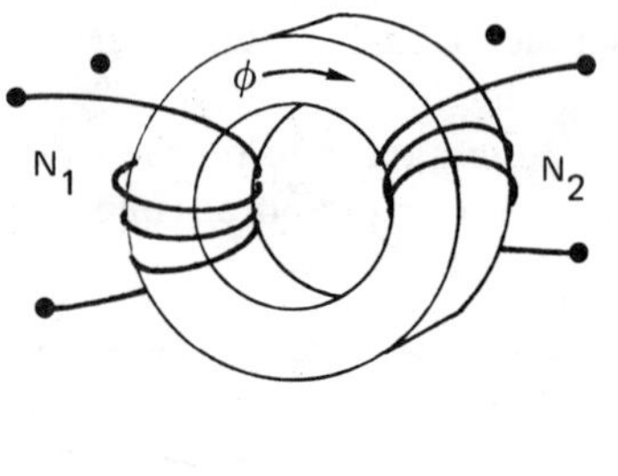

Figure 2.4. Mutually coupled coils having mutual inductance

If the device is shown in a schematic drawing as in figure 2.4, the current and voltage directions are given reference according to the dots: positive currents go into the dotted end of each coil, and positive voltages appear at the dotted terminal. Following the defined directions strictly, the voltage-current relationships are

$$v_1 = L_1 \frac{di_1}{dt} + M \frac{di_2}{dt} \qquad 2.15$$

$$v_2 = M \frac{di_1}{dt} + L_2 \frac{di_2}{dt} \qquad 2.16$$

In a system involving mutual inductance, M, the total stored energy is

$$W = \frac{1}{2} L_1 i_1^2 + \frac{1}{2} L_2 i_2^2 + M i_1 i_2 \qquad 2.17$$

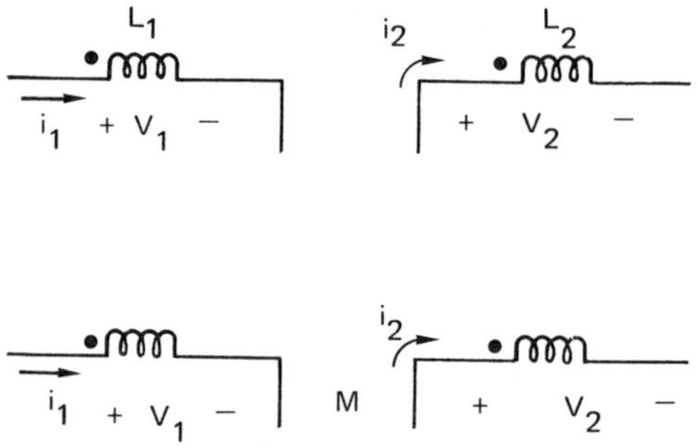

Figure 2.5 Circuit symbols and directions for mutually coupled coils

The energy cannot change instantaneously, but it is permissible for both currents to change abruptly. The mutual inductance is always less than the geometric mean of the two self-inductances L_1 and L_2 because the mutual flux is always less than the total flux. The ratio of the mutual inductance to the geometric mean is called the *coupling coefficient, k.*

$$k = \frac{M}{\sqrt{L_1 L_2}} \qquad 2.18$$

An *ideal transformer* is a simplified model of coupled coils assuming $k = 1$ and a high permeability to make the inductances insignificant in the circuit calculations. All flux from one coil is assumed to link all the turns of the other coil so that the voltage per turn of each coil is identical at each instant in time.

$$\frac{v_1}{N_1} = \frac{v_2}{N_2} = \frac{v_3}{N_3} = \cdots \qquad 2.19$$

Also, the algebraic sum of the *magnetomotive force* terms is zero. (Positive currents are defined by the dots.)

$$N_1 i_1 + N_2 i_2 + N_3 i_3 + \cdots = 0 \qquad 2.20$$

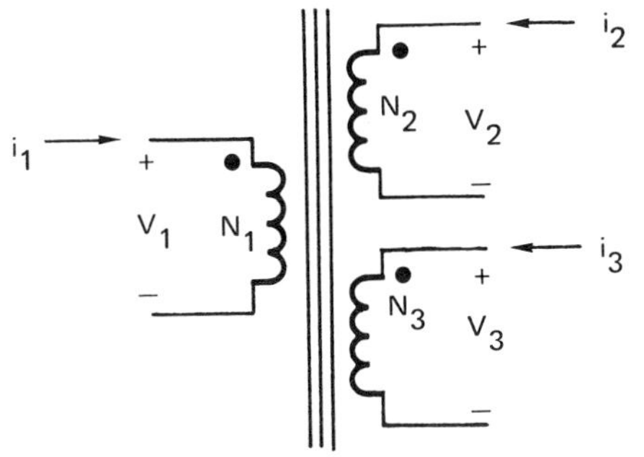

Figure 2.6 Circuit symbols and directions for an ideal transformer

An ideal transformer is lossless and stores no energy. Its functions are to change voltage levels and match impedance in circuits, as well as to provide equal and opposite waveforms in some applications. The ideal transformer model adequately describes the behavior of a transformer in many circuits. This is not true in power circuit problems where efficiency, regulation, or power factor are being determined.

Example 2.4

A transformer consists of a primary winding with 500 turns and two secondary windings of 125 turns and 36 turns. The 125 turn secondary winding has 60 ohms connected to its terminals, and the 36 turn secondary winding has 3 ohms connected to its terminals. If the primary winding is connected to a 120 volt 60 Hz source, determine the current rating for each winding.

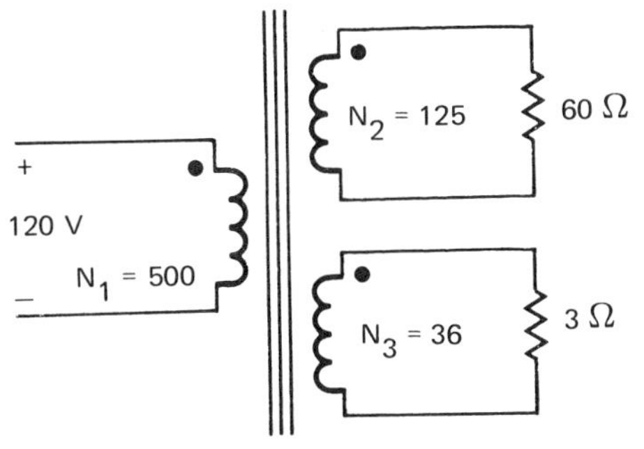

From equation 2.19, the secondary voltages are

$$v_2 = \frac{N_2 v_1}{N_1} = \frac{125 \times 120}{500}$$
$$= 30 \text{ volts}$$
$$v_3 = \frac{N_3 v_1}{N_1} = \frac{36 \times 120}{500}$$
$$= 8.64 \text{ volts}$$

The currents in the secondaries can then be found from the resistances.

$$i_2 = -\frac{v_2}{R_2} = \frac{-30}{60}$$
$$= -0.5 \text{ amps}$$
$$i_3 = -\frac{v_3}{R_3} = \frac{-8.64}{3}$$
$$= -2.88 \text{ amps}$$

The negative signs on the currents are necessary because of the defined current directions (i.e., into the dotted ends of the coils).

From equation 2.20, the primary current is

$$N_1 i_1 + N_2 i_2 + N_3 i_3 = 500 \times i_1 - 125 \times 0.5 - 36 \times 2.88$$
$$= 0$$
$$i_1 = .332 \text{ amps}$$

The minimum current ratings are

332 mA	for winding #1
500 mA	for winding #2
2.88 A	for winding #3

2 LINEAR SOURCE MODELS

Electrical power sources are typically nonlinear, but for purposes of circuit analysis they are modeled as linear devices. The model includes a constant or ideal voltage or current source in combination with a linear impedance. Such sources are classified as *independent sources* to distinguish them from the *dependent sources* of active devices which are not actually sources of power.

A third source type is the *transducer* which produces a voltage or current signal dependent on non-electric quantities. *Thermocouples* are transducers which produce voltages as functions of temperature; *photodiodes* are transducers which provide currents according to the level of illumination; *microphones* are transducers which provide voltages as functions of sound intensity.

Independent sources may be modeled by a linear approximation to measured values. For example, a voltage-current characteristic curve for a direct current source (e.g., a battery or generator) is shown in example 2.5. A straight line approximation to this characteristic curve can be obtained by analytic geometry.

Example 2.5

Obtain an expression for the characteristic curve shown over the range from 8 to 12 amps. Obtain linear circuit models from the derived expressions.

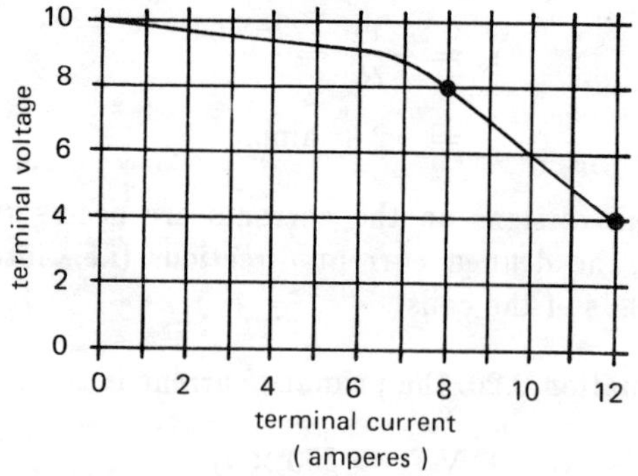

The two-point method is used by taking the pairs (8,8) and (4,12).

$$\frac{I-8}{V-8} = \frac{12-8}{4-8} = -1$$

Either voltage or current can be chosen as the dependent variable.

$$V = -I + 16 \qquad (A)$$

$$I = -V + 16 \qquad (B)$$

Equation A is a loop voltage equation where the terminal voltage V is the sum of two terms, one independent of the current and the other dependent on the current. The dependent term is modeled as a 1 ohm resistance in series with the 16 volt source. Equation B is a node current equation where the terminal current I is the difference in two terms. One term is constant at 16 amps and is modeled as a 16 amp source. The other is dependent on the terminal voltage V and is modeled as a shunt resistance of 1 ohm across the terminals.

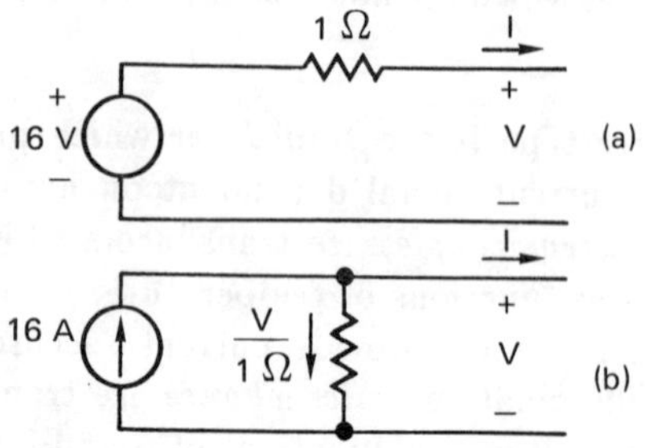

Source characteristics are sometimes given in terms of *open circuit* or *no-load* voltage and the current with some known load resistance or impedance. These two parameters can be used to establish a linear model.

Example 2.6

A d.c. generator has a no-load voltage of 125 volts and a full-load voltage of 120 volts at its rated current of 20 amps. Obtain a linear circuit model for the generator.

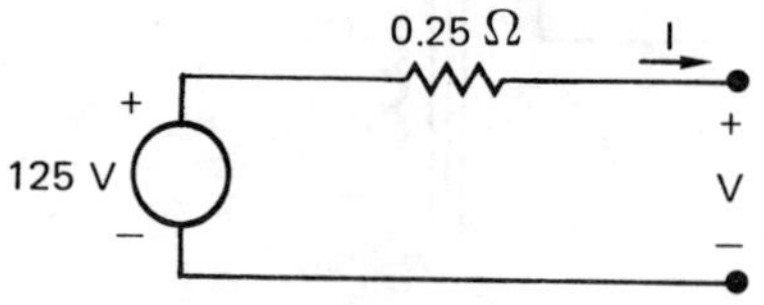

The two current-voltage pairs are (0,125) and (20,120).

$$\frac{I-0}{V-125} = \frac{20-0}{120-125} = -4$$

Solving for V as a function of I produces

$$V = 125 - .25I$$

The generator can be modeled as a voltage source of 125 volts in series with a resistance of .25 ohms.

The percentage change of voltage from the no-load to full-load condition is called the *voltage regulation*. This is a generator parameter often given together with full-load voltage and current.

$$\text{voltage regulation} = \frac{V_{\text{no-load}} - V_{\text{full-load}}}{V_{\text{full-load}}} \qquad 2.21$$

Example 2.7

Find the percentage regulation for the source of example 2.6.

$$\text{regulation} = \frac{125-120}{120} \times 100 = 4.17\%$$

Piecewise linear models of nonlinear elements can be obtained by a method similar to that used in obtaining linear models of power sources. However, the results are only applicable over a limited range which may not be known in advance. Furthermore, in many cases the working of a circuit depends on transitions from one nonlinear region to another. Power sources, on the other hand, usually remain within the linear range described by the equivalent circuit.

Example 2.8

A typical silicon diode is approximated by a *forward voltage drop* of 0.6 volts at negligible current and 1.5

volts at the rated current. Obtain an equivalent circuit for a silicon diode rated 10 watts at 5 amps.

Power is $V_{rated} \times I_{rated}$, so the voltage at rated current is $V_{rated} = \frac{10}{5} = 2$ volts. Then the two (V,I) pairs are (.6,0) and (2,5). The two-point equation is

$$\frac{V-.6}{I-0} = \frac{2-.6}{5-0} = .28$$

Solving for V yields

$$V = .6 + .28I$$

The equivalent circuit is shown. This model is valid only when the current is positive and the voltage exceeds .6 volts.

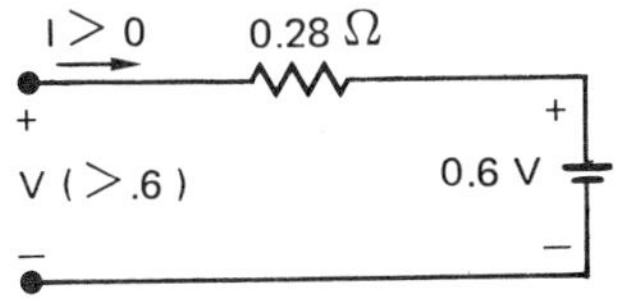

A *zener diode* is typically a silicon diode operated in the reverse direction. It essentially operates as an open circuit for low reverse voltages (defined as the positive direction) and has a voltage known as the *zener* or *avalanche voltage* at which the current abruptly increases when further voltage is applied. Zener diode characteristices may be specified graphically but are more frequently specified by the breakdown (zener) voltage and by the rated current and power.

Example 2.9

Obtain an equivalent circuit for a 6-volt zener diode with a rated current of .75 amps and rated power of 5 watts.

Taking the *reverse voltage* and *reverse current* as positive, the voltage at rated power is

$$V_{rated} = \frac{5}{.75} = 6.67 \text{ volts}$$

The two (V, I) pairs are then (6,0) and (6.67,.75). The two-point method then yields

$$\frac{V-6}{I-0} = \frac{6.67-6}{.75-0} = .889$$

The V-I equation is

$$V = 6 + .889I$$

The equivalent circuit is shown.

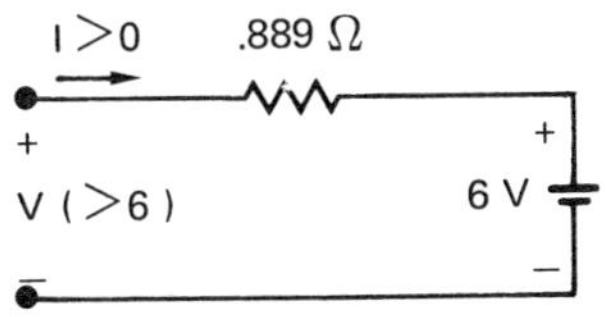

It should be noted that the equivalent circuits for nonlinear circuit elements are different from circuits for power sources in the assigned current directions and are, therefore, drawn in a different manner.

Dependent or *controlled sources* which depend on circuit quantities (e.g., voltage or current) are necessary in the analysis of linear *active circuits* such as transistor amplifiers. Controlled sources are typically devices which have three or more terminals. An example is the two-winding transformer modeled by equations 2.15 and 2.16. These equations can be treated as two loop equations, which then produce an equivalent circuit as shown in figure 2.7. The voltage generated in loop 1 is due to the current in loop 2, and the voltage generated in loop 2 is due to the current in loop 1. Such a circuit is said to be *bi-lateral.*

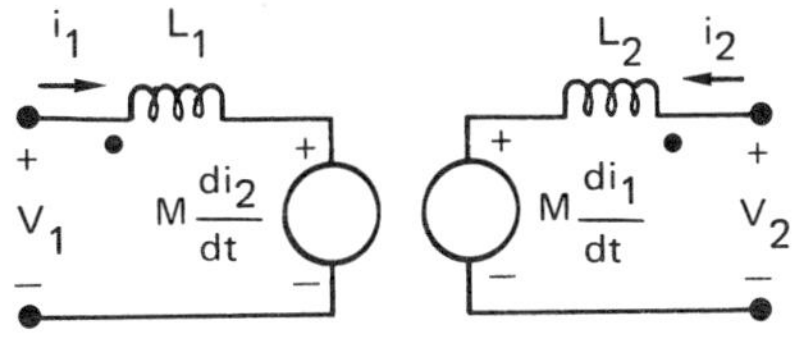

Figure 2.7 Equivalent circuit model for a mutually coupled circuit using controlled sources

It is unusual to represent circuits with controlled sources as shown in figure 2.7. It is simpler to represent them as coupled circuits with the coupling included as common circuit elements. The most common instance is of *unilateral devices* (e.g., transistors) in which the collector circuit only mildly affects the base circuit, while the behavior of the collector circuit is principally controlled by the base circuit. In these cases, the circuit can be described by a linear relationship in matrix form. The matrix terms can be impedance, admittance, or mixed parameters.

A *three-terminal device* can be described by the matrix equation 2.22. The voltages are measured with respect to the third terminal.

$$\begin{bmatrix} v_1 \\ v_2 \end{bmatrix} = \begin{bmatrix} z_i & z_r \\ z_f & z_o \end{bmatrix} \times \begin{bmatrix} i_1 \\ i_2 \end{bmatrix} \qquad 2.22$$

The subscripts on the *z-parameters* stand for *input, reverse, forward,* and *output.* The convention is that terminal 1 is the input and terminal 2 is the output. Equation 2.22 is a voltage relationship. It can be implemented in two series circuits with an impedance and a current controlled voltage source as shown in figure 2.8.

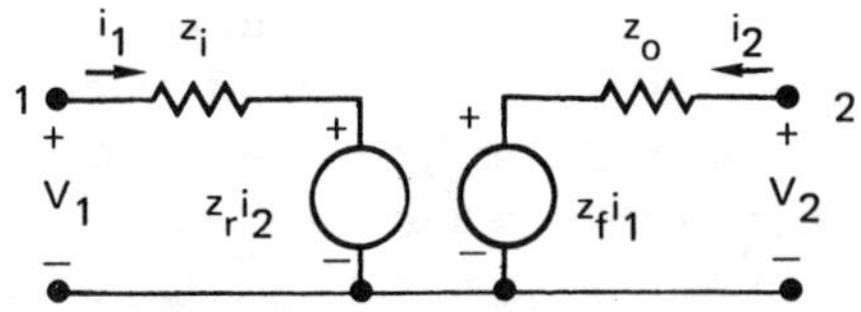

Figure 2.8. Equivalent circuit of controlled voltage sources

The z-parameters result in current-controlled voltage sources in series with impedances. Equation 2.22 can also be written in terms of currents, which then results in voltage-controlled current sources in parallel with admittances. This is the *y-parameter* formulation.

Example 2.10

A three-terminal device has the following y-parameters: $y_i = .02$, $y_r = -.0002$, $y_f = .95$, $y_o = .0001$. Obtain an equivalent circuit for this device.

The y-parameter equations can be written as equation 2.22. Voltages and currents are interchanged, and the y-parameters are substituted for the z-parameters with the same subscripts.

$$i_1 = .02v_1 - .0002v_2$$

$$i_2 = .95v_1 + .0001v_2$$

These are node current equations. Each equation defines the connections to one node, and the voltages are measured with respect to the reference node. The equivalent circuit is shown. Admittances are in *mhos.*

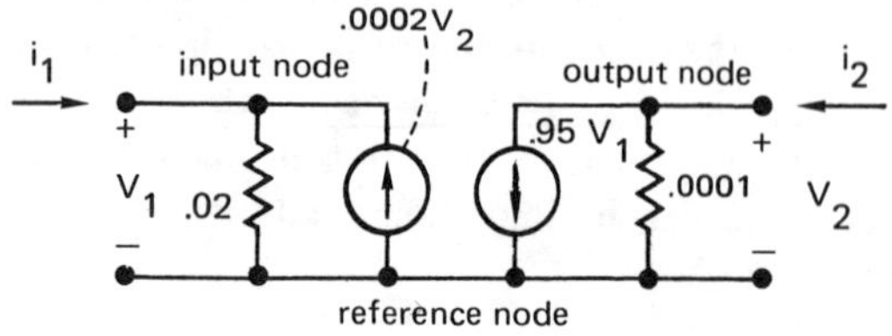

Junction transistors are usually specified by a mixed set of small-signal parameters. These are called *hybrid parameters* because the equivalent circuit consists of a series input circuit and a parallel output circuit. Since the hybrid parameters are usually associated with a common emitter configuration, they carry an additonal subscript *e*. The parameters are h_{ie}, h_{re}, h_{fe}, and h_{oe}. h_{re} is typically negligible and is ignored. The useful equations are

$$v_{be} = h_{ie} i_b \tag{2.23}$$

$$i_c = h_{fe} i_b + h_{oe} v_{ce} \tag{2.24}$$

The equivalent circuit corresponding to equations 2.23 and 2.24 is shown in figure 2.9. This circuit is the most useful of all transistor equivalent circuits. It will be used extensively in the chapter covering linear electronics.

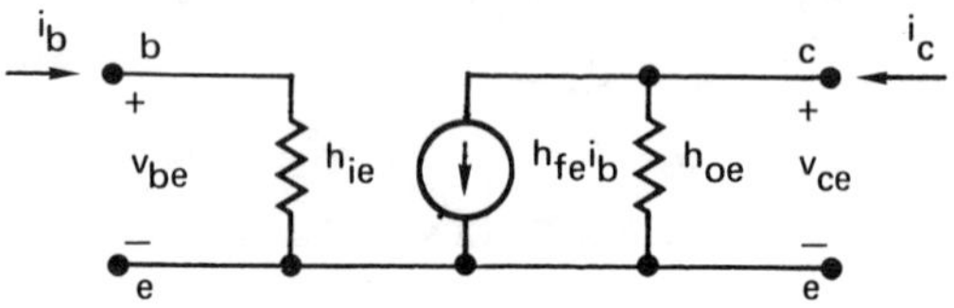

Figure 2.9 Hybrid equivalent circuit for a junction transistor in the common emmiter configuration

Transducers can usually be treated as controlled sources, with the controlling variable being the non-electric quantity converted to an electric signal. Such devices have an output circuit similar to those of three-terminal devices, but a circuit is not usually included in the input.

A silicon *photo-diode* is constructed so that the PN junction is exposed to visible light. Typical characteristic curves of the photoelectric effect are shown in figure 2.10. The curves show that with the diode reverse biased, the current will be roughly proportional to the illumination. The current is approximately constant for any voltage above a certain level. Applying a high enough voltage, therefore, will cause the photodiode to behave as an essentially linear controlled current source.

In cases where the curves are not available, it is usually adequate to use a single point (I_{light}, i) where I_{light} is the illumination in footcandles (fc) and i is the current. The *dark current* (the current for no illumination) is assumed to be zero. In cases where this is not a reasonable assumption, the dark current must be known. In cases of high illumination levels, it is possible to ignore the constant current generator i_0.

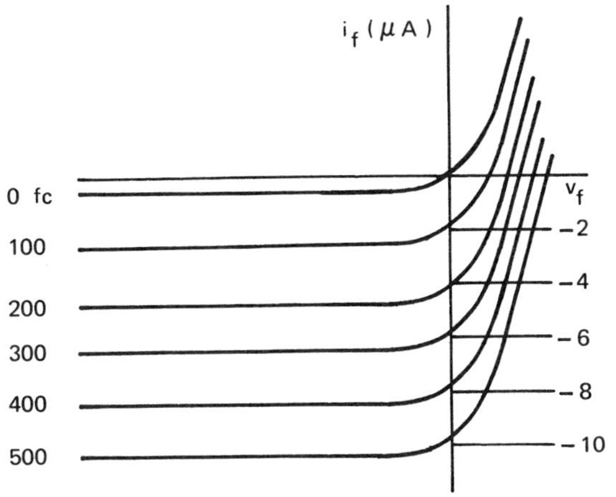

Figure 2.10 Typical photodiode characteristics

Example 2.11

For the curves shown in figure 2.10, construct the equivalent circuit.

The dark current is about $0.7\mu A$, and the current is about $10.3\mu A$ at 500 fc. The two-point method is used to find an equation of current versus illumination.

$$\frac{i - .7}{I_L - 0} = \frac{10.3 - .7}{500 - 0} = .0192$$

$$i = .7 + .0192 I_{light}$$

The equivalent circuit is shown. The biasing is also shown, but it is not necessary in the equivalent circuit as the voltage is in series with a current source.

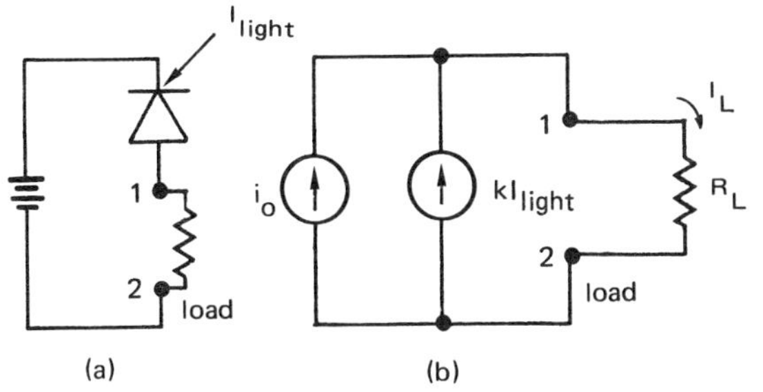

(a) Biasing circuit for photo-diode of figure 2.10
(b) equivalent circuit.

A *phototransistor* has characteristics similar to the photo-diode, but it has a much larger gain (k) because of the current gain of the transistor, β. Thus, the values on the current axis would be approximately β times those on figure 2.10.

Diode characteristics are dependent on temperature, as indicated in figure 2.11. For a temperature-dependent device, behavior can be modeled by a circuit with a temperature dependent voltage source. To model such a device, it is necessary to obtain the voltage-temperature pairs to generate an equation. Then, the voltage is modeled as two sources, one dependent and the other independent of the temperature.

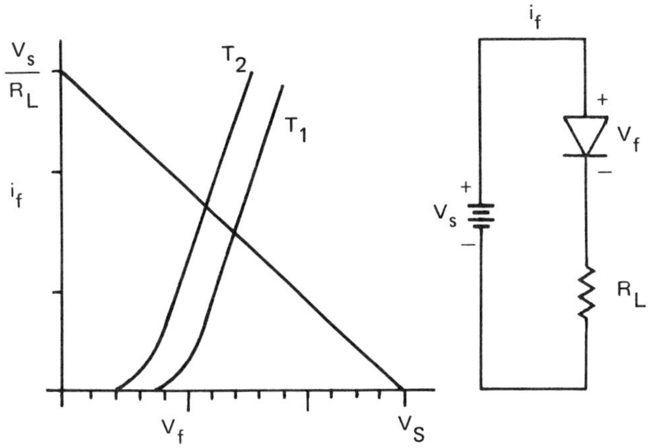

Figure 2.11 Temperature characteristics for a diode

Example 2.12

The characteristics shown in figure 2.11 have voltage divisions of .2 volts and current divisions of 1 amp. Obtain an equivalent circuit for a diode which accounts for the temperature dependence with $T_1 = 20°C$ and $T_2 = 80°C$.

At T_1 the (v_f, i_f) points are (.95, 0) and (1.4, 1.5). The resulting equation is

$$v_{f1} = .95 + .3 i_f$$

As the intercept of the straight line approximation at T_2 is $v_f = .6$, and the slope is essentially the same, the voltage is

$$v_{f2} = .6 + .3 i_f$$

The only difference in the two voltages is the intercept which can be paired with the temperatures as (.95, 20°) and (.6, 80°). Then the two point method gives

$$\frac{V_0 - .95}{T - 20} = \frac{.6 - .95}{80 - 20} = -.0058$$

$$v_f = 1.06 - .0058T + .3 i_f$$

The equivalent circuit is shown.

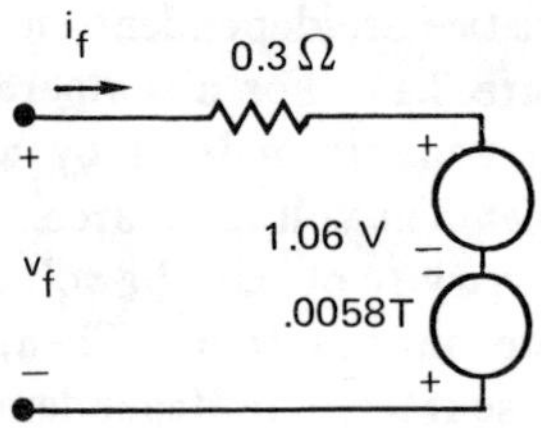

Thermocouples generate a voltage dependent on the difference in temperatures between two junctions of dissimilar metals. This is known as the *Seebeck effect* and is a nonlinear function of the form

$$V_{ab} = C_{ab}(T_h - T_c) \tag{2.25}$$

V_{ab} is the voltage generated by the temperature difference between the hot and cold junctions, C_{ab} is the *Seebeck coefficient*, T_h is the hotter junction's temperature, and T_c is the colder junction's temperature. Equation 2.25 is linear, but the coefficient C_{ab} is nonlinear as indicated in figure 2.12.

Such devices can be considered linear only over the range of temperatures where the coefficient is flat (i.e., from 50° to 100°C). Over that range, the voltage is accurately predicted by equation 2.25. The equivalent circuit is shown in figure 2.13. The resistance of the wires and junctions is assumed negligible.

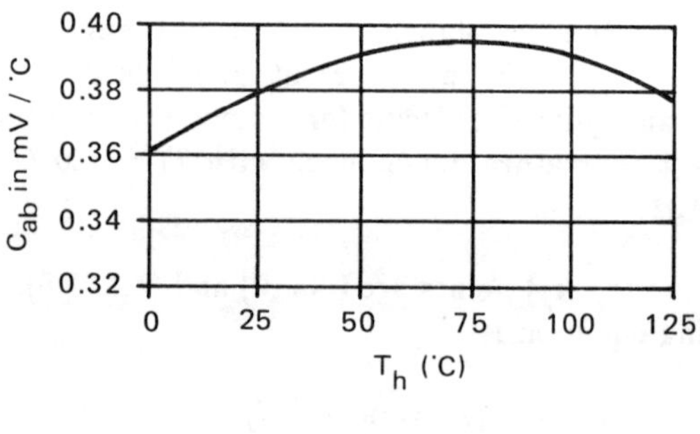

Figure 2.12 Seebeck coefficient versus temperature

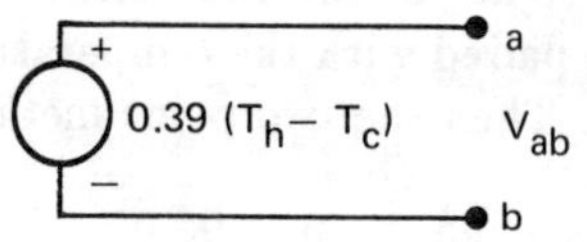

Figure 2.13 Equivalent circuit for thermocouples operating with hot temperature from 50° to 100°C

The *Hall effect* is a consequence of the *Lorentz force equation*

$$\mathbf{F} = q(\mathbf{E} + \mathbf{v} \times \mathbf{B}) \tag{2.26}$$

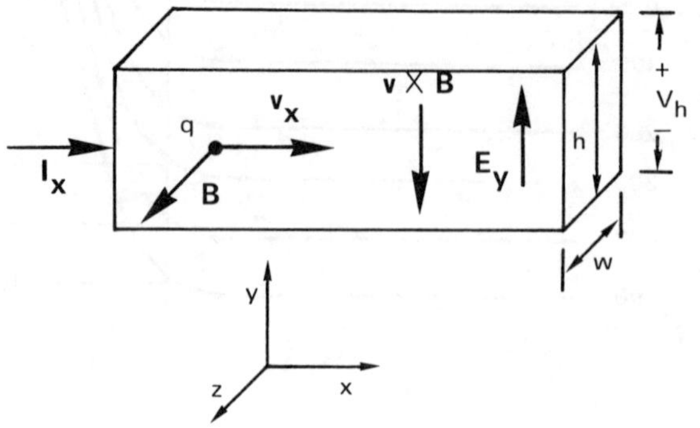

Figure 2.14 Geometry for the calculation of the Hall voltage in a conductor

F is the force on a charge q moving at velocity **v** in an electric field **E** and magnetic field with flux density **B** (**F**, **E**, **v**, and **B** are vectors). In a conductor with the current flow in the x-direction and with the magnetic flux density in the z-direction as shown in figure 2.14, the force is zero. The electric field must be in the direction shown.

$$-E_y = v_x B_z \tag{2.27}$$

The voltage drop from the top to the bottom of the conductor is $E_y h$. The current flowing is $nqv_x A$, where n is the number of charge carriers per unit volume, q is the charge per carrier, and A is the cross-sectional area of the conductor wh. Then, the voltage generated is

$$V_{\text{Hall}} = \left(\frac{I}{W}\right)\left(\frac{1}{nq}\right)B_z \tag{2.28}$$

Example 2.13

Copper has an electron density of 10^{28} electrons per cubic meter. The charge on each electron is 1.6×10^{-19} coulombs. For a copper conductor with width 5 mm, height 5 mm, and length 10 cm, determine the Hall voltage for a current of 1 amp when the magnetic field intensity is $1 \frac{\text{amp}}{m}$. $B = \mu_0 H$, and $\mu_0 = 4\pi \times 10^{-7}$.

With $H = 1 \frac{\text{amp}}{m}$, $B = 4\pi \times 10^{-7}$.

$$V_{\text{Hall}} = \frac{1 \times 4\pi \times 10^{-7}}{(5 \times 10^{-3})(10^{28})(1.6 \times 10^{-19})}$$

$$= 1.57 \times 10^{-13} \text{volts}$$

3 TWO TERMINAL NETWORKS

Many system design problems include fixed portions of networks which are to be incorporated but not changed. Each fixed portion can usually be described by the characteristics seen from its terminals. The fixed portions of the networks can be simplified to Thevenin or Norton equivalents. This simplification is always possible when the system is linear or piecewise linear.

Thevenin's theorem states that any circuit composed of linear impedances, power sources, and linear controlled sources can be replaced with a single voltage source and a series impedance. The resulting voltage source is called the *Thevenin equivalent voltage source.* The resulting impedance is called the *Thevenin equivalent impedance.*

The Thevenin equivalent circuit is found in the following manner.

- *Step 1*: With the terminals of the circuit open circuited, determine the voltage across the terminals.
- *Step 2*: With the terminals short-circuited, determine the short-circuit current.

The open-circuit voltage is the Thevenin equivalent voltage source, and the ratio of the open-circuit voltage to the short-circuit current is the Thevenin equivalent impedance.

A short cut to the Thevenin equivalent voltage is to find the impedance looking into the terminals with all power sources altered: voltage sources are short-circuited and current sources are open-circuited. In this process, all controlled sources are retained as is.

Norton's theorem is the dual of Thevenin's theorem. Under the same linear constraints, the circuit can be replaced by a Norton equivalent current source and an equivalent impedance in parallel with the current source. The *Norton equivalent current* is the short circuit current. The impedance is the same as found in the Thevenin equivalent circuit.

Example 2.14

A simplified small-signal amplifier circuit is shown. The voltage source v_{in} is a variable signal source and can be considered to be an ideal voltage source. The transistor has parameters $h_{ie} = 500$, $h_{re} = 0$, $h_{fe} = 50$, and $h_{oe} = .0001$. Determine the Thevenin equivalent circuit (as seen from terminals a and b) in terms of the unknown turns N_1 and N_2 on the right-hand transformer.

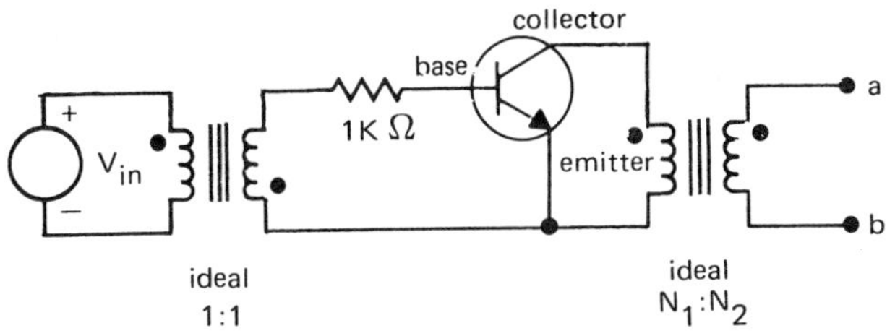

Method 1: The straightforward method involves finding the open-circuit voltage and the short-circuit current. The ideal transformer relations are

$$\frac{V_1}{N_1} = \frac{V_2}{N_2}$$

$$N_1 i_1 + N_2 i_2 = 0$$

All quantities are referred to the dots as previously discussed. For the input transformer, the voltage at the dot on the secondary is the same as at the dot on the primary with a phase reversal. The current is driven into the dot on the primary side, so the current is out of the dot on the secondary side.

The equivalent transistor model contains a resistance of 500 ohms for h_{ie}, a controlled current source $50i_b$ for h_{fe} and a parallel resistance of 10K ohms for h_{oe}.

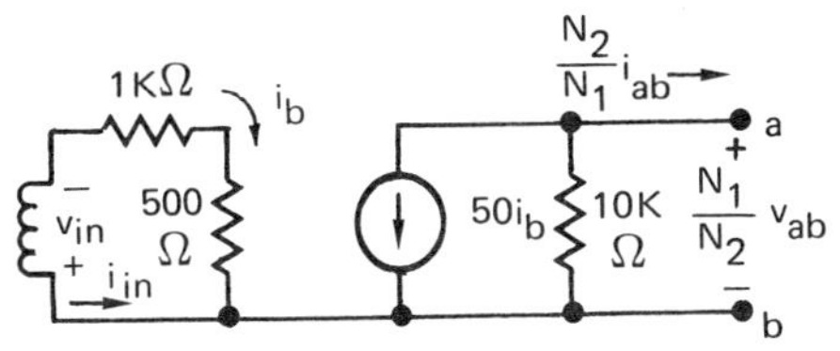

First, determine the open circuit voltage across the output terminals. $i_{ab} = 0$, and all currents from the current source go through the 10K resistor.

$$\frac{N_1 v_{ab}}{N_2} = -50 i_b \times 10^4$$

From the input loop, it is apparent that

$$i_b = \frac{-v_{in}}{1500}$$

Then, by substituting this into the previous expression containing i_b, the open circuit voltage is

$$v_{ab}\Big|_{\text{o.c.}} = \left(\frac{N_2}{N_1}\right) 333.3 v_{\text{in}}$$

In the short-circuit condition, v_{ab} is zero. The 10K collector resistor is shorted out causing all of the current source current to flow through the short circuit.

The short-circuit current in the primary of the output transformer is

$$\left(\frac{N_2}{N_1}\right) i_{ab}\big|_{s.c.} = -50 i_b$$

Since i_b has been determined to be $\frac{-v_{in}}{1500}$,

$$i_{ab}\big|_{s.c.} = \left(\frac{N_1}{N_2}\right)\frac{v_{in}}{30}$$

The Thevenin equivalent impedance is found by taking the ratio of the open-circuit voltage to the short-circuit current.

$$Z_{\text{Thevenin}} = 10^4 \left(\frac{N_2}{N_1}\right)^2$$

The Thevenin equivalent voltage source is $v_{ab}\big|_{o.c.}$

$$v_{\text{Thevenin}} = 333.3\left(\frac{N_2}{N_1}\right) v_{in}$$

Method 2: With this method, the open circuit voltage is used as the voltage source, but the Thevenin equivalent impedance is found by looking into the output terminals with all power and signal sources short-circuited and all controlled sources operating. (If there were current sources, they would be open circuited.) The impedance is found by assuming a voltage on the output terminals v_{ab}, finding the current that would result, and then taking the ratio of voltage to current.

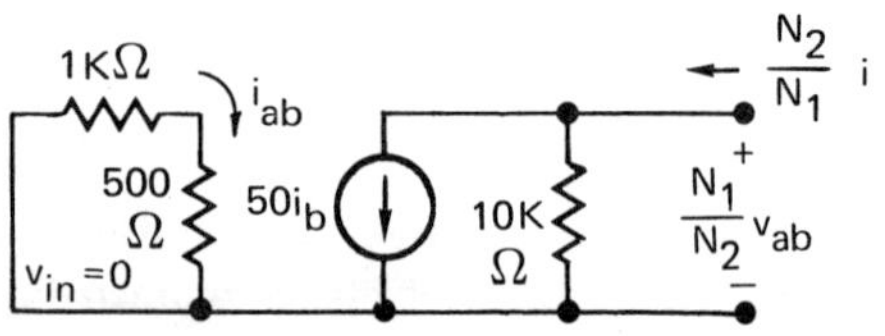

With the signal source short-circuited, the voltage v_{in} is zero, and so the current from the controlled current source will also be zero since i_b is zero. The current $\frac{N_2 i}{N_1}$ is

$$\frac{N_1 v_{ab}}{N_2 \times 10^4}$$

The Thevenin equivalent impedance is

$$Z_{\text{Thevenin}} = \frac{v_{ab}}{i} = \frac{v_{ab} \times 10^4}{(\frac{N_1}{N_2})^2 v_{ab}}$$

$$= \left(\frac{N_2}{N_1}\right)^2 \times 10^4$$

The Maximum Power Transfer Theorem says that maximum power will be transferred from a fixed network to an impedance which is the complex conjugate of the Thevenin equivalent impedance of the fixed network. In example 2.15, the impedance is a pure resistance with no imaginary part, so the complex conjugate is the same as the Thevenin equivalent impedance.

Example 2.15

A 60 Hz alternator generates a single-phase voltage of 120 volts (rms) and has a synchronous reactance of 0.25 ohms. The alternator feeds a line which is essentially resistive with an impedance of 0.1 ohms. The line is terminated with a 2:1 step-up transformer. What should the secondary impedance of the transformer be in order to have maximum power transfer to the load?

Maximum power will be transferred when the Thevenin source impedance seen by the load is the complex conjugate of the load. With the source fixed in this example, the maximum power transfer theorem describes how to obtain maximum power transfer. The real part of the load impedance must be the same as the real part of the source, and an imaginary impedance which is the negative of the imaginary part of the source impedance as seen from the load, must be added.

Since there is a transformer, the source impedance can be transferred to the load side and matched, or the load impedance can be transferred to the source side and matched. The impedance attached to the secondary is transformed to the primary side by multiplying by the square of the turns ratio. This is the ratio of turns on the winding looked into to the turns on the winding the impedance is attached to. Thus, if you look into N_{in} turns, and an impedance Z is attached to N_z turns, the impedance seen is

$$\left(\frac{N_{in}}{N_z}\right)^2 Z$$

In this example, the primary impedance is $.1 + j.25$. The synchronous reactance is an inductive reactance appearing in series with the alternator. Since the turns ratio is 2:1 as seen from the secondary side (step-up means the secondary has more turns than the primary), the impedance seen from the load (secondary) side is

$$2^2 \times (.1 + j.25) = .4 + j1$$

The load impedance for maximum power transfer is the complex conjugate

$$Z_{\text{load}}\Big|_{\text{max power transfer}} = .4 - j1 \text{ ohms}$$

The *superposition theorem* states that the responses to several signals or sources can be calculated separately (setting all other independent sources to zero), and then added algebraically to obtain the total response. This theorem is valid only for linear circuits. The superposition theorem is of most value when several signal sources of different frequencies are applied to a circuit.

Example 2.16

The circuit shown has $Z_1 = 1\mu f$ and $Z_2 = 200$ mh. The voltage source is represented by the partial Fourier series

$$v_s = 20\cos 1000t - 5\cos 4000t$$

Determine the voltage across the 100 ohm resistor.

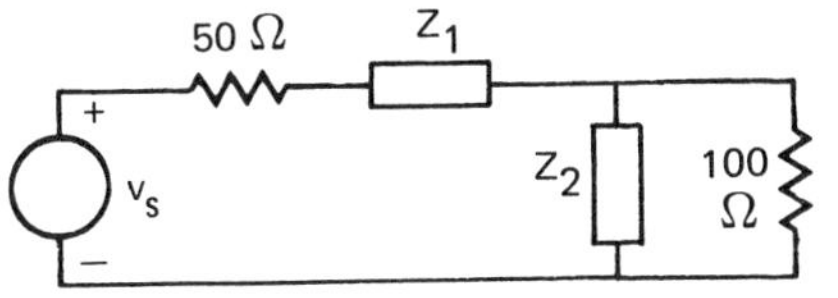

Working with the 1000 $\frac{\text{rad}}{\text{sec}}$ source first, the impedances are $Z_1 = -j10^3$, and $Z_2 = j200$ ohms at this frequency. The Thevenin equivalent circuit seen from the 100 ohm resistor is then a voltage source

$$V_{\text{Th}} = v_s \frac{Z_2}{50 + Z_1 + Z_2} = \frac{j200}{50 - j800} v_s$$
$$= -0.250 v_s \angle -3.58°$$

The Thevenin equivalent impedance seen from the 100 ohm resistance is Z_2 in parallel with $(50 + Z_1)$.

$$Z_{\text{Th}} = \frac{(50 + Z_1) \times Z_2}{50 + Z_1 + Z_2}$$
$$= \frac{(50 - j1000) \times j200}{50 - j800}$$
$$= 3.1 + j249.8$$

The voltage across the 100 ohm resistance can be obtained for the 1000 $\frac{\text{rad}}{\text{sec}}$ component of v_s.

$$v_{100}(1000) = \frac{100}{100 + Z_{\text{Th}}} V_{\text{Th}}$$
$$= \frac{100 \times (-.250 v_s) \angle -3.58°}{103.1 + j249.8}$$
$$= -.0925 v_s \angle -71.15°$$

Expressed trigonometrically, this voltage is

$$1.85\cos(1000t + 108.85°)$$

For the component of v_s at 4000 $\frac{\text{rad}}{\text{sec}}$ the impedances are $Z_1 = -j250$, and $Z_2 = j800$. For variety, the two loop equations are written in matrix form.

$$\begin{bmatrix} v_s \\ 0 \end{bmatrix} = \begin{bmatrix} (50 + j550) & -j800 \\ -j800 & (100 + j800) \end{bmatrix} \begin{bmatrix} i_1 \\ i_2 \end{bmatrix}$$

i_1 is the clockwise left-hand loop current, and i_2 is the clockwise right-hand loop current. This is solved for i_2 with the determinant method explained in chapter 1, as $100i_2$ is the voltage desired:

$$i_2 = .00354 v_S \angle 65.1°$$

The voltage across the 100 ohm resistor is

$$v_{100}(4000) = .354 v_S \angle 65.1°$$

This can be expressed trigonometrically as

$$1.77\cos(4000t + 65.1°)$$

Then applying the superposition theorem,

$$v_{100} = v_{100}(1000) + v_{100}(4000)$$
$$= 1.85\cos(1000t + 108.85°)$$
$$+ 1.77\cos(4000t + 65.1°)$$

4 TWO-PORT NETWORKS

Two-port networks such as transformers and transistors may be either 3-terminal or 4-terminal devices. They must be linear or piecewise linear to be treated under the theory presented in this section. A representation of such a network is shown in figure 2.15 with the current directions and voltage polarities. There are four variables: v_1, i_1, v_2 and i_2. The port with the subscript 1 is called the *input port* and the corresponding terminals are the *input terminals*. The terminals with subscript 2 are called *output terminals* and make up the *output port*.

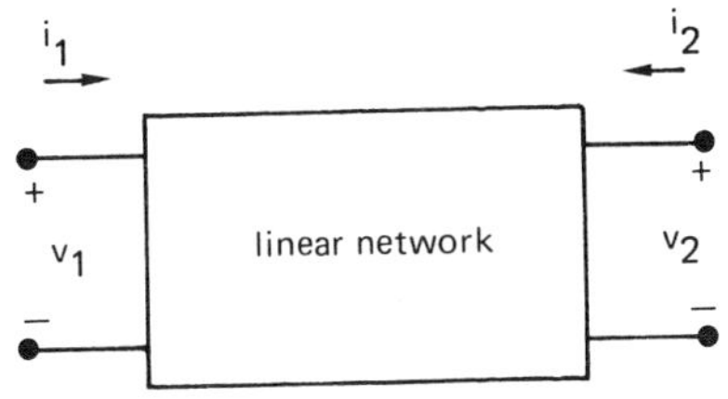

Figure 2.15. Two-port network showing defined directions

Two-port network theory is particularly important in circuits, such as amplifiers, which employ feedback. The four port variables can be arranged into dependent and independent pairs. For a linear network, the distinction is arbitrary. The result is a set of parameters, such as the z-parameters used to describe transistors. In example 2.16, the independent variables were the currents, and the dependent variables were the voltages.

By choosing voltage as the independent variable, *y-parameters* are obtained. By choosing the input current and the output voltage as independent variables, the *h-parameters* are obtained. Generally, the parameters are identified by double subscripts in standard matrix notation with the first subscript corresponding to the row and the second subscript to the column. Thus, the z-parameters are written

$$\begin{bmatrix} v_1 \\ v_2 \end{bmatrix} = \begin{bmatrix} z_{11} & z_{12} \\ z_{21} & z_{22} \end{bmatrix} \begin{bmatrix} i_1 \\ i_2 \end{bmatrix} \qquad 2.29$$

The z-parameters are particularly useful when the parameters of two networks are known and when the input and output currents are the same for the two networks. This is illustrated in figure 2.16 for a common emitter transistor with an emitter resistor. The two sets of z-parameters are combined by addition since the input and output voltages are the sum of the dependent variables. The upper network is indicated by unprimed variables and parameters; the lower network is indicated by primed variables. Equation 2.29 applies to the upper network. By priming all quantities, it also applies to the lower network.

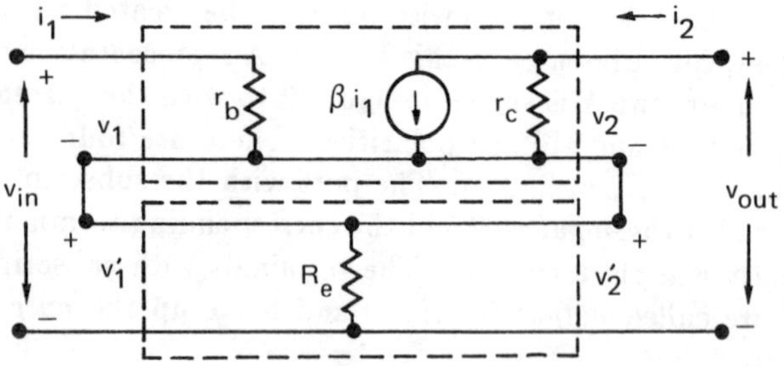

Figure 2.16. Combination of 2-port networks with common currents

Example 2.17

Determine the z-parameters for the upper and the lower networks shown in figure 2.16, and find the z-parameters for the entire circuit.

For the upper network, the equations are

$$v_1 = r_b i_1$$
$$i_2 = \beta i_1 + \frac{v_2}{r_c}$$

The first equation is already in the z-parameter form. The second equation must be manipulated to get v_2 as the dependent variable.

$$v_2 = -\beta r_c i_1 + r_c i_2$$

Thus the z matrix for the upper network is

$$\mathbf{Z} = \begin{bmatrix} r_b & 0 \\ -\beta r_c & r_c \end{bmatrix}$$

For the lower network,

$$v_1' = (i_1 + i_2) R_e$$
$$v_2' = (i_1 + i_2) R_e$$

Note that $i_1' = i_1$ and $i_2' = i_2$. Thus the $\mathbf{Z}'$ matrix for the lower network is

$$\mathbf{Z}' = \begin{bmatrix} R_e & R_e \\ R_e & R_e \end{bmatrix}$$

Since $v_{in} = v_1 + v_1'$ and $v_{out} = v_2 + v_2'$, the z matrixes can be added to give the z-parameters for the entire circuit.

$$\begin{bmatrix} v_{in} \\ v_{out} \end{bmatrix} = \begin{bmatrix} r_b + R_e & R_e \\ -\beta r_c + R_e & r_c + R_e \end{bmatrix} \begin{bmatrix} i_1 \\ i_2 \end{bmatrix}$$

With the z-parameters, the input currents are the same for both circuits. The output currents are also the same. These currents are the independent variables. The dependent voltage variables add together, permitting the addition of corresponding parameters.

This method can be used in parallel circuits for three other combinations of input and output variables: for y-parameters when the terminal voltages are the same, for h-parameters when the input currents are the same and the output voltages are the same, and for *g-parameters* when the input voltages are the same and the output currents are the same.

$$\begin{bmatrix} i_1 \\ v_2 \end{bmatrix} = \begin{bmatrix} g_{11} & g_{12} \\ g_{21} & g_{22} \end{bmatrix} \begin{bmatrix} v_1 \\ i_2 \end{bmatrix}$$

The corresponding g-parameter circuit model is shown in figure 2.17.

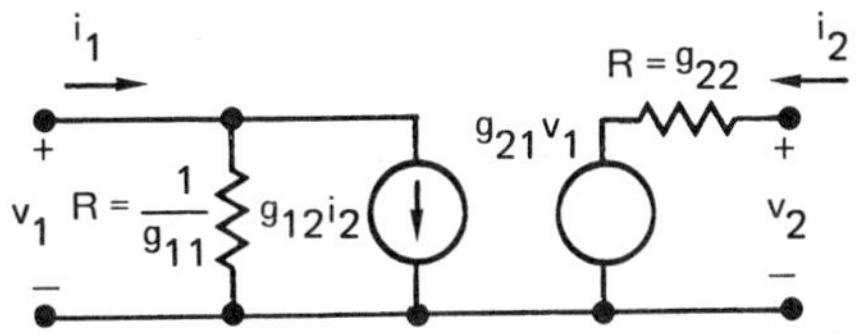

Figure 2.17 g-parameter circuit model

It is also possible to choose independent variables which permit *cascading* of networks. The parameters are then called *chain parameters*. The dependent variables are input voltage and current. The independent variables are the output voltage and the negative of the output current. The output current is negated to cause it to be the same as the following network's input current.

$$\begin{bmatrix} v_1 \\ i_1 \end{bmatrix} = \begin{bmatrix} A & B \\ C & D \end{bmatrix} \begin{bmatrix} v_2 \\ -i_2 \end{bmatrix} \qquad 2.30$$

5 NETWORK ANALYSIS

This section shows how linear network equations can be generated by inspection, bypassing the tedious process of writing Kirchhoff law relationships.

Node voltage equations are generated by taking all voltages with respect to a reference node and writing the Kirchhoff current law equations in terms of those node voltages. The process can be quickly accomplished by the following steps.

Step 1: Use Norton's theorem to convert all voltage sources with series impedances to current sources with shunt impedances. Voltage sources without series impedances between two nodes make one of the nodes a *dependent node*. The voltage at a dependent node is the voltage at the other node plus the value of the voltage source. Voltage at dependent nodes is known.

Step 2: Assign voltage variables to all nodes except the reference node, including the dependent nodes.

Step 3: Set up an empty square *admittance matrix* with the same number of rows and columns as the number of independent and dependent nodes, but not including the reference node.

Step 4: Fill out the square matrix. At each diagonal position (where the row and column have the same number) insert the sum of the admittances connected directly to the node with that number. At other positions, place the negative of the sum of admittances which directly connect the nodes having the same subscripts as the row and column of the position.

Step 5: To the right of the square matrix construct the column matrix for the node voltages. Enter them in order. Include the dependent node voltages.

Step 6: To the left of the square matrix, construct the column matrix for the sources. Enter the sum of the current sources feeding directly into the node with the same numeric subscript as the row. Current sources feeding directly out of a node are negative. Include controlled (dependent) sources as well as independent sources.

The above procedure is adequate whenever there are no dependent sources and no dependent node voltages.

Example 2.18

For the network shown, obtain the network matrix equation in terms of the node voltages.

Number the nodes from left to right. The reference node is the line across the bottom of the diagram.

Step 1: Convert all voltage sources with series impedances. The 5 volt source in series with the 10 ohm resistance converts to a 0.5 amp current source connected between the same two nodes.

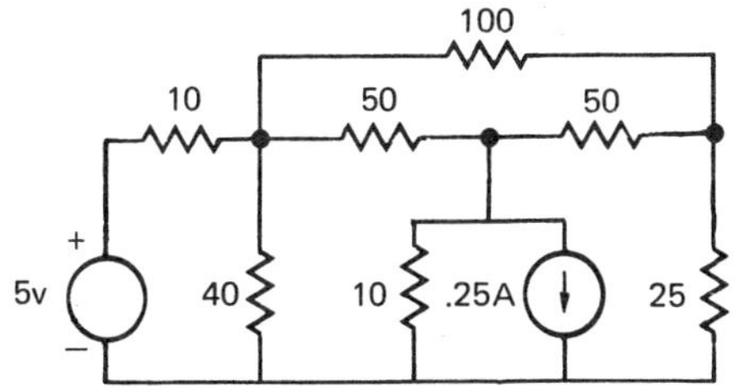

Step 2: Assign the voltage subscripts at the nodes. The results of step 1 and step 2 are shown. Resistance has been converted to conductance (admittance) in the figure shown.

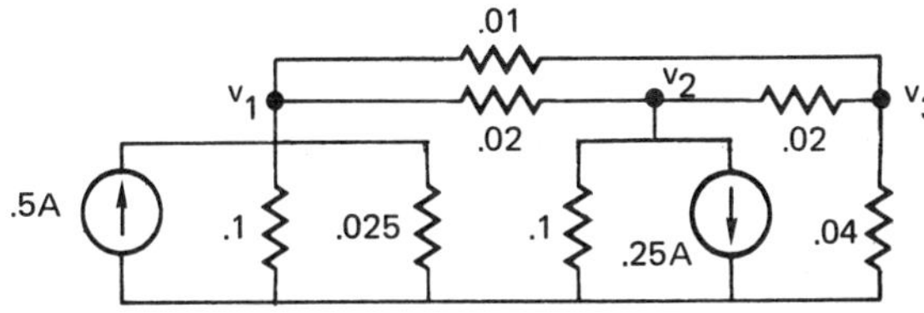

Step 3: This step calls for a 3 × 3 admittance matrix.

$$\begin{array}{c|c} & \text{columns} \\ & 1 \quad 2 \quad 3 \\ \text{rows} \begin{array}{c}1\\2\\3\end{array} & \begin{bmatrix} & & \\ & & \\ & & \end{bmatrix}\end{array}$$

Step 4: Fill out the admittances matrix. Work first along the main diagonal with the node self-admittance y_{11}. (Take the connected admittances counterclockwise.)

$$y_{11} = .1 + .025 + .02 + .01$$
$$= .155 \qquad (node\,1)$$
$$y_{22} = .02 + .1 + .02 = .14 \qquad (node\,2)$$
$$y_{33} = .01 + .02 + .04 = .07 \qquad (node\,3)$$

The negatives of the interconnecting admittances are placed in the corresponding rows and columns

$$y_{12} = y_{21} = .02$$
$$y_{13} = y_{31} = .01$$
$$y_{23} = y_{32} = .02$$

The result is then

$$\begin{bmatrix} .155 & -.02 & -.01 \\ -.02 & .14 & -.02 \\ -.01 & -.02 & .07 \end{bmatrix}$$

Step 5: The single column node voltage matrix is

$$\begin{bmatrix} v_1 \\ v_2 \\ v_3 \end{bmatrix}$$

Step 6: Step 6 calls for a one column, three row source matrix.

row 1: .5A enters node 1.
row 2: −.25A enters node 2. (.25A leaves node 2)
row 3: there are no sources connected to node 3.

$$\begin{bmatrix} .5 \\ -.25 \\ .0 \end{bmatrix}$$

The final result is

$$\begin{bmatrix} .5 \\ -.25 \\ .0 \end{bmatrix} = \begin{bmatrix} .155 & -.02 & -.01 \\ -.02 & .14 & -.02 \\ -.01 & -.02 & .07 \end{bmatrix} \begin{bmatrix} v_1 \\ v_2 \\ v_3 \end{bmatrix}$$

If there are controlled sources in the network, an additional step is required.

Step 7: When *controlled sources* are present, they are placed in the column source matrix as if they were independent sources. It is then necessary to obtain the controlling variable in terms of the independent node voltages.

For example, suppose the .25A independent source had been a controlled source whose value was one quarter of the current flowing from left to right in the left-hand 50 ohm resistance. It would then be necessary to express that current in terms of the node voltages $\frac{(v_1-v_2)}{50}$. The current source would then be given a value of

$$.25 \times \frac{(v_1 - v_2)}{50} = .005(v_1 - v_2)$$

The source matrix would be

$$\begin{bmatrix} .5 \\ -.005(v_1 - v_2) \\ 0 \end{bmatrix}$$

The row in the source matrix that contains the independent variables must be modified. For any independent variable appearing in the column source matrix, subtract the coefficient of that variable from the admittance term in the same row and in the column having the same subscript as that variable.

The controlled source of $-.005(v_1 - v_2)$, for example, appears in the source matrix in the second row. The coefficient of v_1 is −.005, and that quantity must be subtracted from the term in the first column of the second row (i.e., −.02). The coefficient of v_2 is +.005, so that quantity must be subtracted from the term in the second column and the second row (i.e., .14). This also removes the controlled source from the source matrix. The resulting matrix equation for the example is

$$\begin{bmatrix} .5 \\ 0 \\ 0 \end{bmatrix} = \begin{bmatrix} .155 & -.02 & -.01 \\ -.015 & .135 & -.02 \\ -.01 & -.02 & .07 \end{bmatrix} \begin{bmatrix} v_1 \\ v_2 \\ v_3 \end{bmatrix}$$

The presence of *dependent nodes* requires an additional step. Dependent nodes can only be approximated in practice, as there is some impedance to any real source.

However, some highly regulated power supplies and flat compounded d.c. generators approach being impedanceless sources, which produces dependent nodes.

The method of treating such nodes is to include them within the original admittance matrix equations. Rather than being given a distinct name, however, they are carried in the independent variable (node voltage) column matrix as an adjacent node voltage plus the intervening voltage. The treatment involves a step called *node merging*.

Step 8: When dependent nodes are present, the admittances of those node voltages must be (1) added to the admittance of the node voltages that appears in their place, and (2) the product of the internode voltage sources and the admittances must be subtracted from the source terms for those rows.

As an example, suppose there were a dependent node in example 2.18 from a 5 volt source connected from node 1(+) to node 2(−). A term would then appear in the second row of the matrix equation.

$$\begin{bmatrix} .5 \\ -.25 \\ 0 \end{bmatrix} = \begin{bmatrix} .155 & -.02 & -.01 \\ -.02 & .14 & -.02 \\ -.01 & -.02 & .07 \end{bmatrix} \begin{bmatrix} v_1 \\ v_1 - 5 \\ v_3 \end{bmatrix}$$

The $v_1 - 5$ term results from the 5 volt impedanceless source between nodes 1 and 2, with v_2 being 5 volts less than v_1. The terms in the second column of the admittance matrix are the coefficients of the voltage at the node 2. The equation of the first row is

$$.5 = .155v_1 - .02(v_1 - 5) - .01v_3$$

The two coefficients of v_1 are .155 and −.02, which must be added in order to obtain the correct coefficient. Then, as v_1's coefficients are in column 1 and v_2's coefficients are in column 2, the column 2 values must be added to the column 1 values and the result put in column 1. This step removes v_1 from the second row position in the node voltage column matrix.

$$\begin{bmatrix} .5 \\ -.25 \\ 0 \end{bmatrix} = \begin{bmatrix} .135 & -.02 & -.01 \\ .12 & .14 & -.02 \\ -.03 & -.02 & .07 \end{bmatrix} \begin{bmatrix} v_1 \\ -5 \\ v_3 \end{bmatrix}$$

The second column is composed of the coefficients of −5, which belong on the source side of the matrix. For the row 1 equation, this is done by multiplying the known voltage (in this case −5) by its coefficients and subtracting the result from the source term in the same row.

$$\begin{bmatrix} .5 - (-5 \times (-.02)) \\ -.25 - (-5 \times .14) \\ 0 - (-5 \times (-.02)) \end{bmatrix} = \begin{bmatrix} .135 & -.01 \\ .12 & -.02 \\ -.03 & .07 \end{bmatrix} \begin{bmatrix} v_1 \\ v_3 \end{bmatrix}$$

To complete the node merging, it is necessary to add the rows of the two nodes that have been merged. In this case, nodes 1 and 2 were merged, so rows 1 and 2 must be added.

$$\begin{bmatrix} .4 + .45 \\ -.1 \end{bmatrix} = \begin{bmatrix} .135 + .12 & -.01 + (-.02) \\ -.03 & .15 \end{bmatrix} \begin{bmatrix} v_1 \\ v_3 \end{bmatrix}$$

$$\begin{bmatrix} .85 \\ -.1 \end{bmatrix} = \begin{bmatrix} .255 & -.03 \\ -.03 & .07 \end{bmatrix} \begin{bmatrix} v_1 \\ v_3 \end{bmatrix}$$

If you began with a symmetrical matrix, as you should have, you will finish with a symmetrical matrix.

Loop current analysis can also be often carried out by inspection. Loop current analysis, however, is restricted to planar networks, i.e., those which have no crossing wires. The method consists of the following steps:

Step 1: Convert all current sources with parallel impedances to voltage sources with series impedances using Thevenin's theorem. If there are current sources without parallel impedances, node voltage analysis is recommended.

Step 2: Assign loop currents to all meshes (loops) of the network. Assume the directions of all currents to be clockwise. Identify each loop current by a numeric subscript corresponding to the mesh.

Step 3: From left to right, set up a column source matrix with the number of rows equal to the number of loops, a square impedance matrix with the number of rows and columns equal to the number of loops, and a column loop current matrix with the loop currents entered in numerical order.

Step 4: In each row of the source matrix, place the sum of the source voltage *rises* going around the loops. The rows of the source and current matrixes are the same. Voltage rises in the current direction are positive. Include controlled voltage sources in the source matrix.

Step 5: In the diagonal positions of the impedance matrix, place the sum of all the impedances in the loops having the respective numeric subscripts.

Step 6: In non-diagonal positions, place the negative of the impedances common to the two loops with the different subscripts.

Step 7: If there are controlled sources, their voltages will have been assigned to the column source matrix in step 4. It is necessary to obtain the controlling variable in terms of the loop currents. When this is done, the coefficients of the controlling loop currents are subtracted from the term with the column subscript of that current and the row subscript of the impedance matrix.

Example 2.19

Perform a loop current analysis of the circuit shown.

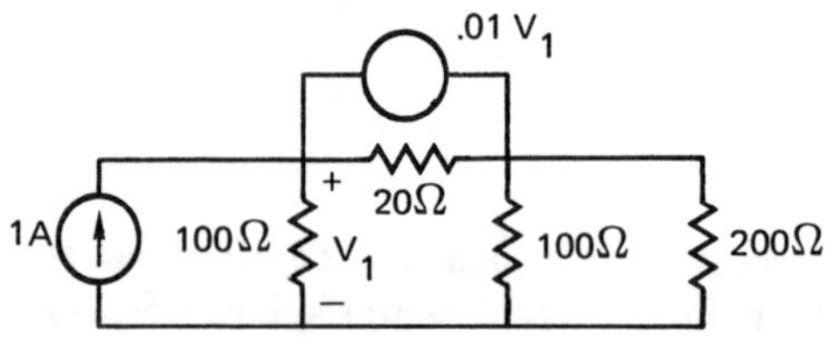

Step 1: The first step is to convert the current sources to voltage sources. The 1 amp independent source on the left has a parallel resistance of 100 ohms. The Thevenin equivalent is a 100 volt source in series with a 100 ohm resistance. The Thevenin equivalent of the controlled current source with a 20 ohm shunting resistance is a controlled voltage source of 20 × $.01v_1$ in series with the 20 ohm resistance.

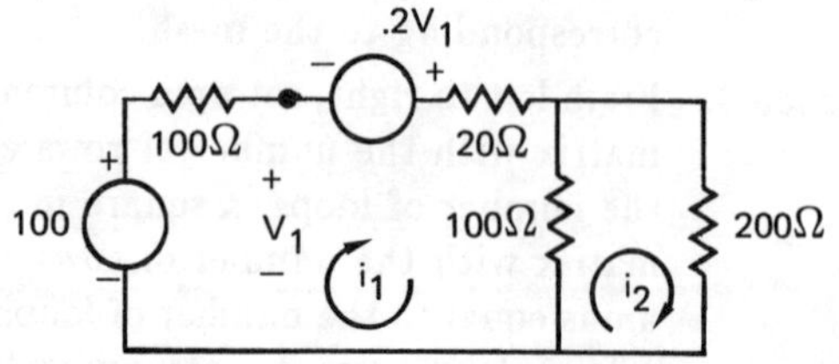

Step 2: The loop currents are chosen in the clockwise direction and are numbered as shown.

Step 3: This step is illustrated here.

$$\begin{bmatrix} \\ \end{bmatrix} = \begin{bmatrix} & \\ & \end{bmatrix}\begin{bmatrix} i_1 \\ i_2 \end{bmatrix}$$

Step 4: Since their polarities are in the current direction, the sources in loop 1 add to $100 + .2v_1$. This is entered in row 1 of the source matrix. Loop 2 has no sources, so enter a zero in row 2.

$$\begin{bmatrix} 100 + .2v_1 \\ 0 \end{bmatrix} = \begin{bmatrix} & \\ & \end{bmatrix}\begin{bmatrix} i_1 \\ i_2 \end{bmatrix}$$

Step 5: Loop 1 has a total of 220 ohms resistance; so enter 220 in the top diagonal position of the impedance matrix. In loop 2, there is a total of 300 ohms, so enter 300 in the lower diagonal position.

Step 6: The impedance common to loops 1 and 2 is 100 ohms, so enter −100 in the non-diagonal positions of the impedance matrix.

The results of steps 5 and 6 are

$$\begin{bmatrix} 100 + .2v_1 \\ 0 \end{bmatrix} = \begin{bmatrix} 220 & -100 \\ -100 & 300 \end{bmatrix}\begin{bmatrix} i_2 \\ i_2 \end{bmatrix}$$

Step 7: The voltage v_1 is the source voltage (100) minus the voltage drop in the 100 ohm resistance in series:

$$v_1 = 100 - 100i_1$$

v_1 is multiplied by .2 to obtain

$$.2v_1 = 20 - 20i_1$$

The constant 20 is added to the 100 in row 1 of the column source matrix. $-20i_1$ must be transferred across the equals sign. This is done by subtracting −20 from the proper coefficient of i_1 in the impedance matrix. The row is 1, and the column of i_1 coefficients is 1 in the impedance matrix. Therefore, the 220 in the impedance matrix has −20 subtracted from it.

$$z_{11} = 220 - (-20) = 240$$

$$\begin{bmatrix} 120 \\ 0 \end{bmatrix} = \begin{bmatrix} 240 & -100 \\ -100 & 220 \end{bmatrix}\begin{bmatrix} i_1 \\ i_2 \end{bmatrix}$$

6 IMPEDANCE ANALYSIS

This section will review the method of evaluating impedances of circuit elements under steady state (d.c. and a.c.) and transient conditions.

D.C. steady state impedance analysis is based on the fact that the voltages and currents in a circuit are constant. This means that the time derivatives of all voltages and currents are zero.

$$\frac{dv}{dt} = 0 \qquad 2.31$$

$$\frac{di}{dt} = 0 \qquad 2.32$$

The d.c. voltage-current relationships are developed here.

A. RESISTANCE:

$$Z_R|_{d.c.} = R \qquad 2.33$$

B. INDUCTANCE:

$$v = L\frac{di}{dt} \qquad 2.34$$

$$\frac{di}{dt} = 0 \qquad 2.35$$

$$v = 0 \qquad 2.36$$

$$Z = \frac{v}{i} \qquad 2.37$$

$$Z_L|_{d.c.} = 0 \text{ (a short-circuit)} \qquad 2.38$$

C. CAPACITANCE

$$i = C\frac{dv}{dt} \qquad 2.39$$

$$\frac{dv}{dt} = 0 \qquad 2.40$$

$$Z_C|_{d.c.} = \infty \text{ (an open circuit)} \qquad 2.41$$

Thus, in steady state analysis of d.c. circuits, inductances are treated as short-circuits and capacitances are treated as open circuits.

A.C. steady state impedances are most easily evaluated by using the *phasor* representation of voltages and currents. Phasor conventions are derived from the definition of a complex number in exponential form.

$$e^{j\omega t} = \cos\omega t + j\sin\omega t \qquad 2.42$$

$$j = \sqrt{-1} \qquad 2.43$$

The real part of $e^{j\omega t}$ is $\cos\omega t$, and the imaginary part is $\sin\omega t$. $e^{j\omega t}$ can be used to represent a general sinusoid rather than specific sine or cosine waveforms. The time derivative of $e^{j\omega t}$ is

$$\frac{d}{dt}e^{j\omega t} = j\omega e^{j\omega t} \qquad 2.44$$

If a sinusoidal current is represented in the exponential form, then

$$i = I_m e^{j\omega t} \qquad 2.45$$

$$\frac{di}{dt} = j\omega I_m e^{j\omega t} = j\omega i \qquad 2.46$$

For a general sinusoidal voltage,

$$v = V_m e^{j\omega t} \qquad 2.47$$

$$\frac{dv}{dt} = j\omega V_m e^{j\omega t} = j\omega v \qquad 2.48$$

Sinusoidal impedances can also be defined using exponential forms.

A. RESISTANCE

$$Z_R|_{a.c.} = R \qquad 2.49$$

B. INDUCTANCE:

$$v = L\frac{di}{dt} = j\omega Li \qquad 2.50$$

$$Z_L|_{a.c.} = \frac{v}{i} = j\omega L \qquad 2.51$$

C. CAPACITANCE:

$$i = C\frac{dv}{dt} = j\omega Cv \qquad 2.52$$

$$Z_C|_{a.c.} = \frac{v}{i} = \frac{1}{j\omega C} \qquad 2.53$$

For *transient analysis* of linear circuits, a more general impedance form is useful. Currents and voltages are given the form e^{st}. s is a complex variable of the form $\sigma + j\omega$; it is the same as the Laplace transform variable. The voltage and current forms are $V_m e^{st}$ and $I_m e^{st}$ respectively, so the time derivatives are

$$\frac{di}{dt} = sI_m e^{st} = si \qquad 2.54$$

$$\frac{dv}{dt} = sV_m e^{st} = sv \qquad 2.55$$

The general forms of the impedances are found from the voltage-current relationships. These impedance forms are important in transient analysis and in system analysis involving stability.

A. RESISTANCE:

$$Z_R = R \qquad 2.56$$

B. INDUCTANCE:

$$v = L\frac{di}{dt} = sLi \qquad 2.57$$

$$Z_L = \frac{v}{i} = sL \qquad 2.58$$

C. CAPACITANCE:

$$i = C\frac{dv}{dt} = sCv \qquad 2.59$$

$$Z_C = \frac{v}{i} = \frac{1}{sC} \qquad 2.60$$

Practice Problems: LINEAR CIRCUIT ANALYSIS

Warmups

1. Using the same diameter nichrome wire as in example 2.1, determine the length needed to obtain a resistance of 1 ohm at 100°C. Determine the resistance at 20°C.

2. For the circuit in example 2.2, assume the switch has been closed for a long time and is opened at $t = 0$. Determine (a) the current through the capacitor at the instant the switch is opened, (b) the time derivative of the capacitor voltage at the instant the switch is opened, and (c) the voltage that will exist across the capacitor after a long time.

3. For the circuit in example 2.3, assume that the switch has been closed for a long time (the final conditions of example 2.3 apply), and the switch is opened at $t = 0$. Determine at the instant the switch opens (a) the current in the inductance, (b) the time rate of change of the current in the inductance, and (c) the voltage across the inductance.

4. Repeat example 2.4 with the 120 volt source connected to winding 2 and the 60 ohm load attached to winding 1.

5. Obtain a linear model of voltage versus current for the characteristic shown in example 2.5 over in the current range of 0 to 6 amps. Obtain the current source and voltage source equivalent circuits.

6. A 5 volt, 25 amp d.c. source has voltage regulation of 1%. Obtain an equivalent circuit for this source.

7. All silicon diodes have a reverse breakdown voltage essentially the same as the avalanche voltage of a zener diode. This results in three distinct regions of operation: one for the normal forward region, one for the avalanche region, and an intermediate region where the device behaves essentially as an open circuit. Obtain the linear models in these three regions for a 1 amp, 2 watt silicon diode with a reverse breakdown voltage of 5.5 volts and a maximum reverse voltage of 5.6 volts.

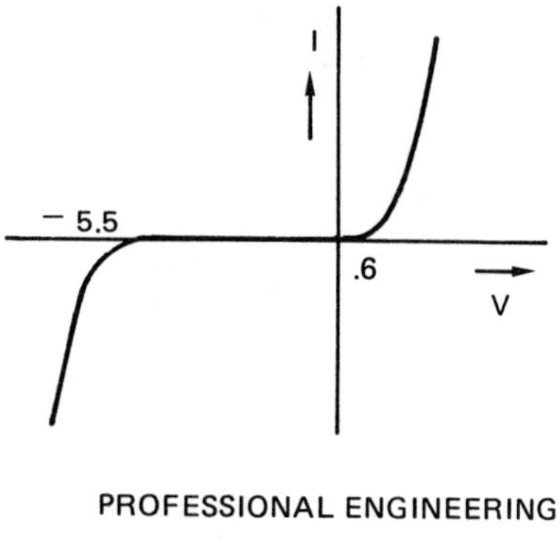

8. A three-terminal device is described by the following z-parameter equations.

$$v_{in} = 250 i_{in} + 5 i_{out}$$
$$v_{out} = -100 i_{in} + 25 i_{out}$$

Obtain an equivalent circuit for this device.

9. Common base transistor configurations are often described in terms of *common base y-parameters*: y_{ib}, y_{rb}, y_{fb}, and y_{ob}. (The *b*'s in the subscripts indicate that the parameters were obtained from a common base configuration with the emitter as the input terminal and the collector the output terminal.) A common circuit model for the transistor used in this configuration is shown. Obtain the y-parameters for this comon base circuit.

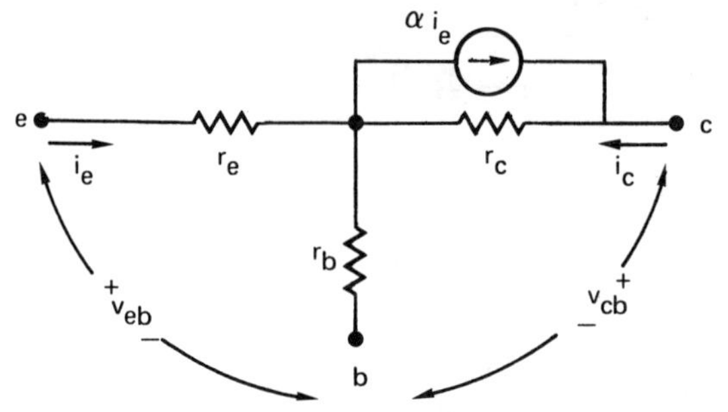

10. For a photodiode with characteristics shown in figure 2.10 and used in the circuit for example 2.11 with a 5 volt source and a 10,000 ohm load, determine the load voltage as a function of illumination level. Over what range of illumination levels will this result be valid? Assume the curves extend to −5 volts on the third quadrant of the characteristic curves.

11. Repeat example 2.12 for $T_1 = 0°C$ and $T_2 = 100°C$.

12. Use the dimensions given in example 2.13, but assume the conductor is an n-type semiconductor with 10^{19} electrons per cubic meter. The current is .1 amp. Obtain an equivalent circuit for a transducer using the Hall voltage to measure magnetic field intensity. Assume that the voltage pickups are capacitive so that there is no connection of the output to the current circuit.

13. For the circuit of example 2.14, obtain the Norton equivalent circuit.

14. Consider the circuit of example 2.14 with an 8 ohm speaker attached to terminals a-b. Show that the necessary turns ratio $\left(\frac{N_1}{N_2}\right)$ to provide maximum power transfer is 35.36.

15. The elements Z_1 and Z_2 are interchanged in example 2.16, and the voltage v_s takes on the value of

$$20 + 5\cos 2236t \text{ volts.}$$

Find the voltage across the 100 ohm resistor.

16. The circuit of example 2.17 has $r_b = 200$ ohms, $\beta = 50$, $r_c = 2500$ ohms and $R_e = 10$ ohms. Find the voltage gain $\frac{v_{out}}{v_{in}}$ when a 1000 ohm load is placed across the v_{out} terminals. Note that the output current i_2 is

$$\frac{-v_{out}}{1000}$$

for this load.

17. Using y-parameters, obtain the total circuit y-parameters for the circuits indicated by the dashed lines. Hint: first find the y-parameters of the two indicated two-port networks, then combine them to obtain the total network y-parameters.

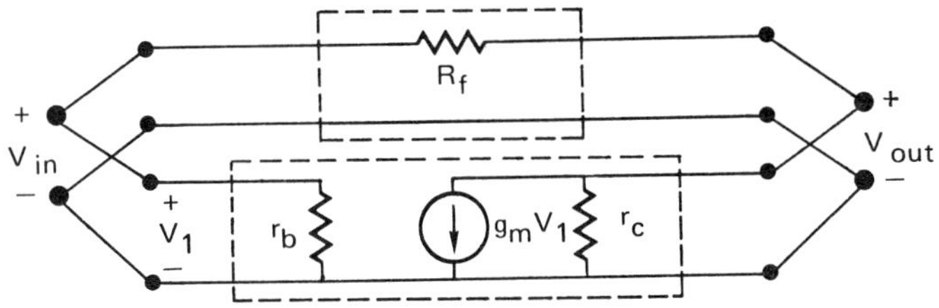

Concentrates

1. In the circuit shown in example 2.16, the voltage source is operating at a frequency of 2000 rad/sec. Specify the impedances Z_1 and Z_2 so that maximum power is transferred to the 100 ohm load resistance. (Hint: Z_1 and Z_2 must be reactive or else they will absorb some of the power.) In specifying Z_1 and Z_2, give their values in microfarads and/or millihenries. There are two possible sets of answers.

2. For the circuit shown, obtain the node voltage equations in matrix form. Solve these equations for v_1 and v_2. (The actual output is $v_1 - v_2$.) Obtain the voltage gain v_{out}/v_s for this emitter-coupled amplifier.

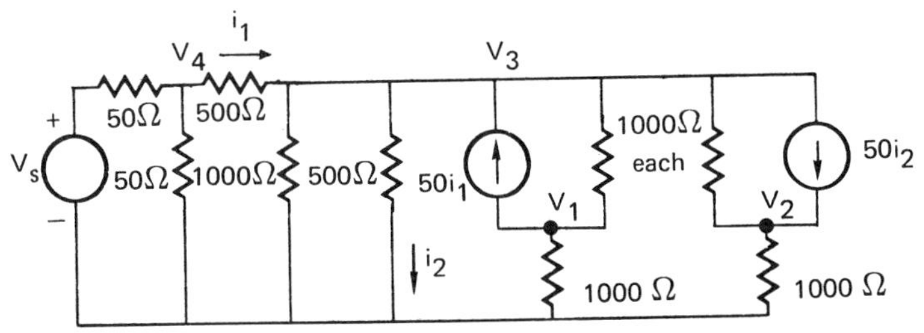

3. For the circuit shown, obtain the loop current equations. Solve for the current in the 200 ohm resistance connected to the secondary of the ideal transformer.

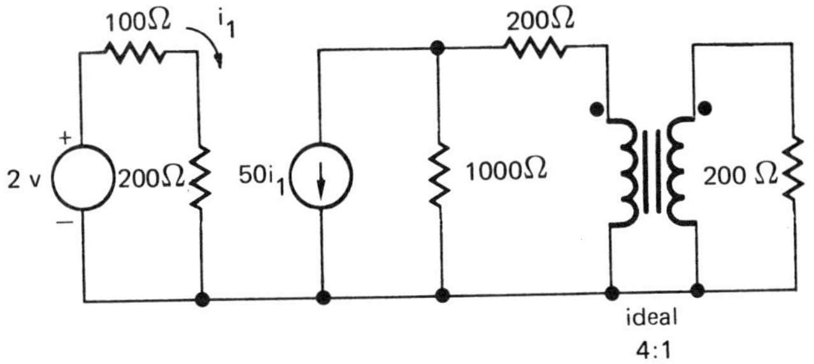

4. A 60 hz source has its voltage measured under various loads, with the results shown below:

Load	Voltmeter Reading
open circuit	120 volts rms
60 Ω resistance	72 volts rms
60 Ω pure capacitance	360 volts rms
60 Ω pure inductance	51.43 volts rms

Determine the voltage for a load of 120 ohms pure capacitance.

5. A bridge network is used to measure the temperature of a chemical bath.

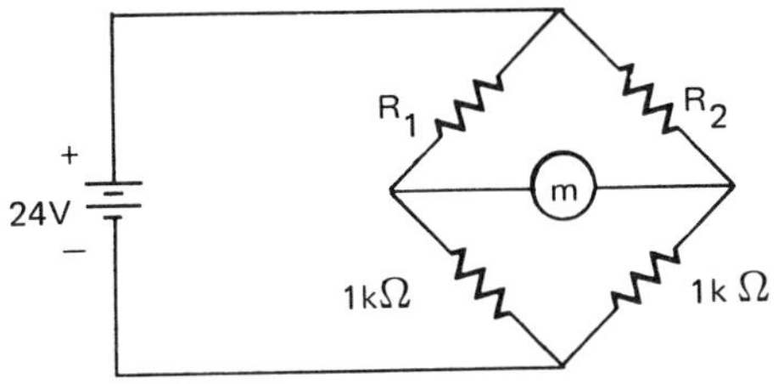

R_1 and R_2 are identical thermistors with negative temperature coefficients of resistance of $-4\%/^0C$. At 25^0C they are both 1500 ohms. The meter (m) has an internal resistance of 800 ohms. Resistance R_2 is held at 25^0C. (a) Determine the meter current when R_1 is at 20^0C. (b) Determine the meter current when R_1 is at 30^0C.

6. An oscillator circuit is shown. Determine the minimum value of K such that $e_0 = E$. Also, find the frequency at which $e_0 = E$.

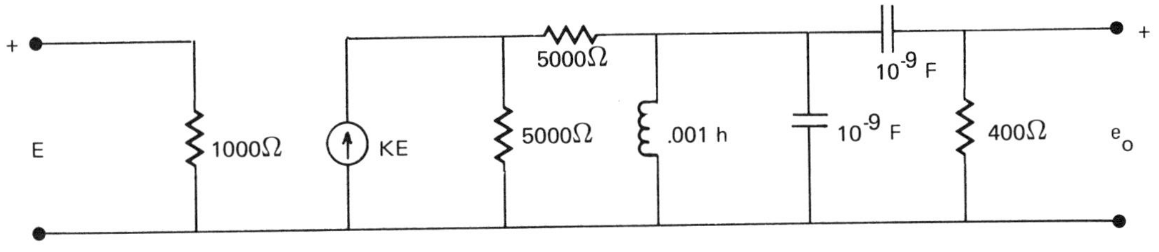

Figure for the circuit in problem 6

RESERVED FOR FUTURE USE

3 MEASUREMENTS OF SIGNAL WAVEFORMS

PART 1: PERIODIC SIGNALS

Nomenclature

a_0	Fourier coefficient	-
a_n	Fourier coefficient	-
b_n	Fourier coefficient	-
B	susceptance	mhos
$\mathbf{f}$	a function	-
f	frequency or function value	Hz,-
$\mathbf{F}$	a phasor function	-
G	conductance	mhos
I	phasor current	amps
I^*	complex conjugate of I	amps
j	imaginary operator, $\sqrt{-1}$	-
n	an integer	-
p	instantaneous power	watts
PF	power factor	-
Q	reactive power	volt-amperes reactive
R	resistance	ohms
RMS	root-mean-square value	-
S	apparent power	volt-amperes
t	time	seconds
T	period of periodic function	seconds
v	instantaneous voltage	volts
V	phasor voltage	volts
X	reactance	ohms
Y	admittance	mhos
Z	impedance	ohms

Symbols

ϕ	an angle	degrees or radians
θ	an angle	degrees or radians
ω	radian frequency	radians/second

Subscripts

C	capacitance or capacitive
f	fundamental
$f.s.$	full-scale
L	inductance or inductive
p	peak value
s	source
thr	threshold

A periodic signal is one which repeats a large number of times. Examples are constant currents and voltages, sinusoids, square waves, pulse trains, and triangular and sawtooth waveforms. Periodic signals can be investigated with Fourier series analysis.

Several different parameters are useful in describing periodic signals, depending on the particular waveform and the detail desired. Parameters which describe periodic signals include the average value, root-mean-square (RMS) value, period or time between repetitions, frequency content or spectrum, and peak value (or some other significant measure of the *size* of the waveform).[1]

1 AVERAGE VALUES

The *average value* of a periodic waveform is the first (zero-frequency) term of the Fourier series described in chapter 1. If $\mathbf{f}(t)$ is a periodic function which repeats itself each T seconds, then the average value of $\mathbf{f}(t)$, f_{avg}, is given by equation 3.1. (t_1 is some convenient time for the evaluation of the integral.)

$$f_{\text{avg}} = \frac{1}{T}\int_{t_1}^{t_1+T} \mathbf{f}(t)\,dt \qquad 3.1$$

[1]The RMS value is also known as the *effective* and *heating value*.

Since the act of integration can be interpreted as finding the area under a curve, equation 3.1 divides the net area of the waveform by the period T. This is illustrated in figure 3.1. For the function shown, the left-most shaded area is a positive area, and the shaded area on the right is a negative area. The negative area is subtracted from the positive area, and the average value is the net remaining area divided by the period, T.

$$f_{\text{avg}} = \frac{\text{positive area} - \text{negative area}}{T} \qquad 3.2$$

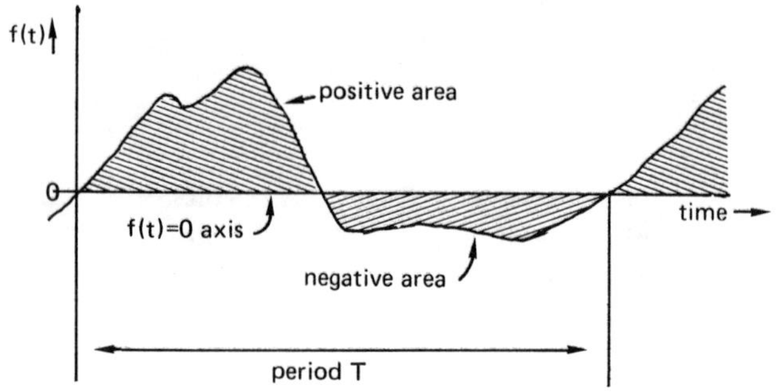

Figure 3.1. Areas used to find the average value.

Example 3.1

A *constant function* (i.e., $f(t)$=constant) has an arbitrary period. It can be thought of as having any convenient period. In this example, the function is a constant voltage of 3 volts. A period of T is arbitrarily marked on the time axis. From equation 3.1, the average voltage is

$$v_{\text{avg}} = \frac{1}{T}\int_{t=0}^{T} 3\,dt = \left.\frac{3t}{T}\right|_{t=0}^{t=T}$$
$$= \frac{3(T-0)}{T} = 3 \text{ volts}$$

Using equation 3.2, it is seen that the positive area (shaded) is $3 \times T = 3T$. Dividing by T, the value of 3 is obtained. The obvious result of this example is that the average value of a constant function is that constant.

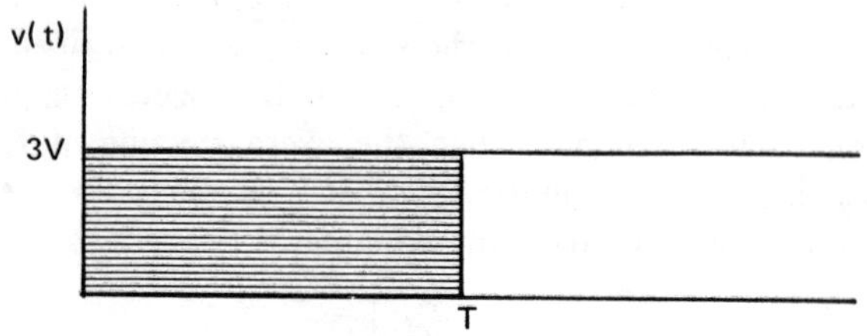

The average value of a *pure sinusoid* is zero, since it has equal positive and negative areas as shown in figure 3.2.

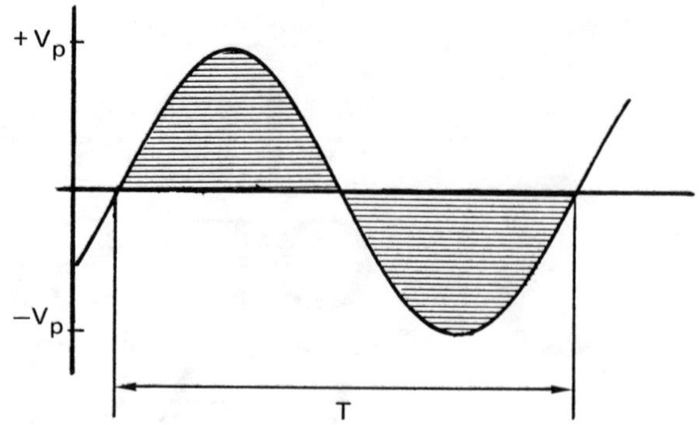

Figure 3.2 A pure sinusoid

A *uniform pulse train* with a pulse width of ΔT seconds and a period of T seconds is shown in figure 3.3. The maximum voltage is V_{max} and the minimum is V_{min}. The net area is

$$V_{\text{min}}T + (V_{\text{max}} - V_{\text{min}})\Delta T$$

This, when divided by the period T (equation 3.2), gives an average value of

$$V_{\text{min}} + (V_{\text{max}} - V_{\text{min}})\frac{\Delta T}{T}$$

Using integration and equation 3.1,

$$v_{\text{avg}} = \frac{1}{T}\int_{t=0}^{\Delta T} V_{\text{max}}\,dt + \frac{1}{T}\int_{t=\Delta T}^{T} V_{\text{min}}\,dt$$
$$= V_{\text{max}}\frac{\Delta T}{T} + V_{\text{min}}\frac{T-\Delta T}{T} \qquad 3.3$$

This is identical to the value found by the area method.

If $V_{\text{min}} = 0$, then

$$v_{\text{avg}} = V_{\text{max}}\frac{\Delta T}{T} \qquad 3.4$$

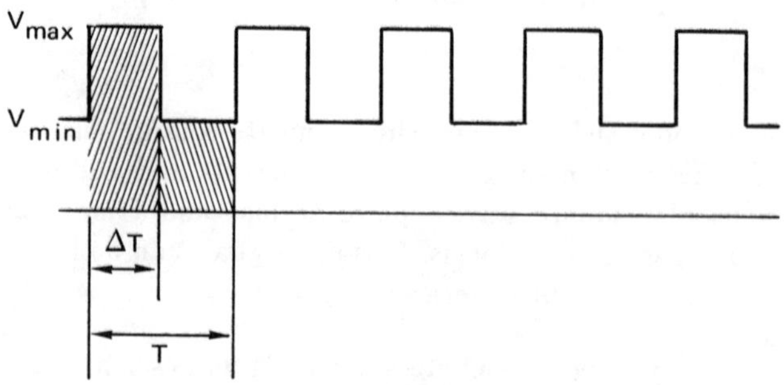

Figure 3.3 A uniform pulse train

If $V_{\text{min}} = -V_{\text{max}}$, and $\Delta T = \frac{T}{2}$ (a square wave), $v_{\text{avg}} = 0$.

A *triangular waveform* as shown in figure 3.4 with $t = 0$ taken at the minimum and $t = T$ at the following minimum. The average value can be found by the area method (equation 3.2).

$$v_{avg} = \left(\frac{1}{2T}\right)(V_{max} - V_{min})T + \left(\frac{1}{T}\right)V_{min}T$$
$$= \frac{1}{2}(V_{max} + V_{min}) \qquad 3.5$$

Equation 3.5 applies to all forms of triangular waves, regardless of whether V_{min} is positive or negative.

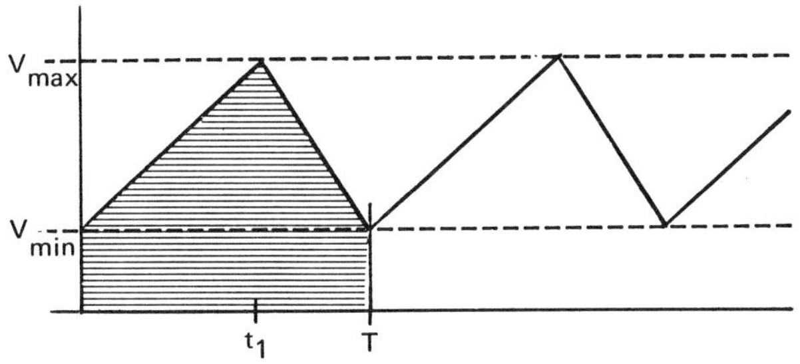

Figure 3.4 A triangular wave

For a *rectified sinusoid*, one half of the wave has been eliminated.[2] For purposes of analysis, it is usually assumed that the negative part has been eliminated. An ideally rectified sinusoid is shown in figure 3.5. During the interval $0 < t < \frac{T}{2}$, the voltage can be represented as

$$v(t) = \begin{cases} V_p \sin \omega t & (t < \frac{T}{2}) \\ 0 & (\frac{T}{2} < t < T) \end{cases} \qquad 3.6$$

$$\omega = \frac{2\pi}{T} \qquad 3.7$$

Equation 3.1 can be used to find the average value. The following calculation recognizes that $\omega T = 2\pi$.

$$v_{avg} = \frac{1}{T}\int_{t=0}^{\frac{T}{2}} V_p \sin \omega t \, dt + \frac{1}{T}\int_{t=\frac{T}{2}}^{T} 0 \, dt$$
$$= \frac{V_p}{\omega T}(-\cos \omega t)\Big|_{t=0}^{\frac{T}{2}}$$
$$= \frac{V_p}{\pi} \qquad 3.8$$

[2]Rectified sinusoids are also known as *half-wave sinusoids*.

For a full-wave rectified sinusoid, the period is half as long. The average is $\frac{2V_p}{\pi}$.

Real *rectifier diodes* will not pass current until a threshold voltage is reached. The rectified voltage is the difference between the sinusoid and the threshold value. This is illustrated in figure 3.6 for full-wave rectification. The first graph shows the ideal rectified waveform and the threshold level; the second graph shows the net voltage.

Because of symmetry, half the average value can be found from equation 3.1 by integrating from $t = 0$ to $t = \frac{T}{2}$. The waveform is represented by equation 3.9.

$$v(t) = \begin{cases} V_p \cos \frac{\pi}{T}t - V_{thr} & \text{for } 0 < t < t_1 \\ 0 & \text{for } t_1 < t < \frac{T}{2} \end{cases} \qquad 3.9$$

At $t = t_1$,

$$V_p \cos \frac{\pi}{T}t_1 = V_{thr} \qquad 3.10$$

$$\frac{1}{2}v_{avg} = \frac{1}{T}\int_{t=0}^{t_1} (V_p \cos \frac{\pi}{T}t - V_{thr})dt$$
$$= \frac{1}{T}\left(\frac{T}{\pi}\right)V_p \sin \frac{\pi}{T}t_1 - V_{thr}\left(\frac{t_1}{T}\right) \qquad 3.11$$

From equation 3.10,

$$\sin\left(\frac{\pi}{T}\right)t_1 = \sqrt{1 - \left(\frac{V_{thr}}{V_p}\right)^2} \qquad 3.12$$

$$\frac{t_1}{T} = \frac{1}{\pi}\arccos\left(\frac{V_{thr}}{V_p}\right) \qquad 3.13$$

$$v_{avg} = \frac{2}{\pi}V_p\sqrt{1 - \left(\frac{V_{thr}}{V_p}\right)^2} - \frac{2}{\pi}V_{thr}\arccos\left(\frac{V_{thr}}{V_p}\right) \qquad 3.14$$

The arccos in equation 3.14 must be expressed in radians. If it is assumed that V_{thr} is negligible (i.e., diodes are ideal), the following errors are generated.

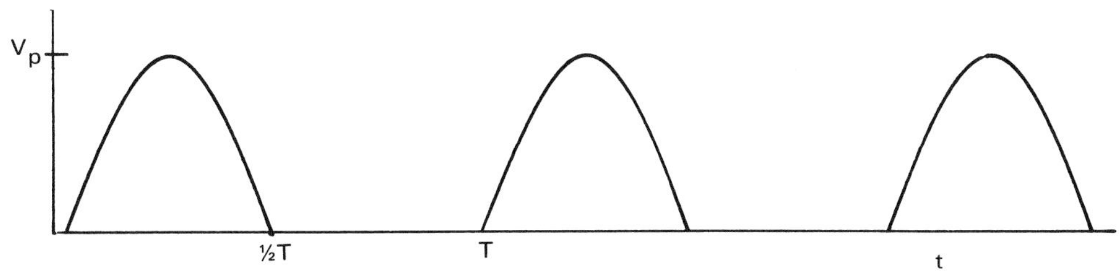

Figure 3.5 A rectified sinusoid

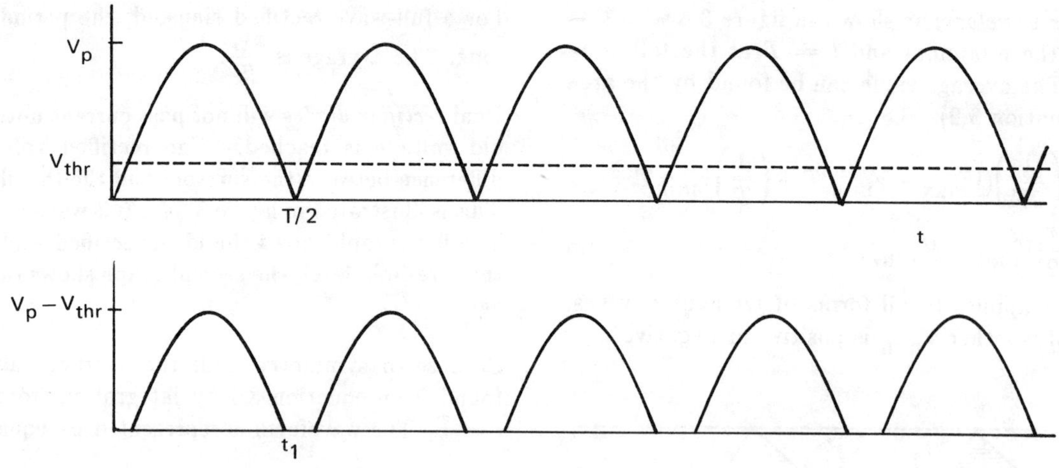

Figure 3.6. Determination of the average value of a sinusoid after full-wave rectification with non-ideal diodes.

1% error for $\frac{V_{thr}}{V_p} = 0.0064$

2% error for $\frac{V_{thr}}{V_p} = 0.0128$

5% error for $\frac{V_{thr}}{V_p} = 0.0321$

10% error for $\frac{V_{thr}}{V_p} = 0.0650$

2 RMS VALUES

The *effective value* of a steady-state voltage or current waveform can be calculated from a single cycle, since all cycles are identical. The effective voltage or current value, when combined with the circuit resistance, will determine the average power for a cycle. For a single complete cycle, the effective value is calculated from the average of the square of the signal. The effective value is, then, the square Root of the Mean of the Square of the signal, hence the name RMS.

$$f_{RMS}^2 = \frac{1}{T}\int_{t_1}^{t_1+T} f^2(t)\,dt \qquad 3.15$$

T in equation 3.15 is the period of the repeating function $f(t)$. t_1 is a convenient time to begin the evaluation of the integral.

Example 3.2

Calculate the RMS of a constant 3 volt signal.

Using equation 3.15, the RMS value is

$$v_{RMS}^2 = \frac{1}{T}\int_0^T 3^2\,dt = \left(\frac{1}{T}\right)9(T-0)$$
$$= 9$$
$$v_{RMS} = \sqrt{9} = 3 \text{ volts}$$

For a pure *sinusoidal voltage,*

$$v(t) = V_p \cos(\omega t + \phi) \qquad 3.16$$

V_p is the peak value or amplitude, $\omega = \frac{2\pi}{T}$ is the radian frequency, and ϕ is the arbitrary phase angle. From equation 3.15,

$$v_{RMS}^2 = \frac{1}{T}\int_0^T V_p^2 \cos^2(\omega t + \phi)dt \qquad 3.17$$

$$\cos^2 x = \frac{1}{2}(1 + \cos 2x) \qquad 3.18$$

$$v_{RMS}^2 = \frac{1}{T}\int_0^T \frac{1}{2}V_p^2 dt + \frac{1}{T}\int_0^T \frac{1}{2}V_p^2 \cos(2\omega t + 2\phi)dt$$
$$= \frac{1}{2}V_p^2 + 0 \qquad 3.19$$

$$v_{RMS} = \sqrt{\left(\frac{1}{2}\right)}V_p = .707V_p \qquad 3.20$$

The RMS value of a *uniform pulse train* as shown in figure 3.3 is

$$v_{RMS}^2 = \frac{1}{T}\int_0^{\Delta T} V_{max}^2 dt + \frac{1}{T}\int_{\Delta T}^T V_{min}^2 dt$$
$$= V_{max}^2 \frac{\Delta T}{T} + V_{min}^2 \frac{T - \Delta T}{T}$$
$$= \left(V_{max}^2 - V_{min}^2\right)\frac{\Delta T}{T} + V_{min}^2 \qquad 3.21$$

If $V_{min} = 0$, then

$$v_{RMS} = V_{max}\sqrt{\frac{\Delta T}{T}} \qquad 3.22$$

If the pulse train is a square wave with $V_{min} = -V_{max}$, then $v_{RMS} = V_{max}$.

For simplification, the *triangular wave* shown in figure 3.4 is redefined in figure 3.7.

$$v(t) = \begin{cases} \dfrac{-(V_{max} - V_{min})}{t_a} t + V_{min} (-t_a < t < 0) \\ \dfrac{V_{max} - V_{min}}{t_b} t + V_{min} (0 < t < t_b) \end{cases} \qquad 3.23$$

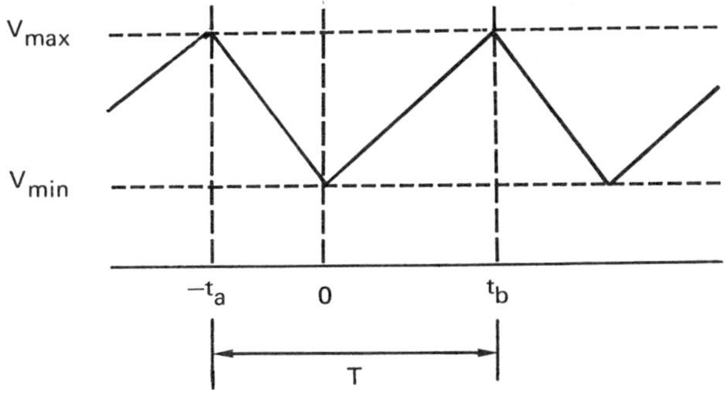

Figure 3.7 Triangular wave redefined

The application of equation 3.15 to figure 3.7 is

$$v_{RMS}^2 = \frac{1}{T}\int_{-t_a}^{0}\left(-\frac{V_{max} - V_{min}}{t_a}t + V_{min}\right)^2 dt + \frac{1}{T}\int_{0}^{t_b}\left(\frac{V_{max} - V_{min}}{t_b}t + V_{min}\right)^2 dt$$

$$= \frac{1}{3}(V_{max}^2 + V_{min}^2 + V_{max}V_{min}) \qquad 3.24$$

If $V_{min} = 0$ or if $V_{min} = -V_{max}$, then

$$v_{RMS} = \frac{V_{max}}{\sqrt{3}} \qquad 3.25$$

For the *half-wave rectified sinusoid* shown in figure 3.5, equation 3.15 becomes

$$v_{RMS}^2 = \frac{1}{T}\int_0^{\frac{T}{2}} V_p^2 \sin^2 \omega t \, dt$$

$$= \frac{V_p^2}{T}\int_0^{\frac{T}{2}} \frac{1}{2}(1 - \cos 2\omega t)dt$$

$$= \frac{1}{4}V_p^2 \qquad 3.26$$

Thus, $v_{RMS} = \frac{1}{2}V_p$.

A full wave rectified sinusoid has the same RMS value as a pure sinusoid. The RMS value of a sinusoid after full-wave rectification using diodes with a threshold can also be calculated. From figure 3.6, the application of equation 3.15 is

$$\frac{1}{2}v_{RMS}^2 = \frac{1}{T}\int_0^{t_1} (V_p \cos \omega t - V_{thr})^2 dt$$

$$= \frac{1}{T}\int_0^{t_1}\left(\frac{1}{2}V_p^2(1 + \cos 2\omega t) - 2V_pV_{thr}\cos \omega t + V_{thr}^2\right)dt$$

$$= \frac{1}{T}\left(\frac{1}{2}V_p^2 t_1 + \frac{V_p^2}{4\omega}\sin 2\omega t_1 - 2\frac{V_pV_{thr}}{\omega}\sin \omega t_1 + V_{thr}^2 t_1\right) \qquad 3.27$$

However,

$$\omega T = \pi \qquad 3.28$$

$$\sin 2\omega t_1 = 2 \sin \omega t_1 \cos \omega t_1 \qquad 3.29$$

Therefore,

$$\frac{1}{2}v_{RMS}^2 = \left(\frac{1}{2}V_p^2 + V_{thr}^2\right)\frac{t_1}{T} + \frac{1}{\pi}\sin \omega t_1 \left(\frac{1}{2}V_p^2 \cos \omega t_1 - 2V_pV_{thr}\right) \qquad 3.30$$

Substituting equations 3.10, 3.12, and 3.13,

$$v_{RMS}^2 = \frac{1}{2}V_p^2\left(\left(1 + \frac{2V_{thr}^2}{V_p^2}\right)\left(\frac{2}{\pi}\right)\arccos\frac{V_{thr}}{V_p} - \frac{6V_{thr}}{\pi V_p}\sqrt{1 - \left(\frac{V_{thr}}{V_p}\right)^2}\right) \qquad 3.31$$

If

$$\frac{V_{thr}}{V_p} = 0$$

then

$$\arccos 0 = \frac{\pi}{2}$$

Then

$$v_{RMS}^2 = \frac{1}{2}V_p^2$$

This is the same as for a complete sinusoid.

3 PARAMETERS OF FOURIER SERIES

All periodic functions can be represented by a *Fourier series.* This series, as discussed in chapter 1, can be written in the form

$$f(t) = \frac{1}{2}a_0 + \sum_{n=1}^{\infty} a_n \cos\frac{2\pi n}{T}t + \sum_{n=1}^{\infty} b_n \sin\frac{2\pi n}{T}t \quad 3.32$$

The *combined form* can also be used.

$$f(t) = \frac{1}{2}a_0 + \sum_{n=1}^{\infty} (a_n^2 + b_n^2)^{\frac{1}{2}} \cos\left(\frac{2\pi n}{T}t - \phi_n\right) \quad 3.33$$

$$\phi_n = \arctan\left(\frac{b_n}{a_n}\right) \quad 3.34$$

The average value of the function represented by the Fourier series is

$$f_{\text{avg}} = \frac{1}{2}a_0$$

The RMS value is

$$f_{\text{RMS}} = \sqrt{\left(\frac{1}{2}a_0\right)^2 + \frac{1}{2}\sum_{n=1}^{\infty}(a_n^2 + b_n^2)} \quad 3.35$$

Example 3.3

A voltage waveform is approximated by the truncated Fourier series $v(t)$. Determine (a) the average value, (b) the fundamental frequency, and (c) the RMS value.

$$v(t) = 5 + 3\cos 100t + .5\cos 300t + .1\cos 600t$$
$$+ 4\sin 100t + .2\sin 200t$$

(a) The value of $\frac{1}{2}a_0$ is 5, therefore, $v_{\text{avg}} = 5$.

(b) The lowest frequency, other than zero, is 100 $\frac{\text{rad}}{\text{sec}}$, so the fundamental frequency is

$$\frac{100}{2\pi} = 15.9\,\text{Hz}$$

(c) From equation 3.35,

$$v_{\text{RMS}}^2 = 5^2 + \frac{1}{2}(3^2 + .5^2 + .1^2 + 4^2 + .2^2)$$
$$= 37.64$$
$$v_{\text{RMS}} = 6.13 \text{ volts}$$

Frequency spectra are generally defined in terms of energy content per unit frequency. For a Fourier series with discrete frequencies, it is more convenient to define energy per second or average power as a function of frequency. Since power is proportional to the square of the RMS signal value and the Fourier series is a series of pure sinusoids, the power spectra can be constructed from the RMS values of each of the sinusoids. If ω_f is the fundamental radian frequency, then

$$P(0) = \left(\frac{1}{2}a_0\right)^2 \quad 3.36$$

$$P(\omega_f) = \frac{1}{2}(a_1^2 + b_1^2) \quad 3.37$$

$$P(2\omega_f) = \frac{1}{2}(a_2^2 + b_2^2) \quad 3.38$$

$$P(n\omega_f) = \frac{1}{2}(a_n^2 + b_n^2) \quad 3.39$$

$$\omega_f = \frac{2\pi}{T} \quad 3.40$$

$P(0)$ is the spectral value at $\omega = 0$; $P(\omega_f)$ is the spectral value at $\omega = \omega_f$; and $P(2\omega_f)$ is the spectral value at $\omega = 2\omega_f$.

The general term is $P(n\omega_f)$, the spectral value at $\omega = n\omega_f$.

Signal spectra for voltage or current differ from power spectra in that the RMS value of each frequency component is used. For example, the RMS values associated with a voltage are

$$v(0) = \sqrt{P(0)} \quad 3.41$$

$$v(\omega_f) = \sqrt{P(\omega_f)} \quad 3.42$$

$$v(2\omega_f) = \sqrt{P(2\omega_f)} \quad 3.43$$

$$v(n\omega_f) = \sqrt{P(n\omega_f)} \quad 3.44$$

Example 3.4

A voltage signal is approximated by the truncated Fourier series

$$v(t) = 2 + 1.5\cos 1000t + .3\sin 2000t$$
$$+ .5\cos 3000t + .1\sin 4000t$$

Plot the power and voltage spectra for this signal.

The power spectrum is found from equations 3.36 through 3.39.

$$P(0) = \left(\frac{1}{2}a_0\right)^2 = 4$$

$$P(1000) = \frac{1}{2}(1.5^2 + 0^2) = 1.125$$

$$P(2000) = \frac{1}{2}(0^2 + 0.3^2) = 0.045$$

$$P(3000) = \frac{1}{2}(0.5^2 + 0^2) = 0.125$$

$$P(4000) = \frac{1}{2}(0^2 + 0.1^2) = 0.005$$

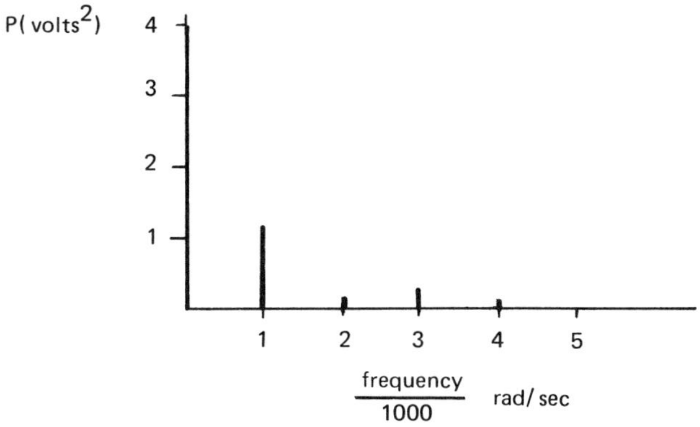

The voltage spectrum is found from equations 3.41 through 3.44.

$$v(0) = \sqrt{4} = 2$$

$$v(1000) = \sqrt{1.125} = 1.061$$

$$v(2000) = \sqrt{.045} = 0.212$$

$$v(3000) = \sqrt{.125} = 0.354$$

$$v(4000) = \sqrt{.005} = 0.071$$

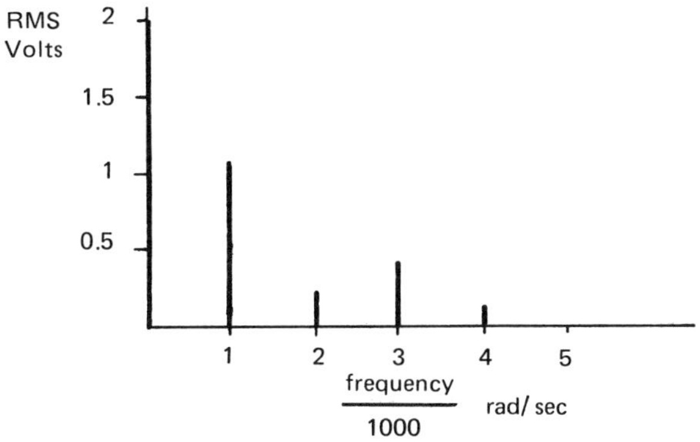

PART 2: MEASUREMENT OF DC AND PERIODIC SIGNALS

4 INTRODUCTION

A variety of instruments is available for the measurement of DC and sinusoidal voltages and currents. Included are several types of electromechanical meter movements, the thermocouple meter, and a variety of electronic instruments.

Electromechanical instruments respond to a voltage or current through the mechanical rotation of a rotor with an attached pointer. These instruments are calibrated so that specific angles of rotation correspond to specific voltage or current levels, and a scale is so marked.

Because of their mechanical inertia, these instruments are normally inadequate in the 5-20 Hz frequency range. Significant inductance limits usefulness to a maximum frequency between 100 and 2500 Hz, depending on the movement type. Such instruments are primarily used in DC or power systems.

Electronic instruments can also employ an electromechanical movement, or they may use a digital readout system. Due to the inclusion of special signal conditioning circuits, electronic meters can be used at much higher frequencies than electromechanical movements

5 PERMANENT MAGNET MOVING COIL INSTRUMENTS

The *permanent magnet moving coil meter movement*, known as the *d'Arsonval movement*, produces an electromagnetic torque proportional to the current flowing in its rotating coil. The pointer is attached to the coil, as is a restraining spring. Coil torque is proportional to the current; the mechanical spring torque is proportional to the angular displacement, but opposite in direction. The pointer will come to rest at the position where the torques are equal and opposite.

The d'Arsonval movement is used primarily to measure average or DC currents and voltages. A similar arrangement is used in data recording systems, where the pointer is replaced with a pen or thermal stylus. Such *pen-recorders* are capable of responding to waveforms with frequencies up to about 15 Hz, although special designs employing compensation can extend the useful range to as much as 100 Hz.

A specific d'Arsonval meter movement is designed to show a full-scale pointer deflection for some specific current, I_{fs}. The usual range of full-scale coil currents is 50 microamps to 10 milliamps. Usually the lower full-scale current requires more turns of wire on the coil, producing a correspondingly higher coil resistance.

Another parameter in ammeter applications is the DC voltage necessary to cause I_{fs} to flow. A standard value of 50 to 100 millivolts (usually 50) is used. This typically requires an additional resistance, called a *swamping resistor*, in series with the coil.

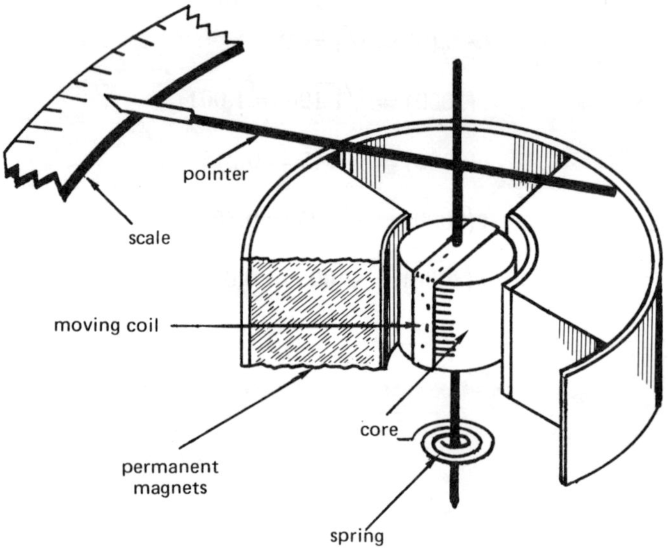

Figure 3.8 d'Arsonval meter movement

A. DC VOLTMETERS

A d'Arsonval meter movement can be incorporated into a *DC voltmeter* (with a full-scale voltage V_{fs}) by connecting a series resistance to limit coil current to I_{fs} when V_{fs} is applied.

$$\frac{R_{\text{coil}} + R_{\text{ext}}}{V_{fs}} = \frac{1}{I_{fs}} \qquad 3.45$$

Equation 3.45 shows that a full-scale swing with a voltage greater than $R_{\text{coil}} I_{fs}$ can be obtained by proper selection of R_{ext}. The quantity $1/I_{fs}$ is fixed for a particular meter movement. It is called the *sensitivity*, with units of $\frac{\text{ohms}}{\text{volt}}$.

Common DC voltmeters ($I_{fs} = 1\,\text{ma}$) have a sensitivity of 1000 ohms per volt, while more expensive models have sensitivities of 5000, 10,000 and 20,000 $\frac{\text{ohms}}{\text{volt}}$.

Accuracy is another design consideration. Movements are available with standard accuracies of 0.5, 1, 2 and 5%. Accuracy is defined as the maximum error relative to the full-scale value, regardless of the current at which that error occurs. In precision instruments a calibration curve is provided to convert a reading from the voltage measured.

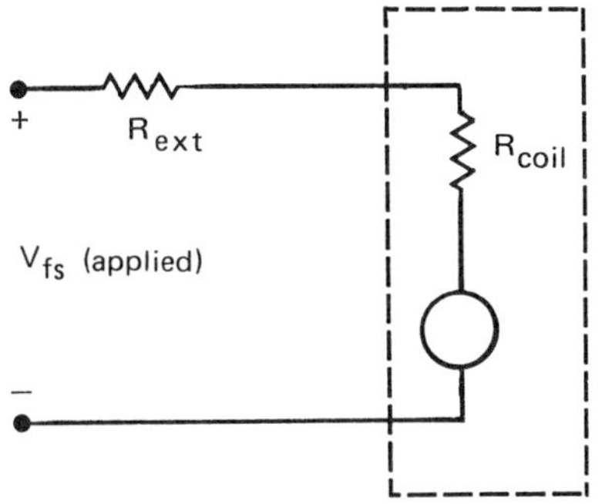

Figure 3.9 DC voltmeter

Example 3.5

A 1000 $\frac{\text{ohm}}{\text{volt}}$, 2% meter movement has a coil resistance of 200 ohms. Specify the external resistance necessary to provide a full-scale reading of 5 volts with an overall accuracy of at least 3%.

Use equation 3.45 with $V_{fs} = 5$ and $R_{coil} = 200$.

$$200 + R_{ext} = (1000)(5)$$
$$R_{ext} = 4800 \text{ ohms (nominal)}$$

To account for the accuracy required, the maximum error introduced by an error in resistance will occur at full-scale voltage. Since the voltage at which maximum error occurs in the meter is unknown, it is assumed to occur at the same value as the resistance error, i.e., at full-scale voltage. Therefore, the maximum error that can be introduced by R_{ext} is 1%. As R_{ext} is much greater than R_{coil}, a 1% tolerance on R_{ext} is the maximum that can be allowed. Therefore, R_{ext} must be a 4800 ohm, 1% resistor.

B. DC AMMETERS

The d'Arsonval movement, when used as an ammeter, is typically specified in terms of the voltage needed to give full-scale pointer deflection. The standard value for full-scale deflection is 50 millivolts, which is more than the coil usually needs to cause I_{fs} to flow. Therefore, an external resistance, called a *swamping resistor*, is added in series with the coil as shown in figure 3.10. The value of the swamping resistance can be found from equation 3.45 with I_{fs} and V_{fs} known.

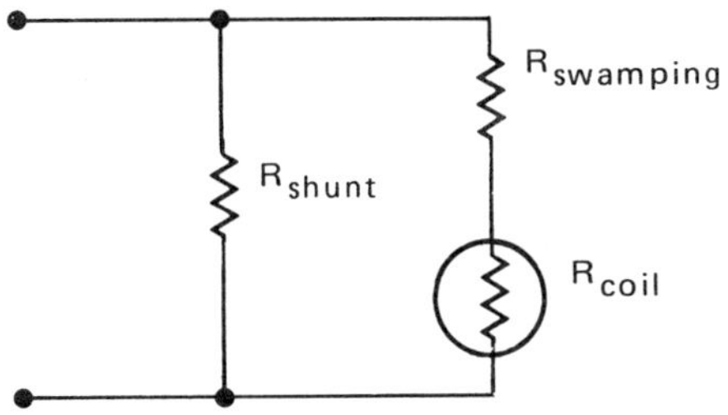

Figure 3.10 Design of a DC ammeter using a swamping resistance and a shunt resistance for a 50 mV system

The standard ammeter is designed so that the shunt will have 50 mV across it, and the coil will have I_{fs} flowing when the sum of the coil and shunt currents are equal to the desired full-scale current.

$$I_{\text{design full-scale}} = \frac{V_{fs}}{R_{\text{shunt}}} + I_{fs(movement)} \qquad 3.46$$

Example 3.6

A 1% d'Arsonval movement has coil and swamping resistances adding to 200 ohms. The full-scale voltage is 50 mV. Determine the full-scale current when no shunt resistance is connected and the shunt resistance and its tolerance to produce a 1 amp full-scale current meter with an overall accuracy of at least 2%.

$$I_{fs} = \frac{50 \times 10^{-3} V}{200 \text{ ohms}} = 2.5\,10^{-4}\, amps$$

From equation 3.46,

$$R_{\text{shunt}} = \frac{50 \times 10^{-3}}{1 - (2.5\,10^{-4})} = 0.05001 \text{ ohms}$$

R_{shunt} must be 0.05 ohms with a tolerance of 1%.

6 THE ELECTRODYNAMOMETER MOVEMENT

Electrodynamometer movements differ from the d'Arsonval movement in that the permanent magnet of the d'Arsonval movement is replaced by a fixed field coil. It has a moving coil, pointer and scale similar to the d'Arsonval, but the scale is non-linear (square-law) in voltmeter or ammeter applications.

The field and moving coils produce magnetic fields which interact when both are carrying currents. The magnetic fields of the two coils try to align with one-another, producing an instantaneous torque on the moving coil that is proportional to the product of the two currents.

The mechanical portion of the system, including inertia and springs, causes the pointer to move into an equilibrium position which is proportional to the average of the product of the two coil currents. Thus, the electrodynamometer movement is an RMS meter when the same current is flowing in both coils.

Electrodynamometers can be used to measure RMS values of voltage and current waveforms in the range of 25 Hz to 2500 Hz. The lower limit is due to the movement's attempt to follow instantaneous values; the upper limit is due to the inductive reactance becoming significant.

The electrodynamometer can be used to measure average values of DC voltages and currents since the average and RMS values are identical. They can also be configured as *wattmeters* over the same frequency range in which AC voltmeters and ammeters are used. They can be calibrated using precisely known DC values, after which they will retain the calibration for AC signals over their entire frequency range. Electrodynamometers, however, generally require more power for operation than comparable d'Arsonval meters.

7 MOVING-IRON INSTRUMENTS

Instruments which make use of the attraction or repulsion of magnetic materials in the presence of a magnetic field are called *moving-iron instruments*. These instruments can be used as voltmeters or ammeters similar to the d'Arsonval movement. They are useful in the measurement of AC signals in the range from 25 to 125 Hz. They are essentially RMS reading (i.e., they use the square-law scale).

They are less useful for DC measurements because of a tendency to acquire significant remnant magnetization. They are typically used to measure voltages and currents in power systems. Sensitivity and power consumption are intermediate to d'Arsonval and electrodynamometer movements.

8 THERMOCOUPLE INSTRUMENTS

When two dissimilar metal wires are joined in a complete circuit and a temperature differential between the two joints is maintained, a current will flow. The current will be proportional to the temperature differential between the two junctions. True RMS measurements of electric currents can be made using this thermocouple effect.

One junction (the *hot junction*) is placed in physical, but not electrical, contact with a wire carrying an unknown electric current. The *cold junction* is connected to a heat sink which remains at the ambient temperature. The current heats the wire and the hot junction. The temperature rise is proportional to the square of unknown current. The current generated by the thermocouple is measured by a d'Arsonval movement properly calibrated with a square-law scale. The result is a true RMS ammeter useful to frequencies up to 100 MHz.

9 RECTIFIER METERS

The linearity and consistency of d'Arsonval movements make them desirable for use in AC measurements. Such use requires *rectification* or *peak detection*. A variety of rectifier arrangements, using from one to four diodes, has long been in use. An undesirable characteristic of such circuits is due to the threshold level of the diodes which causes a serious limitation on the accuracy of measurements at low voltages.

A 1% error results when a diode is used with a threshold 0.64% of the peak value. It is, of course, possible to calibrate the scales to account for the nonlinearity introduced by the diode thresholds, but such calibration is valid only for a single temperature.

Even with ideal diodes (zero or negligible threshold levels), the d'Arsonval movement responds to the average value of the coil current, and it is necessary to calibrate the scale for a specific waveform (presuming RMS values are desired). Rectifier meters are usually designed only for pure sinusoids, using either half-wave or full-wave rectification. The scales are then calibrated to read the RMS value that the corresponding rectified wave would have if it were a pure sinusoid.

A full-wave rectified sinusoid has an average value of $\frac{2V_p}{\pi}$ and an RMS value of $\frac{V_p}{\sqrt{2}}$. The scale of an RMS reading full-wave rectifier d'Arsonval meter will read approximately 1.11 (same as $\frac{\pi}{2\sqrt{2}}$) times the average value of the rectified wave.

A half-wave rectified sinusoid has an average value of $\frac{V_p}{\pi}$, so a meter using half-wave rectification will read approximately 2.22 (same as $\frac{\pi}{\sqrt{2}}$) times the average value of the current flowing in the moving coil.

Example 3.7

A pulse train has $V_{max} = 4.3$ volts, $V_{min} = 0.7$ volts and $\frac{\Delta T}{T} = 0.5$. (Refer to figure 3.3.) Voltage is measured with a full-wave rectifier meter calibrated for RMS values of pure sinusoids. Assume ideal (zero threshold) diodes. Determine the meter reading.

For full-wave rectification the reading will be 1.11 times the average voltage of the waveform. The average value of this wave is

$$\frac{1}{2}(V_{\text{max}} + V_{\text{min}}) = 2.5 \text{ volts}$$

Thus the meter will read $1.11 \times 2.5 = 2.78$ volts.

10 ELECTRONIC ANALOG METERS

Electronic analog meters will eliminate the problems of threshold voltages and other nonlinearities associated with diodes. It is expedient to measure either the peak value of the waveform or the difference between the maximum and minimum values. A scale factor is included to give the correct reading when the waveform is a pure sinusoid. The peak-to-peak measurement

$$\left(V_{\text{max}} - V_{\text{min}}\right)$$

has the advantage of being applicable to a pure sinusoid added to an average value (as is found in amplifier circuits).

Example 3.8

A pulse train with $V_{\text{max}} = 4.3$ volts, $V_{\text{min}} = 0.7$ volts, and $\frac{\Delta T}{T} = 0.5$ is measured with an electronic analog meter using peak-to-peak detection. Determine the meter reading.

The peak-to-peak voltage is $4.3 - .7 = 3.6$ volts. The meter will assume this is $2V_p$.

$$V_{\text{RMS}} = \frac{V_p}{\sqrt{2}} = \frac{3.6}{2\sqrt{2}}$$

$$= 1.27 \text{ volts}$$

11 DIGITAL METERS

Ordinary *digital meters* use circuitry similar to electronic analog meters. Either peak or peak-to-peak values are obtained. One of several schemes is used to display the reading on a numerical display. Accuracy of such meters is typically one-half of the least significant binary digit. They have the same problems with non-sinusoidal signals as the electronic meters calibrated to read RMS values of sinusoids.

Digital meters based on microprocessors can be programmed for any desired parameter of a waveform. In particular, they can be used when a waveform is not a pure sinusoid.

PART 3: POWER WITH SINUSOIDAL CIRCUITS

12 REVIEW OF PHASORS

Phasor representation of steady state sinusoids and the accompanying impedances permits the analysis and design of complicated systems without having to use differential equations. The phasor representation was previously introduced in chapter 2. The *complex plane* is shown in figure 3.11. Sinusoidal voltages, currents, and impedances can be represented by points on the complex plane. The horizontal axis represents real values, and the vertical axis represents imaginary values. Since *complex numbers* contain real and imaginary parts, the plane of points with real and imaginary coordinates is called a complex plane.

Two forms of complex quantities are used: *rectangular* and *polar forms.* A complex value **F** with a real part A and an imaginary part B is written in the rectangular form as

$$\mathbf{F} = A + jB \qquad 3.47$$

$$j = \sqrt{-1} \qquad 3.48$$

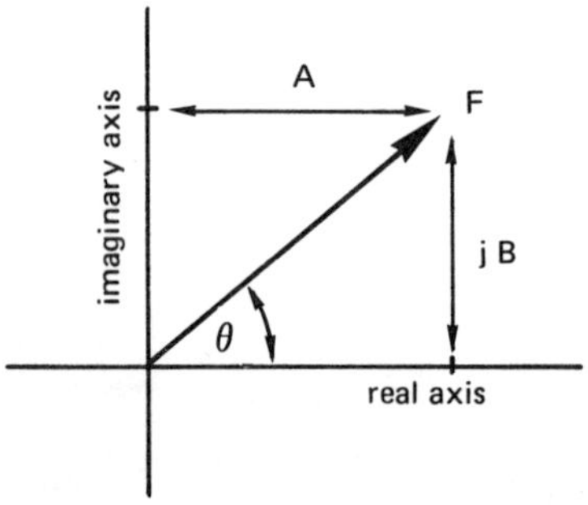

Figure 3.11 the complex plane

F can be written in the polar form as

$$\mathbf{F} = Ce^{j\theta} \qquad 3.49$$

F is a point in the complex plane, but it is typically represented in a vector-like fashion. The vector is directed to the actual point from the origin.

$$C = \sqrt{A^2 + B^2} \qquad 3.50$$

$$A = C\cos\theta \qquad 3.51$$

$$B = C\sin\theta \qquad 3.52$$

$$\tan\theta = \frac{B}{A} \qquad 3.53$$

Phasor representation of sinusoidal voltages and currents is based on the polar form with the angle θ being a function of time.

$$v = V_{\text{max}}\cos(\omega t + \phi) \qquad 3.54$$

Substituting θ for $\omega t + \phi$, v becomes the *real part* of $V_{\text{max}}e^{j(\omega t+\phi)}$. If the function *sine* is used rather than *cosine,* the *imaginary part* of $V_{\text{max}}e^{j\omega t}$ is obtained. Thus, the exponential form of a sinusoid can be used to simplify calculations. The real or imaginary part of the calculated voltage or current is obtained, depending on whether the *cosine* or the *sine* function, respectively, is used. Since

$$e^{j(\omega t+\phi)} = e^{j\omega t}e^{j\phi}$$

the *fixed part* can be separated from the *time variable part* in this notation. These are shown in figure 3.12 for several values of ωt. Notice that the magnitude of $e^{j\omega t}$ remains equal to 1, but the angle increases (rotating counterclockwise) linearly with time. All functions having a component $e^{j\omega t}$ are assumed to rotate counterclockwise with an angular velocity of ω on the complex plane.

Multiplying $e^{j\omega t}$ and $e^{j\phi}$ results in the addition of the exponents: $e^{j(\omega t+\phi)}$. Thus the product gives the points on figure 3.12(b) for times corresponding to those of figure 3.12(a).

Voltage can be represented as

$$v = V_{\text{max}}e^{j\phi}e^{j\omega t} \qquad 3.55$$

In electric circuits involving sinusoids it is more convenient to deal with RMS values of voltage and current rather than with peak values, and with angles in degrees rather than in radians. (Angles in the exponential form must be in radians for mathematical calculations.) Thus, an alternative representation, which is called the *effective value phasor notation* has evolved.

$$v = v_{\text{RMS}}\angle\theta \qquad 3.56$$

$$v_{\text{RMS}} = \frac{V_{\text{max}}}{\sqrt{2}} \qquad 3.57$$

The *reference phasor* is one whose value is known. All other voltages and currents have their positions on the complex plane measured from the reference phasor at the time that the reference phasor is on the positive real axis. The relative angles between phasor voltages and currents will remain constant for sinusoidal steady state operation.

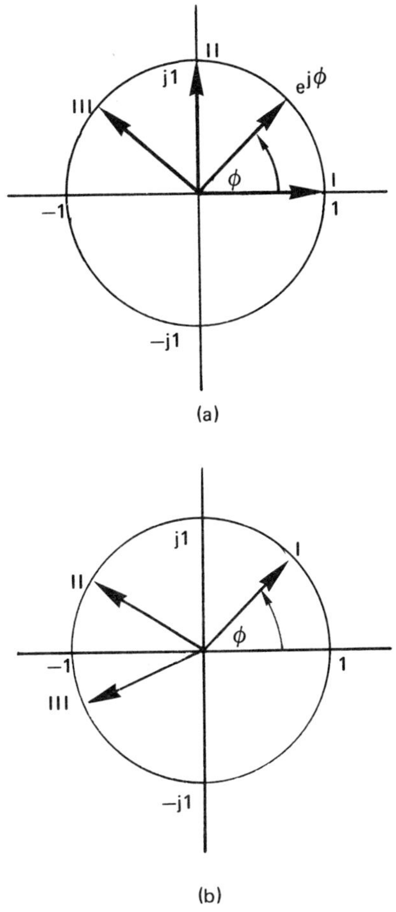

Figure 3.12 (a) $e^{j\phi}$ for all time and $e^{j\omega t}$ for (I) $\omega t = 0$, (II) $\omega t = \frac{\pi}{2}$, and (III) $\omega t = \frac{3\pi}{4}$; and (b) showing $e^{j(\omega t+\phi)}$ for (I) $\omega t = 0$, (II) $\omega t = \frac{\pi}{2}$ and (III) $\omega t = \frac{3\pi}{4}$

13 PHASOR POWER

Power into a pair of terminals is the instantaneous product of the voltage and current. With the voltage at an angle θ and the current at an angle ϕ, the power is

$$p(t) = [V_{max}\cos(\omega t + \theta)] \times [I_{max}\cos(\omega t + \phi)] \quad 3.58$$

Using trigonometric identities, equation 3.58 becomes

$$p(t) = \frac{V_{max}I_{max}}{2}[\cos(\theta - \phi) + \cos(2\omega t + \theta + \phi)] \quad 3.59$$

The first cosine term in the brackets is a constant. The second term is a sinusoidal time function with an average of zero. Thus, the average power is

$$P = V_{RMS}I_{RMS}\cos(\theta - \phi) \quad 3.60$$

The average power can also be found using phasor notation.

$$v = V_{RMS}\angle\theta \quad 3.61$$

$$i = I_{RMS}\angle\phi \quad 3.62$$

$$P = iv = V_{RMS}I_{RMS}\angle(\theta + \phi) \quad 3.63$$

Equation 3.63 is in error, since the power angle should be the difference between the voltage and the current angles as in equation 3.60. This is handled by equation 3.64.

$$P = \mathbf{Re}\{VI^*\} \quad 3.64$$

Re means the real part of the bracketed quantity. I^* is the complex conjugate of the phasor I. The complex conjugate is obtained by negating the angle in polar form. In rectangular form it is obtained by negating the imaginary part.

$$\begin{aligned} VI^* &= (V_{RMS}\angle\theta) \times (I_{RMS}\angle -\phi) \\ &= V_{RMS}I_{RMS}\angle(\theta - \phi) \end{aligned} \quad 3.65$$

The real part of equation 3.65 is

$$P = V_{RMS}I_{RMS}\cos(\theta - \phi) \quad 3.66$$

The *power factor* (abbreviated p.f. or PF) is the quantity $\cos(\theta - \phi)$, which is the same as $\cos(\phi - \theta)$ due to the symmetry of the *cosine* function. The *power factor angle* is the difference between the voltage and current angles ($\theta - \phi$ in this case). It is positive when the voltage *leads* the current (i.e., is counterclockwise from the current on the complex phasor plane) and negative when the voltage *lags* the current (i.e., is clockwise from the current on the complex phasor plane) as indicated in figure 3.13.

The *power triangle* consists of the product VI^* and its real and imaginary components. The components are added on a complex plane, but the plane is not a phasor plane because average rather than instantaneous values are used. In most power problems the voltage is the reference quantity. Then, the power triangle is in the first quadrant when the current lags the voltage (i.e., an inductive circuit) and is in the fourth quadrant when the voltage lags the current (i.e., a capacitive circuit).

Figure 3.14 (a) shows the phasor and power diagrams for an inductive circuit, and figure 3.14 (b) shows the diagrams for a capacitive circuit. The phasors are shown stopped at the instant the reference voltage reaches the positive real axis.

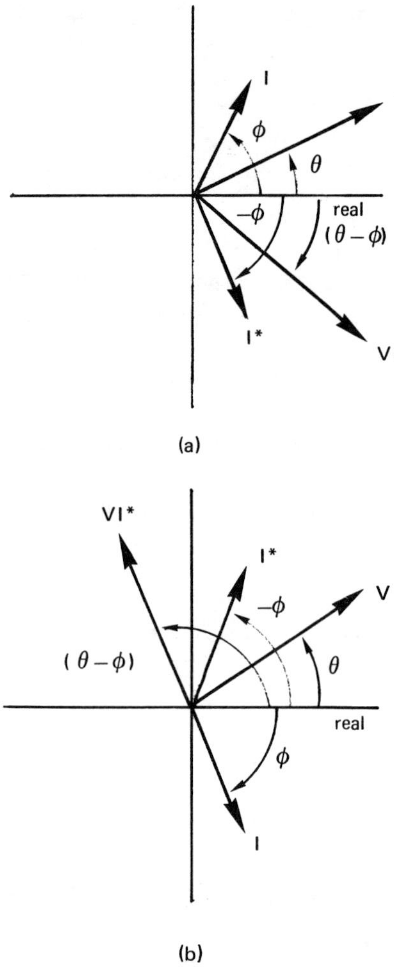

Figure 3.13 Phasor representations of V $\angle\theta$, I $\angle\phi$, $I^* \angle{-\phi}$ and the angle $(\theta - \phi)$ for (a) current leading voltage (capacitive) circuit and (b) voltage leading current (inductive) circuit

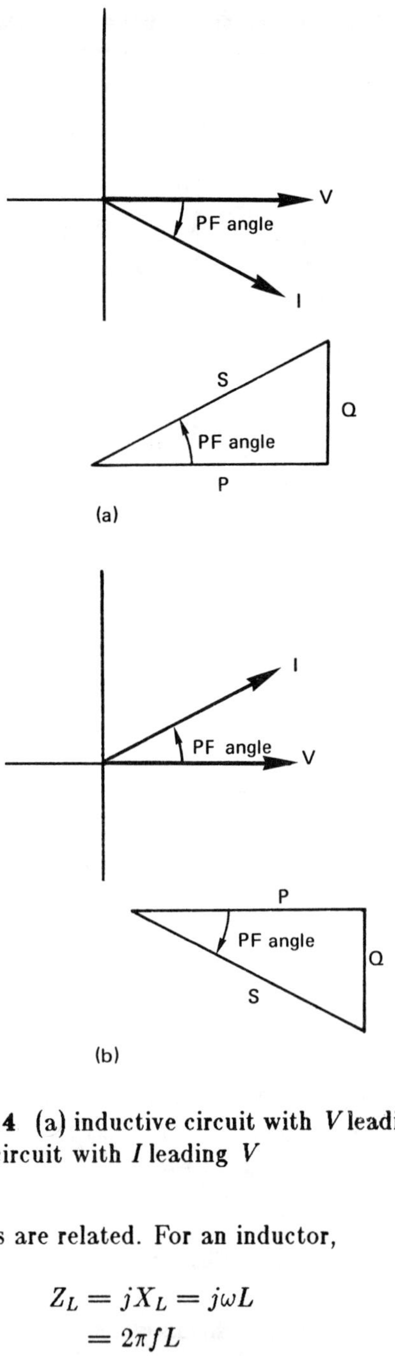

Figure 3.14 (a) inductive circuit with V leading I, (b) capacitive circuit with I leading V

The power triangle is constructed from the *average power* P in watts (W) measured along the real axis; the average *apparent power* S in volt-amperes (VA) as the hypotenuse of the triangle; and the *reactive power* Q in volt-amperes reactive (VAR) in the imaginary direction.

$$S(\text{volt-amps}) = P(\text{watts}) + jQ(\text{volt-amps-reactive}) = VI^* \qquad 3.67$$

$$|S|^2 = |P|^2 + |Q|^2 = I_{\text{RMS}}V_{\text{RMS}} \qquad 3.68$$

$$PF = \frac{|P|}{|S|} \qquad 3.69$$

Reactance is the ratio of voltage to current for an inductor or capacitor in a circuit operating with a sinusoidal voltage. The *impedance* is j times the reactance, symbolized by jX. The negative reciprocal of reactance is *susceptance.*

$$B = \frac{-1}{X} \qquad 3.70$$

These terms are related. For an inductor,

$$Z_L = jX_L = j\omega L = 2\pi f L \qquad 3.71$$

$$Y_L = \frac{1}{Z_L} = \frac{1}{j\omega L} = \frac{-j}{\omega L} = \frac{-j}{2\pi f L} = -jB_L \qquad 3.72$$

For a capacitor,

$$Z_C = \frac{1}{j\omega C} = \frac{-j}{\omega C} = \frac{-j}{2\pi f C} \qquad 3.73$$

$$Y_C = \frac{1}{Z_C} = j\omega C = jB_C \qquad 3.74$$

In the sinusoidal steady state, impedances are considered as a real part plus j times an imaginary part. The real part of the impedance is the *resistive part*, and the imaginary part is the *reactive part*. The same is generally true for admittances, although the real part should be called the *conductive part*, and the imaginary part should be called the *susceptive part*.

X and Y can be written in rectangular form as equations 3.75(a) and 3.75(b).

$$Z = R + jX \text{ (ohms)} \qquad 3.75(a)$$

$$Y = G + jB \text{ (mhos)} \qquad 3.75(b)$$

Example 3.9

In the circuit shown the voltage is 230 volts RMS at 60 Hz.

(a) Obtain the reactance of each reactive element.

(b) Using the current I as a reference, calculate the voltages around the loop, and draw the phasor diagram.

(c) Draw the impedance diagram on the complex impedance plane.

(d) Calculate the current, with its angle given with respect to the voltage V_s.

(e) Determine the power factor.

(f) Determine the power triangle.

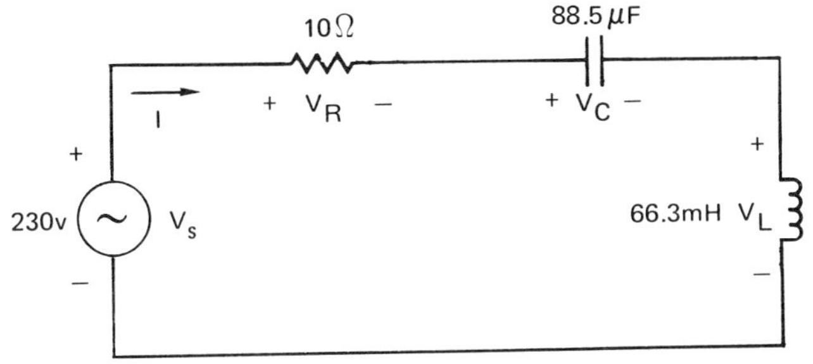

(a)

$$X_C = \frac{-1}{377C} = -29.97$$

$$\approx -30 \text{ ohms}$$

$$X_L = 377L = 24.995$$

$$\approx 25 \text{ ohms}$$

(b)

$$V_R = 10I, \quad V_C = -j30I,$$

and

$$V_L = j25I.$$

$$V_s = V_R + V_C + V_L$$

$$= (10 - j5)I$$

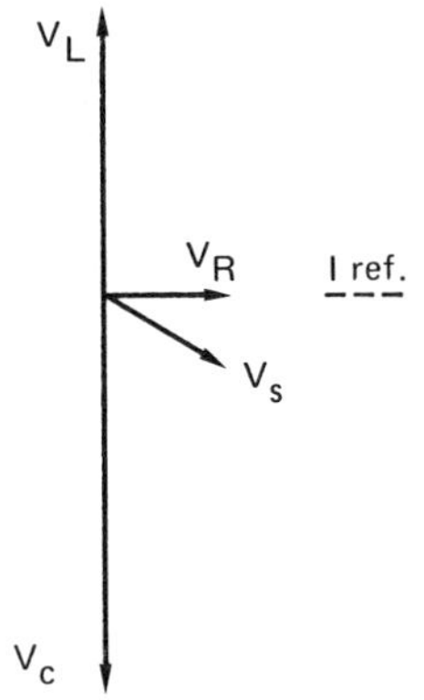

(c)

$$Z_R = 10 \text{ ohms}, \quad Z_C = -j30 \text{ ohms}$$

and

$$Z_L = j25 \text{ ohms}.$$

$$Z_{\text{loop}} = Z_R + Z_C + Z_L$$

$$= 10 - j5$$

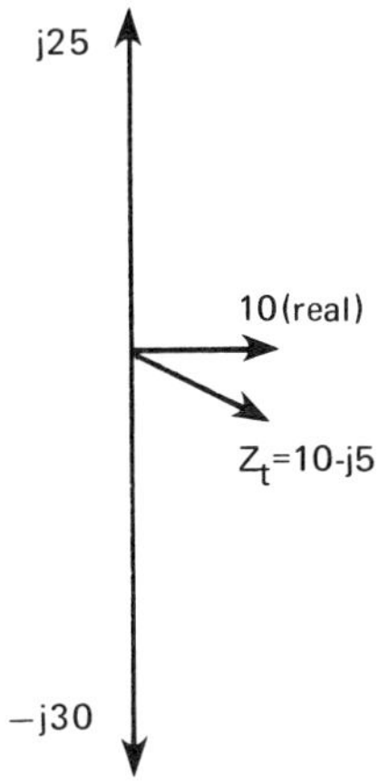

(d)

$$V_s = Z_{\text{total}} I$$

$$230\angle 0^\circ = (10 - j5)I = (11.18\angle{-26.57^\circ})I$$

$$I = \frac{230\angle 0^\circ}{11.18\angle{-26.57^\circ}} = 20.57\angle 26.57^\circ$$

(e)

$$PF = \cos(\angle V - \angle I) = \cos(0 - 26.57^\circ)$$

$$= 0.894$$

(f)

$$|S| = |V| \times |I| = 230 \times 20.57$$

$$= 4731 \text{ VA}$$

$$P = |S| \cos(-26.57^\circ) = 4230 \text{ W}$$

$$Q = |S| \sin(-26.57^\circ) = -2112 \text{ VAR}$$

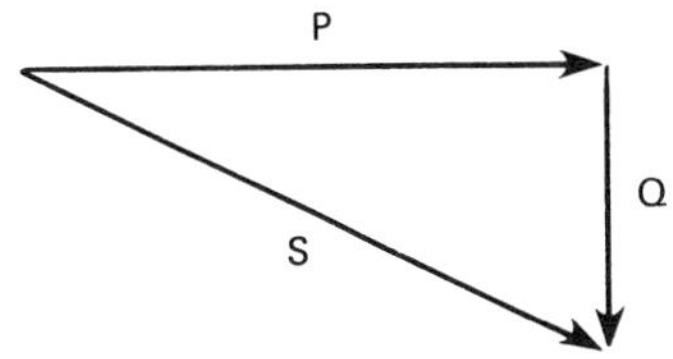

14 POWER FACTOR CORRECTION

Correction in power circuits is generally accomplished by placing a *shunt capacitance* across the load terminals, (i.e., in parallel with the load). A capacitance is usually required to correct the power factor because most power circuits are inductive. The power factor may be specified as *leading* (capacitive) or as *lagging* (inductive).

The power angle is positive for inductive loads involving lagging current and negative for capacitive loads where the current is leading. However, care must be taken when the power factor is known, since the cosines of both positive and negative angles less than 90° are positive. In the absence of a specific information about the leading or lagging nature, it is assumed that the load is inductive.

The power triangle for inductive currents (current lagging voltage) is in the first quadrant, and for capacitive currents (current leading the voltage) the power triangle is in the 4th quadrant.

The fundamental method used in power factor correction is to draw a new line current which cancels all or part of the reactive component already flowing to the original load. The *in-phase component* of the current accounts for the power, P, delivered to the load.

$$\begin{aligned} P &= V \times (\text{component of current in phase with } V) \\ &= VI\cos\theta = VI \times PF \end{aligned} \quad 3.76$$

The component of the current that is *out-of-phase* with V constitutes the reactive component.

$$Q = VI\sin\theta = VI\sqrt{1-PF^2} \quad 3.77$$

The component of current out-of-phase with the voltage is

$$I_{\text{reactive}} = \frac{Q}{V} \quad 3.78$$

The current needed to produce a desired power factor can be determined from the in-phase component of the current, $\frac{P}{V}$, as

$$I_{\text{corrected}} = \left(\frac{P}{V}\right)\frac{\sqrt{1-PF^2}}{PF} \quad 3.79$$

PF is the power factor specified for the corrected system.

The correcting current provides the difference between the new reactive component of the current and the preexisting reactive component of the load current, (Q/V). The *corrected current* is the pre-existing reactive component of the current minus the correcting current.

$$I_{\text{correcting}} = \frac{Q}{V} - \left(\frac{P}{V}\right)\frac{\sqrt{1-PF^2}}{PF} \quad 3.80$$

The reactance necessary to do the correction is

$$\begin{aligned} X_{\text{correcting}} &= \frac{V}{I_{\text{correcting}}} \\ &= \frac{V^2}{Q - \frac{P\sqrt{1-PF^2}}{PF}} \end{aligned} \quad 3.81$$

Q is the initial reactive power in VARS, P is the power (unchanged after the correction takes place), and PF is the desired power factor.

Example 3.10

A load operates at 60 Hz, 440 V, 15 KVA, and 10 KW. Determine the capacitance necessary to correct the power factor to 0.95. Determine the VA rating of the capacitor.

The in-phase component of the load current is

$$\frac{P}{V} = \frac{10,000}{440} = 22.7 \text{ amps}$$

The amplitude of the current is the volt-amperes divided by the voltage.

$$\frac{S}{V} = \frac{15,000}{440} = 34.1 \text{ amps}$$

The reactive component can be determined by trigonometry.

$$\begin{aligned} I^2_{\text{reactive}} &= (34.1)^2 - (22.7)^2 \\ &= 647.5 = (25.5)^2 \end{aligned}$$

The new lagging current is

$$I_{\text{corrected}} = \frac{22.7}{.95}\sqrt{1-(.95)^2} = 7.5$$

The correcting current is

$$I_{\text{correcting}} = 7.5 - 25.5 = -18 \text{ amps (leading)}$$

The reactance necessary to produce such a current is

$$\frac{-440}{18} = -24.4 \text{ ohms (reactive)}$$

As a capacitive reactance is $\frac{1}{2\pi f C}$ and at 60 Hz $2\pi f = 377$,

$$\begin{aligned} C &= \frac{1}{377 \times 24.4} = 1.09 \times 10^{-4} - \mu F \\ &= 109 \mu F \end{aligned}$$

Since the current through the correcting capacitance is 18 amps, the volt-ampere rating is

$$(440)(18) = 7920 \text{ VA (say 8 KVA)}$$

Practice Problems

MEASUREMENTS OF SIGNAL WAVEFORMS

Warmups

1. Using integration, show that the average value of a pure sinusoid is zero.

2. A triangular wave has

$$V_{min} = -V_{max} = -10\,V.$$

After half-wave rectification, find its (a) average value, (b) RMS value.

3. An irregular pulse train repeats itself after 50 mS. The pulses are 1 mS in duration, 2 volts in amplitude, and a total of 10 occur during any 50 mS cycle. Determine the average and RMS values of this waveform.

4. A peak-to-peak detecting AC digital voltmeter shows four significant decimal figures. Determine the resolution (precision) as a percentage of full-scale.

5. A sinusoid has its peak value at $t = 1$ mS, and a period of 10 mS. Express this signal as cosine and sine functions.

6. The voltage $10\cos(100t + 25°)$ is applied to a resistance of 25 ohms and an inductance of 0.5 henrys in parallel. Find the RMS value and the phase angle of the sum of the currents.

7. Construct the power triangle for problem 6.

8. Specify a capacitance in farads to completely correct the power factor of the circuit of problem 6 (i.e., to make $Q = 0$).

Concentrates

1. A voltmeter is being designed to measure voltages in the full-scale ranges of 3, 10, 30 and 100 volts DC. The meter movement to be used has an internal resistance of 50 ohms and a full-scale current of 1 mA. Using a 4-pole, single-throw switch, design the voltmeter.

2. An AC voltmeter consists of a d'Arsonval meter with a full-scale current of 0.125 mA and a series resistance of 10^6 ohms. A full-wave bridge of silicon diodes ($V_{thr} = 0.6$ volts) is used to rectify the AC voltage. The full-scale needle deflection is 50 degrees. Give the scale increments in degrees from 0 volts to 100 volts in 10 volt increments for the RMS value of a pure sinusoid.

3. A train of rectangular pulses is applied to an ideal low-pass filter circuit. The pulse height is 5 volts, and the duration of each pulse is 2 mS. The repetition period is 10 mS. The low-pass filter has a cut-off frequency of 500 Hz. What percentage of the signal power is available at the output of the filter?

4. A full-wave rectifier-type VTVM (vacuum tube voltmeter) is set to an RMS AC scale with a range of 50 volts. The meter is connected to a symmetrical (zero-average) triangular waveform of 100 volts peak-to-peak. What does the meter read?

5. An impedance receives a line current which lags the voltage by 30°. When the voltage across the impedance is 100 volts (RMS), the impedance dissipates 200 watts. Specify the reactance of a capacitance to be placed in parallel with the impedance which would make the line current be in phase with the voltage.

6. A parallel combination of a resistance (10 ohms), capacitance (88.5 microfarads), and inductance (66.3 millihenrys) has 60 Hz, 230 volt (RMS) applied. Obtain the

(a) reactances of C and L.
(b) admittance of each circuit element.
(c) phasor diagram for the currents, using the applied voltage as the reference.
(d) admittance diagram for the circuit, including the total admittance.
(e) input current as a sine function, taking the applied voltage as reference. (Is the circuit inductive or capacitive?)
(f) power factor.
(g) power triangle.

Timed

1. An average-responding meter with an RMS calibrated scale is placed in a full-wave ideal diode bridge. A voltage $v(t) = 6 + 10\cos 1000t$ is measured. Determine the

(a) RMS value.
(b) waveform of the rectified wave.
(c) reading of the meter.

2. A triangular saw-tooth wave has a maximum of 12 volts and a minimum of −4 volts. Determine the meter readings for:

(a) a d'Arsonval voltmeter.
(b) a meter consisting of a d'Arsonval meter and an ideal diode with the meter scale calibrated to read RMS voltage.
(c) a peak-to-peak electronic voltmeter calibrated to read the RMS voltage of a sinusoid.
(d) an electrodynamometer voltmeter.

3. An ammeter is being designed to measure over the five ranges indicated in the accompanying illustration.

The indicating meter is a 1.00 milliampere movement with an internal resistance of 50 ohms. The total resistance ($R_1 + R_2 + R_3 + R_4 + R_5$) is to be 1000 ohms. Specify the resistances R_a and R_1 through R_5.

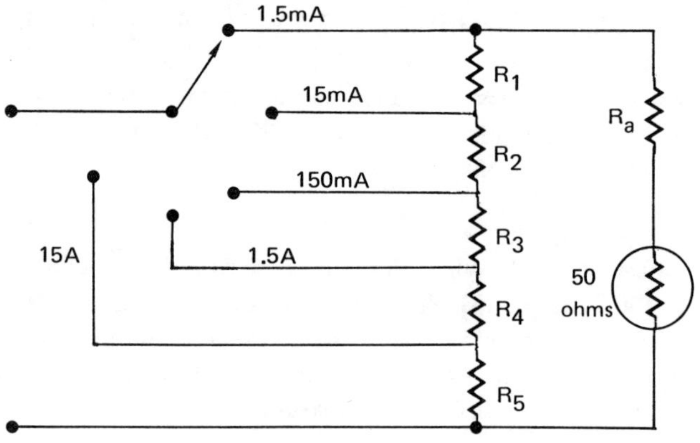

4. A triangular voltage waveform measures 5 volts with a d'Arsonval meter movement. A peak-to-peak reading, RMS calibrated, electronic voltmeter measures the same waveform as 7 volts. What are the maximum and minimum values of this voltage waveform?

5. A load is connected to a voltage of 1320 volts at 60 Hz. The load dissipates 100 KW with a .867 lagging power factor. Specify the capacitance needed to correct the power factor to
   (a) .895 lagging.
   (b) .95 leading.
(Give the voltage and volt-ampere-reactive ratings at 60 Hz for the capacitors.)

Reserved for Future Use

Reserved for Future Use

4 TIME AND FREQUENCY RESPONSE

Nomenclature

C	capacitance	farads
e	base of the natural logarithm	2.718
f	a function of t	-
F	a function of s	-
i	instantaneous current	amps
I	constant current	amps
I(s)	Laplace transformed current (usually subscripted)	amp-sec
j	$\sqrt{-1}$	-
L	inductance	henrys
n	an integer	-
q	a function of t	-
Q	quality factor	-
R	resistance	ohms
s	the Laplace transform variable or a root for natural response	-,rad/sec
t	time	seconds
u(t)	the unit step function	-
v	instantaneous voltage	volts
V	(unsubscripted) voltage	volts
V(s)	Laplace transformed voltage (usually subscripted)	volt-sec
z	impedance	ohms

Symbols

α	exponent	1/sec
β	natural frequency	rad/sec
ζ	damping coefficient	-
τ	time constant	sec
ϕ	an angle	radians
ω	angular frequency	rad/sec
ω_1	lower half power frequency	rad/sec
ω_2	upper half power frequency	rad/sec

Subscripts

C	capacitance
h	high frequency
l	low frequency
L	inductance
m	mid-frequency
n	natural
N	Norton
o	initial value, or resonant
s	source, or signal
ss	steady state
t	terminal
Th	Thevenin

1 INTRODUCTION

This chapter reviews circuit analysis techniques based on the previous chapter. Some basic circuit reduction methods are reviewed by means of examples. The classical solution of differential equations for transient analysis is covered as a review of the physical interpretation of circuit conditions. The application of Laplace transforms expands on the previous material and relates to the classical solutions. Finally, the Laplace transform approach to sinusoidal steady state conditions and a discussion of transfer functions broadens the treatment begun in chapter 3.

2 CIRCUIT REDUCTION TECHNIQUES

A. SOURCE CONVERSION

An electrical source can be modeled either by a voltage source in series with an impedance or by a current source in parallel with an impedance. These models are equivalent to one another in that they have the same terminal voltage-current relationships under the condition that

$$v_{\text{Th}} = i_N Z_{\text{Th}} \qquad 4.1$$

$$i_N = \frac{v_{\text{Th}}}{Z_{\text{Th}}} \qquad 4.2$$

With this equivalency, it is possible to convert a source from one form to the other. This is an essential operation in circuit reduction.

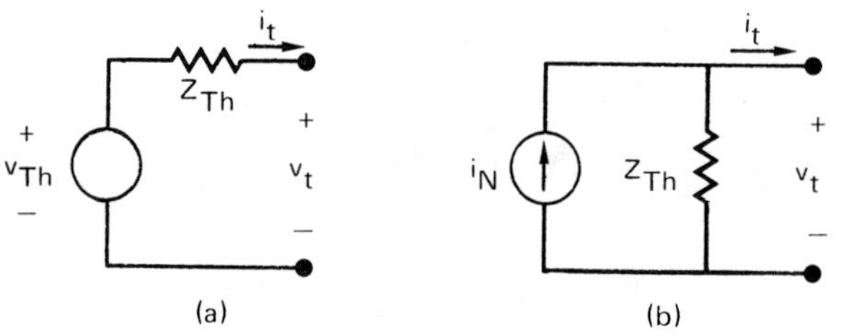

Figure 4.1 (a) Equivalent voltage source, (b) equivalent current source

Example 4.1

A source is modeled as a voltage of $v = 10\cos(20t)$ with an impedance consisting of a 4 ohm resistance and a 0.15 henry inductance in series. Obtain the equivalent current source model.

First, convert the voltage source to phasor form, taking $\cos 20t$ as the reference.

$$v_{Th} = 10\angle 0°$$

Since $\omega = 20$, the inductive reactance is

$$20 \times 0.15 = j3 \text{ ohms}$$

The impedance is $4 + j3$, which should be converted to polar form.

$$Z_{Th} = 4 + j3 = 5\angle 36.87°$$

The value of the current source is found from equation 4.2.

$$I_N = \frac{10\angle 0°}{5\angle 36.87°} = 2\angle -36.87°$$

This can be converted to the time form.

$$i_N = 2\cos(20t - 36.86°)$$

The equivalent source is this current source in parallel with the original impedance as in figure 4.1(b).

There is confusion as to the relative *directions* of the sources in equivalent circuits. A rule is helpful:

- **Rule 4.1** Both the original model and the equivalent must have the same *open-circuit voltage polarity*. Also, both the original model and the equivalent must cause a short-circuit current to flow in the same direction in the short-circuited condition.

B. SERIES AND PARALLEL COMBINATION RULES

The models discussed in section A consist of pure voltage or current sources and impedances. The rules for the combinations of these elements are listed here.

- **Rule 4.2** A pure voltage source model cannot be placed in parallel with another pure voltage source model, as the result would be indeterminant. (This rule also applies to pure current source models in series.)

- **Rule 4.3** A pure voltage source in parallel with a pure current source makes the current source redundant. The voltage is set by the pure voltage source, and the pure voltage source will absorb all the current from the current source.

- **Rule 4.4** A pure current source in series with a voltage source makes the voltage source redundant. The current is set by the current source, and the current source can absorb all the voltage from the voltage source.

- **Rule 4.5** Pure current source models in parallel can be combined algebraically with Kirchhoff's current law.

- **Rule 4.6** Pure voltage sources in series can be combined algebraically with Kirchhoff's voltage law.

- **Rule 4.7** Impedances in series add directly. Impedances in parallel add reciprocally.

C. PARTIAL REDUCTIONS

It is sometimes convenient to reduce a network with a number of sources to a simpler circuit by making combinations described in the previous section. This is done to check analytic results, as each step of the reduction can be easily checked. It is particularly useful when the impedances involved are purely resistive.

Parallel voltage sources

Combining two voltage sources in parallel is best accomplished by first converting both to equivalent current sources. Then, the resulting parallel current sources can be added algebraically, and the parallel impedances can be added reciprocally. It will then be possible to convert the resulting current source to its equivalent voltage source.

Example 4.2

Two voltage sources as shown are to be combined into a single source model.

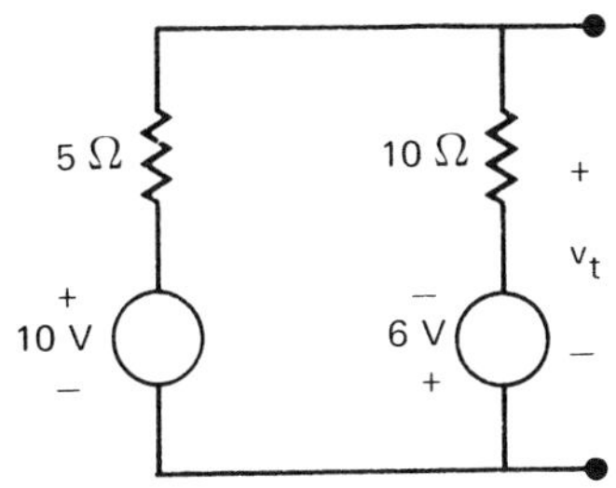

(a)

First, each voltage source is converted to an equivalent current source.

$$i_{N1} = \frac{10}{5} = 2\,(\text{upward})$$

$$i_{N2} = \frac{6}{10} = .6\,(\text{downward})$$

$$= -.6\,(\text{upward})$$

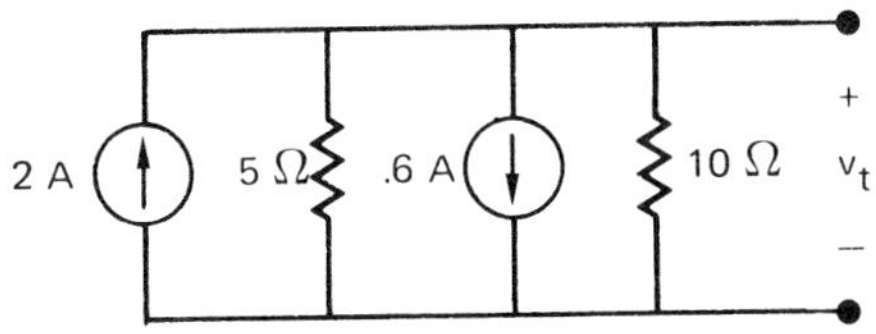

(b)

These two current sources are combined in parallel with the upward direction taken as positive. The result is 1.4 amp.

The two parallel resistances are combined reciprocally to give

$$\frac{1}{R} = \frac{1}{5} + \frac{1}{10} = \frac{3}{10}$$

The combined impedance is $\frac{10}{3}$ ohms.

The reduced source is shown. It could be converted to a voltage source.

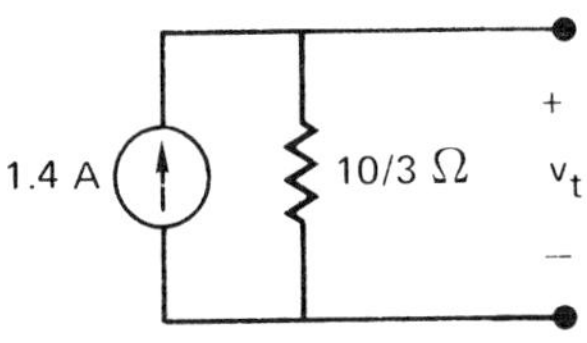

(c)

Series current sources

Current source models in series first should be converted to equivalent voltage sources and then combined by adding the equivalent voltage source values algebraically. The impedances should be added directly.

Example 4.3

Two current source models are shown. These are to be combined into a single source model.

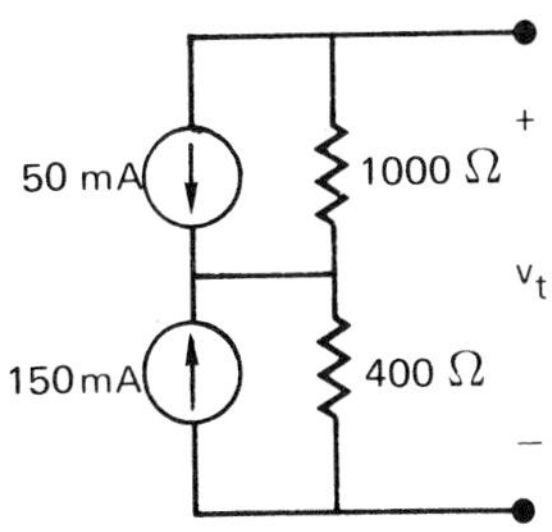

(a)

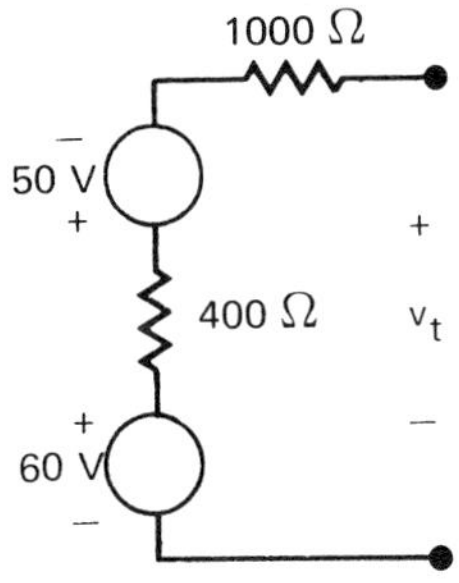

(b)

First, each current source is converted to an equivalent voltage source.

$$v_{\text{Th}1} = .05 \times 1000 = 50\,(\text{down})$$

$$v_{\text{Th}2} = .15 \times 400 = 60\,(\text{up})$$

These equivalent voltage sources are added algebraically to give 10 volts (up). The resistances are added directly to give a total of 1400 ohms.

The resulting reduced source is then as shown.

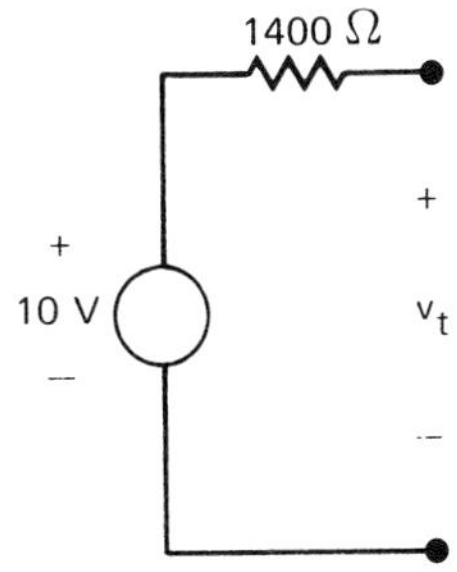

(c)

Iterative method of circuit reduction

The techniques of reducing parallel voltage source models and series current source models can be extended to an iterative method of circuit reduction where an entire network is reduced to a single source and impedance. This is useful with phasor analysis or where capacitance and/or inductance is present in a complex network and the transient voltage or current is of interest. It is also useful in the analysis of circuits containing nonlinear or active elements.

Example 4.4

The network shown has capacitance and inductance which are neither in series nor in parallel. The voltage source v_s is a signal source model, and it is a variable.

Simplify the circuit as much as possible and find the voltage across the capacitance.

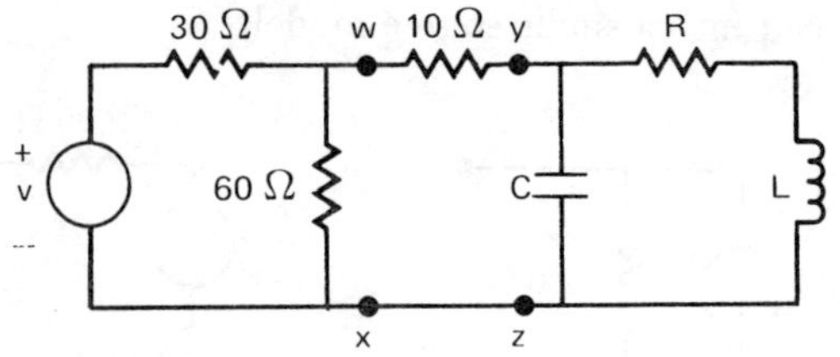

(a)

Note the nodes marked w, x, y, and z. These will be used for intermediate steps in the iterative procedure.

The portion of the network to the left of nodes w-x is first converted to an equivalent current source model. The 30 and 60 ohm resistances are then combined in parallel. The circuit to the left of w-x is shown.

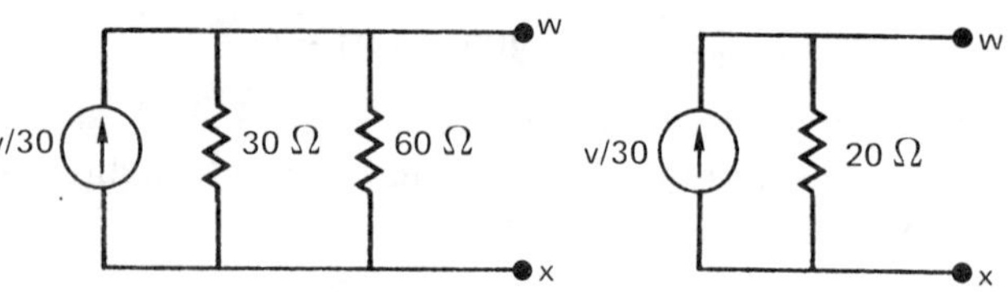

(b) (c)

The current source model to the left of w-x is next converted to an equivalent voltage source. The 20 ohm resistance can be combined in series with the 10 ohm resistance between w and y to further reduce the circuit.

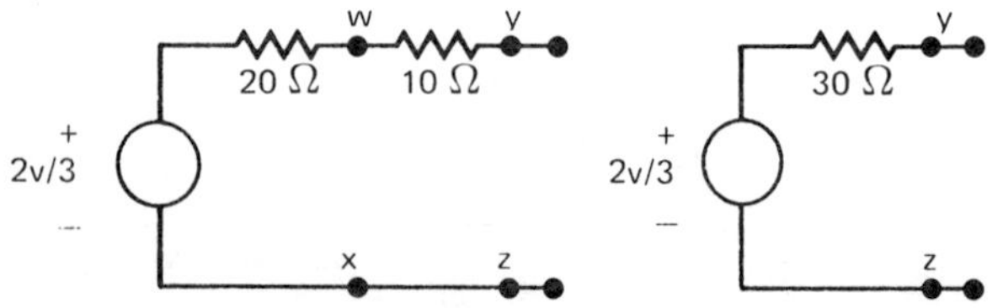

(d) (e)

Finally, the voltage source model is converted to an equivalent current source model. This circuit form is used to establish the circuit equations.

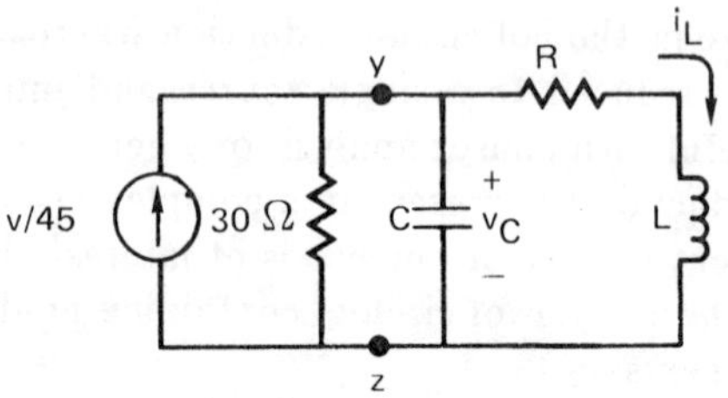

The circuit equations are written in terms of the capacitance voltage and inductance current—the states of the two energy storage elements. The current summation at the node above the capacitance C is

$$\frac{v_S}{45} = \frac{v_C}{30} + C\left(\frac{dv_C}{dt}\right) + i_L$$

The voltage equation around the right-hand loop is

$$v_C = Ri_L + L\left(\frac{di_L}{dt}\right)$$

These two equations are solved most easily by methods of Laplace transformations as outlined in chapter 1. If a sinusoidal analysis was required, the impedance of the inductance and resistance in the right-hand branch would be combined in parallel with the 30 ohm resistance to further simplify the circuit.

D. WYE-DELTA AND DELTA-WYE CONVERSION

It is sometimes convenient (particularly in power networks) to make a more complex conversion involving three terminals. The basic conversions are called *wye-delta* and *delta-wye* transformations. They are alternatively known as the *tee-pi* and *pi-tee* transformations, respectively.

Two forms of a circuit are shown in figure 4.2. Note that for the wye form, the impedances are single subscripted according to the terminal to which they are connected. The delta impedances are double subscripted according to the two terminals to which they are connected.

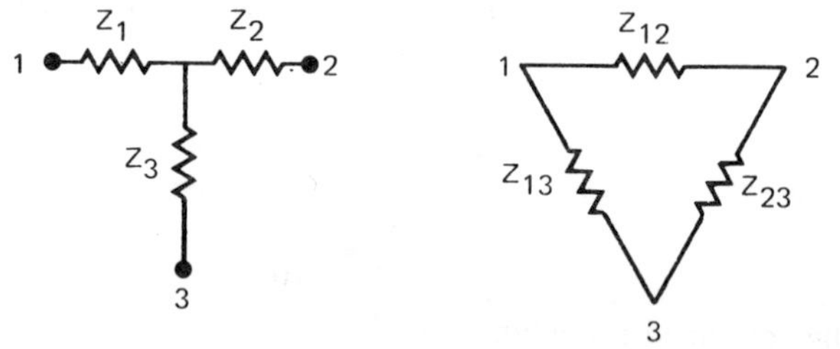

(a) (b)

Figure 4.2 (a) Wye (tee) network, (b) delta (pi) network

These networks are equivalent as long as equations 4.3, 4.4, and 4.5 are valid.

$$Z_{12} = \frac{Z_1Z_2 + Z_1Z_3 + Z_2Z_3}{Z_3} \qquad 4.3$$

$$Z_{13} = \frac{Z_1Z_2 + Z_1Z_3 + Z_2Z_3}{Z_2} \qquad 4.4$$

$$Z_{23} = \frac{Z_1Z_2 + Z_1Z_3 + Z_2Z_3}{Z_1} \qquad 4.5$$

Equations 4.3, 4.4, and 4.5 can be used to convert a wye (tee) circuit to a delta (pi) circuit. In order to convert in the other direction, equations 4.6, 4.7, and 4.8 are needed.

$$Z_1 = \frac{Z_{12}Z_{13}}{Z_{12} + Z_{13} + Z_{23}} \qquad 4.6$$

$$Z_2 = \frac{Z_{12}Z_{23}}{Z_{12} + Z_{13} + Z_{23}} \qquad 4.7$$

$$Z_3 = \frac{Z_{13}Z_{23}}{Z_{12} + Z_{13} + Z_{23}} \qquad 4.8$$

Example 4.5

In example 4.4, a wye configuration, repeated here for convenience, consisted of 30, 60, and 10 ohm resistances. (Note the terminal designations.) Convert this configuration to a delta structure.

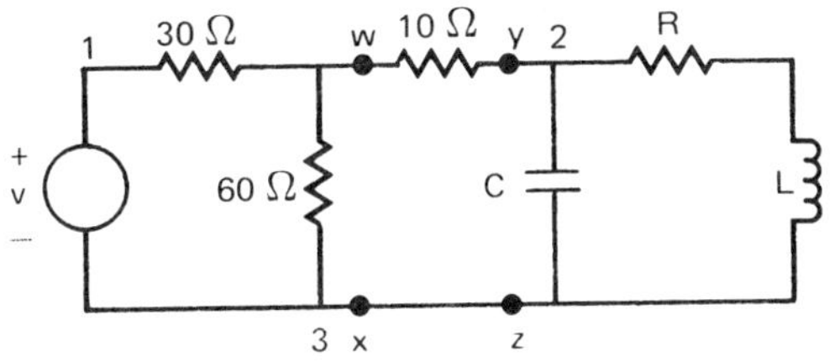

$$Z_{12} = \frac{30 \times 10 + 30 \times 60 + 10 \times 60}{60}$$

$$= \frac{2700}{60} = 45 \text{ ohms}$$

$$Z_{13} = \frac{30 \times 10 + 30 \times 60 + 10 \times 60}{10}$$

$$= \frac{2700}{10} = 270 \text{ ohms}$$

$$Z_{23} = \frac{30 \times 10 + 30 \times 60 + 10 \times 60}{30}$$

$$= \frac{2700}{30} = 90 \text{ ohms}$$

The circuit is next shown after the delta structure is substituted into the circuit to the left of terminals y-z.

The 270 ohm resistance in parallel with the source v_s is redundant. It can be removed.

Then, the source model consisting of v_s and the 45 ohm resistance is converted to the equivalent current model as shown. The parallel resistances of 45 and 90 ohms are combined into 30 ohms. The resultant circuit is identical to the solution of example 4.4.

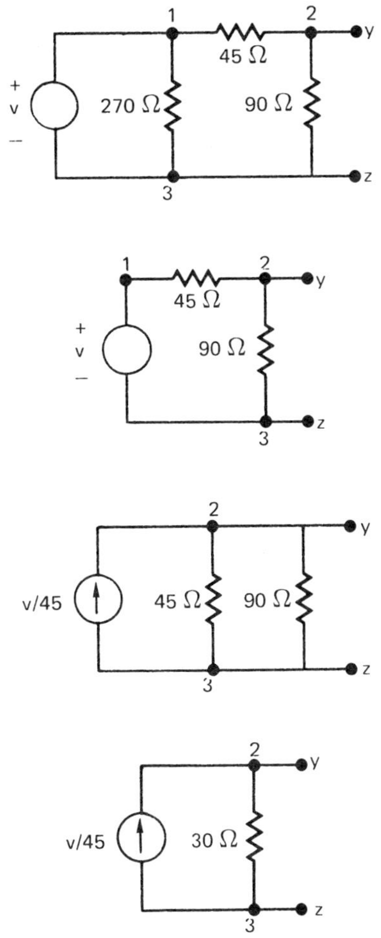

E. REDUNDANT IMPEDANCES

It was shown in example 4.5 that a resistance placed across an ideal voltage source is redundant and may be removed. As far as the remainder of the circuit is concerned, an impedance across an ideal voltage source (which theoretically can supply an unlimited amount of current) will not affect any other voltage or current in the circuit.

This is true except for the current through the source, which will be in error by the voltage divided by the redundant impedance. This source current usually is of no concern in circuit analysis problems.

An impedance in series with an ideal current source is likewise redundant and can be removed. The current source provides the same current to the remainder of the circuit regardless of whether the series impedance is present or not. The voltage across either the ideal current source or the series combination of the ideal current source and the impedance will be determined by the remainder of the circuit.

F. VOLTAGE DIVIDERS

A *voltage divider* circuit is one where a source is in series with two impedances, and the voltage of interest is across one of the two impedances. The voltage divides between the two impedances proportionally to their values. This is an important technique in finding the Thevenin equivalent of a partial circuit. It shortens the number of steps in circuit reduction problems.

Figure 4.3 shows a basic voltage divider circuit. A simple analysis will show that the voltage across Z_2 is

$$v_2 = v_1\left(\frac{Z_2}{Z_1 + Z_2}\right) \qquad 4.9$$

This is also the Thevenin equivalent voltage looking into the terminals across Z_2, (i.e., into terminals a-b as seen from the right). The Thevenin equivalent impedance looking into those same terminals is the parallel combination of Z_1 and Z_2.

$$Z_{\text{Th}} = \frac{Z_1 \times Z_2}{Z_1 + Z_2} \qquad 4.10$$

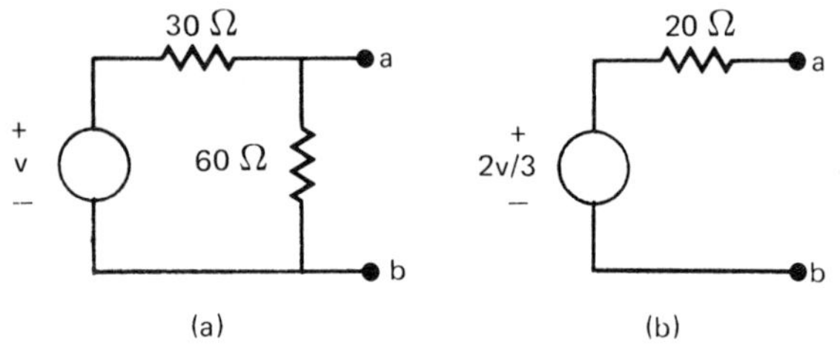

Figure 4.3 (a) circuit, (b) Thevenin equivalent

Example 4.6

The circuit shown in figure 4.3(a) is part of the circuit analyzed in examples 4.4 and 4.5. The Thevenin equivalent voltage is found from equation 4.9.

$$v_{\text{Th}} = \frac{60}{60 + 30} v_s = \frac{2v_s}{3}$$

The Thevenin equivalent impedance is found from equation 4.10.

$$Z_{\text{Th}} = \frac{30 \times 60}{30 + 60} = 20 \text{ ohms}$$

This coincides with the values found in examples 4.4 and 4.5.

3 FIRST-ORDER TRANSIENT ANALYSIS

A. INTRODUCTION

There are three classes of transient problems based on the nature of the energy sources—D.C. switching transients, A.C. switching transients, and pulse transients. When switches (or other devices performing switching functions) operate, they change the circuit topology, either connecting or disconnecting different parts of the networks: hence the name *switching transients*. *Pulse transients* do not involve topology changes in the circuit, as only the voltage or current waveforms of independent sources are changed.

The energy storage elements in electric circuits are inductors and capacitors. A network which contains only one energy storage element can be described by a first-order differential equation. The use of Thevenin or Norton equivalent circuits will result in a standard form which can be readily solved.

A *capacitive first-order circuit* is described by a differential equation using either the capacitor current or the voltage as the dependent variable. The voltage is most conveniently used because it results in a true differential equation. (The capacitor current equation is an integral equation.) The capacitor voltage is also preferred because the energy stored is a function of the capacitor voltage.

$$W = \frac{1}{2} C\, v_C^2 \qquad 4.11$$

After the determination of the Thevenin equivalent voltage source and resistance, the capacitance voltage is described by the equation

$$v_{\text{Th}} = i_{\text{Th}} R_{\text{Th}} + v_C \qquad 4.12$$

As i_{Th} is the capacitance current, it can be determined from the capacitor's voltage-current relationship.

$$i_{\text{Th}} = C \frac{dv_C}{dt} \qquad 4.13$$

Substituted into the previous equation, this produces

$$v_{\text{Th}} = R_{\text{Th}} C \frac{dv_C}{dt} + v_C \qquad 4.14$$

An *inductive first-order circuit* is best described in terms of the current for the same reasons that the capacitor voltage was chosen. The inductance current is the state variable, and the describing equation in terms of the current is

$$v_{\text{Th}} = R_{\text{Th}} i_L + L \frac{di_L}{dt} \qquad 4.15$$

The energy stored in an inductor is

$$W = \frac{1}{2} L i_L^2 \qquad 4.16$$

B. FIRST-ORDER SOLUTIONS

The solution of first-order differential equations contains a *natural response* term in exponential form and a *forced response* term with the same form as the forcing function. The forcing function is v_{Th} in equations 4.14 and 4.15. The general form for the first order differential equation is

$$f(t) = b\frac{dq}{dt} + cq \qquad 4.17$$

$f(t)$ is the forcing function and is a general function of time, b and c are constants, and q is the quantity which is to be determined. The solution is

$$q(t) = Ae^{-t/\tau} + \text{a function of } f(t) \qquad 4.18$$

τ is the *time constant* and is equal to

$$\tau = \frac{b}{c} \qquad 4.19$$

For a first-order R-C circuit, the time constant is $R_{\text{Th}}C$.

For a first-order R-L circuit, the time constant is $\frac{L}{R_{\text{Th}}}$.

The classical solution of a differential equation is to first determine the *steady state value* and then to adjust the natural response coefficient (i.e., A in equation 4.18) to match known conditions at some time—typically at the onset of the transient.

C. FIRST-ORDER SWITCHING TRANSIENTS

First-order switching transients result from the closing or opening of a switch. A switching, whether by actual switches, relay contacts, or nonlinear device operation (such as a saturating or cut-off transistor), results in a change of the circuit topology. The simplest case is where a D.C. power supply is switched into or out of a network. Figure 4.4 is an illustration.

Example 4.7

The network shown in figure 4.4 has $V = 10$ volts, $R_1 = 10$ ohms, $R_2 = 10$ ohms, and an inductance of .15 henrys. Determine the steady state inductance current with the switch (a) closed, and (b) open.

With the switch closed, figure 4.4 (b) applies. $v_{\text{Th}} = 5$ volts, $R_{\text{Th}} = 5$ ohms. The circuit equation is

$$5 = 5i + 0.15\frac{di}{dt}$$

For a D.C. circuit, the steady state current is constant. Therefore,

$$\frac{di}{dt} = 0$$
$$5 = 5i + 0$$
$$i = 1 \text{ amp}$$

With the switch open, figure 4.4(c) applies.

$$v_{\text{Th}} = 0$$
$$R_{\text{Th}} = 10$$

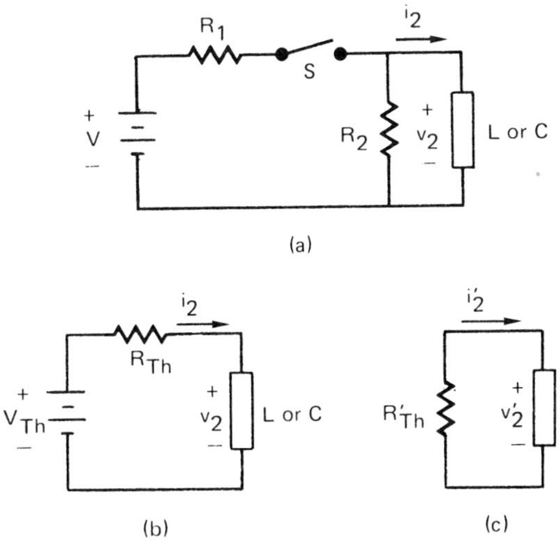

Figure 4.4 (a) First-order network with one energy storage element, (b) Thevenin equivalent when switch is closed, (c) Thevenin equivalent circuit when switch is open

The system equation is given by equation 4.15.

$$0 = 10i + 0.15\frac{di}{dt}$$

The steady state D.C. current is constant, so $\frac{di}{dt} = 0$.

$$0 = 10i + 0$$
$$i = 0$$

Example 4.8

The network shown in figure 4.4 has $V = 10$ volts, $R_1 = 10$ ohms, $R_2 = 10$ ohms, and a capacitance of 1 μF. Determine the steady state capacitance voltage with the switch (a) closed, and (b) open.

The Thevenin values are the same as in example 4.7. From equation 4.14,

$$5 = 5 \times 10^{-6}\frac{dv_C}{dt} + v_C$$

For D.C. steady state, all voltages must be constant.

$$\frac{dv_C}{dt} = 0$$
$$5 = 0 + v_C$$
$$v_C = 5 \text{ volts}$$

With the switch open, figure 4.4(c) applies. Using $v_{\text{Th}} = 0$, equation 4.14 yields

$$0 = 10^{-5}\frac{dv_C}{dt} + v_C$$

Again,

$$\frac{dv_C}{dt} = 0$$
$$0 = 0 + v_C$$
$$v_C = 0$$

The *sinusoidal steady state* case is approached similarly. However, the forcing function is a sinusoid.

$$v_{\text{Th}} = V_m \cos(\omega t + \theta) \qquad 4.20$$

Thus, equation 4.17 becomes

$$V_m \cos(\omega t + \theta) = b\frac{dq}{dt} + cq \qquad 4.21$$

Example 4.9

Figure 4.4 has $v = 10\cos(1000t + 60°)$ volts, $R_1 = 10$ ohms, $R_2 = 10$ ohms, and an inductance of .15 henrys. Determine the steady state current with the switch (a) closed, and (b) open.

With the switch closed, the Thevenin resistance is 5 ohms.

$$X_L = 1000 \times .15 = 150$$

The phasor impedance is $5+j150$. The Thevenin equivalent voltage in phasor form is $V_{\text{Th}} = 5\angle 60°$. The phasor impedance is $Z = 150\angle 88°$.

$$i = \frac{V_{\text{Th}}}{Z} = .033\angle{-28.1°}$$
$$i = .033\cos(1000t - 28.1°)$$

When the switch is opened, there will be no steady-state current.

Example 4.10

Figure 4.4 has $v = 10\cos(1000t + 60°)$ volts, $R_1 = 10$ ohms, $R_2 = 10$ ohms, and a capacitance of 1 μF. Determine the steady state capacitance voltage with the switch (a) closed, and (b) open.

Equation 4.14 can be put into phasor form.

$$v_{\text{Th}} = (j\omega R_{\text{Th}}C + 1)v_C$$
$$= 5\angle 60°$$
$$\omega R_{\text{Th}}C = 10^3 \times 5 \times 10^{-6} = 5 \times 10^{-3}$$

So,

$$1 + j\omega R_{\text{Th}}C = 1\angle .3°$$

and,

$$v_C = 5\angle 59.7°$$

This result can be transformed to the time form.

$$v_C = 5\cos(1000t + 59.7°)$$

The steady state voltage for the switch open is again zero.

Once the steady state solution for the switching circuit is known, the *total solution* is obtained by matching conditions at the switching time for the two circuits. The conditions that must match are the inductance current (for inductive circuits) and the capacitance voltage (for capacitive circuits).

Stored energy cannot change instantaneously, so an inductance current must remain the same during switching. Similarly, a capacitance voltage must remain the same during switching.

The usual procedure is to determine inductance current or capacitance voltage just prior to switching and to set that value just after switching. This information is used in equations 4.22 and 4.23.

$$i_L(t) = Ae^{\frac{-t}{T}} + I_{L,\text{steady state}} \qquad 4.22$$
$$v_C(t) = Ae^{\frac{-t}{T}} + V_{C,\text{steady state}} \qquad 4.23$$

When the time of switching (t_S) and the steady-state values before and after switching are known, the transient coefficient A of equations 4.22 and 4.23 can be found. This is illustrated in examples 4.11 and 4.12.

Example 4.11

The network in figure 4.4 has $V = 10$ volts, $R_1 = 10$ ohms, $R_2 = 10$ ohms, and an inductance of .15 henrys as in example 4.7. This network has had the switch open for a long time, and the switch is closed at $t = 0$. Find i_L after the switch closes.

Having the switch open for a long but unspecified time implies that steady-state conditions have been reached. Thus, $i_L = 0$. This also will be the initial value, $i_L(0^+)$.

The steady-state current with a closed switch was found to be 1 amp in example 4.7. The time constant is

$$\tau = \frac{L}{R} = \frac{.15}{5}$$
$$= .03$$

We can write equation 4.22 as

$$i_L(t) = Ae^{-t/.03} + 1$$

At $t = 0^+$, $i_L(0^+) = 0$ and $e^0 = 1$, so

$$0 = A + 1$$
$$A = -1$$

Thus for $t > 0$,

$$i_L(t) = 1 - e^{-t/.03}$$

Example 4.12

The network in figure 4.4 has values $V = 10$ volts, $R_1 = 10$ ohms, $R_2 = 10$ ohms, and capacitance of 1000 μF. The switch has been closed for a long time, and it is opened at $t = 0$. Determine the capacitance voltage as a function of time after the switch opens.

The initial voltage is 5 volts at the instant the switch is opened. The final voltage is zero. From equation 4.23 at $t = 0$ with $v_C(0) = 5$, A is found to be 5.

The time constant is

$$\tau = R_{\text{Th}}C = 5 \times 10^{-6}$$
$$v_C(t) = 5e^{-2\times10^5 t}$$

For sinusoidal forcing functions, the procedure is similar. However, the precise instant of switching must be given in order to derive an exact solution.

Example 4.13

As in example 4.9, the network of figure 4.4 has values $V = 10\cos(1000t + 60°)$, $R_1 = 10$ ohms, $R_2 = 10$ ohms, and an inductance of .15 henrys. The switch has been closed for a long time and is opened at $t = 0$. Determine the inductance current after the switch opens.

The steady state current was found in example 4.9 to be $.033\cos(1000t - 28.1°)$ for the switch closed. As the switch opens at $t = 0$, the initial inductance current will be

$$i(0^+) = .033\cos(0 + 28°) = .029$$

The final current will be zero. From equation 4.22, at $t = 0$,

$$e^{-t/\tau} = 1$$

So $A = i(0^+)$.

$$i(t) = .029e^{-t/.03}$$

Example 4.14

As in example 4.9, the network of figure 4.4 has values $V = 10\cos(1000t+60°)$, $R_1 = 10$ ohms, $R_2 = 10$ ohms, and an inductance of .15 henrys. The switch has been open for a long time and is closed at $t = 0$. Determine the inductance current after the switch opens.

For example 4.9, the initial current in the inductance is 0. The steady state current is $.033\cos(1000t - 30°)$. From equation 4.22,

$$i(t) = Ae^{-t/.03} + .033\cos(1000t - 28.1°)$$

At $t = 0^+$, $i(0^+) = 0$. Since $e^0 = 1$,

$$0 = A + .033\cos(-28.1°)$$

or $A = -.029$.

$$i(t) = -.029e^{-t/.03} + .033\cos(1000t - 28.1°)$$

D. FIRST-ORDER PULSE TRANSIENTS

First order pulse transients are not caused by switching changes in network topology. Rather, the transients are caused by the pulses from the sources. Pulses are represented by *unit steps* superimposed on one another as shown in figure 4.5.

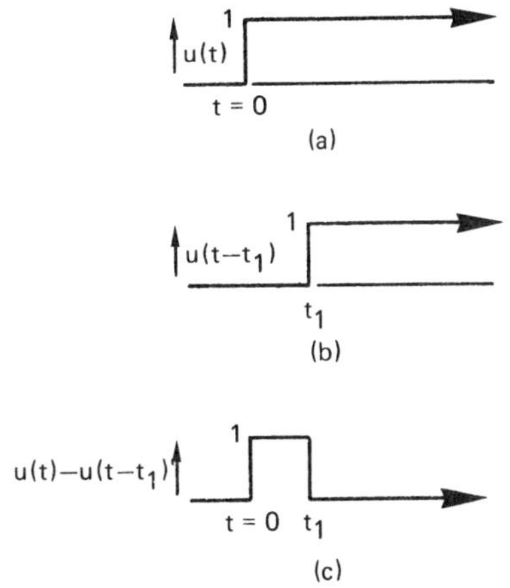

Figure 4.5 (a) unit step $u(t)$, (b) unit step $u(t - t_1)$, (c) pulse $u(t) - u(t - t_1)$

The unit step $u(t - t_1)$ is defined by equation 4.24.

$$u(t - t_1) = 0 \quad t < t_1$$
$$= 1 \quad t > t_1 \qquad 4.24$$

Any pulse train can be constructed from a series of unit steps. A single pulse beginning at $t = 0$ and ending at

$t = t_1$ is shown in figure 4.5(c). Because the networks are linear, it is convenient to determine the response to each unit step and apply superposition by adding the solutions. The response to a unit step is called the *step response*. Since the network is linear, the step responses are all the same but are only translated in time by the appropriate amount.

The response of an energy storage network to a unit step at time t_1 is determined in terms of the state variable of the energy storage device—inductance current or capacitance voltage. If other quantities are required, they can be obtained by use of circuit relations.

The unit step is zero for all time up to t_1, so there is no initial inductance current or capacitor voltage. Then, the response to the unit step at $t = t_1$ is

$$v_C(t) = u(t - t_1)\left[1 - e^{-(t-t_1)/\tau}\right] \qquad 4.25$$

$$i_L(t) = \frac{1}{R_{\text{Th}}} u(t - t_1)\left[1 - e^{-(t-t_1)/\tau}\right] \qquad 4.26$$

τ is $R_{\text{Th}}C$ for the capacitance circuit and $\frac{L}{R_{\text{Th}}}$ for the inductance circuit.

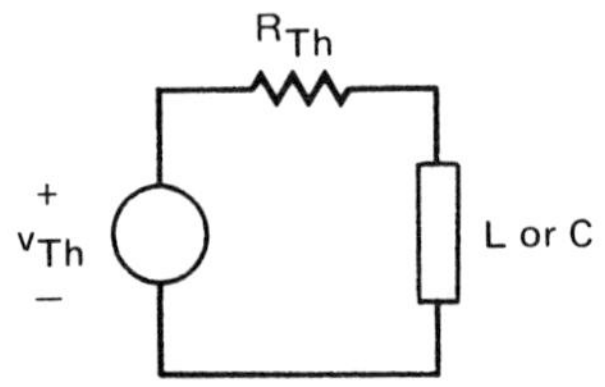

Figure 4.6 Single energy storage network

Example 4.15

A capacitance of 100 μF sees a Thevenin resistance of 1000 ohms and a step voltage which changes from 0 to 10 volts at $t = 1$. Find the capacitance voltage.

The voltage required is not limited to $t > 1$.

$$RC = \tau = 10^3 \times 100 \times 10^{-6} = 10^{-1}$$

From equation 4.25,

$$v_C(t) = 10u(t-1)\left[1 - e^{10(t-1)}\right]$$

The factor of 10 is necessary to scale the unit step to step of height 10.

Example 4.16

The pulse source in example 4.15 puts out two 10 volt pulses 10 mS in length and separated by 20 mS. Determine the capacitance voltage.

The voltage source can be represented as

$$v = 10\,u(t) - 10\,u(t - .01) + 10\,u(t - .03) - 10\,u(t - .04)$$

The response to a general step of 10 volts is 10 times equation 4.25. Using superposition, the response is

$$\begin{aligned} v_C(t) &= 10\,u(t)[1 - e^{-10t}] \\ &\quad - 10\,u(t - .01)\left[1 - e^{-10(t-.01)}\right] \\ &\quad + 10\,u(t - .03)\left[1 - e^{-10(t-.03)}\right] \\ &\quad - 10\,u(t - .04)\left[1 - e^{-10(t-.04)}\right] \end{aligned}$$

4 SECOND-ORDER TRANSIENT ANALYSIS

A. INTRODUCTION

Networks containing two independent energy storage elements require a second order differential equation to describe them. Such a differential equation can be written in terms of any circuit variable. However, inductance current and capacitance voltage are preferred because they relate directly to the initial conditions.

The general form of a second-order differential equation is

$$f(t) = a\frac{d^2q}{dt^2} + b\frac{dq}{dt} + cq \qquad 4.27$$

There are three possible cases, depending on the magnitudes of a, b and c.

- CASE I: Overdamped $b^2 > 4ac$

Overdamped transients die out quickly without oscillation. The solution will have two transient exponential parts and a steady-state part. The form of the solution will be

$$q(t) = Ae^{s_1t} + Be^{s_2t} + q_{ss} \qquad 4.28$$

$$s_1 = \frac{-b + \sqrt{b^2 - 4ac}}{2a} \qquad 4.29$$

$$s_2 = \frac{-b - \sqrt{b^2 - 4ac}}{2a} \qquad 4.30$$

q_{ss} is the steady-state value of the variable q.

The coefficients A and B in equation 4.28 are found from the initial and steady state values of q and $\frac{dq}{dt}$. Assuming the transient is initiated at $t = 0^+$, the initial values $q(0^+)$ and $q'(0^+)$ as well as the values of q_{ss} and q'_{ss} at $t = 0^+$ must be determined.

$$q(0^+) = A + B + q_{ss}(0^+) \qquad 4.31$$

$$q'(0^+) = s_1 A + s_2 B + q'_{ss}(0^+) \qquad 4.32$$

The solution of equations 4.31 and 4.32 determines A and B.

$$A = \frac{1}{2}\left[1 + \frac{b}{\sqrt{b^2 - 4ac}}\right]\left[q(0^+) - q_{ss}(0^+)\right] + \left[\frac{a}{\sqrt{b^2 - 4ac}}\right]\left[q'(0^+) - q'_{ss}(0^+)\right] \qquad 4.33$$

$$B = \frac{1}{2}\left[1 - \frac{b}{\sqrt{b^2 - 4ac}}\right]\left[q(0^+) - q_{ss}(0^+)\right] - \left[\frac{a}{\sqrt{b^2 - 4ac}}\right]\left[q'(0^+) - q'_{ss}(0^+)\right] \qquad 4.34$$

Example 4.17

A system is described by the differential equation

$$12 = \frac{d^2 v}{dt^2} + 5\frac{dv}{dt} + 4v$$

The initial conditions are $v = -5$ and $\frac{dv}{dt} = 2$. Determine the voltage v for $t > 0$.

In steady state, dv/dt and d^2/dt^2 are both zero, so the forcing function is a constant 12. The steady state value is

$$v_{ss} = \frac{12}{4} = 3$$

$$v'_{ss} = 0$$

From equations 4.29 and 4.30

$$s_1 = \frac{-5 + \sqrt{25 - 16}}{2} = \frac{-5 + 3}{2} = -1$$

$$s_2 = \frac{-5 - \sqrt{25 - 16}}{2} = \frac{-5 - 3}{2} = -4$$

From equations 4.33 and 4.34,

$$A = \frac{1}{2}\left[1 + \frac{5}{3}\right]\left[-5 - 3\right] + \frac{1}{3}\left[2 - 0\right] = -10$$

$$B = \frac{1}{2}\left[1 - \frac{5}{3}\right]\left[-5 - 3\right] - \frac{1}{3}\left[2 - 0\right] = 2$$

Then, from equation 4.28

$$v(t) = -10e^{-t} + 2e^{-4t} + 3$$

Example 4.18

A system is described by

$$12\cos 2t = \frac{d^2 v}{dt^2} + 5\frac{dv}{dt} + 4v$$

Initial conditions are $v = -5$ and $\frac{dv}{dt} = 2$ at $t = 0$. Determine v for $t > 0$.

Assume a steady state solution of $v_{ss} = A\cos(2t + \phi)$. After successive differentiating,

$$12\cos 2t = -4A\cos(2t + \phi) - 10A\sin(2t + \phi) + 4A\cos(2t + \phi)$$

From this,

$$A = 1.2$$

$$\phi = \frac{-\pi}{2}$$

$$v_{ss} = 1.2\sin 2t$$

For the natural response, assume

$$v_n = B\,e^{s_1 t} + C\,e^{s_2 t}$$

Substitute e^{st} into the describing equation (while ignoring $12\cos 2t$, as it is taken care of by the steady state response).

$$0 = \left[s^2 + 5s + 4\right]e^{st}$$

Then, $s_1 = -1$, and $s_2 = -4$.

$$v(t) = 1.2\sin 2t + Be^{-t} + Ce^{-4t}$$

$$\frac{dv}{dt} = 2.4\cos 2t - Be^{-t} - 4Ce^{-4t}$$

These equations are evaluated at $t = 0$ and matched with initial conditions.

$$-5 = 0 + B + C$$

$$2 = 2.4 = B - 4C$$

Or, $B = 6.8$ and $C = 1.8$.

$$v(t) = 1.2\sin 2t - 6.8e^{-t} + 1.8e^{-4t}$$

•CASE II: Critically Damped $b^2 = 4ac$

A second order system needs to have two transient components. With only one time constant, the solution is

$$q(t) = Ae^{st} + Bte^{st} + q_{ss} \qquad 4.35$$

$$s = \frac{-b}{2a} \qquad 4.36$$

The initial values at $t = 0^+$ are

$$q(0^+) = A + 0 + q_{ss}(0^+) \qquad 4.37$$

$$q'(0^+) = sA + B + q'_{ss}(0^+) \qquad 4.38$$

Equations 4.37 and 4.38 can be solved to yield A and B.

$$A = q(0^+) - q_{ss}(0^+) \qquad 4.39$$

$$B = q'(0^+) - q'_{ss}(0^+) + \frac{b}{2a}A \qquad 4.40$$

Example 4.19

A system is described by the differential equation

$$20 = \frac{d^2i}{dt^2} + 4\frac{di}{dt} + 4i$$

Initial conditions are $i = 0$ and $\frac{di}{dt} = 4$ at $t = 0$. Determine i for $t > 0$. The steady state current is

$$i_{ss} = \frac{20}{4} = 5$$

The initial current was given as $i_0 = 0$. From equation 4.36,

$$s = \frac{-4}{2} = -2$$

From equations 4.39 and 4.40,

$$A = 0 - 5 = -5$$

$$B = 4 - 0 + 2(-5) = -6$$

Then,

$$i(t) = 5(1 - e^{-2t}) - 6te^{-2t}$$

Example 4.20

Repeat example 4.19 with $f(t) = 20 \sin 4t$ as the forcing function.

Assume $i_{ss} = A\sin(4t + \phi)$ and substitute into the defining equation.

$$20 \sin 4t = A[-16 \sin(4t + \phi) + 16 \cos(4t + \phi) + 4 \sin(4t + \phi)]$$

This implies $A = -1$ and $\phi = 53.13°$ (alternatively, $A = 1$ and $\phi = -126.87°$).

$$20 \sin 4t = -20A \sin(4t + \phi - 53.13°)$$

This system is critically damped with $s = -2$ (see equation 4.35).

$$i(t) = -\sin(4t + 53.13°) + Be^{-2t} + Cte^{-2t}$$

At $t = 0$, $i = 0$, so

$$0 = -\sin 53.13° + B$$

From this, B is found to be 0.8.

$$\frac{di}{dt} = -4\cos(4t + 53.13°) - 2Be^{-2t} + Ce^{-2t} - 2Cte^{-2t}$$

At $t = 0$, $di/dt = 4$, so

$$4 = -4 \cos 53.13° - 2B + C$$

This determines $C = 8$. Therefore,

$$i(t) = -\sin(4t + 53.13°) + 0.8e^{-2t} + 8te^{-2t}$$

•CASE III: Underdamped $b^2 < 4ac$

In this case, there is an oscillatory transient conveniently represented by

$$q = Ae^{-\alpha t}\cos\beta t + Be^{-\alpha t}\sin\beta t + q_{ss} \qquad 4.41$$

$$\alpha = \frac{b}{2a} \qquad 4.42$$

$$\beta = \frac{\sqrt{4ac - b^2}}{2a} \qquad 4.43$$

and, for a constant $f(t)$

$$q_{ss} = \frac{f(t)}{c} \qquad 4.44$$

The initial conditions result in

$$q(0^+) = A + q_{ss}(0^+) \qquad 4.45$$

$$q'(0^+) = -\alpha A + \beta B + q'_{ss}(0^+) \qquad 4.46$$

Solving for A and B,

$$A = q(0^+) - q_{ss}(0^+) \qquad 4.47$$

$$B = \frac{\alpha}{\beta}A + \frac{1}{\beta}[q'(0^+) - q'_{ss}(0^+)] \qquad 4.48$$

An alternative solution has the form

$$q = Ce^{-\alpha t}\cos(\beta t + \phi) + q_{ss} \qquad 4.49$$

$$C = \sqrt{A^2 + B^2} \qquad 4.50$$

$$\tan\phi = \frac{B}{A} \qquad 4.51$$

Example 4.21

A system is described by

$$100 = 5\frac{d^2v}{dt^2} + 6\frac{dv}{dt} + 5v$$

At the time of switching, the derivative of the voltage is 3, and the voltage is 25. Determine the voltage after the switching at $t = 0$.

In this case, a is 5, b is 6, and c is 5. Then,

$$b^2 - 4ac = -64$$

So the system is underdamped.

From equations 4.42 and 4.43,

$$\alpha = \frac{6}{2 \times 5} = .6$$

$$\beta = \frac{\sqrt{64}}{2 \times 5} = .8$$

The form of $v(t)$ is

$$v(t) = Ae^{-.6t}\cos .8t + Be^{-.6t}\sin .8t + v_{ss}$$

v_{ss} is found from the original equation.

$$v_{ss} = \frac{100}{5} = 20$$

A is found from equation 4.47.

$$A = 25 - 20 = 5$$

B is found from equation 4.48.

$$B = .75(5) + 1.25(3 - 0) = 7.5$$

Then, $v(t)$ is

$$v(t) = 5e^{-.6t}\cos .8t + 7.5e^{-.6t}\sin .8t + 20$$

In the alternative form of equation 4.49, $v(t)$ is

$$v(t) = 20 + 9.01e^{-.6t}\cos(.8t - 56.31°)$$

Example 4.22

The forcing function in example 4.21 is replaced by $120\cos 2t$. Determine $v(t)$ for $t > 0$.

A different approach is used to find the steady state response. Assume

$$v_{ss} = M\cos 2t + N\sin 2t$$

This is then differentiated and used in the describing equation.

$$\begin{aligned}120\cos 2t = &[-20M + 12N + 5M]\cos 2t\\ &+ [-20N - 12M + 5N]\sin 2t\end{aligned}$$

By equating coefficients of the sine and cosine terms,

$$\begin{aligned}\text{from cos:}\quad 120 &= [-15M + 12N]\\ \text{from sin:}\quad 0 &= [-15N - 12M]\end{aligned}$$

Solving for M and N,

$$v_{ss} = -4.9\cos 2t + 3.9\sin 2t$$

The derivative is

$$\frac{dv_{ss}}{dt} = 7.8\cos 2t + 9.8\sin 2t$$

The values at $t = 0$ are

$$v_{ss}(0^+) = -4.9$$

$$\frac{dv_{ss}(0^+)}{dt} = 7.8$$

From equations 4.47 and 4.48,

$$A = 25 - (-4.9) = 29.9$$

$$B = \frac{.6}{.8}\Big[29.9\Big] + \frac{1}{8}\Big[3 - 7.8\Big] = 16.4$$

Then, $v(t)$ for $t > 0$ is

$$\begin{aligned}v(t) = &29.9e^{-.6t}\cos .8t + 16.4e^{-.6t}\sin .8t\\ &- 4.9\cos 2t + 3.9\sin 2t\end{aligned}$$

This can be simplified by combining terms of the same frequencies.

$$\begin{aligned}v(t) = &34.1e^{-.6t}\cos(.8t - 28.7°)\\ &- 6.2\cos(2t + 38.5°)\end{aligned}$$

B. STEP RESPONSE OF SECOND-ORDER SYSTEMS

The *step response of second order systems* for zero initial conditions is useful in the analysis of pulse transients. The final condition with a step input is a constant.

$$\frac{f(t)}{c} = q_{ss} \qquad 4.52$$

For a unit step, the final condition is $\frac{1}{c}$. Thus, for zero initial conditions the unit step response is found by setting $f(t) = 1$ in equation 4.27.

•CASE I: Overdamped — $b^2 > 4ac$

$$s_1 = \frac{-b + \sqrt{b^2 - 4ac}}{2a} \qquad 4.53$$

$$s_2 = \frac{-b - \sqrt{b^2 - 4ac}}{2a} \qquad 4.54$$

$$q_{ss} = \frac{1}{c} \qquad 4.55$$

Then, from equations 4.33 and 4.43,

$$q(t) = -\frac{1}{2c}\left[1 + \frac{b}{\sqrt{b^2 - 4ac}}\right]e^{s_1 t} - \frac{1}{2c}\left[1 - \frac{b}{\sqrt{b^2 - 4ac}}\right]e^{s_2 t} + \frac{1}{c} \qquad 4.56$$

•CASE II: Critically Damped — $b^2 = 4ac$

From equations 4.39 and 4.40,

$$q(t) = -\frac{1}{c}e^{-st} - \frac{b}{2ac}te^{-st} + \frac{1}{c} \qquad 4.57$$

$$s = \frac{-b}{2a} \qquad 4.58$$

•CASE III: Underdamped — $b^2 < 4ac$

$$\alpha = \frac{-b}{2a} \qquad 4.59$$

$$\beta = \frac{\sqrt{4ac - b^2}}{2a} \qquad 4.60$$

From equations 4.47 and 4.48,

$$q(t) = \frac{1}{c} - \frac{1}{c}e^{\alpha t}\cos\beta t - \frac{\alpha}{\beta c}e^{-\alpha t}\sin\beta t \qquad 4.61$$

Of course, the angles must be in radians.

Example 4.23

A network has an output current described by

$$.2V_S = 20\frac{d^2 i}{dt^2} + 15\frac{di}{dt} + 6i$$

V_S is a ten volt pulse lasting .1 second. Determine the current 0.2 second after the pulse has ended.

$b^2 - 4ac = 225 - 480$, so the system is underdamped.

$$\alpha = \frac{15}{2 \times 20} = .375$$

$$\beta = \frac{\sqrt{4ac - b^2}}{2a} = \frac{\sqrt{255}}{40} = .4$$

The unit step response is found from equation 4.61.

$$i(t) = \frac{1}{6}\left[1 - e^{-.375t}\cos .4t - .939e^{-.375t}\sin .4t\right]u(t)$$

The pulse is represented by

$$V_s = 10u(t) - 10u(t - .1)$$

$$.2V_s = 2[u(t) - u(t - .1)]$$

The response to the second step is

$$i_2(t) = \frac{2}{6}\left[1 - e^{-.375(t-.1)}\cos .4(t - .1) - .939e^{-.375(t-.1)}\sin .4(t - .1)\right]u(t - .1)$$

The total current is $i_1 + i_2$. At $t = .3$ (.2 after the pulse), the current is

$$i(.3) = \frac{1}{3}\left[1 - e^{-.375\times.3}\cos(.4 \times .3) - .939e^{-.375\times.3}\sin(.4 \times .3)\right] - \frac{1}{3}\left[1 - e^{-.375\times.2}\cos(.4 \times .2) - .939e^{-.375\times.2}\sin(.4 \times .2)\right]$$

$$= .0023$$

C. INITIAL CONDITIONS IN SECOND-ORDER SYSTEMS

Initial conditions in second-order systems must be found from the conditions just prior to switching. A method of finding them is to isolate the portion of the network containing the two energy storage elements. Then, the Thevenin equivalent of the remaining portion is found.

The energy stored in a capacitance or inductance cannot change during the switching. When a capacitance voltage and or inductance current is known prior to switching, the initial conditions are then known for those elements.

Series LC circuits, as indicated in figure 4.7, will have zero steady state current. The capacitance blocks the D.C. current and eventually absorbs the entire Thevenin equivalent current. With initial current I_0 and initial voltage V_0 across the capacitance, the circuit differential equation is

$$V_{\text{Th}} = R_{\text{Th}} i + L\frac{di}{dt} + v_C \qquad 4.62$$

$$i = C\frac{dv_C}{dt} \qquad 4.63$$

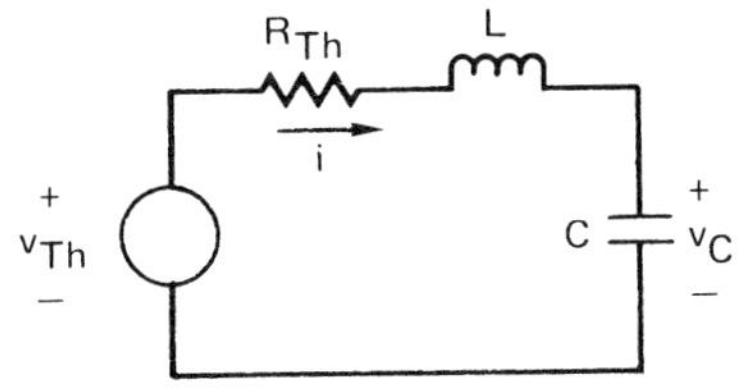

Figure 4.7 Series L-C and Thevenin equivalent circuit.

The differential equation in terms of v_C is

$$v_{Th} = LC\frac{d^2v_C}{dt^2} + R_{Th}C\frac{dv_C}{dt} + v_C \qquad 4.64$$

The initial conditions result in $v_C(0^+) = V_0$ and

$$C\frac{dv_C(0^+)}{dt} = I_0 \qquad 4.65$$

$$\frac{dv_C(0^+)}{dt} = \frac{I_0}{C} \qquad 4.66$$

Example 4.24

A series combination of 1 henry and 10 μF is connected to a Thevenin equivalent circuit having a 10 volt source in series with a 500 ohm resistance. The initial current is 0.01 amps, and the drop across the capacitance is initially 5 volts. Determine v_C for $t > 0$.

Calculate the products for use with equation 4.64.

$$LC = 1 \times 10^{-5} = 10^{-5}$$

$$R_{Th}C = 500 \times 10^{-5} = 5 \times 10^{-3}$$

The differential equation is

$$10 = 10^{-5}\frac{d^2v_C}{dt^2} + 5 \times 10^{-3}\frac{dv_C}{dt} + v_C$$

Using $a = 10^{-5}$, $b = 5 \times 10^{-3}$, and $c = 1$,

$$b^2 - 4ac = -1.5 \times 10^{-5}$$

This is an underdamped circuit. From equations 4.53 and 4.54, the roots are

$$s_1, s_2 = \frac{-5 \times 10^{-3} \pm j3.87 \times 10^{-3}}{2 \times 10^{-5}}$$

$$= -250 \pm j194$$

Therefore, the form of $v_C(t)$ is

$$v_C(t) = Ae^{-250t}\cos 194t + Be^{-250t}\sin 194t + 10$$

From equation 4.66,

$$\frac{dv_C(0^+)}{dt} = \frac{.01}{10^{-5}} = 10^3$$

Next, simultaneously solve for A and B.

$$v_C(0^+) = 5 = A + 10$$

$$\frac{dv_C(0^+)}{dt} = 10^3 = -250A + 194B$$

So, $A = -5$, and $B = -1.29$. The capacitance voltage is

$$v_C(t) = 10 - 5e^{-250t}\cos 194t - 1.29e^{-250t}\sin 194t$$

A *parallel LC circuit* will have a D.C. steady state response of zero volts because of the inductance. The network shown in figure 4.8 is described by the differential equation

$$\frac{v_{Th}}{R_{Th}} = \frac{v_C}{R_{Th}} + C\frac{dv_C}{dt} + i_L \qquad 4.67$$

$$v_C = v_L = L\frac{di_L}{dt} \qquad 4.68$$

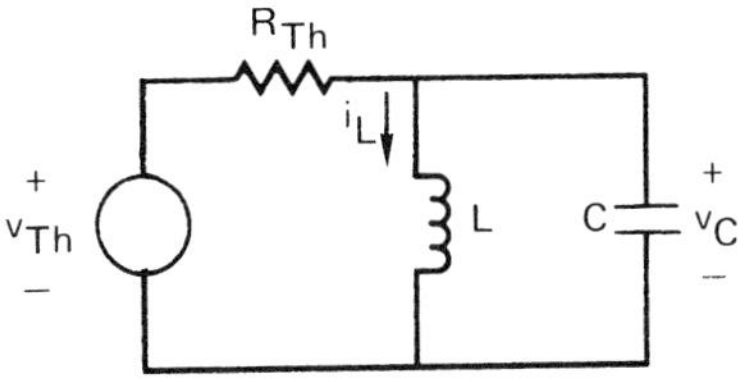

Figure 4.8 Parallel LC and Thevenin equivalent circuit

By use of equation 4.68, the differential equation can also be written in terms of the inductance current.

$$\frac{v_{Th}}{R_{Th}} = \frac{L}{R_{Th}}\frac{di_L}{dt} + LC\frac{d^2i_L}{dt^2} + i_L \qquad 4.69$$

$$\frac{di_L(0^+)}{dt} = \frac{v_C(0^+)}{L} \qquad 4.70$$

Example 4.25

The parallel LC circuit shown in figure 4.8 has an inductance current $i_L(0^+) = .5$ amp and capacitance voltage $v_C(0^+) = 5$ volts. $v_{Th} = 10$ volts, $R_{Th} = 500$ ohms, $L = .1$ henry, and $C = 10\ \mu F$. Determine the inductance current after $t = 0$.

The differential equation is given by equation 4.69.

$$\frac{10}{500} = \frac{.1}{500}\frac{di_L}{dt} + 10^{-6}\frac{d^2 i_L}{dt^2} + i_L$$

or

$$.02 = 10^{-6}\frac{d^2 i_L}{dt^2} + 2 \times 10^{-4}\frac{di_L}{dt} + i_L$$

$$b^2 - 4ac = 4 \times 10^{-8} - 4 \times 10^{-6} = -3.96 \times 10^{-6}$$

Thus, the circuit is underdamped. The roots are

$$s_1, s_2 = \frac{-2 \times 10^{-4} \pm j1.99 \times 10^{-3}}{2 \times 10^{-6}}$$
$$= -100 \pm j995$$

The solution for i_L is

$$i_L(t) = Ae^{-100t}\cos 995t + Be^{-100t}\sin 995t + .02$$
$$i_L(0^+) = .5 = A + .02$$

So, $A = .48$

Differentiating the expression for $i_L(t)$,

$$\frac{di_L(t)}{dt} = -100[Ae^{-100t}\cos 995t + Be^{-100t}\sin 995t] + 995[-Ae^{-100t}\sin 995t + Be^{-100t}\cos 995t]$$

At $t = 0$,

$$\frac{di_L(0^+)}{dt} = -100A + 995B$$
$$= \frac{5}{.1} = 50$$

This can be solved for B since A is known.

$$B = \frac{50 + 100A}{995} = .098$$

Thus, the inductance current is

$$i_L(t) = .02 + .48e^{-100t}\cos 995t + .098e^{-100t}\sin 995t$$

5 LAPLACE TRANSFORM ANALYSIS

A. INTRODUCTION

Laplace transforms are a convenient general form for system analysis. Laplace transforms can be applied either by taking the transform of a differential equation which has been obtained in the classical way, or by analyzing a transformed circuit.

The first step in the classical approach is to obtain the system differential equations. The next step is to obtain the Laplace transform of the differential equations and of all voltage and current sources. The transformed equations are then solved algebraically.

With the circuit transformation method, circuit analysis is performed on transformed impedances with initial condition sources for each energy storage element. This results in the same transformed equations obtained from the classical approach.

Regardless of which approach is taken, the transformed equations are broken into their partial fractions to be inversely transformed. This is particularly true for transient analysis of second and higher order systems. For steady state sinusoidal analysis, the equations reduce to phasor form, and graphical representations of system characteristics can be used.

The transformation of classical differential equations is the most direct method in problems where the desired quantity is an energy storage device variable. In these cases, the sources and resistances can be reduced simply and are either in a Thevenin form (with a voltage source and series resistance) or in a Norton form (with an equivalent current source and parallel resistance). The Laplace transforms can be applied directly to each variable.

$$\mathcal{L}\{f(t)\} = F(s) \qquad 4.71$$

$$\mathcal{L}\left\{\frac{df(t)}{dt}\right\} = sF(s) - f(0^+) \qquad 4.72$$

$f(t)$ can be any function of time, and $f(0^+)$ is the function value at $t = 0^+$. State variables are usually capacitance voltage and inductance current. These are energy storage variables which cannot change value instantaneously. When the functions are state variables, values cannot change instantaneously. In that case,

$$f(0) = f(0^+) \qquad 4.73$$

B. CAPACITANCE

The voltage across a capacitance is a state variable. It is preferred as one of the variables in system equations. The capacitance voltage does not change instantaneously, so that $v_C(0^+) = v_C(0)$. The capacitance voltage $v_C(t)$ and the capacitive current, $C(dv_C/dt)$, are transformed according to equations 4.74 and 4.75.

$$\mathcal{L}\{v_C(t)\} = V_C(s) \qquad 4.74$$

$$\mathcal{L}\{i_C(t)\} = \mathcal{L}\{C\frac{dv_C(t)}{dt}\}$$
$$= sCV_C(s) - Cv_C(0) \qquad 4.75$$

Equation 4.75 is a key to the transformed circuit. In a transformed circuit, the variables are the Laplace transforms of the voltage and current. The elements are impedances and transformed sources. The terminals of a transformed capacitance must behave according to equation 4.75.

Such a circuit element could be modeled with two current sources as indicated in figure 4.9. However, recognizing that the current through the left-hand source is a function of the terminal voltage, it can be replaced by an admittance sC or by an impedance $1/sC$. This equivalent transformed circuit element accounts for the initial conditions by the parallel current source $Cv_C(0)$.

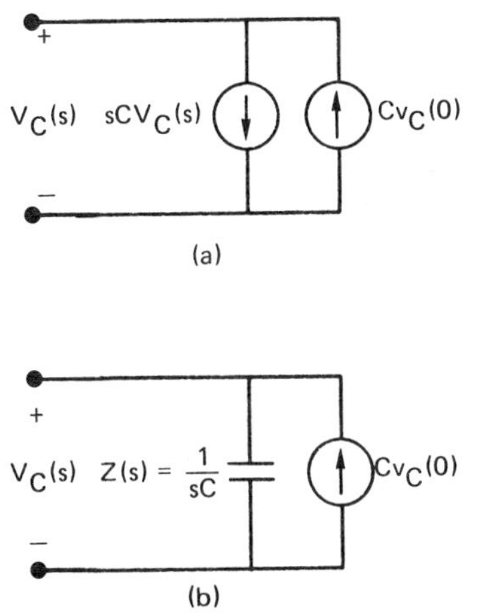

Figure 4.9 Transformed equivalent capacitance circuit: (a) two source model, (b) impedance and source model

The equivalent circuit of figure 4.9(b) will be used in general Laplace analysis in this chapter.

C. INDUCTANCE

Inductance current is similarly a state variable. It is preferred as one of the variables when an inductance is a part of a circuit. The inductance current $i_L(t)$ and the inductance voltage, $L(di_L/dt)$, are transformed by equations 4.76 and 4.77.

$$\mathcal{L}\{i_L(t)\} = I_L(s) \qquad 4.76$$

$$\mathcal{L}\{v_L(t)\} = \mathcal{L}\left\{L\frac{di_L(t)}{dt}\right\}$$

$$= sLI_L(s) - Li_L(0) \qquad 4.77$$

The terminals of a transformed inductance must behave according to equation 4.77. Such a circuit element could be modeled with two voltage sources as indicated in figure 4.10.

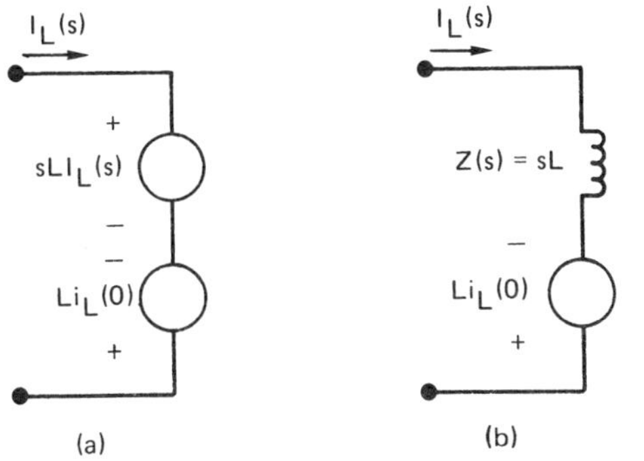

Figure 4.10 Transformed equivalent inductance circuit: (a) two source model, (b) impedance and source model

The voltage across the upper source is a function of the terminal current. It can be replaced with an impedance sL as in figure 4.10(b). This equivalent transformed circuit element accounts for the initial conditions with the series current source $Li_L(0)$.

The equivalent circuit of figure 4.10(b) will be used in general Laplace analysis in this chapter.

D. FIRST ORDER SYSTEMS

Consider the Laplace transformation of a differential equation such as that found in example 4.4. The first step in the analysis of any circuit is to obtain the system differential equations by using Kirchhoff's voltage and current laws. For first-order systems where the state variable is the required quantity, the circuit seen from the terminals of the energy storage device is reduced to a Thevenin or Norton equivalent. The circuit equation is easily written, as illustrated in the following examples.

For a capacitance, the Thevenin form is preferred, since the final value of the capacitance voltage must be the Thevenin equivalent voltage source's final value. This provides a quick check on the resulting expression for the capacitance voltage.

Example 4.26

The reduced circuit shown represents the circuit after the switching or source changes. The circuit equation is Kirchhoff's voltage law around the single loop. The current flowing is

$$i(t) = C\frac{dv_C(t)}{dt}$$

The differential equation is

$$v_{\text{Th}}(t) = RC\frac{dv_C(t)}{dt} + v_C(t)$$

Equation 4.75 is used to transform the system equation term by term.

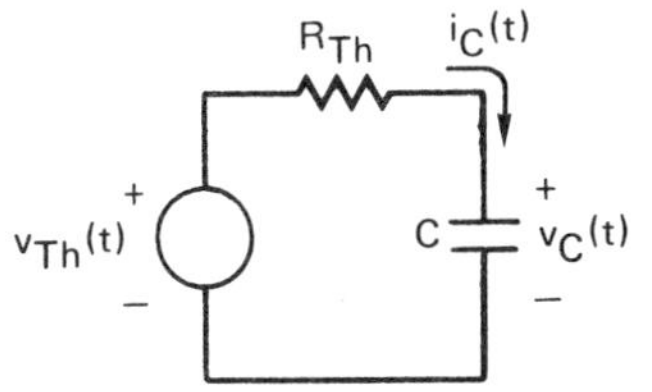

$$V_{\text{Th}}(s) = RC[sV_C(s) - v_C(0)] + V_C(s)$$

This is solved for $V_C(s)$.

$$V_C(s) = \frac{V_{\text{Th}}(s) + RCv_C(0)}{sCR + 1}$$

This resulting equation for $V_C(s)$ is solved for $V_C(t)$ by the methods of chapter 1 after obtaining the transform for the source voltage.

For an inductance current in a first order system, the Norton equivalent source is preferred, since the final value of the inductance current will be identical to the final value of the Norton current source.

Example 4.27

A reduced circuit is shown, and the inductance current is desired.

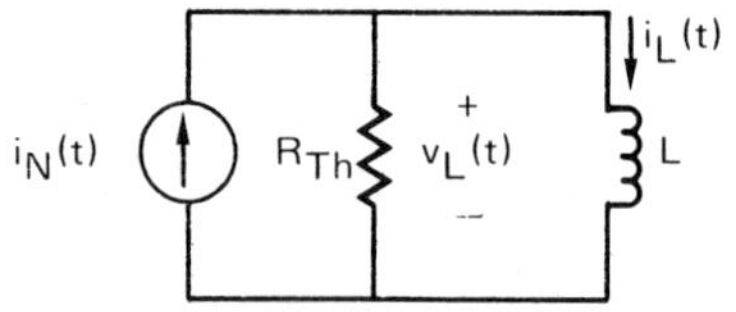

The circuit equation is Kirchhoff's current law taken at the upper node. The current through the resistance is the inductance voltage divided by the resistance.

$$i_N(t) = \frac{L}{R}\frac{di_L(t)}{dt} + i_L(t)$$

This is transformed with equations 4.76 and 4.77.

$$I_N(s) = \frac{L}{R}\Big[sI_L(s) - i_L(0)\Big] + I_L(s) \qquad 4.78$$

This can be solved for $I_L(s)$.

$$I_L(s) = \frac{RI_N(s) + Li_L(0)}{sL + R}$$

$I_L(t)$ can be found from $I_L(s)$ by using the inverse transformation method introduced in chapter 1.

The last two examples resulted in identical forms of the Laplace transformed expression for the state variable. In general, any first-order system will have a similar form. However, where an initial condition term is not a state variable, it will have to be evaluated at $t = 0^+$ as illustrated in section 4.

To use the convenience of state variables in circuits where the desired quantities are not state variables, circuit transformations prior to circuit analysis are required. Then, any ordinary method of circuit analysis, including network reduction, can be used. For network reduction, the impedances can be treated as resistances and combined. Initial condition sources are treated as any other independent source.

Example 4.28

The circuit shown has the switch closed at $t = 0$ after being open for a long time. Obtain the Laplace transform expression for the voltage across the 100 ohm resistance.

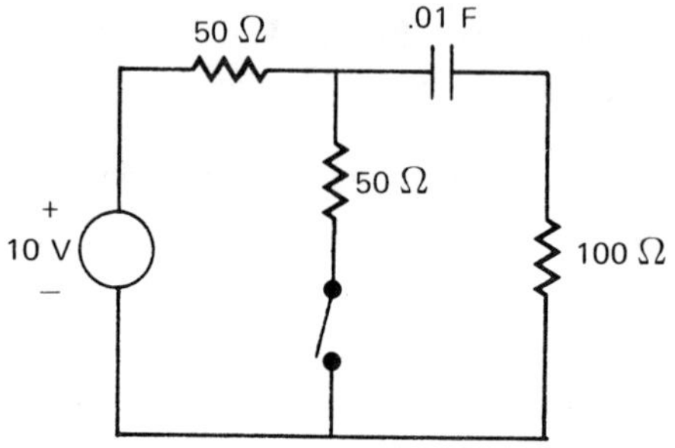

The initial voltage across the capacitance is $v_C(0) = 10$ volts.

It is necessary to convert the voltage divider consisting of the 10 volt source and the two 50 ohm resistances to the Thevenin equivalent and its transformation. These two steps result in the transformed circuit shown.

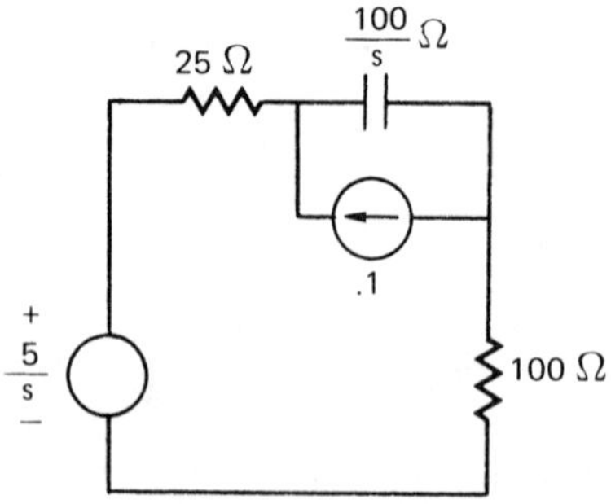

Next, the parallel impedance $(s/100)$ and the initial condition current source of .1 amp are converted to a Thevenin equivalent, yielding the circuit shown.

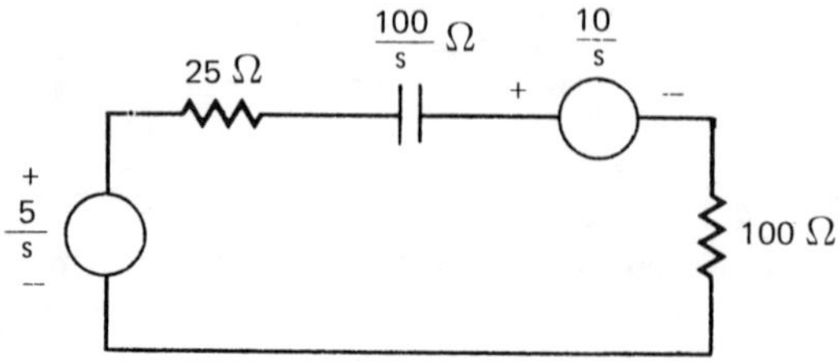

As the voltage across the 100 ohm resistance is desired, solve for its current and multiply by 100. Using Kirchoff's voltage law around the loop yields

$$\frac{5}{s}-\frac{10}{s}=125I(s)+\frac{100}{s}I(s)$$

This is solved for $I(s)$.

$$I(s)=\frac{\frac{-5}{s}}{\frac{100}{s}+125}=\frac{-0.04}{s+0.8}$$

The expression for $I(s)$ is transformed back to the time domain by using Laplace transform pairs.

$$i(t)=-0.04e^{-0.8t}$$
$$v_{100}(t)=100i(t)=-4e^{-0.8t}$$

E. SECOND-ORDER SYSTEMS

A second-order system contains two energy storage devices. When the desired quantity is one of the state variables, the method of transforming the system differential equations (written in terms of the state variables) is appropriate. This is demonstrated in example 4.29.

Example 4.29

Begin with the reduced circuit shown. (This circuit was previously used in example 4.4.)

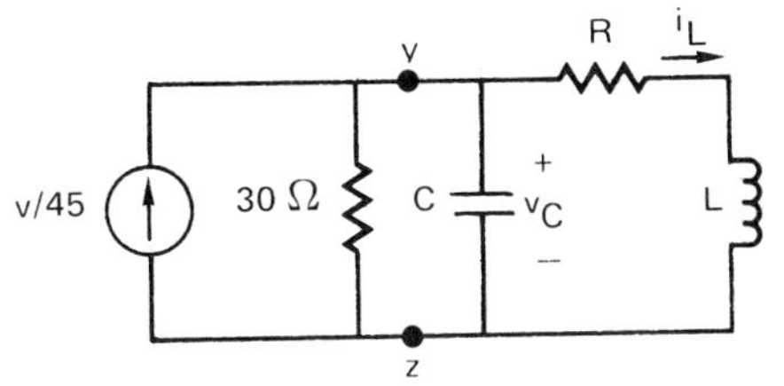

The circuit equations were determined previously in example 4.4. They were specified in terms of the capacitance voltage and inductance current, the variables most easily handled by the Laplace transformation.

$$\frac{v_s}{45}=\frac{v_C}{30}+C\frac{dv_C}{dt}+i_L$$
$$v_C=Ri_L+L\frac{di_L}{dt}$$

These equations are next transformed term by term using

$$\mathcal{L}\{v_s(t)\}=V_s(s)$$

The transformed circuit equations are

$$\frac{V_s(s)}{45}=\frac{V_C(s)}{30}+sCV_C(s)-Cv_C(0)+I_L(s)$$
$$V_C(s)=RI_L(s)+sLI_L(s)-Li_L(0)$$

The capacitance voltage is needed. Solve for the inductance current in terms of the capacitance voltage in the first equation. Substituting into the second equation,

$$V_C(s)=\frac{(R+sL)\left(\frac{V_S(s)}{45}+Cv_C(0)\right)-Li_L(0)}{s^2LC+s\left(\frac{L}{30}\right)+1+\frac{R}{30}}$$

This will allow the methods of chapter 1 to be used.

When the desired variables include more than the state variables, circuit transformation is recommended.

Example 4.30

Consider the circuit and initial conditions previously used in example 4.29. Determine the response of the inductance current to a 5 volt step voltage at $t=0$.

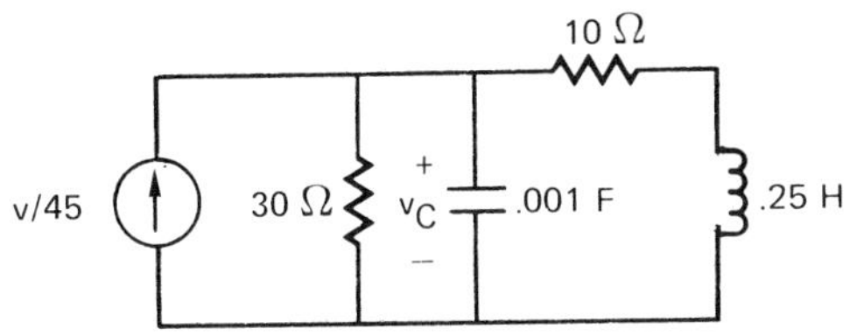

First, transform the circuit according to the models of sections B and C. The initial values of the state variables are 8 volts (dropping across the capacitor) and 40 mA of current downward in the inductance. The current source for the capacitance's transformed circuit is

$$(.001\text{ farads})\times(8\text{ volts})=.008\text{ amp-sec}$$

The voltage source for the inductance's transformed circuit is

$$(.25\text{ henry})\times(.04\text{ amp})=.01\text{ volt-sec}$$

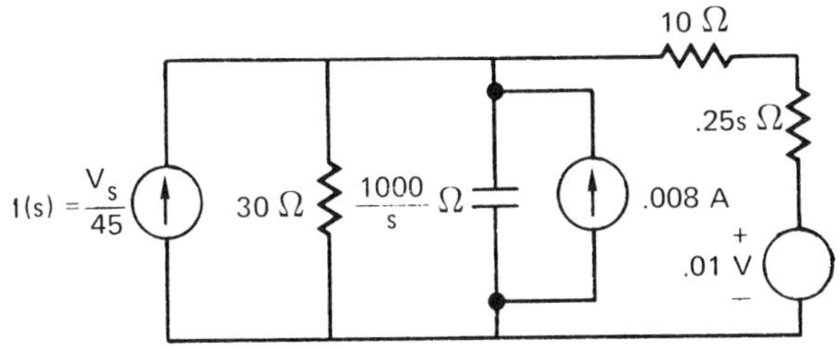

The parallel elements to the left of the 10 ohm resistance are combined into a Norton equivalent. The parallel current sources add directly, giving a total of $I_S(s)+.008$. The 30 ohm resistance and the parallel capacitance impedance of $1000/s$ combine to $1000/(s+33.3)$.

Next, the Norton source is converted to a Thevenin equivalent voltage source by multiplying by the impedance of the parallel combination of 30 ohms and $1000/s$ ohms and by substituting $V_S/45$ for $I_S(s)$.

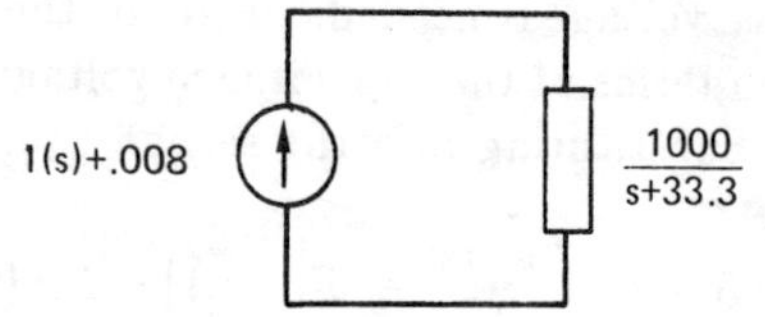

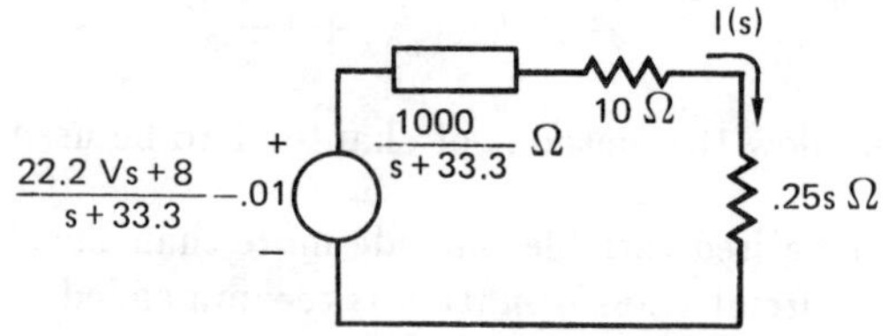

The Kirchoff voltage law equation is written in terms of the inductance current.

$$\frac{22.2V_s+8}{s+33.3}-.01=\left[\frac{1000}{s+33.3}+10+.25s\right]I(s)$$

This is solved for $I(s)$.

$$I(s)=\frac{88.9V_s+30.7-.04s}{s^2+73.3s+5333}$$

The function $v_s(t)$ is a 5 volt step. Therefore, $V_s=5/s$. The denominator polynomial factors to $(s+36.7)^2+63.2^2$ (to put it into the proper form for transformation). Thus, the transformed current is

$$I(s)=\frac{\frac{444}{s}+30.7-.04s}{(s+36.7)^2+63.2^2}$$

F. HIGHER-ORDER SYSTEMS

Higher-order systems are best analyzed by transforming the circuit (as was done in the last example). Higher-order systems are encountered in control systems (covered in chapter 11), and in sinusoidally-excited circuits.

6 SINUSOIDAL ANALYSIS

A. INTRODUCTION

The phasor analysis of sinusoids in chapter 3 dealt with fixed frequency sinusoids. However, a Laplace analysis can be used to study the effects of *variable frequencies*, including *spectral inputs* which have multiple frequencies.

In a sinusoidal steady state analysis, neither initial conditions nor natural transient response terms are significant. They will die out, leaving only the response to the sinusoidal signal. For sinusoidal analysis, several steps used in the previous sections can be skipped.

In particular, the Laplace transform method leads to a partial fraction expansion involving all the factors of the denominator polynomial. Terms appear for the natural response as well as for the signal. The natural response terms will include the effects of initial conditions, but the initial conditions will not affect the signal terms.

The partial fraction expansion terms for the signal (*forced response*) are dependent only on the characteristic terms of the system. These can be expressed as a transfer function dependent only on impedances. The solution procedure obtains only the partial fraction expansion for the signal term. This is the residue for the signal term. For this case, it is sufficient to replace s by $j\omega$.

B. TRANSFER FUNCTIONS

A *transfer function* is the Laplace transform of a ratio of variables. The variables can be impedance, admittance, current ratio, or voltage ratio. Transfer functions assume that the initial conditions are zero or do not matter. For example, in example 4.29 the expression obtained was

$$V_C(s)=\frac{(R+sL)\left(\frac{V_S(s)}{45}+Cv_C(0)\right)-Li_L(0)}{s^2LC+s\left(\frac{L}{30}+RC\right)+1+\frac{R}{30}}$$

The transfer function V_C/V_s can be obtained by eliminating the term $Cv_C(0)-Li_L(0)$ and dividing both sides by $V_s(s)$.

$$G(s)=\frac{V_C(s)}{V_S(s)}$$

$$=\frac{\frac{R+sL}{45}}{s^2LC+s(\frac{L}{30}+RC)+1+\frac{R}{30}}$$

This is equivalent to removing the initial condition sources from the transformed circuit, solving for a variable as a function of the signal, and taking the ratio of the response to the signal. Transfer functions are used for transient and sinusoidal steady state analysis and design. The transient application has been covered in the previous sections, so the frequency response application is now considered.

In *frequency response analysis*, the transfer function is evaluated in the *frequency domain* rather than in the time domain. In essence, the operation requires the evaluation of the residue due to the signal in the partial fraction expansion. This is done by replacing s by $j\omega$ in the transfer function. The result is a complex number, which is normally expressed in polar form with magnitude and angle. If the time response to a sinusoidal signal is desired, the magnitude of the transfer function can be multiplied by the signal magnitude and the angle of the transfer function added to the signal angle.

C. FIRST-ORDER RESPONSE

The capacitance and inductance transformations for sinusoidal steady state analysis use only the impedances. Initial condition sources are omitted. The capacitance initial condition current source is replaced by an open circuit and the inductance initial condition voltage source is replaced by a short circuit.

Example 4.31

Consider the circuit of example 4.28 with the 10 volt source replaced by an arbitrary sinusoidal source $V_s(s)$. The complete circuit for sinusoidal analysis is shown. This circuit omits the initial condition source and has included the voltage divider effect of the source and resistance.

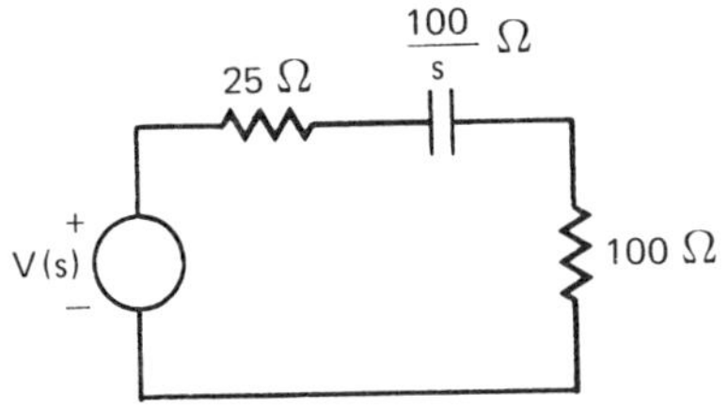

The Kirchoff voltage law equation is

$$V_S(s) = \left[125 + \frac{100}{s}\right] I(s)$$

Solving for $I(s)$,

$$I(s) = \frac{sV_S(s)}{125s + 100}$$

The transfer function is

$$G(s) = \frac{I(s)}{V_S(s)} = \frac{s}{125s + 100}$$

The final step is to substitute $j\omega$ for s and obtain the magnitude and angle of the transfer function as a function of the source frequency ω.

$$G(j\omega) = \frac{j\omega}{125j\omega + 100}$$

$$|G(j\omega)| = \frac{j\omega}{25(|5j\omega + 4|)}$$

$$|G(j\omega)| = \frac{\omega}{25\sqrt{25\omega^2 + 16}}$$

The $j\omega$ term in the numerator $G(j\omega)$ gives a constant 90° angle, so

$$\angle G(j\omega) = 90^\circ - \arctan\left(\frac{125}{100}\right) = 90^\circ - \arctan(1.25)$$

The result obtained here is identical to that which would have been obtained by phasor methods. Therefore, it is not necessary to use Laplace transforms to obtain general sinusoidal results.

Information on the frequency response of a system is typically given by plots of magnitude and angle (called *phase*) as a function of frequency. Most common is the Bode[1] plot, a plot of magnitude in decibels and angle in degrees versus the logarithm of frequency. The advantage of using decibels to represent magnitude is that decibels are a logarithmic measure. The logarithm of a product of two terms is the sum of the logarithms of the separate terms, while the logarithm of a ratio of terms is the logarithm of the numerator term minus the logarithm of the denominator term.

$$\log(xy) = \log(x) + \log(y) \qquad 4.79$$

$$\log\left(\frac{x}{y}\right) = \log(x) - \log(y) \qquad 4.80$$

The *decibel* is defined as 10 times a power ratio, but it has been loosely adopted for transfer functions in general. It can be used for ratios of voltage or current by taking the square of the ratio and ignoring the fact that the two quantities are not acting on the same impedance. Thus, the decibel measure of voltages or currents can be written as equations 4.81 and 4.82.

$$G_{I|db} = 10\log\left|\left(\frac{I_2}{I_1}\right)\right|^2 = 20\log\left|\left(\frac{I_2}{I_1}\right)\right| \qquad 4.81$$

$$G_{V|db} = 10\log\left|\left(\frac{V_2}{V_1}\right)\right|^2 = 20\log\left|\left(\frac{V_2}{V_1}\right)\right| \qquad 4.82$$

It is generally assumed that the decibel measure is 20 times the logarithm of the transfer function.

$$G_{|db} = 20\log(|G(j\omega)|) \qquad 4.83$$

Example 4.32

Use equation 4.83 to solve example 4.31.

The transfer function is

$$G(j\omega) = \frac{j\omega}{25(4 + j5\omega)}$$

From equations 4.80 and 4.83,

$$\begin{aligned} G(j\omega)_{|db} &= 20\log|j\omega| - 20\log 25 - 20\log|4 + j5\omega| \\ &= 20\log|\omega| - 20\log 25 - 20\log\sqrt{16 + 25\omega^2} \\ &= 20\log|\omega| - 20\log 25 - 10\log(16 + 25\omega^2) \end{aligned}$$

D. SECOND-ORDER RESPONSE

The method for solving second-order systems is similar to first-order methods. The major difference is that

[1]Pronounced "bow′-dee".

critically damped and underdamped responses can occur. For first-order systems, the denominator polynomial of a transfer function is limited to *simple poles.* This means that there can be only a single factor of the form $s + a$.

In second-order systems, the denominator must be factored into its *poles.* These can be two simple pole factors such as $(s+a)(s+b)$, repeated simple poles $(s+a)^2$, or a pair of complex conjugate poles.

Denominators of transfer functions can be expressed in several convenient forms depending on the problem type.

$$s^2 + 2\varsigma\omega_n s + \omega_n^2 \quad \text{(control systems)} \qquad 4.84$$

$$s^2 + \frac{\omega_o}{Q}s + \omega_o^2 \quad \text{(resonance analysis)} \qquad 4.85$$

$$(s+\alpha)^2 + \beta^2 \quad \text{(transient analysis)} \qquad 4.86$$

The relationships between the parameters are

$$\omega_o^2 = \omega_n^2 = \alpha^2 + \beta^2 \qquad 4.87$$

$$\varsigma = \frac{1}{2Q} \qquad 4.88$$

$$\alpha = \frac{\omega_o}{2Q} \qquad 4.89$$

Complex poles occur only when $\varsigma < 1$. This is equivalent to $Q > .5$ and β being a real number.

Of particular interest is behavior at or near resonance. The contribution of a complex pole-pair to the frequency response is typically examined at the resonance frequency and at the two *half-power frequencies.* Considering only the complex pole pair, resonance occurs where the phase shift due to these poles is 90° or alternatively, where the magnitude of the response is a minimum.

The minimum of the magnitude occurs where $\omega = \omega_o$, which is also the value of ω for which the phase is 90°. The *half-power points* occur at frequencies where the magnitude is $\sqrt{2}$ times the minimum value. This can be calculated from any of the expressions above, but equation 4.85 is used here. First, define G as

$$G = \frac{1}{(j\omega)^2 + \frac{j\omega_o\omega}{Q} + \omega_o^2} = \frac{1}{\omega_o^2 - \omega^2 + \frac{j\omega_o\omega}{Q}} \qquad 4.90$$

At resonance, the magnitude is found by taking $\omega = \omega_o$.

$$|G_o| = \frac{Q}{\omega_o^2} \qquad 4.91$$

$$\angle G_o = -90° \qquad 4.92$$

At the half-power frequencies, a quadratic equation must be solved.

$$|G|^2 = \frac{1}{|\omega_o^2 - \omega^2 + \frac{j\omega_o\omega}{Q}|^2} = \frac{1}{(\omega_o^2 - \omega^2)^2 + \frac{\omega^2\omega_o^2}{Q^2}} = \frac{Q^2}{2\omega_o^4} \qquad 4.93$$

Example 4.33

Determine the bandwidth between the half-power points for the transfer function $G(s)$.

$$G(s) = \frac{50}{s^2 + 50s + 5000}$$

First, find the resonant frequency from equation 4.85.

$$\omega_o = \sqrt{5000} = 70.7$$

The quality factor is

$$Q = \frac{\omega_o}{50} = 1.414$$

Using equation 4.93,

$$\frac{1}{(70.7^2 - \omega^2)^2 + \frac{70.7^2\omega^2}{1.414^2}} = \frac{1.414^2}{2 \times 70.7^4}$$

By making the substitution

$$x = \left(\frac{\omega}{70.7}\right)^2$$

and dividing the denominator by 70.7^4,

$$(1-x)^2 + \frac{x}{2} = 1$$

The roots are 0 and 1.5. The half-power frequencies are

$$\omega_1 = 0$$

$$\omega_2^2 = 70.7^2 \times 1.5$$

$$\omega_2 = 86.6 \text{ rad/sec}$$

Note that for a system with a Q of 1.414, the low half-power frequency is 0. For smaller values of Q, the lower half-power frequency will be negative, which means that there is no lower half-power frequency.

The *bandwidth* is the upper half-power frequency minus the lower, which is 86.6 rad/sec.

For Q higher than about 5, the bandwidth $\omega_2 - \omega_1$ is approximately

$$\text{bandwidth} = \frac{\omega_o}{Q} \qquad 4.94$$

E. HIGHER-ORDER RESPONSE

Higher-order systems can have combinations of simple poles and complex poles according to the number of energy storage devices in the system. There is one pole for each energy storage device (i.e., one pole for each state variable). The major problem is determining the poles, a matter of factoring a denominator polynomial of order three or higher.

The analysis requires a circuit transformation to Laplace impedances as described in the preceding sections. Once a transfer function is obtained, the task is to factor the polynomials of the numerator and denominator for purposes of obtaining graphical plots or other information. A method to obtain the roots for the third-order case, is given here.

In a *third-order system,* there must be at least one real root. Upon finding one, that value can be factored out to leave a second-degree equation whose roots can be found by the usual methods. Consider the function of equation 4.95.

$$F = s^3 + as^2 + bs + c \qquad 4.95$$

If a known value of s will make this function zero, one of the roots will be known. This can be done in a random fashion, converging on the root by recognizing that when two trials result in values of F with opposite signs, the root lies between the corresponding values of s. This process can be simplified by making good starting guesses. The function itself gives some clues to the range of possible values for the roots. The good starting guesses are found from the adjacent terms in the function.

For functions of the form of equation 4.95, high, mid-, and low frequency approximations can be found.

$$s_{\text{high}} = -a \qquad 4.96$$

$$s_{\text{mid}} = -\frac{b}{a} \qquad 4.97$$

$$s_{\text{low}} = -\frac{c}{b} \qquad 4.98$$

The procedure is to first evaluate F for these three values. At least one of these will result in a negative F and at least one in a positive F. The two closest values of s which give opposite signs for F are used to determine the next guess, either by intuition, simple averaging, or linear interpolation.[2]

Example 4.34

The denominator of a transfer function is

$$F = s^3 + 2s^2 + 6s + 1$$

Determine a root.

First evaluate the starting values of s.

$$s_{\text{high}} = -2$$

$$F = (-2)^3 + 2(-2)^2 + 6(-2) + 1 = -13$$

$$s_{\text{mid}} = \frac{-6}{2} = -3$$

$$F = (-3)^3 + 2(-3)^2 + 6(-3) + 1 = -26$$

$$s_{\text{low}} = \frac{-1}{6}$$

$$F = \left(\frac{-1}{6}\right)^3 + 2\left(\frac{-1}{6}\right)^2 + 6\left(\frac{-1}{6}\right) + 1 = +.0509$$

It can be seen that a root must exist between $s = \frac{-1}{6}$ and $s = -2$. As F is near 0 at $s = \frac{-1}{6}$, intuition is used to choose the next guess near this value, say at $\frac{-1}{5} = -.2$.

$$F = (-.2)^3 + 2(-.2)^2 + 6(-.2) + 1 = -.128$$

This indicates the root is between $s = \frac{-1}{6}$ and $s = \frac{-1}{5}$. The next guess is $s = -1.8$, which results in $F = -.021$. $s = -.17$ results in $F = +.033$, indicating the root lies between $-.17$ and $-.18$. At $s = -.175$, $F = +.006$. This process is continued to the desired accuracy. The root is reasonably approximated by $-.176$. After long division by the factor $(s + .176)$, the function is

$$F = (s + 0.176)(s^2 + .182s + 5.679)$$

The remaining second order roots are complex. They can be found by factoring the quadratic term.

[2]This method is an extension of the Bisection Method of root determination.

Practice Problems:

TIME AND FREQUENCY RESPONSE

Warmups

1. For the circuit shown, determine the steady state inductance current i_L for (a) the switch closed, and (b) the switch open.

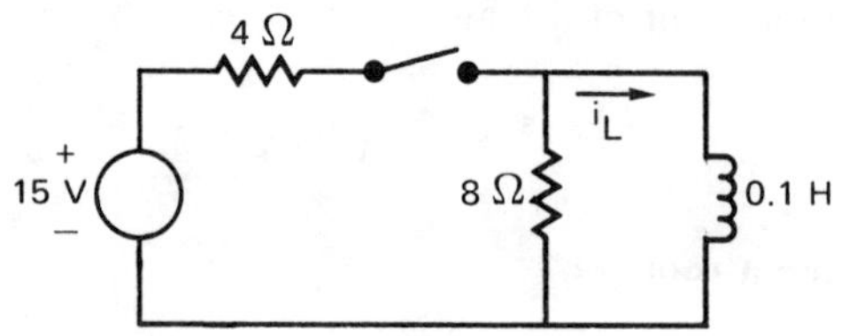

2. For the circuit shown, determine the steady state capacitance voltage v_C for (a) switch closed, and (b) the switch open.

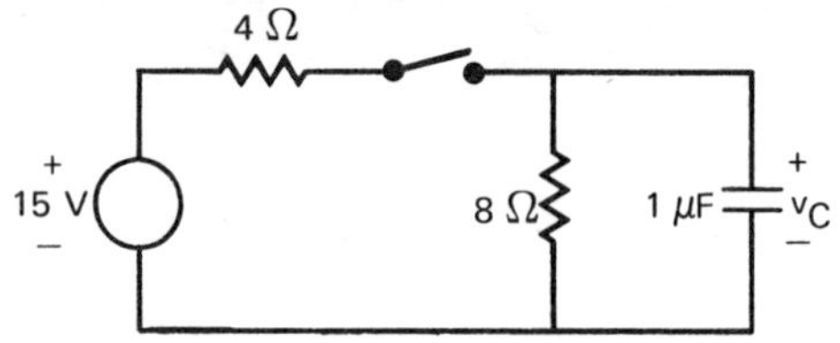

3. For the circuit shown determine the steady state inductance current i_L for (a) the switch closed, and (b) the switch open.

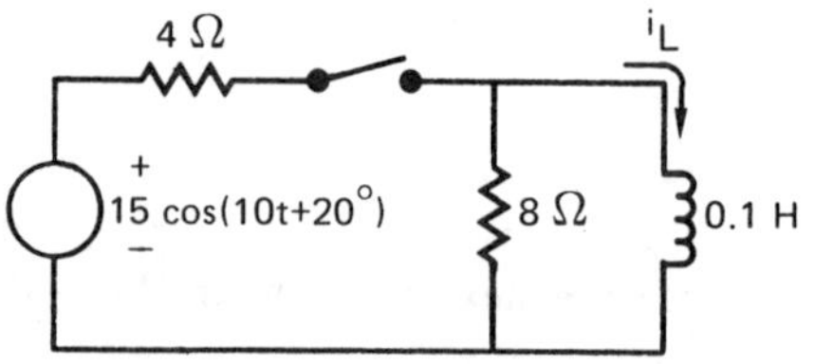

4. The circuit in problem 2 has had the switch closed for a long time. At $t = 0$, the switch opens. Determine the capacitance voltage $v_C(t)$ for $t > 0$.

5. The circuit in problem 2 has had the switch open for a long time. At $t = 0$, the switch closes. Determine the capacitance voltage $v_C(t)$ for $t > 0$.

6. A system is described by the equation

$$12 \sin 2t = \frac{d^2 v}{dt^2} + 5\frac{dv}{dt} + 4v$$

The initial conditions on the voltage are

$$v(0) = -5$$
$$\frac{dv(0)}{dt} = 2$$

Determine $v(t)$ for $t > 0$.

7. For the circuit shown, the current source is

$$i(t) = .01 \cos 500t$$

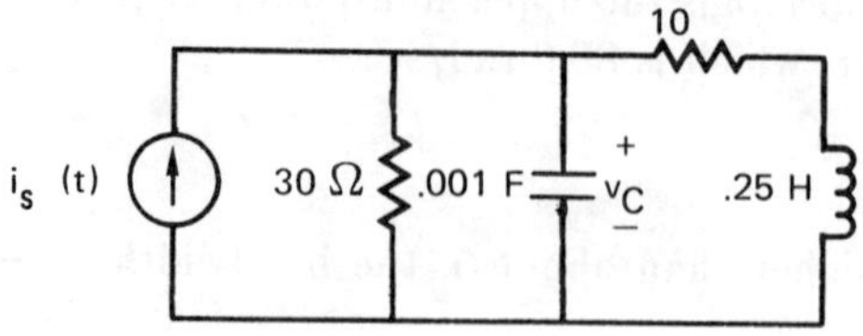

Find the transfer function

$$G(s) = \frac{V_C(s)}{I_s(s)}$$

8. For a series RLC circuit of 10 ohms, .5 H, and 100 μF, determine the resonant frequency, the quality factor, and the bandwidth.

9. For a parallel RLC circuit of 10 ohms, .5 H, and 100 μF, determine the resonant frequency, the quality factor, and the bandwidth.

10. The switch of the circuit shown has been open for a long time and is closed at $t = 0$.

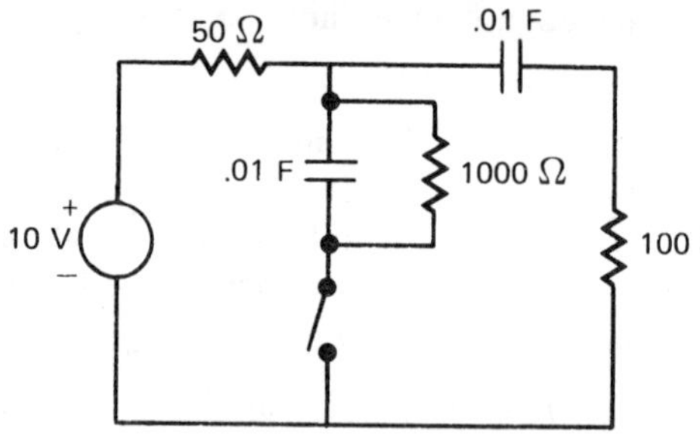

Determine the currents flowing in the two capacitances at the instant the switch closes. Give the magnitudes and directions.

Concentrates

1. A system is described by the differential equation

$$20 \sin 4t = \frac{d^2 i}{dt^2} + 4\frac{di}{dt} + 4i$$

The initial conditions are

$$i(0) = 0$$
$$\frac{di(0)}{dt} = 4$$

Determine $i(t)$ for $t > 0$.

2. The circuit shown has had the switch open for a long time. It closes at $t = 0$. Determine the values of (a) the capacitance current at $t = 0^+$, and (b) the inductance voltage at $t = 0^+$.

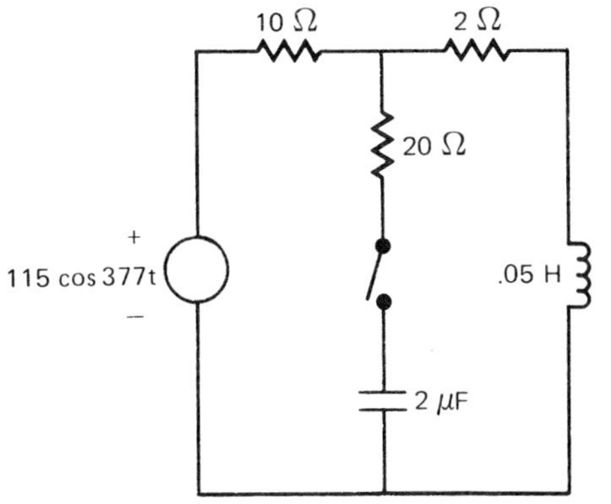

3. The circuit has an initial capacitance voltage of 20 volts and initial inductance current of 1 amp at $t = 0$. Determine the capacitance voltage for $t > 0$.

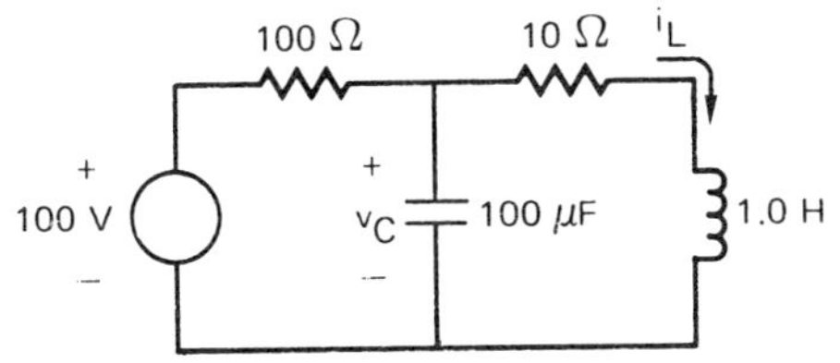

4. The circuit shown has the 1 μF capacitance charged to 100 volts and the 2 μF capacitance uncharged before the switch is closed. Determine (a) the energy stored in each capacitance at the instant the switch closes, (b) the energy stored in each capacitance a long time after the switch closes, and (c) the energy dissipated in the resistance.

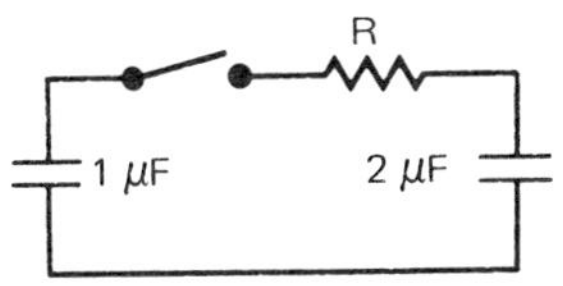

Timed

1. The circuit shown is a *phase-lead compensator*. Together with the amplifier, it is to provide a voltage ratio (gain) of 1 and a phase lead of ϕ_d at a frequency of ω_d. The amplifier has a voltage gain of K.

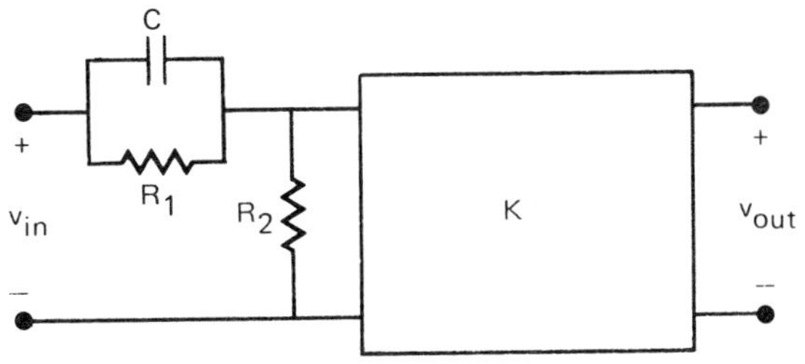

Determine the values of R_1, R_2, and C in terms of K, ϕ_d and ω_d. Discuss the limitations on the angle and the desirable ranges of values.

2. For the circuit, shown find (a) the system transfer function, (b) the response to a unit step at the input, and (c) the response to a unit pulse with a duration of 1 millisecond.

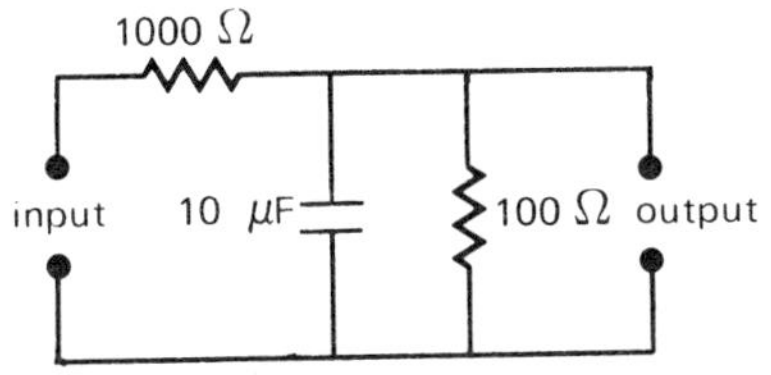

3. The circuit shown has had the switch closed for a long time. It is opened at $t = 0$. For the instant just after the switch opens ($t = 0^+$), determine the current i_1, the current i_2, the voltage v, and the time rate of change of the current i_3.

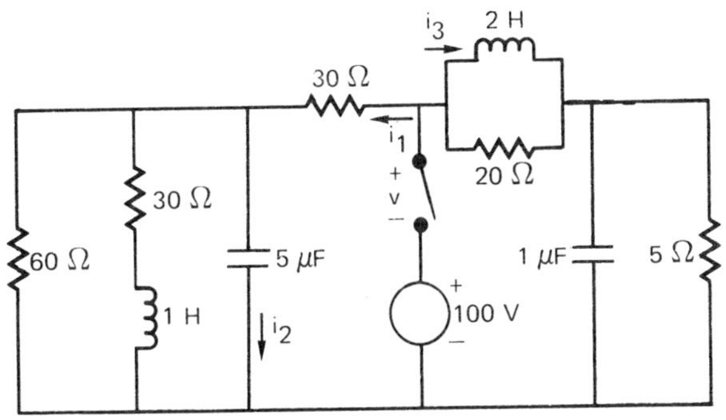

RESERVED FOR FUTURE USE

5 POWER SYSTEMS

Nomenclature

a	transformer turns ratio	-
$\hat{a}$	$.5(-1+j\sqrt{3})$	-
A	an ABCD parameter	volts/volt
$AWG\#$	American Wire Gage number	-
B	(subscripted) susceptance	mhos
B	(unsubscripted) ABCD parameter	ohms
C	an ABCD parameter	mhos
D	an ABCD parameter	amps/amp
e	an equivalent voltage source	per unit
E_g	back EMF	volts
EMF	electromotive force	volts
G	conductance	mhos
I	current	amps rms
j	$\sqrt{-1}$	-
L	inductance	henrys
MMF	magnetomotive force	amps/meter
N	number of turns	-
P	power	watts
PF	power factor	-
$P.U.$	per unit	-
Q	reactive power	vars
R	resistance	ohms
S	apparent power	volt-amps
t	time	sec
v	instantaneous voltage	volts
V	voltage	volts
X	reactance	ohms
Y	admittance	mhos
Z	impedance	ohms

Symbols

η	efficiency	-
Δ	delta connection	
δ	angle	degrees or radians
θ	angle	degrees or radians
ω	frequency	rad/sec

Subscripts

a	phase a
b	phase b
c	phase c, core
d	direct axis
g	generator
l	line
L	load
m	motor
n	neutral
oc	open circuit
p	primary
ps	primary to secondary
pt	primary to tertiary
q	quadrature axis
s	secondary
sc	short circuit
t	tertiary
Th	Thevenin
Y	wye connection

1 THREE-PHASE CIRCUITS

Three-phase power is widely used. In addition to economic considerations, a three-phase generator operating under a balanced load condition produces a constant (rather than pulsating) torque. This causes less bearing wear than a single-phase generator.

Three-phase power is delivered by three-wire or four-wire systems. A mixture of the two can also be used with intervening transformers making the change from three- to four-wire arrangements. A four-wire system is made up of three power lines and a neutral. Three-wire systems use only the three power lines. Voltages and currents in the 3-phase systems are essentially sinusoidal. In a balanced system, there are 120 electrical degrees between the three phases, regardless of whether the voltages are measured line-to-line or line-to-neutral.

As sinusoids are negative for half of each cycle, it is necessary to establish reference conventions for voltage polarities and current directions. In single-phase circuits, the voltage peak can be taken as a reference, and all other voltages then can be referred to that reference.

In a three-phase system, there are three references, possessing a fixed relationship between them. These references are either the peaks of the generator or bus voltages, and they are referred to by the letters a, b and c. In general, voltages are given in terms of node pair voltage drops (e.g., the *voltage drop* from node a to node b is designated as V_{ab}). This voltage is instantaneously positive whenever there is a voltage drop from node a to node b, and instantaneously negative whenever the voltage drop is from node b to node a.

Voltage reference polarities are used in phasor diagrams, in the fixed relationship to other voltages established by the generator or bus voltages. *Current directions* are given in double subscript notation. A current flowing from node a through a circuit element to node b would be designated as I_{ab}. This current would be instantaneously positive when the current was flowing from node a to node b, and would be instantaneously negative whenever the current flow was from node b toward node a. The *neutral node,* when present, is given the letter subscript n.

A current designated $I_{aa'}$ is from node a to node a'. The current $I_{a'a}$ is from node a' to node a. Thus, $I_{a'a} = -I_{aa'}$.

Three-phase voltage generators produce three voltages which are 120 degrees apart in phase. The three sources can be connected with a wye (Y) connection or in a delta (Δ) configuration, as shown in figure 5.1.

The voltage at node a (dropping to the neutral in the wye connection, or dropping to node b in the delta connection) is usually taken as the phasor reference. It is parallel to the positive-real axis in phasor diagrams such as figure 5.2. For the *wye connection,* there are two possible phase sequences of the voltages V_{an}, V_{bn}, and V_{cn}. These two arrangements are called *phase sequence abc* and *phase sequence acb.* All other combinations are identical to one of these, although perhaps displaced in time. Such time displacement is of no consequence when steady state operation is considered. The instantaneous voltages for the two sequences are defined by the following equations 5.1 through 5.6.

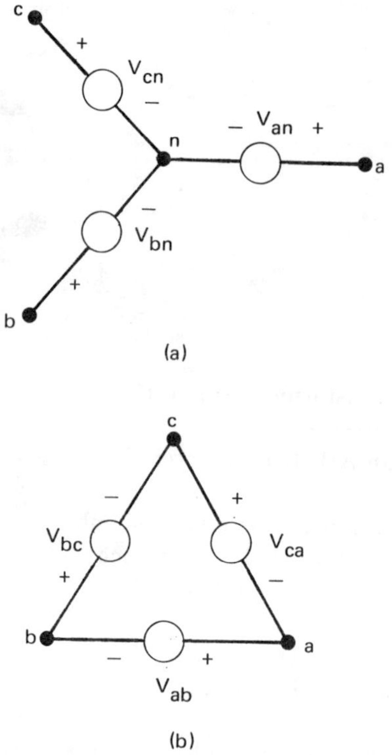

Figure 5.1 Wye and delta three-phase generator connections.

- PHASE SEQUENCE abc (wye)

$$V_{an} = V_m \cos \omega t \qquad 5.1$$

$$V_{bn} = V_m \cos(\omega t - 120^\circ) \qquad 5.2$$

$$V_{cn} = V_m \cos(\omega t - 240^\circ)$$

$$= V_m \cos(\omega t + 120^\circ) \qquad 5.3$$

- PHASE SEQUENCE acb (wye)

$$V_{an} = V_m \cos \omega t \qquad 5.4$$

$$V_{cn} = V_m \cos(\omega t - 120^\circ) \qquad 5.5$$

$$V_{bn} = V_m \cos(\omega t - 240^\circ)$$

$$= V_m \cos(\omega t + 120^\circ) \qquad 5.6$$

The phasor diagrams for the two phase sequences are shown in figure 5.2.

Note that the nodes are indicated in figure 5.2. It is also possible to add the phasor vectors in string diagrams as in figure 5.3. In loop analysis of voltages, however, it is more convenient for the nodes and phasor diagrams to have the same nodal relations. An example of this is shown in figure 5.4 where a loop from node a to node b and then to node n and back to node a is indicated. This phasor diagram shows the node and voltage relations simultaneously. The vector operation can be written in equation form, keeping in mind that the quantities are vectors and the subtraction is vector subtraction.

$$V_{ab} = V_{an} - V_{bn} \quad 5.7$$

The relationship between the *line-to-line voltage* and the *line-to-neutral voltage* is

$$V_{ll} = \sqrt{3}V_{ln} \quad 5.8$$

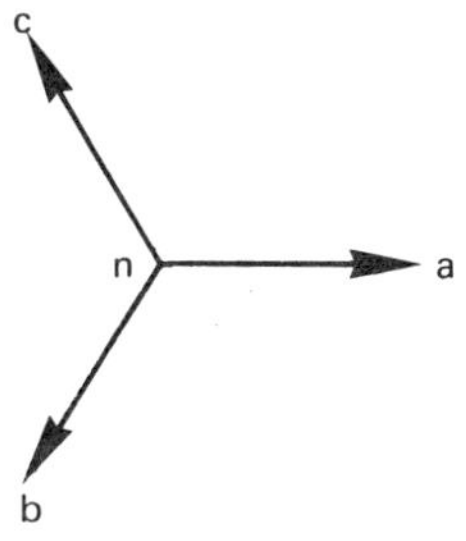

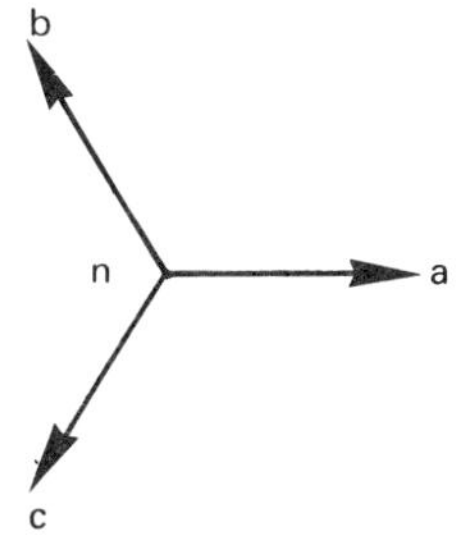

Figure 5.2 Phasor diagrams for abc and acb wye connected generators.

- PHASE SEQUENCE abc

$$V_{ab} = \sqrt{3}V_{an}\angle 30^{\circ} \quad 5.9$$
$$V_{bc} = \sqrt{3}V_{bn}\angle 30^{\circ} \quad 5.10$$
$$V_{ca} = \sqrt{3}V_{cn}\angle 30^{\circ} \quad 5.11$$

- PHASE SEQUENCE acb

$$V_{ab} = \sqrt{3}V_{an}\angle{-30^{\circ}} \quad 5.12$$
$$V_{bc} = \sqrt{3}V_{bn}\angle{-30^{\circ}} \quad 5.13$$
$$V_{ca} = \sqrt{3}V_{cn}\angle{-30^{\circ}} \quad 5.14$$

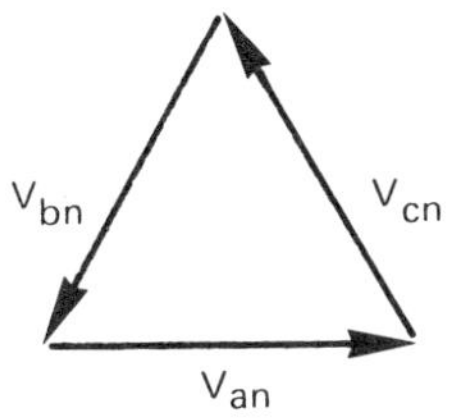

Figure 5.3 The phasor sum of line-to-neutral voltages for a phase sequence abc.

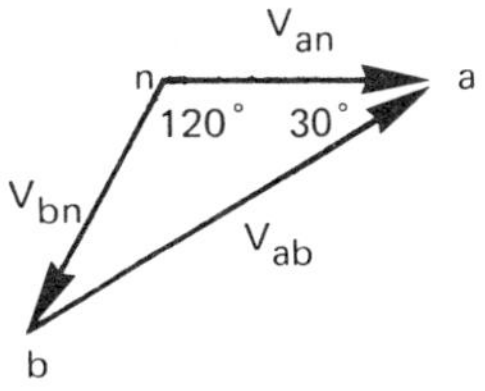

Figure 5.4 The line-to-line voltage from node *a* to node *b* for a phase sequence abc.

The line-to-neutral voltages are usually referred to as the *phase voltages*. If the voltages V_{ab}, V_{bc} and V_{ca} are added together as phasors, the sum is zero. This permits the connection of generators in the delta connection, provided that the phasing is correct. To determine if the phasing is correct before closing the circuit, a voltage measurement can be made between the last two terminals being connected. If the voltage is zero, the phasing is correct and the circuit can be closed. If there is a substantial voltage (twice the voltage for one winding or phase) then the last coil has its terminals reversed.

There also are two ways to connect the three generator windings in a *delta configuration*. One connection will result in the phase sequence *abc* while the other will result in the phase sequence *acb*. These are indicated in figure 5.5. Taking the voltage referenced to node *a* as the reference waveform, the equations 5.15 through 5.20 can be derived. V_{ll} is an effective (rms) value.

- PHASE SEQUENCE abc (delta)

$$V_{ab} = \sqrt{2}V_{ll}\cos\omega t \quad 5.15$$
$$V_{bc} = \sqrt{2}V_{ll}\cos(\omega t - 120^{\circ}) \quad 5.16$$
$$V_{ca} = \sqrt{2}V_{ll}\cos(\omega t - 240^{\circ}) \quad 5.17$$

- PHASE SEQUENCE acb (delta)

$$V_{ab} = \sqrt{2}V_{ll}\cos\omega t \quad 5.18$$
$$V_{bc} = \sqrt{2}V_{ll}\cos(\omega t - 240^{\circ}) \quad 5.19$$
$$V_{ca} = \sqrt{2}V_{ll}\cos(\omega t - 120^{\circ}) \quad 5.20$$

A *balanced wye-wye connection* consists of wye connected generators and wye connected loads consisting of three identical impedances as shown in figure 5.6. The neutral wire is indicated by a dashed line. In a balanced condition, the instantaneous sum of the currents at nodes n and n' are each zero, so no neutral current will flow. However the neutral should be retained to permit small amounts of unbalance.

With a balanced condition, it is possible to carry out an analysis on only one of the circuits. Such analysis is called *single-line analysis*. Having obtained the results of an analysis for one line, the real and apparent powers of the three-phase system are simply three times the single-line powers. Working with the a node, the phase currents are all given by equation 5.21.

$$I_{aa'} = I_{a'n} = \frac{V_{an}}{Z} \qquad 5.21$$

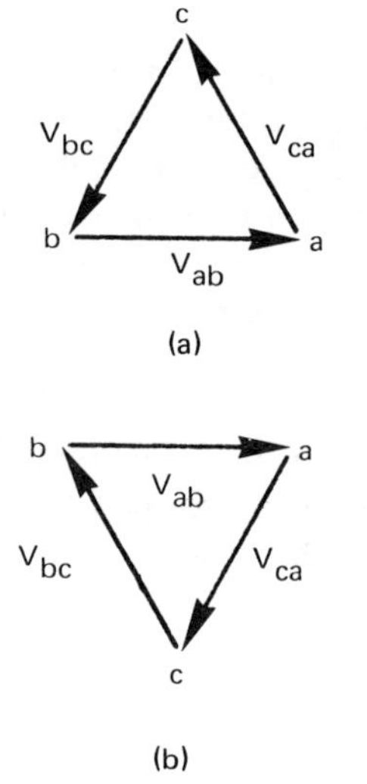

Figure 5.5 Phasor diagrams for abc and acb delta connected generators.

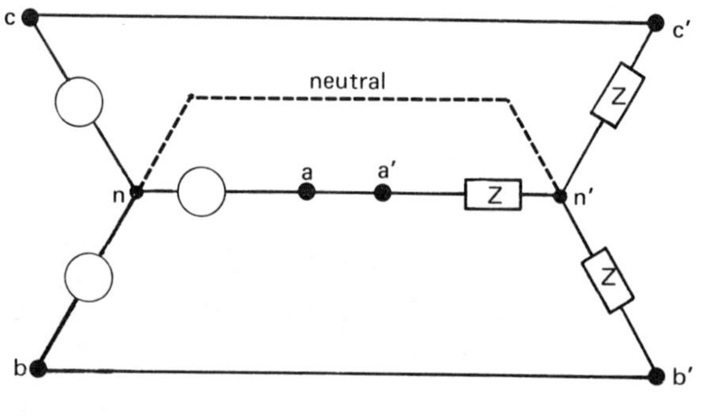

Figure 5.6 Wye-wye balanced circuit

Example 5.1

A three-phase, 6920 volt (line-to-line) generator is connected to a balanced 24 kW wye-connected load with a power factor of .8. Determine (a) the line current, (b) the generator KVA rating, and (c) the compensating reactance to bring the PF to unity.

The voltage must be converted to line-to-neutral with equation 5.8.

$$V_{ln} = \frac{6920}{\sqrt{3}} = 4000 \text{ volts}$$

The real power per phase is one-third of the generator power or 8 kW. The apparent power per phase is the real power per phase divided by the power factor. The line current (same as the phase current) is the apparent power per phase divided by the phase voltage.

$$I_{\text{line}} = \frac{P_{\text{phase}}}{PF \times V_{ln}} = \frac{8000}{(.8)4000} = 2.5 \text{ amps}$$

The generator KVA rating is three times the phase KVA.

$$S_{\text{phase}} = \frac{8}{.8} = 10 \text{ KVA}$$

$$\text{KVA}_{\text{rated}} = 3 \times 10 = 30 \text{ KVA}$$

The total line current is 2.5 amps. The in-phase component is

$$\frac{8000}{4000} = 2 \text{ amps}$$

The out-of-phase component of the current is

$$\sqrt{2.5^2 - 2^2} = 1.5 \text{ amps}$$

The compensating reactance must take an equal and opposite reactive current, so the reactance is

$$X_{\text{compensating}} = \frac{4000}{1.5} = 2667 \text{ ohms (reactive)}$$

The *balanced wye-delta connection* consists of a delta connected generator and wye connected loads. Since the system is balanced, it is possible to treat it as a wye-wye circuit after obtaining the line-to-neutral phase voltages by dividing the line-to-line voltage by $\sqrt{3}$, as was done in example 5.1.

The *balanced delta-delta connection* consists of a delta connected source and delta connected equal impedances as shown figure in 5.7. The current in each branch of the load is the line-to-line voltage divided by the impedance. It is, therefore, more useful to deal with admittances. The *phase currents* are the currents through the impedances, so that the line current is $\sqrt{3}$ times the phase current in magnitude. The phase currents are

$$I_{a'b'} = V_{ab}Y \qquad 5.22$$

$$I_{b'c'} = V_{bc}Y \qquad 5.23$$

$$I_{c'a'} = V_{ca}Y \qquad 5.24$$

For an abc phase sequence, the line currents can be calculated as follows.

$$I_{aa'} = I_{a'b'} - I_{c'a'} = (V_{ab} - V_{ca})Y \qquad 5.25$$

$$V_{ab} - V_{ca} = \sqrt{3}V_{ab}\angle{-30°} \qquad 5.26$$

$$I_{aa'} = \sqrt{3}I_{a'b'}\angle{-30°} \qquad 5.27$$

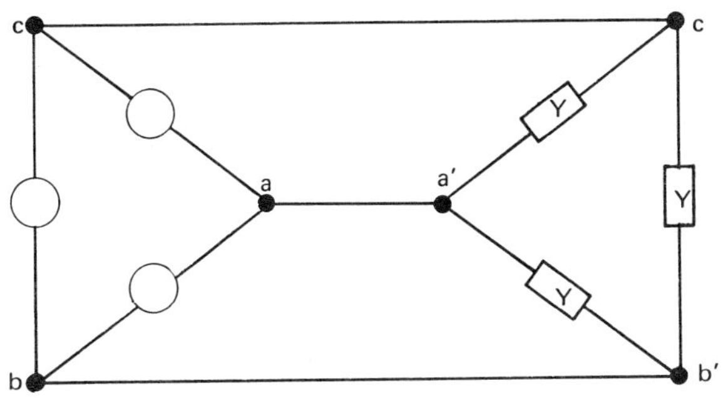

Figure 5.7 Delta-delta balanced circuit

The *power* is the sum of the powers in the load admittances.

$$P = 3|V_{ab}^2| \times \text{Real}\{Y\} \tag{5.28}$$

$$S = 3|V_{ab}^2| \times |Y| \tag{5.29}$$

$$Q = 3|V_{ab}^2| \times \text{Imaginary}\{Y\} \tag{5.30}$$

Example 5.2

A 138 KV (rms) line-to-line three-phase system delivers 70.7 MVA to a balanced delta load with a power factor of .707. Determine (a) the line current, and (b) the reactance necessary to achieve unity power factor.

Since the voltage is rms, the apparent power to each phase is

$$S_{\text{phase}} = \frac{70.7}{3} = 23.567\,\text{MVA}$$

The phase and line currents are

$$I_{\text{phase}} = \frac{S_{\text{phase}}}{V_{\text{line}}} = 170.8\,\text{amps}$$

$$I_{\text{line}} = \sqrt{3}I_{\text{phase}} = 295.8\,\text{amps}$$

The reactive component of the phase current is

$$I_{\text{phase, reactive}} = I_{\text{phase}}\sqrt{1-(PF)^2}$$
$$= 170.8\sqrt{1-.5} = 120.8\,\text{amps}$$

The compensating reactance is

$$|X| = \frac{V_{\text{phase}}}{I_{\text{phase, reactive}}}$$
$$= \frac{138 \times 10^3}{120.8} = 1142.8\,\text{ohms (reactive)}$$

The *balanced wye-delta connection* consists of a wye connected generator and a delta connected load. The neutral from the generator cannot be connected to the load so the generator should be converted to an equivalent delta connection. This is accomplished by multiplying the line-to-neutral voltage by $\sqrt{3}$ and treating the problem as delta-delta.

2 POWER TRANFORMERS

Power transformers are employed to step up voltage levels from generators. Power can then be transmitted at a low current (and therefore with less loss) using small conductors. At the end of the transmission line, a step-down transformer reduces the voltage to a useable level.

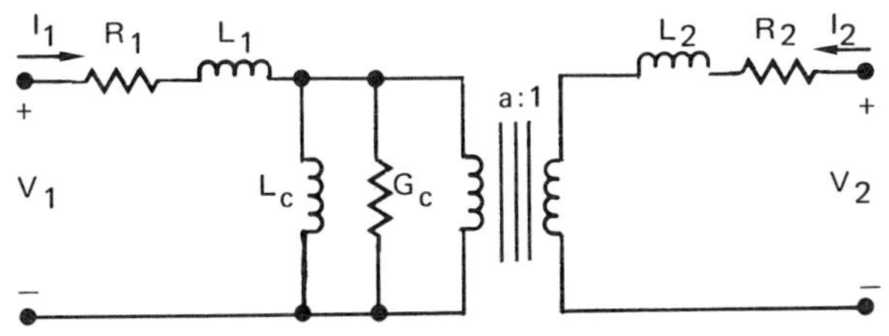

Figure 5.8 Circuit model for a power transformer

A two-winding transformer model is shown in figure 5.8. The common model parameters are:

- R_1 primary winding resistance
- L_1 primary winding leakage inductance
- L_c core inductance
- G_c core conductance
- a_{ps} primary to secondary turns ratio N_p/N_s
- L_2 secondary winding leakage inductance
- R_2 secondary winding resistance

Transformer parameter measurements require open-circuit and short-circuit tests. These can be performed on the primary or secondary winding. The results provide approximate values for the parameters shown in figure 5.8.

A. OPEN CIRCUIT TEST

Open circuit tests are always performed at the rated voltage of the winding to which the voltage is applied. This is usually the primary winding.

For transformers with a high voltage primary and low voltage secondary, measurements usually are made on the secondary side. Then, core impedances are placed on the secondary side of the equivalent circuit shown in figure 5.8.

With the secondary winding open-circuited (except for a high impedance voltmeter), the primary winding impedance is negligible. A measurement of the primary voltage, current, and power will determine the parameters G_c and L_c or X_c. With the voltage drop across R_1 and L_1 negligible, the admittance is

$$Y_{\text{open circuit}} = G_c - jB_c \quad 5.31$$

The susceptance is

$$B_c = \frac{1}{X_c} = \frac{1}{\omega L_c} \quad 5.32$$

$$S_{oc} = V_{1,oc}^2 Y_c = V_{1,oc}^2 G_c - jV_{1,oc}^2 B_c \quad 5.33$$

The open circuit power is

$$P_{oc} = V_{oc}^2 G_c \quad 5.34$$

Therefore,

$$G_c = \frac{P_{oc}}{V_{1,oc}^2} \quad 5.35$$

Furthermore,

$$Q_{oc}^2 = S_{oc}^2 - P_{oc}^2 \quad 5.36$$

$$Q_{oc} = V_{1,oc}^2 B_c \quad 5.37$$

Therefore,

$$B_c = \frac{\sqrt{S_{oc}^2 - P_{oc}^2}}{V_1^2} \quad 5.38$$

The ratio of the primary to secondary voltages with no secondary current is approximately the turns ratio.

$$a_{ps} = \frac{V_{1,oc}}{V_{2,oc}}\Big|_{I_2=0} \quad 5.39$$

Example 5.3

An open circuit test is performed on a 440 volt transformer winding. The results are $V_1 = 440$ volts, $P = 100$ watts, $I = 1$ amp, and $V_2 = 1100$ volts. Find the transformer parameters.

$$S = (440)(1) = 440 \text{ VA}$$

$$Q^2 = 440^2 - 100^2 = 428^2$$

$$G_c = \frac{100}{440^2} = .00052 \text{ mhos}$$

$$B_c = \frac{-428}{440^2} = -.00221 \text{ mhos}$$

$$a_{ps} = \frac{440}{1100} = 0.4$$

B. SHORT-CIRCUIT TESTS

The short-circuit tests are used to determine the winding impedances. Short circuit tests are always conducted at the rated current of the winding to which the voltage is applied. The secondary winding is short-circuited, causing the secondary winding impedance to essentially short out the core admittance. The effective circuit is the primary winding impedance in series with the transformed secondary winding impedance.

$$Z_{sc} = R_1 + jX_1 + a_{ps}^2(R_2 + jX_2) \quad 5.40$$

The primary voltage, V_{sc}, current, I_{sc}, and power, P_{sc}, are next measured.

$$\begin{aligned} V_{sc}I_{sc} &= I_{sc}^2 Z \\ &= I_{sc}^2(R_1 + a_{ps}^2 R_2) \\ &\quad + jI_{sc}^2(X_1 + a_{ps}^2 X_2) \end{aligned} \quad 5.41$$

However,

$$P_{sc} = I_{sc}^2(R_1 + a_{ps}^2 R_2) \quad 5.42$$

Therefore,

$$R_1 + a_{ps}^2 R_2 = \frac{P_{sc}}{I_{sc}^2} \quad 5.43$$

Furthermore,

$$Q_{sc}^2 = S_{sc}^2 - P_{sc}^2 \quad 5.44$$

$$Q_{sc} = I_{sc}^2(X_1 + a_{ps}^2 X_2) \quad 5.45$$

Therefore,

$$X_1 + a_{ps}^2 X_2 = \frac{\sqrt{V_{sc}^2 I_{sc}^2 - P_{sc}^2}}{I_{sc}^2} \quad 5.46$$

For reasons of efficiency, power transformers are typically designed so that $R_1 = a_{ps}^2 R_2$ and $X_1 = a_{ps}^2 X_2$. If this is not the case, then further tests are needed to obtain the equivalent circuit parameters. When these relationships are valid, then

$$R_1 = a_{ps}^2 R^2 = \frac{P_{sc}}{2I_{sc}^2} \quad 5.47$$

$$X_1 = a_{ps}^2 X_2 = \frac{Q_{sc}}{2I_{sc}^2} \quad 5.48$$

It is also possible to determine the turns ratio by measuring the short circuit currents in the primary and secondary windings.

$$a_{ps} = \frac{I_{2,sc}}{I_{1,sc}} \quad 5.49$$

Example 5.4

A transformer rated at 15 KVA and 1320 primary volts is subjected to a short circuit test. The secondary current is measured as 56.8 amps, the primary voltage is 100 volts, and the power is 750 watts. Determine the equivalent circuit parameters. The rated current is the volt-amp rating divided by the voltage rating.

$$I_{\text{rated}} = \frac{15}{1.32} = 11.36 \text{ amps}$$

The apparent power and reactive powers are

$$S = 11.36 \times 100 = 1136 \text{ VA}$$

$$Q = \sqrt{(1136)^2 - (750)^2} = 853 \text{ VAR}$$

From equation 5.47,

$$\begin{aligned} R_1 &= a_{ps}^2 R_2 \\ &= \frac{750}{2 \times (11.36)^2} = 2.9 \text{ ohms} \end{aligned}$$

From equation 5.43,

$$X_1 = a_{ps}^2 X_2$$

$$= \frac{853}{(2 \times 11.36^2)} = 3.3 \text{ ohms}$$

and from

$$a = \frac{I_{2,sc}}{I_{1,sc}}$$

$$a_{ps} = \frac{56.8}{11.36} = 5$$

C. ABCD PARAMETERS

ABCD Parameters are convenient in power transmission and distribution problems. The ABCD or *chain parameters* are of the form

$$V_1 = AV_2 - BI_2 \qquad 5.50$$

$$I_1 = CV_2 - DI_2 \qquad 5.51$$

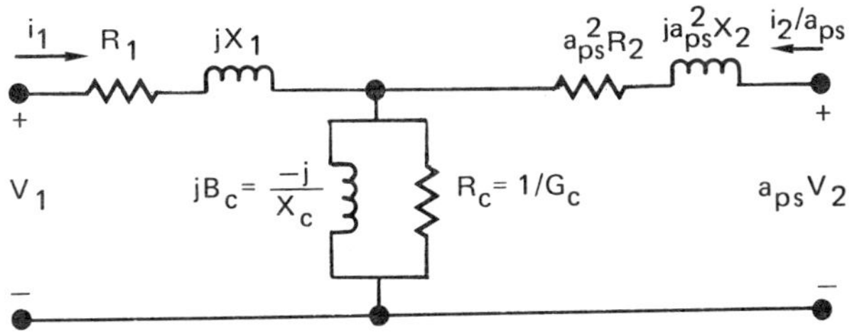

Figure 5.9 Transformer circuit with all quantities referred to the primary winding.

Circuit analysis of the transformer with all quantities referred to the primary winding, as in figure 5.9, yields equations 5.52 and 5.53.

$$V_1 = [1 + Z_1Y_c]a_{ps}V_2 - \frac{[Z_1 + a_{ps}^2 Z_2(1 + Z_1Y_c)]i_2}{a_{ps}} \qquad 5.52$$

$$i_1 = [Y_c]a_{ps}V_2 - \frac{[1 + a_{ps}^2 Z_2 Y_c]i_2}{a_{ps}} \qquad 5.53$$

$$Z_1 = R_1 + jX_1 \qquad 5.54$$

$$Z_2 = R_2 + jX_2 \qquad 5.55$$

$$Y_c = G_c - jB_c \qquad 5.56$$

For the usual case of $a_{ps}^2 Z_2 = Z_1$,

$$A = 1 + Z_1Y_c \qquad 5.57$$

$$B = Z_1(2 + Z_1Y_c) \qquad 5.58$$

$$C = Y_c \qquad 5.59$$

$$D = A \qquad 5.60$$

Example 5.5

A transformer has $Z_1 = .01 + j.02$ and $Y_c = .0002 - j.001$. Determine the ABCD parameters referred to the primary. Assume $a_{ps}^2 Z_2 = Z_1$. Then,

$$Z_1Y_c = 2.2 \times 10^{-5} - j6 \times 10^{-6}$$

For all practical purposes this is zero.

$$A = 1$$

$$B = 2Z_1 = .02 + j.04$$

$$C = Y_c = .0002 - j.001$$

$$D = A = 1$$

A transformer with a load impedance attached to the secondary behaves according to $V_2 = -I_2Z_L$. If ABCD parameters are used, then

$$V_1 = \frac{(Aa_{ps}^2 Z_L + B)I_2}{a_{ps}} \qquad 5.61$$

$$I_1 = \frac{-(Ca_{ps}^2 Z_L + D)I_2}{a_{ps}} \qquad 5.62$$

Solving for I_1 and I_2 in terms of V_1 and V_2,

$$I_2 = \frac{-a_{ps}V_1}{Aa_{ps}^2 Z_L + B} \qquad 5.63$$

$$I_1 = \frac{V_1(Ca_{ps}^2 Z_L + D)}{Aa_{ps}^2 Z_L + B} \qquad 5.64$$

Example 5.6

The transformer of example 5.5 has a load of 0.5 ohms referred to the primary. Determine the primary input impedance.

Substitute the values from example 5.5 into equation 5.64.

$$\frac{I_1}{V_1} = \frac{(2 \times 10^{-4} - j10^{-3}) \times .5 + 1}{0.5 + (.01 + j.02)}$$

$$= \frac{1.0001 - j.0005}{.51 + j.02}$$

$$Z_{\text{in}} = \frac{.51 + j.02}{1.0001 - j.005} = .51 + j.023$$

ABCD parameters are particularly useful for chaining two-port networks together. Chaining is applicable to cascaded amplifiers as well as power distribution systems. For two cascaded networks, the resulting ABCD parameters are

$$A = A_1A_2 + B_1C_2 \qquad 5.65$$

$$B = A_1B_2 + B_1D_2 \qquad 5.66$$

$$C = A_2C_1 + C_2D_1 \qquad 5.67$$

$$D = B_2C_1 + D_1D_2 \qquad 5.68$$

ABCD constants for common network forms are given in figure 5.10.

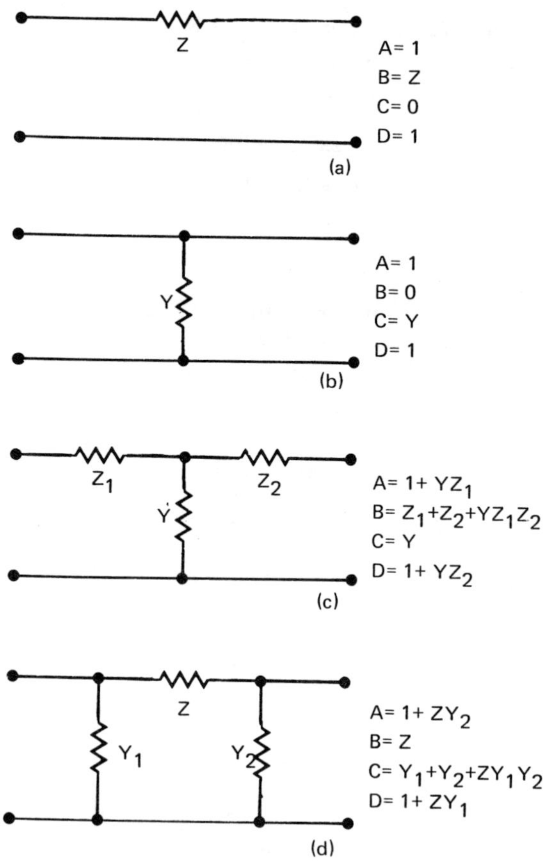

Figure 5.10 ABCD parameters for other 2-port networks: series impedance, shunt admittance, unbalanced TEE network, and unbalanced PI network.

3 THE PER-UNIT SYSTEM

In power systems, voltage, current, VA rating and impedance are usually expressed in a *per-unit system,* which is percent expressed as a decimal. The usual *bases* of the system are the KVA and KV ratings. The base current and impedance can then be calculated. For single-phase systems and per-phase values, the per-phase kilovoltamperes and line-to-neutral kilovolts can be used as the bases.

$$\text{base current} = \frac{\text{base } KVA_{1\phi}}{\text{base } KV_{ln}} \qquad 5.69$$

$$\text{base impedance} = \frac{\text{base} V_{ln}}{\text{base current}} \qquad 5.70$$

$$\text{base } KW_{1\phi} = \text{base } KVA_{1\phi} \qquad 5.71$$

$$\text{base } KVAR_{1\phi} = \text{base } KVA_{1\phi} \qquad 5.72$$

The quantities expressed in per-unit notation are

$$\text{current } (\textbf{P.U.}) = \frac{\text{actual current}}{\text{base current}} \qquad 5.73$$

$$\text{voltage } (\textbf{P.U.}) = \frac{\text{actual voltage}}{\text{base voltage}} \qquad 5.74$$

$$\text{impedance } (\textbf{P.U.}) = \frac{\text{actual impedance}}{\text{base impedance}} \qquad 5.75$$

$$\text{power } (\textbf{P.U.}) = \frac{\text{actual power}}{\text{base power}} \qquad 5.76$$

As three-phase power systems are usually specified in terms of total KVA and line-to-line voltages,

$$KVA_{1\phi} = \frac{KVA}{3} \qquad 5.77$$

$$V_{ln} = \frac{V}{\sqrt{3}} \qquad 5.78$$

Within any part of a system, all impedances must be expressed in the same per-unit system. Thus, it may be necessary to convert from another per-unit system. For instance, if the resistance and reactance of a device are given as per-unit quantities, the base is the rated KVA and KV of the device.

$$\text{base } Z = \frac{\text{base } V_{ln}}{\text{base } I} = \frac{(\text{base } V_{ln})^2}{\text{base } VA_{1\phi}} \qquad 5.79$$

The actual impedance is

$$Z = Z\,(\textbf{P.U.}) \times \frac{(\text{base } V_{ln})^2}{\text{base } VA_{1\phi}} \qquad 5.80$$

Equation 5.81 can be used to convert to a new per unit system.

$$Z\,(\text{new } \textbf{P.U.}) = Z\,(\textbf{P.U.}) \times \left(\frac{\text{base } V_{ln}}{\text{new base } V_{ln}}\right)^2 \times \frac{\text{new base } VA_{1\phi}}{\text{base } VA_{1\phi}} \qquad 5.81$$

Example 5.7

A 3000 KVA, 150 KV machine has a per-unit impedance of $.1 + j.6$. Determine the actual impedance of the machine. Express the per-unit impedance in a system which has 15 MVA and 150 KV bases.

On the basis of 3000 KVA and 150 KV, the per-phase bases are 1000 KVA and $\frac{150}{\sqrt{3}}KV = 86.6KV$. Then, the base current is $\frac{1000}{86.6} = 11.55$A, and the base impedance is

$$Z_{\text{base}} = \frac{86,600}{11.55} = 7500 \text{ ohms}$$

Then,

$$Z = 7500(.1 + j.6) = 750 + j4500 \text{ ohms}$$

In the new system, base $KVA = 5000$ and base $KV = \frac{150}{\sqrt{3}} = 86.6$, so that equation 5.69 gives

$$\text{base } I = \frac{5000}{86.6} = 57.74$$

The base impedance for the new per-unit system is

$$\text{new base } Z = \frac{86,600}{57.74} = 1500 \text{ ohms}$$

Therefore,

$$\text{new } Z\,(\mathbf{P.U.}) = \frac{750 + j4500}{1500} = .5 + j3\,\mathbf{P.U.}$$

The per-unit system has special advantages in evaluating performance of transformers, particularly two-winding transformers, as the per-unit impedances remain the same for bases taken from the transformer ratings.

Example 5.8

A 13.8 KV: 440 V, 50 KVA single-phase transformer has a leakage reactance of 300 ohms referred to the 13.8 KV side. Determine the **P.U.** value of the leakage reactance (a) for the high-voltage base and (b) for the low-voltage base.

The high-voltage has a KVA base of 50 KVA and a voltage base of 13.8 KV. Thus the current base is $\frac{50\,KVA}{13.8\,KV} = 3.62\ A$, and an impedance base of $\frac{13.8KV}{3.62} = 3808.8$ ohms. Then,

$$X\,(\mathbf{P.U.}) = \frac{300}{3808.8} = .079$$

For the low-voltage side, the leakage reactance is adjusted by the square of the turns ratio.

$$X_l\,(\text{secondary}) = \left(\frac{N_2}{N_1}\right)^2 X_l\,(\text{referred to primary})$$
$$= \left(\frac{.440}{13.8}\right)^2 \times 300 = .305 \text{ ohms}$$

The bases for the low-voltage side are 50 KVA and 0.44 KV. From equation 5.69, the current base is

$$\text{base } I = \frac{50,000}{440} = 113.6 \text{ amps}$$

From equation 5.70, the impedance base is

$$\text{base } Z = \frac{440}{113.6} = 3.872$$

Then

$$X_l\,(\mathbf{P.U.}) = \frac{.305}{3.872} = .079$$

Three-winding transformers present a special problem. The transformer KVA rating applies to the primary winding, while the secondary and tertiary windings have ratings which add up to the primary KVA ratings. The tests required are three partial short circuit tests.

TEST 1: Test 1 is performed from the primary side with shorted secondary and open tertiary windings.

$$Z_{SC_1} = R_1 + jX_1 = R_p + jX_p + a_{ps}^2(R_s + jX_s) \quad 5.82$$

The subscript s indicates quantities in the secondary circuit, and the subscript p indicates quantities in the primary circuit.

TEST 2: Test 2 is performed from the primary side with open secondary and shorted tertiary windings.

$$Z_{SC_2} = R_2 + jX_2 = R_p + jX_p + a_{pt}^2(R_t + jX_t) \quad 5.83$$

The subscript t indicates quantities in the tertiary circuit.

TEST 3: Test 3 is performed from the secondary side with open primary and shorted tertiary windings.

$$Z_{SC_3} = R_3 + jX_3 = R_s + jX_s + \frac{a_{pt}^2}{a_{ps}^2}(R_t + jX_t) \quad 5.84$$

Multiplying through by the square of the primary to secondary turns ratio, (a_{ps}^2), equation 5.84 becomes

$$a_{ps}^2 Z_{SC_3} = a_{ps}^2 R_3 + ja_{ps}^2 X_3 \quad 5.85$$
$$= a_{ps}^2(R_s + jX_s) + a_{pt}^2(R_t + jX_t)$$

Equations 5.86 through 5.91 can be obtained by separating real and imaginary parts in equations 5.82, 5.83, and 5.84.

$$R_p = \frac{R_1 + R_2 - a_{ps}^2 R_3}{2} \quad 5.86$$

$$X_p = \frac{X_1 + X_2 - a_{ps}^2 X_3}{2} \quad 5.87$$

$$a_{ps}^2 R_s = \frac{R_1 + a_{ps}^2 R_3 - R_2}{2} \quad 5.88$$

$$a_{ps}^2 X_s = \frac{X_1 + a_{ps}^2 X_3 - X_2}{2} \quad 5.89$$

$$a_{pt}^2 R_t = \frac{R_2 + a_{ps}^2 R_3 - R_1}{2} \quad 5.90$$

$$a_{pt}^2 R_t = \frac{X_2 + a_{ps}^2 X_3 - X_1}{2} \quad 5.91$$

The primary base is used as base for problems involving feeding through a 3-winding transformer.

Example 5.9

A three-winding, single-phase transformer is rated 6.8 KV, 50 KVA; 440 V, 30 KVA; and 1380 V, 20 KVA. Short circuit tests are performed as follows:

I. On the 6.8 KV winding, 440 V winding short-circuited and 1380 V winding open circuited:

$$Z_I = 100 + j180$$

II. On the 6.8 KV winding, 440 V winding open and 1380 V winding shorted:

$$Z_{II} = 120 + j200$$

III. On the 1380 V winding, 6.8 KV winding open and 440 V winding shorted:

$$Z_{III} = 5 + j6.6$$

Find the per-unit impedances in the primary base.

Notice that the third test is reversed from that described in the procedure (if the 440 V winding is the secondary). To overcome this problem, designate the 1380 volt winding as the secondary, so the two first tests must be reversed to use equations 5.82 through 5.91.

$$Z_1 = R_1 + jX_1 = 120 + j200$$
$$Z_2 = R_2 + jX_2 = 100 + j180$$

The turns ratio from primary to secondary is

$$a_{ps} = \frac{6.8}{1.38}$$
$$a_{ps}^2 = 24.28$$
$$a_{ps}^2 Z_3 = 121 + j160$$

Thus from equations 5.86 through 5.91,

$$R_p = \frac{120 + 100 - 121}{2} = 49.5 \text{ ohms}$$

$$X_p = \frac{200 + 180 - 160}{2} = 110 \text{ ohms}$$

$$a_{ps}^2 R_s = \frac{120 - 100 + 121}{2} = 70.5 \text{ ohms}$$

$$a_{ps}^2 X_s = \frac{200 - 180 + 160}{2} = 90 \text{ ohms}$$

$$a_{ps}^2 R_t = \frac{-120 + 100 + 121}{2} = 50.5 \text{ ohms}$$

$$a_{pt}^2 X_t = \frac{-200 + 180 + 160}{2} = 70 \text{ ohms}$$

The base impedance is

$$\text{base } Z = \frac{(6800)^2}{50,000} = 925 \text{ ohms}$$

Then,

$$Z_p = .054 + j.119 \textbf{ P.U.} \text{ (6.8 KV winding)}$$
$$a_{ps}^2 Z_s = .076 + j.097 \textbf{ P.U.} \text{ (1.38 KV winding)}$$
$$a_{pt}^2 Z_t = .055 + j.076 \textbf{ P.U.} \text{ (440 V winding)}$$

4 ONE-LINE DIAGRAMS

The normal operation of three-phase power systems is balanced. Since balanced three-phase systems are analyzed on a per-phase basis, it is redundant to draw more than one phase. The *one-line diagram* is a simplification which omits even the neutral, leaving one line for each three-phase circuit. Certain symbols are in standard use to describe elements of power systems. Some of these are shown in figure 5.11.

Locations of circuit breakers and fuses are not shown in *load studies*. In *stability studies* of transients due to faults, the locations and characteristics of fuses and circuit breakers (as well as control relays) are essential, and they must be included on the one-line diagram.

Figure 5.12 shows a one-line diagram of a portion of a hypothetical power system. This diagram begins at the top with the *generation level*. Two synchronous generators are shown connected to a *transmission line*. The generation voltage is 6.9 KV, which is in the usual range of power generation. The generators are connected with intervening *circuit breakers* to the low-voltage sides of wye-delta transformers.

The transmission voltage level is indicated by the usual ranges. The generators are shown with their neutrals connected through surge current limiting impedances.

This limits the maximum current in case of a ground fault in the generator circuit. One or more additional generators can be connected to each power bus.

There may also be local loads from the generation stations. The transmission lines interconnect the generators and lead to subtransmission distribution buses. *Subtransmission systems* are typically in a *grid form*, with interconnections between input buses, and with many paths to any distribution substation. There are usually at least two switchable inputs to a substation input bus to allow for isolation of faults while continuing service.

Each distribution substation feeds a number of distribution transformers, either in a radial, tree, or loop pattern, or via a grid system. The *distribution transformer* serves secondary mains from which individual three-phase and single-phase feeders bring the service to consumers.

Very large users may have their own substation. Intermediate size users may have a separate distribution transformer. Small users typically share a distribution transformer.

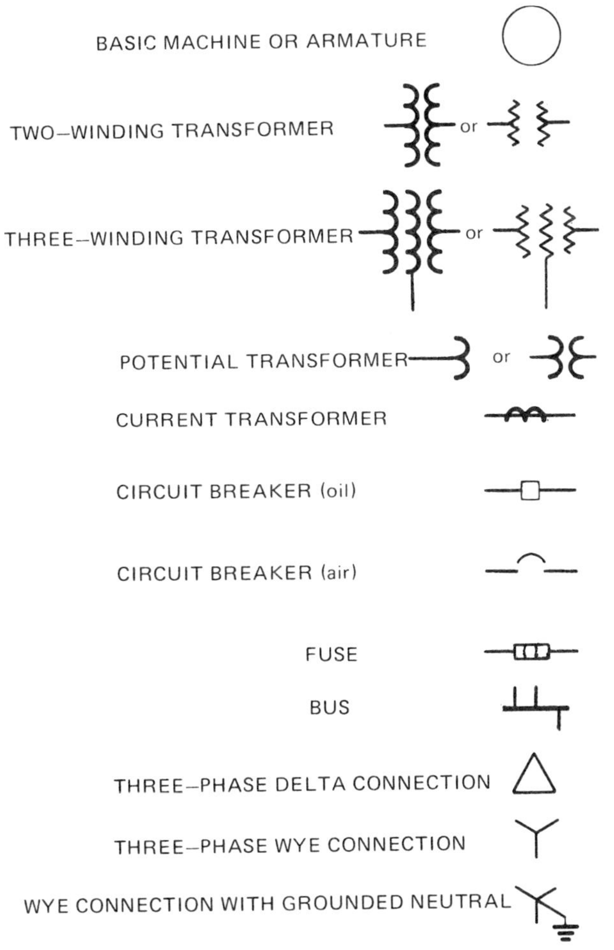

Figure 5.11 Symbols for One-Line Diagrams

The frequency of the power is rigidly controlled by varying the prime-mover speeds. Power transferred from one generator bus to another is controlled by varying the relative phase angle between the two voltages via the rotor current magnitude. The voltage levels are largely controlled by the use of taps on the substation transformers. These taps allow minor changes in the primary to secondary turns ratios on the transformers.

5 DISTRIBUTION TRANSFORMERS

Distribution transformers are able to serve either balanced three-phase loads or unbalanced single-phase loads. However, the overall distribution system remains balanced through a statistical process of combining single-phase and other unbalances. Nevertheless, the load at an individual distribution transformer may be unbalanced, requiring more complex analysis than the single-line method.

For most line regulation calculations, power-factor and primary voltage loading of the distribution transformer can be considered to be constant. This situation implies an *infinite bus* because the change in loading from an individual distribution transformer does not change the primary voltage. In such calculations, it is necessary to include only the series portion of the transformer equivalent impedance (i.e., the leakage reactance and the winding resistance of the combined primary and secondary) to account for the transformer.

A. TYPICAL PER-UNIT VALUES

Typical per-unit values for *distribution transformer impedances* can be estimated from the following formulas where such information is not known. The estimates will be within 25% of actual values for practical machines.

- *single-phase distribution transformers*:

$$Z = [.0252 - .0029 ln(KVA)] + j[.010 + .0042 ln(KVA)] \ \mathbf{P.U.} \qquad 5.92$$

- *three-phase transformers (per-phase)*:

$$Z = [.0364 - .0045 ln(KVA)] + j[.026 + .0018 ln(KVA)] \ \mathbf{P.U.} \qquad 5.93$$

These formulas illustrate the essentially inductive nature of transformer impedances. Most single-phase transformers are 2400 volts to 120/240 volts with ratings up to 100 KVA. Three-phase distribution transformers are typically 2400/4160 volts to 240/480 volts with ratings up to 400 KVA.

B. VOLTAGE REGULATION

Voltage regulation is conveniently calculated using per-unit quantities because the primary voltage of a transformer (referred to the secondary winding) is considered constant (i.e., an infinite bus). The difference between the primary voltage and the output of the transformer in per-unit values is the voltage regulation of the transformer when the output voltage is 1 **P.U.**.

Example 5.10

A 50 KVA, single-phase transformer has a 440 volt secondary voltage at rated KVA and 80% lagging power-

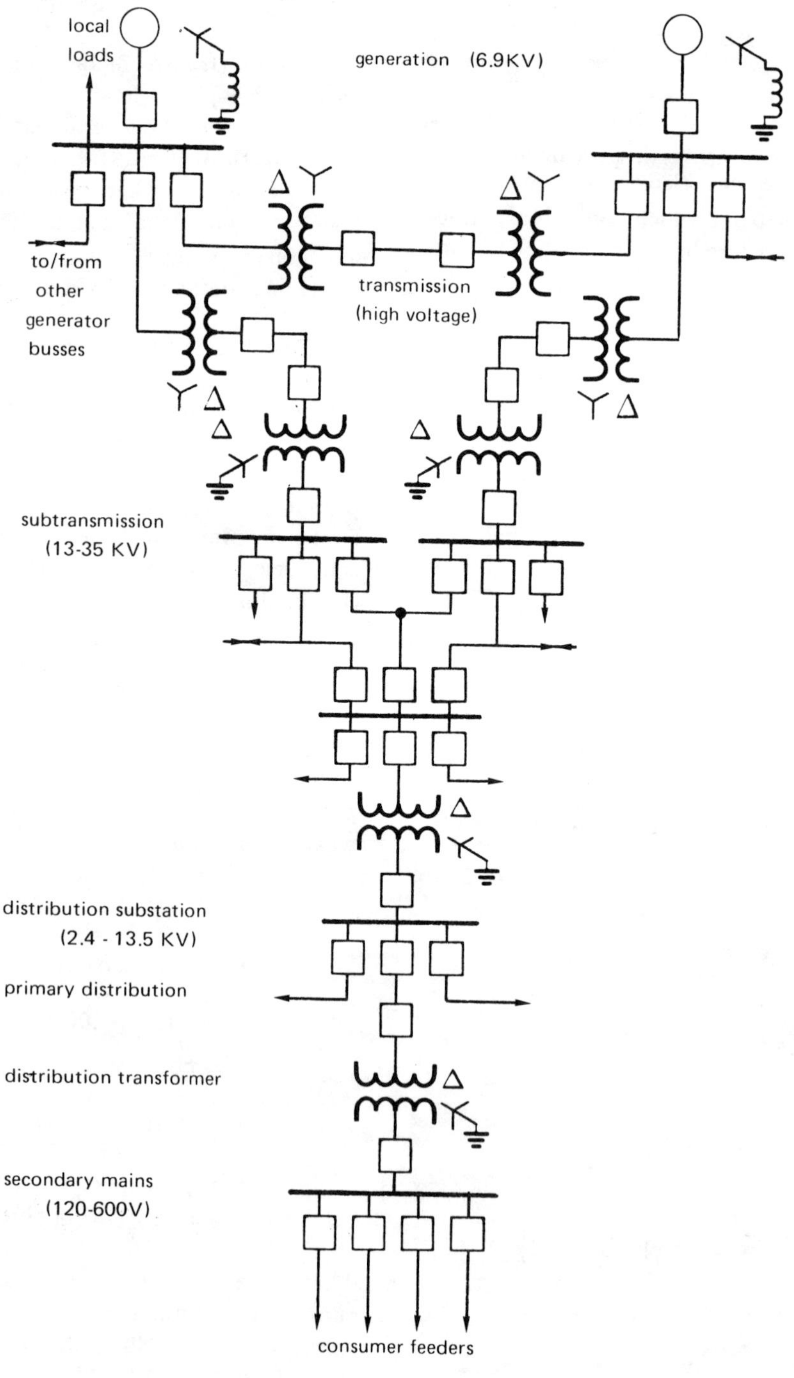

Figure 5.12 Partial one-line diagram of a hypothetical power system.

factor. Determine the half-load output voltage at the same power-factor. Calculate the voltage regulations going from no-load to full-load, and going from half-load to full-load.

The primary voltage of the transformer is considered constant. Then, from equation 5.92 the winding impedance is

$$Z_{\text{winding}} = .014 + j.026\,\textbf{P.U.} = .030\angle 62.0^\circ\,\textbf{P.U.}$$

Since $\arccos(0.8) = 36.9^\circ$, the current under full-load is $1\angle{-36.9^\circ}$. Taking the transformer output voltage as 1 **P.U.**, the primary voltage referred to the secondary is

$$V'_p = |1 + .030\angle 62.0^\circ \times 1\angle{-36.9^\circ}| = 1.027\,\textbf{P.U.}$$

To find the voltage of the transformer at half load, assume that it is 1 **P.U.** and calculate what the primary voltage would be. Then, adjust proportionally from the given information.

$$V''_p = |1 + .030\angle 62.0^\circ \times .5\angle{-36.9^\circ}| = 1.014\,\textbf{P.U.}$$

$$V_{\text{out}} = \frac{1.027}{1.014} \times 1 = 1.013 \text{ for } 50\% \text{ output}$$

The voltage regulation from half to full-load at $PF = .8$ is $1.01 - 1 = .013\,(1.3\%)$. And from the first result, the regulation from no-load to full-load is $1.027 - 1 = .027\,(2.7\%)$.

C. POWER FACTOR CORRECTION

Power factor correction is sometimes used to improve the power-carrying capacity of a transformer or a feeder line from a secondary main. The current capacity of a line is determined by its size and acceptable temperature rise. The power carried by a line under small power factors is less than at high power factors because the current amplitude (not the phase) determines the capacity of feeder lines and transformers.

Example 5.11

A feeder from a distribution transformer is 500 feet long. It consists of three AWG# 4/0 wires having a resistance of .052 ohms per 1000 feet and current capacity of 225 amps. The three-phase load is balanced and draws 120 KW at 440 volts with a 60% lagging power factor. Determine the KVA rating of capacitors capable of correcting the power factor enough to allow rated current to flow. The uncorrected current is

$$|I| = \frac{\frac{120{,}000}{3}}{.6 \times \frac{440}{\sqrt{3}}} = 262.5 \text{ amps per phase}$$

Referenced to the phase voltage, the current is

$$I = 157.5 - j210 \text{ amps}$$

This must be corrected to a magnitude of 225 while retaining the real part of 157.5. The corrected current is

$$I_{\text{corrected}} = 157.5 - j160.7 \text{ amps}$$

The leading current required through the capacitors is

$$210 - 160.7 = 49.3 \text{ amps}$$

Thus, the required per-phase rating of the capacitors is

$$\text{rating} = \frac{254 \times 49.3}{1000} = 12.5\,\text{KVA}$$

When the power factor of a transformer output is unity (or just slightly leading), the voltage regulation is minimized. This is because the voltage drop across the winding impedance is essentially in *quadrature* (90° out of time phase) with the primary voltage.

Example 5.12

A transformer rated 150 KVA, three-phase, supplies 100 KVA at a 50% lagging power factor. The transformer's winding impedance is 2% resistance and 5% reactance, based on 150 KVA. Assume 100% voltage. If the power factor is corrected to 90%, what improvement in voltage regulation will be achieved?

At $\frac{2}{3}$ of the rated KVA and 100% of the voltage, the current will be $\frac{2}{3}$ **P.U.**.

$$I = .667\angle{-60^\circ} = .333 - j.577$$

Correction to 90% PF with the same in-phase component of current will result in a current of

$$I_{\text{corrected}} = .333 - j.333\tan(\arccos .9)$$
$$I_{\text{corrected}} = .333 - j.161 = .37\angle{-25.8^\circ}$$

The uncorrected voltage drop across the winding impedance is

$$\begin{aligned} V &= [.02 + j.05] \times .667\angle{-60^\circ} \\ &= .054\angle 68.2^\circ \times .667\angle{-60^\circ} \\ &= .0357 + j.005 \end{aligned}$$

This is virtually in phase with the terminal voltage, so that the primary voltage is $1.0357 + j.005$ with a magnitude of 1.0357 **P.U.**. This corresponds to a voltage regulation of 3.57%.

With a 90% power factor, the voltage across the transformer's winding impedance is

$$.054\angle 68.2^\circ \times .37\angle -25.8^\circ = .0148 + j.0135$$

Thus, the primary voltage referred to the secondary is $1.0148 + j.0135$. This has a magnitude of 1.0149, corresponding to a voltage regulation of 1.49%. The improvement in voltage regulation is

$$3.57\% - 1.49\% = 2.08\%$$

D. INDUCTION MOTOR STARTING

Induction motor starting across secondary mains can cause undesirable *flicker* in lighting circuits and momentary undervoltage in the feeders. The usual starting current for induction motors is the motor power factor multiplied by six times the rated current. Upon starting, the motor is essentially inductive, as the winding reactances are on the order of five times the resistances. For larger distribution transformers, the reactance of the windings is so much greater than the resistance that only the reactance of the transformer winding needs to be included.

On starting, the transformer-induction motor combination appears as two inductances in series forming a voltage divider. The drop across the winding impedance is essentially in phase with the primary voltage, causing the worst case of voltage regulation. The drop in terminal voltage is estimated algebraically (rather than vectorially) from the drop across the winding impedance.

Example 5.13

A 250 KVA, three-phase, distribution transformer has a 440 volt secondary rating and a 4% inductive reactance based on its own ratings. What size induction motor can be started directly on the line if the dip in secondary main voltage is limited to 2%? Assume a machine efficiency of 90% and a PF of .9 running. State your answer in horsepower.

Use the transformer base of 250 KVA and 254 volts line-to-neutral. The starting current is approximated as

$$I_{\text{starting}} = 6 \times .9 \times I_{\text{rated}} = 5.4\, I_{\text{rated}}$$

This current can cause a .02 **P.U.** drop in voltage across the .04 **P.U.** reactance of the transformer winding.

$$5.4 I_{\text{rated}} \times .04 = .02\,\textbf{P.U.}$$

$$I_{\text{rated}} = .0926\,\textbf{P.U.}$$

To obtain current in amps, use the ratings of the transformer.

$$I_{\text{rated}} = .0926 I_{\text{base}}$$

$$= .0926 \times \frac{\frac{250\,KVA}{3}}{.254\,KV} = 30.4\text{ amps}$$

The KVA rating of the motor is

$$30.4A \times 254V \times 3\text{ phases} = 23.16\,KVA$$

With a 90% efficiency and a PF of .9, the rated horsepower of the motor is

$$\text{rated horsepower} = 23.16\,KVA \times \frac{1\text{ hp}}{.746\,KW} \times \frac{.9\,KW}{KVA} \times .9 = 25.15\text{ hp}$$

6 THREE-PHASE TRANSFORMER CONNECTIONS

Voltage ratios for three-phase transformers specify the line-to-line voltages regardless of the transformer configuration. However, when turns ratios are given, the primary and secondary winding configurations must be known in order to determine the line-to-line voltages. For either a wye primary with wye secondary configuration or a delta primary with delta secondary (i.e., wye-wye or delta-delta), the turns and voltage ratios are the same. For a wye primary with delta secondary (wye-delta), the secondary voltage is

$$V_s = \frac{N_s}{\sqrt{3}N_p} V_p \qquad 5.94$$

For a delta primary with wye secondary (delta-wye),

$$V_s = \frac{\sqrt{3}N_s}{N_p} V_p \qquad 5.95$$

Distribution transformers with three-phase secondaries are usually wye connected with a grounded neutral. This has the effect of maintaining the primary voltage referred to the secondary (V_p') when unbalanced loads are applied.

The nonlinear nature of transformers designed for minimum weight and cost causes a *third-harmonic* voltage to be generated. A delta connected winding causes the currents from third-harmonic voltages to circulate around the delta, being essentially shorted out, and thereby not interfering with the line currents or voltages. Thus, it is essential that all transformers have at least one delta winding. Where wye-wye is required, a tertiary winding should be connected in delta and left open (no load) to suppress the third-harmonic effects.

An *open delta* or *vee transformer* is sometimes used with only two phases being connected in an incomplete delta. This is typically accomplished with two single-phase transformers (not necessarily equally rated). When an accurate analysis is required, a complete circuit diagram including the winding impedances of the two transformers is required. If less accuracy is required, the winding impedances can be ignored and two single-line analyses performed.

Example 5.14

An open-delta, three-phase transformer consisting of two single-phase transformers is operating with a balanced three-phase load of 50 KVA, 440 volts and a lagging power factor of 0.8. Determine the required ratings of the two transformers.

Take the line-neutral voltage of the **a** phase as reference. Assuming an **abc** phase sequence with the **c-a** leg of the transformer missing, the voltages are

$$\begin{aligned} V_{an} &= 254\angle 0^\circ \\ V_{ab} &= 440\angle 30^\circ \\ V_{cn} &= 254\angle 120^\circ \\ V_{cb} &= 440\angle 90^\circ \end{aligned}$$

Neglecting the winding impedances, the currents are

$$\begin{aligned} I_a &= \frac{\frac{50,000}{3}}{254}\angle -\arccos^{-1} .8 \\ &= 65.6\angle -36.9^\circ \text{ amps} \\ I_c &= 65.6\angle -36.9 + 120^\circ \\ &= 65.6\angle 83.1^\circ \text{ amps} \end{aligned}$$

With the **c-a** leg of the transformer missing, I_a is the current in the **a-b** leg, and I_c is the current in the **c-b** leg. If asterisks are used to represent complex conjugates, then the ratings of the two legs are

$$\begin{aligned} S_{a-b} &= V_{ab} \times I_a^* = .44\angle 30 \times (65.6\angle -36.9)^* \\ &= .44\angle 30 \times 65.6\angle 36.9^\circ \\ &= 28.87\angle 66.9^\circ \, KVA \\ S_{c-b} &= V_{cb} \times I_c^* = .44\angle 90 \times 65.6\angle -83.1^\circ \\ &= 28.87\angle 6.9^\circ \, KVA \end{aligned}$$

Unbalanced loads can add to the complexity of problems with open delta transformers.

Example 5.15

The transformer in example 5.14 has an additional single-phase load of 10 KW resistive connected across the **c-a** leg. What should be the required ratings for the transformers?

A component of current in phase with V_{ca} is added to the currents previously found. On the **c** line this current will be in phase with V_{ca}, while on the **a** line it will be 180° out of phase with V_{ca} (i.e., in phase with V_{ac}).

$$|I_{1\phi}| = \frac{10,000}{440} = 22.73 \text{ amps}$$

This current is at the angle of V_{ca}, which leads V_{ab} by 120°. It is, therefore, at 150° with respect to V_{an}, the reference taken in example 5.14. Adding this current vectorially to the currents found in example 5.14,

$$\begin{aligned} I_c &= 65.6\angle 83.1^\circ + 22.73\angle 150^\circ = 77.40\angle 98.8^\circ \text{ amps} \\ I_a &= 65.6\angle -36.9^\circ + 22.73\angle -30^\circ = 88.21\angle -35.1^\circ \end{aligned}$$

The KVA for each leg is

$$\begin{aligned} S_{c-b} &= V_{cb} I_c{}^* = .440\angle 90^\circ \times 77.40\angle -98.8^\circ \\ &= 34.06\angle -8.8 \text{ KVA} \\ S_{a-b} &= .440\angle 30^\circ \times 88.21\angle 35.1^\circ \\ &= 38.8\angle 65.1^\circ \text{ KVA} \end{aligned}$$

7 VOLTAGE CONTROL

Voltage level is largely controlled at the distribution substation by tap-changes made under load. The taps can be on the primary or secondary sides or on both. The switching under load is done by an auto-transformer. The auto-transformer is initially connected between the present and previous tap, with its secondary adjusted to the present tap end. The other end of the auto-transformer is switched to the next tap on the transformer. Then the output (secondary) is moved by small steps to the next tap end. This establishes a new turns-ratio without interrupting the current flow.

It is sometimes necessary to bank two or more transformers. Placing two transformers in parallel will increase the current carrying capability and reliability of a system. When tap changes are made on banked transformers, very large currents in both the primary and secondary circuits can circulate. Only the leakage reactance of the parallel transformers is present to limit this circulating current.

Example 5.16

Two 1000 KVA transformers are connected in parallel. Taps are available in 2% increments above and below the nominal rated voltage. The reactances of the transformers are 10% (.1 **P.U.**). If the taps are changed consecutively rather than concurrently, what will the circulating current be during the tap changes?

The per-unit voltage between taps is .02. When one transformer has completed the tap change and the other has not yet begun, there is a .02 **P.U.** voltage around the loop consisting of the two secondaries. This voltage across the two .1 **P.U.** reactances in series will cause a circulating current of $\frac{.02}{.2}$ = .1 **P.U.**. The current is essentially in quadrature with the load current (assuming that the load PF is high) so that the increase in current caused by a 10% component added in quadrature to the rated current is only about .5% even though the component is 10%.

That is,

$$1 + j.1 = 1.005\angle 5.7°$$

Banked transformers are often not identical since systems develop over time and new capacity is added. This can lead to uneven load distribution if proper design is not observed.

Example 5.17

Two parallel transformers have 8% reactance each. One is rated 25,000 KVA and the other at 15,000 KVA. For a 30,000 KVA load with a lagging PF of .8, what fraction of the load will each transformer carry?

First put both transformers on the same base. The 25,000 KVA base is chosen (a combined base of 40,000 KVA would also be a reasonable choice). The reactance of the smaller transformer is then converted to the 25,000 KVA base.

$$X_{2,25000\,\mathrm{KVA}} = \frac{25,000}{15,000} \times .08 = .133\,\mathbf{P.U.}$$

The circuit can now be modeled by two voltage sources with identical open circuit voltages of 1 **P.U.**. The series reactances are .08 and .133, respectively. The outputs are tied together. The result is a current division independent of the load.

$$I_{25,000\,\mathrm{KVA}} = \frac{.133}{.133 + .08} \times I_{\mathrm{load}} = .625\,I_{\mathrm{load}}$$

$$I_{15,000\,\mathrm{KVA}} = \frac{.08}{.133 + .08} \times I_{\mathrm{load}} = .375\,I_{\mathrm{load}}$$

As the output voltages are identical, each transformer carries the same percentage of load power as it does the load current.

$$S_{25,000\,\mathrm{KVA}} = .625 \times 30,000 = 18,750\,\mathrm{KVA}$$
$$S_{15,000\,\mathrm{KVA}} = .375 \times 30,000 = 11,250\,\mathrm{KVA}$$

Note that the fraction carried by each transformer is proportional to its KVA rating. This is a result of the per-unit impedance being the same for both.

8 REPRESENTATIVE VALUES OF POWER DEVICE PARAMETERS

The per-unit values of impendances and other parameters of power devices lie in a reasonably restricted range. This restricted range is dependent on the base quantities of KVA and KV ratings and on power factors. Distribution transformer winding impedances were given previously. Power transformers are designed so that per-unit resistance decreases similarly to the distribution transformers. However, the resistance is so low at high KVA levels as to be negligible in all but efficiency studies.

When the resistance is needed, an estimated linear interpolation is reasonable. With a 1000 KVA (three-phase) or a 500 KVA (single-phase) transformer, the per-unit resistance is 1%; for a 10 MVA (three-phase) or a 5 MVA (single-phase) transformer, the per unit value is about .5%.

The leakage reactance of power transformers is approximately proportional to the high-voltage winding rating in KV. Reasonable estimates are given by equations 5.96 and 5.97.

- forced air-cooled transformers

$$X = .05 + 10^{-4}KV\,\mathbf{P.U.} \qquad 5.96$$

- forced oil-cooled transformers

$$X = .08 + 2 \times 10^{-4}KV\,\mathbf{P.U.} \qquad 5.97$$

The *magnetizing current* is approximately 1% of the rated current.

Synchronous generators have synchronous reactances of approximately 10% (.1 **P.U.**) and winding resistances on the order of 1% (.01 **P.U.**). Synchronous capacitors have synchronous reactances of approximately 14% (.14 **P.U.**) and resistances of 1% or less.

Synchronous motors present a problem in terms of the KVA rating because motor name-plate information is in horsepower. If the efficiency and running power factor are given, then the reactance and resistance values for the synchronous generators are used after determining the rated KVA from equation 5.98.

$$\mathrm{KVA}_{\mathrm{rated}} = \frac{(\mathrm{horsepower})(.746)}{\eta(PF)} \qquad 5.98$$

If neither efficiency nor power factor is given, the KVA rating is estimated to be the same as the horsepower. If unity power factor is given, the efficiency is assumed to be such that the KVA rating is .85 times the horsepower rating. If only the power factor is given as 80%, the KVA rating is 1.1 times the horsepower rating.

The starting current for induction motors is estimated as

$$I_{\text{starting}} = 6 \times P.F. \times I_{\text{rated}} \qquad 5.99$$

This is equivalent to a winding reactance of $\frac{1}{6 \times PF}$ **P.U.** based on the machine's KVA. However, the KVA rating should be taken as the horsepower rating if no other information is given. (This implies approximately 75% efficiency.) The *running power-factor* (typically around 85%) is the power factor when running at the rated current, voltage and horsepower. It can be taken as essentially the $\frac{\text{rotor resistance}}{\text{slip}}$. If the slip is not given, it can be assumed to be about 3%.

Example 5.18

A 50 HP, three-phase induction motor has a running power factor of 85%. Estimate its per-unit equivalent circuit parameters X_e, r_s, and $a^2 r_r$.

The starting current will be

$$6 \times PF = 6 \times .85 = 5.1\,\textbf{P.U.}$$

The equivalent reactance is dominant in starting, so its value is estimated as

$$\frac{1}{5.1} = .196\,\textbf{P.U.}$$

The PF angle is $\cos^{-1} .85 = 31.98°$, and its tangent is .620. This is the ratio of X_e to $(r_s + \frac{a^2 r_r}{\text{slip}})$.

$$r_s + \frac{a^2 r_r}{\text{slip}} = \frac{X_e}{.620} = .316\,\textbf{P.U.}$$

r_s is usually 1.5 to 2 times $a^2 r_r$. Since the slip was not given, it is assumed to be .03. Then

$$\left[2 + \frac{1}{.03}\right] a^2 r_r = .316$$

$$a^2 r_r = .009\,\textbf{P.U.}$$

$$r_s = .018\,\textbf{P.U.}$$

$$X_e = .196\,\textbf{P.U.}$$

9 FAULT CALCULATIONS

In power systems, faults are expected. System reliability and preservation require that protection of equipment be a part of the initial design.

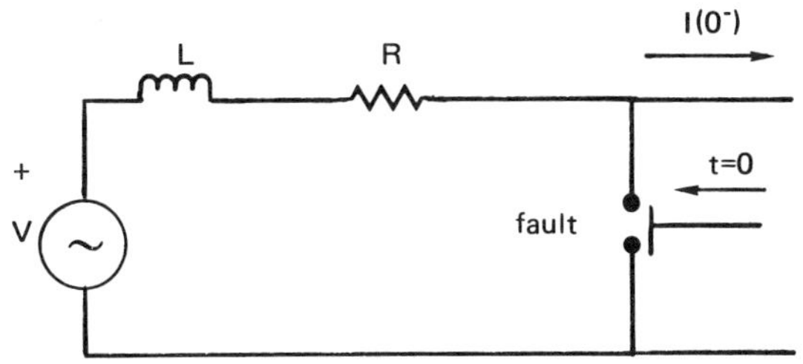

Figure 5.13 Model for an R-L circuit with a fault

Figure 5.13 shows a model for a system which experiences a fault. If the voltage is sinusoidal, the current is given by equation 5.101.

$$v = V_{\max} \sin(\omega t + \delta) \qquad 5.100$$

$$i = \frac{V_{\max}}{|Z|}\left[\sin(\omega t + \delta - \theta) - e^{\frac{-Rt}{L}} \sin(\delta - \theta)\right] + I(0^-)e^{\frac{-Rt}{L}} \qquad 5.101$$

$$|Z| = \sqrt{R^2 + (\omega L)^2} \qquad 5.102$$

$$\tan\theta = \frac{\omega L}{R} \qquad 5.103$$

δ is an arbitrary phase angle and $I(0^-)$ was the current flowing just prior to the fault occurence.

The time at which the fault occurs is crucial in determining the *transient component* (i.e., the coefficient of $e^{\frac{-Rt}{L}}$). The worse case of the unidirectional component of fault current occurs when the pre-fault current is at its maximum value and $\sin(\delta - \theta) = 1$. Then, the two components add to equation 5.104.

$$|I_{\text{unidirectional, max}}| = \frac{V_{\max}}{|Z|} + \sqrt{2} I_{\text{RMS (pre-fault)}} \qquad 5.104$$

In most power systems, the fault current is much higher than the pre-fault current, so that the pre-fault current can be ignored.

$e^{\frac{-Rt}{L}}$ can be written as $e^{\frac{-R(\omega t)}{X}}$ in order to study the behavior of the unidirectional component (called the *d.c. component*). For most power systems, the short-circuit PF is between .1 and .25, which also corresponds to a $\frac{R}{X}$ ratio of .1 to .25. As one cycle corresponds to $\omega t = 2\pi$, the d.c. component diminishes to $\frac{1}{e}$ of its initial value in $\frac{X}{2\pi R}$ cycles. For $\frac{X}{R} = 4$ it is less than 1 cycle. Even for $\frac{X}{R} = 50$, the time constant is eight cycles. For most purposes, the unidirectional component can be ignored. Thus, the fault current after a few cycles can be expressed by equation 5.105.

$$I \approx \frac{V_{\text{max}}}{|Z|} \sin(\omega t + \delta - \theta) \qquad 5.105$$

The RMS value of equation 5.105 is

$$I_{\text{RMS}} = \frac{V_{\text{RMS}}}{|Z|} \qquad 5.106$$

Synchronous generators have synchronous reactances dependent on the armature current. In addition, the phase angle of the armature current has an effect on the synchronous reactance. This is due to the non-uniformity of the rotor (most pronounced in salient pole machines) and the non-uniformity of the stator structure.

At high (unity) power factors, the armature current is essentially in phase with the terminal voltage, and the armature reaction voltage is leading the generator voltage by about 45°. Within the air gap, the reluctance is somewhat higher for the armature reaction MMF than for the rotor MMF. Therefore, a lower synchronous reactance is experienced. With small lagging power factors, the armature sees approximately the same (lower) reluctances as the rotor, and thus has higher synchronous reactances.

This can be taken into account by a *two-reactance theory*. The two reactances are called the *direct* and the *quadrature reactances*. The current is divided into two components — one in phase with the generated voltage E_g, and the other lagging E_g by 90°. The 90° lagging current component flows into the same reluctance as the rotor current (direct component). The current in phase with E_g acts into the reluctance that is 90° (electrically) from the rotor current (quadrature component). The quadrature reactance is always somewhat less than the direct reactance, even in non-salient machines. Thus, there are two reactances, X_d and X_q, and the armature reactance voltage jI_aX_s consists of two components.

$$jI_aX_s = jI_{a,\text{direct}}X_d + jI_{a,\text{quadrature}}X_q \qquad 5.107$$

$I_{a,\text{direct}}$ is the component of I_a which lags E_g by 90°. $I_{a,\text{quadrature}}$ is the component in phase with E_g.

For fault conditions, the power factor is low. The reactance to be used is X_d. Typical reactance values are given in table 5.1.

The reactance increases with current because the reluctance increases due to saturation. Under transient conditions a lower reactance is produced. The transient reactance is given by the primed quantities X'_d and X'_q. Table 5.1 gives values of X'_d useful in fault calculations.

During the initial part of the fault transient, a few cycles are required to establish the transient reactance. This period is called the *subtransient time* interval. During this interval, the effective reactance is changing from a considerably lower value to the transient reactances X'_d and X'_q. The initial reactances are called *subtransient reactances* and are identified by double primes X''_d and X''_q. Table 5.1 gives values for X''_d applicable to the pre-transient fault currents.

The generator voltage, E_g, just prior to a fault can be calculated from the synchronous reactance. However, this value will not be correct for the calculation of fault currents because of the requirement for continuity of energy. The sudden change of current in the armature winding results in a lower armature reactance, but the current through the leakage reactance must remain the same at the instant of the fault. (See equation 5.101 for $t = 0$.) Thus, a change in E_g to $E_{g'}$ for transient analysis and to $E_{g''}$ is necessary.

These voltages are calculated from the load condition just prior to the occurrence of the fault. However, the transient reactance X'_d or subtransient reactance X''_d are used as appropriate substitutes for the synchronous reactance X_s.

When a synchronous motor is in a system, it must be considered to be a generator for fault calculations. A value for the generated voltage is calculated on the basis of the transient reactance or the subtransient reactance.

Circuit breakers for specific applications are rated according to the current they can interrupt and the voltage they can withstand after interruption. The usual rating specification is in KVA or MVA. This rating is the product of the interrupt current capacity and the line-to-line KV rating.

Example 5.19

A synchronous generator and a synchronous motor are both rated at 15 MVA, 13.9 KV with 25% subtransient reactances. They are connected through a line having $.3 + j.4$ **P.U.** impedance on the 15 MVA, 13.9 KV base. The motor is running at 12 MVA, 13 KV, 0.8 P.F. lagging. (a) Determine the equivalent generated voltages for the generator ($E_{g''g}$) and the motor ($E_{g''m}$) under the expectation of a fault. (b) Determine the subtransient fault current of the generator and the motor for a three-phase fault at the terminals of the motor. (c) Determine the MVA rating of circuit breakers to protect the machines under these conditions. The fault is on the bus side of the circuit breakers.

The synchronous motor will act as a generator into the fault while it gradually loses speed. The subtransient time interval is too short for the motor to have lost significant speed, so it can be treated as a synchronous generator for this interval.

The terminal voltage is $\frac{13}{13.9}$ or 0.935 per unit. This is taken as the reference (at 0°). The current is

$$|I_L| = \frac{\frac{12\,MVA}{13\,KV}}{\frac{15\,MVA}{13.9\,KV}} = .855\,\textbf{P.U.}$$

$$I_L = .855\angle{-36.9^\circ} = .684 - j.513\,\textbf{P.U.}$$

With a 25% subtransient reactance, the necessary E_{gm} is

$$E_{g''m} = V_t - I_L(jx''_d) = .935 - (.685 - j.513) \times (j.25)$$
$$= 807 - j.171 = .825\angle{-12^\circ}\,\textbf{P.U.}$$

The impedance from the motor terminals is

$$Z = Z_{\text{line}} + j.25 = .3 + j.65 = .716\angle 65.2^\circ$$

Then, the required value of $E_{g''g}$ is

$$E_{g''g} = V_t + I_L Z = .935 + (.716\angle 65.2^\circ) \times (.855\angle{-36.9^\circ})$$
$$= 1.47 + j.29 = 1.5\angle 11.1^\circ\,\textbf{P.U.}$$

When the fault occurs at the motor terminals, the motor current will be $\frac{E_{g''m}}{jx''_d}$.

Table 5.1	X_d (unsat.)	X_q rated current	X'_d rated voltage	X''_d rated voltage	X_2 rated current	X_0 rated current (note 1)
two-pole turbine generators	1.20 / 0.95–1.45	1.16 / 0.92–1.42	0.15 / 0.12–0.21	0.09 / 0.07–0.14	$= X''_d$	0.03 / 0.01–0.08
Four-pole turbine generators	1.20 / 1.00–1.45	1.16 / 0.92–1.42	0.23 / 0.20–0.28	0.14 / 0.12–0.17	$= X''_d$	0.08 / .015–0.14
Salient-pole generators and motors (with dampers)	1.25 / 0.60–1.50	0.70 / 0.40–0.80	0.30 / 0.20–0.50 *note 2*	0.20 / 0.13–0.32 *note 2*	0.20 / 0.13–0.32 *note 2*	0.18 / 0.03–0.23
Salient-pole generators (without dampers)	1.25 / 0.60–1.50	0.70 / 0.40–0.80	0.30 / 0.20–0.50 *note 2*	0.30 / 0.20–0.50 *note 2*	0.48 / 0.35–0.50	0.19 / 0.03–0.24
Capacitors (air-cooled)	1.85 / 1.25–2.20	1.15 / 0.95–1.30	0.40 / 0.36–0.50	0.27 / 0.19–0.30	0.26 / 0.18–0.40	0.12 / .025–0.15
Capacitors (hydrogen-cooled at ½ psi) KVA rating	2.20 / 1.50–2.65	1.35 / 1.10–1.55	0.48 / 0.36–0.60	0.32 / 0.23–0.36	0.31 / 0.22–0.48	0.14 / 0.03–0.18

note 1 X_0 varies so critically with armature winding pitch that an average value can hardly be given. Variations are from 0.1 to 0.7 of X''_d. Low limit is for two-thirds pitch windings.

note 2 High-speed units tend to have low reactance, and low-speed units have high reactance.

Table 5.1 Typical constants of three-phase synchronous machines. Typical values are given above the bars, and the ranges given below the bars. From *Electrical Transmission and Distribution Reference Book*, by permission of the Westinghouse Electric Corporation.

$$I_m = \frac{.825\angle{-12^\circ}}{.25\angle 90^\circ} = 3.3\angle{-102^\circ}\ \textbf{P.U.}$$

The generator current will be $\frac{E_{g''g}}{Z}$.

$$I_g = \frac{1.5\angle 11.1^\circ}{.716\angle 65.2^\circ} = 2.09\angle{-54.1^\circ}\ \textbf{P.U.}$$

The total fault current is

$$3.3\angle{-102^\circ} + 2.09\angle{-54.1^\circ} = 4.95\angle{-83.7^\circ}$$

The three-phase breaker rating can be found from **P.U.** current, the base KVA, and base KV.

$$\text{KVA rating of breaker (3-phase)} = \text{I } (\textbf{P.U.}) \times \text{base KVA}$$

The rating of the motor circuit breaker is

$$3.3 \times 15\ MVA = 49.5\ MVA$$
at 13.9 KV three-phase rating

The rating for the generator is

$$2.09 \times 15\ MVA = 31.35\ MVA$$
at 13.9 KV three-phase rating

Asymmetrical faults occur when there is a single line-to-ground (or line-to-neutral) fault, a line-to-line fault, or a two-line-to-ground (or neutral) fault. The three-phase fault previously discussed permits the use of per-phase calculations because of the symmetry.

When an asymmetrical fault occurs, the analysis can be performed laboriously on an unbalanced three-phase basis, or by use of *symmetrical components*. The use of symmetrical components is convenient because fault currents can be determined in terms of the pre-fault conditions and the associated *sequence impedances.* A partial development of symmetrical components is included here.

An unbalanced three-phase problem can be solved by using three sets of balanced voltage and current components. These sets consist of a positive sequence (normally *abc*), a negative sequence (normally *acb*), and a zero sequence. With an unbalanced condition, each phase current can be represented by three components: positive sequence, negative sequence and zero sequence. These are illustrated in figure 5.14.

The *positive sequence* has been used previously. It includes balanced three-phase generators, transformers, and motors.

The *negative sequence* can be considered either as a mirror image of the positive sequence with phasor rotation in the opposite direction, or as an order reversal in phase sequence (e.g., *acb* if the positive sequence is *abc*, or *abc* if the positive sequence is *acb*). The latter method is preferred.

The *zero sequence* is a set of equal voltages or currents which have the same time phase in each of the three legs. The zero sequence can occur only in four wire systems (e.g., wye connections with a neutral wire or a grounded neutral).

It is possible to calculate the three sequence components from the phasor diagram if the magnitudes and angles are known. However, we are interested primarily in determining fault currents when only the generator voltages are known. For this procedure, the equivalent circuits for each phase sequence are needed.

Symmetrical components are mathematical concepts used to account for the imbalance of the circuit. They are useful in determining all currents in a system when a fault occurs. It is possible, however, to use some of the theory to determine the fault currents alone, thus specifying breaker requirements. In order to do this, it is necessary to determine the equivalent circuits for the three sequences, so as to determine the sequence impedances.

Positive sequence impedances are those which have been used in balanced three-phase analyses up to this point. The generators are positive sequence by definition, so they appear only in positive sequence equivalent circuits. The positive sequence circuit is established with the neutral as the reference bus. The impedance of a generator can be the subtransient reactance, the transient reactance, or the synchronous reactance depending on the analysis required. For fault analysis, the subtransient reactance is used. The equivalent generator voltage as calculated in example 5.19 is appropriate.

For rotating machines, the *negative sequence impedance's* contribution to the armature reaction is a field rotating in the opposite direction from the rotor field. Thus, the reactance is different from that of the positive sequence. The *negative sequence reactance* is the average of the subtransient direct and the quadrature reactances.

$$X = \frac{1}{2}(X''_d + X''_q) \qquad 5.108$$

Zero sequence impedances are quite different from positive sequence impedances. In machines, such as synchronous motors, the three components of the zero sequence currents do not form a rotating field. The only part of the machine impedance they see is the leakage reactance, which is given in table 5.1 as X_0.

On a transmission line, the zero sequence currents are equal, in phase with one another, and travel parallel in the three lines. The result is a single line over ground the plane or a single-phase line if there is a ground wire present. The series reactance is increased by a factor of 2 to 3.5 over the positive sequence reactance.

The resistance of a zero sequence line is not changed from the positive sequence value, as it is a function of frequency and not of spacing. The only capacitance seen by zero sequence currents is the capacitance to ground. This produces an increase in capacitive reactance by about the same factor as the increase in inductive reactance.

A zero sequence current flowing in one line of a three-phase network must, by definition, also be flowing in the other two lines. This has drastic effects on the impedance of three-phase loads. The symmetry of the zero sequence circuit requires that the zero sequence voltage from any line to the reference be the same as from any other line to the reference at that point. Thus, the zero sequence voltages across any delta-connected load must be zero, so that zero sequence currents cannot flow in a delta-connected load. Similarly, a wye-connected load with a floating neutral cannot support zero-sequence currents, as current must sum to zero at the floating neutral.

Only a wye-connected load with the grounded neutral, either directly or through an impedance Z_n, permits zero-sequence currents. The impedance seen from the line side of such a load to ground is the load impedance Z_L plus three times the neutral to reference impedance. The factor 3 is required because the three zero sequence components must flow through Z_n. Z_n will also include *ground impedance* from the actual grounding point and the zero sequence reference point or *reference bus*.

Transformers are the most difficult for which to determine zero sequence impedances. The key point for transformers is that primary current can flow only when secondary current flows, and vice versa. When one side

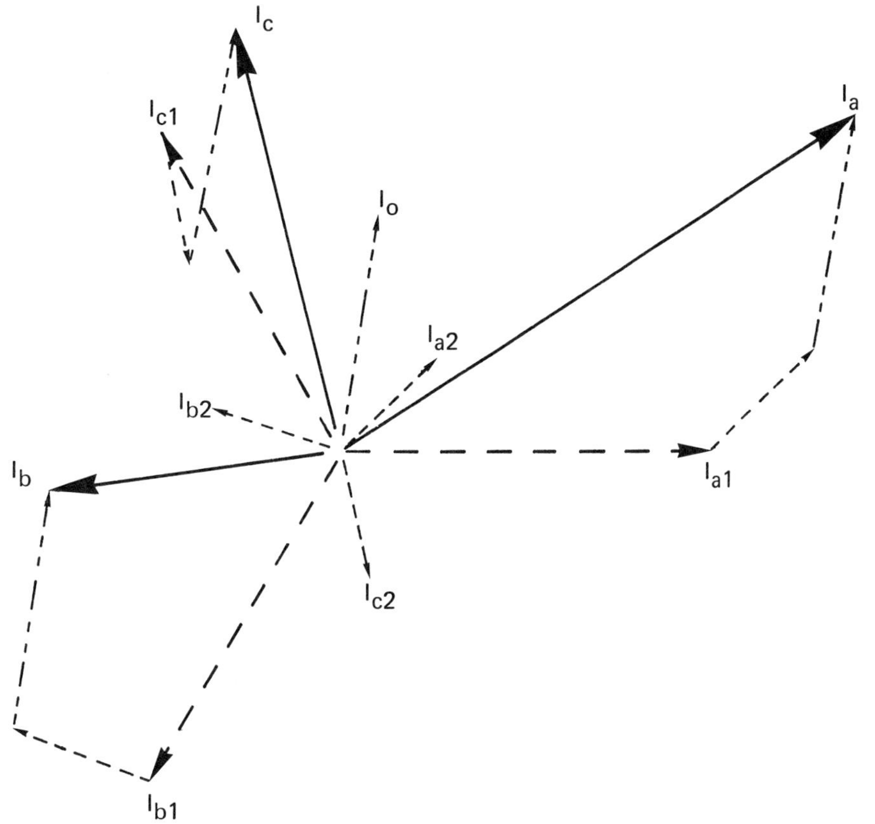

Figure 5.14 Phasor diagram showing symmetrical components for unbalanced currents in a three-phase system.

of a two-winding transformer contains a wye-connected winding with a floating neutral, it is an open circuit to zero-sequence currents. Only transformers with grounded wye connections on both primary and secondary windings permit a connection from primary to secondary. In this situation, the impedance is the the same as for the positive sequence circuit.

Only a delta-connected transformer secondary winding can act as a current path for zero sequence currents. In such a case, the same current flows around the closed path of the delta satisfying the requirement of equal zero sequence currents in each phase. However, the equal phase to reference bus on a delta connected winding requires that there be zero voltage from line to line in the delta-connected secondary. This will appear as a short circuit to a grounded wye-connected primary. Conversely, a delta-connected primary will appear as a short circuit to a grounded wye-connected secondary.

Figure 5.15 shows zero sequence impedances for various circuit configurations.

Sequence networks are obtained by the correct use of the elements outlined previously. To illustrate the process, consider the simple distribution system shown in figure 5.16(a). The positive sequence network, as previously determined for balanced networks, is shown in figure 5.16(b).

The negative sequence network is similar to the positive sequence network, but it omits all positive sequence generators and replaces X''_d by $\frac{1}{2}(X''_d + X''_q)$. This is shown in figure 5.16(c).

The zero sequence network is more difficult to obtain. Care must be taken in observing ground paths. The generator has a ground path which carries $3I_0$, so for any individual phase it appears as $3Z_n$, as shown in figure 5.16(d). The wye-connected primary of transformer T_1 is ungrounded and results in an open circuit at point A. The secondary of T_1 and primary of T_2 are both delta-connected, and they isolate the transmission line from the zero sequence network.

The delta connection of T_2 and the grounded secondary wye connection effectively ties the transformer impedances to ground. The grounded wye connections of the motor and load put them in parallel as shown.

Symmetrical components are most conveniently represented with the phasor operator **a**, which is a unit vector at an angle of 120°(similar to j, which is a unit vector at 90°).

$$\mathbf{a} = 1\angle 120° = -.5 + j.866 \qquad 5.109$$

$$\mathbf{a}^2 = 1\angle 240° = -.5 - j.866 \qquad 5.110$$

$$\mathbf{a}^3 = 1\angle 360° = 1 \qquad 5.111$$

$$\mathbf{a}^4 = a \qquad 5.112$$

$$\mathbf{a}^5 = a^2 \qquad 5.113$$

$$\mathbf{a}^6 = a^3 \qquad 5.114$$

Analysis of the symmetrical components in terms of actual circuit values makes use of equations 5.115 through 5.118.

$$I_a = I_{a0} + I_{a1} + I_{a2} \qquad 5.115$$

$$I_b = I_{b0} + I_{b1} + I_{b2} \qquad 5.116$$

$$I_c = I_{c0} + I_{c1} + I_{c2} \qquad 5.117$$

$$I_{a0} = I_{b0} = I_{c0} = \frac{1}{3}(I_a + I_b + I_c) \qquad 5.118$$

For an *abc* sequence the positive sequence components are

$$I_{b1} = \mathbf{a}^2 I_{a1} \qquad 5.119$$

$$I_{c1} = \mathbf{a} I_{a1} \qquad 5.120$$

The negative sequence components are

$$I_{b2} = \mathbf{a} I_{a2} \qquad 5.121$$

$$I_{c2} = \mathbf{a}^2 I_{a2} \qquad 5.122$$

Rearranging equations 5.115 through 5.122 results in equations 5.123 through 5.126.

$$I_{a0} = \frac{1}{3}(I_a + I_b + I_c) \qquad 5.123$$

$$I_{a1} = \frac{1}{3}(I_a + \mathbf{a}I_b + \mathbf{a}^2 I_c) \qquad 5.124$$

$$I_{a2} = \frac{1}{3}(I_a + \mathbf{a}^2 I_b + \mathbf{a}I_c) \qquad 5.125$$

A similar development for sequence to phase voltage conversion results in equations 5.126 through 5.128. The zero sequence voltage reference is ground. The positive and negative sequence voltages are referenced to neutral.

$$V_{a0} = \frac{1}{3}(V_{an} + V_{bn} + V_{cn}) \qquad 5.126$$

$$V_{a1} = \frac{1}{3}(V_{an} + \mathbf{a}V_{bn} + \mathbf{a}^2 V_{cn}) \qquad 5.127$$

$$V_{a2} = \frac{1}{3}(V_{an} + \mathbf{a}^2 V_{bn} + \mathbf{a}V_{cn}) \qquad 5.128$$

Symmetrical three-phase faults can be calculated since it is known that the load is balanced and that $I_b = \mathbf{a}^2 I_a$ and $I_c = \mathbf{a} I_a$.

$$I_{a0} = \frac{1}{3}(1 + \mathbf{a}^2 + \mathbf{a})I_a = 0 \qquad 5.129$$

$$I_{a1} = \frac{1}{3}(1 + \mathbf{a}^3 + \mathbf{a}^3)I_a = I_a \qquad 5.130$$

$$I_{a2} = \frac{1}{3}(1 + \mathbf{a}^4 + \mathbf{a}^2)I_a = 0 \qquad 5.131$$

Thus balanced (positive sequence only) analysis can be done.

Example 5.20

For the power system shown in figure 5.16, determine the fault current for a three-phase fault at the load end (point C) of the transmission line. The load bus is at 90% voltage, the motor is running at 0.8 leading power factor, and the current is 0.5 **P.U.**

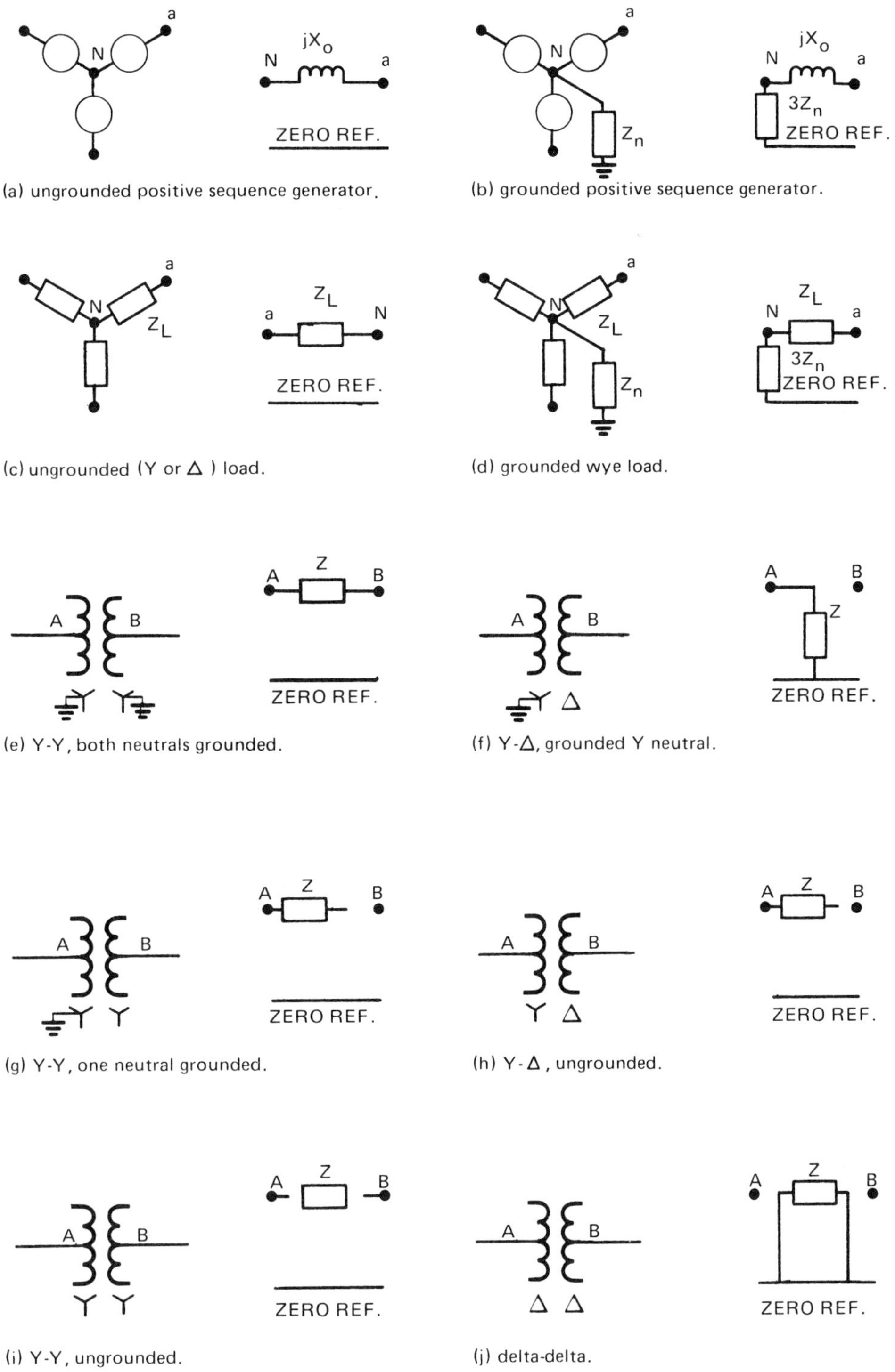

Figure 5.15 Zero sequence impedances

Since only the fault current is required, it is sufficient to obtain the Thevenin equivalent circuit, using the appropriate subtransient reactances, just prior to defaulting.

With the 90% voltage at the load bus taken as reference, the three-phase load takes a current of

$$\left(\frac{.9}{2}\right)\angle{-36.9^\circ} = .45\angle{-36.9^\circ}$$

The motor is operating with a power factor of 0.8 leading, with a current of $.5\angle 36.9^\circ$. The total bus current is the sum of the load and motor currents.

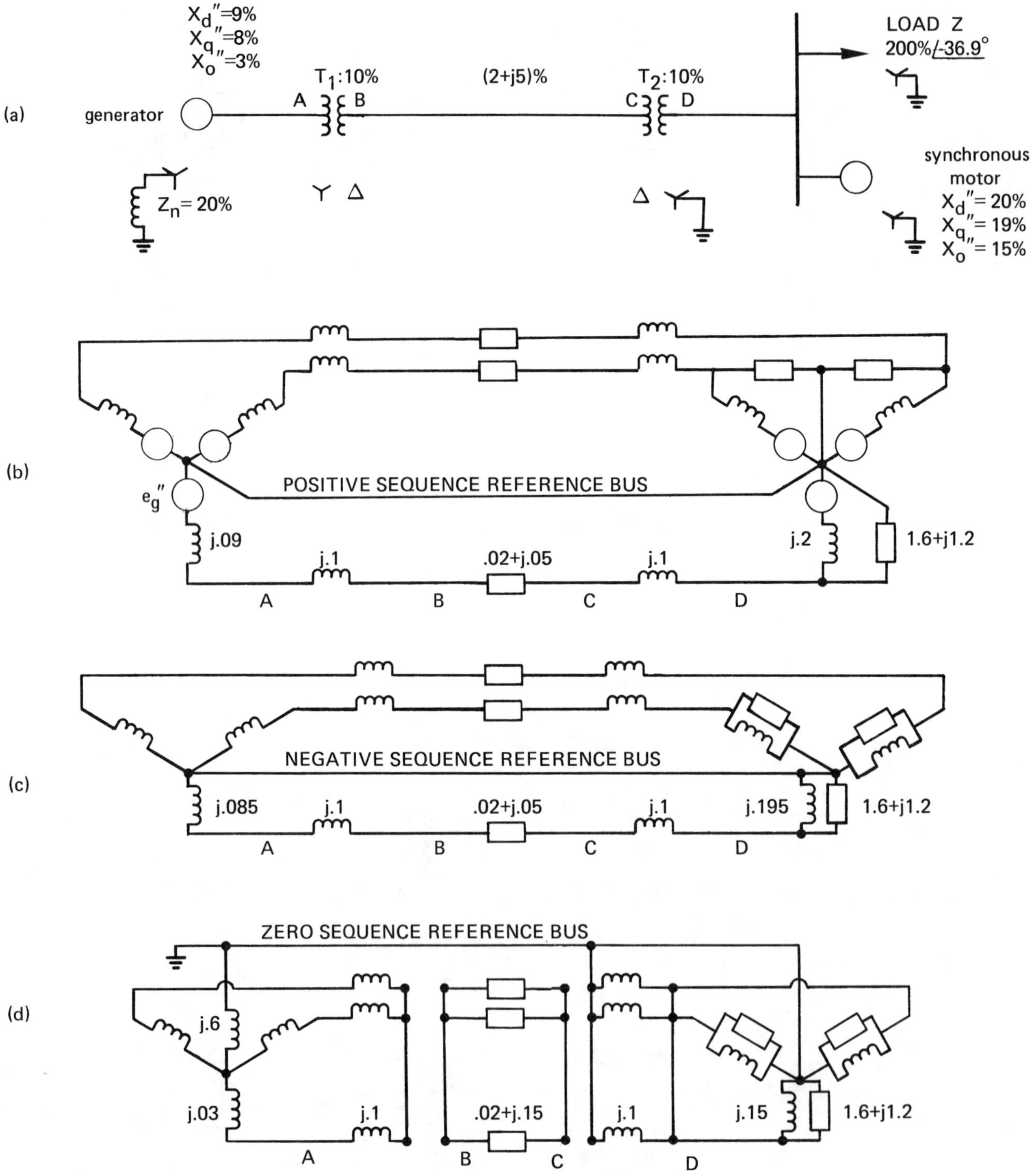

Figure 5.16 A simple distribution system with corresponding sequence networks

$$.45\angle{-36.9^\circ} + .5\angle 36.9^\circ = .76\angle 2.3^\circ$$
$$= .76 + j.03$$

The voltage rise across the 10% transformer reactance is

$$j.1 \times (.76 + j.03) = -.003 + j.076$$

Therefore, the pre-fault voltage is

$$.9 + (-.003 + j.076) = 0.9\angle 4.8^\circ \textbf{ P.U.}$$

The impedance seen from the fault is to the left of point C. $(.02+j.05)+(j.1)+j.09$ is in parallel with the impedance to the right. On the right is the $j.1$ **P.U.** for the transformer. Also, the parallel combination of $j.2$ from the motor and $2\angle{-36.9}$ produces an impedance of $.018 + j.211$. The parallel combination is $0.113\angle 85.2^\circ$ **P.U.**.

The fault current is the Thevenin equivalent pre-fault voltage divided by the impedance seen from the point of the fault to the neutral.

$$|I_{\text{fault}}| = \frac{.9}{.113} = 7.96 \textbf{ P.U.}$$

(When the load has no effect, it is necessary to treat the symmetrical fault as was done in example 5.19.)

Single-phase faults to ground require finding the Thevenin equivalent network for the fault assuming that any type of fault can occur there. The pre-fault circuit must be found for each of the sequence circuits. Equations 5.123 through 5.128 can be used to determine the fault currents.

The Thevenin equivalent circuits are found by determining pre-fault voltage at the fault and impedance from fault point to reference bus for each of the sequence circuits.

For a single-phase fault from the a phase to ground, for example, the currents through the three phases of Thevenin equivalents are I_a, $I_b = 0$, and $I_c = 0$. I_a is unknown and must be determined.

From equations 5.123 through 5.125, the single phase to ground currents are

$$I_{a0} = \frac{I_a}{3} \qquad 5.132$$

$$I_{a1} = \frac{I_a}{3} \qquad 5.133$$

$$I_{a2} = \frac{I_a}{3} \qquad 5.134$$

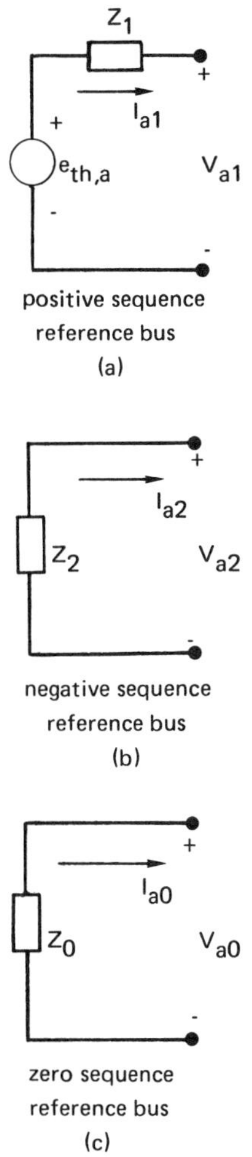

Figure 5.17(a) Thevenin equivalent networks for sequence networks as seen from the fault point.

This makes I_{a0}, I_{a1}, and I_{a2} equal, so the three circuits shown in figure 5.17 and 5.17(a) are in series. With line a shorted, $V_a = 0$, so the sum of the voltages V_{a1}, V_{a2}, and V_{a0} is zero. The resulting circuit must be as in figure 5.17. From this it is seen that the fault current is

$$I_{a1} = \frac{e_{Th,a}}{Z_0 + Z_1 + Z_2} \qquad 5.135$$

$$I_a = 3I_{a1} \qquad 5.136$$

Example 5.21

Determine the current in a fault from line to ground at point C for the circuit and conditions in example 5.20.

The Thevenin equivalent for the positive sequence circuit has been calculated.

$$e_{a,Th} = .9\angle 4.8^\circ$$
$$Z_1 = .113\angle 85.2^\circ$$

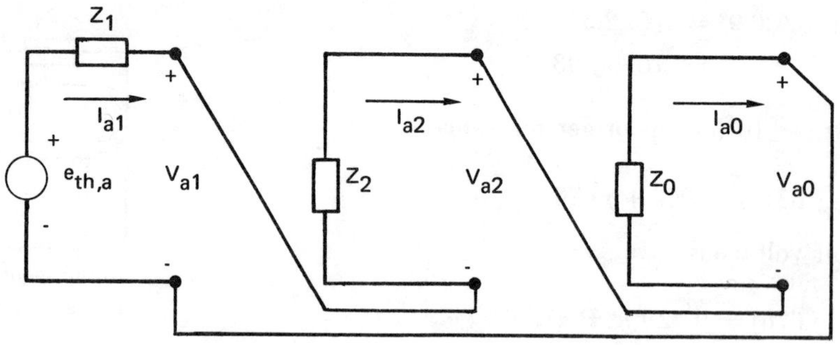

Figure 5.17(b) Phase *a* circuit for a line to ground fault on phase *a*.

The Thevenin equivalent impedance for the negative sequence to the left of point C is

$$.02 + j.05 + j.1 + j.085 = .02 + j.235 = .236\angle 85.1^\circ$$

To the right, $j.1$ is in series with the parallel combination of $2\angle -36.9^\circ$ and $j.195$ on a per-unit basis. This parallel combination is $.206\angle 82.3^\circ$. When added to $j.1$, the total to the right is $.305\angle 84.8^\circ$. The Thevenin impedance is the sum of the parallel impedances to the left and the right.

$$Z_2 = \frac{1}{\frac{1}{.305\angle 84.1^\circ} + \frac{1}{.236\angle 85.1^\circ}}$$

$$= .133\angle 85^\circ$$

In figure 5.16(c) the impedance from point C to the zero-sequence reference is infinite because there is no zero-sequence path to ground: $Z_0 = \infty$.

Thus, a single-line to ground fault at C causes no fault current.

A *line-to-line fault* will require working with all three phases of the Thevenin equivalent network, as indicated in figure 5.18. The application of equations 5.123 and 5.128 requires the following equations.

$$I_a = 0 \qquad 5.137$$

$$I_c = -I_b \qquad 5.138$$

$$V_b = V_c \qquad 5.139$$

$$3I_{a0} = 0 + I_b - I_b = 0 \qquad 5.140$$

$$3I_{a1} = 0 + \mathbf{a}^2 I_b - \mathbf{a}^2 I_b = (\mathbf{a} - \mathbf{a}^2) I_b \qquad 5.141$$

$$3I_{a2} = 0 + \mathbf{a}^2 I_b - \mathbf{a} I_b = (\mathbf{a}^2 - \mathbf{a}) I_b \qquad 5.142$$

Therefore,

$$I_{a2} = -I_{a1} \qquad 5.143$$

Also,

$$3V_{a0} = V_a + V_b + V_b = V_a + 2V_b \qquad 5.144$$

$$3V_{a1} = V_a + \mathbf{a}V_b + \mathbf{a}^2 V_b = V_a + (\mathbf{a} + \mathbf{a}^2) V_b \qquad 5.145$$

$$3V_{a2} = V_a + \mathbf{a}^2 V_b + \mathbf{a} V_b = V_a + (\mathbf{a}^2 + \mathbf{a}) V_b \qquad 5.146$$

Therefore,

$$V_{a1} = V_{a2} \qquad 5.147$$

These conditions are fulfilled by the interconnection of the Thevenin equivalent sequence circuits shown in figure 5.17(b). For this equivalent circuit,

$$I_{a1} = \frac{e_{Th,a}}{Z_1 + Z_2} \qquad 5.148$$

$$I_b = I_{a0} + \mathbf{a}^2 I_{a1} + \mathbf{a} I_{a2} \qquad 5.149$$

With $I_{a2} = -I_{a1}$ and $I_{a0} = 0$,

$$I_b = I_{\text{fault}} = -j\sqrt{3} I_{a1} \qquad 5.150$$

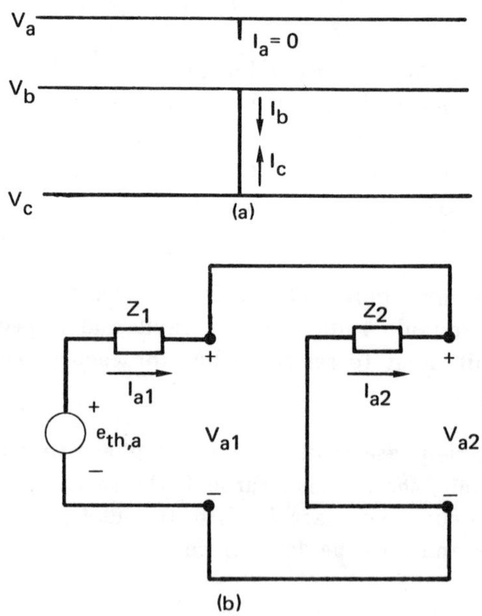

Figure 5.18 Line-to-line fault and equivalent circuit for sequence currents.

Example 5.22

Determine the fault current for a line to line fault at point C of the circuit shown in example 5.20 under the load conditions of example 5.20. Also, determine the line to neutral voltages at point C.

The Thevenin equivalent circuits were found for point C in examples 5.20 and 5.21.

$$e_{Th,a} = .9\angle 4.8°$$
$$Z_1 = .113\angle 85.2°$$
$$Z_2 = .133\angle 85°$$

Then, from equation 5.148

$$I_{a1} = 3.66\angle{-80.3°}$$
$$V_{a1} = I_{a1}Z_2 = .487\angle 4.7°$$

Furthermore, it is known that $I_{a0} = 0$, $I_{a2} = -I_{a1}$, $V_{a0} = 0$, and $V_{a2} = V_{a1}$. Therefore,

$$I_b = j\sqrt{3}I_{a1} = 6.34\angle 9.7°$$
$$V_a = V_{a0} + V_{a1} + V_{a2} = 2V_{a1}$$
$$= .974\angle 4.7°$$
$$V_b = V_{a0} + \mathbf{a}V_{a1} + \mathbf{a}^2V_{a2} = (\mathbf{a} + \mathbf{a}^2)V_{a1} = -V_{a1}$$
$$= .487\angle{-175.2°}$$

It is possible to calculate the actual (per-unit) line currents with these values.

A two-phase to ground fault from 5.19 follows the relations

$$V_b = V_c = 0 \qquad 5.151$$
$$I_a = 0 \qquad 5.152$$

From equations 5.123 through 5.128

$$V_{a0} = V_{a1} = V_{a2}$$
$$= \frac{V_a}{3} \qquad 5.153$$

Using the Thevenin equivalent circuits

$$I_{a0} = \frac{-V_{a1}}{Z_0} \qquad 5.154$$
$$I_{a2} = \frac{-V_{a1}}{Z_2} \qquad 5.155$$
$$I_{a1} = \frac{e_{Th,a} - V_{a1}}{Z_1} \qquad 5.156$$

Thus,

$$I_a = \frac{e_{Th,a}}{Z_1} - V_{a1}\left(\frac{1}{Z_0} + \frac{1}{Z_1} + \frac{1}{Z_2}\right)$$
$$= 0 \qquad 5.157$$

$$V_{a1} = \frac{e_{Th,a}Z_0Z_2}{Z_0Z_1 + Z_1Z_2 + Z_0Z_2} \qquad 5.158$$

Now substitute $e_{Th,a} - I_{a1}Z_1$ for V_{a1}.

$$I_{a1} = \frac{e_{Th,a}}{Z_1 + \frac{Z_0Z_2}{Z_0+Z_2}} \qquad 5.159$$
$$I_{a2} = \frac{e_{Th,a}Z_0}{Z_0Z_1 + Z_0Z_2 + Z_1Z_2} \qquad 5.160$$
$$I_{a0} = \frac{e_{Th,a}Z_1}{Z_0Z_1 + Z_0Z_2 + Z_1Z_2} \qquad 5.161$$

This implies that the three Thevenin equivalent circuits are in parallel as shown in figure 5.19.

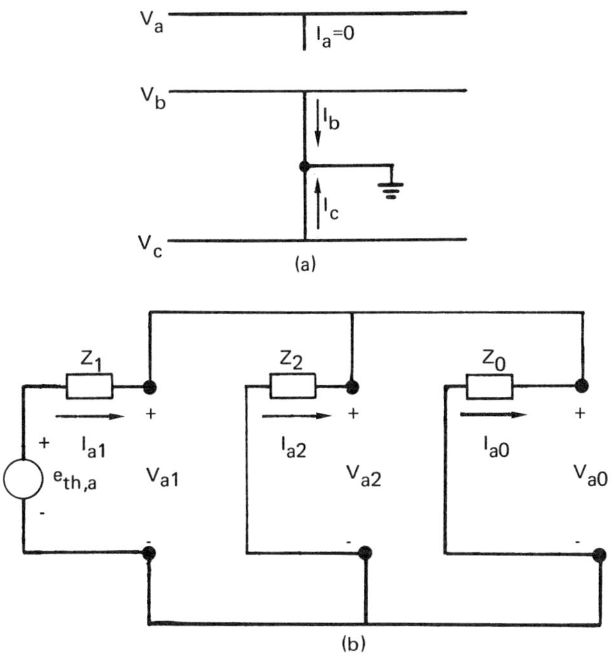

Figure 5.19 Double line-to-ground fault with corresponding sequence circuit for phase a.

The fault currents are found from equations 5.162, 5.163, and 5.164.

$$I_a = 0 \qquad 5.162$$
$$I_b = I_{a0} + \mathbf{a}^2I_{a1} + \mathbf{a}I_{a2} \qquad 5.163$$
$$I_c = I_{a0} + \mathbf{a}I_{a1} + \mathbf{a}I_{a2} \qquad 5.164$$

Since $V_{a2} = 3V_{a1}$, the line currents can also be found.

Practice Problems: POWER SYSTEMS

Warm-ups

1. A 230 KV (line-to-line), three-phase system delivers 200 MVA to a balanced load with a 0.9 power factor (lagging). Determine the line current and the reactance necessary to achieve unity power factor.

2. A 13.8 KV single-phase transformer is subjected to an open circuit test with the following results: $P_{\text{in}} = 900\ W$, $I_{\text{in}} = 0.2\ A$, $V_{\text{out}} = 460\ V$ (secondary). Determine the values of the equivalent circuit parameters which can be found from this test.

3. A 32 KVA single-phase transformer is rated at 1.5 KV on the primary. A 60 Hz short circuit test on the primary indicates $P_{\text{in}} = 1$ KW, $I_{\text{in}} = 20$ A, $I_{\text{secondary}} = 100$ A and $V_{\text{primary}} = 80$ V. Determine the winding resistances and the leakage inductances.

4. A three-phase synchronous motor is rated 10 KVA, 440 V and has a synchronous reactance of 0.8 **P.U.** It is used in a 100 KVA, 440 V system. Determine the actual synchronous reactance, and the per-unit value of the synchronous reactance in the system base.

5. A single-phase transformer is rated 6.8 KV: 115 KV at 500 KVA. A short circuit test on the high-voltage side at rated current indicates $P_{\text{sc}} = 435$ W and $V_{\text{sc}} = 2.5$ KV. Determine the per unit impedance of the transformer.

6. A 50 KVA single-phase transformer has a 440 V secondary at rated KVA and unity power factor. Determine the half-load output voltage at the same power factor for per unit resistance of 1.4% and reactance of 2.4%. Also determine half-load to full-load and no-load to full-load voltage regulations.

7. Repeat problem 6 for a 0.4 lagging power factor.

8. A three-phase, 100 KVA transformer with phase impedance $.03+j.04$ **P.U.** is operating at full load with a power factor of 0.8 lagging. Determine the change in voltage regulation if the power factor is corrected to unity. Assume rated current in both cases.

9. An open-delta, three-phase system consisting of two single-phase transformers drives a balanced three-phase load of 50 KVA and 440 V, unity power factor. Determine the necessary ratings for the two transformers.

10. Two transformers in parallel are rated at 2.5 MVA, one with 5% reactance and the other with 6% reactance. If tap changes are 2.5% , what current will circulate with one tap difference between the two transformers?

Concentrates

1. A three-phase, 11.5 KV generator drives a 500 KW, 0.866 lagging power factor load. Determine (a) the line current, (b) the necessary generator rating, and (c) the compensating reactance required to bring the power factor to 1.0.

2. A transformer is rated 50 MVA, 115 KV: 13.8 KV. Open and short circuit tests are carried out on the secondary side: $P_{\text{oc}} = 250$ KW, $I_{\text{oc}} = 350$ A, $P_{\text{sc}} = 300$ KW, $V_{\text{sc}} = 1.5$ KV. (a) Determine the equivalent circuit parameters. (b) Obtain the ABCD parameters. (c) For a resistive load of 0.5 ohms referred to the primary, determine the primary input impedance.

3. Two identical transformers are cascaded. Each has $Z_p = .01 + j.02$ and $Y_c = .0002 - j.001$. The secondary winding of the second transformer is loaded with a resistance of 0.1 ohms (referred to the primary of the first transformer). Calculate the input impedance seen at the primary of the first transformer.

4. A three-winding, single-phase transformer has the following winding impedances referred to the primary: $Z_p = 49.5 + j110$; $a_{ps}^2 Z_s = 70.5 + j90$; $a_{pt}^2 Z_t = 50.5 + j70$ ohms. The winding ratings are: primary - 50 KVA, 6.8 KV; secondary - 30 KVA, 440 V; and tertiary - 20 KVA, 1380 V. Determine the percent inpedances in the secondary and tertiary bases.

5. The feeder from a distribution transformer is 500 feet long and consists of three AWG # 4/0 wires having a resistance of $\frac{.053 \text{ ohms}}{1000 \text{ ft}}$ and a current capacity of 225 A. The three-phase motor load is 120 KW at 440 V with a power factor of 0.6 lagging. The transformer is rated 250 KVA, 440 V and has 1% resistance and 5% reactance. Determine the per cent change in load voltage when power factor correction is made to allow 225 A phase currents to flow.

6. (a) Repeat example 5.19 for a motor $PF = 1.0$. (b) Repeat example 5.19 for a motor $PF = 0.8$ leading. (c) Draw conclusions concerning the effect of power factor on three-phase faults.

7. For the power system of figure 5.16, determine the fault current for a three-phase fault at the primary of transformer T_1 (point A). The load bus is at 90% voltage, and the motor is running at 50% current with a power factor of 0.8 leading.

8. Determine the fault current for a single-phase fault-to-ground with condition the same as in problem 7.

9. A three-phase, wye-connected auto-transformer connects a 345 KV substation to a 500 KV substation.

It must be able to transfer 750 MVA in either direction. Determine the voltage, current, and MVA ratings of both common and series windings of the auto-transformers.

10. A 500 HP, 2400 volt, 1200 RPM squirrel cage motor is supplied from the secondary of a 2500 KVA, 13.8 KV delta to 2.4 KV wye transformer having an impedance of 5.5% and an X/R ratio of 7. The impedance of the 13.8 KV system is $5 + j20\%$ on a 100 MVA, 13.8 KV base. Determine the percent voltage drop if the motor is line started.

Timed

1. A 250 KVA, three-phase distribution transformer with a 440 V secondary and a 4% reactance supplies an induction motor rated at 440 V. (a) How large an induction motor (in horsepower) can be started directly across the line if the maximum voltage dip is limited to 3%? Assume 90% efficiency and a running power factor of 85%. (b) If the motor is delta-connected for running but wye-connected for starting, determine the maximum horsepower to meet the 3% restriction on line voltage dip. (c) If a three-phase auto-transformer with 40% taps is available for starting, what would be the largest motor which could be started without changing from wye to delta?

2. An open-delta, three-phase 440 V transformer consists of two single-phase transformers. The *ca* leg is missing. The transformers serve a 50 KVA balanced three-phase load with a 0.8 lagging power factor. Determine the required transformer ratings if (a) a 10 KW resistive single-phase load is added to the *ab* leg, and (b) if a 10 KW resistive load is added to the *bc* leg.

3. Two paralleled transformers, one rated 25 MVA and the other at 15 MVA, are serving a load of 30 MVA with a 0.8 lagging power factor. The 25 MVA transformer has a 10% reactance, and the 15 MVA transformer has a 6% reactance (referred to their own bases). (a) What fraction of the load current will each carry? (b) If the secondary voltage is 35 KV and the 25 MVA transformer is on a 5% overvoltage tap, determine the resulting circulating current. What is the total current for each transformer?

4. Determine the fault current for a line-to-line fault at point *A* of figure 5.16 with the load bus at 90% voltage. The motor is running at 50% current with an 80% lagging power factor. Also determine the line-to-neutral voltages at point *A*.

5. A power system has a line-to-line short at bus number 3 as shown. Find the voltages at busses 1 and 2.

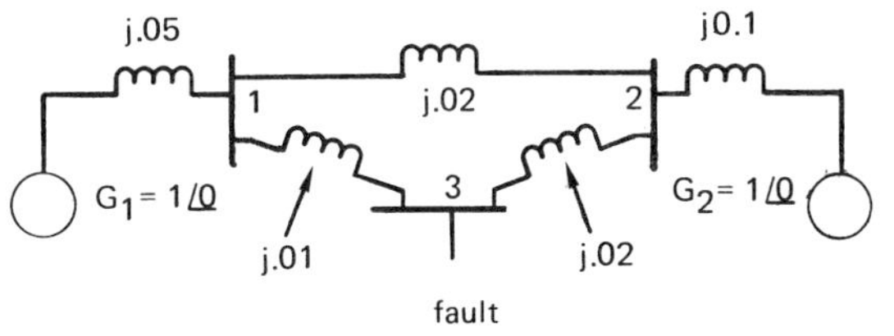

6. A three-phase transformer is fed from a three-phase, 138 KV (line-to-line) system with an impedance of $j3.1$ ohms per line. The transformer impedance base is 6 ohms referred to the 138 KV windings. The winding parameters from the nameplate are: H winding: 138 KV line-to-line, connected in a grounded wye; X winding: 34.5 KV line-to-line, connected in a grounded wye; T winding: 13.8 KV line-to-line connected in delta. The impedance and KVA ratings from winding to winding are: H to X: $j9.77\%$ at 60 MVA (20 MVA/phase); H to T: $j7.46\%$ at 21 MVA (7 MVA/phase); and X to T: $j2.88\%$ at 21 MVA (7 MVA/phase). For faults occurring on the X winding, determine the fault current for (a) a three-phase fault; and (b) a single-phase fault to ground.

7. Two three-phase transformers, each rated 100 MVA and 138 KV:12.47 KV, are operated in parallel in a substation. The leakage reactance is based on their own ratings and are 4% and 6% for transformers 1 and 2 respectively. The winding resistances are negligible. Transformer 1 has a tap-changing accessory which allows the turns ratio to be modified ±10% in 32 equal steps. Transformer 2 has a phase-shifting accessory that can adjust the phase angle continuously over a range of ±3 degrees. Determine the tap changer and phase-angle settings needed to allow the two transformers to carry a load of 200 MVA at 0.8 power factor lagging without either transformer being overloaded.

8. A 30/40/50 MVA substation transformer (115 KV-12 KV) has $Z = 7.5\%$ and $X/R = 15$. The transformer is connected to a 100 MVA, 12.47 KV line with 2% R and 10% X, which in turn is connected to a 34 MW load with an 85% lagging power factor. The load voltage is measured at 12.16 KV. Find the voltage regulation at the load.

9. A transmission line with 8% reactance on a 200 MVA base connects two substations. At the first substation the voltage is $1.03\angle 5°$ **P.U.** and at the second the voltage is $0.98\angle -2.5°$ **P.U.** Find the power and volt-amperes reactive flowing out of the first substation.

10. A 500 V:100 V transformer can be connected as an auto-transformer to obtain a 600 V:500 V ratio (600 V input). There are four possible ways of connecting the auto-transformer, two of which are incorrect. (a) For a wrong connection, what is the output voltage? (b) What are the effects of the wrong connection? (c) Show either correct connection.

RESERVED FOR FUTURE USE

6 TRANSMISSION LINES

Nomenclature

A	area, or ABCD parameter	m^2, or volt/volts
b	normalized susceptance	-
B	ABCD parameter or susceptance	ohms, mhos
C	ABCD parameter or capacitance	mhos, farads
d_S	geometric mean distance	ft
D	ABCD parameter or center-to-center spacing	amps/amp or ft
f	frequency	hz
GMR	geometric mean radius	ft
I	RMS phasor current	amps
$I.D.$	inner dimension or diameter	ft or m
J	current density	$\frac{\text{amps}}{\text{m}^2}$
K_c	capacitive correction factor	-
K_L	inductive correction factor	-
l	line length	miles
L	inductance	henrys
L_l	inductance per unit length	henrys/m
M	mutual inductance	henrys
$O.D.$	outside dimension or diameter	ft or m
P	power	watts
PF	power factor	-
q	charge per unit length	coulombs/m
Q	reactive power	volt-amps reactive
r	radial distance	m or ft
R	resistance	ohms
R_l	resistance per unit length	ohms/m
S	apparent power	volt-amperes
V	RMS phasor voltage	-
x	distance or normalized reactance	m or ft -
X	reactance	ohms
y	normalized admittance, or admittance per unit length	-,or mhos/m
Y	admittance	mhos
z	normalized impedance, or impedance per unit length	-, or ohms/m
z_l	apparent normalized impedance	-
z_o	characteristic impedance	ohms
Z	impedance	ohms

Symbols

α	attenuation constant	nepers/m
β	phase constant	radians/m
γ	propagation constant	radians/m
δ	skin depth	m
ϵ	permittivity	farads/m
μ	permeability	henrys/m
ρ	resistivity	ohm-m
ψ	flux linkage	webers
ω	radian frequency	rad/sec

Subscripts

a	line a
b	line b
c	line c
C	capacitance
cp	compensation point
e	external
eff	effective
fl	full load
i	internal
l	line, or per unit length
L	inductive, or load
n	neutral
nl	no load
o	characteristic value
O	D.C. value
R	receiving
S	sending

1 INTRODUCTION

Power distribution and communication systems both use transmission lines. The behavior of such lines is an important aspect of the overall system performance, and it must be taken into account in system design or analysis. The properties of interest in transmission lines are resistance and inductance along the line, and the capacitances from line-to-line and line-to-ground. The *shunt* or *line-to-line conductance* is usually not significant and is omitted in this chapter.

A variety of length units are used in practice, including mils, inches, feet, miles, meters, and kilometers. Conversions are straight-forward, and table 6.1 can be used when necessary.

Table 6.1

Length Conversion Factors

multiply	by	to obtain
mils	10^{-3}	inches
mils	2.54×10^{-3}	centimeters
mils	2.54×10^{-5}	meters
inches	2.54	centimeters
miles	5280	feet
miles	1609.3	meters
feet	0.3048	meters

The transmission line resistance, inductance, and capacitance are universally specified as *per-unit-length* quantities. The resistance is a frequency dependent quantity because of the *skin effect*, increasing as the square root of the frequency. This, together with the *temperature effect*, presents a problem in the design of conductors for power systems. In communication systems at high frequencies, the usual length of a transmission line is short enough to neglect resistance.

Inductance is made up of two components. The *internal inductance* decreases with frequency because of the skin effect. The *external conductance* is dependent on the geometric arrangement of the conductors within the transmission line. The capacitance also is a function of the geometric arrangement of the conductors.

2 SKIN EFFECT

The *resistance per unit length* of a conductor carrying a constant D.C. current is the resistivity divided by the cross sectional area. Resistance per unit length in this chapter is R_l, but it often is given with no identification.

$$R_{lo} = \frac{\rho}{A} \qquad 6.1$$

The *internal inductance* of a circular wire carrying D.C. current is independent of the radius. For non magnetic materials it can be calculated[1] from basic field principles as

$$L_{loi} = \frac{\mu}{8\pi} = .5 \times 10^{-7} \qquad 6.2$$

[1]Ramo, Whinnery, and Van Duzer, *Fields and Waves in Communication Electronics*, John Wiley and Sons, 1965, p.305.

L_{li} is the internal inductance per unit length, and L_{loi} is the D.C. value of the internal inductance.

The *skin effect* is an electromagnetic phenomenon which determines the time needed for an external field to penetrate a conductor. This time is governed by the resistivity and permeability of the conducting material and the dimensions of the conductor. When the electric field fails to fully penetrate the conductor, the current density is higher near the surface than at the middle. This produces an effective increase of the resistance and a decrease in the internal inductance. For a flat plate conductor, the current density depends on the *skin depth* δ.

$$J = J_{\text{surface}} e^{-\frac{x}{\delta}(1+j)} \qquad 6.3$$

$$\delta = \frac{1}{\sqrt{\frac{\pi f \mu}{\rho}}} \qquad 6.4$$

The impedance per unit length per unit width is

$$Z_{lw} = \frac{\rho}{\delta}(1+j) = (1+j)\sqrt{\pi f \mu \rho} \qquad 6.5$$

Equation 6.5 is interpreted to mean that the effective thickness of the plate is δ. (Obviously, this will be true only if the actual thickness of the plate is somewhat greater than δ.)

For round wires, equation 6.5 is useful only if the skin depth is much less than the radius of the wire (e.g., $\frac{r}{\delta} > 5.5$ for 10% or less error).

Example 6.1

The skin depth for copper is given by $\delta = \frac{.066}{\sqrt{f}}$ (meters). Determine the wire radius (in mils) for which the skin depth equals the radius for frequencies of (a) 60 hz, (b) 10 khz, and (c) 1 Mhz. (A mil is 0.001 inches.)

First convert meters to mils.

$$\delta \text{ in mils} = \delta(m) \times \frac{10^5}{2.54}$$
$$= .066 \times \frac{10^5}{2.54\sqrt{f}}$$

Now substitute the frequencies into the expression for δ.

(a) $f = 60$ hz, $\delta = 336$ mils
(b) $f = 10$ khz, $\delta = 26$ mils
(c) $f = 1$ Mhz, $\delta = 2.6$ mils

At 60 hz, the skin depth of copper is 336 mils (from example 6.1). The largest copper wire is AWG #0000 (larger sizes being stranded). The radius of AWG #0000 is 230 mils, so there can be some effect of the skin depth phenomenon even at 60 hz.

Example 6.2

AWG #20 wire has a diameter of 32 mils. Its D.C. resistance is 10.15 ohms per 1000 feet at 25°C. Determine its resistance per 1000 feet at 25°C at a frequency of 1 Mhz.

At this frequency, the skin depth is 2.6 mils (from example 6.1). The radius is 16 mils. Thus, equation 6.5 could be used for finding $\frac{\rho}{\delta}$. However, the effective cross sectional area can be found by subtracting the area of a circle with a radius one skin depth less than the wire radius from the area of the wire.

$$A_{\text{effective}} = \pi r^2 - \pi (r - \delta)^2 \qquad 6.6$$

As the resistivity remains the same, the ratio of the resistance is inversely proportional to the area.

$$\frac{R_{1Mhz}}{R_O} = \frac{A}{A_{\text{eff}}} = \frac{\pi r^2}{\pi r^2\left[1 - (1 - \frac{\delta}{r})^2\right]} \qquad 6.7$$

$$R_{1Mhz} = \frac{R_O}{1 - \left[1 - \frac{\delta}{r}\right]^2} = \frac{10.15}{1 - \left[1 - \frac{2.6}{16}\right]^2} = 37.5\,\Omega$$

For coaxial cables at high frequencies, the resistance and internal inductance of the line can be found from equation 6.5. The width of the conductor is taken as 2π times the radius for the inner and outer conductors. The thickness is taken as the skin depth.

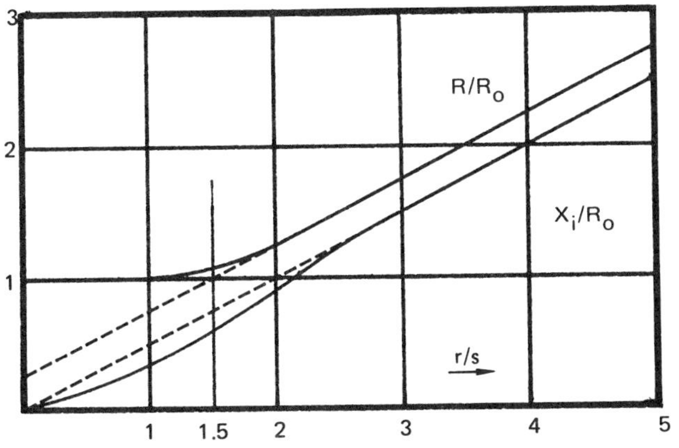

Figure 6.1 Normalized resistance and internal reactance for circular solid wires

Calculation of impedance of round wires at intermediate frequencies is a complicated process involving Bessel function tables. It is usually sufficient to use normalized curves such as figure 6.1.

Over the range of $\frac{r}{\delta} > 2$, the resistance is approximated by

$$\frac{R}{R_O} \approx .25 + .5\frac{r}{\delta} \qquad 6.8$$

The inductive reactance is approximated by equation 6.9 for $\frac{r}{\delta} > 2.5$.

$$\frac{X_i}{R_O} \approx .5025\frac{r}{\delta} \qquad 6.9$$

Example 6.3

Find the internal impedance per 1000 feet for AWG #1 solid copper wire operated at 60 hz. The D.C. resistance is .124 ohms per 1000 feet at 25°C. The wire diameter is 289 mils.

From example 6.1, $\delta = 336$ mils. Then,

$$\frac{r}{\delta} = \frac{289}{(2)(336)} = .43$$

From figure 6.1,

$$\frac{R}{R_O} = 1$$

$$\frac{X_i}{R_O} = .12$$

Therefore,

$$Z_L = R_O(1 + j.12) = .124 + j.0014 \text{ ohms/1000 ft}$$

3 EXTERNAL INDUCTANCE

A pair of parallel solid round wires carrying the same current in opposite directions has an external inductance in henrys per meter of approximately

$$L_{l,e} = \frac{\mu}{\pi}\,ln\left(\frac{D}{r}\right) \qquad 6.10$$

D is the center-to-center spacing of the wires and r is the wire radius.

As the permeability of nonmagnetic materials is essentially $4\pi \times 10^{-7}$, the usual case is that

$$L_{l,e} = \left(4 \times 10^{-7}\right) ln\left(\frac{D}{r}\right) \qquad 6.11$$

A coaxial line has an external inductance of

$$L_{l,e} = \left(2 \times 10^{-7}\right) ln\left(\frac{\text{I.D. of outer conductor}}{\text{O.D. of inner conductor}}\right) \qquad 6.12$$

The *total inductance per unit length* is the sum of the internal and external inductances.

$$L_l = L_i + L_e \qquad 6.13$$

Example 6.4

Determine the total inductive reactance per meter at 10 Mhz for a coaxial line (copper, $\rho = 1.8 \times 10^{-8}$ ohm-m) with an inner conductor diameter of 0.2 cm and an outer conductor inner radius of 1 cm.

The internal impedance of the line is $.147(1+j)$ ohms, (as determined in warm-up problem #1), so the internal reactance is the real part of .147 ohms. From equation 6.12, the external inductance is

$$L_{l,e} = (2 \times 10^{-7})\, ln\left(\frac{2}{.2}\right) = 4.605 \times 10^{-7}\,\frac{H}{m}$$

$$X_e = 2\pi f L_e = 28.935\,\frac{\text{ohms}}{m}$$

The total inductive reactance per unit length is given by equation 6.13

$$X_L = .147 + 28.935 = 29.082\,\frac{\text{ohms}}{m}$$

4 CAPACITANCE

The capacitance in farads/meter of a pair of parallel solid round wires is approximately

$$C_l = \frac{2\pi\epsilon}{ln\left[\frac{(D-r_1)(D-r_2)}{r_1 r_2}\right]} \qquad 6.14$$

D is the center-to-center spacing of the wires. r_1 and r_2 are the radii of the two wires. The permittivity is composed of the permittivity of free space ($\epsilon_o = 1/(36\pi \times 10^9)$) and the permittivity ratio ($\epsilon_r = 1$ for air).

$$\epsilon = \epsilon_r \epsilon_o \qquad 6.15$$

For a coaxial line, the capacitance in farads/meter is

$$C_l = \frac{2\pi\epsilon}{ln\left[\frac{\text{I.D. of outer conductor}}{\text{O.D. of inner conductor}}\right]}. \qquad 6.16$$

Example 6.5

Determine the capacitive reactance at 1 Mhz per 1000 feet for a pair of AWG #20 wires separated by 1 cm (center-to-center).

AWG #20 wire has a diameter of 32 mils (a radius of .04064 cm). For air, $\epsilon = \epsilon_o = \frac{1}{36\pi \times 10^9}$.

$$2\pi\epsilon = \frac{10^{-9}}{18}$$

$$C_l = \frac{10^{-9}}{18\, ln\frac{(1-.04)^2}{.04^2}} = 8.740 \times 10^{-12}\, F/m$$

Converting farads per meter to farads per 1000 feet,

$$C = 2.678 \times 10^{-9}\, F/1000\,\text{ft}$$

Then,

$$X_C = \frac{1}{2\pi f C} = 59.43\,\text{ohms}/1000\,\text{ft}$$

5 POWER TRANSMISSION LINES

Power transmission lines differ from the lines previously discussed in several ways. They are typically made of aluminum strands with a core of steel or aluminum alloy wires for higher tensile strength. Also, they occur in sets of three or four lines for three-phase power transmission.

Calculating the line parameters of resistance, inductance, and capacitance is complicated and unnecessary in most cases since these parameters are tabled for standard conductors. Table 6.2 gives the electrical properties of the common standard conductors which have code names for easier identification. The inductive and capacitive reactances are per-conductor values based on a 1 foot center-to-center spacing. For a single-phase, two-wire system with 1 foot spacing, the inductive and capacitive reactances per mile would be doubled.

The GMR is the *geometric mean radius* of the conductor. It takes into account the layers and the number of individual conductors per layer. The GMR is the radius adjusted for the internal inductance of the conductor. Therefore, the entire inductance (internal and external) can be found from one calculation.

For a solid round wire (when skin effect is not important), the GMR is the actual radius multiplied by $e^{-.25}$. For example, Bluebird conductor in table 6.2 has an outside diameter of 1.762 inches, so its radius is .881 inches. If it were a solid wire, the GMR would have been $.881 \times e^{-.25} = .6861$ inches or .0572 feet. This is close to the tabulated GMR .0586 ft.

The inductive reactance in ohms per mile is given by equation 6.17, where D is the line-to-line spacing.

$$X_L = (2.022 \times 10^{-3}) f \times ln\left(\frac{D}{GMR}\right) \qquad 6.17$$

Equation 6.17 can be broken into two parts.

$$X_L = (2.022 \times 10^{-3}) f\left[ln\left(\frac{1}{GMR}\right) + ln(D)\right] \qquad 6.18$$

A multiplicative correction factor is needed for line-to-line spacings of other than one foot.

$$K_L = 1 + \frac{ln(D)}{ln(\frac{1}{GMR})} \qquad 6.19$$

Both D and GMR must be in feet to use equations 6.17 through 6.19.

CODE NAME	aluminum area in (c. mils)	strands (Al/St)	O.D. (in.)	$R_{D.C.}$ 20°C (mΩ/ft.)	$R_{A.C.}$, 60 hz 20°C (Ω/mi.)	$R_{A.C.}$, 60 hz 50°C (Ω/mi.)	GMR (ft.)	1-ft. spacing reactance per conductor @ 60 hz X_L (Ω/mi.)	1-ft. spacing reactance per conductor @ 60 hz X_C (MΩ-mi.)
Waxwing	266,800	18/1	0.609	0.0646	0.3488	0.3831	0.0198	0.476	0.1090
Partridge	266,800	26/7	0.642	0.0640	0.3452	0.3792	0.0217	0.465	0.1074
Ostrich	300,000	26/7	0.680	0.0569	0.3070	0.3372	0.0229	0.458	0.1057
Merlin	336,400	18/1	0.684	0.0512	0.2767	0.3037	0.0222	0.462	0.1055
Linnet	336,400	26/7	0.721	0.0507	0.2737	0.3006	0.0243	0.451	0.1040
Oriel	336,400	30/7	0.741	0.0504	0.2719	0.2987	0.0255	0.445	0.1032
Chickadee	397,500	18/1	0.743	0.0433	0.2342	0.2572	0.0241	0.452	0.1031
Ibis	397,500	26/7	0.783	0.0430	0.2323	0.2551	0.0264	0.441	0.1015
Pelican	477,000	18/1	0.814	0.0361	0.1957	0.2148	0.0264	0.441	0.1004
Flicker	477,000	24/7	0.846	0.0359	0.1943	0.2134	0.0284	0.432	0.0992
Hawk	477,000	26/7	0.858	0.0357	0.1931	0.2120	0.0289	0.430	0.0988
Hen	477,000	30/7	0.883	0.0355	0.1919	0.2107	0.0304	0.424	0.0980
Osprey	556,500	18/1	0.879	0.0309	0.1679	0.1843	0.0284	0.432	0.0981
Parakeet	556,500	24/7	0.914	0.0308	0.1669	0.1832	0.0306	0.423	0.0969
Dove	556,500	26/7	0.927	0.0307	0.1663	0.1826	0.0314	0.420	0.0965
Rook	636,000	24/7	0.977	0.0269	0.1461	0.1603	0.0327	0.415	0.0950
Grosbeak	636,000	26/7	0.990	0.0268	0.1454	0.1596	0.0335	0.412	0.0946
Drake	795,000	26/7	1.108	0.0215	0.1172	0.1284	0.0373	0.399	0.0912
Tern	795,000	45/7	1.063	0.0217	0.1188	0.1302	0.0352	0.406	0.0925
Rail	954,000	45/7	1.165	0.0181	0.0977	0.1092	0.0386	0.395	0.0897
Cardinal	954,000	54/7	1.196	0.0180	0.0988	0.1082	0.0402	0.390	0.0890
Ortolan	1,033,500	45/7	1.213	0.0167	0.0924	0.1011	0.0402	0.390	0.0885
Bluejay	1,113,000	45/7	1.259	0.0155	0.0861	0.0941	0.0415	0.386	0.0874
Finch	1,113,000	54/19	1.293	0.0155	0.0856	0.0937	0.0436	0.380	0.0866
Bittern	1,272,000	45/7	1.382	0.0136	0.0762	0.0832	0.0444	0.378	0.0855
Bobolink	1,431,000	45/7	1.427	0.0121	0.0684	0.0746	0.0470	0.371	0.0837
Plover	1,431,000	54/19	1.465	0.0120	0.0673	0.0735	0.0494	0.365	0.0829
Lapwing	1,590,000	45/7	1.502	0.0109	0.0623	0.0678	0.0498	0.364	0.0822
Falcon	1,590,000	54/19	1.545	0.0108	0.0612	0.0667	0.0523	0.358	0.0814
Bluebird	2,156,000	84/19	1.762	0.0080	0.0476	0.0515	0.0586	0.344	0.0776

Table 6.2 Electrical characteristics of the multilayer, steel-reinforced Bare Aluminum Conductors. From *Aluminum Electrical Conductor Handbook*, New York, September 1971. Used with permission of the Aluminum Association.

Example 6.6

Determine the inductive reactance of Bluebird conductor when the line-to-line spacing is 20 feet.

The correction factor is

$$K_L = 1 + \frac{ln\,20}{ln(\frac{1}{.0586})} = 2.056$$

The one foot inductive reactance from table 6.2 is .344. Therefore,

$$X_L = 2.05 \times 0.344 = 0.707\,\Omega/\text{mile}$$

The capacitive reactance times miles per conductor in air is approximately

$$X_C = \frac{1.781 \times 10^6}{f}\,ln\left(\frac{D}{r}\right) \qquad 6.20$$

Equation 6.20 can be corrected for values other than $D = 1$ by the multiplicative factor K_C.

$$K_C = 1 + \frac{ln\,D}{ln(\frac{1}{r})} \qquad 6.21$$

r is the actual conductor radius, not the *GMR*. D and r both must be in feet.

The value of X_C from table 6.2 can be used with K_C to find the actual capacitive reactance

$$X_C = X_{C,\text{table}} \times K_C \qquad 6.22$$

Example 6.7

Find the total shunt reactance per mile of a Bluebird conductor pair with 15 feet center-to-center separation.

The table value is .0776 MΩ-mi. The correction factor is

$$K_C = 1 + \frac{ln\,15}{ln(\frac{24}{1.762})} = 2.0369$$

The reactance per line is

$$X_C = (2.0369)(.0776) = .1581\ M\Omega\text{-mi.}$$

With multiple conductors, the reactances add.

$$X_{\text{total}} = (2)(.1581) = .3162\ M\Omega\text{-mi.}$$

6 THREE-PHASE TRANSMISSION LINES

For three-wire, three-phase transmission lines, the sum of the currents is instantaneously zero, as is the sum of the line-to-line voltages. This permits certain simplifications.

A. INDUCTANCE

Inductance is determined from geometry, considering the flux linking each conductor. The flux linking two conductors a and b separated by a distance r_{ab} is

$$\psi_{ab} = \frac{\mu_o I_b}{2\pi}\int_{r_{ab}}^{\infty}\frac{dr}{r} \qquad 6.23$$

Equation 6.23 is an infinite integral, and a limiting process can be used. Rather than integrating to infinity, integrate to a large value, say A.

$$\psi_{ab} = \frac{\mu_o I_b}{2\pi}\int_{r_{ab}}^{A}\frac{dr}{r} = \frac{\mu_o I_b}{2\pi}\,ln\left(\frac{A}{r_{ab}}\right) \qquad 6.24$$

For a three-wire system such as is shown in figure 6.2, the flux linking conductor a is

$$\psi_a = \frac{\mu_o}{2\pi}\left[I_a\,ln\left(\frac{A}{GMR}\right) + I_b\,ln\left(\frac{A}{r_{ab}}\right) + I_c\,ln\left(\frac{A}{r_{ac}}\right)\right] \qquad 6.25$$

Since $ln(a/b) = ln\,a - ln\,b$, equation 6.25 can be rearranged.

$$\psi_a = \frac{\mu_o}{2\pi}\Big[(I_a + I_b + I_c)\,ln(A) - I_a\,ln(GMR) - I_b\,ln(r_{ab}) - I_c\,ln(r_{ac})\Big] \qquad 6.26$$

Since $I_a + I_b + I_c = 0$, the first term of equation 6.26 vanishes as long as the radius A is large. Then, the distance to all the conductors is approximately the same. Then,

$$\psi_a = \frac{\mu_o}{2\pi}\Big[-I_a\,ln(GMR) - I_b\,ln(r_{ab}) - I_c\,ln(r_{ac})\Big] \qquad 6.27$$

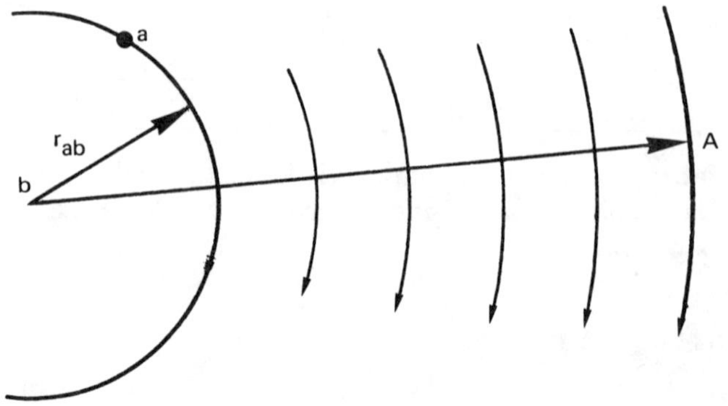

Figure 6.2 Flux linking conductor a due to conductor b, limited to a maximum radius A

The *geometric mean* of the three distances separating the conductors is

$$d_S = \sqrt[3]{r_{ab}r_{bc}r_{ca}} \qquad 6.28$$

Furthermore,

$$ln(r_{ab}) = ln\frac{r_{ab}r_{bc}r_{ca}}{r_{bc}r_{ca}} = ln\frac{d_S^3}{r_{bc}r_{ca}}$$
$$= ln\frac{d_S^2}{r_{bc}r_{ca}} + ln\ d_S \qquad 6.29$$

Equation 6.27 can be rearranged,

$$\psi_a = \frac{\mu_o}{2\pi}\Bigg[- I_a\ ln(GMR) - I_b\ ln(d_S) - I_c\ ln(d_S)$$
$$- I_b\ ln\left(\frac{d_S^2}{r_{bc}r_{ca}}\right) - I_c\ ln\left(\frac{d_S^2}{r_{bc}r_{ab}}\right)\Bigg] \qquad 6.30$$

Also, $I_b + I_c = -I_a$. So, the first three terms of the right-hand side of equation 6.30 can be combined.

$$\psi_a = \left[\frac{\mu_o}{2\pi}ln\left(\frac{d_S}{GMR}\right)\right]I_a + \left[\frac{\mu_o}{2\pi}ln\left(\frac{r_{ac}r_{bc}}{d_S^2}\right)\right]I_b$$
$$+ \left[\frac{\mu_o}{2\pi}ln\left(\frac{r_{ab}r_{bc}}{d_S^2}\right)\right]I_c \qquad 6.31$$

Equation 6.31 is the sum of the self and mutual inductive terms.

$$\psi_a = L_a I_a + M_{ab}I_b + M_{ac}I_c \qquad 6.32$$

When the three wires are identical (as they typically are), the three self inductances are the same.

$$L = \frac{\mu_o}{2\pi}ln\left(\frac{d_S}{GMR}\right)$$
$$= (2 \times 10^{-7})\,ln\left(\frac{d_S}{GMR}\right) \qquad 6.33$$

The mutual inductances are calculated from equation 6.34.

$$M = \frac{\mu_o}{2\pi}\ ln\left(\frac{r_{bc}r_{ac}}{d_S^2}\right)$$
$$= (2 \times 10^{-7})\,ln\left(\frac{r_{bc}r_{ac}}{d_S^2}\right) \qquad 6.34$$

When the three conductors are arranged as an equilateral triangle, the geometric mean distance is the same as each of the sides, and $ln(1) = 0$. Then, the mutual inductance terms vanish. The self-inductance found in equation 6.33 is called *phase inductance*. It is the same as the per-conductor value given in table 6.2. The reactances are not doubled as was done in the two-conductor, single-phase system.

Example 6.8

A three-conductor transmission line has Waxwing conductors arranged in a plane, with each of the conductors separated by 1 foot. Find the self and mutual inductances per meter of conductor.

$$d_S^3 = \sqrt[3]{1 \times 1 \times 2} = \sqrt[3]{2}$$

From table 6.2, $GMR = .0198$.

$$L = (2 \times 10^{-7})\,ln\left[\frac{\sqrt[3]{2}}{.0198}\right] = 8.306 \times 10^{-7}\ \text{H/m}$$

$$M_{12} = (2 \times 10^{-7})\,ln\left[\frac{1 \times 2}{2^{2/3}}\right] = 4.62 \times 10^{-8}\ \text{H/m}$$

$$M_{13} = (2 \times 10^{-7})\,ln\left[\frac{1 \times 1}{2^{2/3}}\right] = -9.24 \times 10^{-8}\ \text{H/m}$$

$$M_{23} = M_{12}$$

B. CAPACITANCE

The *capacitance* of a three-phase, three-conductor system is also determined from the geometry. For two conductors with line-charges q_a and q_b coulombs per meter, the voltage drop from line a to line b is determined by the line integral of equation 6.35.

$$V_{ab} = \frac{q_a}{2\pi\epsilon}\oint_{r_a}^{r_{ab}}\frac{dr}{r} + \frac{q_b}{2\pi\epsilon}\oint_{r_{ab}}^{r_a}\frac{dr}{r} \qquad 6.35$$

r_{ab} is the center-to-center separation of line a from line b. The integration of equation 6.35 is

$$V_{ab} = \frac{q_a}{2\pi\epsilon}ln\left(\frac{r_{ab}}{r_a}\right) + \frac{q_b}{2\pi\epsilon}ln\left(\frac{r_a}{r_{ab}}\right) \qquad 6.36$$

If a third line with different distances is included, it will contribute a drop. The total voltage drop from line a to line b becomes

$$V_{ab} = \frac{q_a}{2\pi\epsilon}ln\left(\frac{r_{ab}}{r_a}\right) + \frac{q_b}{2\pi\epsilon}ln\left(\frac{r_a}{r_{ab}}\right)$$
$$+ \frac{q_c}{2\pi\epsilon}ln\left(\frac{r_{cb}}{r_{ac}}\right) \qquad 6.37$$

Consideration of unequal spacing is complicated and lengthy. In most cases, the approximation of equal spacing is adequate. Thus, the last term of equation 6.37 will vanish because $ln(1) = 0$.

The voltage drop from line a to line c is

$$V_{ac} = \frac{q_a}{2\pi\epsilon}ln\left(\frac{d}{r}\right) + \frac{q_c}{2\pi\epsilon}ln\left(\frac{r}{d}\right) \qquad 6.38$$

r is now the radius of each of the three identical conductors, and d is the separation between each pair. Equation 6.36 can be rewritten in the same form as equation 6.38.

$$V_{ab} = \frac{q_a}{2\pi\epsilon} ln\left(\frac{d}{r}\right) + \frac{q_b}{2\pi\epsilon} ln\left(\frac{r}{d}\right) \qquad 6.39$$

Adding equations 6.38 and 6.39,

$$V_{ab} + V_{ac} = \frac{2q_a}{2\pi\epsilon} ln\left(\frac{d}{r}\right) + \frac{q_b + q_c}{2\pi\epsilon} ln\left(\frac{r}{d}\right) \qquad 6.40$$

However, the sum of the charges is approximately zero. (It is exactly zero only if the lines are equilaterally spaced and the ground is an infinite distance away.) Therefore, $q_b + q_c \approx -q_a$.

$$V_{ab} + V_{ac} = \frac{3q_a}{2\pi\epsilon} ln\left(\frac{d}{r}\right) \qquad 6.41$$

The phasor relations indicate that $V_{ab} + V_{ac} = 3\,V_{an}$, where V_{an} is the voltage drop to a hypothetical neutral. Therefore, the charge is

$$q_a = \frac{2\pi\epsilon}{ln(\frac{d}{r})} V_{an} = C_{an} V_{an} \qquad 6.42$$

C_{an} is the phase capacitance, which is the per-conductor capacitance listed in table 6.2. The capacitive reactance is $1/2\pi f C$, so with air as the dielectric, $\epsilon = 10^{-9}/36\pi$. The capacitive reactance times miles at 60 hz is

$$X_C = 29{,}668\ ln\left(\frac{d}{r}\right)\ \Omega\text{-mi} \qquad 6.43$$

The actual capacitive reactance is obtained by dividing X_C by the number of miles of conductor.

7 POWER TRANSMISSION LINE REPRESENTATION

A. INTRODUCTION

Three-phase transmission lines are normally operated as balanced as possible. However, the spacing is not equilateral. Transposing the positions of the phases as the transmission line travels across the countryside will tend to balance the inductance and capacitance per phase, but this is seldom done. Nevertheless, calculations which assume equilateral spacing and transposition closely approximate the equilateral cases in the previous sections, as long as the geometric mean distance between pairs d_3 from equation 6.28 is used for the effective spacing.

Under these assumptions, it is possible to analyze balanced three-phase systems by considering only one phase and the hypothetical neutral (which contributes no resistance, inductance, or capacitance to the system). Seriously unbalanced lines, however, must be evaluated with *symmetrical components.*

As three-phase systems are analyzed using one phase only, it is necessary to perform a delta-wye transformation to obtain the equivalent impedance-to-neutral when the actual load is a delta configuration.

B. DELTA-WYE TRANSFORMATION

The *delta-wye transformation* for unbalanced loads is given by equation 6.44.

$$Z_{an} = \frac{Z_{ab} Z_{ca}}{Z_{ab} + Z_{bc} + Z_{ca}} \qquad 6.44$$

For a balanced load, all the delta-connected impedances are equal. The equivalent line-to-neutral impedance is one-third of the line-to-line impedance.

$$Z_{ln} = \frac{Z_{ll}}{3} \qquad 6.45$$

C. SHORT TRANSMISSION LINES

Short transmission lines at 60 hz are under 50 miles in length. In such lines, the shunt capacitance is not important because it is much greater than the usual impedances of the loads (and incomparably larger than the line impedance). In such cases, it is necessary to determine only the resistance and inductance for the line.

$$Z_l = \left[\frac{R}{\text{mile}} + \frac{jX_L}{\text{mile}}(\text{corrected})\right] \times [\text{length in miles}] \qquad 6.46$$

A circuit model for a short transmission line is shown in figure 6.3. The ABCD parameters for this circuit result in equations 6.47 and 6.48.

$$V_S = V_R + Z\,I_R \qquad 6.47$$

$$I_S = I_R \qquad 6.48$$

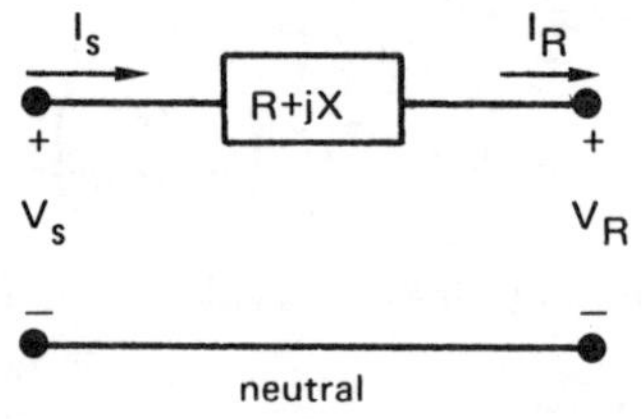

Figure 6.3 One phase of a short transmission line

One of the principal concerns in transmission line design is the effect of power factor on the voltage regulation. The *voltage regulation* (or simply *regulation*) in percent is

$$\text{regulation} = \frac{|V_{R,nl}| - |V_{R,fl}|}{|V_{R,fl}|} \times 100 \qquad 6.49$$

$V_{R,nl}$ is the voltage at the receiving end of the transmission line when the load is disconnected (i.e., no load current is flowing); $V_{R,fl}$ is the voltage at the receiving end of the transmission line when full line current is flowing.

Example 6.9

A 25 mile Flicker conductor transmission line has 7 feet between the conductors. The full-load current is 225 amps per phase. The line-to-neutral voltage at the sending end is 11 kV per phase, and the power factor of the load is 0.9 lagging. Determine the voltage regulation in percent. Assume $T = 20°$ C.

From table 6.2, resistance per mile is .1943 ohms; X_L@ 1 foot spacing is .432 ohms/mi, and $GMR = .0284$ feet. The correction factor for 7 feet spacing is

$$K_L = 1 + \frac{ln 7}{ln(1/.0284)} = 1.5464$$

$$X_L = 1.5464 \times .432 \text{ ohms/mi} = .669 \text{ ohms/mi}$$

For 25 miles of line,

$$Z = (.1943 + j.669) \times 25 = 4.86 + j16.7$$
$$= 17.87\angle 74.2°$$

With a power factor of 0.9 lagging, the current at the receiving end lags the receiving end voltage by 25.8°. If the receiving end voltage is taken as a reference, then

$$I_R = 225\angle{-25.8°}$$

The line voltage drop is

$$I_R Z = 225\angle{-25.8°} \times 17.87\angle 74.2° = 4021\angle 48.4°$$

From equation 6.47,

$$V_R\angle 0° = V_S\angle\theta - I_R Z = 11\angle\theta - 4.021\angle 48.4° \text{ kV}$$

θ is the angle of the sending end voltage with respect to the receiving end voltage V_R. This equation is rewritten in rectangular form.

$$V_R = 11\cos\theta + j11\sin\theta - 2.67 - j3.01$$

Since there is no imaginary part on the left-hand side of this equation, the right-hand side's imaginary part must be zero. Then,

$$\sin\theta = \frac{3.01}{11}$$

$$\cos\theta = .9619$$

$$V_R = 11 \times .9619 - 2.67 = 7.91 \text{ kV}$$
$$= V_{R,fl}$$

The no-load voltage at the receiving end is the same as the voltage at the sending end.

$$V_{R,nl} = 11 \text{ kV}$$

Therefore,

$$\text{regulation} = \frac{11 - 7.91}{7.91} \times 100 = 39.0\%$$

D. MEDIUM-LENGTH TRANSMISSION LINES

Medium length transmission lines are generally considered to be from 50 to 150 miles in length at 60 hz. (It is sometimes permissible to consider a line up to 200 miles long to be a medium length transmission line.) The capacitive reactance cannot be ignored, as significant amounts of shunt current are lost in the transmission process.

There are two circuit models for medium-length lines, the tee and pi approximations as shown in figure 6.4. The *tee approximation* is made by splitting the series impedance in half with the shunt susceptance in the downward leg.

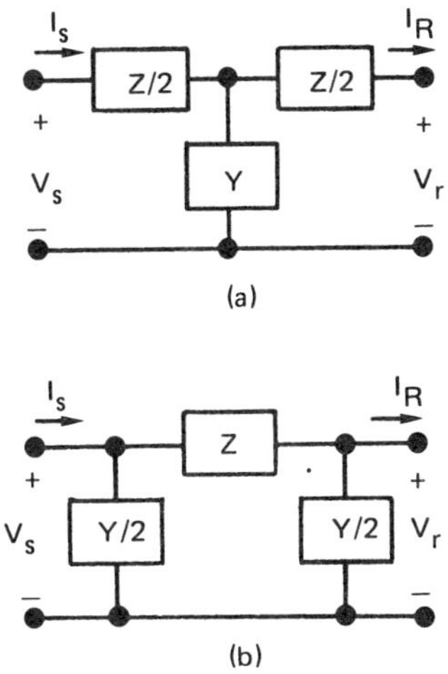

Figure 6.4 One phase of a medium length transmission line: (a) tee equivalent circuit, (b) pi equivalent circuit

$$Y = \frac{-1}{jX_C} = \frac{j}{X_C} = jB_C \qquad 6.50$$

The *pi approximation* is made by splitting the shunt susceptance with the series impedance between the two downward legs. The ABCD parameters for each of these circuit models are given by the following equations.

- tee approximation

$$A = 1 + \frac{YZ}{2} \qquad 6.51$$

$$B = Z\left(1 + \frac{YZ}{4}\right) \qquad 6.52$$

$$C = Y \qquad 6.53$$

$$D = 1 + \frac{YZ}{2} \qquad 6.54$$

•pi approximation

$$A = 1 + \frac{YZ}{2} \qquad 6.55$$

$$B = Z \qquad 6.56$$

$$C = Y\left(1 + \frac{YZ}{4}\right) \qquad 6.57$$

$$D = 1 + \frac{YZ}{2} \qquad 6.58$$

The two approximations are not the same, except when $YZ/4$ is insignificant compared to 1. The pi approximation results in calculations for regulation that are slightly simpler.

$$V_S = \left(1 + \frac{YZ}{2}\right)V_R + Z\,I_R \qquad 6.59$$

$$I_S = Y\left(1 + \frac{YZ}{4}\right)V_R + \left(1 + \frac{YZ}{2}\right)I_R \qquad 6.60$$

The regulation can be found by using equation 6.59.

Example 6.10

A transmission line delivers 7500 kW at 8500 kVA and 190 kV to a balanced load. The line is 100 miles long, and the conductors are Flicker with a separation of 7 feet. Assume the operating temperature is 50°C. Determine the regulation.

The power factor is 7500/8500=.88, assumed to be lagging. The phase voltage is $190/\sqrt{3} = 109.7$ kV, and the phase kVA is 8500/3=2833. The phase current is then $(2833/109.7)\angle{-28.07°}$.

The inductive reactance per mile is the same as in example 6.9.

$$X_L = .669\text{ ohms/mile} \times 100\text{ miles} = 66.9\text{ ohms}$$

$$R = .2134\text{ ohms/mile} \times 100\text{ miles}$$

$$= 21.43\text{ ohms}$$

$$X_C = .0992\text{ megohm-miles}$$

$$K_C = 1.58$$

From equation 6.21

$$K_c = 1 + \frac{ln 7}{ln(\frac{24}{.864})}$$

$$X_C = 1.58 \times (.0992 \times 10^6)/100\text{ miles}$$

$$= 1569\text{ ohms}$$

$$Y = \frac{j}{X_c} = j6.37 \times 10^{-4}\text{ mhos}$$

$$Z = 21.34 + j66.9$$

$$1 + \frac{YZ}{2} = .9787 + j.0136 \approx .9787$$

$$IZ = 1.86\text{ kV}\angle 44.69° = (1.32 + j1.31)\text{ kV}$$

From equation 6.59,

$$V_S = .9787 \times 109.7 + 1.32 + j1.31 = 108.7\text{ kV}$$

At no load with $I_R = 0$,

$$V_{R,nl} = \frac{108.7}{.9787} = 111.1\text{ k V}$$

Since $V_{R,fl}$ was given as 109.7,

$$\text{regulation} = \frac{111.1 - 109.7}{109.7} \times 100 = 1.29\%$$

E. LONG TRANSMISSION LINES

Long transmission lines must be evaluated with distributed parameters. Consider the section of transmission line in figure 6.5: z is the impedance per unit length, and y is the shunt admittance per unit length. (Both z and y are complex quantities.) The impedance for the section is, then, z dx, and the admittance is y dx.

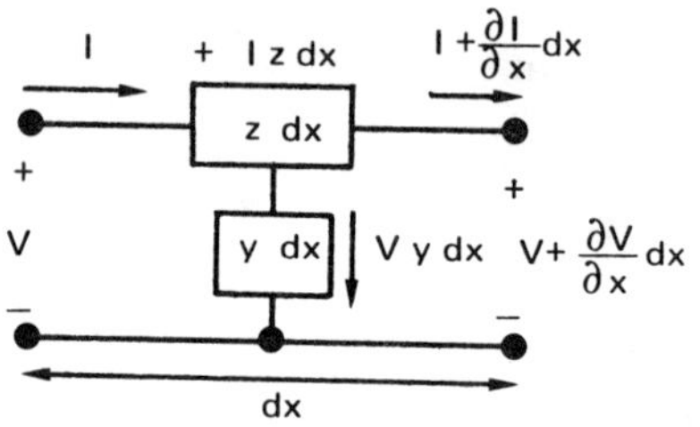

Figure 6.5 Section of a long transmission line

Kirchhoff's voltage law can be written around figure 6.5.

$$V = Iz\,dx + V + \frac{\partial V}{\partial x}dx \qquad 6.61$$

This reduces to

$$\frac{\partial V}{\partial x} = -Iz \qquad 6.62$$

Similarly, Kirchhoff's current law equation can be written for the section.

$$I = Vy\,dx + I + \frac{\partial I}{\partial x}dx \qquad 6.63$$

This reduces to

$$\frac{\partial I}{\partial x} = -Vy \qquad 6.64$$

Differentiating equations 6.62 and 6.64,

$$\frac{\partial^2 V}{\partial x^2} = yzV \qquad 6.65$$

$$\frac{\partial^2 I}{\partial x^2} = yzI \qquad 6.66$$

These differential equations have solutions of the form

$$V = V_1 e^{-\gamma x} + V_2 e^{\gamma x} \qquad 6.67$$

The *propagation constant* is

$$\gamma = \sqrt{yz} \qquad 6.68$$

$$y = jB_C \qquad 6.69$$

$$B_C = \frac{1}{X_C} \qquad 6.70$$

$$z = r + jX_L \qquad 6.71$$

Then,

$$yz = -B_C X_L + jrB_C \qquad 6.72$$

$$\gamma = \alpha + j\beta \qquad 6.73$$

$$|\gamma| = \sqrt{B_C}(r^2 \times X_L^2)^{\frac{1}{4}} \qquad 6.74$$

α is the *attenuation constant.* Common units are nepers/mile.

$$\alpha = |\gamma| \cos\left[\frac{1}{2}\arctan\left(\frac{-r}{X_L}\right)\right] \qquad 6.75$$

β is the *phase constant.* Common units are radians/mile.

$$\beta = |\gamma| \sin\left[\frac{1}{2}\arctan\left(\frac{-r}{X_L}\right)\right] \qquad 6.76$$

The voltage along the transmission line is

$$V = V_1 e^{-\alpha x} e^{-j\beta x} + V_2 e^{\alpha x} e^{j\beta x} \qquad 6.77$$

From equation 6.62,

$$I = \sqrt{\frac{\gamma}{z}} V_1 e^{-\alpha x} e^{-j\beta x} - \sqrt{\frac{\gamma}{z}} V_2 e^{\alpha x} e^{j\beta x} \qquad 6.78$$

These equations represent *traveling waves.* The $e^{-\alpha x} e^{-j\beta x}$ terms are waves traveling in the positive x-direction, and terms with $e^{\alpha x} e^{j\beta x}$ are waves traveling in the negative x-direction.

The *characteristic impedance* is

$$z_o = \sqrt{\frac{z}{y}} \qquad 6.79$$

Using the line parameters,

$$\frac{z}{y} = \frac{r + jX_L}{jB_C} = X_L X_C - jX_c r \qquad 6.80$$

$$|z_o| = \sqrt{X_C}(r^2 + X_L^2)^{\frac{1}{4}} \qquad 6.81$$

$$\angle z_o = \frac{1}{2}\arctan\left(\frac{-r}{X_L}\right) \qquad 6.82$$

The relative values of V_1 and V_2 (or I_1 and I_2) are determined from conditions at the sending and receiving ends of the line.

$$I_1 z_o = V_1 \qquad 6.83$$

$$I_2 z_o = -V_2 \qquad 6.84$$

At the receiving end,

$$I_R = I_1 + I_2 \qquad 6.85$$

$$V_R = V_1 + V_2 \qquad 6.86$$

The origin ($x = 0$) is ordinarily taken at the receiving end of the line. Since $V_R/I_R = Z_L$, equations 6.85 and 6.86 can be combined.

$$V_1 + V_2 = Z_L(I_1 + I_2) \qquad 6.87$$

Introducing equations 6.83 and 6.84,

$$V_2 = \left[\frac{Z_L - z_o}{Z_L + z_o}\right] V_1 \qquad 6.88$$

The ratio of V_2 to V_1 is the *reflection coefficient,* ρ.

$$\rho = \frac{Z_L - z_o}{Z_L + z_o} \qquad 6.89$$

Example 6.11

For Waxwing conductors with 3 foot spacing at 20°C, determine (a) the propagation constant γ in polar form, (b) the attenuation constant α, and (c) the phase constant β.

From table 6.2, $O.D.$ = .609 inch, so radius =.0254 feet. GMR = 0198 ft. R = .3488 ohms/mi, X_L = .476 ohms/mile, and X_C = .109 Mohm-mi.

First, correct the reactance.

$$X_L = \left[1 + \frac{ln\,3}{ln(\frac{1}{.0198})}\right] \times .476 = .6093\,\Omega/\text{mi}$$

$$X_C = \left[1 + \frac{ln\,3}{ln(\frac{1}{.0254})}\right] \times .109 = 0.1416\,\text{Mohm-mi}$$

$$B_C = 7.06 \times 10^{-6}\ \text{mho/mi}$$

$$yz = (j7.06 \times 10^{-6})(.3488 + j.6093)$$

$$= 4.96 \times 10^{-6}\angle 150^\circ$$

The propagation constant is

$$\gamma = \sqrt{yz} = 2.23 \times 10^{-3}\angle 75^\circ$$

$$\gamma = 5.72 \times 10^{-4} + j(2.15 \times 10^{-3})$$

The attenuation and phase constants are

$$\alpha = 5.72 \times 10^{-4}\ \text{nepers/mile}$$

$$\beta = 2.15 \times 10^{-3}\ \text{radians/mile}$$

Example 6.12

Find the characteristic impedance for the line of example 6.11.

$$z_o^2 = \frac{z}{jB_C} = \frac{.3488 + j.6093}{j(7.06 \times 10^{-6})}$$

$$= 9.94 \times 10^4\angle{-29.8^\circ}$$

$$z_o = 315.3\angle{-14.9^\circ}$$

Example 6.13

The line of examples 6.11 and 6.12 has a line-to-neutral (phase) load impedance of 10 ohms. Determine the reflection coefficient.

$$\rho = \frac{Z_L - z_o}{Z_L + z_o} = \frac{10 - 304.6 + j81.1}{10 + 304.6 - j81.1}$$
$$= \frac{305.6\angle 164.6}{325\angle -14.40} = .94\angle -179^\circ$$

The ABCD parameters can be found from V_1, V_2, I_1, and I_2. From equations 6.83 through 6.86,

$$V_1 = \frac{V_R + z_o I_R}{2} \qquad 6.90$$

$$V_2 = \frac{V_R - z_o I_R}{2} \qquad 6.91$$

Substituting into equations 6.77 and 6.78,

$$V = \left[\frac{V_R + I_R z_o}{2}\right] e^{-\gamma x} + \left[\frac{V_R - z_o I_R}{2}\right] e^{\gamma x} \qquad 6.92$$

$$I = \left[\frac{\frac{V_R}{z_o} + I_R}{2}\right] e^{-\gamma x} - \left[\frac{\frac{V_R}{z_o} - I_R}{2}\right] e^{-\gamma x} \qquad 6.93$$

Terms can be combined to obtain equations 6.94 and 6.95.

$$V = V_R\left[\frac{e^{\gamma x} + e^{-\gamma x}}{2}\right] - I_R z_o\left[\frac{e^{\gamma x} - e^{-\gamma x}}{2}\right] \qquad 6.94$$

$$I = \frac{V_R}{z_o}\left[\frac{e^{\gamma x} + e^{-\gamma x}}{2}\right] + I_R\left[\frac{e^{\gamma x} - e^{-\gamma x}}{2}\right] \qquad 6.95$$

The hyperbolic functions are defined by equations 6.96 and 6.97.

$$\cosh A = \frac{e^A + e^{-A}}{2} \qquad 6.96$$

$$\sinh A = \frac{e^A - e^{-A}}{2} \qquad 6.97$$

When A is the complex number $a + jb$,

$$\cosh(a + jb) = \cosh a \cos b + j(\sinh a \sin b) \qquad 6.98$$

$$\sinh(a + jb) = \sinh a \cos b + j(\cosh a \sin b) \qquad 6.99$$

The distance x is defined as being positive to the right, the direction of power flow. Therefore, with V_R and I_R as the dependent variables in equations 6.94 and 6.95 distances toward the sending end are negative, and at the sending end $x = -l$. Thus equations 6.94 and 6.95 become

$$V_S = (\cosh \gamma l)V_R \pm (z_o \sinh \gamma l)I_R \qquad 6.100$$

$$I_S = \left(\frac{1}{z_o}\right)(\sinh \gamma l)V_R + (\cosh \gamma l)I_R \qquad 6.101$$

This defines the ABCD parameters.

$$A = \cosh \gamma l \qquad 6.102$$

$$B = z_o \sinh \gamma l \qquad 6.103$$

$$C = \frac{1}{z_o}\sinh \gamma l \qquad 6.104$$

$$D = \cosh \gamma l \qquad 6.105$$

Example 6.14

The line of examples 6.11 through 6.13 has a length of 1000 miles. Determine the sending end voltage and current, the power loss along the line, and the voltage regulation ($V_R = 110$ kV, $I_R = 200$ amps, and power factor $= .8$).

From example 6.11,

$$\alpha l = .572 \text{ nepers}$$
$$\beta l = 2.15 \text{ radians} = 123.2^\circ$$
$$\cosh \alpha l = 1.168$$
$$\sinh \alpha l = 0.6037$$
$$\cos \beta l = -0.5474$$
$$\sin \beta l = 0.8369$$

From equation 6.98,

$$\cosh \gamma l = .8149\angle 141.6^\circ$$

From equation 6.99

$$\sinh \gamma l = 1.032\angle 108.7^\circ$$

From example 6.12, $z_o = 315.3\angle -14.9^\circ$, and for $PF = .8$, $I_R = 200\angle -36.9^\circ$. Then, from equation 6.100

$$V_S = 115.5\angle 107.5^\circ$$

The voltage regulation is

$$\frac{\frac{115.5}{.8149} - 110}{110} \times 100 = 28.8\%$$

From equation 6.101,

$$I_S = 516.9\angle 117.7^\circ \text{ amps}$$

Then,

$$S_S = P_S + jQ_S$$
$$= V_S I_S^* = 58,747 - j10,586 \text{ kVA}$$
$$S_R = 110 \times 200\angle 36.9 = 17,600 + j13,200 \text{ kVA}$$

Then, the power loss is

$$P = P_S - P_R = 41,147 \text{ kW}$$

8 HIGH FREQUENCY TRANSMISSION LINES

At high frequencies the wavelengths become shorter, so even a few feet of line must be treated as a long line. However, the resistance is negligible, considerably simplifying the calculations. With $R = 0$, the series impedance per unit length is jwL_C. The simplified equations are

$$z = j\omega L \tag{6.106}$$

$$y = j\omega C \tag{6.107}$$

$$\gamma = \sqrt{yz} = j\omega\sqrt{LC}$$

$$= j\beta \tag{6.108}$$

$$z_o = \sqrt{\frac{z}{y}} = \sqrt{\frac{L}{C}} \tag{6.109}$$

$$\cosh j\beta x = \cos \beta x \tag{6.110}$$

$$\sinh j\beta x = j \sin \beta x \tag{6.111}$$

At high frequencies, it is desired to compensate a line to minimize reflected waves. This is done by using a capacitive compensation to make the load impedance appear to be the characteristic impedance. The sending voltage and current are found from equations 6.77, 6.90 and 6.107 through 6.111.

$$V_S = V_1\left[e^{j\beta l} + \rho e^{-j\beta l}\right] \tag{6.112}$$

$$I_S = \frac{V_1}{z_o}\left[e^{j\beta l} - \rho e^{-j\beta l}\right] \tag{6.113}$$

Then,

$$V_S = V_1[(1+\rho)\cos\beta l + j(1-\rho)\sin\beta l] \tag{6.114}$$

$$I_S = \frac{V_1}{z_o}[(1-\rho)\cos\beta l + j(1+\rho)\sin\beta l] \tag{6.115}$$

The *input impedance* is V_S/I_S, which becomes

$$Z_i = z_o\left[\frac{(1+\rho)\cos\beta l + j(1-\rho)\sin\beta l}{(1-\rho)\cos\beta l + j(1+\rho)\sin\beta l}\right] \tag{6.116}$$

Using the definition of the reflection coefficient ρ, the input impedance is

$$Z_i = z_o\left[\frac{Z_L\cos\beta l + jz_o\sin\beta l}{z_o\cos\beta l + jZ_L\sin\beta l}\right] \tag{6.117}$$

Example 6.15

A transmission line has a characteristic impedance of $z_o = 50$ ohms. It has a 100 ohm terminating resistance. Find (a) the reflection coefficient, and (b) the input impedance if the line length is an even number of wavelengths ($\beta l = 2\pi n$).

$$\rho = \frac{Z_L - z_o}{Z_L + z_o}$$

$$= \frac{100-50}{100+50} = .333$$

$$\cos\beta l = 1$$

$$\sin\beta l = 0$$

$$Z_{\text{in}} = z_o\frac{Z_L}{z_o}$$

$$= Z_L = 100 \text{ ohms}$$

The voltage at any point along the line is a function of the distance from the load because of the two traveling waves. One of the measurable variables along the line is the ratio of the maximum (RMS) voltage to the minimum voltage, known as the voltage standing wave ratio VSWR. Where the magnitudes of the two waves add, the total voltage is a maximum. Where they subtract, it is a minimum.

$$\text{VSWR} = \frac{|V_1|+|V_2|}{|V_1|-|V_2|} = \frac{1+|\rho|}{1-|\rho|} \tag{6.118}$$

Example 6.16

Find the VSWR for the line of example 6.15.

$$\text{VSWR} = \frac{1+.333}{1-.333} = 2$$

The current along the line will also exhibit a standing wave pattern. The current is maximum where the voltage magnitudes add.

$$I_{\text{max}} = \frac{V_1(1+|\rho|)}{z_o} \tag{6.119}$$

The current is minimum where they subtract.

$$I_{\text{min}} = \frac{V_1(1-|\rho|)}{z_o} \tag{6.120}$$

From equation 6.112 and 6.113, the maximum current occurs where the voltage minimum occurs and the conditions for the minimum current occur when the voltage amplitudes add. Then, the impedance will have a maximum at the point of maximum voltage.

$$Z_{\text{max}} = z_o \times \text{VSWR} \tag{6.121}$$

$$Z_{\text{min}} = \frac{z_o}{\text{VSWR}} \tag{6.122}$$

9 THE SMITH CHART

The analysis and design of lossless transmission lines is greatly aided by the use of Smith charts.[2] The chart is a polar representation of normalized (with respect to z_o) and constant resistance and reactance curves, as well as the reflection coeficient ρ. The apparent normalized line impedance is obtained by finding the *normalized load impedance.*

$$z_L = \frac{R_L + jX_L}{z_o} = r + jx \tag{6.123}$$

The z_L point is located on the chart at the intersection of the r and x lines. A circle centered at the origin of the chart then is drawn through that value of $r + jx$. This circle is known as the *impedance locus circle.* Next, a radial line is drawn from the origin (center) through the z_L point. Finally, the angle βl is determined, and the normalized impedance a distance l from the load can be determined by following the circle from the load point clockwise through an angle $2\beta l$.

[2] P.H. Smith, *An Improved Transmission-Line Calculator*, Electronics, 17, 130, Jan. 1944.

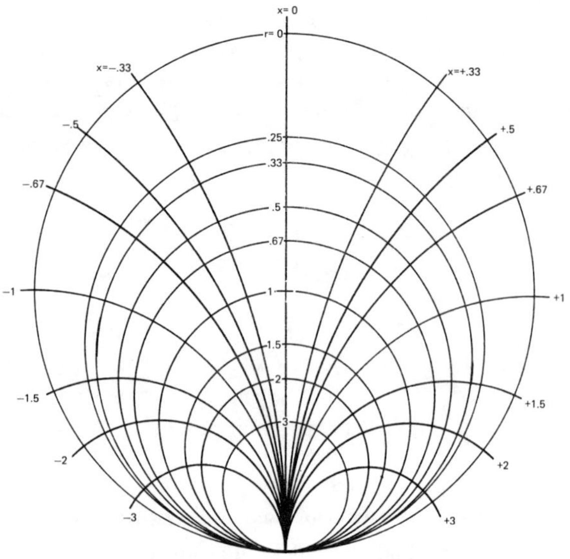

Figure 6.6 Smith Chart

Example 6.17

A transmission line of length $\beta l = 6$ radians (344°) and $z_o = 50$ ohms has a termination load of $75 + j25$ ohms. Determine the input impedance.

Step 1 Calculate the normalized load impedance.

$$\frac{R_L + jX_L}{z_o} = \frac{75 + j25}{50} = 1.5 + j.5$$

Step 2 Plot point $z_L = 1.5 + j.5$ on the Smith chart.

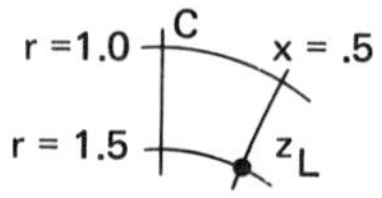

Step 3 Draw a circle with center at $1.0 + j0$ (point C) which passes through point z_L.

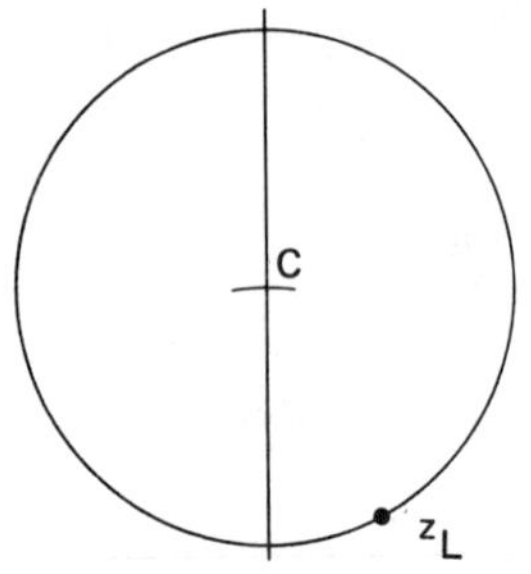

Step 4 Extend a line from point C through point z_L to the edge of the Smith chart, point L (load point).

Step 5 Calculate the angular separation between the source and load points.

$$2 \times 344 = 688°$$

Since this is more than one revolution, subtract the full revolution.

$$688 - 360 = 328°$$

Move clockwise 328° from the load to the generator point.

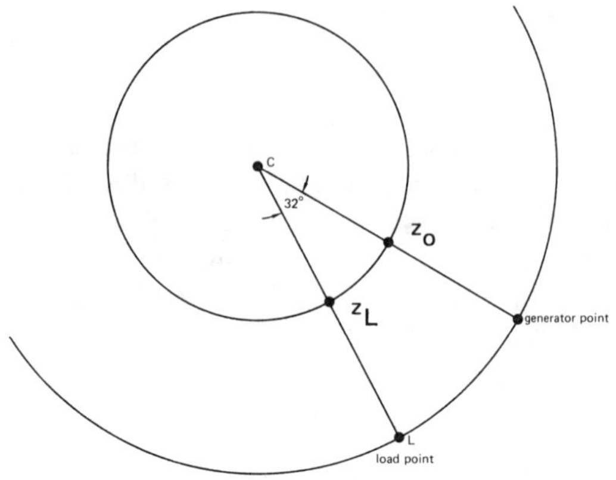

Step 6 Read the impedance seen from the generator as the intersection of the generator line and the impedance circle, point z_L

$$z_L = 1.1 + j.6$$

Step 7 Calculate the input load impedance.

$$z_l = 50(1.1 + j.6) = 55 + j30$$

The intersection of the $r = 1$ circle and $x = 0$ is the origin of the Smith chart. At that point, the impedance circle locus has zero radius, and the impedance will not change with length of line. This is the *matched condition*, which can be achieved by placing a *compensating reactance* a particular distance away from the load (toward the source).

The compensation technique is to set up the impedance circle locus as usual and then traverse the line toward the source a distance which will put the circle on the $r = 1$ circle. At that point, the impedance will be

$$z_{cp} = 1 + jx \qquad 6.124$$

This can be adjusted to $z = 1$ by adding a series reactance of $-x$. As the most satisfactory compensating reactance is a capacitance, the compensation point should be chosen on the right side ($x > 0$) of the Smith chart.

Example 6.18

A transmission line with $z_o = 50$ ohms is terminated with a load impedance of $100 + j100$ ohms. Find the electrical angle (βl) where a compensating capacitor can be inserted in the line to cause the impedance seen at that point to be 50 ohms. Determine the reactance of the capacitor.

Step 1 The normalized load impedance is

$$\frac{100 + j100}{50} = 2 + j2$$

Step 2 Plot the load point z_L.

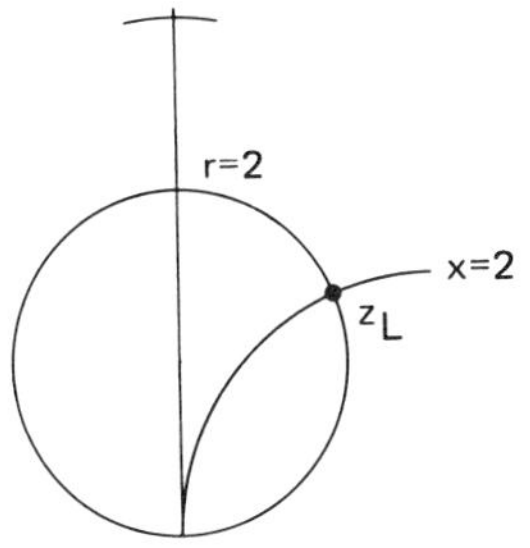

Step 3 Draw a circle through point z_L with center at $1.0 + j0$ (point C). This is the impedance locus circle.

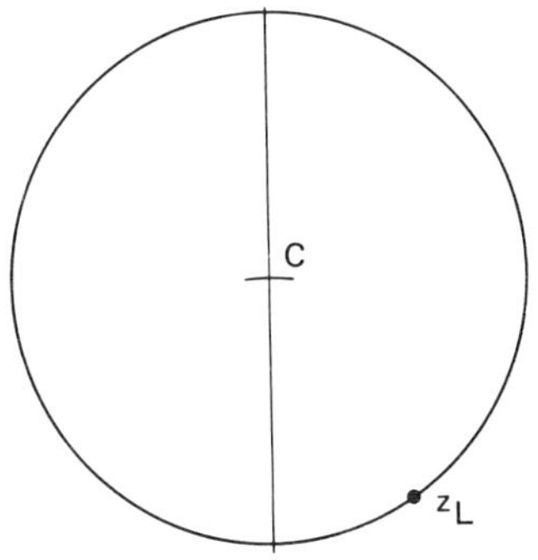

Step 4 Determine the compensation point P, as the intersection of the impedance circle and the $r = 1$ circle. The intersection is at $1 + j1.6$.

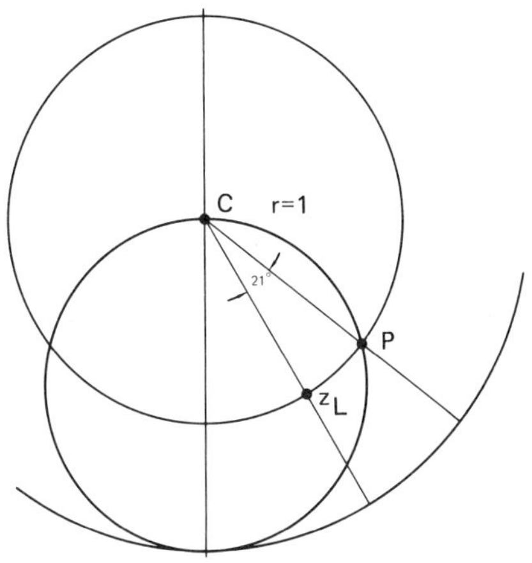

Step 5 Extend lines from point C through z_L and P, and measure the angular difference between them as 21°.

Step 6 Since the direction toward the source is clockwise, it is necessary to travel $360 - 21 = 339°$ around the circle toward the load. The electrical angle along the line is one-half that amount or 169.5°. The distance from the load to the compensation point is 169.5 electrical degrees or 169.5/360 wavelengths. Therefore, $l = .471\lambda$. At this point, the normalized reactance is about 1.6. The compensating reactance is

$$x_c = -1.6 \times 50 = -80 \text{ ohms}$$

The coordinates of the Smith chart are $p\angle\phi$ where p is the magnitude of ρ and ϕ is the angle of ρ. Other information that can be obtained from the Smith chart is $|\rho|$ and the VSWR.

$$\rho = p\angle\phi = u + jv \qquad 6.125$$

On the Smith chart, the loci of r and x values are circles described by equations 6.126 and 6.127.

$$\left[u - \frac{v}{1+r}\right]^2 + v^2 = \frac{1}{(1+r)^2} \qquad 6.126$$

and

$$(u-1)^2 + \left[v - \frac{1}{x}\right]^2 = \frac{1}{x^2} \qquad 6.127$$

Equation 6.126 shows that the circle of $r = 0$ has a unit radius ($p = 1$). Therefore, p is the ratio of the distance from the origin to the point $z = r + jx$. The VSWR is

$$VSWR = \frac{1+|\rho|}{1-|\rho|} = \frac{1+p}{1-p} \qquad 6.128$$

The circle of radius p intersects the $x = 0$ line in two places. $(1+p)$ is the distance from the intersection to the near side of the $r = 0$ circle, and $(1-p)$ is the distance from either intersection to the opposite side of the $r = 0$ circle. Referring to figure 6.7, the length $\overline{OZ}$ divided by the length $\overline{OA}$ is p.

$$p = \frac{\overline{OZ}}{\overline{OA}} \qquad 6.129$$

The lengths $\overline{AB}$ and $\overline{AC}$ give the VSWR.

$$\text{VSWR} = \frac{\overline{AB}}{\overline{BC}} \qquad 6.130$$

The reciprocal of a complex value can be found from the Smith chart by following a circle 180° through the value to a point diametrically opposed to the original point. This method can be used to determine an admittance from an impedance and vice versa. Thus, for a coaxial line where a shunt capacitance is more suitable than a series capacitance for compensation, the load point for admittance is on the same P circle as the impedance 180° from the load point.

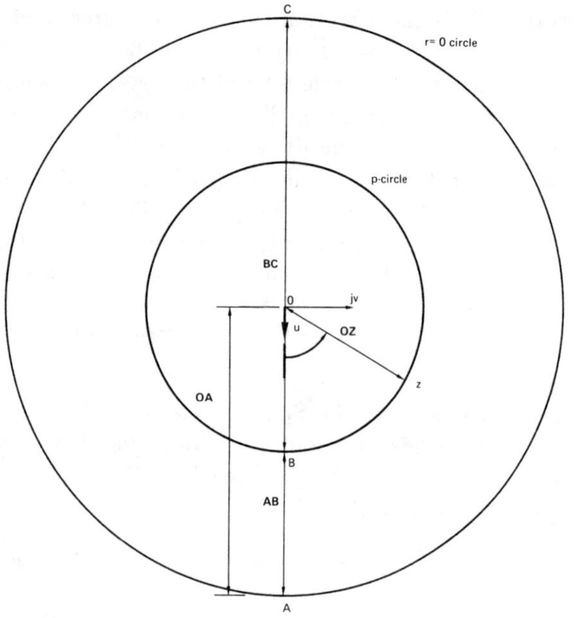

Figure 6.7 Lengths for determining the reflection coefficient and the VSWR

Example 6.19

The transmission line of example 6.18 is to be compensated with a shunt capacitance. Determine the value and position (in wavelengths) of the compensating capacitance.

The Smith chart is the same as example 6.18, except the load is 180° away ($y = g + jb = .25 - j.25$). The compensation point is where the circle intersects $g = 1$ (also $r = 1$).

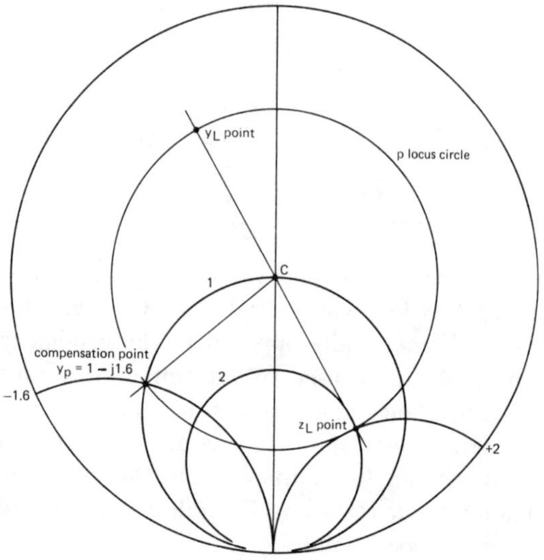

However, the compensation point using a capacitance must be a point where the susceptance b is negative. Therefore, the compensation point is at the intersection of the $g = 1$ circle with an angle of 264°. The admittance at this point is $y = 1 - j1.6$.

The electrical angle along the line is 264/2. Therefore,

$$l = \left(\frac{264}{720}\right)\lambda = .337\lambda$$

The susceptance is

$$B = \frac{b}{z_o} = \frac{1.6}{50}$$

$$X_C = \frac{-50}{1.6} = -31.25\,\text{ohms}$$

Shorted and open transmission lines with lengths of $\lambda/4$ can be used as shunt compensation. Such a line has essentially zero resistance, so the impedance and admittance are on the $r = 0$ circle. The values of x or b can then be set by the length of the line. For a shorted line, the termination reactance is zero. This is at the top of the Smith chart where $r = 0$ and $x = 0$ coincide. The susceptance at this point is infinite (the line is shorted). On the other hand, an open line has infinite reactance at the load end. This point is at the bottom of the Smith chart (180° from $x = 0$).

In the case of shorted and open lines, it is possible to place the compensation at the first clockwise intersection from the load. This will result in different compensation points for both examples, as inductive compensation (positive reactance or negative susceptance) is available.

Example 6.20

Determine lengths of a 72 ohm line which, when placed (a) in series, and (b) in parallel, will compensate the 50 ohm line in examples 6.18 and 6.19.

(a) The angle to the first intersection of the $r = 1$ circle is 83°. So, the distance is

$$l = \left(\frac{83}{720}\right)\lambda = .115\lambda$$

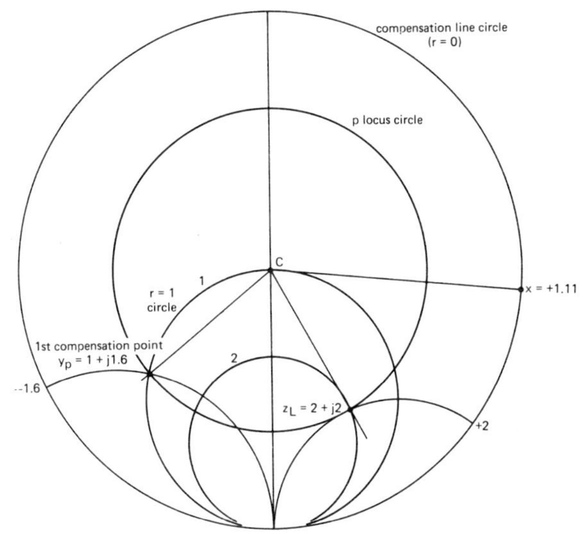

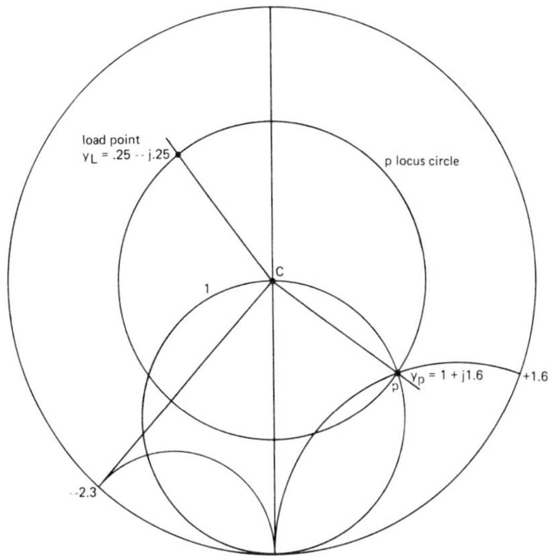

The normalized impedance at that point is -1.6, so the actual compensating reactance needed is

$$-1.6 \times 80 = -80 \text{ ohms}$$

With a 72 ohm line, the normalized reactance is $80/72 = -1.11$ ohms. This requires a series reactance of $+1.11$ ohms for compensation at that point.

The shortest compensating line is one that will begin at the top of a Smith chart because length increases in a clockwise direction. For a reactance of 1.11, the angle will be 94° (just below the $x = +1$ intersection with the $r = 0$ circle). So, the compensating line length is

$$l_{cp} = \left(\frac{94}{720}\right)\lambda = .131\lambda$$

This is a shorted length because the termination is at $x = 0$.

(b) From example 6.19, the first clockwise intersection from the load point $Y_L = .25 - j.25$ occurs at $y = 1 + j1.6$. It is again capacitive ($b = 1.6$), and the angle is 160°. Therefore, the length is

$$l = \left(\frac{160}{720}\right)\lambda = .222\lambda$$

Since $b = 1.6$, $B = 1.6/50 = .032$. Then, in the 72 ohm normalization $b_C = .032 \times 72 = -2.3$.

This is a negative value. The shortest line will be at the bottom of the Smith chart at its termination. Since $b = \infty$, this is a shorted line. The point at $g = 0$ and $b = -2.3$ is between the $x = -2$ and $x = -3$ intersection with the $r = 0$ circle ($g = 0$). The angle from the bottom is measured as approximately 45°. Thus,

$$l_{cp} = \left(\frac{45}{720}\right)\lambda = .0625\lambda$$

Practice Problems

TRANSMISSION LINES

Warmups

1. A coaxial line with an inner conductor diameter of .20 cm and an outer conductor with an inner radius of 1 cm is operated at 10 Mhz. Determine the internal impedance per meter of length ($\rho_{cu} = 1.8 \times 10^{-8}$ ohm-meters).

2. Find the internal impedance per 1000 feet of AWG #1 solid copper wire at 10 khz.

3. Determine the total inductive reactance per 1000 feet for a pair of AWG #20 wires separated by 1 cm (center to center) and operating at 1 Mhz.

4. Determine the capacitance per meter for a coaxial line with an inner conductor diameter of .2 cm and inner diameter of the outer conductor of 2 cm.

5. Determine the inductive reactance of a Bluebird conductor when the line-to-line spacing is 10 feet.

6. Find the capacitive reactance of a 2 mile single-phase Bluebird line operating at 60 hz with a spacing of 20 feet between centers.

7. Verify that equation 6.43 gives the correct values of per-conductor capacitance for several conductors in table 6.2. (Hint: You must convert the given conductor diameter to the conductor radius in feet. Assume that the spacing is 1 foot.)

8. An unbalanced three-phase load is delta-connected with $R_{ab} = 2$ ohms, $R_{bc} = 2.5$ ohms, and $R_{ca} = 1.8$ ohms. Determine the wye equivalent resistances.

9. In applying capacitive reactance for power factor correction, compare capacitor ratings for wye-connection and delta-connection. (Ratings are in volt-amperes reactive and rated voltage.)

Concentrates

1. A three-conductor transmission line has the conductors arranged as three of the vertices of a square which is 10 feet on a side. The conductors are Partridge, and the frequency is 60 hz. Determine the per-phase inductive reactance X_L and the mutual reactances between phases X_{ab}, X_{ac}, and X_{bc}. (The order is not specified.)

2. A balanced three-phase power system has line impedances $3+j4$ and line-to-line load impedances of $20+j15$ ohms. The load power is 30 kW. (a) Determine the source voltage, power, and volt-amperes. (b) Determine the source voltage, power and volt-amperes if the load power factor is corrected to 1.

3. A 25 mile length of transmission line consisting of Flicker conductors has spacing of 7 feet between the conductors. The full load current is 225 amps per phase, the voltage at the sending end of the line is 11 kV per phase (line-to-neutral), and the power factor of the load is 0.9 leading. Determine the voltage regulation. Assume $T = 20°\text{C}$.

4. For Falcon conductors spaced 20 feet apart at $T = 50°\text{C}$, determine (a) the propagation constant γ, (b) the attenuation constant α, (c) the phase constant β, (d) the characteristic impedance, and (e) the reflection coefficient for a load of 100 ohms per phase.

5. A transmission line with a characteristic impedance of 50 ohms and a termination resistance of 70 ohms has $\beta l = \frac{\pi}{2}$. Determine (a) the reflection coefficient, (b) the input impedance, and (c) the VSWR.

6. A transmission line with $z_o = 50$ ohms is terminated with a load impedance of $25-j25$ ohms. Find the electrical angle (βl) where a compensating capacitor can be inserted in the line to cause the impedance seen at that point to be 50 ohms. Also, determine the reactance of the capacitor.

7. Given a coaxial cable with $z_o = 50$ ohms, determine the length of cable (in wavelengths) necessary to produce (a) a reactance of -25 ohms, and (b) a reactance of $+50$ ohms. (The stubs may be either open or shorted.)

Timed

1. A transmission line delivers 7500 kW at 8500 kVA, and 86.6 kV to a balanced load. The line is 100 miles long, and the conductors are Flicker with a separation of 7 feet. Assume the operating temperature is 50°C. Determine the regulation.

2. A transmission line with $z_o = 72$ ohms and $\beta = .5$ rad/m is 7.5 meters long. It is terminated with a pure 50 ohm resistance. Determine the impedance seen from the source.

3. A transmission line with $z_o = 100$ ohms is terminated with $Z_L = 25 + j25$ ohms. This is to be compensated with a 72 ohm line section. Determine the length of the compensating line section and specify its placement from the load on the 100 ohm line for (a) series compensation, and (b) parallel compensation.

4. A transmission line with $z_o = 50$ ohms and $Z_L = 25-j25$ ohms is to be compensated with a shunt capaci-

tor. Specify the capacitance of the capacitor and its placement from the load in wavelengths.

5. A 200 mile transmission line carrying three-phase power at 60 hz has approximate pi parameters of $Z_{\text{series}} = 40 + j160$ and $Y_{\text{shunt}} = 0.001063j$. The receiving end voltage is 220 kV with a load power of 50 MW at a power factor of 0.9 lagging. Determine the sending end voltage.

6. A 60 hz, three-phase transmission line is 100 miles long. It delivers 75 MW at 132 kV with a lagging power factor of 0.92. The line constants are 0.1 ohms/mile, 0.6 inductive ohms/mile, and 0.15 capacitive MΩ-mile. Using the pi approximation, determine the sending end voltage.

Appendix A

COPPER-WIRE TABLE, STANDARD ANNEALED COPPER
American Wire Gage

gage no.	diameter, mils at 20°C	cross section at 20°C circular mils	sq. in	ohms per 1,000 ft at 20°C	lb. per 1,000 ft.
0000	460.0	211,600.0	0.1662	0.04901	640.5
000	409.6	167,800.0	0.1318	0.06180	507.9
00	364.8	133,100.0	0.1045	0.07793	402.8
0	324.9	105,500.0	0.08289	0.09827	319.5
1	289.3	83,690.0	0.06573	0.1239	253.3
2	257.6	66,370.0	0.05213	0.1563	200.9
3	229.4	52,640.0	0.04134	0.1970	159.3
4	204.3	41,740.0	0.03278	0.2485	126.4
5	181.9	33,100.0	0.02600	0.3133	100.2
6	162.0	26,250.0	0.02062	0.3951	79.46
7	144.3	20,820.0	0.01635	0.4982	63.02
8	128.5	16,510.0	0.01297	0.6282	49.98
9	114.4	13,090.0	0.01028	0.7921	39.63
10	101.9	10,380.0	0.008155	0.9989	31.43
11	90.74	8,234.0	0.006467	1.260	24.92
12	80.81	6,530.0	0.005129	1.588	19.77
13	71.96	5,178.0	0.004067	2.003	15.68
14	64.08	4,107.0	0.003225	2.525	12.43
15	57.07	3,257.0	0.002558	3.184	9.858
16	50.82	2,583.0	0.002028	4.016	7.818
17	45.26	2,048.0	0.001609	5.064	6.200
18	40.30	1,624.0	0.001276	6.385	4.917
19	35.89	1,288.0	0.001012	8.051	3.899
20	31.96	1,022.0	0.0008023	10.15	3.092
21	28.46	810.1	0.0006363	12.80	2.452
22	25.35	642.4	0.0005046	16.14	1.945
23	22.57	509.5	0.0004002	20.36	1.542
24	20.10	404.0	0.0003173	25.67	1.223
25	17.90	320.4	0.0002517	32.37	0.9699
26	15.94	254.1	0.0001996	40.81	0.7692
27	14.20	201.5	0.0001583	51.47	0.6100
28	12.64	159.8	0.0001255	64.90	0.4837
29	11.26	126.7	0.00009953	81.83	0.3836
30	10.03	100.5	0.00007894	103.2	0.3042
31	8.928	79.7	0.00006260	130.1	0.2413
32	7.950	63.21	0.00004964	164.1	0.1913
33	7.080	50.13	0.00003937	206.9	0.1517
34	6.305	39.75	0.00003122	260.9	0.1203
35	5.615	31.52	0.00002476	329.0	0.09542
36	5.000	25.00	0.00001964	414.8	0.07568
37	4.453	19.83	0.00001557	523.1	0.06001
38	3.965	15.72	0.00001235	659.6	0.04759
39	3.531	12.47	0.000009793	831.8	0.03774
40	3.145	9.888	0.000007766	1049.0	0.02993

Appendix B

Approximate Dielectric Constants
(20°C and 1 atmosphere)

material	ϵ_r
Acetone	21.3
Air	1.00059
Alcohol	16 – 31
Amber	2.9
Asbestos paper	2.7
Asphalt	2.7
Bakelite	3.5 – 10
Benzene	2.284
Carbon dioxide	1.001
Carbon tetrachloride	2.238
Castor oil	4.7
Diamond	16.5
Glass	5 – 10
Glycerine	56.2
Hydrogen	1.003
Lucite	3.4
Marble	8.3
Methanol	22
Mica	2.5 – 8
Mineral oil	2.24
Mylar	2.8 – 3.5
Olive oil	3.11
Paper	2.0 – 2.6
Paper (kraft)	3.5
Paraffin	1.9 – 2.5
Polyethylene	2.25
Polystyrene	2.6
Porcelain	5.7 – 6.8
Quartz	5
Rock	$\approx$ 5
Rubber	2.3 – 5.0
Shellac	2.7 – 3.7
Silicon oil	2.2 – 2.7
Slate	6.6 – 7.4
Sulfur	3.6 – 4.2
Teflon	2.0 – 2.2
Vacuum	1.000
Water	80.37
Wood	2.5 – 7.7

Appendix C

The Smith Chart

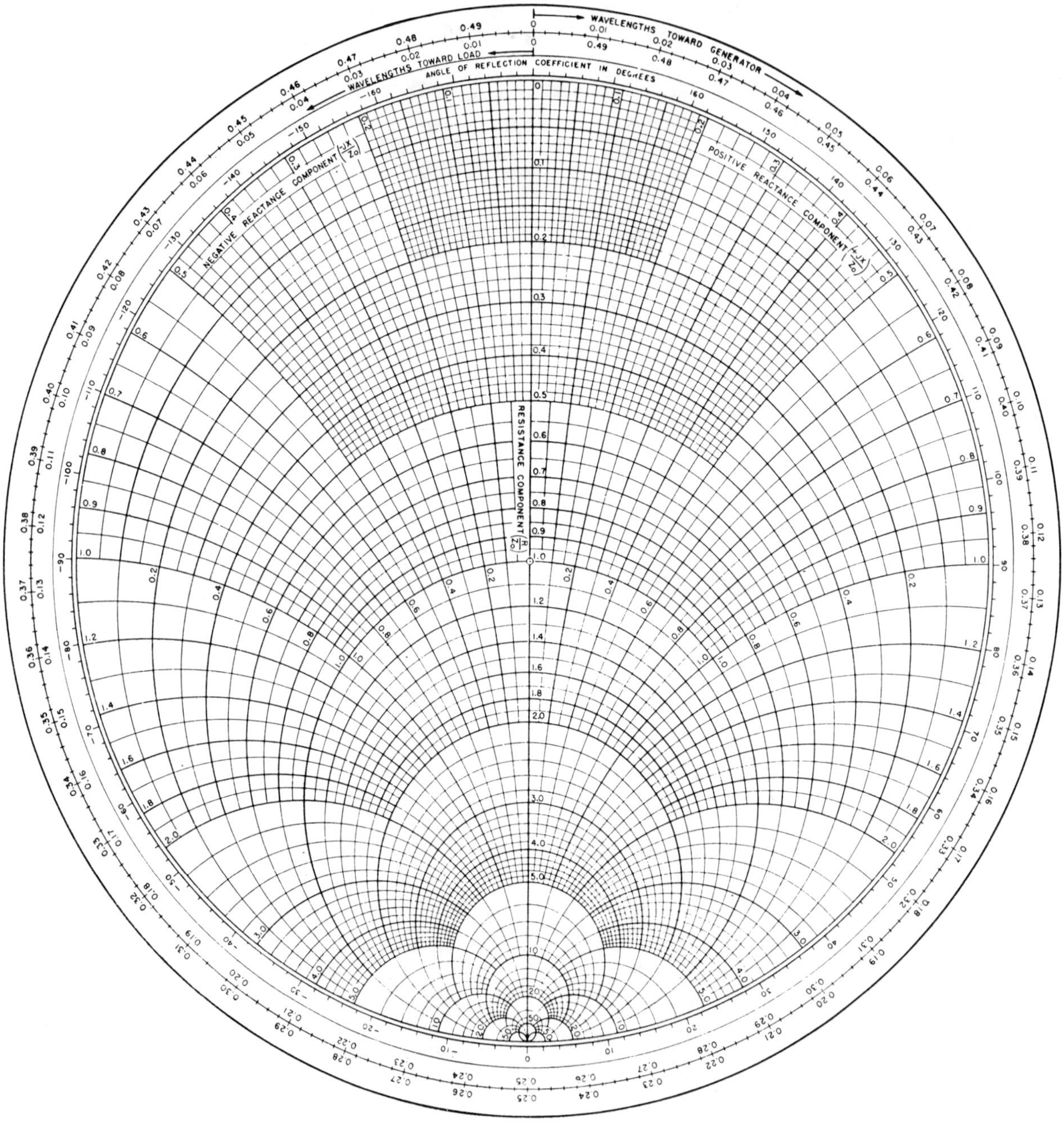

RESERVED FOR FUTURE USE

RESERVED FOR FUTURE USE

7 ROTATING MACHINES

Nomenclature

a	equivalent turns ratio	-
a_p	number of parallel current paths	-
A_r	area of rotor winding	m^2
b	susceptance	mhos
B	magnetic flux density	teslas
E_g	EMF or generated voltage	volts
EMF	electromotive force	volts
f	frequency	hz
f_v	coefficient of viscous friction	N-m-sec/rad
g	gap length, or conductance	m, or mhos
I	D.C. or RMS current	amps
I^*	complex conjugate of current	amps
J	mass moment of inertia	N-m-sec^2/rad
K_T	torque constant	-
K_v	voltage constant	-
K_w	winding constant	-
K_Φ	flux constant	-
l	a length	m
L_a	armature inductance	H
MMF	magnetomotive force	amp-turns
n	rotary speed	RPM
N_f	number of shunt filed winding turns	-
N_r	number of rotor winding turns	-
N_s	number of series field winding turns	-
p	number of poles	-
P	power	W
PF	power factor	-
q	number of phases	-
Q	reactive power	volt-amps reactive
r	resistance	ohms
R	resistance, or radius	ohms, or m
s	slip	-
S	apparent power	volt-amps
t	time	sec
T	torque	N-m
V	voltage	volts
x	reactance	ohms
X_s	synchronous reactance	ohms
Y	admittance	mhos
z_c	the number of armature conductors	-
Z	impedance	ohms

Symbols

δ	torque angle	degrees or rad
θ	an angle	degrees or rad
μ_o	$4\pi \times 10^{-7}$	webers/amp-turn-meter
Φ	magnetic flux	webers
ψ	power factor angle	degrees or rad
Ω	mechanical angular velocity	rad/sec
ω	electrical angular frequency	rad/sec

Subscripts

a	armature
an	from line a to neutral
aa'	from point a to point a'
$b.r.$	blocked rotor
c	core, or conductors
e	equivalent
f	field
g	gap
in	input
L	load
$l-l$	line to line
$l-n$	line to neutral
m	magnetizing or mechanical
n	normalized, normalizing
o	operating point
p	parallel
r	rotor, or rotational
R	rated
s	stator, shaft, or synchronous
T	torque
v	voltage

1 REVIEW OF THE FUNDAMENTALS OF D.C. MACHINES

A circuit model of D.C. motors and generators is shown in figure 7.1.

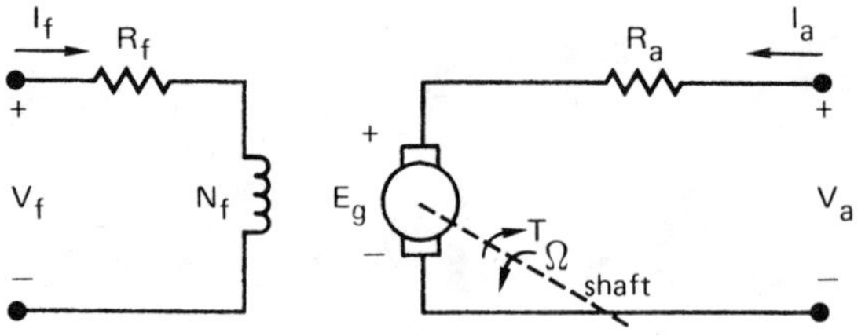

Figure 7.1 D.C. motor circuit model. (Reverse T and I_a for a generator)

The *generated voltage* E_g is proportional to the shaft speed Ω and the field flux, Φ. The proportionality constant K_v is approximately equal to the number of series turns between the commutator brushes. (When the speed is given in revolutions per minute, the constant K_v must be multiplied by $30/\pi$.)

$$E_g = K_v \Phi \Omega \qquad 7.1$$

$$K_v \approx N_r \qquad 7.2$$

Electro-mechanical energy conversion can be described by electrical and mechanical expressions. Mechanical power is usually given in horsepower, while torque is usually given in foot-pounds. Appendix 7.2 gives important conversion factors for energy and power variables.

$$P_{\text{shaft}} = E_g I_a = T\Omega \qquad 7.3$$

From equations 7.1 and 7.3, the relationship between armature current and torque is

$$T = K_T \Phi I_a \qquad 7.4$$

In the MKS system $K_T = K_v \approx N_r$. If mixed units are used, then adjustments have to be made to K_T. If the torque is given in foot-pounds, then

$$(T \text{ in N-m}) \times .7376 = T \text{ in ft-lb} \qquad 7.5$$

The relationship between the flux Φ and the magnetomotive force MMF is frequently given by a *saturation curve* of generated voltage E_g and the field current I_f for constant shaft speed. The saturation curve is a scaled version of the *magnetization curve*. It is obtained by exciting the field winding with a known current and then measuring the *open circuit armature voltage* (E_g in figure 7.1 with no load current flowing).

In this chapter, the saturation curve is scaled by dividing it by the generated voltage E_g and speed Ω, and normalizing it to some convenient value of E_g/Ω. That value of E_g is known as E_n.

The horizontal axis of the magnetization curve is frequently the field current I_f. This is scaled from the MMF. In the case of only one current contributing to the field, MMF is $N_f I_f$. In this chapter, the horizontal axis is scaled to some convenient value of I_f known as I_n.

The resulting characteristic is a *normalized magnetization curve* for the machine, usually referred to as the *saturation curve* when it is given in terms of the field current and generated voltage at constant speed. The normalized curve is used in this section to demonstrate the basic principles, since it avoids confusion about the changes of speed and the compounding problem presented in the next section. A normalized saturation curve is shown in figure 7.2.

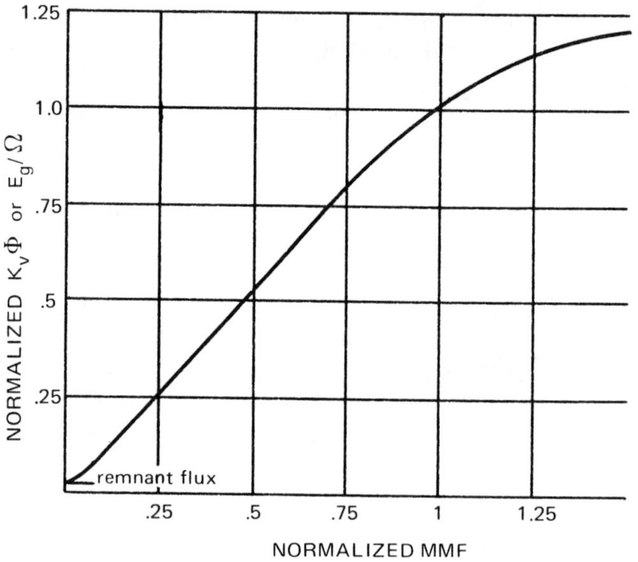

Figure 7.2 Saturation curve

Actual curves for various load currents are better than the saturation curve for determining the behavior of D.C. machines. Actual curves show the effects of *armature reaction*, a phenomenon which occurs because the armature current produces a component of field flux which affects the stator poles, reducing the field flux. This effect is compensated in most machines by the use of *interpoles* in the *quadrature plane* of the field. (The rotor field component is in the quadrature plane, i.e., at 90° to the stator field.) The armature current passes through the windings of the compensating poles to cancel the flux contributed by the armature current. This chapter assumes that compensating interpoles are used so that the saturation curve is adequate.

2 D.C. GENERATORS

A. INTRODUCTION

A voltage applied to the field winding of a D.C. machine results in current and flux. The flux produces a voltage potential in each rotor (armature) conductor that passes through the flux. The result is summarized by equation 7.1. The field winding can be excited externally (resulting in *separately-excited generation*) or by use of the generated voltage itself to supply the field current (resulting in *self-excited generation*).

The self-excited system results in decreased terminal voltage V_a as the load current increases. This is due to the armature resistance R_a. The terminal voltage reduction is particularly detrimental in the self-excited generator because it results in a decrease in field current, causing an even lower generated voltage.

It is sometimes necessary to supply a series field winding. The load (armature) current supplies additional MMF to counter the IR loss in terminal voltage. This technique is called *compounding,* as it uses both shunt and series windings.

B. SEPARATELY-EXCITED D.C. GENERATORS

Separately-excited D.C. generators can be analyzed with equations 7.1 through 7.5. A magnetization curve or mathematical relationship between flux and field current also is required.

Example 7.1

A separately-excited D.C. generator is rated at 10 kW at 240 volts D.C. while turning at 2800 RPM with an input power of 15 horsepower. It is proposed to operate this machine at 3600 RPM and at an output of 300 volts. Determine the ratings of the machine under the new conditions. Determine the field current necessary if the generated voltage at 2800 RPM is given by

$$E_g = 100I_f - I_f^2$$

The armature resistance is $R_a = .5$ ohms.

With a speed increase, the generated voltage will increase proportionally. This assumes a constant flux in equation 7.1.

$$V_{\text{rated}} = 240 \times \frac{3600}{2800} = 308\,\text{volts}$$

The current rating will not change since it is set by the wire size. Then, the power rating will increase since power is proportional to voltage.

$$P_{\text{rated}} = 10\,\text{kW} \times \frac{3600}{2800} = 12.9\,\text{kW}$$

The input horsepower will increase, but not quite proportionally, since part of the input power must supply $I_a^2R_a$ losses. The $I_a^2R_a$ loss is determined from the load current at full load.

$$I_a = \frac{10{,}000}{240} = 41.67\,\text{amps}$$

The armature resistance loss is

$$I_a^2R_a = (41.67)^2(.5) = 868.2\,\text{watts}$$

$$\frac{(868.2)(1.341)}{1000} = 1.16\,\text{hp}$$

Assuming that the remainder of the 15 hp input was due to $T\,\Omega$, the increase in speed will increase this part of the input proportionally.

$$(15 - 1.16) \times \left(\frac{3600}{2800}\right) = 17.79\,\text{hp}$$

This must be added to the $I_a^2R_a$ loss.

$$\text{HP rating} = 17.79 + 1.16 = 18.95\,\text{hp}$$

At full load with a terminal voltage of 300 volts and a current of 41.67 amps through R_a, the generated voltage must be

$$E_g = 300 + (41.67)(.5) = 320.8\,\text{volts}$$

I_f is then found by solving the quadratic equation

$$100I_f = \left(\frac{2800}{3600}\right)320.8 + I_f^2$$

$$I_f = 2.56\,\text{amps}$$

C. SELF-EXCITED GENERATORS

Self-excited D.C. generators use the terminal voltage V_a to supply the field current. The circuit model for a self-excited generator is shown in figure 7.3.

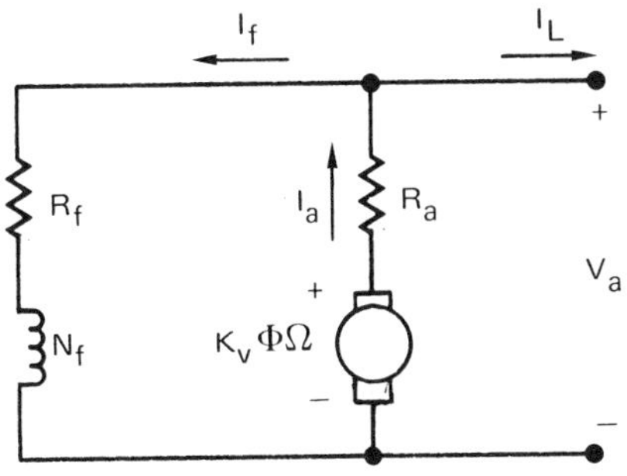

Figure 7.3 Self-excited D.C. generator circuit model

Self-excited D.C. generators behave according to equation 7.6.

$$E_g = K_v\Phi\Omega = V_a + R_aI_L + R_aI_f \qquad 7.6$$

To determine I_f, the values of V_a, I_a and speed must be known. The unknowns in equation 7.6 are then E_g and I_f, as V_a, I_L, and R_a are given. The usual way to solve for two unknowns is to find another equation. However, the only other information available is the saturation curve.

The saturation curve can be used if equation 7.6 is normalized the same way that the magnetization curve is normalized. As equation 7.6 gives voltage, it is normalized to a value of $K_v\Phi\Omega/\Omega$ taken from the magnetization curve data. This implies that some reference point on the curve corresponds to given ratios of E_g/Ω and MMF/N_f. This reference point is given the normalized values of $E_g/\Omega = 1.0$ and $MMF_n = 1.0$.

If the curve is a saturation curve at a given speed, then this normalization is not necessary, and equation 7.6 can be used directly, provided that the curve speed is the same as the problem speed. Otherwise a factor of $\Omega_{\text{design}}/\Omega_{\text{data}}$ must be used on the saturation curve's vertical axis.

Then, divide equation 7.6 by the normalizing voltage E_n.

$$\frac{E_g}{E_n} = \frac{K_v\Phi\Omega}{E_n} = \frac{\frac{E_g}{\Omega}}{\frac{E_n}{\Omega}}$$

$$= \frac{K_v\Phi}{\frac{E_n}{\Omega}}$$

$$= \frac{V_a + R_a I_L}{E_n} + R_a\left(\frac{I_n}{E_n}\right) \times \frac{I_f}{I_n} \qquad 7.7$$

The variables of the normalized magnetization curve become E_g/E_n and I_f/I_n. Equation 7.7 is a straight line. The intersection of the straight line with the magnetization curve gives the required value of I_f/I_n from which I_f is obtained. Then, the required value of R_f is V_a/I_f.

$$R_{f,\text{required}} = \frac{V_{a,\text{specified}}}{I_f} \qquad 7.8$$

Example 7.2

A D.C. generator rated 10 kW, 15 hp, 260 volts, and 2800 RPM has the normalized magnetization curve shown. The normalization reference point is at E_g = 260 volts, I_f = 2.75 amps. Determine the value of R_f necessary to self-excite the generator with V_a = 240 volts at I_L = 41.667 amps if R_a = 0.48 ohms.

Use equation 7.7.

$$\frac{E_g}{260} = \frac{240 + .48(41.667)}{260} + .48\left(\frac{2.75}{260}\right)\frac{I_f}{I_n}$$

$$= 1 + (5.076 \times 10^{-3})\frac{I_f}{I_n}$$

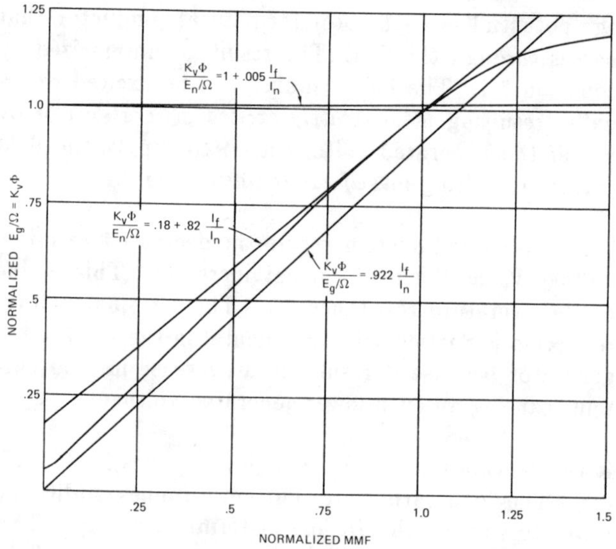

The slope of this curve is small, so it approximately intersects the magnetization curve at the point (1,1). For a more accurate determination of I_f/I_n, take a straight-line approximation to the magnetization curve at that point.

$$\frac{E_g}{260} = 0.18 + 0.82\left(\frac{I_f}{2.75}\right)$$

This intersects the straight line of the previous equation at I_f = 2.767 amps. Then,

$$R_f = \frac{240}{2.767} = 86.7 \text{ ohms}$$

Regulation of the generator is determined from equation 7.6 when the load current is zero.

$$E_g = I_f(R_a + R_f) \qquad 7.9$$

Normalizing this to the magnetization curve coordinates,

$$\frac{E_g}{E_n} = \left(\frac{I_n}{E_n}\right)(R_a + R_f)\left(\frac{I_f}{I_n}\right) \qquad 7.10$$

This is a straight line through the origin. The intersection of this line with the saturation curve will give the no-load value of E_g/E_n.

$$V_a = E_g - I_f R_a \qquad 7.11$$

$$V_a = \frac{E_g}{1 + \frac{R_a}{R_f}} \qquad 7.12$$

Example 7.3

Find the voltage regulation of the self-excited generator in example 7.2.

Substituting into equation 7.10,

$$\frac{E_g}{260} = \frac{2.75}{260}(.48 + 86.7)\frac{I_f}{2.75}$$

This intersects the magnetization curve at $E_g/260 = 1.15$ and $I_f = 2.75 \times 1.22$.

Then,

$$V_a = 299 - .48 \times 2.75 \times 1.22 = 297.4 \text{ volts}$$

So, the voltage regulation is

$$\frac{297.4 - 240}{240} \times 100 = 24.6\%$$

D. COMPOUND GENERATORS

Compound generators are a solution to the voltage regulation problem which can occur for small armature resistances. The load or armature current is used to provide extra MMF to the field to counter the effects of the armature resistance and reaction. The range of usefulness of this method is limited. The characteristics of the generator (i.e., the plot of terminal voltage versus load current) will remain nonlinear. This is illustrated in figure 7.4(c).

There are two possible arrangements of the compound winding in the circuit model. Figure 7.4(a) shows the *short-shunt connection,* where the main field winding shunts the generator. Only the load current flows through the series compounding winding. Figure 7.4(b) shows the *long-shunt connection,* where the shunt winding comes after the series connection. The compensating winding carries the armature current in this case.

Compounding is used when loads are sensitive to voltage changes. (Incandescent lights are sensitive in this manner.) Flat compounding can be used when the load is near the generator, and there is negligible loss in the feeder lines. Over-compounding is necessary when there is a long feeder line with significant resistance.

This chapter uses the short-shunt configuration for computational ease. The results for the long-shunt connection are essentially the same.

With two different currents contributing to the MMF, it is necessary to modify the horizontal coordinates of the saturation curve. The MMF for a short-shunt is

$$MMF = N_f I_f + N_S I_L \qquad 7.13$$

The normalization of the horizontal coordinate of the magnetization curve then becomes

$$\frac{I_f + \left(\frac{N_S}{N_f}\right) I_L}{I_n}$$

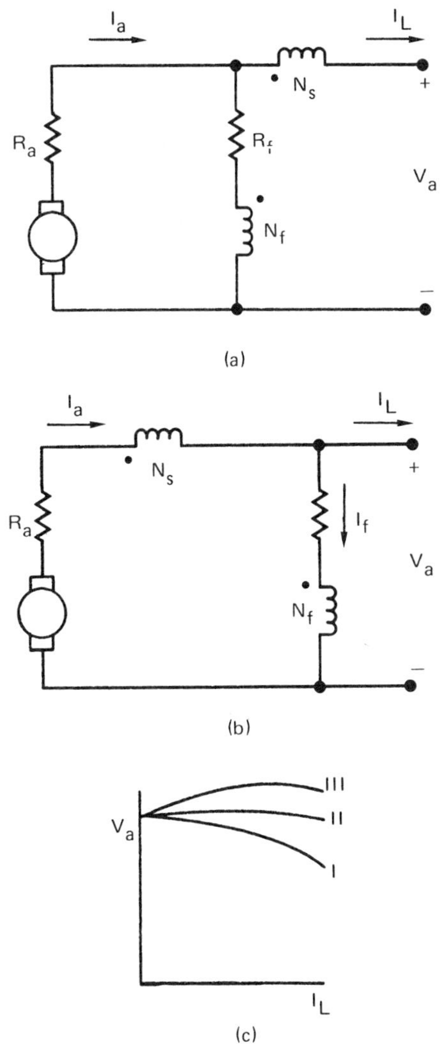

Figure 7.4 Compounded generators: (a) short-shunt, (b) long-shunt, (c) the voltage current characteristic: I uncompounded, II flat-compounded, III over-compounded

Repeating equation 7.6,

$$E_g = V_a + I_L R_a + I_f R_a \qquad 7.14$$

However, $I_f = V_a/R_f$, so

$$E_g = V_a\left(1 + \frac{R_a}{R_f}\right) + I_L R_a \qquad 7.15$$

If the V_a versus I_L line is to be flat at some value of I_L, it is necessary that $\partial V_a/\partial I_L = 0$.

Taking the partial derivative of equation 7.15 with respect to I_L and setting it to zero,

$$\frac{\partial E_g}{\partial I_L} = R_a \qquad 7.16$$

E_g is also a function of the MMF as shown in figure 7.2. Describing this relationship as an unspecified function $E_g/E_n = \mathbf{F}(MMF_n)$, the partial derivative of that function with respect to I_L is

$$\frac{\partial(MMF_n)}{\partial I_L} = \frac{\partial}{\partial I_L}\frac{\left[I_f + \left(\frac{N_S}{N_f}\right)I_L\right]}{I_n} = \frac{1}{I_n}\left(\frac{N_S}{N_f}\right) \quad 7.17$$

$$\begin{aligned}\frac{\partial \mathbf{F}(MMF_n)}{\partial I_L} &= \frac{\partial\left(\frac{E_g}{E_n}\right)}{\partial I_L}\\ &= \frac{\partial \mathbf{F}(MMF_n)}{\partial(MMF_n)} \times \frac{\partial(MMF_n)}{\partial I_L}\\ &= \frac{1}{I_n}\left(\frac{N_S}{N_f}\right)\frac{\partial \mathbf{F}(MMF_n)}{\partial(MMF_n)} \quad 7.18\end{aligned}$$

However, the partial derivitive of $\mathbf{F}(MMF_n)$ with respect to MMF_n is the slope of the magnetization curve at the operating point (call this point **OPS**). Then,

$$\frac{\partial E_g}{\partial I_L} = \frac{N_S}{N_f} \times \frac{E_n}{I_n} \times \mathbf{OPS} = R_a \quad 7.19$$

Equation 7.19 requires trial and error because the operating point and I_f are both unknown. However, if the design follows a selection of R_f which gives the proper output voltage at the specified load current, the range of the operating point will be known. The approximate value of I_f will be somewhat less than would be required for self-excited operation. The operating point should be near the no-load voltage for the separately-excited case when the field current has been adjusted to give the required voltage at the specified load current. Then, using that approximate value of **OPS** in equation 7.18, the ratio of N_S to N_f is

$$\frac{N_S}{N_f} = \frac{R_a \times \frac{I_n}{E_n}}{\mathbf{OPS}} \quad 7.20$$

From this value, it is possible to determine normalized MMF_n from equation 7.17.

$$I_f = I_n \times MMF_n - \left(\frac{N_S}{N_f}\right)I_L \quad 7.21$$

Equation 7.21 can be substituted into equation 7.7.

$$\frac{E_g}{E_n} = \frac{V_a + R_a I_L\left(1 - \frac{N_S}{N_f}\right)}{E_n} + R_a\left(\frac{I_n}{E_n}\right)MMF_n \quad 7.22$$

Equation 7.22 will determine the necessary operating point. If it is sufficiently close to the operating point chosen, I_f and R_f can be determined. If the operating point is inadequate, use the operating point found as a new starting point. Repeat the process from equation 7.19 onward and iterate until the results are adequate.

Example 7.4

The generator of example 7.2 is to be flat-compounded. The flatness in the characteristic will occur at a load current of 30 amps for a voltage $V_a = 240$ volts. Determine the N_S/N_f ratio and the value of R_f to accomplish this.

Begin by assuming that the field current will be about 2.5 amps. The $I_a R_a$ drop is, then

$$32.5 \times .48 = 15.6$$

The operating point will be

$$E_g = 240 + 15.6 = 255.6$$

The normalized value of E_g/E_n is 255.6/260=.98.

From the magnetization curve in example 7.2, the slope at the normalization point is about .83. From equation 7.19,

$$\frac{N_S}{N_f} = \frac{2.75 \times .48}{260 \times .83} = .006$$

Substituting into equation 7.22,

$$\frac{E_g}{260} = .940 + .005\, MMF_n$$

This expression gives .945 for $MMF_n \approx 1$. This is different from the original assumption, so another iteration is performed, starting with the operating point of $E_g/E_n = .945$. Here the slope is about .85, and the value of N_S/N_f is 5.97×10^{-3}, which is not very different. The recalculated operating point is .945 again, so the value of N_f/N_s is taken as .006. The normalized MMF_n at the operating point is .93 and $MMF_n \times 2.75$ is 2.56. Thus,

$$I_f = 2.56 - .006 \times 30 = 2.38\,\text{amps}$$

$$R_f = \frac{240}{2.38} = 100.8\,\text{ohms}$$

E. ADVANCED D.C. GENERATORS FUNDAMENTALS

The power flow path to generators is from the mechanical source through the shaft. The input power is used to overcome mechanical or rotational losses as well as to produce the electric power $E_g I_a$. The mechanical losses can be assumed constant unless otherwise stated. From the shaft, power goes to the load, giving up parts to the $I_a^2 R_a$ and $I_f^2 R_f$ losses. The power equation can be written as the difference between mechanical and rotational terms.

$$\begin{aligned}P_m - P_r &= T\Omega = E_g I_a\\ &= I_a^2 R_a + I_f^2 R_f + V_a I_L \quad 7.23\end{aligned}$$

The usual practice is to use unnormalized saturation curves. In such cases, the normalizing factors E_n and I_n are both 1. Also, speed is normally given in RPM with n being the variable for speed. Some machines have more than two poles. The voltage equation which incorporates these changes is

$$E_g = \left(\frac{pZ_c}{60a_p}\right)\Phi n = K_v \Phi n \qquad 7.24$$

$$K_v = \frac{pZ_c}{60a_p} \qquad 7.25$$

p is the number of poles, Z_c is the number of armature conductors, and a_p is the number of parallel paths.

The torque equation is

$$T = \left(\frac{pZ_c}{2\pi a_p}\right)\Phi I_a = K_T \Phi I_a \qquad 7.26$$

$$K_T = \frac{pZ_c}{2\pi a_p} = \frac{60K_v}{2\pi} \qquad 7.27$$

3 D.C. MOTORS

A. INTRODUCTION

The major difference between D.C. generators and D.C. motors is that generators are more efficiently constructed. The fundamental behavior is the same, and the same equations apply. The directions of the armature current and the shaft torque are reversed, corresponding to the reversed power flow.

Motor equivalent circuits are the same as for generators, although a D.C. series motor is somewhat different. Motor performance is given by speed-torque curves.

Representative speed-torque curves for the four types of motor are shown in figure 7.5.

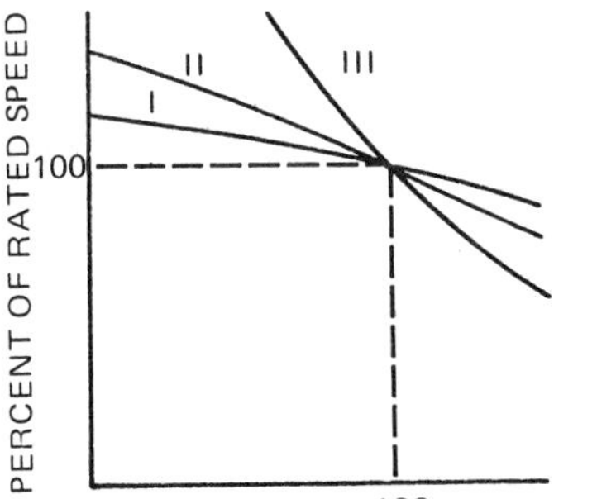

Figure 7.5 Speed-torque characteristics for D.C. motors: (I) shunt and separately excited, (II) compounded, (III) series

B. SEPARATELY-EXCITED MOTORS

Separately-excited D.C. motors are usually small servo type motors of fractional horsepower. The field is maintained at a constant current, and the speed and direction of rotation is controlled by the armature voltage (which is often a power amplifier). The application of the equations for steady state operation is straight forward.

Analysis of transient behavior requires knowing the inductances of the motor windings L_f and L_a and knowing the mass moment of inertia of the shaft system and load, J. In motor control systems, the load must be specified in terms of torque-speed relations, accounting for both friction and windage effects. This results, for the simplest case, in the load being described by the differential equation 7.28.

$$T_L = J\frac{d\Omega}{dt} + f_v\Omega \qquad 7.28$$

For constant separate-excitation, Φ and $K_v\Phi$ are constant terms. The system equations are

$$V_a = I_a R_a + K_v \Phi \Omega \qquad 7.29$$

$$T = K_T I_a = \frac{K_T \Phi}{R_a}(V_a - K_v \Phi \Omega) \qquad 7.30$$

$$T = J\frac{d\Omega}{dt} + f_v\Omega \qquad 7.31$$

By eliminating T from the last two equations,

$$\frac{d\Omega}{dt} = \alpha\Omega + \frac{K_T\Phi}{R_a J}V_a \qquad 7.32$$

$$\alpha = \frac{f_v}{J} + \frac{K_T K_v \Phi^2}{J R_a} \qquad 7.33$$

This first-order differential equation can be used in sinusoidal analysis by replacing d/dt with $j\omega$. For transient analysis, ordinary methods can be used.

C. D.C. SHUNT MOTORS

D.C. shunt motors are essentially constant speed devices. They are used where starting torque requirements are moderate. Equations 7.28 through 7.30 apply.

The equations which govern D.C. motor operation are

$$P_{in} = V_a I_a + V_a I_f = V_a I_a + \frac{V_a^2}{R_f} \qquad 7.34$$

$$V_a I_a = E_g I_a + I_a^2 R_a \qquad 7.35$$

$$E_g I_a = T\Omega = \text{rotational losses} + T_L\Omega \qquad 7.36$$

Example 7.5

A 10 hp, 1200 RPM shunt motor has a rated voltage of 230 volts and current of 42 amps. Its field current is 2 amps and its armature resistance is 0.25 ohms. The motor actually is operating at 1200 RPM with a viscous load torque of 35 ft-lb when the resistance of the armature circuit is suddenly increased to 1.0 ohm. What are the input and output powers after the motor reaches steady state? Assume that the rotational losses remain constant.

For convenience, first convert to MKS units. The initial speed is

$$\Omega = 1200 \times \frac{2\pi}{60} = 125.66 \text{ rad/sec}$$

The load torque is

$$T = 35 \times \frac{746}{550} = 47.47\ N\text{-m}$$

Thus, the load power is

$$P_L = 125.66 \times 47.47 = 5965\ W$$

The rated input power is

$$230 \times 42 = 9660\ W$$

Of this, 2 amps flow in the field circuit, accounting for 460 watts. This leaves 9200 watts for the armature circuit, along with 40 amps of armature current.

The armature resistance accounts for a 10 volt (40 amps $\times$ .25Ω) drop, leaving 220 volts for the back EMF, E_g. The armature $I_a^2R_a$ loss is 400 watts (10 volts $\times$ 40 amps) leaving an air-gap power of

$$220 \times 40 = 8800\ W$$

The power lost in the mechanical system is the difference between the air-gap power and the load power.

$$\begin{aligned} \text{mechanical loss} &= P_{\text{in}} - P_L \\ &= 8800 - (10 \times 746) = 1340\ W \end{aligned}$$

The initial power flow diagram is

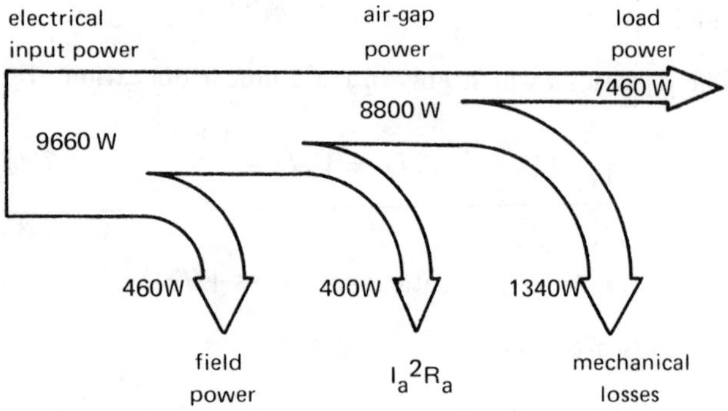

After the change of armature resistance R_a, the field power will remain at 460 watts, and the mechanical loss will remain at 1340 watts. The $I_a^2R_a$ loss and the load power will adjust to satisfy the system equations.

The air-gap power will be E_gI_a. I_a is found from the armature circuit relation.

$$\begin{aligned} 230 &= I_aR_a + E_g \\ I_a &= \frac{230 - E_g}{1} \\ E_gI_a &= 230E_g - E_g^2 \end{aligned}$$

For constant I_f, $K_v\Phi = K$, so

$$E_g = K\Omega$$

The original situation had $E_g = 220$ volts when Ω was 125.66 rad/sec. K was, then

$$K = \frac{220}{125.66} = 1.75$$

The air-gap power expressed in terms of the speed is

$$E_gI_a = 230K\Omega - K^2\Omega^2 = 402.5\Omega - 3.065\Omega^2$$

The load power can be expressed as $f_v\Omega^2$ for a viscous load. From $T = f_v\Omega$ (equation 7.31 with constant speed) and the original conditions of torque and rotational speed

$$\begin{aligned} f_v &= \frac{T}{\Omega} = \frac{47.47}{125.66} \\ &= .378\ \frac{N\text{-sec}}{\text{m-rad}} \end{aligned}$$

The load power is, then, $.378\Omega^2$. Thus, the air-gap power must balance the mechanical loss and the load power.

$$E_gI_a = 1340 + .378\Omega^2$$

Simultaneously solving the electrical and the mechanical expressions for air-gap power produces two solutions of 113.5 and 3.42 rad/sec. As the machine is slowing from 125.66 rad/sec, it will first satisfy the 113.5 solution. Then,

$$\begin{aligned} E_g &= 1.75 \times 113.5 = 198.6 \text{ volts} \\ I_a &= 230 - 198.6 = 31.4 \text{ amps} \end{aligned}$$

The total input current is

$$I_{\text{in}} = 31.4 + I_f = 31.4 + 2 = 33.4 \text{ amps}$$

The input power is

$$P_{\text{in}} = 33.4 \times 230 = 7676\ W$$

The output power is $f_v\Omega^2$, so

$$P_{\text{out}} = .378 \times (113.5)^2 = 4869\ W$$

Starting D.C. shunt motors larger than 2 hp is a problem. Current limiting is provided by *starting resistors* in series with the armature winding. The *controller* consists of a number of relays and armature series resistances. A review of relay and switch symbols is given in figure 7.6

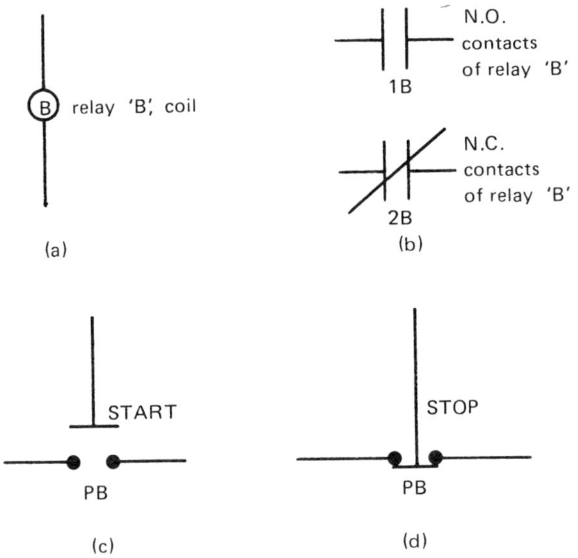

Figure 7.6 Relay and push button symbols

A controller for a large D.C. compound motor is shown in figure 7.7. The resistors R_1 and R_2 are the *current limiting resistors.* Until the start button is pressed, the contacts 1A and 2A are open; when the start button is pushed, the main relay M is energized, which closes contacts M_a. M_a maintains the energizing of coil M. Line voltage is then supplied to the armature circuit. R_1 and R_2 limit the current as the motor starts.

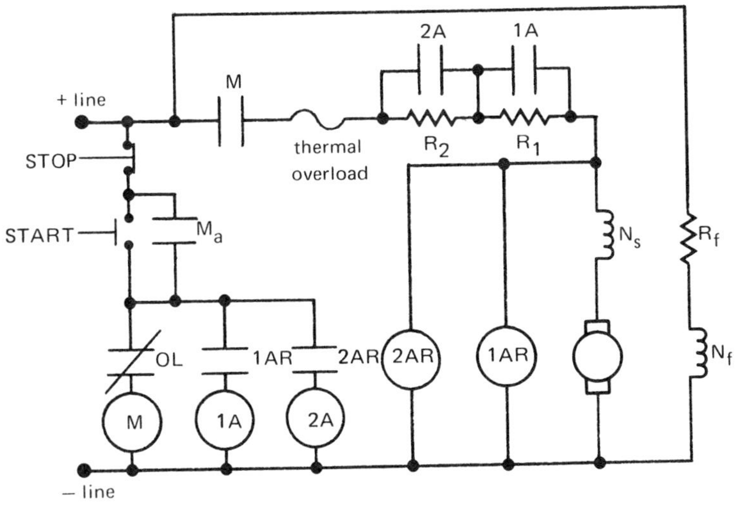

Figure 7.7 Counter-EMF magnetic controller

As the armature counter-EMF and motor speed build up to about half their rated values, *accelerating relays* 1AR and 2AR (which are *light-current relays*) remain un-energized. 1AR pulls in at the half rated voltage, which in turn energizes the coil of 1A to close contact 1A and bypass resistance R_1. This allows more current to flow in the armature circuit and further accelerates the motor.

Relay 2AR pulls in at about 80% of rated voltage and speed. This energizes relay coil 2A through contacts 2AR, which in turn closes contacts 2A and shunts R_2. Full line voltage is then applied to the armature.

The *overload contact* is a *thermal limit switch* sensing the temperature of the armature.

Example 7.6

A 250 volt D.C. motor starter consists of series resistors in the armature circuit and a corresponding set of relays. The rated armature current is 125 amps, and the armature circuit resistance is 0.15 ohms. The starting current is to be limited to 150% of rated current, and each relay coil is to pull in when the current reaches 100% of rated current. Determine the values and wattage of the resistors, as well as the pull-in voltage of each relay coil.

The first resistor is found from the line voltage applied at zero speed.

$$R_{x1} + R_a = \frac{250}{125 \times 1.5} = 1.333$$

$$R_{x1} = 1.333 - .15 = 1.1833 \text{ ohms}$$

For a current of 125 amps to flow through R_{x1}, the armature voltage must be

$$V_a = 250 - (125 \times 1.1833) = 102.1 \text{ volts}$$

This is the pull-in voltage of the first relay and is also the armature voltage at the time the first relay closes. E_g at this time is

$$E_g = 250 - (125 \times 1.33) = 83.33$$

For 150% of rated current at this value of E_g (which will occur when contact relay CR-1 closes),

$$R_{x2} + R_a = \frac{250 - 83.33}{1.5 \times 125} = .89$$

$$R_{x2} = .98 - .15 = .74$$

To have 125 amps through R_{x2},

$$E_g = 250 - (125 \times .89) = 138.75$$

$$V_a = 250 - (125 \times .74)$$

$$= 157.6 = V_{CR2}$$

V_{CR2} is the pull-in voltage of the second relay.

To obtain 150% current with this value of E_g, the resistance must be

$$R_{x3} + R_a = \frac{250 - 138.75}{1.5 \times 125} = .59$$
$$R_{x3} = .59 - .15 = .44$$

Continuing in a similar fashion,

$$E_g = 250 - (125 \times .59) = 176.25$$
$$V_a = 250 - (125 \times .44)$$
$$= 195 = V_{CR3}$$

$$R_{x4} + R_a = \frac{250 - 176.25}{1.5 \times 125} = .39$$
$$R_{x4} = .39 - .15 = .24$$
$$E_g = 250 - (125 \times .39) = 201.25$$
$$V_a = 250 - (125 \times .24)$$
$$= 220 = V_{CR4}$$

$$R_{x5} + R_a = \frac{250 - 201.25}{1.5 \times 125} = .26$$
$$R_{x5} = .26 - .15 = .11$$
$$E_g = 250 - (125 \times .26) = 217.5$$
$$V_a = 250 - (125 \times .11)$$
$$= 236 = V_{CR5}$$

$$R_{x6} + R_a = \frac{250 - 217.5}{1.5 \times 125} = .17$$
$$R_{x6} = .17 - .15 = .02$$
$$E_g = 250 - (125 \times .17) = 228.75$$
$$V_a = 250 - (125 \times .02)$$
$$= 247.5 = V_{CR6}$$

For this value of E_g, it is impossible to obtain 150% rated current, so no further external resistance is needed.

The required resistances are

$$R_6 = R_{x6} = .02 \text{ ohms}$$
$$R_5 = R_{x5} - R_{x6} = .11 - .02$$
$$= .09 \text{ ohms}$$
$$R_4 = R_{x4} - R_{x5} = .24 - .11$$
$$= .13 \text{ ohms}$$
$$R_3 = R_{x3} - R_{x4} = .44 - .24$$
$$= .20 \text{ ohms}$$
$$R_2 = R_{x2} - R_{x3} = .74 - .44$$
$$= .30 \text{ ohms}$$
$$R_1 = R_{x1} - R_{x2} = 1.18 - .74$$
$$= .44 \text{ ohms}$$

D. COMPOUND D.C. MOTORS

Compound D.C. motors can be constructed so that the MMF of the series winding either aids (*cumulative compounding*) the shunt winding MMF or opposes (*differential compounding*) the shunt winding's MMF. The shunt winding can be in parallel with both the armature and the series winding as shown in figure 7.8 (*long shunt*), or it may parallel only the armature, with the series winding being between the parallel combination and the power source (*short shunt*). Cumulative compounding is used to increase starting torque, and differential compounding is used to improve speed regulation. Referring to figure 7.8(b), the MMF for the differentially-compounded motor is

$$MMF = N_f I_f - N_S I_a \qquad 7.37$$

The speed is derived from equation 7.38.

$$\Omega = \frac{V_a - (R_S + R_a) I_a}{K_v \phi} \qquad 7.38$$

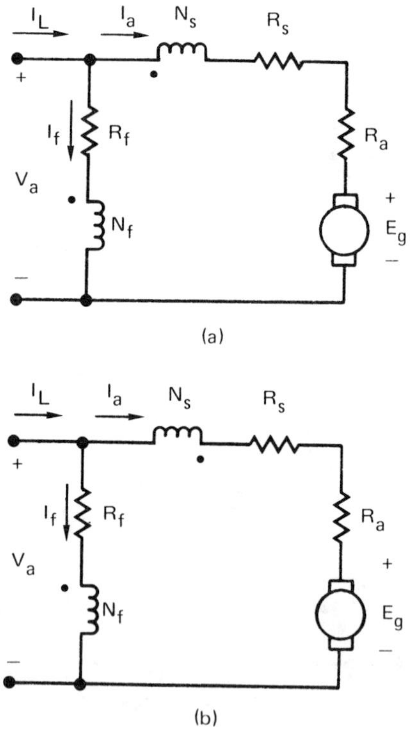

Figure 7.8 Compounded D.C.motors (long shunt): (a) cumulative compounding, (b) differential compounding

Small performance changes around an operating point, indicated by the subscript *o*, can be predicted from equation 7.39.

$$\Omega_o + \Delta\Omega = \frac{V_a - (R_a + R_S)I_{ao} + \Delta I_a)}{K_v(\Phi_o + \Delta\Phi)} \qquad 7.39$$

At constant speed, $\Delta\Omega = 0$, so it is necessary that

$$K_v \Omega_o \Delta\Phi = -(R_a + R_S)\Delta I_a \qquad 7.40$$

Equation 7.40 can be written in partial differential form as

$$\frac{\partial}{\partial I_a}[K_v \Omega_o \Phi] = -(R_a + R_S) \qquad 7.41$$

Recognizing that $E_g = K_v \Omega \Phi$, the chain rule for partial differential equations results in

$$\frac{\partial}{\partial I_a}[K_v \Omega \Phi] = \frac{\partial}{\partial MMF} E_g \times \frac{\partial}{\partial I_a} MMF \qquad 7.42$$

Equations 7.41 and 7.42 can be considered equivalent when Ω is constant. From equation 7.37,

$$\frac{\partial}{\partial I_a} MMF = -N_S \qquad 7.43$$

For an unnormalized curve where the MMF is given in terms of the field current I_f, the actual variable on the abscissa is $I_f \pm (N_S/N_f)I_a$. (The negative sign is used with differential compounding.) For such a curve,

$$\left(\frac{\partial E_g}{\partial MMF}\right)_{\text{unnormalized}} = \frac{\textbf{OPS}}{N_f} \qquad 7.44$$

For a normalized curve, with E_n and $N_f I_n$ as the operating point,

$$\left(\frac{\partial E_g}{\partial MMF}\right)_{\text{normalized}} = \frac{N_f I_n}{E_n}\left(\frac{\partial E_{gn}}{\partial MMF_n}\right) \qquad 7.45$$

$$\frac{\partial E_{gn}}{\partial MMF_n} = \textbf{OPS} \qquad 7.46$$

For constant speed around an operating point, with an unnormalized saturation curve,

$$\left(\frac{N_S}{N_f}\right) = \frac{R_a + R_S}{\textbf{OPS}} \qquad 7.47$$

With a normalized saturation curve,

$$\left(\frac{N_S}{N_f}\right) = \frac{(R_a + R_S)\frac{I_n}{E_n}}{\textbf{OPS}} \qquad 7.48$$

This is the same as equation 7.20 with the exception that R_S was previously ignored.

The only difference in the design between a generator and a motor is that the generator is short-shunted while the motor is long-shunted. In the motor, a *flat speed-torque* characteristic is obtained for differential compounding at the operating point. With cumulative compounding, a higher starting torque is obtained with a resulting degradation of speed regulation. Therefore, differential compounding will result in lower starting torque.

E. D.C. SERIES MOTORS

D.C. series motors (*traction motors*) are used in variable speed applications where the starting torque is of primary importance. Such motors always are connected directly to the load shaft. The series motor has only a series field winding, which carries the armature current. Thus, the field is approximately proportional to the armature current since $MMF = N_S I_a$. The basic relationships for this motor are

$$V_a = E_g + I_a R_a \qquad 7.49$$

$$E_g = K_v \Phi \Omega \qquad 7.50$$

$$MMF = N_S I_a \qquad 7.51$$

$$T = K_T \Phi I_a \qquad 7.52$$

In MKS units $K_T = K_v$. (When n in RPM is used rather than Ω in radius/sec, then $K_T = 60K_v/2\pi$. Torque is still in newton-meters, however.)

A reasonable approach is to take the slope of the magnetization curve through the origin and the maximum current point and set

$$L_a = \frac{K_v \Phi_{\text{max}}}{I_{a,\text{max}}} \qquad 7.53$$

Then,

$$K_v \Phi = L_a I_a \qquad 7.54$$

When working entirely in MKS units, the equations are

$$E_g = L_a I_a \Omega \qquad 7.55$$

$$T = L_a I_a^2 \qquad 7.56$$

$$V_a = (L_a \Omega + R_a) I_a \qquad 7.57$$

$$\Omega = \frac{V_a}{L_a I_a} - \frac{R_a}{L_a} \qquad 7.58$$

The rated power output is

$$P_R = T_R \Omega_R \qquad 7.59$$

P_R will be given for particular Ω_R and $I_{a,R}$. From these values, the armature inductance is

$$L_a = \frac{P_R}{\Omega_R I_{a,R}^2} \qquad 7.60$$

The starting torque can be estimated from equations 7.56 through 7.58.

Example 7.7

A D.C. series motor is rated at 900 volts, 200 amps, and 225 hp at a speed of 1500 RPM. The armature circuit resistance, including the resistance of the series winding, is .001 ohms. If the starting current is limited to six times the rated current, determine the starting torque in foot-pounds.

The torque at 200 amps is

$$T_{200} = \frac{225\,\text{hp} \times 746\,\text{W/hp}}{1500\,\text{rpm} \times \frac{2\pi}{60}\frac{\text{rad/sec}}{\text{rpm}}} \times \frac{550\ \text{ft-lb}}{746\ N\text{-m}}$$

$$= \frac{16,500}{1500\pi} \times 225 = 789\ \text{ft-lb}$$

However,

$$T = KI_a^2$$

So,

$$\frac{T_2}{T_1} = \frac{I_2^2}{I_1^2}$$

Since

$$I_1 = 6I_2$$

$$T_2 = (6)^2 \times 789 = 28,404\ \text{ft-lb}$$

4 INDUCTION MOTORS

A. INTRODUCTION

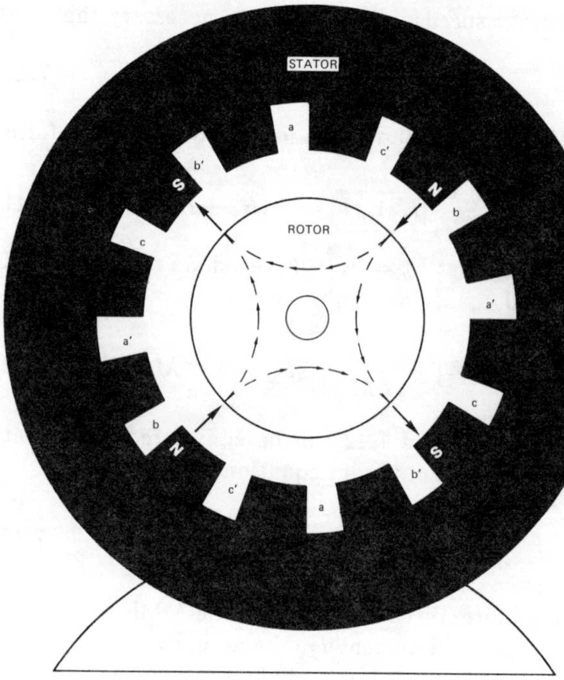

Figure 7.9 A four-pole, three-phase winding

The three-phase induction motor consists of a set of three field windings on the stator, producing a rotating field. The rotating field turns through 360 electrical degrees in each cycle for the applied voltage. The number of field poles is determined by the layout of stator windings. For one complete three-phase winding set on a stator, there are two poles produced by the stator currents. With two complete sets of windings, as indicated in figure 7.9, there are four poles generated by the stator currents. For purposes of systems analysis, the stator presents a balanced load to the line and is, therefore, considered to be wye-connected.

In figure 7.9, phase sequence *acb* results in a field rotating in the clockwise direction. Phase sequence *abc* produces in a counterclockwise field rotation. (In actual motors, the phase windings are distributed over a whole pole-span, but they are shown concentrated in figure 7.9 to clarify the concept.)

As the poles move through a complete set of windings during an electrical cycle, the mechanical speed of rotation of the stator field is the synchronous speed, Ω_S, which is the electrical frequency ω divided by the number of complete sets of windings $p/2$.

$$\Omega_S = \frac{2\omega}{p} \qquad 7.61$$

The rotor has windings with the same distribution as the stator. They may be brought out via slip rings and brushes if speed control is necessary, or internally shorted in a squirrel-cage configuration. The source of excitation for the rotor currents is the rotating stator field, which induces rotor EMF by the relative motion between the stator field and the rotor. The EMF induced in a conductor by a constant amplitude, sinusoidally-distributed flux moving with respect to the conductor at a frequency Ω_c is

$$\text{RMS EMF per conductor} = \frac{\Omega_c K_w \Phi}{\sqrt{2}} \qquad 7.62$$

Φ in equation 7.62 is the *flux per pole* from the rotating field, and K_w is the *winding factor* for the particular conductor's winding. K_w depends on the spatial distribution of windings. It is specific to particular machines but it varies from .65 to .86. Thus, for a stator field with N_S conductors per phase and with stationary windings so that the relative speed is the synchronous speed of the rotating field ($\Omega_{cs} = \omega$), the RMS EMF per phase of the stator is

$$E_S = \frac{N_S \omega K_{ws} \Phi_p}{\sqrt{2}} = \sqrt{2}\pi f N_S K_{wS} \Phi_p$$
$$= 4.44 f N_S K_{wS} \Phi_p \qquad 7.63$$

For the rotor, the relative speed of the field is

$$\omega_r = \frac{p}{2}(\Omega_S - \Omega_r) = \omega - \frac{p\Omega_r}{2} \qquad 7.64$$

Slip is defined as

$$s = \frac{\Omega_S - \Omega_r}{\Omega_S} \qquad 7.65$$

Equation 7.65 can be substituted into equation 7.64.

$$\omega_r = \frac{p}{2}\Omega_S \times s = s\omega \qquad 7.66$$

Then, the RMS EMF induced in a rotor phase with N_r conductors is

$$E_r = 4.44 s f N_r K_{wr} \Phi_p \tag{7.67}$$

The frequency of the rotor current is sf. The RMS rotor EMF can be rewritten in terms of the stator EMF.

$$E_r = \left[\frac{sN_r K_{wr}}{N_S K_{wS}}\right] E_S \tag{7.68}$$

The rotor circuit has an impedance of $r + js\omega L_r$, so the resulting per-phase RMS voltage is

$$E_r = s\left[\frac{N_r K_{wr}}{N_S K_{wS}}\right] E_S = rI_r + js\omega L_r I_r \tag{7.69}$$

Dividing equation 7.69 by the slip s and defining a, the RMS rotor current is

$$I_r = \frac{\frac{E_S}{a}}{\frac{r}{s} + jx} \tag{7.70}$$

$$a = \frac{N_S K_{wS}}{N_r K_{wr}} \tag{7.71}$$

$x = \omega L_r$ is the reactance of the rotor at the stator frequency. Thus, a circuit represented by equation 7.71 is equivalent to a transformer secondary with a turns ratio of a.

The power loss per phase in the rotor circuit is

$$P_r = I_r^2 r_r \tag{7.72}$$

From an analysis of the stator circuit, the power loss per phase appears to be $I_r^2 r_r / s$. This is greater than equation 7.72. The difference is the gap power, P_g.

$$P_g \Big|_{\text{phase}} = I_r^2 r_r \left[\frac{1}{s} - 1\right] = I_r^2 r_r \left[\frac{1-s}{s}\right] \tag{7.73}$$

For q phases, the gap power transferred to the shaft is

$$\begin{aligned} P_g &= q I_a^2 r_r \left[\frac{1-s}{s}\right] \\ &= T\Omega_r = T\Omega_S (1-s) \end{aligned} \tag{7.74}$$

The resistance in ohms associated with the mechanical load is R_m.

$$R_m = r_r \left[\frac{1-s}{s}\right] \tag{7.75}$$

An induction motor can be represented on a per-phase basis as a transformer with the equivalent circuit shown in figure 7.10. (The per-phase basis assumes a wye connection unless otherwise stated.) The power flow through the circuit must be multiplied by q, the number of phases, in order to obtain the total power.

An analysis of figure 7.10(a) shows that the absolute value of the current i'_r is

$$|i'_r| = \sqrt{\frac{|V_{an}|^2}{\left(r_S + a^2 \frac{r_r}{s}\right)^2 + x_e^2}} \tag{7.76}$$

From equation 7.74, the torque per phase is

$$\frac{T}{q} = \frac{|V_{an}|^2}{\Omega_S} \left[\frac{\frac{a^2 r_r}{s}}{\left(r_S + \frac{a^2 r_r}{s}\right)^2 + x_e^2}\right] \tag{7.77}$$

The starting torque is obtained when $s = 1$.

$$T_{\text{starting}} = \frac{q|V_{an}|^2}{\Omega_S} \left[\frac{a^2 r_r}{(r_S + a^2 r_r)^2 + x_e^2}\right] \tag{7.78}$$

The slip at which maximum torque occurs can be found by differentiating equation 7.77 with respect to s and setting the result to zero.

$$s_{\text{maximum torque}} = \frac{a^2 r_r}{\sqrt{r_S^2 + x_e^2}} \tag{7.79}$$

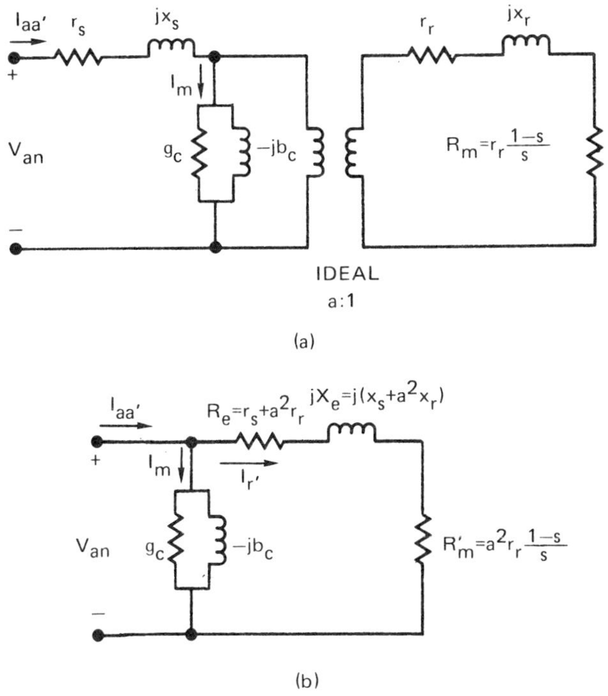

Figure 7.10 Phase models of an induction motor: (a) physical model, (b) practical approximate model

Substituting equation 7.79 into equation 7.77 gives the maximum torque.

$$T_{\text{max}} = \frac{q|V_{an}|^2}{\Omega_S} \left[\frac{1}{r_S + \sqrt{r_S^2 + x_e^2}}\right] \tag{7.80}$$

An external three-phase resistance can be connected to the rotor to adjust the point of maximum torque. It will also adjust the running speed for a given load and the starting torque. This is illustrated in figure 7.12 for shorted rotor windings and for an external resistance R. The slope of the torque-speed curve where $\Omega_r = \Omega_S$ $(s = 0)$ is

$$\frac{dT}{ds}\Big|_{s=0} = \frac{-q|V_{an}|^2}{\Omega_S}\left[\frac{1}{a^2(r_r + R)}\right] \qquad 7.81$$

Power can be evaluated with the aid of figure 7.10(b). The input power per phase is

$$P_{\text{in, phase}} = |V_{an}|^2 g_c + |I'_r|^2\left(r_S + \frac{a^2 r_r}{s}\right) \qquad 7.82$$

Of the power in equation 7.82, the output power transferred to the shaft is

$$P_{\text{shaft, phase}} = \frac{P_L + P_{\text{rotational}}}{q}$$

$$= |I'_r|^2 a^2 r_r\left(\frac{1-s}{s}\right) \qquad 7.83$$

Part of the shaft power goes into rotational losses, and part goes to the load.

$$P_{\text{shaft}} = P_{\text{rotational losses}} + P_{\text{output}} \qquad 7.84$$

Example 7.8

A 10 hp (nameplate), 6 pole, 60 hz, three-phase induction motor delivers 9.9 hp with an input of 9200 watts. The core loss is 450 watts, the stator copper loss is 650 watts, and the rotational loss is 150 watts. What is the motor speed?

From equation 7.82 for three phases,

$$P_{\text{in}} - P_c - P_S = 9200 - 450 - 650 = 8100 \text{ watts}$$

From equation 7.83 for three phases,

$$P_L + P_{\text{rotational}} = 9.9 \times 746 + 150 = 7535.4 \text{ watts}$$

Then,

$$1 - s = \frac{7535.4}{8100} = 0.93$$

The synchronous speed is

$$n_S = \frac{60f}{\frac{p}{2}} = \frac{(60)(60)}{\frac{6}{2}}$$

$$= 1200 \text{ rpm}$$

The motor speed is derived from equation 7.65.

$$n_r = n_S(1 - s) = 1116 \text{RPM}$$

B. INDUCTION MOTOR TESTS

Induction motor tests are comparable to open and short circuit tests on transformers. A no-load test will determine the core characteristics and the rotational losses. When unloaded, slip is usually a fraction of a percent, and the *transformed rotor resistance* $a^2 r_r/s$ becomes very large. Figure 7.10(b) would then be dominated by the core conductance and susceptance.

$$g_c = \frac{P_{\text{no-load}}}{|V_{an}|^2 q} \qquad 7.85$$

$$b_c = \frac{Q_{\text{no-load}}}{|V_{an}|^2 q} \qquad 7.86$$

The no-load test does not distinguish between core losses and rotational losses. It is common to combine those losses into the g_c term or to let g_c be zero and combine the losses with the rotational losses. There will be a slight difference in results depending on which assumption is made.

The equivalent of a transformer short circuit test is the *blocked rotor test*. The rotor is locked to prevent rotation. Under this condition, $s = 1$. The effective input impedance is

$$Z = (r_S + a^2 r_r) + j(x_S + a^2 x_r) \qquad 7.87$$

The primary use of the blocked rotor test is to determine X_e.

$$R_e = r_S + a^2 r_r = \frac{P_{b.r.}}{I_{an}^2 q} \qquad 7.88$$

$$X_e = x_S + a^2 x_r = \frac{Q_{b.r.}}{I_{an}^2 q} \qquad 7.89$$

To obtain $a^2 r_r$ and r_s from R_e, a third test must be made. The third test is to measure input and output power and the slip. By combining with the results of the no-load and blocked rotor tests, it is possible to calculate $a^2 r_r$ from equation 7.83.

Example 7.9

The results of a no-load test on a three-phase induction motor are: $V_{\text{line-to-line}} = 440$ RMS volts, line current= 4 amps, input power= 900 watts. Determine g_c and b_c.

$$V_{an}^2 = \frac{(440)^2}{3}$$

$$= \left(\frac{440}{\sqrt{3}}\right)^2 = 64{,}533$$

From equation 7.85,

$$g_c = \frac{\frac{900}{3}}{64{,}533} = .00465 \text{ mhos}$$

The apparent power per phase is

$$\frac{S}{q} = \frac{440}{\sqrt{3}} \times 4 = 1016 \text{ VA per phase}$$

$$\frac{P}{q} = 300 \text{ watts per phase}$$

$$\frac{Q}{q} = \sqrt{1016^2 - 300^2} = 971 \text{ VAR per phase}$$

From equation 7.86, on a per-phase basis (i.e., $q = 1$),

$$b_c = \frac{Q}{|V_{an}|^2} = -.0150 \text{ mhos reactive}$$

Example 7.10

The motor of example 7.9 is subjected to a blocked rotor test with the following results: $P_{\text{in}} = 100$ watts, $I_{\text{line}} = 12$ amps, $V_{\text{line-to-line}} = 20$ volts. Determine R_e and X_e.

From equation 7.88,

$$R_e = \frac{100}{(12)^2(3)} = .231 \text{ ohms}$$

$$\frac{S}{q} = \frac{V_{\text{line-to-line}}}{\sqrt{3}} \times I_{\text{line}}$$

$$= \left(\frac{20}{\sqrt{3}}\right) \times 12 = 138 \text{ VA per phase}$$

$$\frac{Q}{q} = \sqrt{138^2 - 33.3^2} = 134.5 \text{ VAR}$$

From equation 7.89,

$$X_e = \frac{134.5}{(12)^2} = .934 \text{ ohms reactive}$$

Example 7.11

The motor of examples 7.9 and 7.10 has a slip of .05 with an output power of 10 hp and an input power of 9.3 kW. Determine r_s and a^2r_r.

The output power is given by equation 7.83 since the rotational losses in the no-load test were lumped into g_c. Then, from equation 7.74 or 7.83,

$$I_r'^2 a^2 r_r = \frac{.05}{.95}\left(\frac{10 \times 746}{3}\right) = 130.88$$

$$I_r'^2\left(\frac{a^2 r_r}{s}\right) = \frac{130.88}{.05} = 2617.5 \text{ watts}$$

The input power per phase is 3.1 kW. From example 7.9, $g_c V_{an}^2 = 300$ watts. Then,

$$P_{\text{in}} - g_c V_{an}^2 - I_r'^2 r_s = I_r'^2\left(\frac{a^2 r_r}{s}\right)$$

$$I_r'^2 r_s = 3100 - 300 - 2617.5 = 182.5$$

Then,

$$\frac{I_r'^2 a^2 r_r}{I_r'^2 r_s} = \frac{a^2 r_r}{r_s} = \frac{130.88}{182.5}$$

$$= .717$$

From example 7.10, $R_e = .231$

$$R_e = r_s + a^2 r_r = r_s(1 + .717)$$

$$= .231$$

So,

$$r_s = .135 \text{ ohms}$$

$$a^2 r_r = .135 \times .717 = .096 \text{ ohms}$$

C. STARTING INDUCTION MOTORS

Large starting currents can cause serious drops on branch and feeder circuits. These drops can interfere with other equipment, or the currents drawn may trip over-current protection devices. Some loads cannot be started abruptly, so the starting torque may have to be limited as well.

One control method starts with the windings in a wye configuration and switches them to a delta configuration after starting.

Example 7.12

Assume the motor of examples 7.9 through 7.11 is delta connected for all the previous measurements. Determine the starting torque for delta and wye connections if the synchronous speed is 900 rpm.

The actual impedance attached from line-to-line is three times the phase impedance. At starting, the admittance for the line-to-neutral phase was found to be

$$g_c = .00465$$

$$b_c = .0150$$

$$R_e = .231$$

$$X_e = .934$$

$$Y_{\text{in}} = g_c - jb_c + \frac{1}{R_e + jX_e} = .254 - j1.024$$

Therefore,

$$Z_{l-n} = Z_{\text{in}} = \frac{1}{Y_{\text{in}}}$$

$$= .228 + j.920$$

$$Z_{l-l} = 3\, Z_{l-n} = .685 + j2.760$$

If this is the actual line-to-line phase impedance, and if it is initially connected in a wye configuration, the currents flowing will be one-third those calculated previously. The initial current flowing would then be

$$I_{\text{line}} = \frac{254}{|Z_{l-l}|} = 88 \text{ amps}$$

With a delta connection, the line current would be 264 amps. The starting torque given by equation 7.78 is only one-third of the delta- connected torque.

From equation 7.78 with a delta connection,

$$T_{\text{starting}} = \frac{3(440)^2}{\left(\frac{2\pi}{60}\right)900} \times \frac{3a^2r_r}{(3R_e)^2 + (3X_e)^2}$$

$$= \frac{3(400)^2}{30\pi} \times \frac{a^2r_r}{3[R_e^2 + X_e^2]}$$

With a wye connection, $V_{l-n} = 254$ volts.

$$T_{\text{starting}} = \frac{3(254)^2}{30\pi} \times \frac{3a^2r_r}{(3R_e)^2 + (3X_e)^2}$$

$$= \frac{V^2_{\text{phase}}}{30\pi} \times \frac{a^2r_r}{R_e^2 + X_e^2}$$

Since it is known that $a^2r_r = .096$, $R_e = .231$, and $X_e = .934$,

$$\frac{a^2r_r}{R_e^2 + X_e^2} = .1037$$

$$T_{\text{starting}} = \frac{V^2_{\text{phase}}}{30\pi} \times .1037 = 0.0011V^2_{\text{phase}}$$

For $V_{\text{phase}} = V_{l-l} = 440$, $T_{\text{starting}} = 213$ N-m.

For $V_{\text{phase}} = V_{l-n} = 440/\sqrt{3}$, $T_{\text{starting}} = 71$ N-m.

Another type of controller is the *auto-transformer* with voltage taps that are switched from a low voltage at starting to higher values as the speed increases. There are three auto-transformers in wye connections. All are switched together. They effectively control the line-to-neutral voltage. The lowest voltage tap is for starting, and one or more taps may involve intermediate voltage levels.

Example 7.13

The maximum line current for the motor of examples 7.9 through 7.11 has been set at 180 amps. The motor is to be controlled by a tapped three-phase auto-transformer (wye connection). On starting, the initial current is to be limited to 180 amps when the voltage of tap #1 is applied. When the current drops to 150 amps the voltage is to be switched to tap #2 to supply sufficient voltage to draw 180 amps. Subsequent taps will be switched in when the current drops to 140, 130, 120 amps, etc. At each switching, the initial current is limited to 180 amps. Switching will continue until the voltage necessary to draw 180 amps is greater than the line-to-neutral voltage of 250 volts.

Specify the tap voltage and the switching speed in RPM for a synchronous speed of 900 RPM. Ignore g_c and b_c in the calculations.

The voltage necessary is found from figure 7.10(b).

$$V_{\text{tap}} = I_{\text{max}}\sqrt{\left(\frac{a^2r_r}{s} + r_s\right)^2 + X_e^2}$$

$$= 180\sqrt{\left(\frac{.096}{s} + .135\right)^2 + .934^2}$$

The slip at which the next tap-change should occur is found from the same equation.

$$s_{\text{switching}} = \frac{a^2r_r}{-r_s + \sqrt{\left(\frac{V_{\text{tap}}}{I_{\text{min}}}\right)^2 - X_e^2}}$$

$$= \frac{.096}{-.135 + \sqrt{\left(\frac{V_{\text{tap}}}{I_{\text{min}}}\right)^2 - .934^2}}$$

The switching speed is

$$n_{\text{switching}} = n_s(1 - s_{\text{switching}})$$

Initially, the machine is stopped, so $s = 1$.

$$V_1 = 180\sqrt{(.096 + .135)^2 + .934^2} = 173 \text{ volts}$$

$$s_1 = \frac{.096}{-.135 + \sqrt{\left(\frac{173}{150}\right)^2 - .934^2}} = 0.177$$

$$n_1 = 900(1 - .177) = 740 \text{ rpm}$$

$$V_2 = 180\sqrt{\left(\frac{.096}{.177} + .135\right)^2 + .934^2} = 208 \text{ volts}$$

$$s_2 = \frac{.096}{-.135 + \sqrt{\left(\frac{208}{140}\right)^2 - .934^2}} = .094$$

$$n_2 = 900(1 - .094) = 815 \text{ RPM}$$

$$V_3 = 180\sqrt{\left(\frac{.096}{.094} + .135\right)^2 + .934^2} = 267 \text{ volts}$$

This excedes the line-to-neutral voltage, so the final voltage is 254 volts.

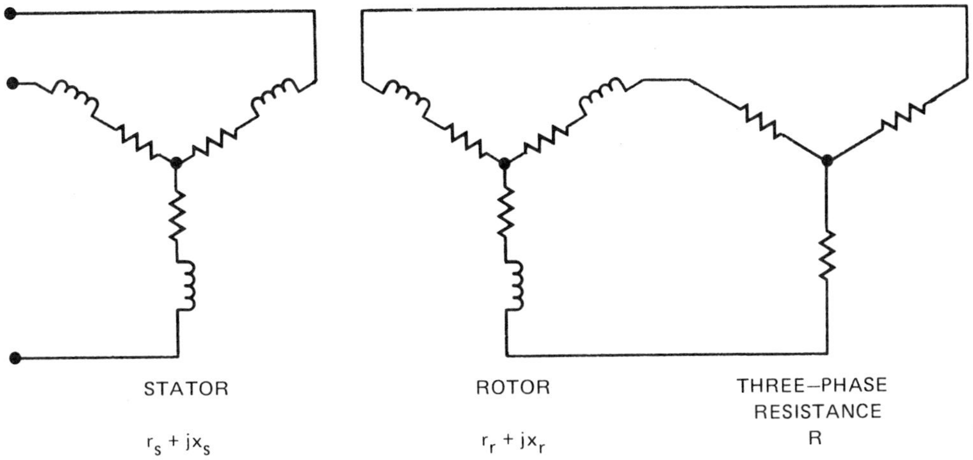

Figure 7.11 External rotor resistance connections

A third method of current control uses series resistors, as were used to start D.C. motors. In such cases, the series resistance is added to R_e. Calculations are similar to those of example 7.13. The procedure for torque limitation is not covered here.

Wound rotor induction motors can be controlled by an external three-phase rotor resistance. This resistance can be controlled by relays to limit starting current torque as well as running speed. In this case, a^2r_r is replaced by $a^2(r_r + R_r)$ where R_r is the external rotor resistance. Figures 7.11 and 7.12 illustrate this method of control.

5 SYNCHRONOUS MACHINES

A. INTRODUCTION

Three-phase synchronous machines have stator windings constructed similarly to three-phase induction machines. However, while the stator is the field for an induction machine, the stator of a synchronous machine is known as an *armature*. The synchronous machine field is supplied by a D.C. current in the rotor. The rotor will consist of *salient poles* for machines operating at 1800 rpm or less. A *cylindrical rotor* is used for higher speeds.

The number of poles in the machine is related to the ratio of electrical frequency to mechanical frequency.

$$p = 2\left[\frac{\text{electrical frequency}}{\text{mechanical frequency}}\right] = \frac{2\omega}{\Omega}$$
$$= \frac{2 \times 60f}{n} \qquad 7.90$$

In synchronous machines, the slip is zero, so rotor currents are not induced by the stator currents. Instead, the *rotor flux* is produced by a D.C. current, and the flux rotates with the rotor at the synchronous speed.

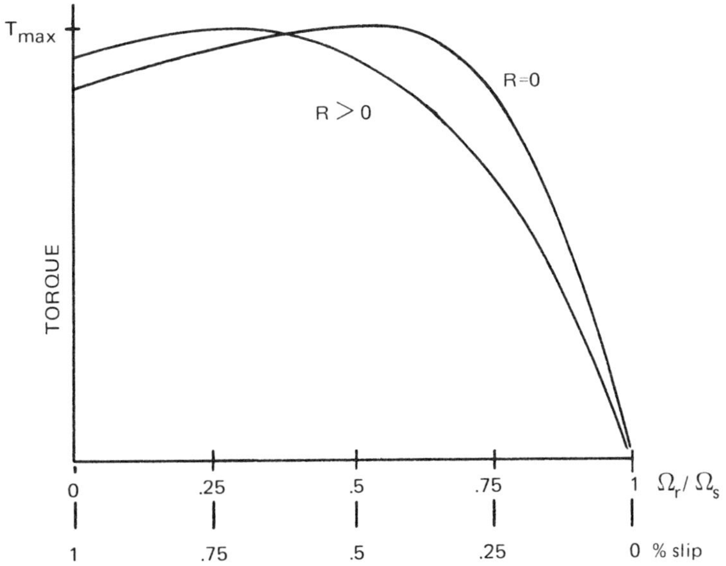

Figure 7.12 Performance curves for two external rotor resistances

The rotor flux is approximated by equation 7.91.

$$\Phi_r = \left[\frac{\mu_o N_r A_r}{gp}\right] I_r \qquad 7.91$$

N_r is the number of rotor winding turns per pole of the rotor, g is the *gap length* (the separation of the rotor from the stator), p is the number of poles, and I_r is the rotor current.

It is assumed that the flux produced by the rotor current is sinusoidally distributed around the rotor or stator. With respect to the rotor, the flux density will complete its positive portion (i.e., the region where the flux density is radially outward) over the span of one pole. Using the electrical angle, the flux density can be written as a sinusoid. (The point of maximum outward flux density is assumed to occur at $\theta_r = 0$.)

$$B(\theta_r) = B_{\max} \cos\theta_r \qquad 7.92$$

If the length of the rotor is l_r and the radius is R_r, the flux from one pole is found by integrating equation 7.92 over the electrical angle $-\pi/2$ to $+\pi/2$.

$$\Phi_r = 2 l_r R_r B_{\max} = B_{\max} A_r \qquad 7.93$$

A_r in equation 7.93 is the area of the rotor seen perpendicular to its axis.

$$A_r = 2 l_r R_r \qquad 7.94$$

The rotor turns at the synchronous speed ω_S, so the rotor flux density seen from the stator will appear as

$$B_r(\theta_S, t) = B_{\max} \cos(\omega_S t - \theta_S) \qquad 7.95$$

Equation 7.95 assumes counterclockwise rotation of the rotor. θ_S is the angle of the stator at which the flux density is observed, measured counterclockwise from where the rotor flux density is maximum at $t = 0$ (i.e., *the stator reference point*).

The EMF induced in a single stator turn located θ_S counterclockwise from the stator reference can be determined.

$$EMF_{\text{turn}} = 2 B_{\max} A_r \omega_S \sin(\omega_S t - \theta_S) \qquad 7.96$$

A three-phase stator winding has the phases separated by 120 *electrical degrees* (240/p *mechanical degrees*), and the windings are distributed symmetrically about their respective centers. Thus, the voltages induced will not be simply the number of turns per phase multiplied by equation 7.96 but will be somewhat less because the individual components of each phase voltage are slightly out-of-phase. For a perfectly sinusoidal distribution of the phase windings, the effective number of turns is the actual number times $2/\pi$, or about $.64N_S$, where N_S is the number of turns per phase.

Then, the RMS EMF induced in the stator winding due to the rotor current is

$$EMF_{rms} = \sqrt{2} K_\Phi I_r \sin(\omega_S t - \theta_S) \qquad 7.97$$

θ_S takes on values of 0°, 120°, and 240° for an ABC sequence for phases A, B, and C, respectively, and K_Φ is an ideal constant given by equation 7.98.

$$K_\Phi = \frac{\sqrt{2} N_r N_S \mu_o A_r \omega_S}{\pi g p} \qquad 7.98$$

Then, the RMS value of the induced voltage per phase is

$$EMF_{rms} = E_g = K_\Phi I_r \qquad 7.99$$

Equation 7.99 shows that the generated voltage is proportional to the rotor current. This is valid because the air-gap between the rotor and stator is fairly large, minimizing the effects of nonlinear rotor and stator reluctances. Equation 7.99 cannot be used when I_r is very large, and magnetization data must then be considered.

When a balanced three-phase current flows in the stator windings, a rotating component of flux is induced, as in the induction machine. In contrast with the induction machine, however, there is no induced voltage in the rotor windings because the rotor is traveling at synchronous speed in unison with the stator field. The stator currents induce an EMF in the stator winding because of the self-inductance of the windings. This EMF corresponds to the rotating stator field, which can be thought of as being in-phase with the stator current. The stator EMF will lead the stator current by approximately 90°, because the inductive reactance is much larger than the winding resistance.

In addition, there is a component of stator generated EMF which is due to leakage inductance, not linking the other stator windings via the air-gap. Thus, the total self-generated EMF of the stator winding can be accounted for by an equivalent stator reactance called the *synchronous reactance* X_S. Winding resistance is typically ignored except when power flow and efficiency are being evaluated.

The RMS terminal voltage of the stator winding is, therefore, made up of two phasor components. One is due to the stator current, and the other is due to the rotor current. (The subscript a in equation 7.100 designates *armature*, as the stator is considered to be the armature in a synchronous machine.)

$$\mathbf{V}_a = jX_S \mathbf{I}_a + \mathbf{E}_g \qquad 7.100$$

Equation 7.100 is a phasor equation, and the power into the machine is positive whenever $\mathbf{I}_a$ is within 90° of $\mathbf{V}_a$. In such a case, the machine will be acting as a motor. If, on the other hand, the current is more than 90° out of phase (positive or negative), the power flow is out of the terminals, and the machine will be acting as a generator.

Equation 7.100 represents a machine acting as a motor with the per-phase circuit shown in figure 7.13. The generated voltage $\mathbf{E}_g$ lags behind the applied voltage $\mathbf{V}_a$ according to the phasor diagram figure 7.13(b). Taking the applied phase voltage as the phasor reference, the generated RMS voltage due to the rotor current is

$$E_g = |E_g| \cos\delta - |E_g| \sin\delta \qquad 7.101$$

δ is the angle by which $\mathbf{E}_g$ lags $\mathbf{V}_a$. Then, equation 7.100 can be rewritten to give the RMS stator current.

$$I_a = \frac{|E_g| \sin\delta}{X_S} - j\frac{V_a - |E_g| \cos\delta}{X_S} \qquad 7.102$$

The input power to the motor is the terminal voltage $|V_a|$ multiplied by the real part of the terminal current (when the terminal voltage is taken as reference). The apparent power per phase is the terminal voltage multiplied by the complex conjugate of the RMS terminal current. Therefore, for a **motor**,

$$S = P + jQ = q[V_a I_a^*]$$
$$= q\left[\frac{V_a |E_g| \sin\delta}{X_S} + j\frac{V_a^2 - V_a |E_g| \cos\delta}{X_S}\right] \qquad 7.103$$

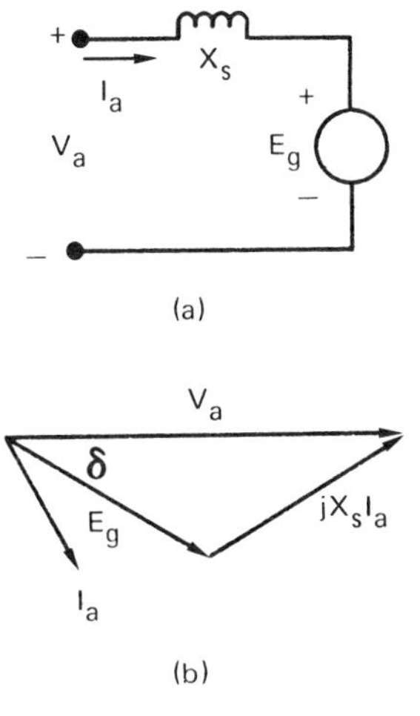

Figure 7.13 (a) Per-phase equivalent circuit for a synchronous motor, (b) phasor diagram

Assuming all circuit losses are negligible, all electrical power is converted to *air-gap mechanical power*, which is the product of the *air-gap torque* and the mechanical angular velocity: $T\,\Omega$. Then, **motor** torque can be written as equation 7.104 where Ω_S is the mechanical synchronous speed.

$$T_{\text{air-gap}} = q\left[\frac{V_a |E_g| \sin\delta}{X_S \Omega_S}\right] \qquad 7.104$$

For a synchronous **generator**, the generated voltage causes a current to flow out of the terminals. So, equation 7.100 is modified to account for the reversal of the current direction. The RMS EMF is

$$\mathbf{E}_g = j\mathbf{I}_a X_S + \mathbf{V}_a \qquad 7.105$$

Equation 7.105 represents the per-phase circuit indicated in figure 7.14. Taking the terminal voltage V_a as the reference and letting δ be the angle by which $\mathbf{E}_g$ leads $\mathbf{V}_a$, the EMF voltage for a **generator** is

$$E_g = |E_g| \cos\delta + j|E_g| \sin\delta \qquad 7.106$$

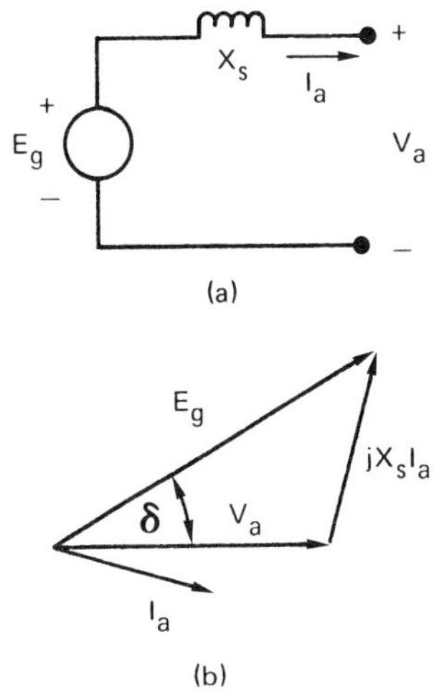

Figure 7.14 (a) Per-phase equivalent circuit for a synchronous generator, (b) phasor diagram

The apparent power for a **generator** can be calculated from equations 7.105 and 7.106. The air-gap torque will be the product of the generated voltage and the complex conjugate of the terminal current $E_g I_a^*$.

$$\frac{S}{q} = \frac{|E_g| V_a \sin\delta}{X_S} + j\frac{|E_g|^2 - |E_g| V_a \cos\delta}{X_S} \qquad 7.107$$

The air-gap torque is the ratio of power to angular velocity, so equation 7.104 can be used for the generator.

The ideal torque-power angle relationship is a sinusoidal curve as shown in figure 7.15. However, there is a second harmonic of the angle in salient pole machines. This harmonic produces higher torque and a non-sinusoidal relationship. Only pure sinusoidal performance is covered in this chapter.

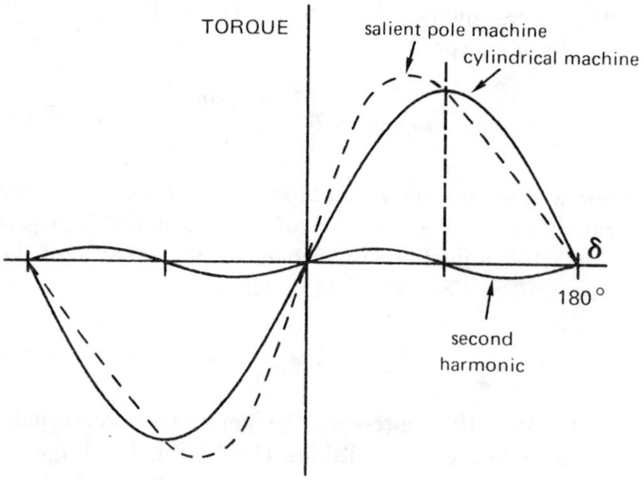

Figure 7.15 Torque-power angle relationship

B. SYNCHRONOUS MOTORS

Synchronous motors generally are started under no-load by means of a squirrel-cage rotor winding. This brings the motor up to nearly synchronous speed, at which time the rotor current is applied. After several oscillations, the rotor locks to the synchronous speed.

Under constant rotor current, the rotor generated voltage remains constant. The terminal voltage is usually constant (except when line regulation is a factor), so the power is related to the sine of the power angle δ. In many synchronous motor problems, the information given includes input power, input volt-amperes, power factor, and a suitable number of other parameters. For this reason, it is convenient to express the apparent power in several forms.

Two additional angles are defined in figure 7.16. θ is the usual *power factor angle*. It is positive when the terminal current leads the terminal voltage. ψ is the power factor angle between the generated voltage $\mathbf{E_g}$ and the terminal current $\mathbf{I_a}$. It also is positive when the current leads the voltage.

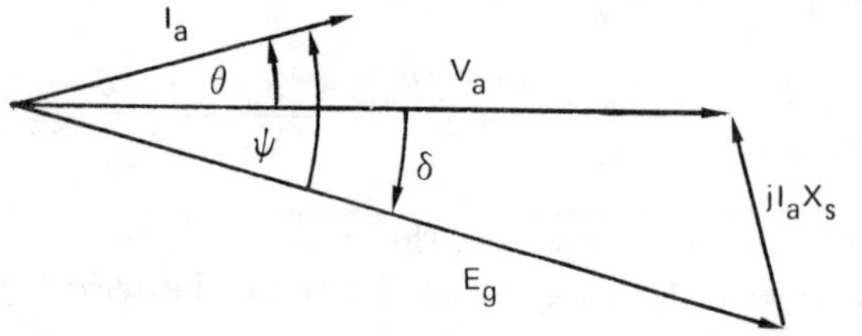

Figure 7.16 Angles for a synchronous motor

The apparent power can be rewritten from equation 7.103 as

$$S = P + jQ = V_a I_a^* = E_g I_a^* + jI_a I_a^* X_S \qquad 7.108$$

Using the angles defined in figure 7.16,

$$S = V_a|I_a|\cos\theta + jV_a|I_a|\sin\theta \qquad 7.109$$

$$S = |E_g||I_a|\cos\psi + j\left[|E_g||I_a|\sin\psi + |I_a|^2 X_S\right] \qquad 7.110$$

Using figure 7.16 and equations 7.109 and 7.110, the following useful formulas can be derived.

$$|I_a X_S|^2 = \frac{|E_g|^2}{2} + \left(\frac{Q}{q}\right)X_S \pm \sqrt{\frac{|E_g|^4}{4} + \frac{|E_g|^2 Q X_S}{q} - \frac{(PX_S)^2}{q^2}} \qquad 7.111$$

$$|E_g|^2 = |I_a|^2 X_S^2 + \frac{P^2 + Q^2}{q^2|I_a|^2} - \frac{2QX_S}{q} \qquad 7.112$$

$$X_S = \frac{Q}{q|I_a|^2} \pm \sqrt{\frac{|E_g|^2}{|I_a|^2} - \frac{P^2}{q^2|I_a|^4}} \qquad 7.113$$

$$|I_a X_S|^2 = |V_a|^2 + |E_g|^2 - 2|E_g||V_a|\cos\delta \qquad 7.114$$

Example 7.14

A three-phase synchronous machine with a line-to-line voltage of 440 volts and a synchronous speed of 900 RPM operates with a power of 9 kW and a lagging power factor of 0.8. The synchronous reactance per phase is 10 ohms. Determine (a) the torque angle δ in electrical degrees and (b) the generated voltage E_g. If the overall machine efficiency is 90%, determine (c) the torque delivered to the output shaft.

The per-phase current is

$$I_a = \frac{\frac{(9)(1000)}{3}}{254 \times .8}\angle -\arccos(.8) = 14.76\angle -36.9°$$

The per-phase reactive power is

$$\frac{Q}{q} = \frac{9\,kW}{3 \times (.8)}\sin 36.9° = 2.25\,\text{KVAR}$$

From equation 7.112,

$$E_g = 203.31\ \text{volts}$$

From equation 7.114,

$$\cos\delta = \frac{-I_a^2 X_S^2 + V_a^2 + E_g^2}{2E_g V_a} = .8140$$

$$\delta = \arccos(.8140) = -35.51°$$

The input power less the losses is the load power.

$$P_L = .9 \times 9\,kW = 8.1\,kW$$

$$T = \frac{P_L}{\Omega} = \frac{8100}{900 \times \frac{2\pi}{60}} = 85.9\,N\text{-m}$$

C. POWER FACTOR CORRECTION

Power factor correction of synchronous motors can be accomplished by adjusting the rotor current, since E_g is approximately proportional to the rotor current. The phasor diagram of figure 7.16 illustrates the principle involved.

For a leading PF (positive θ), the voltages are related by figure 7.16.

$$|E_g|^2 = |V_a|^2 + 2|V_a||I_aX_S|\sin\theta + |I_aX_a|^2 \qquad 7.115$$

The input power per phase is

$$\frac{P}{q} = |V_a||I_S|\cos\theta \qquad 7.116$$

Substituting equation 7.116 into equation 7.115, the generated voltage with a leading power factor angle is

$$|E_g|^2 = |V_a|^2 + 2X_S\left(\frac{P}{q}\right)\tan\theta + \frac{X_S^2(\frac{P}{q})^2}{|V_a|^2}\sec^2\theta \qquad 7.117$$

$$\sec^2\theta = \frac{1}{PF^2} \qquad 7.118$$

Example 7.15

The machine in example 7.14 was operating with a rotor current of 5 amps. It is desired to continue carrying the same load but to provide 5 KVAR of power factor correction to the line. Determine the required rotor current to accomplish this.

E_g can be calculated from equation 7.117. θ is taken as negative so that $\tan\theta$ is negative and $\sec^2\theta$ is positive.

From example 7.14,

$$E_g = 203.31 \text{ volts}$$

From equation 7.99,

$$E_g = K_rI_r$$

So,

$$K_r = \frac{203.31}{5} = 40.66$$

The power factor is corrected to

$$PF = \frac{9}{\sqrt{5^2+9^2}} = .874$$

$$\theta = \arccos(.874) = 29.05°$$

Again, from equation 7.117,

$$|E_g|^2 = 116,117.93$$

$$|E_g| = 340.75 \text{ volts}$$

Then, equation 7.99 gives the rotor current

$$I_r = \frac{340.75}{40.66} = 8.38 \text{ amps}$$

D. SYNCHRONOUS CAPACITORS

Synchronous capacitors are machines run with no load (only rotational losses). The power angle δ is nearly zero, so that $\mathbf{V}_a$ and $\mathbf{E}_g$ are in phase. At no-load, the vector relationship is

$$j\mathbf{I}_aX_S = \mathbf{V}_a - \mathbf{E}_g \qquad 7.119$$

$\mathbf{I}_a$ leads $\mathbf{V}_a$ by 90° if $|E_g| > |V_a|$. $\mathbf{I}_a$ lags $\mathbf{V}_a$ by 90° if $|E_g| < |V_a|$.

The magnitude of I_a is given by equation 7.119. Thus, for a synchronous capacitor,

$$\frac{Q}{q} = V_aI_a = \frac{E_gV_a - V_a^2}{X_S} \qquad 7.120$$

Example 7.16

The machine of examples 7.14 and 7.15 is to provide 10 KVAR power factor correction when running lightly loaded. Determine the rotor current necessary.

From equation 7.120,

$$E_g = \frac{X_S(\frac{Q}{3})}{V_a} + V_a = \frac{10\times10^4}{3\times254} + 254 = 385.23 \text{ volts}$$

Then, by direct proportion,

$$I_r = 5\times\frac{385.23}{203.25} = 9.48 \text{ amps}$$

E. SYNCHRONOUS GENERATORS

Synchronous generators are typically tied into a grid. It is assumed that the terminal voltage and phase are set by the grid. The individual machines are then constrained to that operating frequency and phase. If the prime mover of the synchronous generator does not continuously supply positive torque, the machine will behave as a motor. Thus, the power angle δ must be positive, with E_g leading V_a.

If current direction is reversed from that of the motor, all of the previous equations can be used with that change.

Figure 7.17 shows the current and voltage vectors for a synchronous generator. The *over-excited case*

($|E_g|\cos\delta > V_a$) is where there is a lagging power factor. The *under-excited case* ($|E_g|\cos\delta < V_a$) is where the power factor is leading. For simplicity, describe $\mathbf{E}_g$ as a vector $E\cos\delta + jE\sin\delta$ and $\mathbf{I}_a$ as $I\cos\theta + jI\sin\theta$, where $E = |\mathbf{E}_g|$ and $I = |\mathbf{I}_a|$.

As the terminal voltage $\mathbf{V}_a$ is set by the grid, it is given a phase angle of zero. For generator action, $\mathbf{E}_g$ must lead $\mathbf{V}_a$. Then, the following equations describe the terminal quantities.

$$\mathbf{I}_a = \frac{E_g - V_a}{jX_S} = \frac{E\sin\delta}{X_S} + j\frac{V_a - E\cos\delta}{X_S} \quad 7.121$$

$$(IX)^2 = V_a^2 + E^2 - 2V_aE\cos\delta \quad 7.122$$

$$E^2 = V_a^2 + (IX)^2 - 2V_aIX\sin\theta \quad 7.123$$

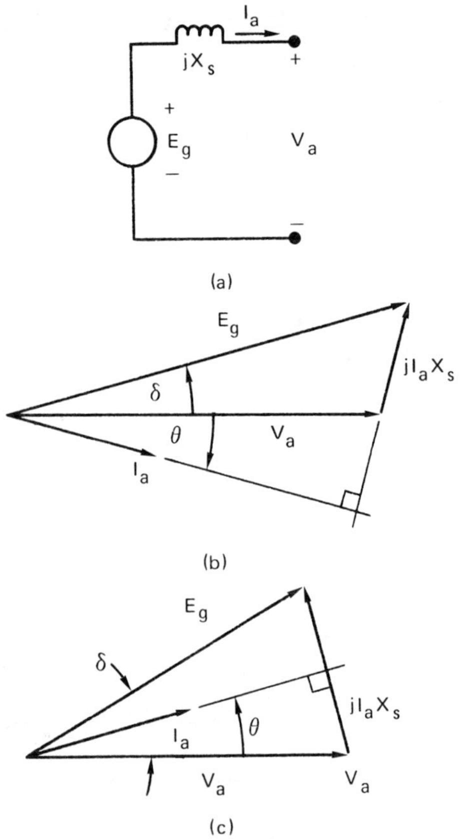

Figure 7.17 Synchronous generator (a) per-phase model, (b) over-excited with lagging power factor, (c) under-excited with leading power factor

As E is roughly proportional to the rotor current I_r, it is possible to use equations 7.122 and 7.123 when there are changes in power, power factor, or rotor current. Since $S = P + jQ = V_aI_a^*$,

$$P = V_aI\cos\theta = \left(\frac{V_aE}{X_S}\right)\sin\delta \quad 7.124$$

$$Q = -V_aI\sin\theta = \frac{V_aE\cos\delta - V_a^2}{X_S} \quad 7.125$$

Example 7.17

A 50 MVA, 13.8 kV, 3600 rpm, 60 hz grid-connected generator with X_S = 3.8 ohms reactive (per phase) operates with a lagging P.F. of 0.88 at rated voltage and current. Determine the new power factor if the rotor current is decreased by 5%.

As the prime-mover's power has not changed, it is assumed that the output power is constant. Thus $E\sin\delta = IX_S\cos\theta$ is constant. The phase voltage is

$$\frac{13.8}{\sqrt{3}} = 7.967$$

The rated current per phase is

$$I_{\text{phase}} = \frac{\frac{50 \times 10^6}{3}}{7.967 \times 10^3} = 2092$$

$$I = 1841 - j994$$

$$IX\sin\theta = 3776$$

Using equation 7.123 and noting that θ and $\sin\theta$ are negative,

$$E^2 = (7.967)^2 + (3.8 \times 2.092)^2 - 2 \times 7.967 \times (-3.776)$$
$$= 186.833 \text{ (in thousands)}$$

Therefore,

$$E_1 = 13.669 \text{ kV}$$

Since the generated voltage is proportional to I_r,

$$E_2 = .95E_1 = 12.986 \text{ kV}$$

The output power $E\sin\delta$ is constant.

$$I_1X\cos\theta_1 = 6.996 \text{ kV}$$

Referring to figure 7.17,

$$\sin\delta_2 = \frac{6.996}{12.986} = .5387$$

$$\cos\delta_2 = \cos[\arcsin(.5387)] = .8425$$

Therefore,

$$E_2\cos\delta_2 = 10.941 \text{ kV}$$

Then, from equation 7.122

$$I_2X = \sqrt{(7.967)^2 + (12.986)^2 - 2 \times 7.967 \times 10.941}$$
$$= 7.6014 \text{ kV}$$

$$\cos\theta_2 = PF_2 = \frac{6.996}{7.6014}$$
$$= .92$$

Example 7.18

The machine of example 7.17 is to provide a lagging reactive power of 40 MVAR and still operate at its rated current. What rotor current is necessary to accomplish this?

In this case, I remains constant, and the input power varies.

$$\sin\theta_2 = \frac{40}{50} = .8$$

From equation 7.123,

$$E = \sqrt{(7.967)^2 + (3.8\cdot 2.092)^2 + 2\cdot 7.967\cdot 2.092\cdot 3.8\cdot .8}$$
$$= 15.1\,\text{kV}$$

Since the original value from example 7.17 was 13.667 kV, the percent increase in rotor current is

$$\frac{15.1 - 13.667}{13.667}\times 100 = 10.05\%$$

E is proportional to rotor current, and it will also experience a 10.05% increase.

Example 7.19

The machine in example 7.17 has its prime-mover power decreased by 25%. If the rotor current decreases by the same percentage, what will the new power and power factor be?

Since the input power drops by 25%, the output power also will drop by 25%. However, V_a will remain constant at 13.8 kV.

$$I_2X_S\cos\theta_2 = .75[I_1X_S\cos\theta_1] = 5.247\,\text{kV}$$

Similarly,

$$E_2\sin\delta_2 = .75[E_1\sin\delta_1]$$

However, the rotor current drops by 25%. Therefore,

$$E_2 = .75E_1$$

From equation 7.99, E is proportional to I_r.

$$\sin\delta_2 = \sin\delta_1$$
$$\cos\delta_2 = \cos\delta_1$$

Since $E_1 = 13.669$, equation 7.124 results in

$$I_1X_S\cos\theta_1 = E_1\sin\delta_1$$
$$6.995 = 13.669\sin\delta_1$$
$$\delta_1 = 30.7°$$
$$\cos\delta_1 = .86$$
$$E_1\cos\delta_1 = 13.669\times .86 = 11.75$$
$$E_2\cos\delta_2 = .75\times 11.75 = 8.814$$

Thus from equation 7.122,

$$I_2X_S = \sqrt{(7.967)^2 + (.75\cdot 13.669)^2 - 2\cdot 7.967\cdot 8.814}$$
$$= 5.304\,\text{kV}$$

However,

$$I_2X_S\cos\theta_2 = .75\times 5.247$$

Therefore,

$$PF_2 = \cos\theta_2 = \frac{.75\times 5.247}{5.304}$$
$$= .74$$

F. POWER FACTOR RELATIONS FOR SYNCHRONOUS GENERATORS

A three-phase generator operating with constant load torque has

$$T_{\text{shaft}} = 3V_aE_g\sin\delta \qquad 7.126$$

E_g is approximately proportional to the rotor current I_r. Figure 7.18(a) shows the phasor diagram with a dashed line for the locus of $\mathbf{E}_g$ at constant load. The line perpendicular to $\mathbf{V}_a$ is the minimum value of $jX_S\mathbf{I}_a$. The angle θ is the power factor angle. From trigonometric considerations,

$$PF = \cos\theta = \frac{jX_S\,I_{a,\text{min}}}{jX_SI_a} = \frac{I_{a,\text{min}}}{I_a} \qquad 7.127$$

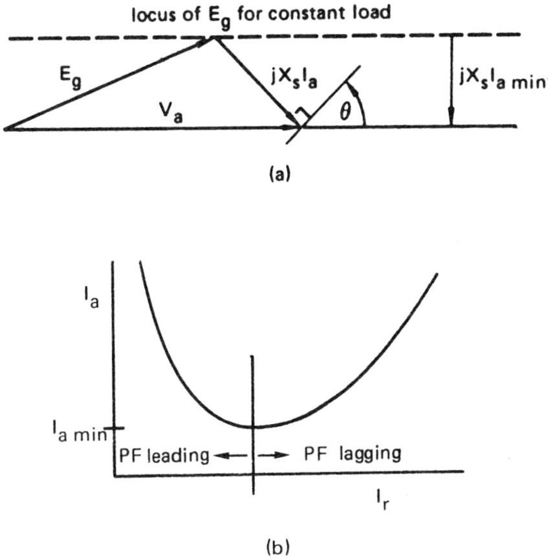

Figure 7.18 (a) Generator vector diagram, (b) vee curve for constant load

With E_g proportional to I_r, it is common to plot I_a versus I_r at constant load. This results in the so-called *vee curve* as shown in figure 7.18(b).

Practice Problems

ROTATING MACHINES

Warmups

1. A D.C. generator has a 250 volt output at 50 amp load current and 260 volts at no-load. Estimate the armature resistance.

2. A separately-excited D.C. generator is rated 10 kW, 240 volts, 2800 rpm for input power of 15 hp. No-load voltage is 260 volts. Determine the armature resistance and the mechanical power loss at 2800 rpm.

3. A D.C. shunt motor is running at rated values when the line voltage suddenly drops by 10%. Determine the resulting changes in input power and motor speed if (a) the load torque is constant, and (b) the load torque is proportional to speed.

4. A D.C. series motor is running at rated speed, voltage, and current when the load torque increases by 25%. Determine the effects of this change, assuming that the line voltage remains the same.

5. A 4-pole, 60 hz, three-phase induction motor is drawing 25 kW under load and running at 1700 rpm. The no-load losses have been measured at 850 watts. The stator copper losses have been measured at 550 watts. Determine the hp and efficiency of this machine.

6. A no-load test on a 440 volt (line-to-line) three-phase induction motor yields 1200 watts and 2500 VA. Determine the per-phase parameters r_c $(1/g_c)$ and x_Φ $(-1/b_c)$.

7. A blocked rotor test on a 440 volt (line-to-line) three-phase induction motor finds a line current of 21 amps, $P_{\text{in}} = 375$ watts, and power factor of 0.37. Find the per-phase values of R_e and X_e.

8. The motor of problems 6 and 7 is run at a slip of 0.03 with an output of 20 hp and an input power of 16.95 kW. Determine r_S and a^2r_r on a per-phase basis.

9. A three-phase, 440 volt (line-to-line) synchronous motor has $X_S = 20$ ohms and armature resistance of 0.5 ohms per phase. The synchronous speed is 1800 rpm. The no-load field current is adjusted for minimum armature current at $I_f = 5$ amps D.C., at which time the input power is 900 watts. This machine is to provide 10 KVAR of power factor correction when running lightly loaded. Determine the rotor current required.

10. A 50 MVA, 13.8 kV, 3600 rpm synchronous motor with per-phase reactance X_S of 2.8 ohms at 60 hz is operating into a grid. The power factor is .88 lagging at rated voltage and current. Determine the new power factor if the rotor current is increased by 10%.

Concentrates

1. A D.C. generator is rated 10 kW, 240 volts, 2800 rpm with 15 hp. Its armature resistance is 1.0 ohm, and it has a no-load voltage of 260 volts at $I_f = 2.75$ amps and 2800 rpm. Using the magnetization curve of figure 7.2 normalized to the no-load values, determine the shunt field resistance necessary to self-excite this generator at rated voltage and current at 2800 rpm.

2. A D.C. generator rated 10 kW, 240 volts, 2800 rpm, and 15 hp has an armature resistance of 1.0 ohm, a self-excitation shunt field resistance of 77 ohms, and follows the normalized magnetization curve of figure 7.2 for $E_n = 260$ volts and $MMF_n = 2.75$ amps. Find the voltage regulation if the normalized $K_V\Phi$ is 1.2 for normalized $MMF > 1.5$ in figure 7.2.

3. A D.C. shunt motor takes 200 volts and 25 amps at 1000 rpm and 5 hp. It has an armature resistance of 0.5 ohms and a shunt field resistance of 100 ohms. Determine (a) the rotational losses, and (b) the no-load speed at 200 volts.

4. A 500 volt, 330 amp, 200 hp D.C. series motor is rated at 900 rpm. With $R_a = 0.08$ ohms, find the starting current and torque at rated voltage.

5. A 440 volt (line-to-line) induction motor has per-phase parameters $X_e = .712$ ohms, $R_e = .283$ ohms, and $r_S = .126$ ohms when it is connected in a delta configuration. The synchronous speed is 1200 rpm. Determine the starting current and the starting torque if (a) the stator windings are connected in the delta configuration, and (b) if the stator windings are connected in the wye configuration.

6. A 440 volt (line-to-line) induction motor with synchronous speed of 1200 rpm has per-phase parameters $X_e = .934$ ohms, $R_e = .231$ ohms, and $r_S = .135$ ohms. The starting of this motor is to be controlled through a tapped auto-transformer which is in a wye-connection. The maximum line current is to be limited to 150 amps, and switching to the next higher tap is to take place when the line current drops to 100 amps. Specify the voltage taps and the switching speeds for the starting transformer.

7. A three-phase, 440 volt (line-to-line), 60 hz, 3600 rpm synchronous motor with per-phase X_S of 20 ohms and armature resistance of 0.5 ohms has its field current adjusted for minimum armature current at $I_f = 5$

amps, at which time the input power is 900 watts. For a rated current $I_a = 12$ amps, determine (a) the power angle, (b) the power factor angle, (c) the load power, (d) the efficiency, and (e) the new rotor current necessary to provide 5 KVAR of power factor correction at these operating conditions.

Timed

1. A D.C. generator rated 10 kW, 240 volts, 2800 rpm, 15 hp has $R_a = 0.48$ ohms. It has a no-load armature voltage of 260 volts at 2800 rpm when a separate excitation of 2.75 amps is applied to the field winding. Use the magnetization curve of figure 7.2 normalized to the no-load values to determine the ratio of series field turns to shunt field turns, N_S/N_f, and the shunt field resistance necessary for flat compounding at rated values.

2. A D.C. shunt motor is rated 7.5 hp, 60 amps, 1000 rpm, 120 volts. Assume the load torque is constant regardless of speed. Give circuit diagrams with parameter values to change the speed to (a) 1150 rpm, and (b) 750 rpm.

3. Design a starting circuit for the armature of a 250 volts, 125 amps, 1150 rpm, 35 hp D.C. shunt motor, assuming the line voltage of 250 volts is always connected to the shunt field winding circuit. The armature resistance is 0.15 ohms, and the maximum starting current is 250% of rated current.

4. A 200 hp squirrel-cage (induction) motor is rated at 460 volts (line-to-line) and 1760 rpm. At no-load, it requires 4.5 kW at 460 volts, 135 amps and turns at 1798 rpm. Under locked rotor conditions, it draws 240 amps at 78 volts and produces a torque of 17.2 ft-lb. The motor is started on an 80% tap. Determine (a) the starting torque, and (b) the starting current.

5. A 60 hz, 480 volt, 50 KVA alternator with rated field current of 20 amps is driven at a constant 30 hp while the following measurements are taken:

I_r (amps)	2	3	5.5	8	12	15	20	23	27
I_a (amps)	60	45	30	21	15	18	32	45	60

For the same load, determine (a) the rotor current necessary for an 80% lagging power factor, and (b) the rotor current necessary to obtain $I_a = 40$ amps with a leading power factor, and (c) the resulting power factor.

Appendix 7.1

MAGNETIC CONVERSION FACTORS

MMF

MKS: amp-turns

cgs: 1 gilbert=0.4π amp-turns

FLUX ϕ

MKS: webers (volt-sec)

cgs: 1 maxwell =1 line=10^{-8} webers

MAGNETIC FIELD INTENSITY H

MKS: amp-turns/m

cgs: 1 oersted=$(250/\pi)$ amp-turns/m

MAGNETIC FLUX DENSITY B

MKS: 1 tesla=1 weber/m^2

cgs: 1 gauss=10^{-4} teslas

Appendix 7.2

Power and Energy Conversions

multiply	by	to get
BTU	3.929×10^{-4}	hp-hrs
BTU	778.3	ft-lbf
BTU	2.930×10^{-4}	kw-hrs
BTU	1.0×10^{-5}	therms
BTU/hr	.2161	ft-lbf/sec
BTU/hr	3.929×10^{-4}	hp
BTU/hr	.2930	watts
ft-lbf	1.285×10^{-3}	BTU
ft-lbf	3.766×10^{-7}	kw-hrs
ft-lbf	5.051×10^{-7}	hp-hrs
ft-lbf/sec	4.6272	BTU/hr
ft-lbf/sec	1.818×10^{-3}	hp
ft-lbf/sec	1.356×10^{-3}	kw
hp	2545.0	BTU/hr
hp	550	ft-lbf/sec
hp	.7457	kw
hp-hr	2545.0	BTU
hp-hr	1.976×10^{6}	ft-lbf
hp-hr	.7457	kw-hrs
kw	1.341	hp
kw	3412.9	BTU/hr
kw	737.6	ft-lbf/sec

RESERVED FOR FUTURE USE

8 LINEAR AMPLIFIERS

Nomenclature

$A(f)$	gain spectrum	-
A_I	current gain	-
A_V	voltage gain	-
BW	bandwidth	hz
C_d	diffusion capacitance	F
C_t	transition capacitance	F
$CMMR$	common mode rejection ratio	db
e^2	spectral voltage squared density	V^2/hz
f	functions	-
F	noise figure	-
G	power gain	-
GBW	gain-bandwidth product	rad/sec
$G(f)$	power gain spectrum	-
g_m	transconductance	mhos
h_{ib}	common base input impedance	ohms
h_{ie}	common emitter input impedance	ohms
h_{fb}	common base current gain	-
h_{fe}	common emitter current gain	-
h_{ob}	common base output admittance	mhos
h_{oe}	common emitter output impedance	mhos
h_{rb}	common base feedback voltage gain	-
h_{re}	common emitter feedback voltage gain	-
I	current	amps
I_{co}	collector reverse saturation current	amps
I_{DSS}	maximum JFET drain current	amps
I^2	spectral current squared density	amps^2/hz
k	(with T) Boltzmann's constant	1.38×10^{-23} Joules/°K
k	a constant	-
N	noise power	W
p	spectral power density	W/hz
q	charge of an electron	1.609×10^{-19} coulombs
r	resistance	ohms
r_m	transresistance	ohms
R	resistance	ohms
S	signal power	W
$S(f)$	spectral voltage squared density	V/hz
S_v	stability factor	mA/°C
T	temperature	°K
V	voltage	V
$V_{GS(\text{off})}$	turn-off gate-source voltage for depletion-mode FETs	V
V_p	pinchoff voltage	V
Y	admittance	mhos
Z_{in}	input impedance	ohms
Z_{out}	output impedance	ohms

Symbols

α_o	common base D.C. current gain	-
β	common emitter current gain	-
ω	frequency	rad/sec
ω_o	GBW product of operational amplifier	rad/sec
ω_{cf}	corner frequency	rad/sec

Subscripts

a	amplifier
b	base
BR	breakdown
c	collector
cc	collector supply
cf	corner frequency
cm	common mode
d	differential or diode
D	drain

DD	drain supply
e	equivalent or emitter
f	forward
fb	feedback
G	gate
in	input
kz	keep-alive
L	load
max	maximum
min	minimum
o	output
out	output
pn	from P-type to N-type
Q	quiescent or bias point value
r	reverse
s	saturation
S	source
z	zener or avalanche

1 IDEAL OPERATIONAL AMPLIFIER

A. INTRODUCTION

An ideal *operational amplifier* (*op-amp*) is an *integrator* (amplifier with infinite D.C. gain) with a very large input impedance, a very small output impedance, a gain bandwidth of ω_o, and a differential input. This can be modeled by the circuit shown in figure 8.1(a). Here the terminals are identified as N for the normal (non-inverting) terminal and I for the inverting terminal. The frequency response curve is a straight line with a slope of -20 db/decade, crossing the zero db axis line at ω_o rad/sec.

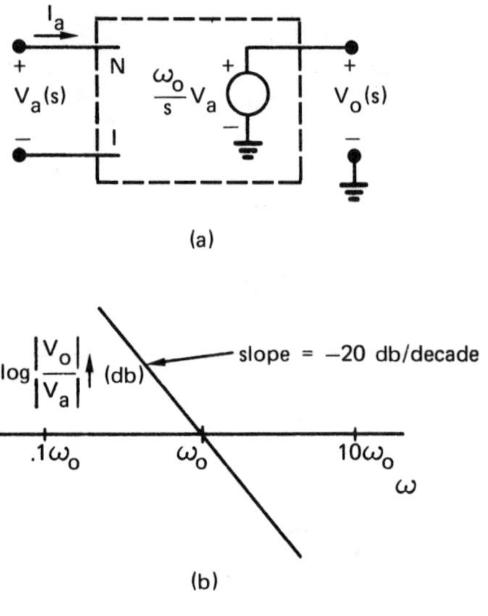

Figure 8.1 (a) Circuit model for ideal operational amplifier, (b) frequency characteristic

B. IDEAL OPERATIONAL AMPLIFIER WITH POSITIVE FEEDBACK

The ideal operational amplifier is supplied with *positive feedback* through connections as shown in figure 8.2. Analysis follows Kirchhoff's current law with current summing at the N terminal.

$$I_{fb} + I_{in} = I_a = 0 \qquad 8.1$$

From Kirchhoff's voltage law,

$$V_{in} = I_{in}R_{in} + \frac{S}{\omega_o}V_o \qquad 8.2$$

$$V_{out} = I_{fb}R_{fb} + \frac{s}{\omega_o}V_o \qquad 8.3$$

Then,

$$V_{out} = \frac{\left(\frac{R_{fb}}{R_{in}}\right)V_{in}}{\frac{s}{\omega_o}\left(1+\frac{R_{fb}}{R_{in}}\right)-1} \qquad 8.4$$

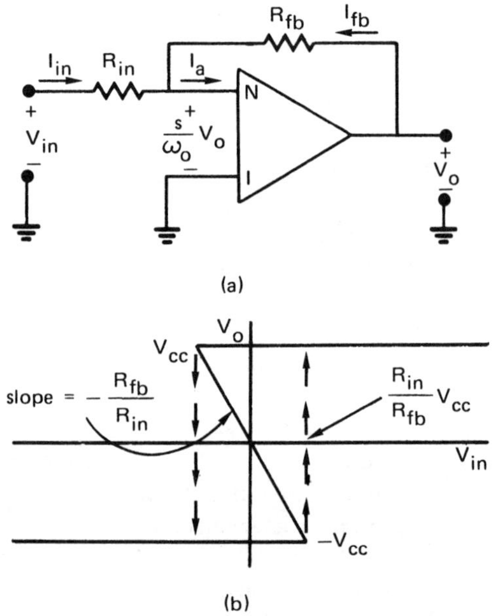

Figure 8.2 (a) Operational amplifier with positive feedback, (b) characteristic

Consider an impulse input of $v_{in}(t) = \delta(t)$. Then $V_{in}(s) = 1$ and

$$V_{out}(s) = \frac{\frac{R_{fb}}{R_{in}}}{\frac{s}{\omega_o}\left(1+\frac{R_{fb}}{R_{in}}\right)-1} \qquad 8.5$$

Taking the Laplace inverse transform, we obtain

$$V_{out}(t) = \frac{-R_f\omega_o}{R_f + R_{in}}e^{\omega_{cf}t} \qquad 8.6$$

$$\omega_{cf} = \frac{\omega_o R_{in}}{R_{in} + R_f} \qquad 8.7$$

This is an unstable situation. The output voltage grows exponentially. An actual amplifier would be limited in output by the D.C. voltage sources. For an operational amplifier, there are typically two equal and opposite D.C. voltages: V_{cc} and $-V_{cc}$. Operation will be limited by those values, and a characteristic for D.C. gain will be similar to that in figure 8.2(b).

Because of the instability (as indicated by the negative slope and the triple values for V_o for single values of V_{in}), a *hysteretic behavior* results, with the dashed arrow paths followed in steady state operation. Such hysteretic behavior can be valuable in applications where *noise immunity* is desired.

C. INVERTING OPERATIONAL AMPLIFIER WITH NEGATIVE FEEDBACK

The circuit for an *inverting operational amplifier* is shown in figure 8.3. Following the same steps taken for the operational amplifier with positive feedback, the input-output voltage relation in s-notation is

$$\frac{V_o}{V_{\text{in}}} = \frac{\frac{-R_{fb}}{R_{\text{in}}}}{\frac{s}{\omega_{cf}} + 1} \qquad 8.8$$

$$\omega_{cf} = \frac{\omega_o R_{\text{in}}}{R_{\text{in}} + R_{fb}} \qquad 8.9$$

Thus, the D.C. ($s = 0$) gain is $-R_{fb}/R_{\text{in}}$, and the *corner frequency* (-3 *db* frequency) is ω_{cf}.

This circuit is the most commonly used form of the operational amplifier, as precision resistances can be used to set the gain precisely.

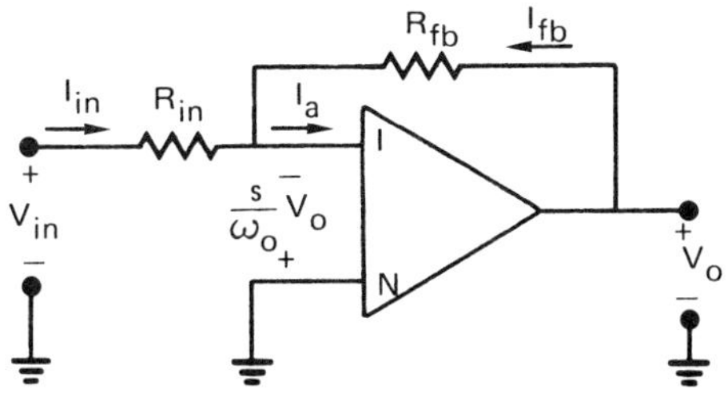

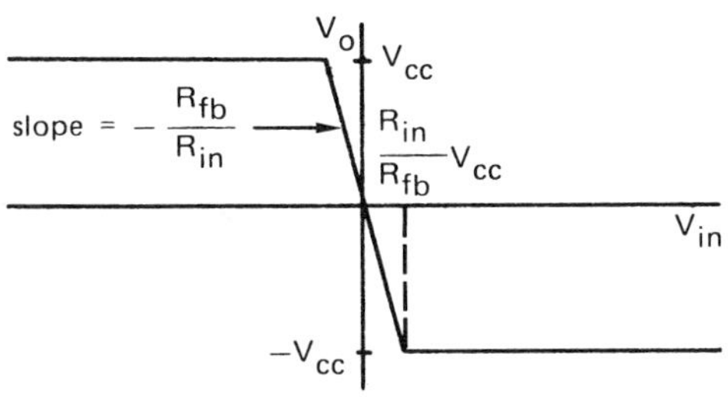

Figure 8.3 (a) Inverting operational amplifier with negative feedback, (b) voltage transfer characteristic

D. NON-INVERTING OPERATIONAL AMPLIFIER WITH NEGATIVE FEEDBACK

A *non-inverting* version of the operational amplifier can be obtained with the connections shown in figure 8.4. Note that the feedback is to the inverting terminal *I*.

The current flowing in the lower circuit toward the *N* terminal is zero, so that $V_N = V_{\text{in}}$. Then, the voltage at the *I* terminal is

$$V_I = V_{\text{in}} - \left(\frac{s}{\omega_o}\right) V_o \qquad 8.10$$

Then,

$$I_{\text{in}} = \frac{\left(\frac{s}{\omega_o}\right) V_o - V_{\text{in}}}{R_{\text{in}}} \qquad 8.11$$

$$I_{fb} = \frac{V_o + \left(\frac{s}{\omega_o}\right) V_o - V_{\text{in}}}{R_{fb}} \qquad 8.12$$

And, as $I_a = 0$,

$$I_{\text{in}} + I_{fb} = 0 \qquad 8.13$$

Then,

$$V_o = \left[\frac{\frac{R_{fb}}{R_{\text{in}}} + 1}{\frac{s}{\omega_{cf}} + 1}\right] V_{\text{in}} \qquad 8.14$$

$$\omega_{cf} = \frac{\omega_o R_{\text{in}}}{R_{\text{in}} + R_f} \qquad 8.15$$

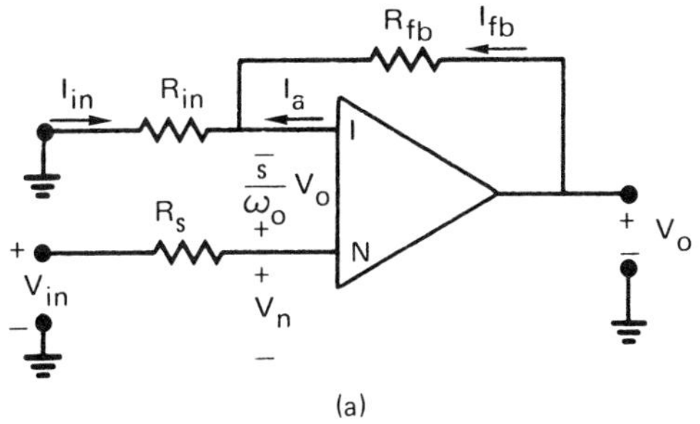

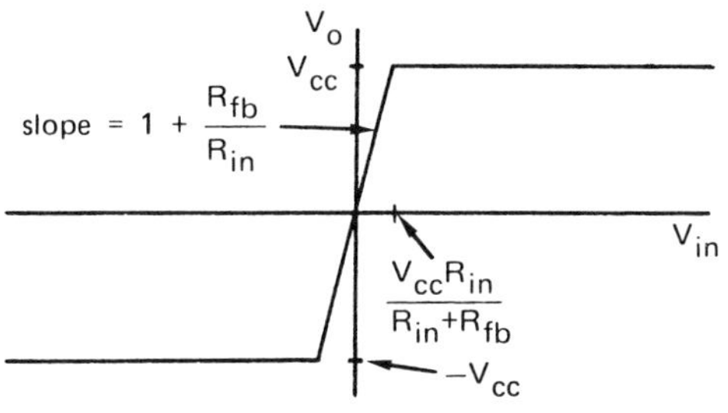

Figure 8.4 (a) Non-inverting operational amplifier with negative feedback, (b) voltage transfer characteristic

With the output limited to the range $-V_{cc}$ to $+V_{cc}$, the D.C. voltage transfer characteristic will be similar to that shown in figure 8.4(b).

The 3-db bandwidth is again ω_c rad/sec.

E. VOLTAGE FOLLOWER

A *voltage follower* circuit, with a gain of 1 and a bandwidth of ω_o, is obtained with the circuit shown in figure 8.4(a) using $R_f = 0$ and $R_{in} = \infty$. In that case, $V_{out} = V_{in}$ and $\omega_c = \omega_o$. Such a circuit is useful when a low power input signal is required to provide significant current to a load. This is a type of impedance matching. A good deal of current gain is provided, so the operational amplifier can drive the load when the signal source cannot.

F. OPERATIONAL AMPLIFIER WITH MULTIPLE INPUTS

The apparent input impedance of an operational amplifier with feedback is quite low and is essentially independent of the actual input impedance. The input terminals appear to be shorted, so that any input impedance essentially is at zero volts (if the non-inverting input is grounded).

However, the input current effectively flows in the feedback impedance. Therefore, the current flowing in any input impedance essentially will be the voltage applied to the input impedance divided by the impedance. Then, for an operational amplifier with multiple inputs as shown in figure 8.5, the output voltage will be

$$V_{out} = -\sum_{i=1}^{n} \frac{Z_{fb}}{Z_i} V_i \qquad 8.16$$

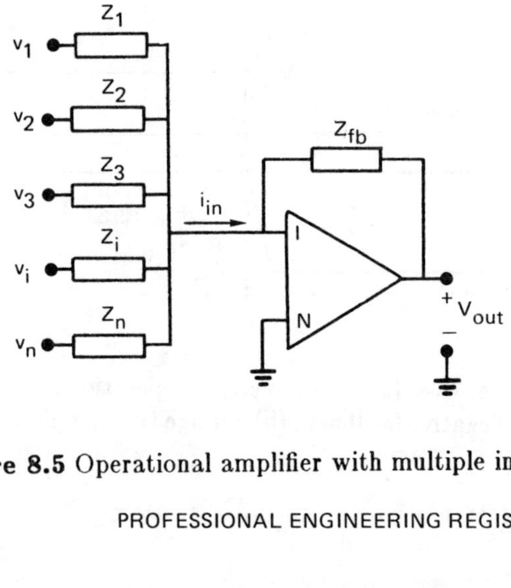

Figure 8.5 Operational amplifier with multiple inputs

2 ACTIVE FILTERS WITH INVERTING OPERATIONAL AMPLIFIERS

A. INTRODUCTION

Two forms of active filters are shown in figure 8.6. Analysis of figure 8.6(a) yields

$$\frac{V_o}{V_{in}} = \frac{-Y_{in}}{\frac{s}{\omega_o}(Y_{fb} + Y_{in}) + Y_{fb}} \qquad 8.17$$

For frequencies much less than the *gain-bandwidth product* ω_o, equation 8.17 reduces to $-Y_{in}/Y_{fb}$.

$$\frac{V_o}{V_{in}} = -\frac{Y_{in}}{Y_f} \quad \text{for} \quad \frac{s}{\omega_o} << 1 \qquad 8.18$$

In any case, it is necessary that Y_{fb} be much greater than $(Y_{in} + Y_{fb})s/\omega_o$.

Analysis of the circuit in figure 8.6(b) yields

$$\frac{V_o}{V_{in}} = \frac{-Y_1Y_3}{Y_5(Y_1 + Y_2 + Y_3 + Y_4)(1 + \frac{s}{\omega_o}) + \frac{s}{\omega_o}Y_3(Y_1 + Y_2)} \qquad 8.19$$

Equation 8.19 reduces to equation 8.20 when the frequency is much less than the gain-bandwidth product ω_o.

$$\frac{V_o}{V_{in}} = \frac{-Y_1Y_3}{Y_5(Y_1 + Y_2 + Y_3 + Y_4) + Y_3Y_4} \quad \text{for} \quad \frac{s}{\omega_o} << 1 \qquad 8.20$$

Equations 8.18 and 8.20 are evaluated for various filters in the following sections.

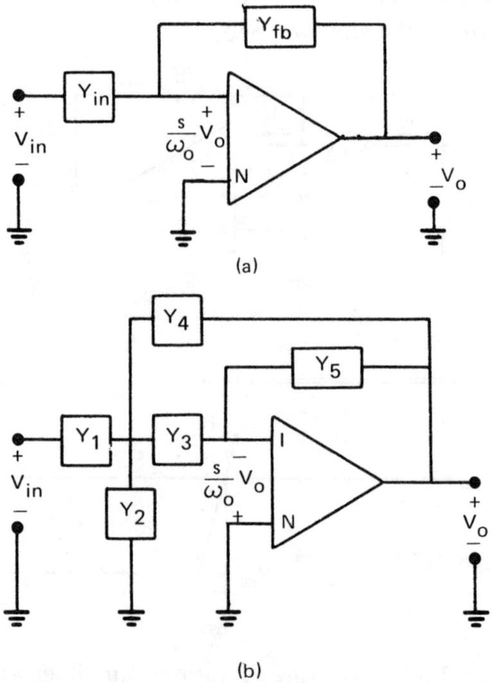

Figure 8.6 Negative feedback filters

B. OPERATIONAL AMPLIFIER INTEGRATOR

The *R-C integrator* shown in figure 8.7(a) behaves according to equation 8.21.

$$\frac{V_o}{V_{\text{in}}} = \frac{1}{sC_oR + 1} \qquad 8.21$$

Operation as an integrator is adequate only for $\omega > 10/RC_o$, as shown in figure 8.7(b).

Analysis of the operational amplifier circuit in figure 8.8(a) yields

$$\frac{V_o}{V_{\text{in}}} = \frac{-1}{sC_{fb}R} \times \frac{1}{1 + \frac{1}{\omega_o RC_{fb}}} \times \frac{1}{1 + \frac{s}{\omega_c}} \qquad 8.22$$

$$\omega_c = \omega_o + \frac{1}{RC_{fb}} \qquad 8.23$$

The circuit has the characteristic indicated in figure 8.8(b). It is adequate as an integrator for frequencies less than $0.1\omega_c$, which is ω_o greater than $1/RC$. As ω_o is on the order of 10^7, this is no real limitation.

For $\omega_o RC_{fb} >> 1$,

$$\frac{V_o}{V_{\text{in}}} = \frac{1}{sC_{fb}R} \qquad 8.24$$

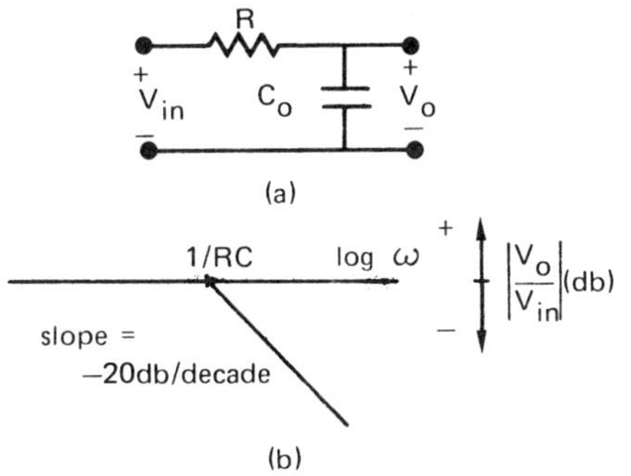

Figure 8.7 (a) Simple RC integrator, (b) frequency characteristic

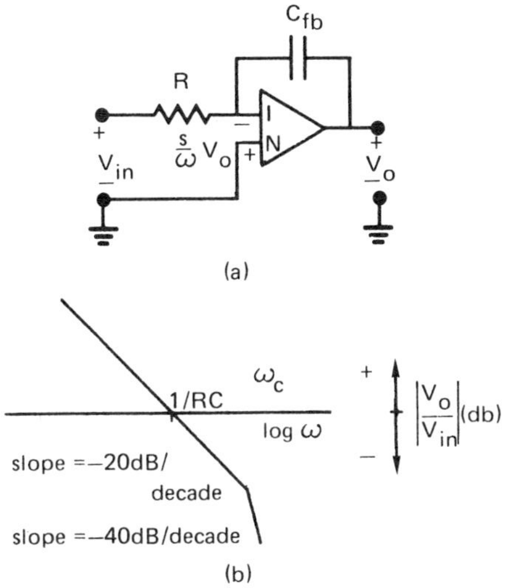

Figure 8.8 (a) Operational amplifier integrator, (b) frequency characteristic

C. SINGLE-POLE OPERATIONAL AMPLIFIER LOW-PASS FILTER

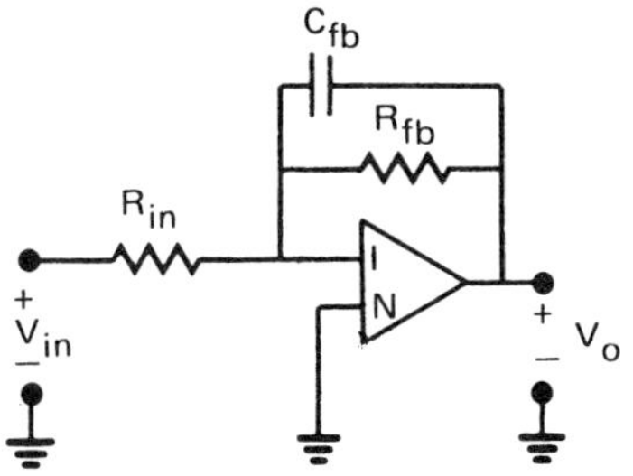

Figure 8.9 Single-pole operational amplifier low-pass filter

For figure 8.9,

$$\frac{V_o}{V_{\text{in}}} = \frac{-\frac{R_{fb}}{R_{\text{in}}}}{sC_{fb}R_{fb} + 1} \qquad 8.25$$

$$\omega_{cf} = \frac{1}{R_{fb}C_{fb}} \qquad 8.26$$

D. SINGLE-POLE OPERATIONAL AMPLIFIER HIGH-PASS FILTER

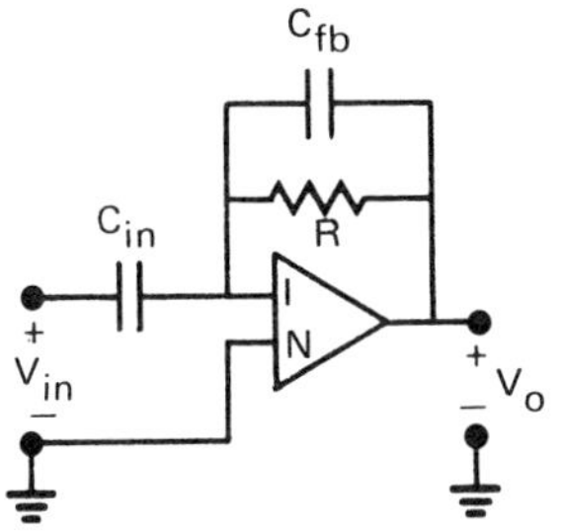

Figure 8.10 Single-pole operator amplifier high-pass filter

For figure 8.10,

$$\frac{V_o}{V_{\text{in}}} = \frac{-sC_{\text{in}}R}{sC_{fb}R + 1} \qquad 8.27$$

$$\omega_{cf} = \frac{1}{RC_{fb}} \qquad 8.28$$

E. LEAD-LAG (OR LAG-LEAD) OPERATIONAL AMPLIFIER FILTER

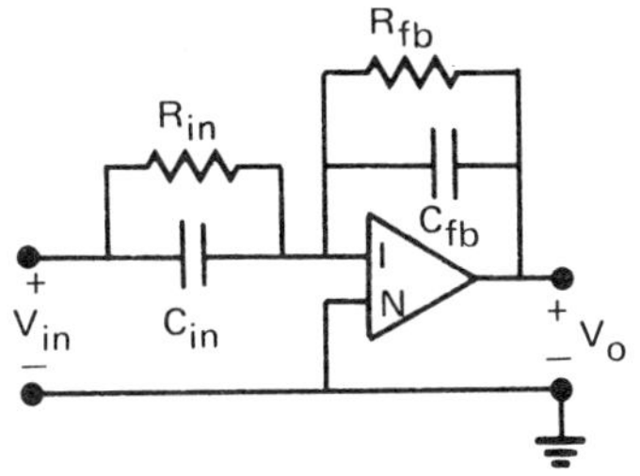

Figure 8.11 Lead-lag (or lag-lead) operator amplifier filter

For figure 8.11,

$$\frac{V_o}{V_{\text{in}}} = -\frac{R_{fb}}{R_{\text{in}}} \times \frac{sC_{\text{in}}R_{\text{in}} + 1}{sC_{fb}R_{fb} + 1} \qquad 8.29$$

F. OPERATIONAL AMPLIFIER BAND-PASS FILTER

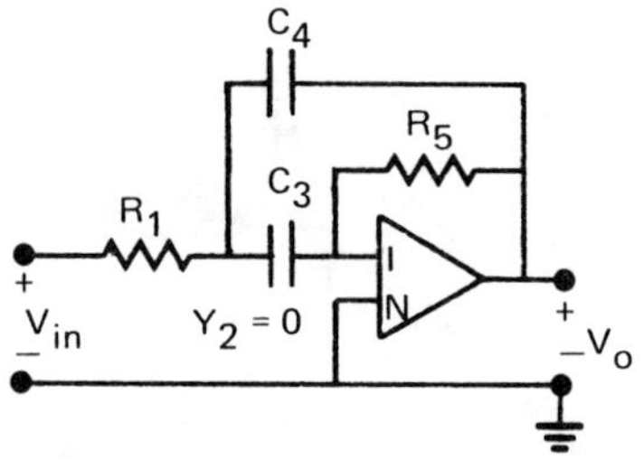

Figure 8.12 Operational amplifier band-pass filter

For figure 8.12,

$$\frac{V_o}{V_{in}} = \frac{-sC_3R_5}{s^2C_3C_4R_1R_5 + s(C_3 + C_4)R_1 + 1} \qquad 8.30$$

G. A TWO-POLE OPERATIONAL AMPLIFIER LOW-PASS FILTER

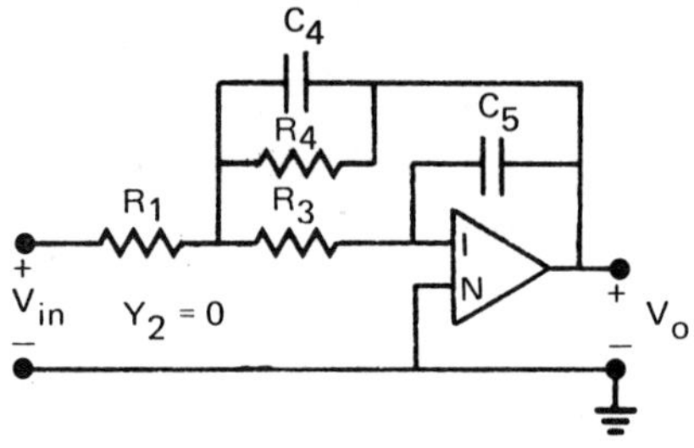

Figure 8.13 A two-pole operational amplifier low-pass filter

For figure 8.13,

$$\frac{V_o}{V_{in}} = \frac{-R_4}{D}$$

$$D = s^2R_1R_3R_4C_4C_5 + s(C_5(R_3R_4 + R_1R_3 + R_1R_4) + C_4R_1R_4) + R_1 \qquad 8.31$$

H. TWO-POLE OPERATIONAL AMPLIFIER HIGH-PASS FILTER

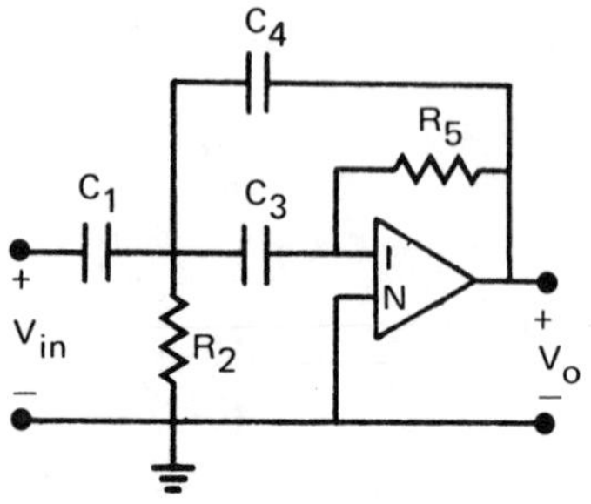

Figure 8.14 Two-pole operational amplifier high-pass filter

$$\frac{V_o}{V_{in}} = \frac{-s^2R_2R_5C_1C_3}{s^2R_2R_5C_3C_4 + sR_2(C_1 + C_3 + C_4) + 1} \qquad 8.32$$

3 ACTIVE FILTERS WITH NON-INVERTING OPERATIONAL AMPLIFIERS

A. INTRODUCTION

The general form of a non-inverting operational amplifier active filter circuit is shown in figure 8.15. The gain formula is

$$\frac{V_o}{V_{in}} = \frac{kY_1Y_3}{Y_5(Y_1 + Y_2 + Y_3 + Y_4) + Y_3(Y_1 + Y_2 + Y_4(1 - k))} \qquad 8.33$$

Specific cases of figure 8.15 and equation 8.33 follow.

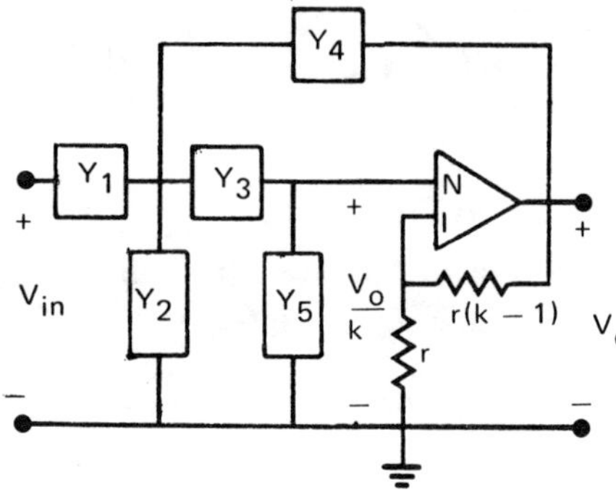

Figure 8.15 Non-inverting amplifier filter with positive feedback

B. LOW-PASS SALLEN AND KEY CIRCUIT

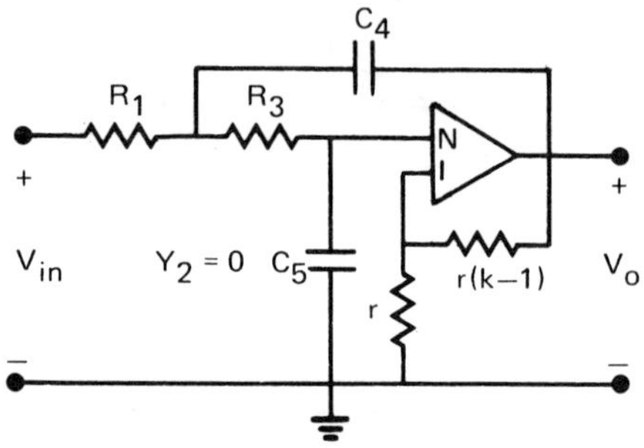

Figure 8.16 Low-pass Sallen and Key circuit

For figure 8.16,

$$\frac{V_o}{V_{in}} = \frac{k}{s^2R_1R_3C_4C_5 + s(C_5(R_1 + R_3) + (1 - k)R_1C_4) + 1} \qquad 8.34$$

C. HIGH-PASS NON-INVERTING OPERATOR AMPLIFIER FILTER

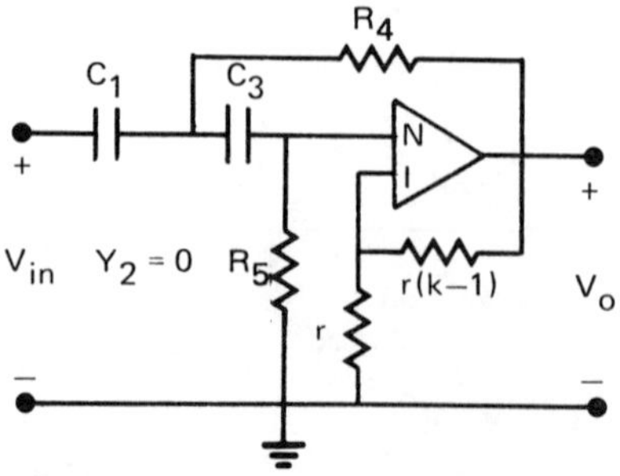

Figure 8.17 High-pass non-inverting operator amplifier filter

For figure 8.17,

$$\frac{V_o}{V_{\text{in}}} = \frac{s^2 k R_4 R_5 C_1 C_3}{s^2 R_4 R_5 C_1 C_3 + s(R_4(C_1 + C_3) + (1 - k)C_3 R_5) + 1} \tag{8.35}$$

D. SALLEN AND KEY BANDPASS CIRCUIT

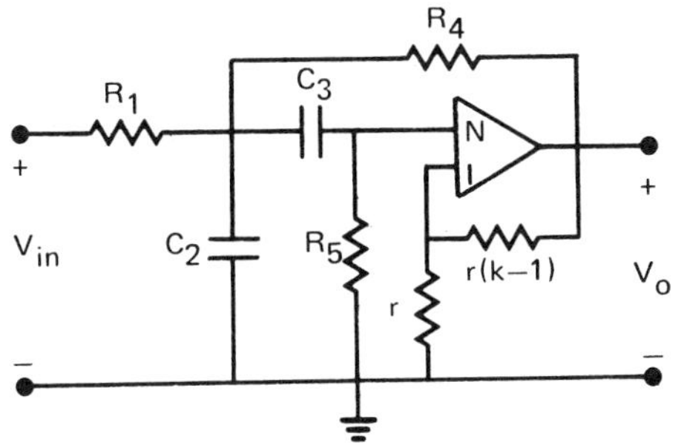

Figure 8.18 Sallen and key bandpass circuit

For figure 8.18,

$$\frac{V_o}{V_{\text{in}}} = \frac{s k R_4 R_5 C_3}{D}$$

$$D = s^2 R_1 R_4 R_5 C_2 C_3 + s(R_1(C_2 + C_3)R_5 + C_3 R_5(R_4 + (1 - k)R_1)) + R_1 + R_4 \tag{8.36}$$

4 P-N JUNCTION CHARACTERISTICS

A. INTRODUCTION

Pure silicon and germanium are poor conductors. They must be doped with a column III element such as boron to become *P-type* (*P* for positive charge carriers) semiconductors, or with a column IV element such as phosphorous to become *N-type* (*N* for negative charge carriers) semiconductors.

When doping is changed from P-type to N-type over a short distance, a *semiconductor junction* (*P-N junction*) is formed. Near this junction (where the effective doping level is zero), a thermally generated *depletion region* exists. Here the charge carriers have diffused away from the junction, and an internal *voltage barrier* exists. See figure 8.19(a).

With the application of an external voltage, the voltage barrier will lower or raise depending on the polarity of the applied voltage. Correspondingly, the depletion layer will narrow or widen depending on the polarity of the externally applied voltage. This is indicated in figure 8.19(b) or (c).

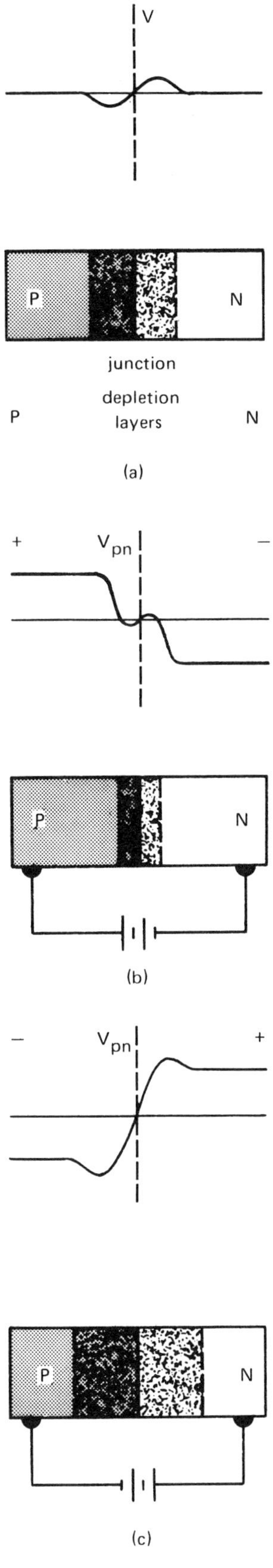

Figure 8.19 P-N junction voltage barrier and depletion regions: (a) no applied voltage, (b) voltage dropping from P to N, and (c) voltage rising from P to N

When the applied junction voltage exceeds the potential barrier, the current flow is approximately exponential with the voltage. The potential barrier is about 0.6 volts for silicon and 0.3 volts for germanium. This corresponds to a *forward-bias condition* as indicated in figure 8.19(b).

In the *reverse-bias condition*, figure 8.19(c), the current is approximately constant, due to *minority carrier conduction* (holes in the N-region and electrons in the P-region) which thermally diffuse to the depletion region and are swept across. These minority carriers are produced from a thermal phenomenon which is constant at a given temperature.

Thus, minority carrier generation is constant, resulting in a constant reverse current. This reverse current is called the *reverse saturation current*, I_S. Over a certain range of applied voltages, the behavior of the junction can be described approximately by the *ideal P-N diode equation.*

$$I_{pn} = I_S\left[e^{qV_{pn}/kT} - 1\right] \qquad 8.37$$

The subscript *pn* indicates that the positive (forward-bias) direction of current is from the P-side to the N-side of the junction, and the positive polarity of voltage is to drop from the P-side to the N- side.

B. AVALANCHE BREAKDOWN OF P-N JUNCTIONS

The width of the depletion region is inversely proportional to the doping level and proportional to the $1/x$ power of the applied voltage V_{np}.

$$\text{depletion width} \propto \frac{1}{\text{doping level}} V_{np}^{1/x} \quad (2 < x < 3) \qquad 8.38$$

When the junction is reverse biased, the depletion region widens. At the same time, however, an electric field proportional to the voltage and inversely proportional to the depletion layer width is established. When the field becomes strong enough, some of the electrons in the atomic bonds are ionized and a current flows.

If the depletion region is very narrow due to intentional high doping density on both sides of the junction, the ionization of the bonding electrons is the major source of current. This ionization is called the *Zener effect.* If the depletion regions are lightly doped so that a wider depletion layer occurs, the ionized electrons collide with other atoms and thereby cause the release of more charge carriers. This is called the *avalanche effect.*

For any P-N junction with reverse bias, there is *reverse breakdown*, usually called *Zener breakdown*. Actual Zener breakdown occurs only for very high doping densities on both sides of the junction and in the range of only a few volts. *Avalanche breakdown* is the more usual mechanism in semiconductors. By changing the doping level, avalanche breakdown can be made to occur over a wide range of voltages. This effect is used in *voltage regulator diodes*, usually misnamed *Zener diodes.*

C. VERY LOW FREQUENCY CHARACTERISTICS OF P-N JUNCTIONS

At very low frequencies and for voltages more positive than the breakdown voltage, the voltage-current characteristics approximate the ideal P-N diode equation. Two methods of approximating equation 8.37 are useful: the incremental resistance method for the forward-biased region, and a piecewise-linear model for general analysis of circuits using P-N junctions in diodes, voltage-regulating diodes, and bipolar junction transistors.

- **Incremental Resistance**

For a junction which is forward-biased and does not change its operating point significantly, an incremental resistance r_d is useful in small-signal analysis. The incremental resistance is found by taking the partial derivative of equation 8.37.

$$\frac{1}{r_d} = \frac{\partial I_{pn}}{\partial V_{pn}} = \frac{q}{kT} I_S e^{qV_{pn}/kT} \qquad 8.39$$

However,

$$I_S e^{-qV_{pn}/kT} = I_{pn} + I_S \qquad 8.40$$

Therefore,

$$\frac{1}{r_d} = \frac{q}{kT}(I_{pn} + I_S) \qquad 8.41$$

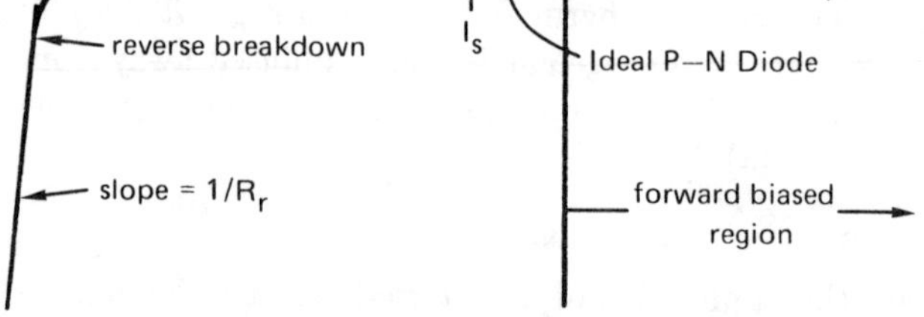

Figure 8.20 Low frequency characteristic of P-N junctions.

As I_S typically is less then 1% of the rated current of a junction, it can be ignored. A reasonable approximation is given by equation 8.42. (It is convenient to recognize that $kT/q = .026$ volts at $300°K$, approximately room temperature.)

$$r_d = \frac{kT}{qI_{pn}} \qquad 8.42$$

- **Piecewise Linear Model**

The piecewise linear model is obtained by finding the best linear approximations to the diode characteristic in the three regions separated by the voltages $-V_r$ and V_f.

•Forward-bias region, $V_{pn} > V_f$

$$I_{pn} = \frac{V_{pn} - V_f}{R_f} \qquad 8.43$$

•Off region, $-V_r > V_{pn} > V_f$

$$I_{pn} = 0 \qquad 8.44$$

•Reverse breakdown region, $V_{pn} > -V_r$

$$I_{pn} = \frac{V_{pn} + V_r}{R_r} \qquad 8.45$$

A single equivalent circuit can be obtained from the definition of an ideal diode. An *ideal diode* will allow any amount of current to flow in the forward direction without forward voltage drop. An ideal diode also will not allow any reverse current to flow, regardless of how much reverse voltage is applied.

Using ideal diodes, the piecewise-linear model is obtained.

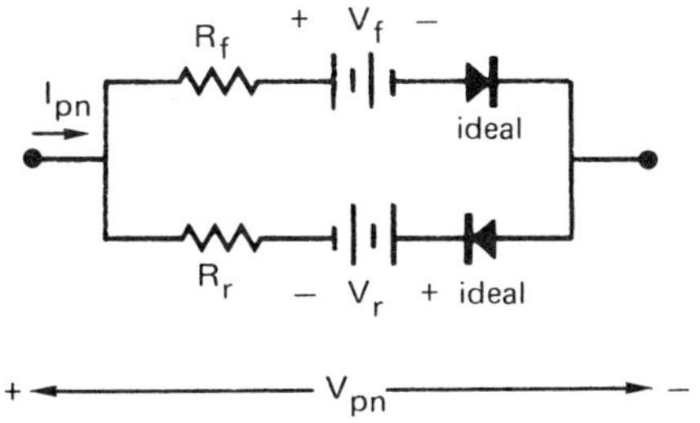

Figure 8.21 Piecewise linear model of real diode

D. JUNCTION CAPACITANCES

The main frequency effects in P-N junctions are two capacitances. The depletion region is related to the depletion layer width. The *diffusion capacitance* is related to charge storage associated with forward current when the forward voltage is large (primarily in bipolar junction transistors).

The *depletion region capacitance* is called the *transition capacitance* C_t, and the *diffusion capacitance* is C_d. The transition capacitance is inversely proportional to the depletion layer width and, therefore, is inversely proportional to $V_{np}^{1/x}$ where x is again between 2 and 3. Transition capacitance is the primary capacitance for reverse bias operation. Both capacitances appear in parallel with the terminals.

The diffusion capacitance is proportional to the forward current, while the incremental resistance is inversely proportional to the forward current. Therefore, the product $r_d C_d$ is constant and is known as the junction cutoff frequency.

$$r_d C_d = \omega_{cf} \qquad 8.46$$

5 FIELD EFFECT TRANSISTORS

A. INTRODUCTION

Field effect transistors (FET's) are divided into junction transistors (JFET's) and isolated gate or metal-oxide-silicon transistors (MOSFET's or IGFET's). Each category is further divided in N-channel and P-channel transistors. MOSFET's are divided into depletion mode and enhancement mode transistors, while JFET's are always depletion mode.

B. JUNCTION FET's (JFET's)

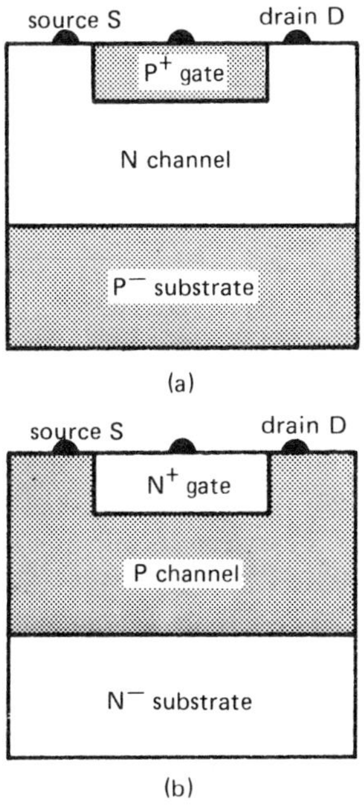

Figure 8.22 Junction FET (a) N-channel, (b) P-channel

Junction FET's are fabricated with three layers: a lightly doped substrate, a moderately doped channel, and a heavily doped gate. Figure 8.22(a) shows an N-channel JFET, and figure 8.22(b) shows the P-channel JFET. Electrical connections are made to the points indicated by *source, drain,* and *gate.* Voltage polarities are referenced to the source as V_{GS} and V_{DS} drops. The channel current is called the *drain current* I_D. It is defined as positive when flowing from the drain to the source.

JFET's are operated with the gate-channel junction reverse-biased. (If they are allowed to have forward bias, they behave as *unijunction transistors,* also called *double-base diodes.*) If the gate is shorted to the source and the drain voltage set so that the gate-channel junction at the drain end is reverse-biased, the N-channel drain will have a positive voltage to reverse bias the junction. The P-channel drain will have a negative voltage to reverse bias the junction at the drain end. This establishes the normal operating conditions for N- and P-channels in depletion mode FET's.

- N-channel: V_{DS} positive, I_D positive, V_{GS} negative
- P-channel: V_{DS} negative, I_D negative, V_{GS} positive

Depletion mode FET's are controlled by a depletion region in the channel. For this reason, the gate is heavily doped and the channel moderately doped. The depletion layer controlled by the gate is thereby practically contained entirely in the channel. Similarly, the substrate is more lightly doped than the channel, so that any depletion region there will occur in the substrate and will not affect the channel. (An alternative construction method uses a second gate between channel and substrate. The sensitivity to gate voltages is thereby doubled.)

C. DEPLETION MODE PINCHOFF

When the gate-source voltage is held at zero and the drain-source voltage is increased in the proper sense, the depletion layer will grow into the channel as indicated in figure 8.23(a). For low values of V_{DS}, the channel appears to be a resistance, but as the voltage is increased, the depletion region extends to the bottom of the channel, and the channel is said to be *pinched off.*

The voltage at which the depletion region completely fills the channel is called the *pinchoff voltage,* V_p. Beyond pinchoff, the channel current is limited to its maximum (I_{DSS}) until avalanche occurs from the drain to the source. (In the saturation region (i.e., where $I_D = I_{DSS}$), a complete pinchoff does not actually occur. A narrow channel remains but continues to narrow so that the resistance is almost proportional to V_{DS}. Therefore, the current remains almost constant until avalanche.) Avalanche also can occur through the gate-drain circuit. This phenomemon should be avoided.

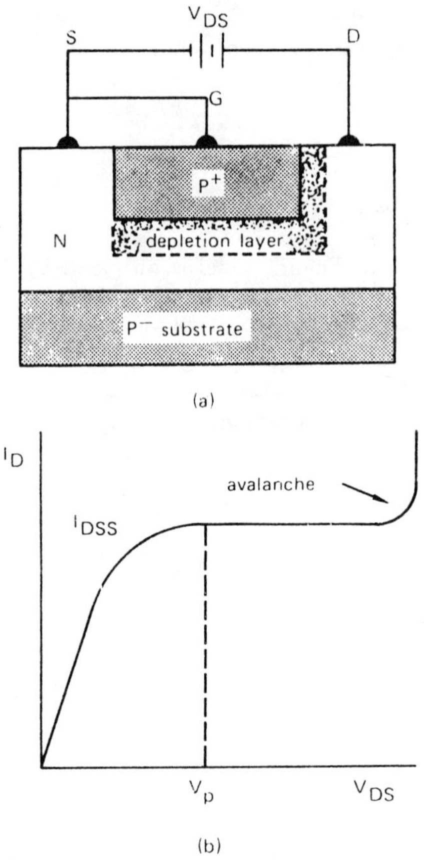

Figure 8.23 Zero biased N-channel JFET (a) depletion layer in the channel, (b) V-I characteristic for drain and source

D. DEPLETION MODE TURN-OFF

If V_{GS} is set to $-V_p$, the entire channel will be pinched off, and virtually no current will flow. This situation is shown in figure 8.24 for a P-channel JFET. This condition is called the *off-condition,* as the channel appears to be an open circuit. The value of V_{GS} is called $V_{GS(OFF)}$, equal to $-V_p$.

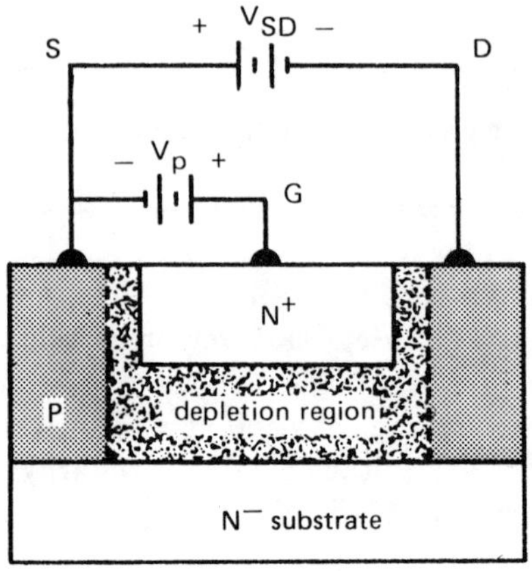

(a)

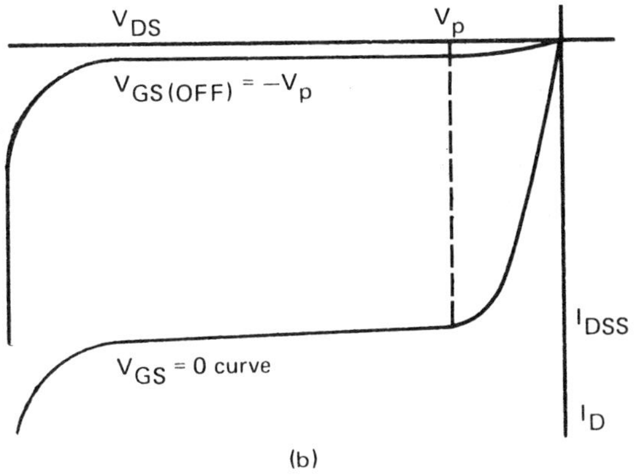

Figure 8.24 Turned-off P-channel JFET (a) depletion region in the channel, (b) V-I characteristic

E. JFET DRAIN-SOURCE CHARACTERISTICS

JFET gate voltages range from zero to $-V_p$ as previously indicated. For gate voltages between these values, characteristic curves show increments of gate-source voltage V_{GS}. These curves are most sensitive in the region around $V_{GS} = 0$. That is, in that region it requires less change in V_{GS} to change I_d than it does for larger V_{GS} values. This is illustrated by the characteristic curves shown in figure 8.25.

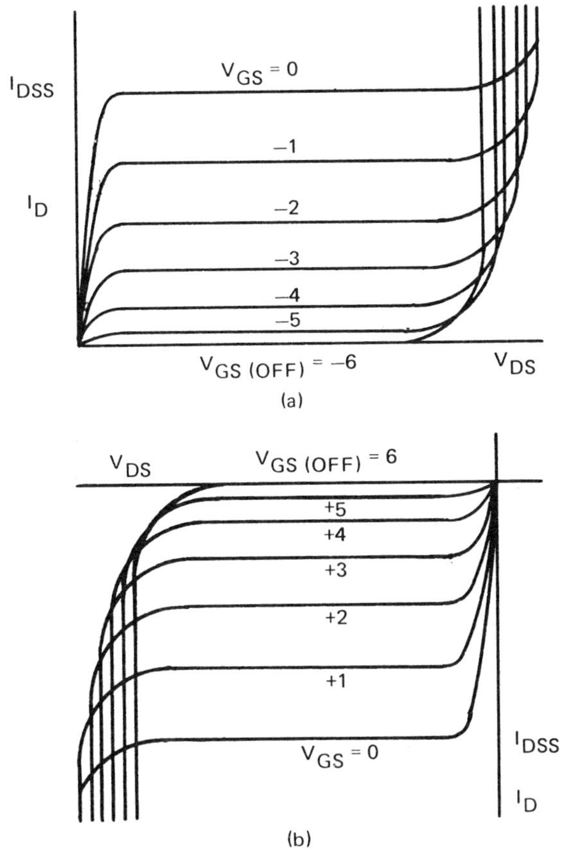

Figure 8.25 JFET drain-source characteristics for $V_{GS(\text{OFF})} = \pm 5$ volts: (a) N-channel JFET, (b) P-channel JFET

In the two characteristic curves shown in figure 8.25, the *pinchoff voltage* on each constant V_{GS} curve is

$$V_p = V_{DS} - V_{GS} = -V_{GS(\text{OFF})} \qquad 8.47$$

The junction FET in saturation can be modeled approximately by equation 8.48.

$$I_D = I_{DSS}\left[1 - \frac{V_{GS}}{V_{GS(\text{OFF})}}\right]^2 \qquad 8.48$$

F. JFET CIRCUIT SYMBOLS

JFET circuit symbols are shown in figure 8.26. (The exact placement of the gate lead is not significant. Various authors and manufacturers may set it higher or lower than shown.) The arrow shown on the gate lead points from the P-region to the N-region. For an N-channel JFET, the arrow points toward the N-channel, away from the P-gate. It points toward the N-gate from a P-channel JFET.

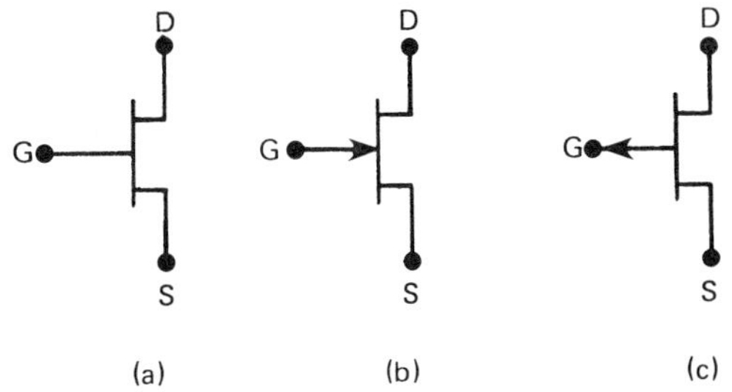

Figure 8.26 JFET symbols: (a) general form showing drain, source, and gate, (b) N-channel, (c) P-channel

6 DEPLETION MODE MOSFET's

A. INTRODUCTION

Depletion mode MOSFET's are similar to JFET's in many respects, with the exception of a reversed bias direction on the gate-source voltage. These devices do not employ a P-N junction between the gate and the channel. Rather they use a capacitance connection. The metallic gate (typically aluminum) is separated from the channel by a thin layer of insulating material (typically silicon dioxide SiO_2). This construction achieves the same effects as the JFET with a simpler manufacturing process, while gaining certain advantages over the JFET.

With the metal gate and intervening insulator, depletion regions are set up in depletion mode MOSFET's, comparable to those in JFET's. The fundamental difference between MOSFET and JFET operation is that there is no bias problem. Therefore, a wider range of gate-source voltages (including positive voltages for N-channel and negative voltages for P-channel MOSFET's) is usable.

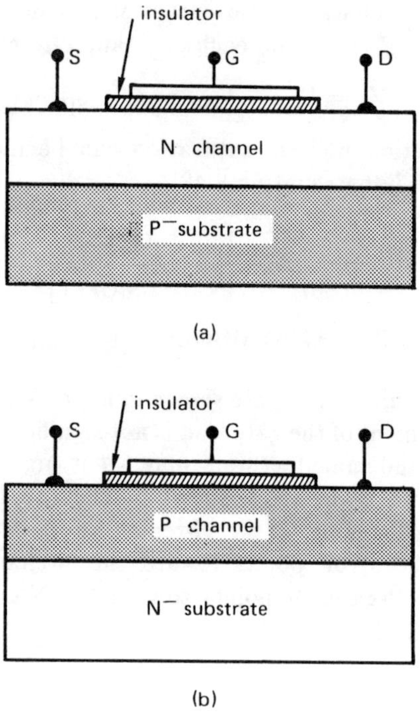

Figure 8.27 Depletion mode MOSFET: (a) N-channel, (b) P-channel

The advantage of JFET's over MOSFET's is in the levels of gate-source voltages which are permissible. Because the JFET is a P-N junction device, static electricity is not a problem. On the other hand, the very thin layers of insulation between the gate and the channel used for MOSFET's create a new problem.

When very high voltages (which can be generated electrostatically) are imposed on the gate lead of a MOSFET, it is possible to cause permanent damage to the gate-channel interface by creating a conductive path. For this reason MOSFET's must be handled with extreme care during fabrication and during installation in circuits. Otherwise, they can be damaged by static electricity produced on the person handling them.

B. CHARACTERISTICS OF DEPLETION MODE MOSFET's

Because MOSFET's do not utilize a P-N junction between the gate and the channel, it is possible to provide bias voltages to the gate with respect to the source of either polarity. Where the gate-source voltage is in the forward-biased region for a P-N junction, there is an accumulation of charge in the channel. This aids the conduction from drain to source (i.e., extra majority carriers are available, so that for any given drain-source voltage a current greater than I_{DSS} of the comparable JFET is possible). In such operation, the MOSFET is said to be in the *enhancement mode*.

Thus, the depletion mode MOSFET is capable of operation in the so-called enhancement mode as well as in the depletion mode utilized by similar JFET's. The result is a somewhat different drain-source characteristic, as shown in figure 8.28.

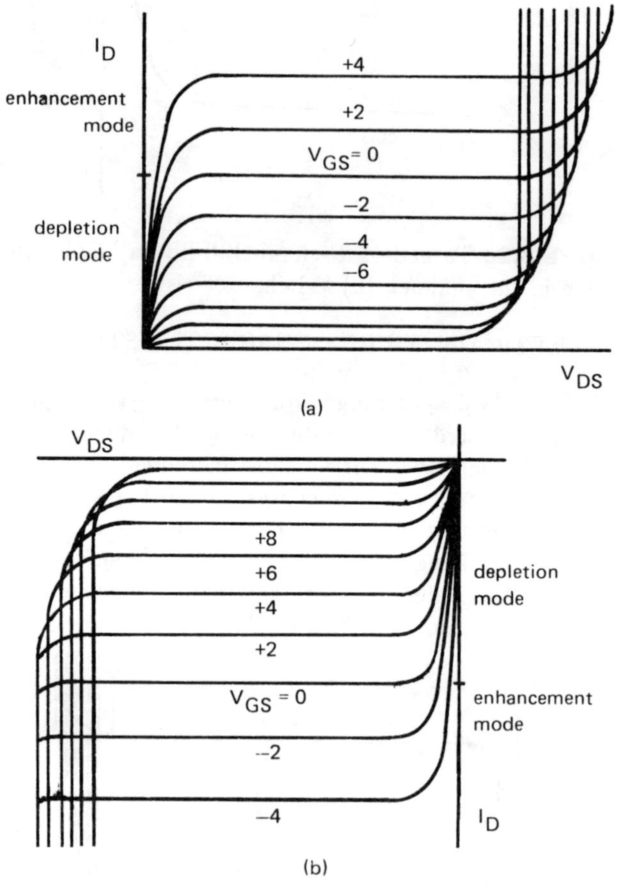

Figure 8.28 Depletion and enhancement mode operation: (a) N-channel, (b) P-channel

C. CIRCUIT SYMBOLS FOR DEPLETION MODE MOSFET's

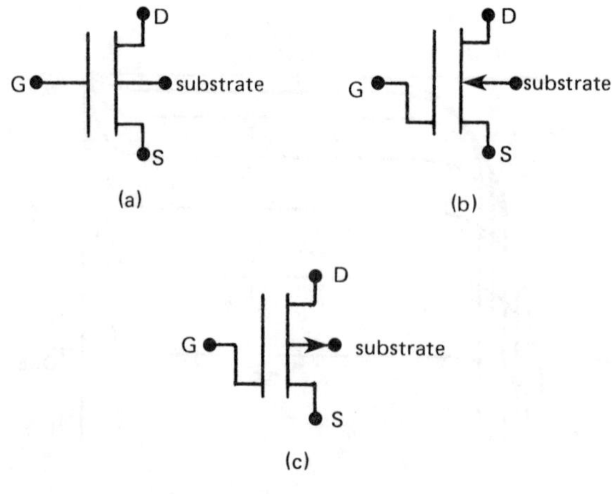

Figure 8.29 Depletion mode MOSFET circuit symbols: (a) general form with drain, gate, source, and substrate lead, (b) N-channel, (c) P-channel

The circuit symbols for depletion mode MOSFET's differ from the JFET symbol in that a space is left between the gate and the channel. This indicates that a capacitance exists by virtue of the insulator. The substrate connection also is included as a separate lead. The arrow identifying the channel as N-type or P-type is shown with the substrate lead. The arrow points toward the N-channel and away from the P-channel.

7 DEPLETION MODE FET BIASING

A. INTRODUCTION

Biasing depletion mode FET's for small-signal amplifier (linear) operation requires choosing an *operating point* in the saturation region.[1] This region is bounded by pinchoff and avalanche breakdown as indicated in figure 8.30(a) for an N-channel FET. For a P-channel FET, the voltages and currents all are reversed from those shown. V_{DSQ}, V_{GSQ}, and I_{DQ} are the operating point values.

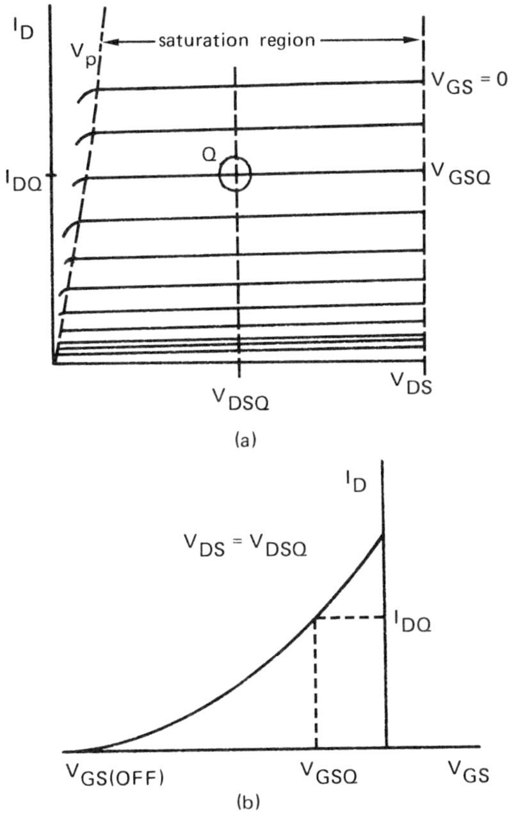

Figure 8.30 (a) N-channel FET operating point, (b) transconductance curve for use in biasing calculations

The most useful curve for determining biasing is the *transconductance curve* shown in figure 8.30(b). This curve is obtained by plotting I_D versus V_{GS} with $V_{DS} = V_{DSQ}$.

[1]The operating point also is known as the *quiescent point* or *Q-point*.

B. A SIMPLE SELF-BIASING CIRCUIT

Figure 8.31 shows a simple self-biasing circuit using an N-channel depletion mode MOSFET. Because no gate current flows, the resistance R_G carries no current, and the gate terminal is at ground potential. The source terminal will be at the voltage $I_D R_S$, so that $V_{GS} = -I_D R_S$.

On the transconductance curve, the relation between I_D and V_{GS} is that of a straight line through the origin with slope $-I/R_S$. The design then requires that

$$R_S = \frac{V_{GSQ}}{I_{DQ}} \qquad 8.49$$

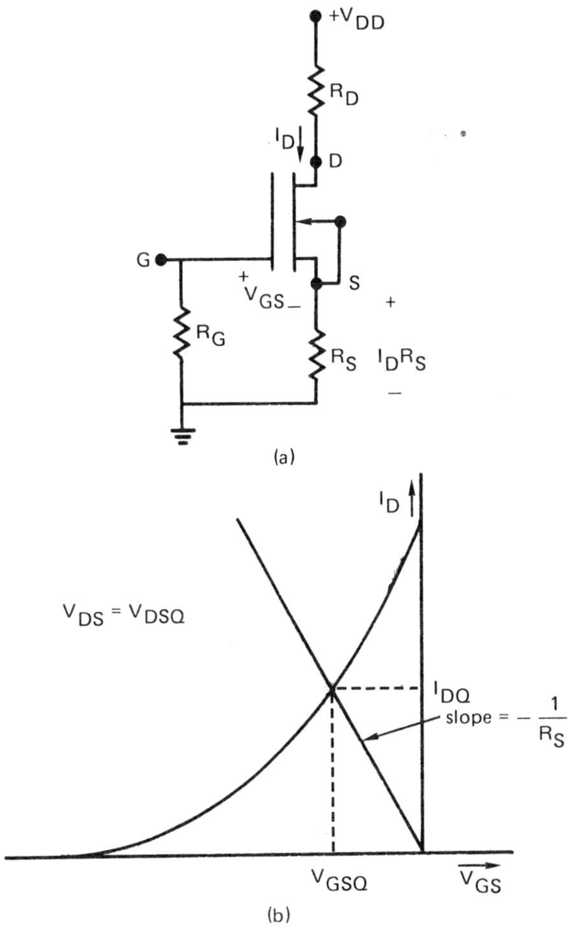

Figure 8.31 (a) Simple self-biasing circuit FET, (b) use of transconductance curve

The design is completed by determining R_D. Kirchhoff's voltage law used around the drain-source circuit yields

$$V_{DD} = I_D R_D + V_{DS} + I_D R_S \qquad 8.50$$

But $I_D R_S = -V_{GS}$, so at the operating point

$$R_D = \frac{V_{DD} - V_{DSQ} + V_{GSQ}}{I_{DQ}} \qquad 8.51$$

The bias method used here is adequate for only one circuit, when device-to-device variations are not important. When many similar circuits are produced, it is necessary to insure that the operating points of V_{DS} and I_{DS} be maintained within narrow limitations.

C. BIASING FOR DEVICE CHARACTERISTIC VARIATIONS

Characteristic curves for two FET's with the same nominal parameters are shown in figure 8.32. These can be plotted together on a single transconductance curve to give a range of variations in operating point. The most important parameter to control for small-signal FET amplifiers is the *incremental transconductance*, g_m. The incremental transconductance is the slope of the transconductance curve at the operating point.

Variations in transconductance can be minimized by choosing operating points on the two limiting curves of figure 8.32(c) which have the same slope.

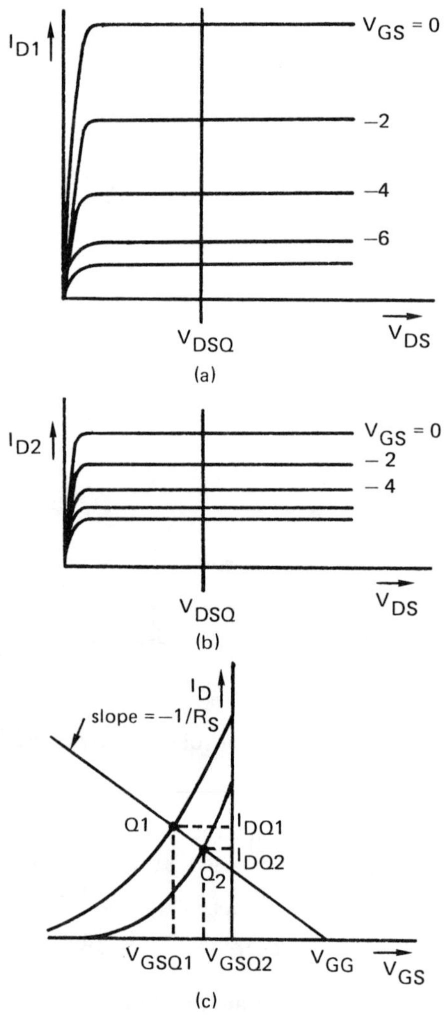

Figure 8.32 Two depletion mode FET's of the same type: (a) highest gain version, (b) lowest gain version, (c) transconductance curves

The *bias load-line* can be calculated from the two-point form for the two operating points.

$$\frac{I_D - I_{DQ2}}{V_{GS} - V_{GSQ2}} = \frac{I_{DQ1} - I_{DQ2}}{V_{GSQ1} - V_{GSQ2}} \qquad 8.52$$

Solving for V_{GS},

$$V_{GS} = V_{GG} - R_S I_D \qquad 8.53$$

$$V_{GG} = V_{GSQ2} + \left[\frac{V_{GSQ2} - V_{GSQ1}}{I_{DQ1} - I_{DQ2}}\right] I_{DSQ2} \qquad 8.54$$

$$R_S = \frac{V_{GSQ2} - V_{GSQ1}}{I_{DQ1} - I_{DQ2}} \qquad 8.55$$

(Note that V_{GSQ1} and V_{GSQ2} are both negative numbers, so that all terms in equations 8.54 and 8.55 are positive.)

Because V_{GG} is positive, it can be supplied by the drain voltage source. A biasing circuit which will accomplish this is shown in figure 8.33.

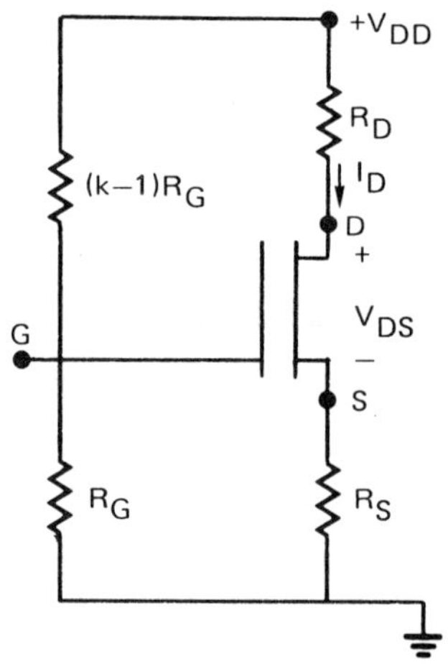

Figure 8.33 Constant g_m bias circuit

Using Kirchhoff's voltage law around the drain-source circuit,

$$R_D = \frac{V_{DD} - V_{DSQ}}{I_{DQ}} - R_S \qquad 8.56$$

$$I_{DQ} = \frac{1}{2}(I_{DQ1} + I_{DQ2}) \qquad 8.57$$

The gate resistance R_G should be large and $V_{GG} = V_{DD}/k$, so

$$k = \frac{V_{DD}}{V_{GG}} \qquad 8.58$$

8 ENHANCEMENT MODE MOSFET's

A. INTRODUCTION

Enhancement mode MOSFET's can be either N-channel or P-channel. They are depletion mode devices with similar external supply requirements. The principle difference between depletion mode and enhancement mode devices is that with no gate-source voltage there is no channel in the enhancement mode device, so the drain current is zero. An enhancement mode (N-channel) MOSFET is shown in figure 8.34(a).

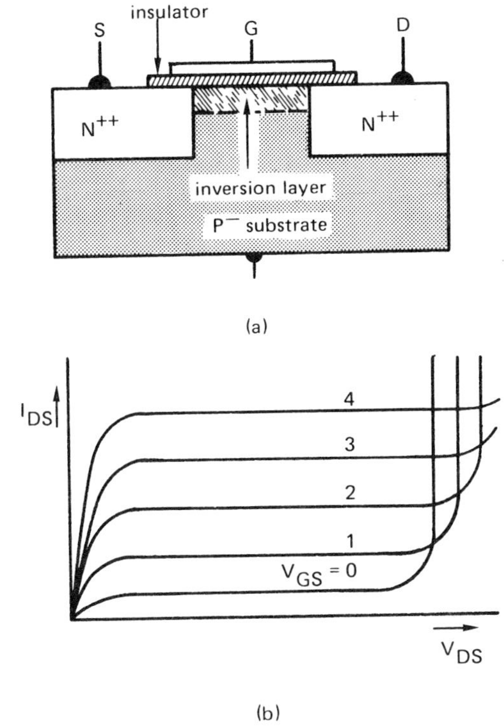

Figure 8.34 (a) Enhancement mode N-channel MOSFET, (b) characteristics

To establish a channel, a positive gate-source voltage depletes positive charge carriers (holes) in the lightly-doped P^- region by attracting electrons to fill those holes, thereby establishing a negative charge in the region near the gate in the P^- region. At the same time, free electrons are attracted from the N^{++} source region. (Note that the gate overlaps the N^{++} source and drain region.) This provides a ready source of electrons to replace any which are attracted to the drain.

The loosely-bound electrons which fill the holes in the depletion region effectively invert the P^- region making it an N-channel. This channel is called an *inversion layer*. Its depth increases with increasing gate-source voltage. The resulting characteristic is shown in figure 8.34(b). Characteristics for P-channel are similar with all voltage and current directions reversed.

B. ENHANCEMENT MODE MOSFET CIRCUIT SYMBOLS

The circuit symbol for the enhancement mode MOSFET is similar to the depletion mode symbol, except that the line representing the channel is divided into three parts: one for the drain, one for the substrate, and one for the source. The placement of the lead on the gate is not significant.

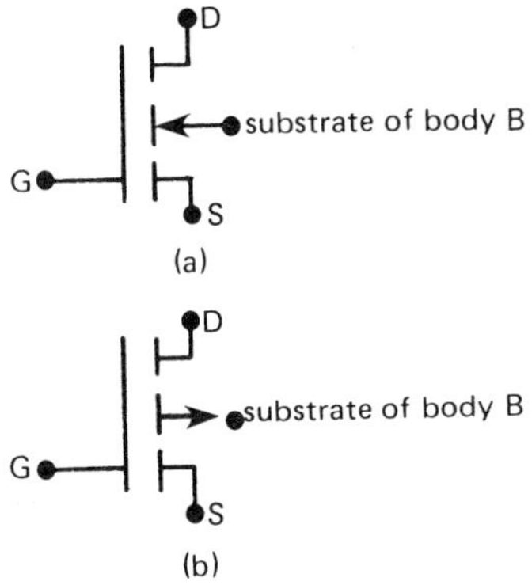

Figure 8.35 Enhancement mode MOSFET: (a) N-channel, (b) P-channel

C. VOLTAGE FEEDBACK BIASING

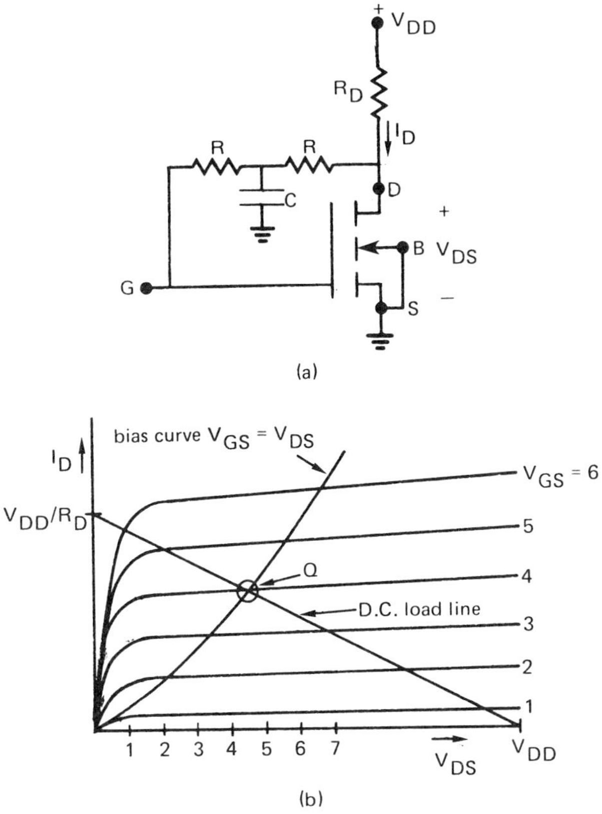

Figure 8.36 (a) voltage feedback biasing, (b) load line and the bias curve

One common method of biasing the enhancement mode MOSFET is shown in figure 8.36. The bias level of the gate is tied directly to the average value of the drain-source voltage. This is known as *voltage feedback biasing*. With the characteristic shown in figure 8.36(b), the locus of points with a gate-source voltage equal to the drain-source voltage is the bias curve. The Kirchhoff voltage law equation for the drain-source circuit is

$$V_{DS} = V_{DD} - I_D R_D \qquad 8.59$$

The D.C. load line intercepts are $V_{DS} = V_{DD}$ and $I_D = V_{DD}/R_D$. The intersection of these two curves gives the operating point, marked by Q on the characteristic curves.

D. CURRENT FEEDBACK BIASING

Another common biasing method for enhancement mode MOSFET's is shown in figure 8.37(a). This is similar to bias stabilization of depletion mode FET's. The transconductance curve is used (as with depletion mode FET's) and is shown in figure 8.37(b).

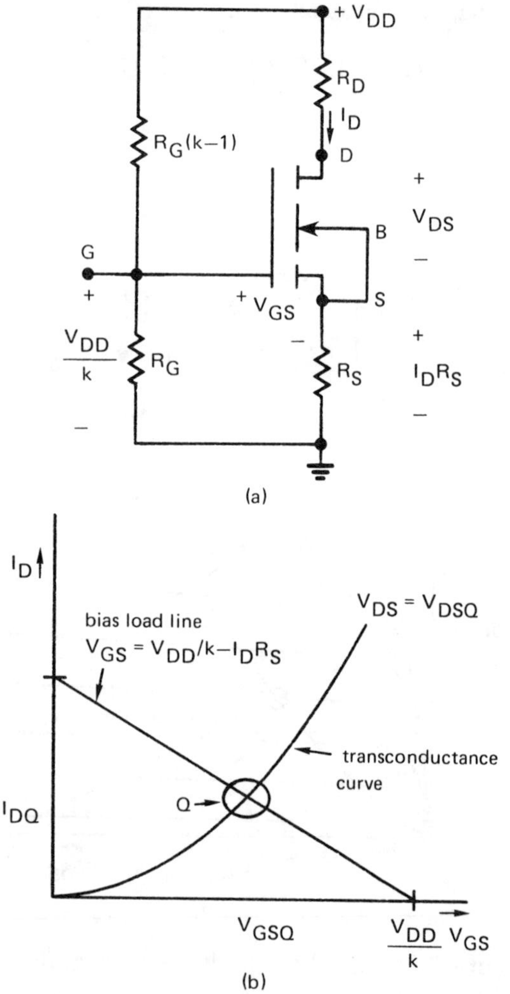

Figure 8.37 (a) Current feedback biasing, (b) transconductance characteristic

The application of Kirchhoff's voltage law around the gate-source circuit generates the bias load line equation.

$$V_{GS} = \frac{V_{DD}}{k} - I_D R_S \qquad 8.60$$

The load line can be plotted on the transconductance curve. The intersection of these two curves is the operating point Q as indicated in figure 8.37(b).

If I_{DQ} is specified and k is selected as 2 for maximum input resistance, then the two-point method can be used for a design problem. (V_{GSQ} is found from the transconductance curve.)

$$\frac{V_{GS} - V_{GSQ}}{I_D - I_{DQ}} = \frac{\frac{V_{DD}}{k} - V_{GSQ}}{0 - I_{DQ}} \qquad 8.61$$

$$V_{GS} = \frac{V_{DD}}{k} - I_D\left[\frac{\frac{V_{DD}}{k} - V_{GSQ}}{I_{DQ}}\right] \qquad 8.62$$

$$R_S = \frac{\frac{V_{DD}}{k} - V_{GSQ}}{I_{DQ}} \qquad 8.63$$

The design is completed by using Kirchhoff's voltage law around the drain-source circuit.

$$R_D = \frac{V_{DD} - V_{DSQ} - I_{DQ}R_S}{I_{DQ}} \qquad 8.64$$

E. VERTICAL ENHANCEMENT MODE MOSFET (VMOS)

All of the MOS devices discussed up to this point are limited to drain currents in the milliampere range. This limitation is due primarily to the relatively long channel. An alternative uses three diffusions: an N^+ drain, a P^- depletion channel, and an N^+ source. After the diffusions are completed, a V-shaped notch is cut in the wafer. An oxide layer placed in the groove is followed by a metal gate layer.

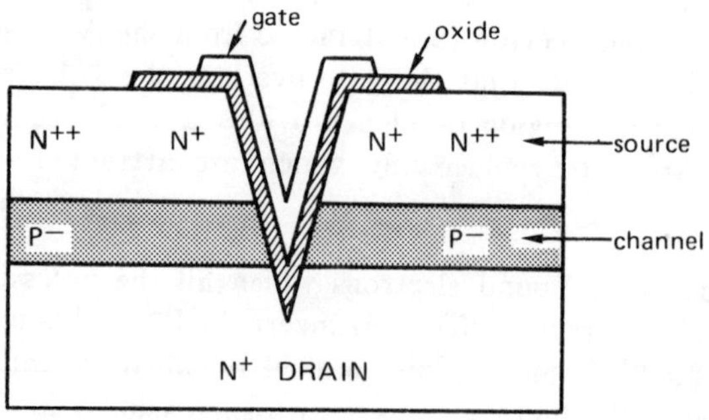

Figure 8.38 Vertical MOSFET (VMOS)

Because the channel can be made short, VMOS transistors are capable of high currents in the range of tens of amperes or more.

The characteristics and control of VMOS circuitry essentially is the same as for the enhancement mode N-channel MOSFET.

9 SMALL-SIGNAL FET PARAMETERS

A. INTRODUCTION

The two main low-frequency parameters are the transconductance and the drain-source resistance. Transconductance is obtained from the slope of the transconductance curve at the operating point. The drain-source resistance is obtained from the slope of the constant V_{GS} characteristic on the I_D versus V_{DS} characteristic curves at the operating point.

$$g_m = \left.\frac{\partial I_D}{\partial V_{GS}}\right|_Q \quad 8.65$$

$$\frac{1}{r_{ds}} = \left.\frac{\partial I_D}{\partial V_{DS}}\right|_Q \quad 8.66$$

This approach produces the same equivalent circuit for field effect transistors, regardless of the type or channel polarity.

B. LOW-FREQUENCY SMALL-SIGNAL FET EQUIVALENT CIRCUIT

For small variations about the operating point, each variable X can be expressed as $X = X_Q + x$. X_Q is the operating point value and x is a small variation in that value. Thus,

$$V_{DS} = V_{DSQ} + v_{ds} \quad 8.67$$

$$V_{GS} = V_{GSQ} + v_{gs} \quad 8.68$$

$$I_D = I_{DQ} + i_d \quad 8.69$$

As the operating values are constant and satisfy Kirchhoff's laws independently of the small variations, they can be omitted from the equations, and the variations can be considered alone. The small variations must satisfy Kirchhoff's laws by themselves.

Thus small-signal circuits do not need to include any constant (D.C.) sources. For any FET, the small-signal equivalent circuit is shown in figure 8.39(a). To find the total voltages or currents for any specific form, the operating point values must be added to the small-signal values.

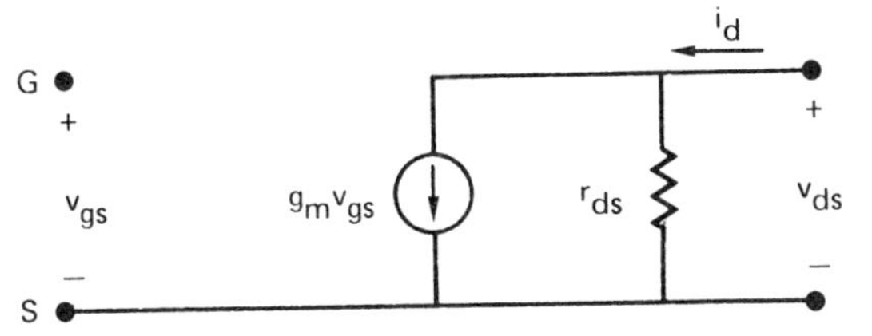

Figure 8.39(a) Low-frequency, small-signal FET equivalent circuit

C. HIGH-FREQUENCY SMALL-SIGNAL FET EQUIVALENT CIRCUIT

The intrinsic design of FET's gives rise to capacitances which are important at high frequencies. These capacitances are associated with depletion and inversion layers in general and are dependent on the voltages associated with them.

The *gate-channel capacitance* C_{ISS} and the *output capacitance* C_{OSS} both are distributed capacitances. They are difficult to assign to either lead along the channel. The input capacitance C_{ISS} usually is divided between the drain and the source to produce capacitances C_{gs} and C_{gd}. The output capacitance includes the *gate-channel capacitance* C_{ISS} and the *channel-substrate capacitance.* Because the substrate often is connected to the source, the channel-substrate capacitance usually is attributed to the drain. The resulting high-frequency, small-signal equivalent circuit is shown in figure 8.39(b).

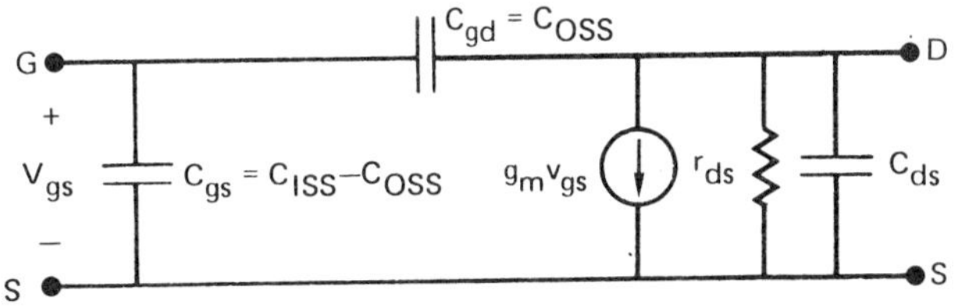

Figure 8.39(b) High-frequency, small-signal FET equivalent circuit

10 BIPOLAR JUNCTION TRANSISTORS

A. INTRODUCTION

Bipolar junction transistors (BJT's) consist of two opposing PN junctions. Both *N-type carriers* (electrons) and *P-type carriers* (holes) carry substantial current. Thus, the devices are called *bipolar.*

The bipolar junction transistor consists of two opposing PN junctions with a narrow intervening layer called the *base.* The other two regions are called the *collector* and the *emitter.* The collector-base junction is biased in the reverse direction while the emitter-base junction is forward-biased for normal amplifier operation.

The bipolar junction transistor characteristics most often provided give the collector current I_c versus the collector – emitter voltage V_{ce}. Such characteristics are called *common emitter characteristics.* Alternatively, the *common base characteristics* are sometimes provided. These are curves of constant I_e (emitter current) with coordinates of the collector-base voltage and the collect current.

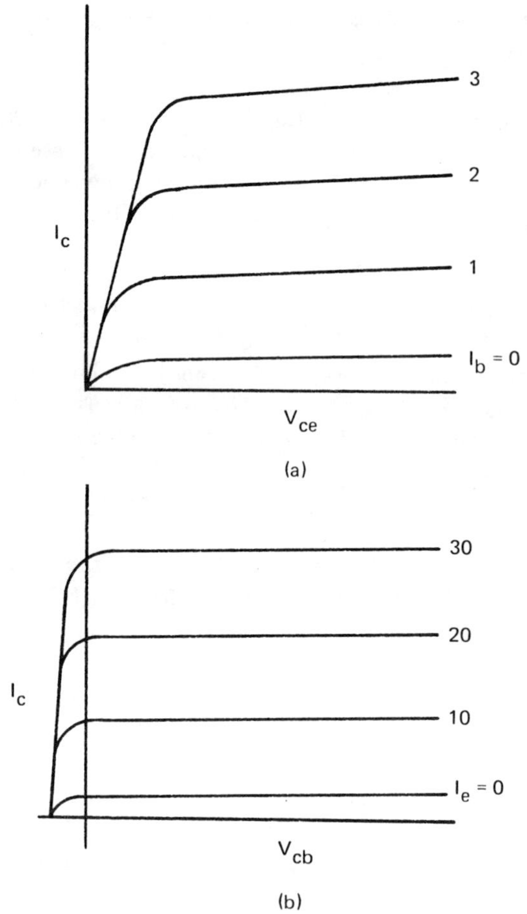

Figure 8.40 NPN bipolar junction transistor characteristics: (a) common emitter, (b) common base

B. BIPOLAR JUNCTION TRANSISTOR BIASING

The operating point Q for a bipolar junction transistor will depend on the D.C. component of current into the third terminal (i.e., the base for common emitter characteristics and the emitter for common base characteristics). The operating point is established by the combination of the load-line (found from the voltage source and resistance connected to the output terminals of the bipolar junction transistor) and the current into the third terminal.

Most bias design uses the common base D.C. parameters. In this case, the collector current is proportional to the emitter current.

$$I_c = I_{co} + \alpha_o I_e \qquad 8.70$$

I_{co} is the collector current (see figure 8.41(b)) when the emitter current is zero, and α_o is the average change in collector current divided by the corresponding change in emitter current.[2] Plotting the load-line on the characteristic curve, the operating point is the intersection of the load line with the characteristic curve corresponding to the D.C. value of the emitter current.

The collector current in the common base configuration is independent of the supply voltage V_S and the Thevenin equivalent resistance R_S is a function of I_e only. The load line sets the relationship between the collector current and the collector-base voltage. The general biasing circuit is shown in figure 8.41.

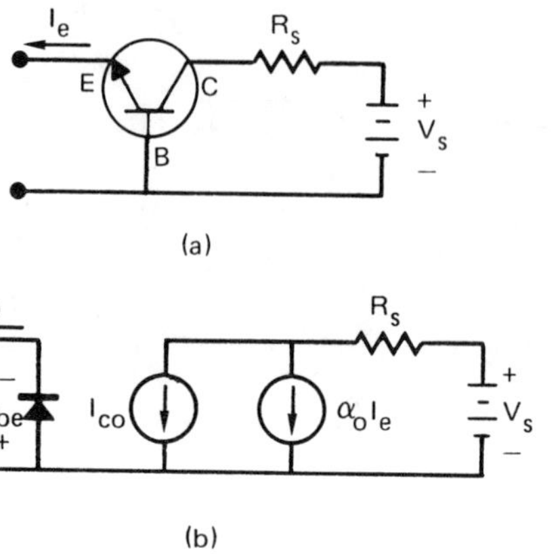

Figure 8.41 (a) Common base biasing circuit, (b) equivalent circuit

The bipolar junction transistor has three terminals, and the currents in each terminal must add to zero to satisfy Kirchhoff's current law. With directions as given in figure 8.42, the current equation is

$$I_b + I_c = I_e \qquad 8.71$$

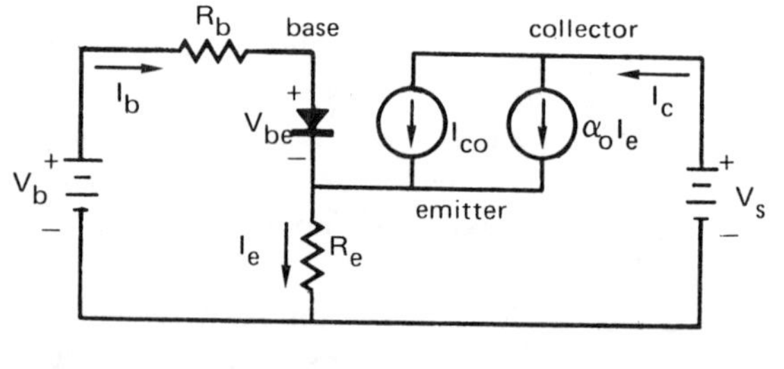

Figure 8.42 Circuit for biasing calculations

From equation 8.71,

$$I_e = \frac{I_c - I_{co}}{\alpha_o} \qquad 8.72$$

Therefore,

$$I_b = \left[\frac{1-\alpha_o}{\alpha_o}\right] I_c - \frac{I_{co}}{\alpha_o} \qquad 8.73$$

Writing Kirchhoff's voltage law around the left-hand loop of figure 8.42,

$$I_c = \frac{\alpha_o(V_b - V_{be}) + I_{co}(R_b + R_e)}{R_b(1-\alpha_o) + R_e} \qquad 8.74$$

[2] α_o sometimes is given the symbol $\alpha_{D.C.}$ in the literature. It is known as the *large signal current gain.*

Example 8.1

A silicon NPN transistor is used in the circuit shown. The transistor parameters are $I_{co} = 10^{-6}$ amps, $\alpha_o = .95$, $R_e = 50\,\Omega$ and $V_{cc} = 24$ volts. Assume $R_1 = R_2$. Determine the value of R_b to provide an operating point of $I_c = 10$ mA. (R_b is the parallel combination of R_1 and R_2 and is written as $R_b = R_1 || R_2$.)

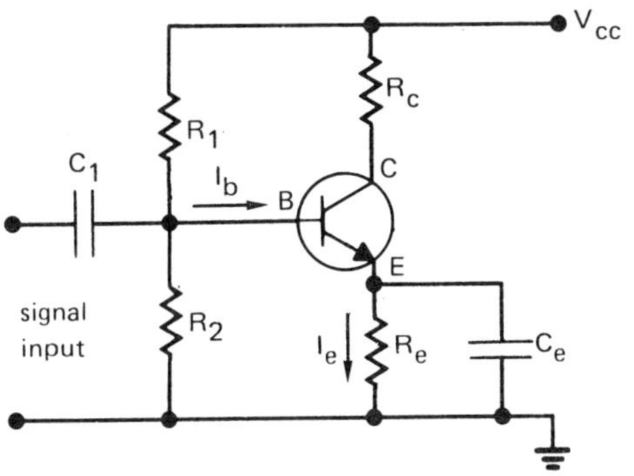

The Thevenin equivalent D.C. circuit for the bias is R_b in series with 12 volts.

$$V_b = 24\left[\frac{R_2}{R_1 + R_2}\right]$$

For a silicon transistor, the base-emitter voltage is the diode drop for a forward-biased diode, on the order of 0.6 to 1.5 volts. The value usually is taken as 0.6 or 0.7 volts. Here it is taken as 0.6 volts. Then solve for R_b from equation 8.74.

$$I_c = .01 = \frac{.95(12 - .6) + 10^{-6}(R_b + 50)}{R_b(1 - .95) + 50}$$

$$R_b = 20.7K\,\Omega$$

C. STABILITY

Bipolar junction transistor stability is fundamental in the design of reliable electronic circuits. The base-emitter junction voltage decreases at a rate of 2.5 mV per degree Celsius. This is the major problem in temperature stabilization of silicon transistors.

$$\frac{\partial V_{be}}{\partial T} = -2.5\,\mathrm{mV}/^\circ C \qquad 8.75$$

The resulting change in collector current is

$$\frac{\partial I_c}{\partial T} = \frac{\partial I_c}{\partial V_{be}} \times \frac{\partial V_{be}}{\partial T} = -.0025\frac{\partial I_c}{\partial V_{be}} \qquad 8.76$$

Using equation 8.73,

$$\frac{\partial I_c}{\partial V_{be}} = \frac{-\alpha_o}{R_b(1 - \alpha_o) + R_e} \qquad 8.77$$

As α_o is approximately 1, a reasonable approximation is

$$S_v = \frac{\partial I_c}{\partial T} = \frac{2.5}{R_e}\alpha_o\ \mathrm{mA}/^\circ\mathrm{C} \qquad 8.78$$

S_v is called the *voltage stability factor*. It is necessary to examine $R_b(1-\alpha_o)$ to determine whether it is negligible compared to R_e. If it is not, then

$$S_v = \frac{2.5\alpha_o}{(1 - \alpha_o)R_b + R_e}\ \mathrm{mA}/^\circ hboxC \qquad 8.79$$

Example 8.2

Determine the change in collector current for a temperature change of 125°C for the circuit of example 8.1.

From equation 8.79,

$$S_v = \frac{2.5\alpha_o}{.05 \times 20{,}700 + 50} = \frac{2.5(.95)}{1085} = .0022\,\mathrm{mA}/^\circ C$$

In this case, $R_b(1 - \alpha_o)$ is not negligible. (See values from example 8.1.)

$$\Delta I_c = S_v \times 125^\circ = 0.275\,\mathrm{mA}$$

D. PARAMETER VARIATION

Interchangeability of bipolar junction transistors is a major concern for circuits which are to be produced in quantity. Parameter variations among transistors of the same type must be taken into account. The primary parameter is the large signal current gain α_o.

The usual method is to design the circuit so that the collector current will remain within a specified range when α_o varies. For a known maximum and minimum α_o, use equation 8.72 but neglect I_{co} because of its insignificant contribution.

$$I_{c,\max} = \frac{\alpha_{\max}(V_b - V_{be})}{(1 - \alpha_{\max})R_b + R_e} \qquad 8.80$$

$$I_{c,\min} = \frac{\alpha_{\min}(V_b - V_{be})}{(1 - \alpha_{\min})R_b + R_e} \qquad 8.81$$

To maintain a specified range independent of the voltages, take the ratio of equations 8.80 and 8.81.

$$k_Q > \frac{I_{c,\max}}{I_{c,\min}} = \frac{(1 - \alpha_{\min})R_b + R_e}{(1 - \alpha_{\max})R_b + R_e} \times \frac{\alpha_{\max}}{\alpha_{\min}} \qquad 8.82$$

k_Q is the specified maximum ratio of the maximum I_c to the minimum at the operating point. This results in the following inequality relating R_b and R_e.

$$\left[\frac{1 - \alpha_{\min}}{\alpha_{\min}} - k_Q\left(\frac{1 - \alpha_{\max}}{\alpha_{\max}}\right)\right]R_b < \left[\frac{k_Q}{\alpha_{\max}} - \frac{1}{\alpha_{\min}}\right]R_e \qquad 8.83$$

Example 8.3

The transistor of example 8.1 has a range of α_o from 0.94 to 0.96. Determine whether R_b and R_e will keep I_c between 9 and 11 mA. If not, redesign the circuit to meet the specifications.

$$k_Q = \frac{11}{9}$$

From equation 8.83,

$$\left[\frac{1-.94}{.94} - \frac{11}{9} \times \frac{1-.96}{.96}\right] R_b < \left[\frac{11}{9 \times .96} - \frac{1}{.94}\right] R_e$$

$$R_e > 0.0616 R_b$$

This is not satisfied when $R_e = 50$ and $R_b = 20.7\text{K}$. The minimum and maximum values of I_c calculated from equation 8.74 are 8.29 and 12.46 mA, respectively, (neglecting I_{co}).

The procedure is to set R_e within the restriction (greater than 0.0616 R_b), so set $R_e = .0625 R_b$. This results in minimum and maximum currents of $87.48/R_b$ and $106.8/R_b$, respectively, which satisfies the range ratio of 11:9. Then, setting the low value at 9 mA,

$$R_b = \frac{87.48}{.009} = 9720\,\Omega$$

E. SMALL-SIGNAL PARAMETERS

Common emitter parameters, such as those shown in figure 8.43, relate variables I_c, V_{ce}, and I_b. There are three nonlinear functional relationships between the characteristic curves. However, I_b is an independent variable, so only the two functions are needed.

$$I_c = \mathbf{f}_1(V_{ce}, I_b) \qquad 8.84$$

$$V_{ce} = \mathbf{f}_2(I_c, I_b) \qquad 8.85$$

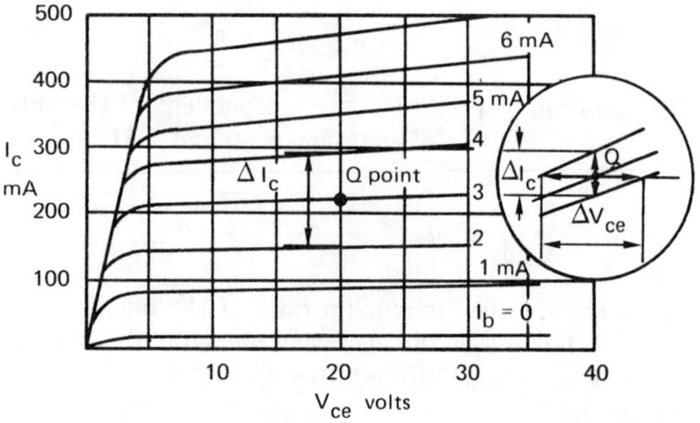

Figure 8.43 Characteristics for determining small-signal parameters

These equations are expanded for small variations. The small change in I_c is represented by i_c and is a variation around the operating Q point. Small changes in V_{ce} are similarly represented as v_{ce}, and small changes in I_b around its bias value are represented by i_b. Thus, for small variations around an operating point,

$$I_c + i_c = \mathbf{f}_1 + \frac{\partial I_c}{\partial V_{ce}} v_{ce} + \frac{\partial I_c}{\partial I_b} i_b \qquad 8.86$$

$$V_{ce} + v_{ce} = \mathbf{f}_2 + \frac{\partial V_{ce}}{\partial I_c} i_c + \frac{\partial V_{ce}}{\partial I_b} i_b \qquad 8.87$$

However, $V_{ce} = \mathbf{f}_2$ and $I_c = \mathbf{f}_1$, so

$$i_c = \frac{\partial I_c}{\partial V_{ce}} v_{ce} + \frac{\partial I_c}{\partial I_b} i_b \qquad 8.88$$

$$v_{ce} = \frac{\partial V_{ce}}{\partial I_c} i_c + \frac{\partial V_{ce}}{\partial I_b} i_b \qquad 8.89$$

These partial derivatives are found from the characteristics. The partial derivative of I_c with respect to V_{ce} is the slope of the constant I_b curve through the operating point.

$$\frac{1}{r_c} = \frac{\partial I_c}{\partial V_{ce}} = \lim_{\Delta V_{ce} \to 0} \frac{\Delta I_c}{\Delta V_{ce}}\Big|_{I_b = \text{constant}} \qquad 8.90$$

The slope of the constant I_b curve is easily evaluated if the curve goes through the actual operating point. In most cases, the operating point will fall between curves, and the slope must be estimated.

The partial derivative of I_c with respect to I_b is more difficult to evaluate. It is estimated by determining the change in I_c from one constant I_b curve below the operating point to the next constant I_b curve above the operating point, both values taken for the operating point value of V_{ce}. Then, this difference is divided by the difference in I_b on two constant I_b curves.

$$\beta = \frac{\partial I_c}{\partial I_b} = \lim_{\Delta I_b \to 0} \frac{\Delta I_c}{\Delta I_b}\Big|_{V_{ce} = \text{constant}} \qquad 8.91$$

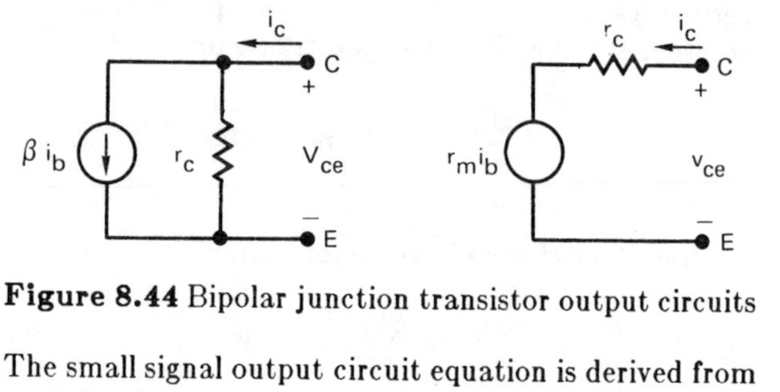

Figure 8.44 Bipolar junction transistor output circuits

The small signal output circuit equation is derived from equation 8.88.

$$i_c = \frac{1}{r_c} v_{ce} + \beta i_b \qquad 8.92$$

The equivalent circuit derived from equation 8.92 is shown in figure 8.44.

r_c can be evaluated in another way from the characteristic curves.

$$r_c = \frac{\partial V_{ce}}{\partial I_c} = \lim_{\Delta I_c \to 0} \frac{\Delta V_{ce}}{\Delta I_c}\Big|_{I_b=\text{constant}} \qquad 8.93$$

This is the inverse of the slope of the constant I_b characteristic curve at the operating point. The partial derivative of V_{ce} with respect to I_b is not easily evaluated for transistors, because of the curve flatness. If the characteristic curves had a steeper slope, as do those shown in the inset of figure 8.43, then the derivative would be taken along the constant I_c value at the operating point.

$$r_m = \frac{\partial V_{ce}}{\partial I_b} = \lim_{\Delta I_b \to 0} \frac{\Delta V_{ce}}{\Delta I_b}\Big|_{I_c=\text{constant}} \qquad 8.94$$

Equations 8.93 and 8.94 can be combined with equation 8.89.

$$v_{ce} = r_c i_c - r_m i_b \qquad 8.95$$

This is represented by the output circuit shown in figure 8.44(b). Note that this is the Thevenin equivalent of the circuit in figure 8.44(a) with $r_m = \beta r_c$. This is valid only if very close values of I_b are available on the curves. Usually, the most satisfactory method of obtaining the circuit shown in figure 8.44(b) is by a Thevenin transformation of the circuit of figure 8.44(a) which has readily obtainable parameters.

Example 8.4

For the operating point given in figure 8.43, obtain the small-signal output parameters.

The slope of the characteristic through the operating point is approximately

$$\frac{1}{r_c} = \frac{.03}{30} = \frac{1}{1000}$$

$$r_c = 1000 \text{ ohms}$$

Using the ΔI_c shown, ΔI_b is $4 - 2 = 2$ mA. The corresponding values of I_c are 280 and 140, so $\Delta I_c = 140$ mA. Thus,

$$\beta = \frac{140}{2} = 70$$

Input characteristics normally are not given on manufacturer specification sheets, but the parameters are. The output parameters are hybrid parameters for the common emitter configuration. Thus, equation 8.92 could be written as

$$i_c = h_{fe} i_b + h_{oe} v_{ce} \qquad 8.96$$

The corresponding input parameters are h_{ie} and h_{re}. h_{re} is negligibly small in the common emitter parameters, so only h_{ie} is necessary. This may be given in the specifications for a specific collector current. It also can be constructed from the *bulk base resistance* $r_{bb'}$ and the *base-spreading resistance* $r_{b'e}$ (also called r'_e).

$$r_{b'e} = \frac{r_e}{1-\alpha} \qquad 8.97a$$

$$r_e = \frac{kT}{qI_e}$$

$$= \frac{.026}{I_e} \text{ at } 300° K \qquad 8.97b$$

Then,

$$h_{ie} = r_{bb'} + r_{b'e} \qquad 8.98$$

$$v_{be} = h_{ie} i_b \qquad 8.99$$

h_{ie} can vary from about 150 ohms to 750 ohms. 500 ohms is a common estimate when the actual value is unknown.

11 SMALL-SIGNAL AMPLIFIERS

A. INTRODUCTION

The analysis of small-signal amplifiers is largely a circuit analysis problem. A number of amplifier parameters, including input impedance, output impedance, voltage gain, current gain, transconductance, and transresistance, can be involved. (The *transconductance* is the ratio of the output current to the input voltage, and the *transresistance* is the ratio of the output voltage to the input current.)

In small-signal analysis, the circuit should be analyzed for its *mid-frequency behavior*. In this case, the reactive elements are not significant (except in the case of tuned circuits), and the capacitors are treated as short circuits. In particular, coupling capacitances along the signal path isolate bias circuits from one stage to the next. These are treated as short circuits, with the provision that the actual impedance must be negligible compared with the impedance seen from the terminals of the capacitance.

Self-biasing resistors also often have bypass capacitors to reduce feedback. These, too, will be considered short circuits to small signals. This results in the shunting resistance being shorted out of the circuit and vanishing from the small-signal equivalent circuit.

Small signal analysis involves small variations about the operating point. All D.C. bias values disappear from the small-signal equivalent circuit with D.C. sources being replaced by their internal resistance. Thus, the small-signal equivalent circuit for the amplifier of figure 8.45(a) is shown in figure 8.45(b).

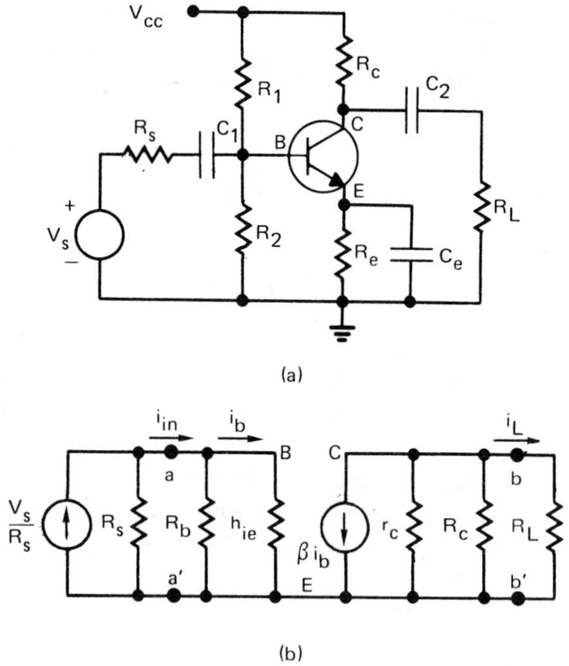

Figure 8.45 Common emitter amplifier: (a) circuit diagram, (b) small-signal equivalent circuit.

B. COMMON EMITTER BIPOLAR JUNCTION TRANSISTOR AMPLIFIER

A common emitter bipolar junction transistor amplifier is shown in figure 8.45 in mid-frequency, small-signal form. The parameters of interest are input and output impedances (resistances in this case) and the voltage and current gains.

Input impedance is the impedance seen from the input from terminals a-a′ in figure 8.45(b)). For the common emitter bipolar junction transistor amplifier shown, the input impedance is

$$Z_{\text{in}} = h_{ie} \| R_1 \| R_2 = h_{ie} \| R_b \qquad 8.100$$

$R_a \| R_b$ means the parallel combination of R_a and R_b.

Output impedance is the impedance seen from the output terminals (b-b′ in figure 8.45(b)) looking back into the output of the amplifier.

$$Z_{\text{out}} = r_c \| R_c \qquad 8.101$$

Voltage gain can be defined as the ratio of the voltage input terminals (i.e., $A_V = v_{bb'}/v_{aa'}$ in figure 8.45(b)). The input current to the base of the transistor is

$$i_b = \frac{v_{aa'}}{h_{ie}} \qquad 8.102$$

The output voltage is

$$v_{bb'} = -\beta i_b [r_c \| R_c \| R_L] \qquad 8.103$$

The voltage gain is

$$A_V = \frac{v_{bb'}}{v_{aa'}} = \frac{-\beta [r_c \| R_c \| R_L]}{h_{ie}} \qquad 8.104$$

The input current at terminals a-a′ is

$$i_{\text{in}} = i_b \left(1 + \frac{h_{ie}}{R_b}\right) \qquad 8.105$$

The output current is $v_{bb'}/R_L$, so the current gain is

$$A_I = \frac{i_L}{i_{\text{in}}}$$

$$= -\frac{\beta [r_c \| R_c \| R_L]}{R_L} \times \left(1 + \frac{h_{ie}}{R_b}\right) \qquad 8.106$$

$$R_b = R_1 \| R_2 \qquad 8.107$$

Example 8.5

The amplifier in figure 8.45 has $R_1 = 6K$, $R_2 = 3K$, $R_c = 1K$, and $R_S = 1K$ ohms. The small-signal parameters are $h_{ie} = 500$ ohms, $\beta = 50$, and $r_c = 5K$ ohms. Determine the value of R_L to give maximum power output.

The *maximum power transfer theorem* requires that the load impedance match the Thevenin equivalent impedance of the driving source. This is the output impedance of the amplifier. From equation 8.101, the output impedance is

$$Z_{\text{out}} = r_c \| R_C = 1K \| 5K$$
$$= 833 \text{ ohms}$$

Thus, the load resistance must be 833 ohms.

C. COMMON COLLECTOR AMPLIFIER

The *common collector amplifier (emitter follower)* is shown in figure 8.46(a). This circuit has a voltage gain of slightly less than 1. It is used primarily for current gain or impedance matching between a low impedance load and a high impedance source. Its small-signal eqivalent circuit requires a more complicated analysis than does the common emitter circuit.

The input voltage is the base-collector voltage, since the collector is effectively grounded to small signals and the output current is the emitter current. A new set of parameters can be defined.

$$v_{bc} = h_{ie} i_b + v_{ec} \qquad 8.108$$

$$i_e = (1 + \beta) i_b + \frac{v_{ec}}{r_c} \qquad 8.109$$

Equations 8.108 and 8.109 define the common collector *hybrid parameters* for a bipolar junction transistor.

$$h_{ic} = h_{ie} \qquad 8.110$$

$$h_{rc} = 1 \qquad 8.111$$

$$h_{fc} = 1 + \beta = 1 + h_{fe} \qquad 8.112$$

$$h_{oc} = \frac{1}{r_c} = h_{oe} \qquad 8.113$$

These parameters could be used to establish a different equivalent circuit, but there is no advantage in doing so. Therefore, the circuit of figure 8.46(b) is used.

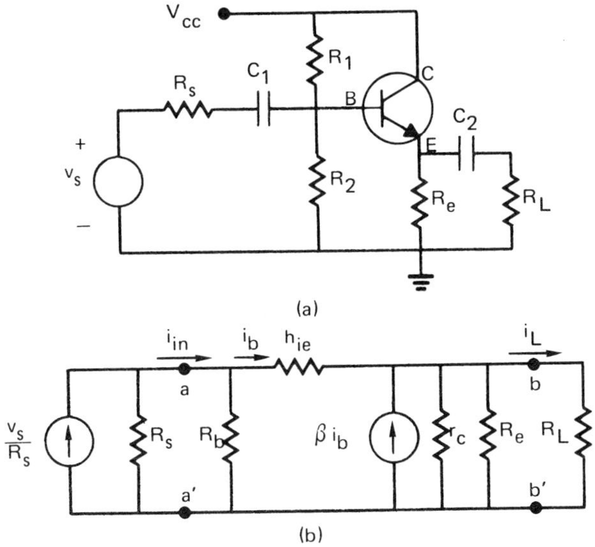

Figure 8.46 Common collector (emitter follower) amplifier: (a) circuit, (b) small-signal equivalent circuit

At terminals a-a',

$$Z_{\text{in}} = R_b \| \left[h_{ie} + (1+\beta)(r_c \| R_e \| R_L) \right] \quad 8.114$$

At terminals b-b'

$$Z_{\text{out}} = R_e \| r_c \| \left[\frac{h_{ie} + R_b \| R_S}{1+\beta} \right] \quad 8.115$$

From terminals a-a' to the load, the current and voltage gains are

$$A_I = \frac{(1+\beta)R_b}{Z_{\text{in}}} \times \frac{r_c \| R_e \| R_L}{R_L} \quad 8.116$$

$$A_V = \frac{1}{1 + \frac{h_{ie}}{(1+\beta)[r_c \| R_e \| R_L]}} \quad 8.117$$

D. THE COMMON BASE AMPLIFIER

The *common base amplifier* is shown in figure 8.47. It has a current gain of less than 1. The change in configuration allows the use of a different set of parameters, the *common base hybrid parameters*. These are calculated from the common emitter parameters.

$$h_{ib} = \frac{h_{ie}}{1 + \beta + \left(\frac{h_{ie}}{r_c} \right)} \approx \frac{h_{ie}}{1+\beta} \quad 8.118$$

$$h_{rb} = \frac{h_{ie}}{(1+\beta)r_c + h_{ie}} \approx 0 \quad 8.119$$

$$h_{fb} = \frac{-(\beta r_c + h_{ie})}{(1+\beta)r_c + h_{ie}} \approx \frac{-\beta}{1+\beta} = -\alpha \quad 8.120$$

$$h_{ob} = \frac{1}{(1+\beta)r_c + h_{ie}} \approx \frac{1}{(1+\beta)r_c} \quad 8.121$$

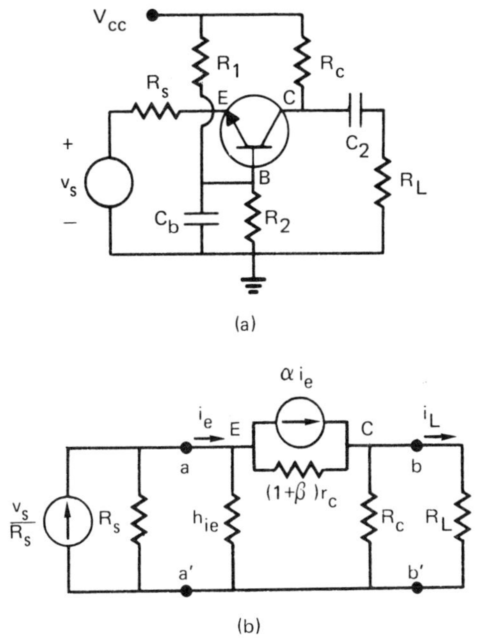

Figure 8.47 Common base amplifier: (a) circuit, (b) small-signal equivalent circuit

Analysis of the small signal equivalent circuit produces the following equations for terminals a-a'.

$$Z_{\text{in}} = \frac{h_{ie}[r_c + R_L \| R_c]}{(1+\beta)r_c + h_{ie} + R_L \| R_c} \approx \frac{h_{ie}}{1+\beta} \quad 8.122$$

For terminals b-b',

$$Z_{\text{out}} = R_c \| \left[(r_c \| h_{ie}) + R_S + \frac{\beta r_c R_S}{h_{ie} + r_c} \right]$$

$$\approx R_c \| [h_{ie} + (1+\beta)R_S] \quad 8.123$$

From terminals a-a' to the load,

$$A_I = \frac{R_c}{R_c + R_L} \times \frac{(1+\beta)r_c + h_{ie}}{(1+\beta)r_c + h_{ie} + R_L \| R_c}$$

$$\approx \frac{\alpha R_c}{R_c + R_L} \quad 8.124$$

$$A_V = \frac{R_c \| R_L}{h_{ie}} \times \frac{\beta r_c + h_{ie}}{r_c + h_{ie}} \approx \frac{\beta (R_c \| R_L)}{h_{ie}} \quad 8.125$$

E. COMMON SOURCE FET AMPLIFIERS

FET amplifiers are simpler to analyze because of the electrical isolation of the gate from the source-drain. For any common source circuit such as the one shown in figure 8.48, the amplifier parameters are

$$Z_{\text{in}} = R_3 + R_1 \| R_2 \quad 8.126$$

$$Z_{\text{out}} = R_d \| r_{ds} \quad 8.127$$

$$A_I = -\frac{g_m Z_{\text{in}}}{R_L} [r_{ds} \| R_d \| R_L] \quad 8.128$$

$$A_V = -g_m [r_{ds} \| R_d \| R_L] \quad 8.129$$

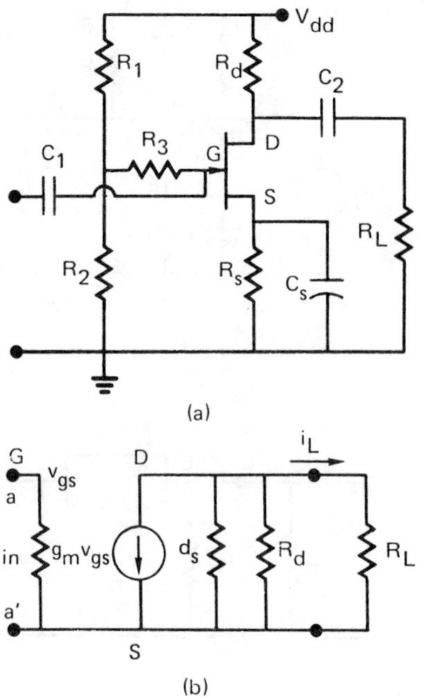

Figure 8.48 Common source FET amplifier: (a) circuit diagram, (b) small-signal equivalent circuit

F. COMMON DRAIN FET AMPLIFIERS

For the common drain (*source follower*) amplifier shown in figure 8.49, the amplifier parameters are

$$Z_{\text{in}} = R_1(1 + \frac{R_{s2}}{R_{s1}} \qquad 8.130$$

Let $R_s = R_{s1} + R_{s2}$

$$Z_{\text{out}} = \frac{1}{g_m + \frac{1}{r_{ds}} + \frac{1}{R_s}} \qquad 8.131$$

$$A_V = \frac{1}{1 + \frac{(R_{ds}\|R_L) + R_s}{gm(r_{ds}\|R_L)R_s}} \qquad 8.132$$

$$A_I = g_m R_1 \frac{r_{ds}}{r_{ds} + R_L} \times \frac{R_{s1} + R_{s2}}{R_{s2} + R(R_L\|r_{ds})(1 + g_m + R_{s1})} \qquad 8.133$$

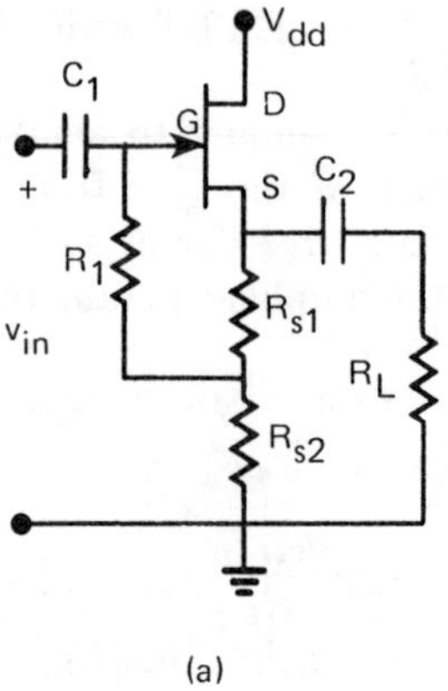

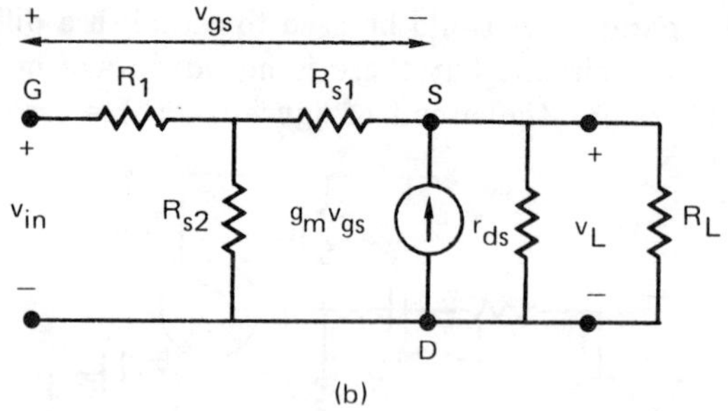

Figure 8.49 Common drain FET amplifier: (a) circuit diagram, (b) small-signal equivalent circuit

12 MULTIPLE TRANSISTOR CIRCUITS

A. INTRODUCTION

To analyze a multiple transistor circuit, it is necessary to obtain the small-signal equivalent circuit. Even with only two transistors, it may be wise to perform an entirely numerical analysis, as symbolic analysis can be unwieldly. The method of analysis may require matrix analysis, or network reduction techniques can be used. The latter is easier to check for errors, since it is a step-by-step process, and each step can be checked separately.

There are several two-transistor amplifiers which are sufficiently common to consider as special cases.

B. DIFFERENTIAL AMPLIFIERS

The *differential amplifier* is a very common circuit with some variation in construction. Figure 8.50(a) shows the configuration of one version of this amplifier. The two input signals supposedly are equal and opposite, but there may be an undesirable common signal which is the same for both inputs. This common signal is the so-called *common mode*. The differential amplifier has very good common mode rejection, particularly when capacitively coupled.

Two different gains are present in the differential amplifier: one involving the differential part of the signal, designated A_d, and the other dealing with the common mode signal and designated A_{cm}. The *common mode rejection ratio* is the ratio A_{cm}/A_d. It is often expressed in decibels.

$$CMRR = 20 \log\left(\frac{A_{cm}}{A_d}\right) \qquad 8.134$$

Example 8.6

The transistors of figure 8.50(a) have $h_{fe} = 50$, $r_c = 3K$ ohms, $h_{ie} = 500$ ohms. The circuit values are $R_b = 100$, $R_e = 300$, $R_{c1} = 500$, and $R_{c2} = R_L = 1000$ ohms. Determine (a) the voltage gain for differential signals, (b) the voltage gain for common mode signals, and (c) the common mode rejection ratio.

The small-signal equivalent network is shown in figure 8.50(b). To use network reduction, it is necessary to preserve voltages and currents which are involved with the control of controlled sources. The currents i_{b1} and i_{b2} are the source controls, and they can be written in terms of V_e and input voltages v_{s1} and v_{s2}. Therefore, it is necessary to preserve these quantities to account for the controlled source variables. The strategy is to determine v_e, from which the output voltage will be found.

The first step is to obtain the numerical expressions for βi_{b1} and βi_{b2}.

$$\beta i_{b1} = \frac{\beta(v_{s1} - v_e)}{R_b + h_{ie}} = \frac{v_{s1} - v_e}{12} \qquad 8.135$$

$$\beta i_{b2} = \frac{v_{s2} - v_e}{12} \qquad 8.136$$

The next step is to reduce the base-emitter circuits to a single Norton equivalent. The controlled sources are converted to Thevenin equivalent circuits and then combined with the series resistances. They are then converted to Norton equivalents as shown in figure 8.50(c).

At this point, control for the controlled current sources involving $-v_e/14$ is dependent on a 14 ohm resistance from the emitter to ground. This leaves only the terms $v_{s1}/14$ and $v_{s2}/14$ which can be combined with the current source. Then, combining the 14 ohm resistances,

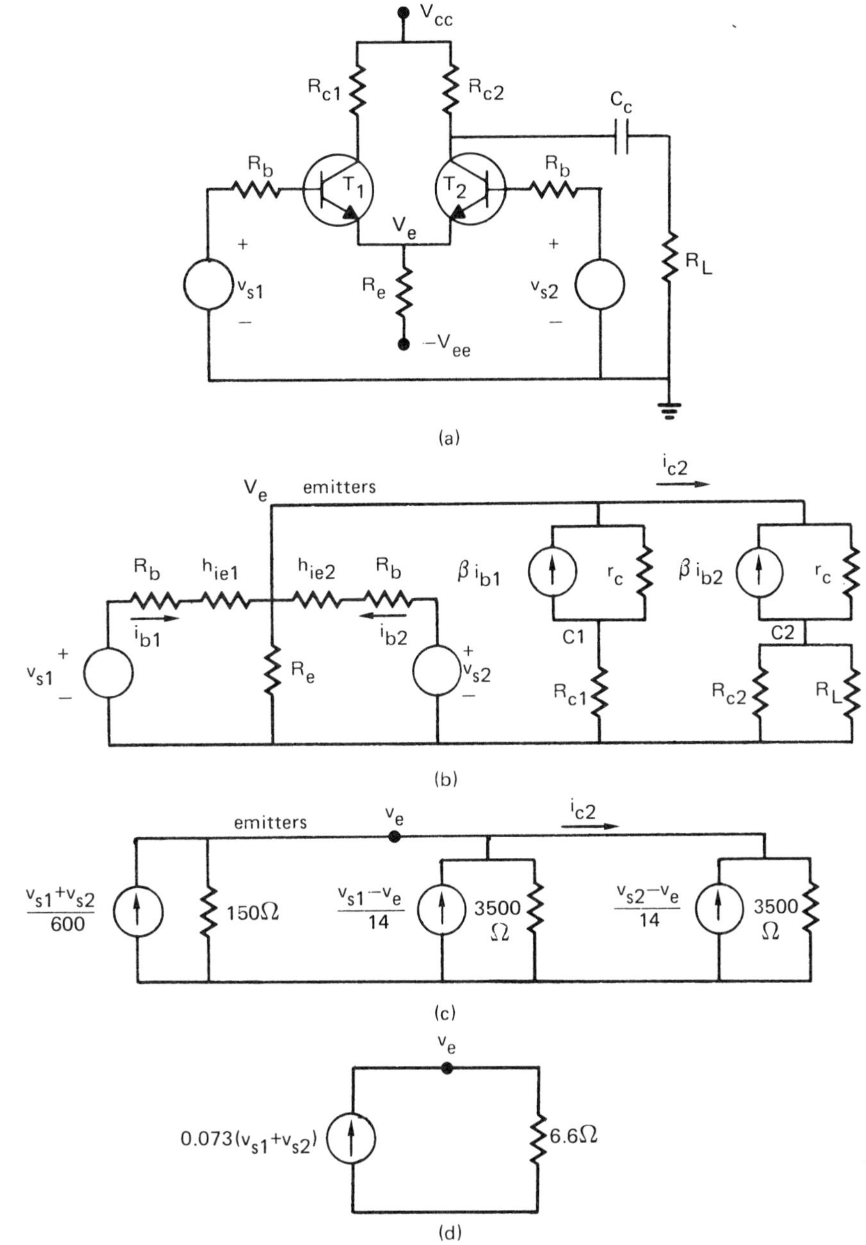

Figure 8.50 Differential amplifier

the 3500 ohm resistances, and the 150 ohm resistance in parallel, the equivalent circuit seen from v_e to ground is given in figure 8.50(d).

The equivalent circuit is described by

$$v_e = 0.048(v_{s1} + v_{s2})$$

Referring to figure 8.50(c),

$$i_{c2} = .487(v_{s1} + v_{s2})\left[\frac{1}{14} + \frac{1}{3500}\right] - \frac{v_{s2}}{14}$$

$$i_{c2} = 0.0349v_{s1} - 0.0365v_{s2}$$

Since $R_L = R_{c2}$, the current to the load is $i_{c2}/2$. So, the load voltage is

$$V_L = 8.725v_{s1} - 9.125v_{s2}$$

In *differential operation*, $v_{s2} = -v_{s1}$, and $v_{in} = v_{s1} - v_{s2}$, so the *differential gain* is

$$A_d = \frac{V_L}{V_{in}} = \frac{8.725 + 9.125}{2} = 8.925$$

For common mode, the input is $v_{in} = v_{s1} = v_{s2}$. The *common mode gain* is

$$A_{cm} = 8.725 - 9.125 = -0.85$$

The common mode rejection ratio is $A_{cm}/A_d = -0.095$ or $CMRR_{db} = -20.4$ *db*.

C. THE DARLINGTON CIRCUIT

The *Darlington circuit* is a pair of transistors (known as *compound transistors*) connected as shown in figure 8.51(a). The increase in gain is approximately the product of the current gains of the two transistors. The input impedance is increased as well. An analysis shows that the three-terminal equations for the pair are

$$v_{be'} = i_{b1}\left[h_{ie1} + \frac{(1+\beta_1)h_{ie2}}{1 + \frac{h_{ie2}}{r_{c1}}}\right] + v_{ce2}\left[\frac{h_{ie2}}{r_{c1} + h_{ie2}}\right] \quad 8.137$$

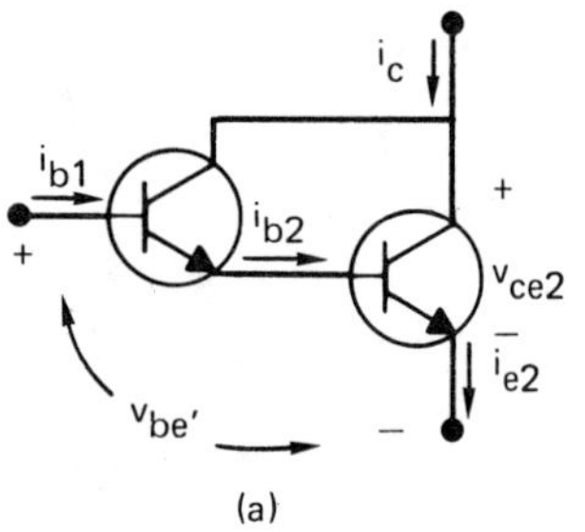

(a)

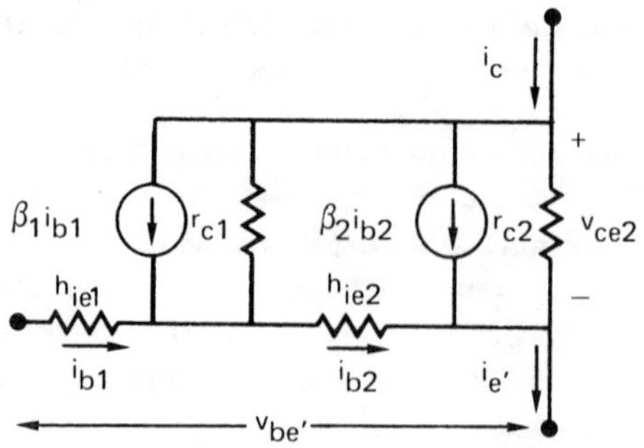

(b)

Figure 8.51 Darlington pair: (a) circuit, (b) small-signal equivalent

$$i_{e'} = i_{b1}\left[\frac{(1+\beta_1)(1+\beta_2)}{1 + \frac{h_{ie2}}{r_{c1}}}\right] + v_{ce2}\left[\frac{1}{r_{c2}} + \frac{1+\beta_2}{r_{c1} + h_{ie2}}\right] \quad 8.138$$

$$i_c = i_{b1}\left[\frac{\beta_1 + \beta_2 + \beta_1\beta_2 - \left(\frac{h_{ie2}}{r_{c1}}\right)}{1 + \frac{h_{ie2}}{r_{c1}}}\right] + v_{ce2}\left[\frac{1}{r_{c2}} + \frac{1+\beta_2}{r_{c1} + h_{ie2}}\right] \quad 8.139$$

These three equations constitute effective hybrid parameters for Darlington pair transistors.

D. CASCODE AMPLIFIERS

The *Cascode amplifier* is another two-transistor circuit which finds use in high frequency circuits because of its superior high frequency response. The input transistor (the lower one in figure 8.52(a)) is a common emitter stage, while the output transistor is in the common base configuration. (The same configurations, with appropriate bias circuits, can be created with FET's or vacuum tubes. For FET's, the input stage is common source, and the output is common gate. For vacuum triode systems, the input stage is common cathode and the output is common grounded grid.) The small-signal equivalent circuit for the bipolar junction transistor Cascode amplifier is shown in figure 8.52(b). For good high-frequency response, R_c should be zero.

The behavior of a Cascode amplifier is not significantly different from that of a common emitter amplifier at mid-frequencies. Its advantage is at high frequencies where improved bandwidth is achieved.

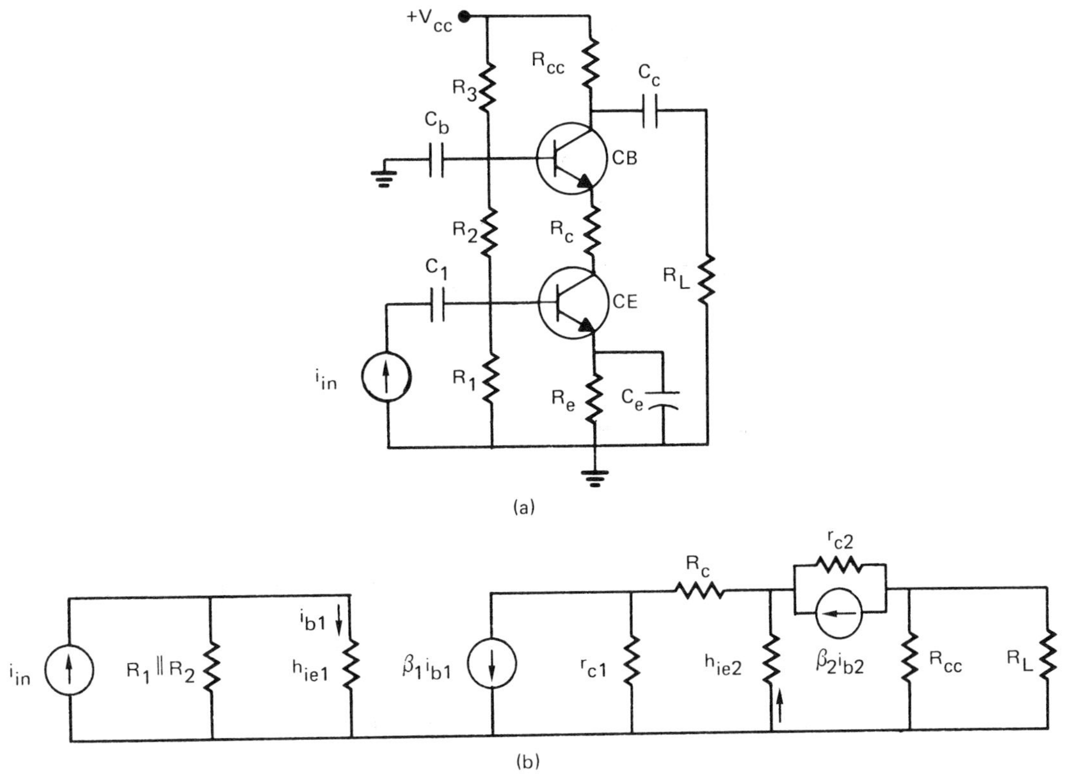

Figure 8.52 Cascode bipolar junction transistor: (a) circuit diagram, (b) small-signal equivalent circuit

13 FREQUENCY EFFECTS

A. LOW FREQUENCY BEHAVIOR

Low frequency behavior usually is influenced by emitter resistance and the bypass capacitors. This is due to the relatively small resistance and the subsequent need for large capacitance. The impedance of a parallel resistance and capacitance is

$$Z_{R\|C} = \frac{R}{1 + j\omega RC} \qquad 8.140$$

The *corner frequency* is

$$\omega_{cf} = \frac{1}{RC} \qquad 8.141$$

The amplitude of the impedance is $R/\sqrt{2}$, or 3 db down from the D.C. value. For $\omega > 10/RC$, the impedance is essentially the capacitive reactance, inversely proportional to the frequency. The circuits is particularly important because significant impedance in that position causes degenerative feedback which lowers the gain of the amplifier.

Coupling capacitors and bypass capacitors both diminish the gain when their impedances become comparable to the resistance they see. The resistance typically seen consists of a parallel combination of the biasing resistances with the input impedance of the amplifying device as well as the output impedance of the previous stage.

Example 8.7

Determine the low frequency limit for the emitter impedance of example 8.5. The bypass capacitance C_e is 100 μF and $R_e = 100$ ohms.

The resistance seen by the capacitance is not simply the emitter resistance R_e. It must be found by determining the effective output impedance as if the amplifier was an emitter follower. The collector circuit is essentially isolated by the large value of r_c. For the common collector amplifier, equation 8.115 indicates that the output impedance included the series connection of h_{ie} and the resistance connected to it, all divided by $1+\beta$. The term of r_c in parallel should be increased to include the load and collector resistances.

$$R_{\|C} = R_e \| (r_c + R_L \| R_c) \| \left[\frac{h_{ie} + R_b \| R_S}{1 + \beta} \right]$$

For this example,

$$R_b = 3K||6K = 2000\,\text{ohms}$$
$$R_S = 1000\,\text{ohms}$$
$$h_{ie} = 500\,\text{ohms}$$
$$\beta = 50$$
$$r_c = 5000\,\text{ohms}$$
$$R_c = 1000\,\text{ohms}$$
$$R_L = 833\,\text{ohms}$$

Then,

$$R_{\|C} = 100||5454||\left[\frac{500 + 1000||2000}{51}\right]$$
$$= 18.6\,\text{ohms}$$

Thus,

$$\frac{1}{RC} = 539\,\text{rad/sec} = 85.7\,\text{hz}$$

B. HIGH FREQUENCY BIPOLAR JUNCTION TRANSISTOR BEHAVIOR

Analysis of *high frequency limitations of bipolar junction transistors* requires a high frequency equivalent circuit. The most general model is the *Hybrid pi* model shown in figure 8.53. The circuit can be simplified by using the *Miller effective impedance.* This causes the capacitance $C_{b'c}$ to be multiplied by $(g_m R'_C + 1)$ and connected from the active base b' to the emitter as shown in figure 8.54.

The *high frequency cutoff* is found by determining the resistance seen by the parallel combination of $C_{b'e}$ and $C_{b'c}(1 + g_m R'_C)$. For a common emitter amplifier, this is the parallel combination of the base biasing resistors and the output impedance of the previous state (R_S) in series with $V_{bb'}$ all in parallel with $r_{b'e}$.

For a cascode amplifier, the Miller effective capacitance is reduced because the first transistor (CE) sees R_C and the low input impedance of the second transistor (CC). With $R_C = 0$, the Miller capacitance is essentially $C_{b'c}$.

Example 8.8

Determine the high frequency cutoff for an emitter follower which has the following parameters:

$$C_{b'e} = 1000\,pF$$
$$C_{b'c} = 25\,pF$$
$$r_{bb'} = 150\,\text{ohms}$$
$$r_{b'e} = 100\,\text{ohms}$$
$$\beta = 50$$
$$R_S||R_b = 200\,\text{ohms}$$
$$R_e = 100\,\text{ohms}$$

As the collector is grounded, the resistance seen from g_m is R_e.

$$g_m = \frac{\beta}{r_e} = .5$$

The Miller capacitance is

$$25(1 + 50) = 1275\,pF$$
$$C_{\text{total}} = 1275 + C_{b'e} = 2275$$

The resistance seen from the capacitance is

$$R_{\|C} = r_e||(r_{bb'} + R_S||R_b + (1 + \beta)R_e)$$
$$= 100||(150 + 200 + 51 \times 100) = 96.7\,\text{ohms}$$

Then, the cutoff frequency is

$$\omega_{cf} = \frac{1}{96.7 \times 2275 \times 10^{-12}} = 4.55 \times 10^6\,\text{rad/sec}$$

$$f_{cf} = \frac{4.55 \times 10^6}{(2\pi)(1000)} = 724\,\text{khz}$$

C. HIGH FREQUENCY EFFECTS ON FET's

The FET at *high frequency* has the hybrid pi equivalent circuit shown in figure 8.55(a). When Miller effective capacitance is taken into effect, figure 8.55(b) represents the equivalent circuit. Finding the high frequency cutoff is similar to the bipolar junction transistor procedure.

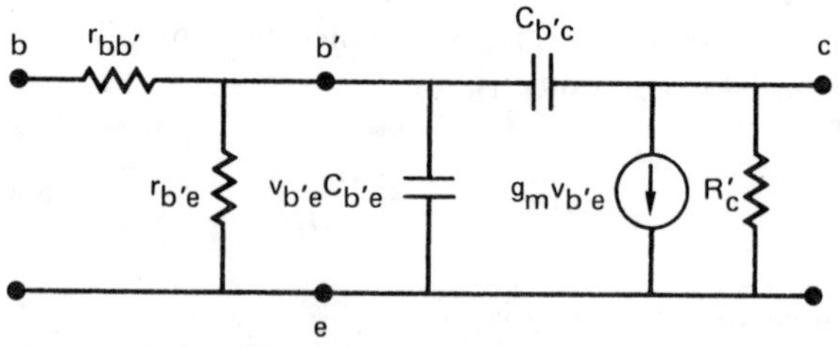

Figure 8.53 Hybrid pi equivalent circuit for bipolar junction transistor at high-frequency

$r_{bb'}$	= base bulk resistance
$r_{b'e}$	= base-spreading resistance $= kT/qI_e$
	$= .026/I_b$ at $300°C$
$C_{b'e}$	= emitter junction capacitance $= 1/(\omega_\beta r_e)$
ω_β	= frequency at which $\beta = 1$
$C_{b'c}$	= collector junction capacitance
R'_c	= the equivalent resistance (including load) across the c-e terminals, including r_c
g_m	$= \beta/r_{b'e}$

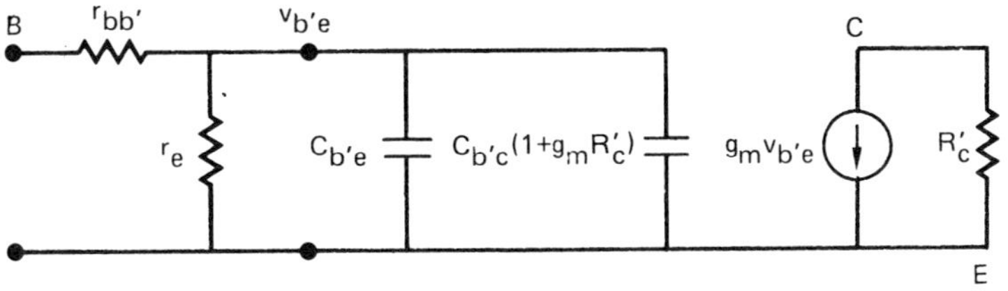

Figure 8.54 Hybrid pi equivalent circuit with Miller capacitance

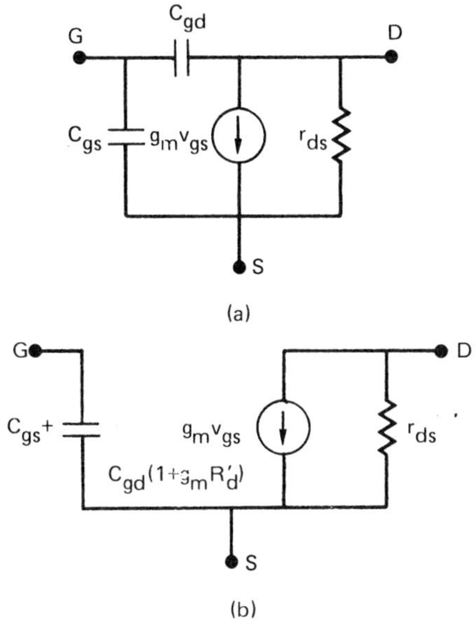

Figure 8.55 High frequency equivalent circuit for FET: (a) hybrid pi circuit, (b) modification with Miller capacitance. (R'_d is the Thevenin equivalent resistance at the current generator terminals)

14 PRACTICAL OPERATIONAL AMPLIFIERS

Operational amplifiers have very high D.C. gain, reasonably high input impedance, and low output impedance. They can, therefore, be treated as essentially pure voltage amplifiers. Typical gain-frequency charcteristics are shown in figure 8.56. The D.C. voltage gain is on the order of 10^5, and the frequency at which the gain goes to 1 (0 db) is on the order of 10^6 or 10^7 rad/sec. The frequency at which the voltage gain is 1 is called ω_o.

As the slope of the gain in figure 8.54(a) is -20 db per decade, the frequency equation is

$$A = \frac{A_o}{1 + j\left(\frac{\omega}{\omega_a}\right)} \qquad 8.142$$

At ω_o, $A = 1$, so that $\omega_a = \omega_o/A_o$. For frequencies above ω_a, the amplifier has an essentially constant *gain-bandwidth product.*

$$GBW = \omega_a A_o = \omega_o \qquad 8.143$$

Alteration of the circuit by feedback will result in a product of gain and bandwidth no greater than $\omega_o A_o$.

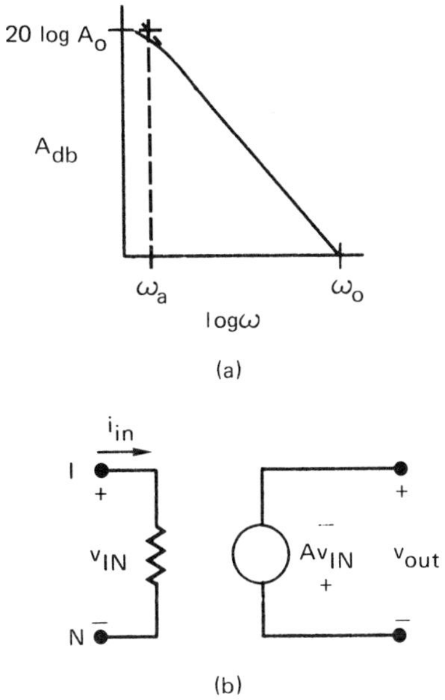

Figure 8.56 Operational amplifier: (a) gain-frequency characteristic, (b) equivalent circuit

The equivalent circuit for the operational amplifier is shown in figure 8.56(b). There are two input terminals, neither of which is grounded. One is the *inverting terminal* and is indicated with an *I*. The other is the *non-inverting terminal,* designated with *N*. A positive voltage applied from *I* to *N* will result in a negative output voltage. For most applications the *N* terminal is grounded, either directly or through a resistance.

The output voltage with respect to ground is

$$V_{out} = \left[\frac{-A_o}{1 + j\left(\frac{\omega}{\omega_o}\right)A_o}\right] V_{in} \qquad 8.144$$

The differential and Laplace forms of equation 8.144 are

$$-v_{in} = \frac{v_{out}}{A_o} + \left(\frac{1}{\omega_o}\right)\frac{dv_{out}}{dt} \qquad 8.145$$

$$-v_{in}(s) = \left[\frac{1}{A_o} + \frac{s}{\omega_o}\right] v_{out}(s) \qquad 8.146$$

15 AMPLIFIER NOISE

A. SHOT NOISE

Shot noise is associated with current. It is an aggregate phenomonon involving numerous individual charge carriers. A current can be calculated from the number of charge carriers passing a cross-section per second.

$$I = q\frac{dn}{dt} \quad 8.147$$

q is the charge per carrier, and dn/dt is the number of carriers passing the cross-section per second. Because of random thermal velocities of the charge carriers, dn/dt fluctuates. If the average current is I, and the total current is $i(t)$, then the noise current, $i_n(t)$ can be defined as

$$i_n(t) = i(t) - I \quad 8.148$$

The resulting spectral density of the noise current, assuming that the arrival of individual charge carriers is entirely random, is constant. That is, all frequencies are equally likely. This can be considered as an aggregate of delta functions in time, which in the frequency domain becomes a constant (Fourier transformation). The amplitude of the spectral density is, then, proportional to the average current.

$$S_{\text{shot}}(f) = 2qI \quad 8.149$$

S_{shot} is the noise shot power spectral density in W/hz. It is constant over all frequencies and is known as *white noise*.

B. JOHNSON OR THERMAL NOISE

Thermal noise is the result of random motions of electrons in conductors, which give rise to random voltage fluctuations across the conductor. This noise voltage exists whether or not there is any current flow in the conductor. The energy of individual electrons in a conductor is random, but the average energy per electron is kT. The resulting noise voltage is related to energy and is expressed in terms of its RMS value squared.

$$S_{\text{thermal}}(f) = 4kTR \quad 8.150$$

S is the spectral power density of the thermal noise in W/hz. R is the resistance of the conductor in ohms.

C. LOW FREQUENCY NOISE

Parameter variations due to changes in temperature over time, aging of components, and any other slow changes (drift) in a system produce low-frequency noise. The spectral distribution is typically $1/f$ in nature, so it would be expected to become infinite at zero frequency. In reality, there is no zero-frequency component actually present.

The lowest frequency which must be considered is $1/T_{\text{on}}$, where T_{on} is the length of time since the device was first (most recently) turned on.

D. INTERFERENCE NOISE

Interference noise is caused by power lines and unrelated signals. These are called *cross-talk* when due to signals. The frequency spectra depend on the source. Power line noise is typically at the power frequency and the frequencies of its harmonics, but it also can involve random bursts from switching of power equipment.

Interference noise can be filtered out if it is not within the spectrum of the desired signal.

E. BANDWIDTH LIMITING OF NOISE

All the noise sources except interference are described by a power frequency spectral distribution. The output noise power from an amplifier will be related to the frequency characteristics of the amplifier and the spectral distribution of the noise sources.

Figure 8.57 shows a thermal noise source (taken as its Thevenin equivalent voltage source) and an amplifier which has a voltage gain frequency response of $A(f)$. The corresponding power gain response is

$$G(f) = |A(f)|^2 \quad 8.151$$

For a voltage-squared input spectrum, S_n, from the noise voltage source, the output voltage-squared value will be

$$\overline{v_o}^2 = \int_{f=0}^{\infty} S_n(f)G(f)df \quad 8.152$$

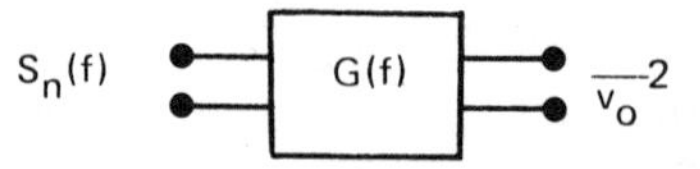

Figure 8.57 Noise into an amplifier

A single-pole, low-pass amplifier corresponding to $A(f)$ will have

$$A(f) = \frac{A_V}{1 + j\left(\frac{f}{f_{cf}}\right)} \quad 8.153$$

$$G(f) = \frac{A_V^2}{1 + \left(\frac{f}{f_{cf}}\right)^2} \quad 8.154$$

Thus, for thermal or shot noise with a constant $S(f) = S$,

$$\overline{v_o}^2 = S \int_0^\infty \frac{A_V^2}{1+\left(\frac{f}{f_{cf}}\right)^2} df$$
$$= A_V S f_{cf} \arctan\left(\frac{f}{f_{cf}}\right)\Big|_{f=0}^{f=\infty}$$
$$= A_V S f_{cf} \frac{\pi}{2} \qquad 8.155$$

For a two-pole, low-pass filter with critical damping,[3]

$$A(f) = \frac{A_V}{\left[1 + j\left(\frac{f}{f_{cf}}\right)\right]^2} \qquad 8.156$$

$$G(f) = \frac{A_V^2}{\left[1 + \left(\frac{f}{f_{cf}}\right)^2\right]^2} \qquad 8.157$$

Again, for thermal or shot noise with $S(f) = S$,

$$\overline{v_o}^2 = S \int_0^\infty \frac{A_V^2}{\left[1+\left(\frac{f}{f_{cf}}\right)^2\right]^2} = A_V S f_{cf} \frac{\pi}{4} \qquad 8.158$$

Bandwidth usually is defined by 3 db points, so that the bandwidth for the single-pole, low-pass amplifier is f_{cf}. The bandwidth for the two-pole filter is not f_{cf}, however. The filter gain is 6 db down at that frequency, and the 3 db frequency is at

$$.6436 f_{cf} = f'_{cf}$$

Scaling this upwards,

$$\frac{\pi}{4} f_{cf} = 1.22 f'_{cf}$$

[3]The power gain G is related to the voltage gain by $G = R_{in} A_v^2 / R_{out}$. Here, it is assumed that $R_{out} = R_{in}$.

The *noise bandwidth* is defined as the frequency range Δf over which the mid-frequency gain A_v holds such that a square filter amplifier would pass noise energy $S A_v^2 \Delta f$. The resulting bandwidths are summarized in table 8.1.

F. NOISE SOURCE EQUIVALENT CIRCUITS

A white noise source may produce thermal noise and shot noise, one of which is dependent on temperature and the other on current level. The thermal noise source is identified as an RMS voltage source, and the shot noise source appears as an RMS current source. The general circuit for a resistive device with both noise sources is shown in figure 8.58. Because the two sources are statistically independent, squaring variables does not produce any cross-products. Thus, the noise input spectrum is the independent sum of the two. From equations 8.149 and 8.150,

$$S(f) = S(f)_{\text{thermal}} + S(f)_{\text{shot}}$$
$$= e_{tn}^2 + R i_{sn}^2 \qquad 8.159$$
$$e_{tn}^2 = 4kTR \qquad 8.160$$
$$R i_{sn} = 2qIR \qquad 8.161$$

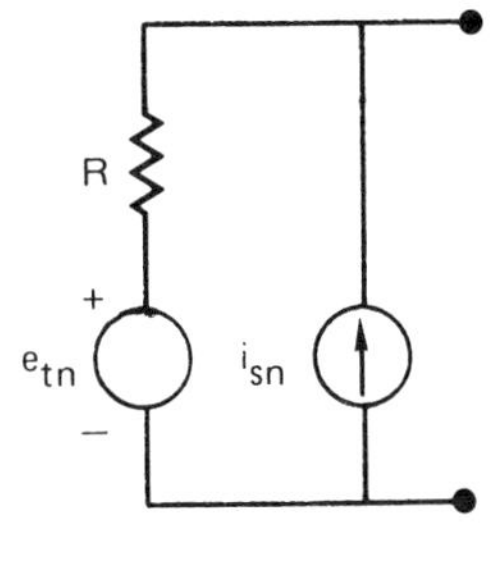

(a)

Table 8.1

Signal and Noise Bandwidths
for Amplifiers

	signal bandwidth (−3 db)	noise bandwidth (Δf)
1-pole, low pass:	f_{cf}	$1.571 f_{cf}$
2-pole, low-pass (critical damping)	$.6436 f_{cf} = BW_s$	$.785 f_{cf} = 1.22 BW_s$
2-pole, band-pass	$f_2 - f_1$	$1.571(f_2 - f_1)$
4-pole, band-pass (critical damping)	$.6436 f_2 - 1.554 f_1 = BW_s$	$1.22 BW_s$

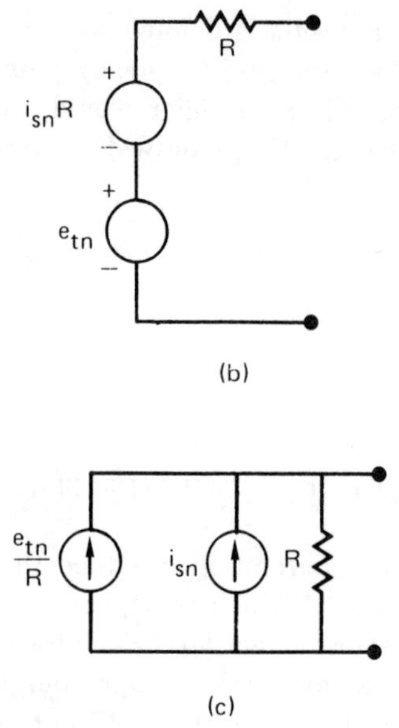

Figure 8.58 Noise source equivalent circuits: (a) thermal and shot noise sources, (b) Thevenin equivalent circuit, (c) the Norton equivalent circuit

G. OPTIMUM SOURCE RESISTANCE

From the usual thermal point of view, it is possible to minimize the effects of internal noise sources in an amplifier by proper selection of the source resistance. Consider the noiseless amplifier shown in figure 8.59(a).

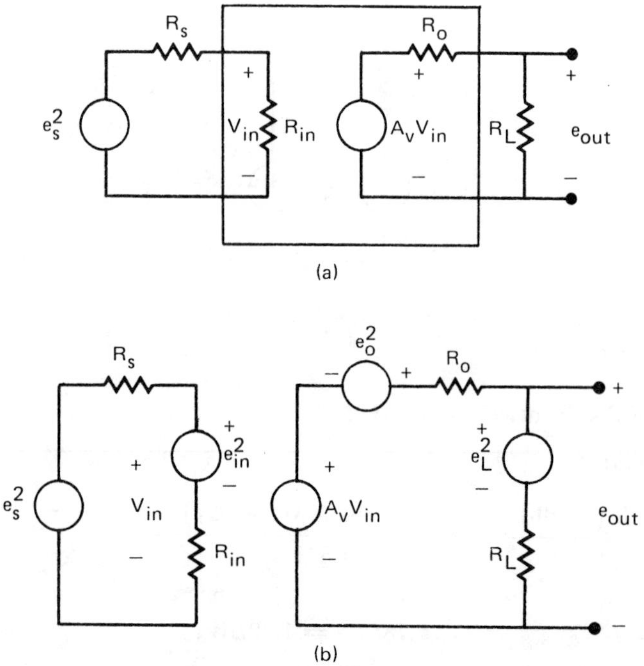

Figure 8.59 (a) Noiseless amplifier, (b) amplifier with internal noise sources

Circuit analysis of the noiseless amplifier results in

$$e_{out} = \frac{A_V R_{in} R_L e_S}{(R_{in} + R_S)(R_L + R_o)} \quad 8.162$$

$$e_{out}^2 = \frac{A_V^2 R_{in}^2 R_L^2 e_S^2}{(R_{in} + R_S)^2 (R_L + R_o)^2} \quad 8.163$$

$$e_S^2 = 4kTR_S \quad 8.164$$

Circuit analysis of the same amplifier with internal noise sources due to the internal resistances and the load resistance results in

$$e_{out}^2 = \frac{A_V^2 R_L^2 (R_{in}^2 e_S^2 + R_S^2 e_{in}^2)}{(R_{in} + R_S)^2 (R_L + R_o)^2} + \frac{R_o^2 e_L^2 + R_L^2 e_o^2}{(R_o + R_L)^2} \quad 8.165$$

The *noise figure* is a measure of the noise generated within an amplifier as compared to the noise that would exist without internally generated noise. It is the ratio of equation 8.165 to equation 8.163.

$$F = 1 + \frac{R_S^2 e_{in}^2}{R_{in}^2 e_S^2} + \frac{(R_S + R_{in})^2 (R_o^2 e_L^2 + R_L^2 e_o^2)}{A_V^2 R_L^2 R_{in}^2 e_S^2} \quad 8.166$$

However,

$$e_{in}^2 = 4kTR_{in} \quad 8.167$$

$$e_o^2 = 4kTR_o \quad 8.168$$

$$e_L^2 = 4kTR_L \quad 8.169$$

Therefore, equation 8.166 reduces to

$$F_{thermal} = 1 + \frac{R_S}{R_{in}} + \frac{R_o (R_S + R_{in})^2 (R_L + R_o)}{A_V^2 R_S R_{in}^2 R_L} \quad 8.170$$

Use equation 8.171 to express F in db.

$$F_{db} = 10 \log(F) \quad 8.171$$

To minimize the noise figure with respect to R_S, the partial derivative of F with respect to R_S is set to zero. As the second partial derivative of F with respect to R_S is always positive, the minimum condition is

$$R_S^2 \Big|_{\min F} = \frac{R_{in}^2 R_o (R_L + R_o)}{A_V^2 R_{in} R_L + R_o (R_L + R_o)} \quad 8.172$$

The noise power is defined as the voltage squared.

$$N = V_{on}^2 = e_S^2 (BW) \left[\frac{A_V^2 R_{in}^2 R_L^2}{(R_{in} + R_S)^2 (R_o + R_L)^2} \right] \quad 8.173$$

$$e_S^2 = 4kTR_S \quad 8.174$$

The voltage squared signal power is needed when calculating the ratio of signal power to noise power, S/N.

$$S = V_{oS}^2 = V_S{}^2 \left[\frac{A_V^2 R_{in}^2 R_L^2}{(R_{in} + R_S)^2 (R_o + R_L)^2} \right] \quad 8.175$$

The ratio of signal power to noise power frequently is given in decibels.

$$\left(\frac{S}{N}\right)_{db} = 10\log\left(\frac{S}{N}\right) \qquad 8.176$$

The noise and signal powers actually are powers only if they appear across a 1 ohm resistance. The actual powers are obtained by dividing S and N by the actual load resistance. However, this does not change the ratio.

Example 8.9

The circuit in figure 8.59 has $R_{\text{in}} = R_o = R_L = 1000$ ohms, and $A_V = 10$. (a) Determine the optimum R_S and the resulting noise figure in decibels. (b) With a noise bandwidth of 10^4 hz and a temperature of $300°K$, determine the noise power voltage at the load. (c) For an RMS input signal of 10^{-5} volts, determine the ratio of output signal to noise.

From equation 8.172, using K as shorthand notation for 1000,

$$R_S^2 = \frac{(1K)^2(1K)(1K+1K)}{(10)^2(1K)(1K)+(1K)(1K+1K)} = \frac{2K^4}{102K^2}$$

$$R_S = 140\,\text{ohms}$$

From equation 8.170,

$$F = 1 + \frac{140}{1K} + \frac{(1K)(140+1K)^2(1K+1K)}{(10)^2(1K)(1K)^2(1K)}$$

$$= 1.166$$

$$F_{db} = 10\log(1.166) = 0.67\,\text{db}$$

Now use the equation 8.173.

$$e_S^2 = 4(1.38\times10^{-23})(300)(140) = 2.32\times10^{-18}$$

$$N = V_{\text{on}}^2$$

$$= (2.32\times10^{-18})(10^4)\left[\frac{(100)(1K)^2(1K)^2}{(140+1K)^2(1K+1K)^2}\right]$$

$$= 4.46\times10^{-13}\,V^2$$

For the signal input of 10^{-5} volts, $V_S^2 = 10^{-10}$. From equation 8.175,

$$S = 1.92\times10^{-9}$$

Therefore, the ratio of signal and noise powers is

$$\frac{S}{N} = \frac{1.92\times10^{-9}}{4.46\times10^{-13}} = 4.31\times10^3$$

$$\left(\frac{S}{N}\right)_{db} = 10\log 4310 = 31.3\,\text{db}$$

H. NOISE POWER TO AN INPUT RESISTANCE

The RMS voltage squared per unit bandwidth (in V^2/hz) for a resistance R_S is

$$e_n^2 = 4kTR_S \qquad 8.177$$

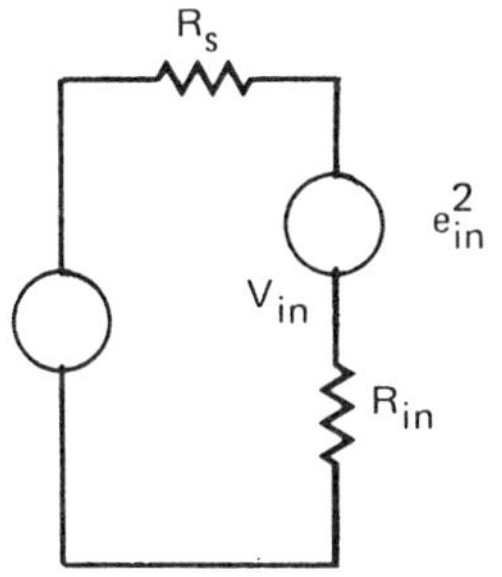

Figure 8.60 Input noise circuit

If this voltage is modeled as in figure 8.60, the voltage across R_{in} is

$$R_{\text{in}} = \frac{e_n R_{\text{in}}}{R_S + R_{\text{in}}} \qquad 8.178$$

With power being voltage squared divided by resistance, the power in W/hz from the resistance R_S which reaches R_{in}) is

$$p_{\text{in}} = \frac{e_n^2 R_{\text{in}}^2}{(R_S + R_{\text{in}})^2 R_{\text{in}}} = \frac{(4kTR_S)R_{\text{in}}}{(R_S + R_{\text{in}})^2} \qquad 8.179$$

In many cases, the input resistance is set equal to the source resistance in order to maximize power transfer. Unfortunately, this also maximizes the noise power input from the source resistance.

The total noise power at the input, including the noise from the input resistance, is minimized by making the source resistance as small as possible. With maximum signal power to the input, the corresponding maximum effect of the input resistance noise occurs as

$$p_{\text{in}} = kT \qquad 8.180$$

I. NOISE FIGURES FOR CASCADED AMPLIFIERS

The noise output from an amplifier with only an input resistance noise is

$$p_o = FGp_{\text{in}} \qquad 8.181$$

Equation 8.180 is used for p_{in} as impedance matching is assumed. This is illustrated for the input stage of figure 8.61.

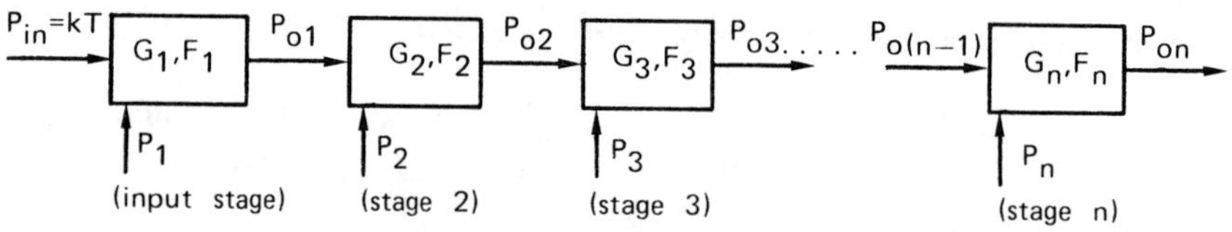

Figure 8.61 Cascaded amplifier stages

The excess noise accounting for F, is modeled by an equivalent internal noise source at the input (indicated by p_1 in figure 8.61). Then,

$$p_{o1} = G_1(kT + p_1) = F_1 G_1 kT \qquad 8.182$$

$$p_1 = kT(F_1 - 1) \qquad 8.183$$

Similarly for the other matched stages,

$$p_2 = kT(F_2 - 1) \qquad 8.184$$

$$p_3 = kT(F_3 - 1) \qquad 8.185$$

$$p_n = kT(F_n - 1) \qquad 8.186$$

Then,

$$\begin{aligned} p_{o2} &= G_2(p_{o1} + p_2) \\ &= G_2 G_1 F_1 kT + G_2(F_2 - 1)kT \end{aligned} \qquad 8.187$$

Since the noiseless contribution is $G_2 G_1 kT = G_2 G_1 p_{\text{in}}$, and the noise figure is $p_{o2}/G_2 G_1 p_{\text{in}}$, an overall noise figure for the first two stages is

$$F_{12} = F_1 + \frac{F_2 - 1}{G_1} \qquad 8.188$$

Similarly for p_{o3} and F_{13},

$$\begin{aligned} p_{o3} &= G_3(p_{o2} + p_3) \\ &= G_3 G_2 G_1 F_1 kT + G_3 G_2(F_2 - 1)kT \\ &\quad + G_3(F_3 - 1)kT \end{aligned} \qquad 8.189$$

$$F_{13} = F_1 + \frac{F_2 - 1}{G_1} + \frac{F_3 - 1}{G_1 G_2} \qquad 8.190$$

It can be seen that the noise figure is obtained by referring all subsequent noise sources to the input of the first stage, dividing by the gains of all previous amplifiers, and then dividing by the input power density. Then, each succeeding stage's contribution to the noise figure is found from

$$F_n = \frac{F_n - 1}{G_1 G_2 G_3 \cdots G_{n-1}} \qquad 8.191$$

Equation 8.191 is applicable to any amplifier components involved in a cascaded system, including antennas, modulators, and mixers. The equivalent noise power density of any source is kT.

J. NOISE TEMPERATURE AND NOISE RESISTANCE

Another method of representing device noise effects is by an equivalent *noise temperature* T_e. The output of the input stage is $G_1(kT_o + kT_e)$ where T_o commonly is taken as room temperature (290°K). Then,

$$F_1 = 1 + \frac{T_e}{T_o} \qquad 8.192$$

$$F_n - 1 = \frac{T_{en}}{T_o} \qquad 8.193$$

Yet another method of representing a device's contribution to noise is by an equivalent *noise resistance*, R_n. For this method, the input resistance of the nth stage must be known.

$$F_1 = 1 + \frac{R_e}{R_S} \qquad 8.194$$

$$F_n - 1 = \frac{R_{en}}{R_{Sn}} \qquad 8.195$$

Appendix 8.1

SEMICONDUCTOR DEVICE SYMBOLS AND CHARACTERISTICS

(A) PN Diode

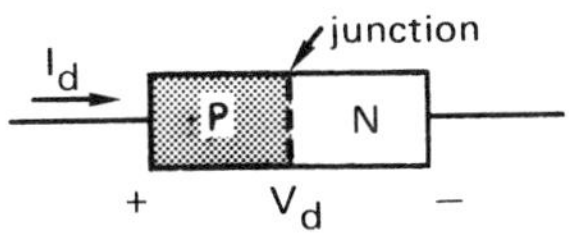

conceptual composition

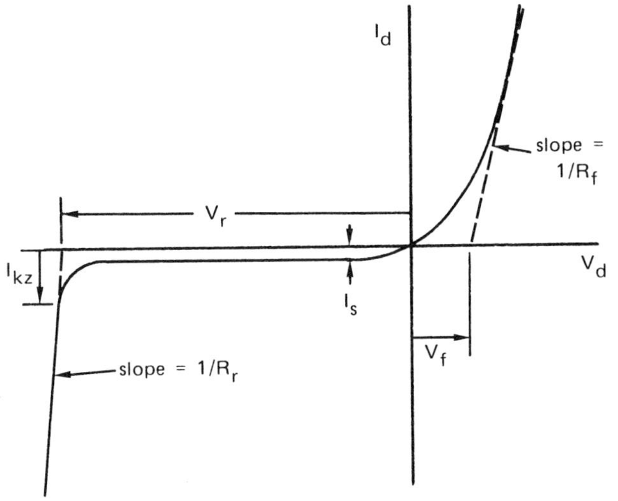

diode characteristic

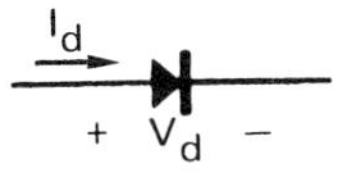

circuit symbol

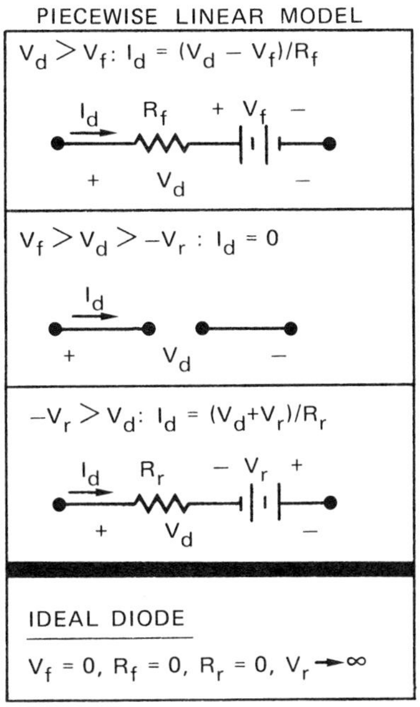

COMPOSITE PIECEWISE LINEAR MODEL

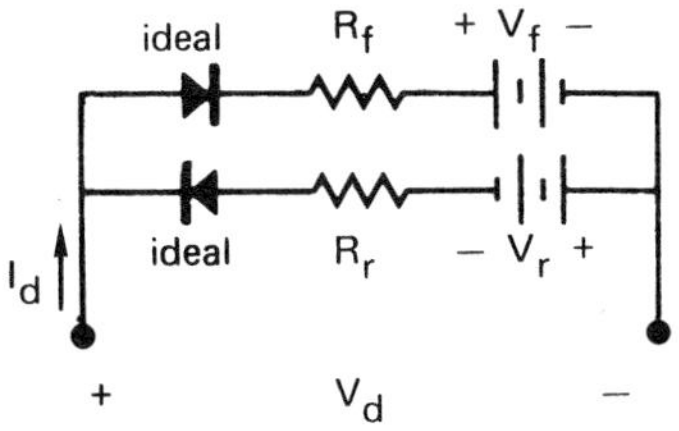

PERFECT DIODE EQUATION:

$$I_d = I_S(e^{-qV_d/kT} - 1)$$
$$\frac{q}{kT} = 40 \text{ at } T = 300°K$$

INCREMENTAL RESISTANCE

$$r_f = \frac{kT}{qI_d} \text{ for } V_d > V_f$$

BIAS DEFINITION

positive bias: $V_d > 0$ (drops from P to N)

forward bias: positive bias

negative bias: $V_d < 0$ (drops from N to P)

reverse bias: negative bias

FORWARD VOLTAGE V_f

Germanium: $V_f = 0.2$ volts

Silicon: $V_f = 0.6$ volts

(B) Zener or Avalanche Diode

The same as the PN diode but with voltage and current directions reversed.

Changes in nomenclature: $V_r = V_z$, $R_r = R_z$

Keep-alive current I_{kz} is the minimum for which characteristic is linear.

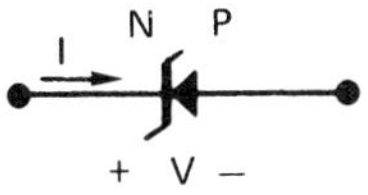

Circuit symbol

(C) NPN Bipolar Junction Transistor (BJT)

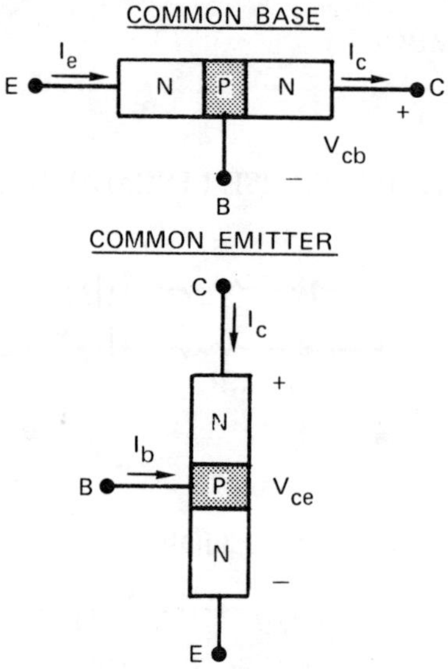

conceptual compositions

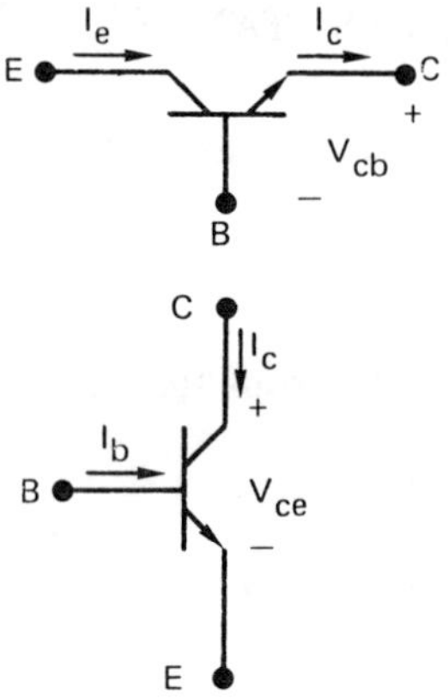

circuit symbols

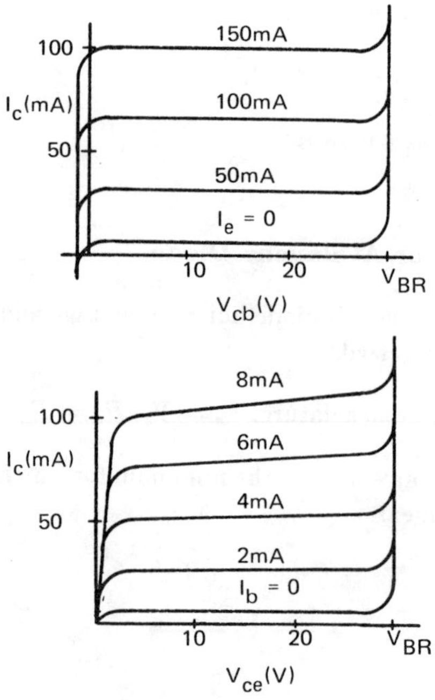

forms of characteristics

Nomenclature

B	base
C	collector
E	emitter
I_b	base current
I_c	collector current
I_e	emitter current
V_{cb}	voltage drop from collector to base
V_{ce}	voltage drop from collector to emitter
V_{BR}	avalanche breakdown voltage

The arrow on the emitter points in the forward bias direction for a PN junction (i.e., from P toward N).

(D) PNP Bipolar Junction Transistor (BJT)

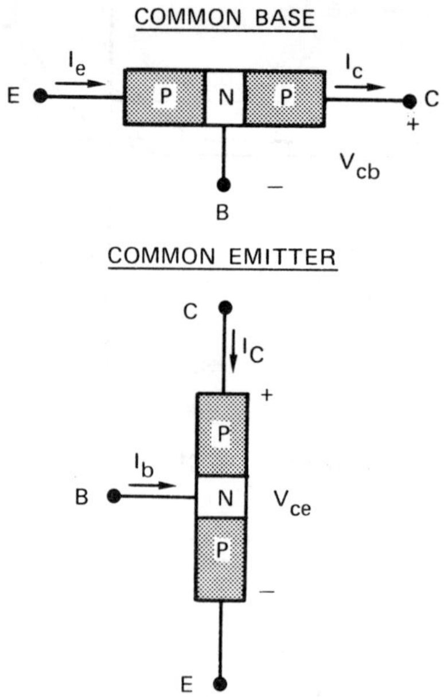

conceptual compositions

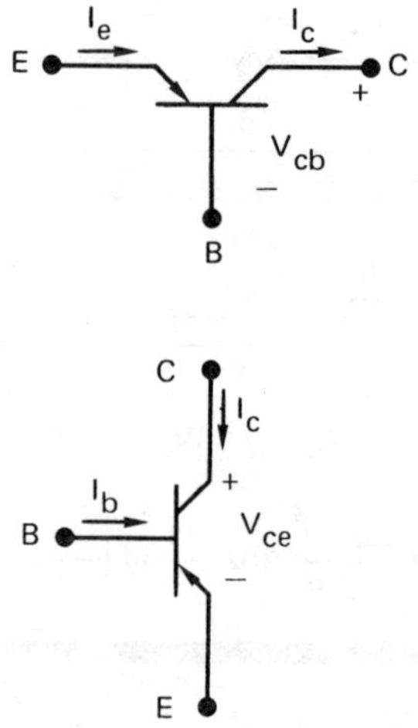

circuit symbols

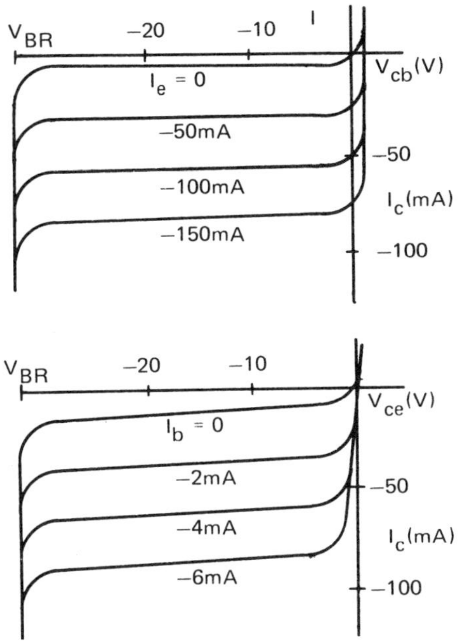

forms of characteristics

Nomenclature

B	base
C	collector
E	emitter
I_b	base current
I_c	collector current
I_e	emitter current
V_{cb}	voltage drop from collector to base
V_{ce}	voltage drop from collector to emitter
V_{BR}	avalanched breakdown voltage

The arrow points in the forward bias direction for a PN junction (i.e., from P to N).

(E) Junction Field Effect Transistors (JFET)

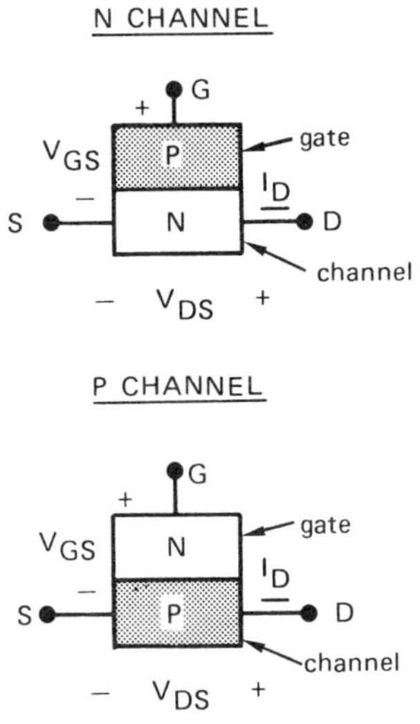

conceptual compositions

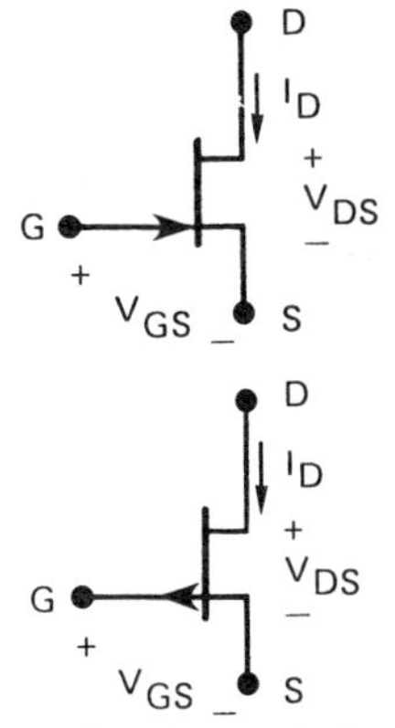

circuit symbols

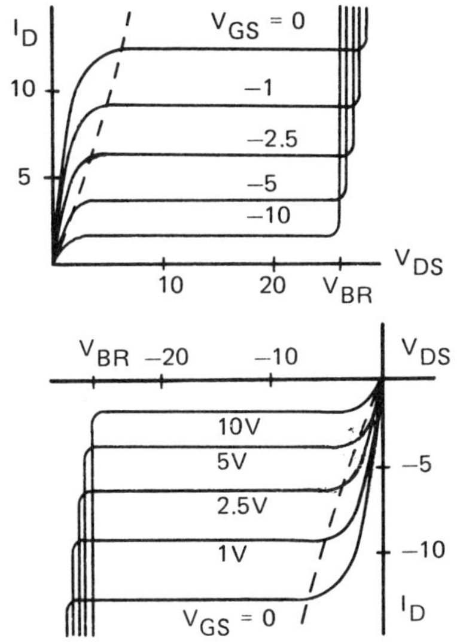

forms of characteristics

Nomenclature

D	drain
G	gate
S	source
I_D	drain current
V_{DS}	voltage drop from drain to source
V_{GS}	voltage drop from gate to source
V_{BR}	avalanche voltage break down

Arrow on the gate lead points from P to N.

(F) Depletion Mode Metal-Oxide-Semiconductor Field Effect Transistors (D-MOSFET)

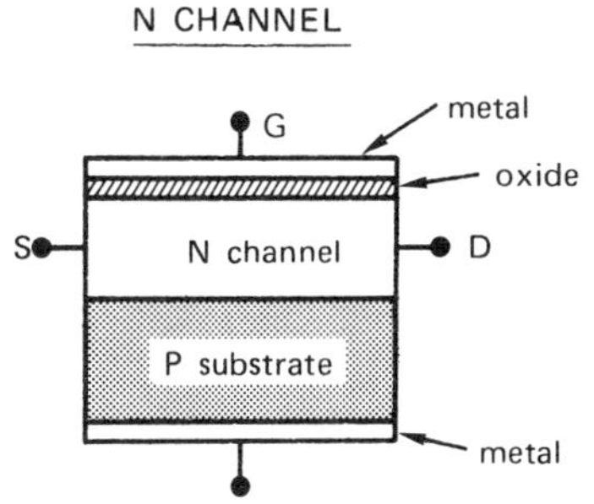

P CHANNEL

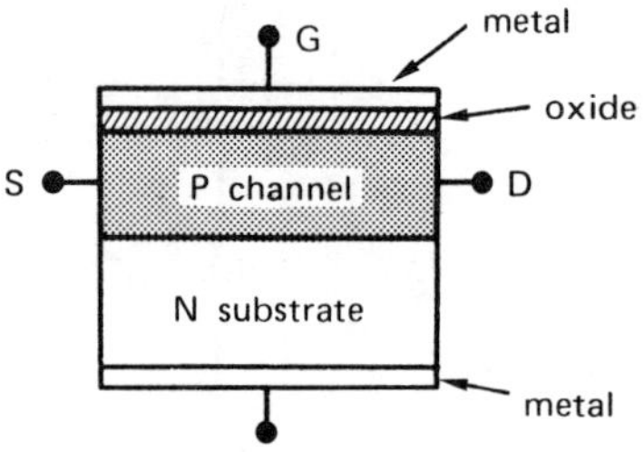

conceptual compositions

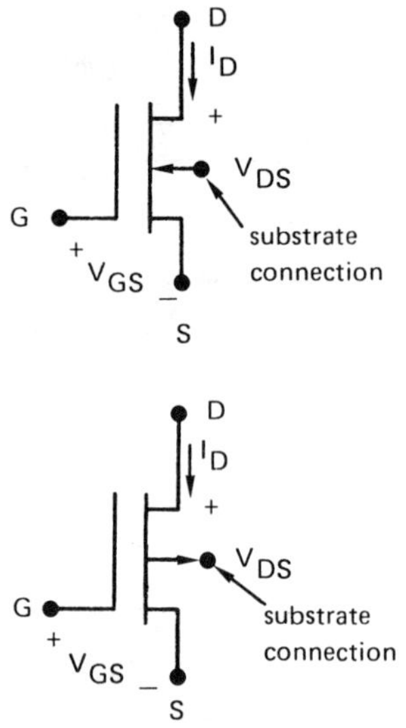

circuit symbols

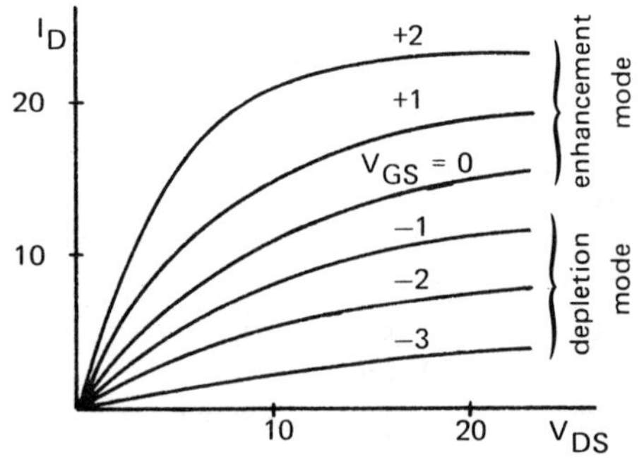

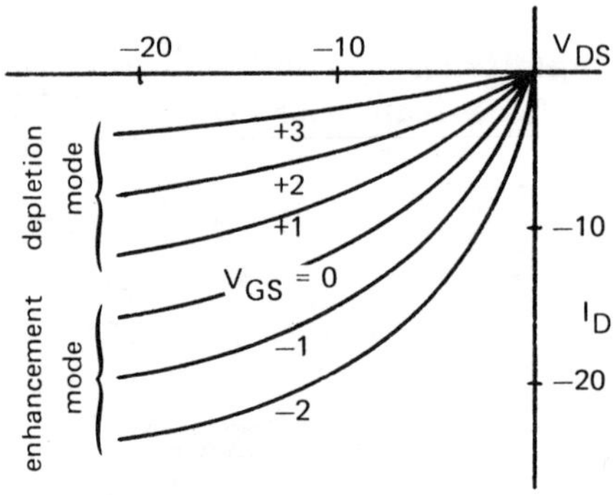

forms of characteristics

Nomenclature

D	drain
G	gate
S	source
V_{DS}	voltage drop from drain to source
V_{GS}	voltage drop from gate to source
I_D	drain current

Arrow on the substrate lead points from P to N.

(G) Enhancement Mode Metal-Oxide-Semiconductor Field Effect Transistors (E-MOSFET)

N CHANNEL

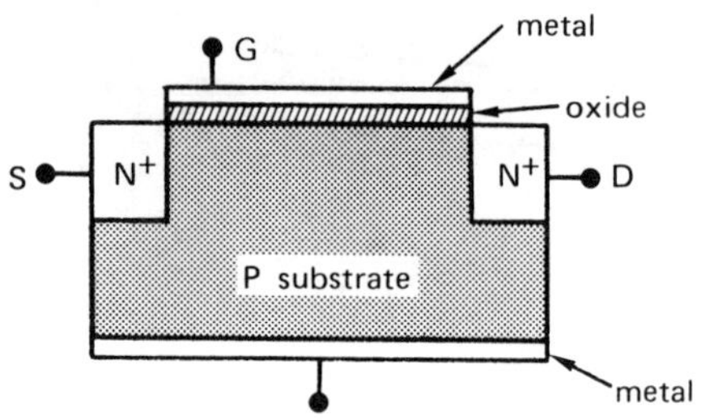

P CHANNEL

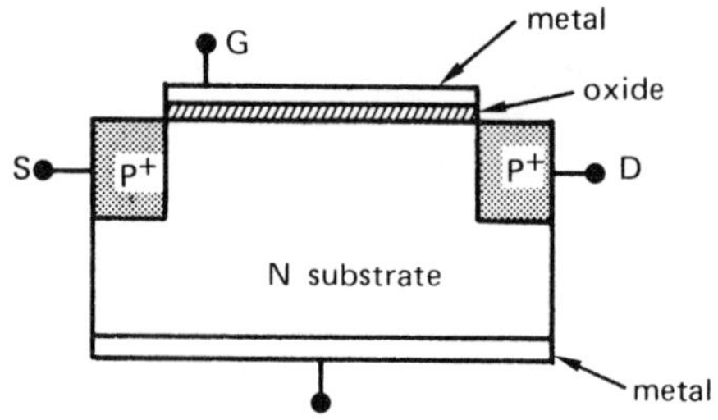

conceptual compositions

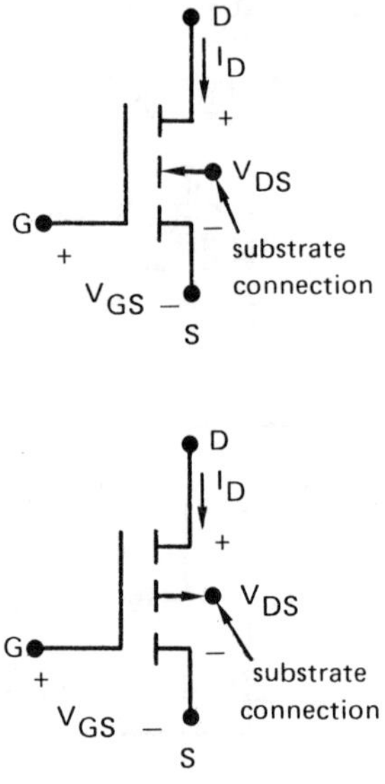

circuit symbols

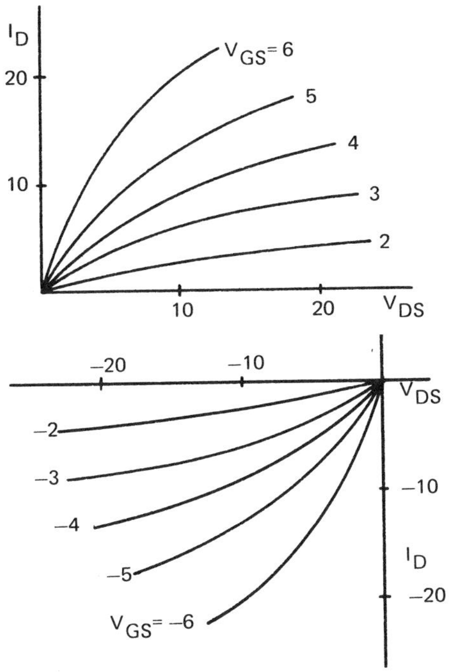

forms of characteristics

Nomenclature

D	drain
G	gate
S	source
V_{DS}	voltage drop from drain to source
V_{GS}	voltage drop from gate to source
I_D	drain current

Arrow on substrate lead points from P to N.

(H) Tunnel Diode

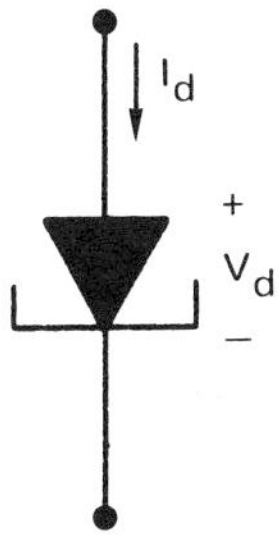

circuit symbol

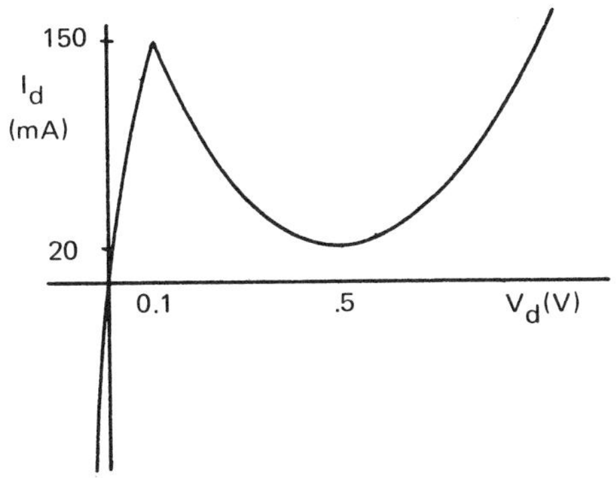

form of characteristic

(I) Unijunction Transistor (UJT) or Double-base Diode

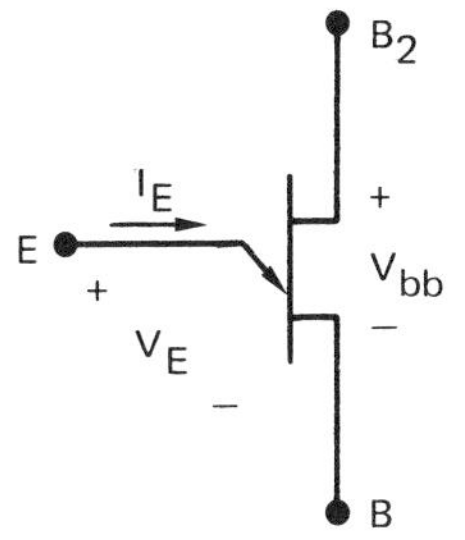

circuit symbol

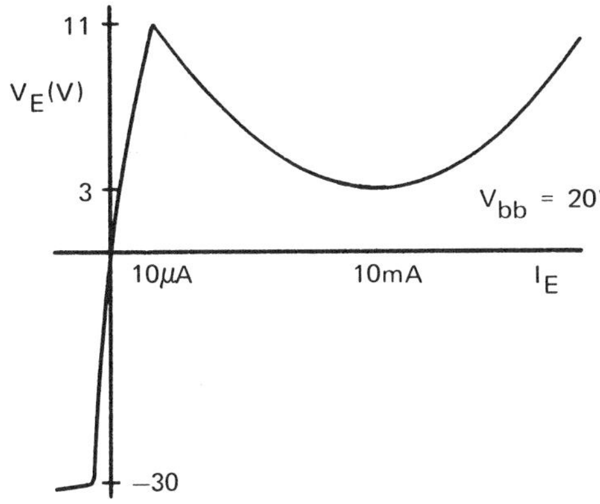

form of characteristic

(J) Current Regulator Diode (see JFET)

JFET equivalent

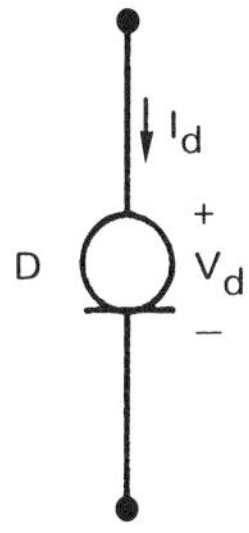

circuit symbol

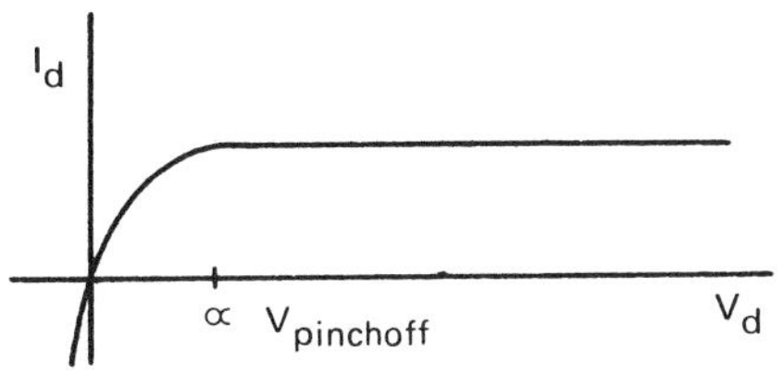

form of characteristic

(K) DIAC

conceptual composition

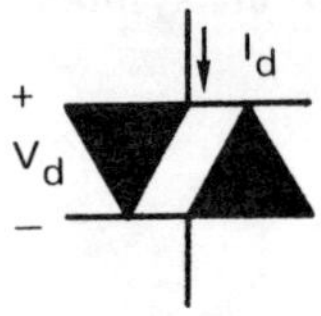

circuit symbol

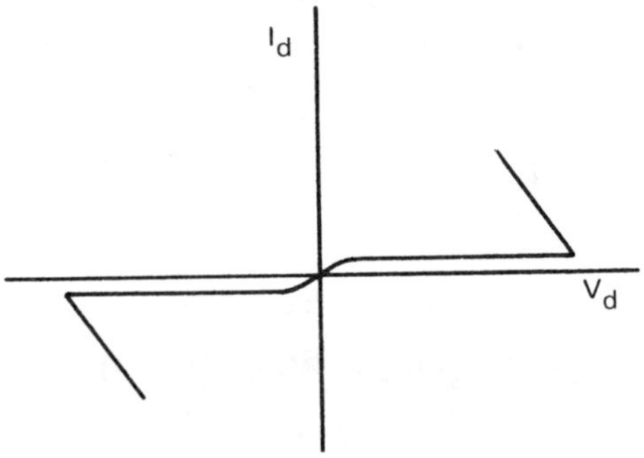

form of characteristic

(L) Silicon Controlled Rectifier (SCR)

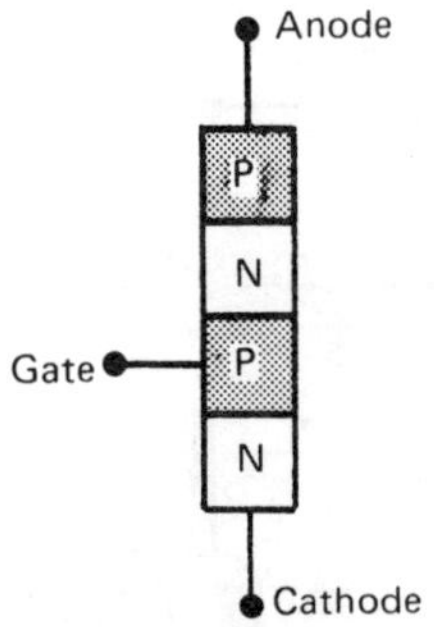

conceptual composition

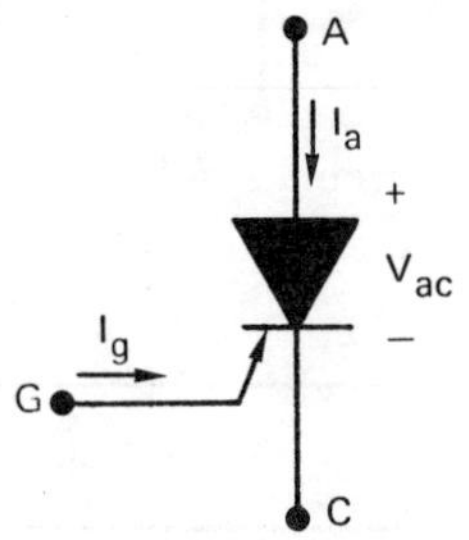

circuit symbol

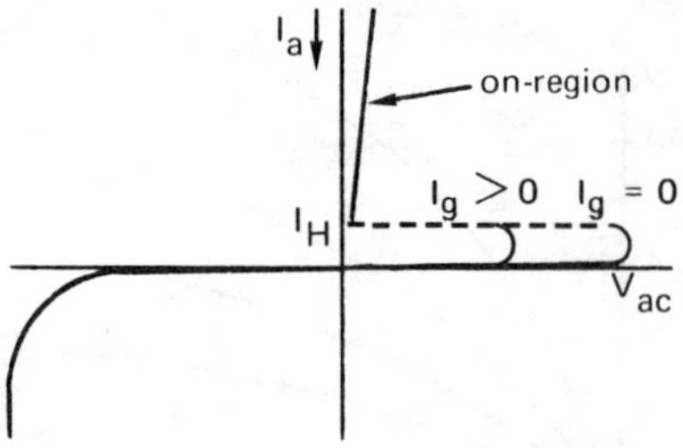

form of characteristic

I_H is the minimum current needed to maintain operation in the on-region

(M) Gate Turn-off Silicon Controlled Rectifier (GTO)

This device is an SCR which can be turned off by reversing the voltage on the gate (making V_{gc} negative). The ordinary SCR cannot be turned off via the gate lead.

(N) Triac

This device is essentially two SCRs connected back-to-back, with the gate leads internally connected.

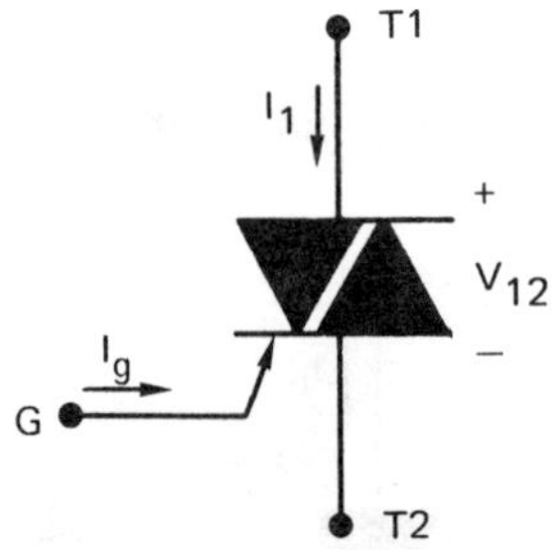

circuit symbol

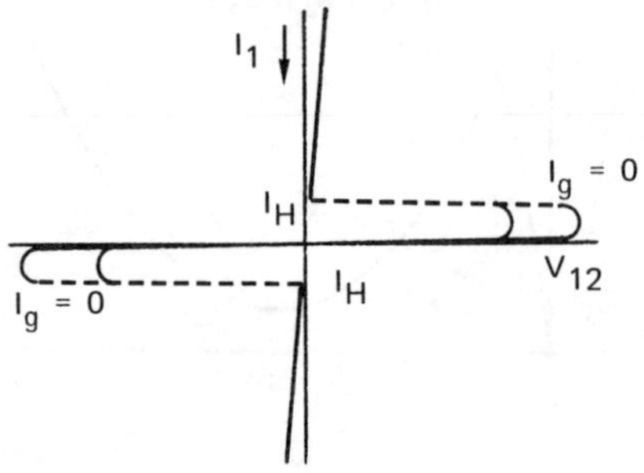

form of characteristic

Practice Problems

LINEAR AMPLIFIERS

Warm-ups

1. An operational amplifier has a gain-bandwidth product of $2\pi \times 10^7$ rad/sec. The amplifier is used in an inverting mode with a feedback resistance of 25,000 ohms and an input resistance 5,000 ohms. Determine the gain and bandwidth of the circuit.

2. The circuit shown is to provide $V_0 = V_1 + V_2$. Determine R_1 and R_2.

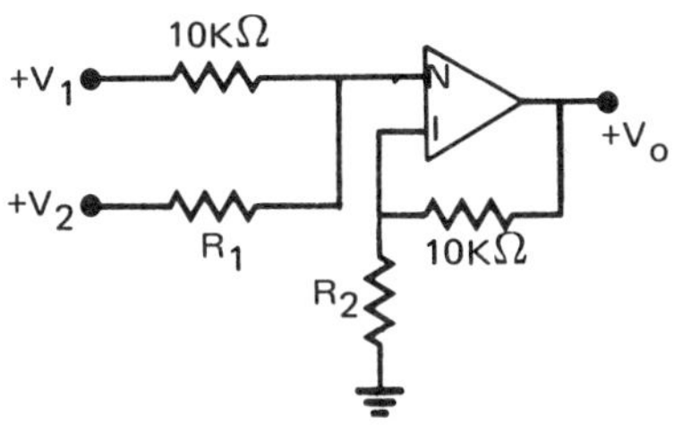

3. A low-pass active filter using one operational amplifier is to have a low frequency gain of 10 and a bandwidth of 30 hz. The only capacitor available is 0.1 μF. Specify the resistances and show the circuit.

4. Design a two-pole high-pass active filter as in figure 8.14. $R_2 = 10K\Omega$ and all capacitors must be powers of 10 less than 10 μF. The poles are to be at 100 rad/sec, and the circuit is to be critically damped.

5. An N-channel JFET has $I_{DSS} = 5$ mA and $V_p = 5$ volts. Estimate I_D at $V_{GS} = -3$ volts.

6. An N-channel JFET has the characteristics shown. As a small-signal amplifier, it is to be self-biased at $V_{DS} = 15$ volts, $I_D = 5$ mA with a drain supply of 30 volts. Specify R_D and R_S.

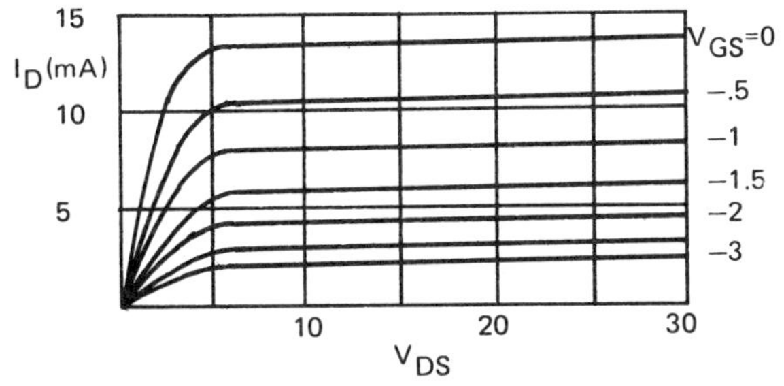

7. An N-channel enhancement mode MOSFET with the given characteristics is to be operated as a source-follower with $R_D = 500$ ohms, and $R_S = 1000$ ohms, a D.C. supply of 16 volts. Determine a bias circuit which will permit an input voltage swing of at least one volt peak-to-peak.

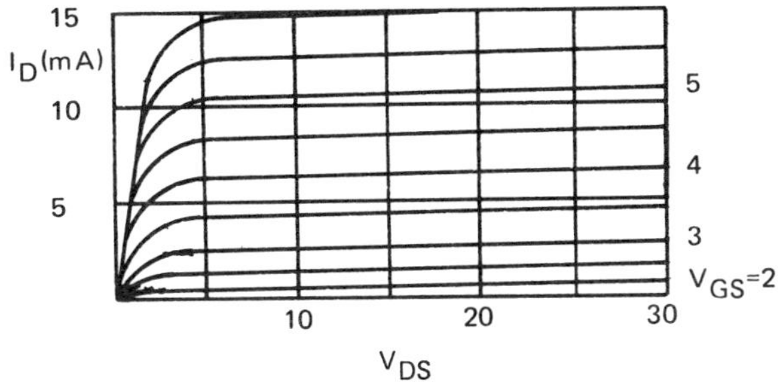

8. The circuit in example 8.1 (p.8-19) has $R_1 = R_2 = 10K\ \Omega$. Determine (a) the value of R_e necessary for an operating point of 15 mA when $\alpha_o = 0.95$, $I_{co} = 10^{-6}$, and $V_{CC} = 24$ volts, and (b) the change in collector current for a temperature change of 125°C and with the parameters found in part (a). Assume silicon construction.

9. A bipolar junction transistor characteristic is shown in figure 8.43. For an operating point of $V_{ce} = 10$ volts and $I_b = 1$ mA, obtain the output circuit for a small-signal amplifier of the form shown in figure 8.44(a).

10. An amplifier in the configuration of figure 8.45 has $R_1 = 6K\ \Omega$, $R_2 = 3K\ \Omega$, $R_e = 1K\ \Omega$, $R_3 = 1K\ \Omega$, and $R_L = 833\ \Omega$. With $h_{ie} = 500\ \Omega$, $\beta = 50$ and $r_c = 5K\ \Omega$, determine the amplifier voltage gain.

11. An amplifier has an input resistance of 50 ohms, a bandwidth of 100 khz and a noise figure of 6 db. Determine (a) the power gain at 293°K if the output noise power is 10^{-12} watts and the input source is matched, (b) the equivalent noise temperature, and (c) the equivalent noise resistance.

Concentrates

1. For the circuit shown, find V_o as a function of V_1, V_2, V_3, and V_4.

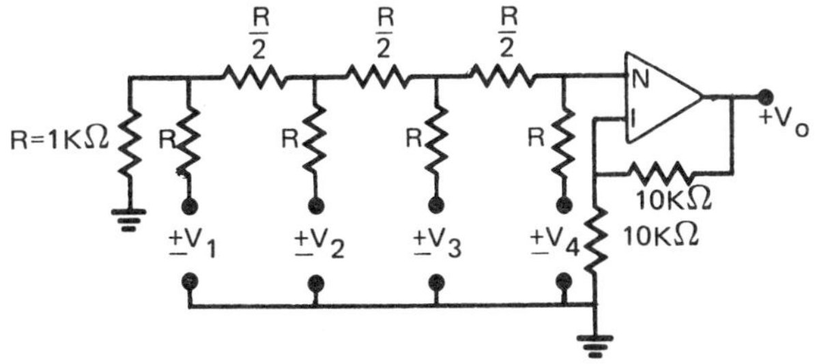

2. A capacitive displacement transducer has a capacitance given by $C_x = A/(0.225d)$ where d is the displacement to be measured. A voltage proportional to the displacement is obtained using the accompanying circuit.

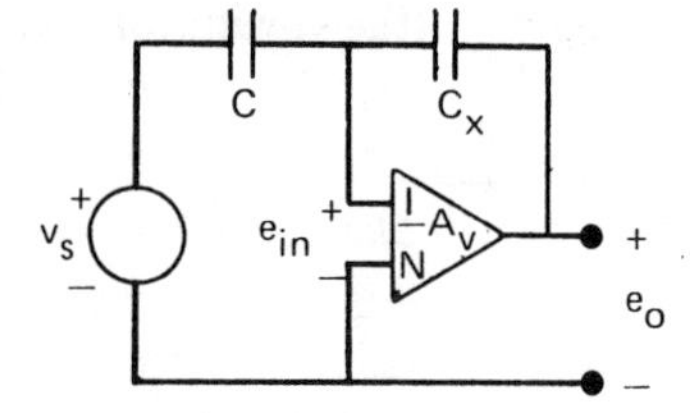

Show that for v_S being a constant amplitude sinusoid at approximately 50 khz and $A \approx 10^8$,

$$e_o = \frac{-(.225)Cv_S d}{A}$$

3. An N-channel JFET has the circuit shown and $C_{OSS} = 2\,pF$, $g_m = 2 \times 10^{-3}$ mhos, $C_{ISS} = 6\,pF$. Determine the voltage gain and high-frequency cutoff (in hz) if (a) $R_S = 5000\,\Omega$, and (b) $R_S = 50\,\Omega$.

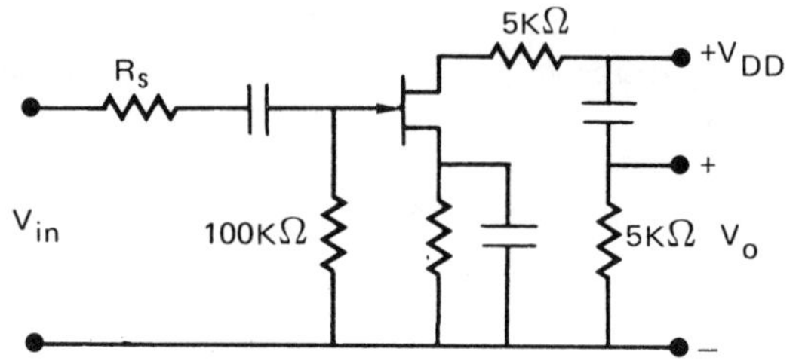

4. A receiving system has a source temperature of 293°, an input stage with noise bandwidth of 1 Ghz, a power gain of 12 db, a noise figure of 6 db, and an input impedance of 72 ohms. The mixer stage has a gain of −6 db and a noise figure of 12 db. The I.F. stage has a power gain of 40 db and a noise figure of 6 db, with a noise bandwidth of 10 Mhz. Determine (a) the overall noise figure, and (b) the minimum detectable input voltage within the input band-pass to permit a signal-to-noise ratio of 1.

5. An emitter follower uses a Darlington pair as shown. $R_1 = R_2 = 220K\,\Omega$, $R_e = 10K\,\Omega$, $R_L = 1K\,\Omega$, $h_{ie1} = 750\,\Omega$, $h_{ie2} = 250\,\Omega$, $h_{fe1} = 100$, $h_{fe2} = 50$, $r_{c1} = 20K\,\Omega$, and $r_{c2} = 10K\,\Omega$.

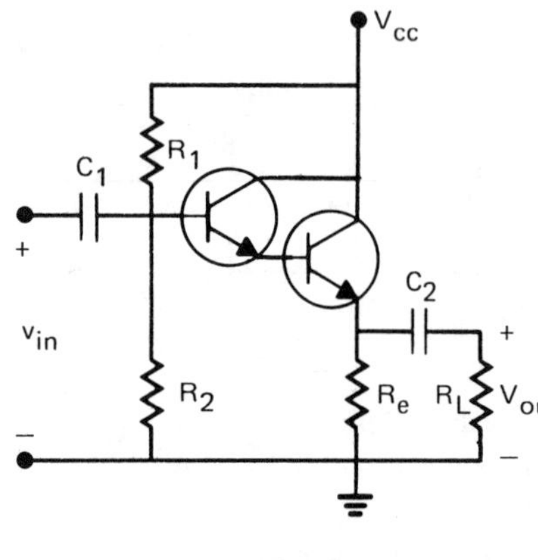

Determine the current gain of this amplifier.

6. For the amplifier shown, $h_{ie} = 500\,\Omega$, $\beta = 39$, and $r_c = 5K\,\Omega$. Determine (a) the low frequency cutoff due to the emitter bypass capacitance, assuming that C_1 is a short circuit, (b) the value of C_1 necessary to insure that the associated corner frequency is less than 10 hz regardless of whether or not the emitter resistance is bypassed, and (c) the frequency response of the voltage gain, assuming C_2 can be neglected. The transistor has high-frequency parameters of $C_{b'c} = 25\,pF$, $C_{b'e} = 1000\,pF$, $r_{bb'} = 400\,\Omega$, and $r_{b'e} = 100\,\Omega$.

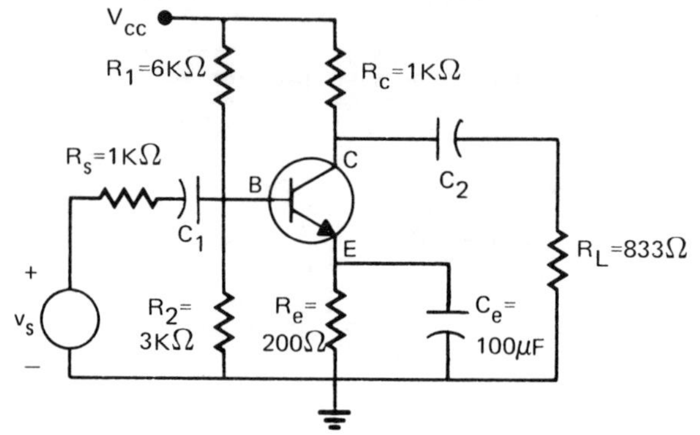

7. For the amplifier circuit and MOSFET characteristics shown, determine the frequency response for V_o/V_{in}. Show a graph of the amplitude response.

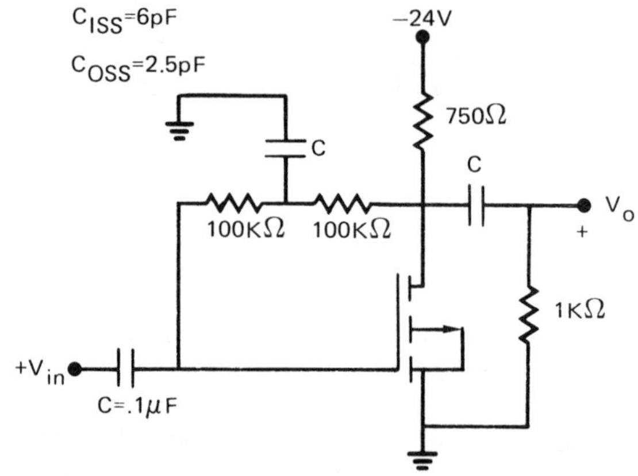

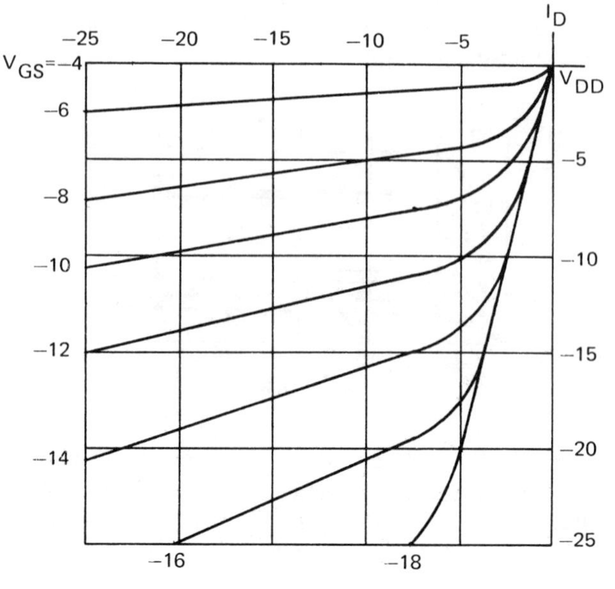

8. Design the bias circuit for the amplifier shown. The desired Q-point is 10 mA and $V_{DS} = 5$ volts, with an input impedance of 100K Ω. Determine the amplifier gain if the resistance R_S is capacitively bypassed.

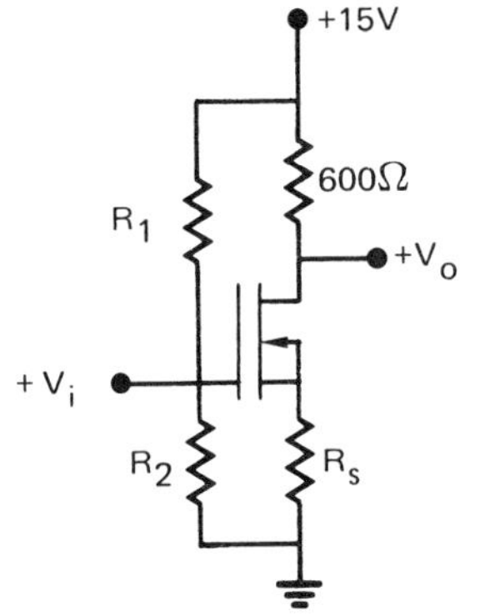

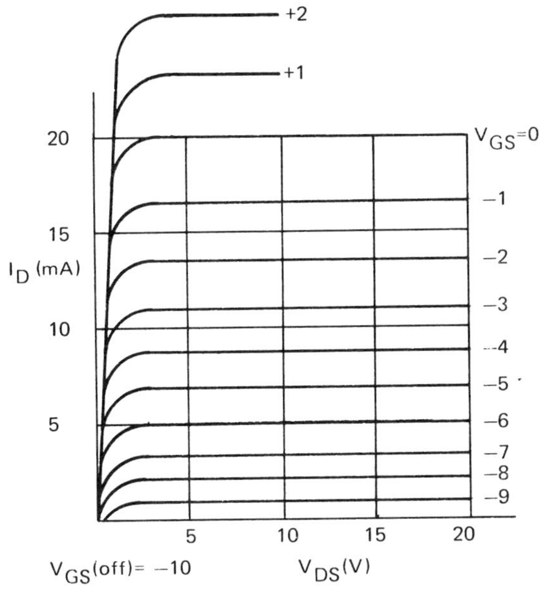

Timed

1. The circuit shown is a notch filter providing at least 30 db of attenuation at 400 hz. It has a 3 db bandwidth of 80 hz. 0.01μF capacitors are available. Determine R_1, R_2, and R_3.

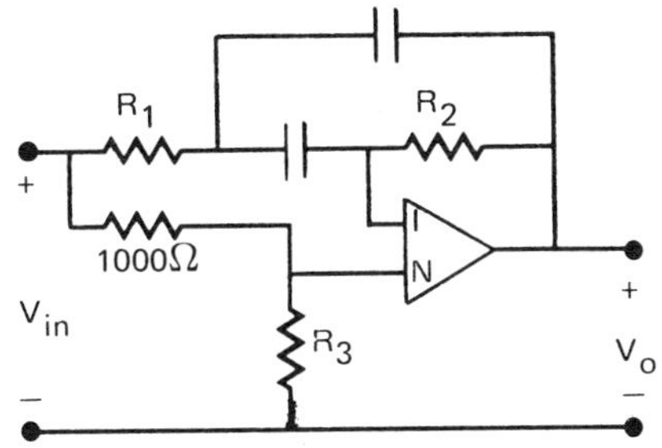

2. The transconductance of the amplifier circuit shown is 5000 μmhos and the drain-source resistance is 20K Ω. Neglect the capacitor effects and determine (a) the voltage gain, and (b) the input impedance.

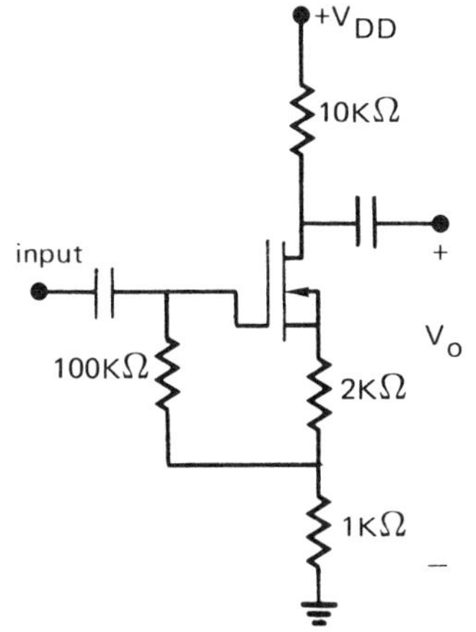

3. The operational amplifiers shown are identical and have a raw GBW of 10^7 rad/sec. Determine the input impedance Z_{in}, and the frequency range over which it appears essentially as a capacitance.

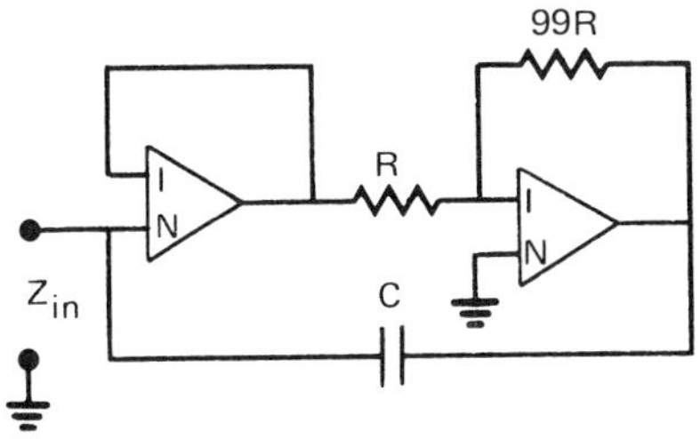

4. The amplifier shown has the following transistor parameters. For T_1: $h_{fe} = 100$, $h_{ie} = 750\,\Omega$, and $r_c = 20K\,\Omega$; for T_2: $h_{fe} = 50$, $h_{ie} = 250\,\Omega$, and $r_c = 10K\,\Omega$. Determine the mid-frequency voltage gain for this amplifier.

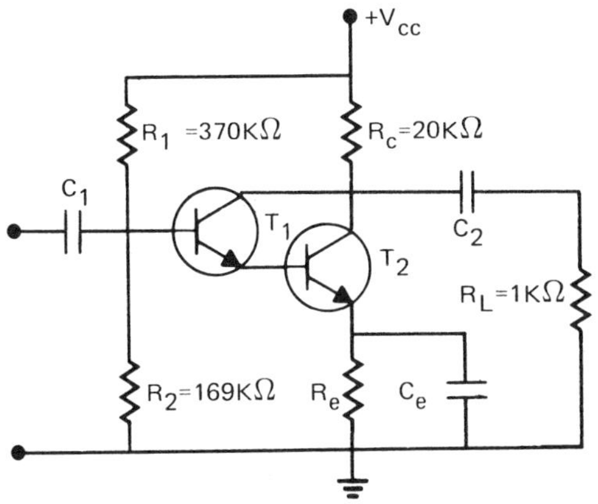

RESERVED FOR FUTURE USE

9 NONLINEAR WAVESHAPING AND SWITCHING

Nomenclature

A_v	amplifier voltage gain	-
C	capacitance	farads
h_{IE}	base-emitter input impedance	ohms
i	current	amps
I	current	amps
I_{CO}	collector reverse saturation current	amps
k	voltage division ratio	-
L	inductance	henrys
N	an integer	-
Q	charge	C
R	resistance	ohms
t	time	sec
T	period of time	sec
v	voltage	volts
V	voltage	volts
V_{bb}	base supply voltage	volts
V_{cc}	collector supply voltage	volts

Symbols

ω	frequency	rad/sec

Subscripts

b	bias
BE	base-emitter
br	breakdown
B	base
BO	breakover
c	control
C	collector or capacitor
d	diode or device
dd	drain supply
D	drain
e	emitter
E	emitter
f	forward, firing, or fall
G	gate
GT	gate turn-on
H	hold
in	input
I	ionization
L	load
o	output
r	reference or rise
s	signal or storage
sat	saturation
S	source
Th	Thevenin
v	voltage
z	Zener
zk	Zener keep-alive

1 DIODE CIRCUITS

A. INTRODUCTION

Semiconductor diodes have characteristics similar to figure 9.1. For *silicon diodes*, the *forward voltage drop* is approximately .6 to .7 volts, while for *germanium diodes*, it varies from .15 to .25 volts. The *reverse voltage breakdown* value is set by the doping levels. (Decreasing doping results in a higher breakdown voltage.) This latter property is used in the regulation of D.C. voltage supplies.

The linear equivalent circuit is derived from the curve parameters shown in figure 9.1(a) and results in the circuit shown in figure 9.1(b).

In figure 9.1(b), the lower branch is active when V_d is less than $-V_{br}$. The upper branch is active when V_d is greater than $+V_f$. When V_d is between $-V_{br}$ and $+V_f$, the device behaves as an open circuit.

B. THE PRECISION DIODE

The equivalent circuit shown in figure 9.1(b) is useful when voltages are greater than V_f, but it is of limited use for voltages near or smaller V_f. For such low-voltage situations, an equivalent circuit more closely approximating the ideal diode is needed. Such an approximation is obtained by using an operational amplifier in conjunction with one or two ordinary diodes.

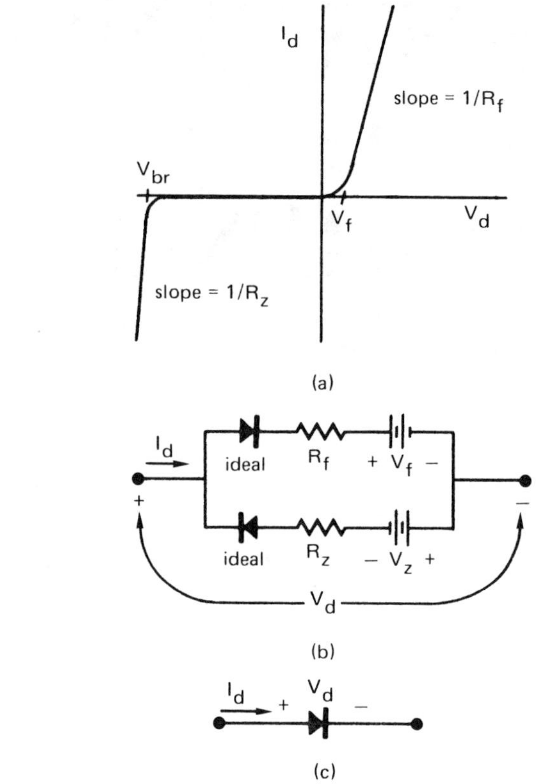

Figure 9.1 (a) Semiconductor diode characteristics, (b) equivalent circuit, and (c) circuit symbol

Consider the circuit of figure 9.2(a). A diode has been inserted between the amplifier output and the feedback point of a voltage follower. For negative inputs, the amplifier output will be saturated at the maximum negative output. The diode will then be reverse-biased. (The diode must have a reverse breakdown greater in magnitude than the maximum negative amplifier voltage.) Then, the output will be disconnected from the input, and the load will see only R_o.

When the input voltage is positive, the amplifier output will be Av_{in} until the diode becomes forward-biased (i.e., until V_d excedes V_f). In this case, the output voltage V_L follows the input with the small offset V_f/A_v. (A_v is the raw gain of the operational amplifier.) This circuit also suppresses the forward resistance and any source resistance associated with V_{in}.

For comparison, consider the diode circuit of figure 9.2(b). Circuit analysis yields equation 9.1.

$$V_L = \frac{R_o V_{\text{in}}}{R_f + R_o} - \frac{R_o V_f}{R_f + R_o} \qquad 9.1$$

R_f and V_f are the linear equivalent circuit parameters for the diode from figure 9.1 when the diode is forward-biased. In cases where the source (V_{in}) has an associated resistance, it should be added to R_f.

The equivalent circuit for when the diode of figure 9.2(a) is forward biased is shown in figure 9.2(c). The circuit analysis yields equation 9.2.

$$V_L = \frac{R_o V_{\text{in}}}{R_o + \frac{R_o + R_f}{A_v}} - \frac{R_o \frac{V_f}{A_v}}{R_o + \frac{R_o + R_f}{A_v}} \qquad 9.2$$

Since A_v is on the order of 10^5, the effective source resistance is $(R_o + R_f) \times 10^{-5}$. This is negligible in the denominator compared to R_o. Thus, the resulting transfer equation is

$$V_L = V_{\text{in}} - \frac{V_f}{A_v} \qquad 9.3$$

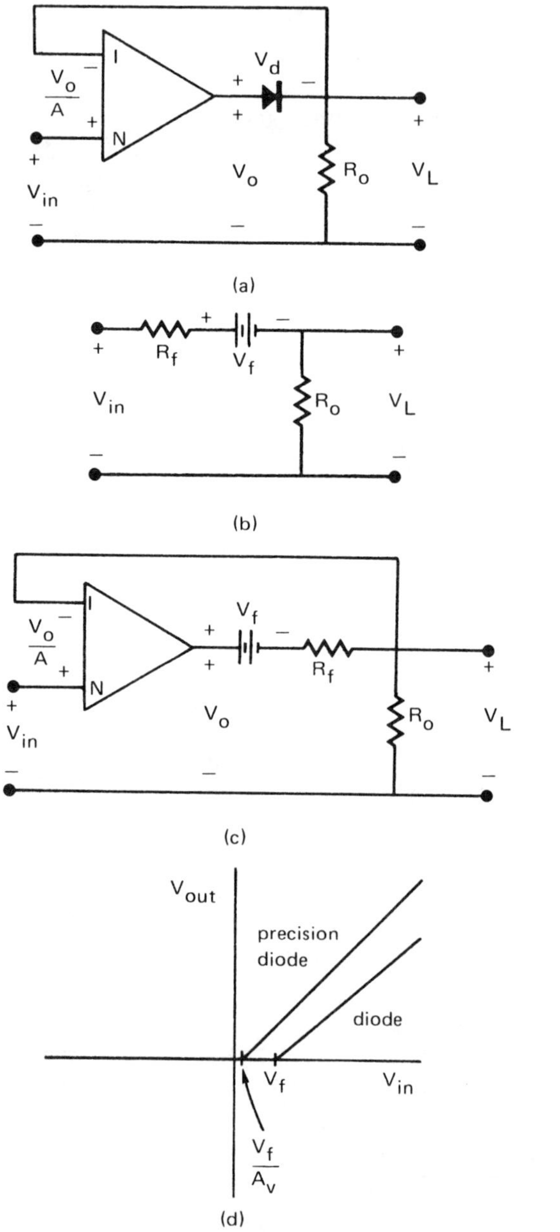

Figure 9.2 (a) Simple precision diode circuit, (b) ordinary diode circuit, (c) equivalent circuit of forward-biased precision diode, and (d) input-output voltage characteristics

A *precision diode* will be seen from the output as the voltage V_{in} in series with a small battery. The equivalent V_f for this precision diode would be about 6 microvolts for a silicon diode, assuming an operational amplifier raw gain of 10^5. The result is a rectifier which can be used for millivolt level signals. Precision and ordinary diode characteristics are shown in figure 9.2(d).

C. THE CLIPPER-LIMITER CIRCUIT

The diode can perform clipping and limiting functions when used in the circuits shown in figure 9.3. Analysis of the first circuit produces equations 9.4 and 9.5.

$$V_L = \frac{R(V_r - V_f)}{R_f + R} + \frac{R_f V_S}{R_f + R} \text{ for } V_S < V_r - V_f \qquad 9.4$$

$$= V_S \text{ for } V_S > V_r - V_f \qquad 9.5$$

The resulting voltage characteristics are shown. V_r can be either negative or positive. If it is negative, the circuit is considered a *clipper*. If V_r is positive, the circuit is a *limiter*.

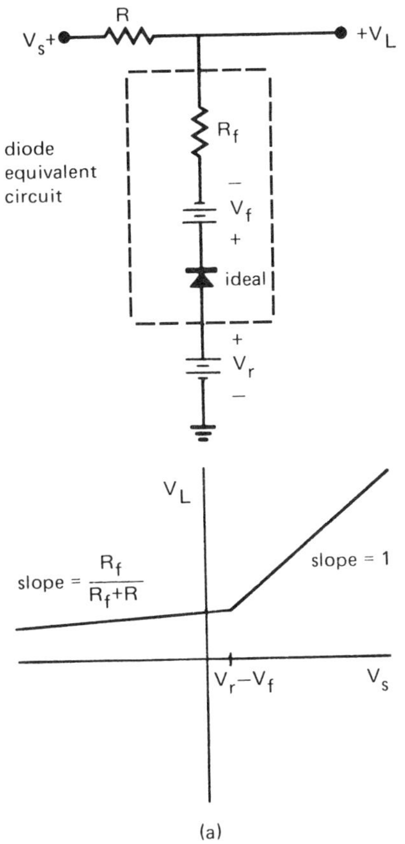

(a)

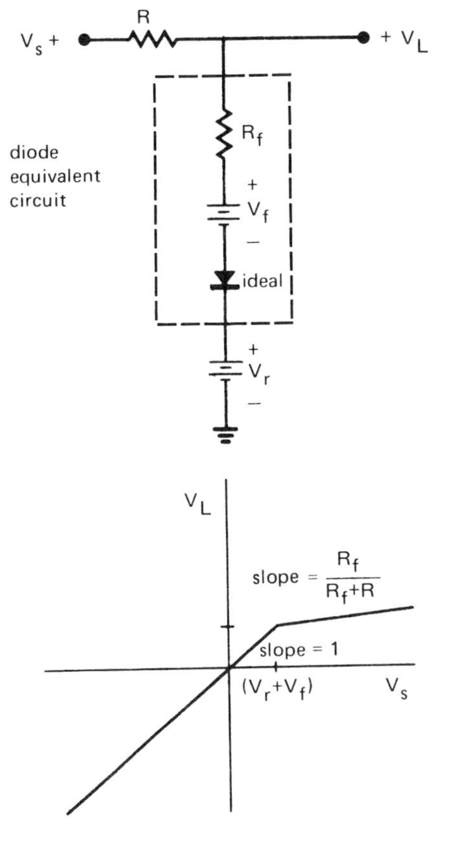

(b)

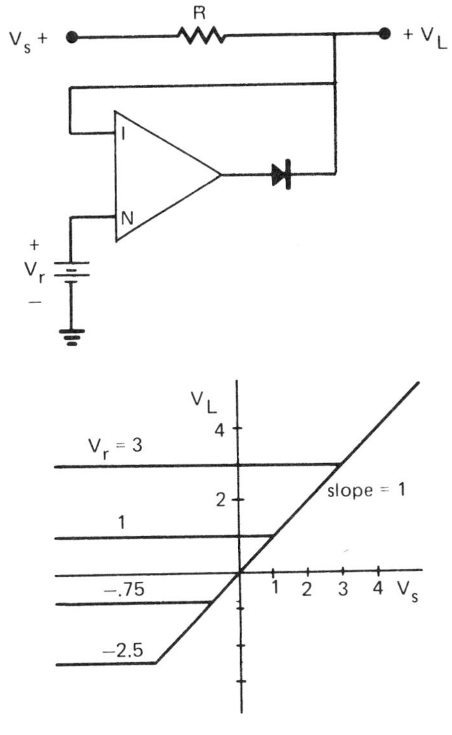

(c)

Figure 9.3 (a) Limiter circuit, (b) clipper circuit, (c) precision limiter/clipper circuit, and (d) precision limiter/clipper circuit sensitive to negative voltages

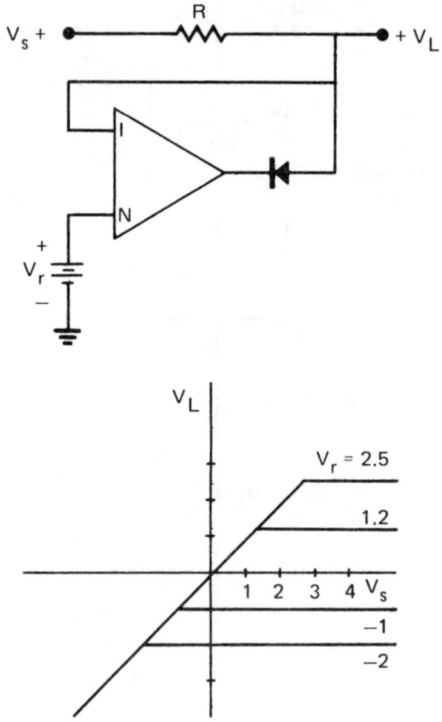

(d)

Figure 9.3 (a) Limiter circuit, (b) clipper circuit, (c) precision limiter/clipper circuit, and (d) precision limiter/clipper circuit sensitive to negative voltages

The circuit of figure 9.3(b) is sensitive to negative values of V_S. The ideal circuit would have $R_f/(R_f + R)$ be equal to zero, and the effect of V_f would be negligible. This can be accomplished in practice by using the precision diode circuit. This is illustrated in figures 9.3(c) and 9.3(d). R_f and V_f are negligible for the two circuits.

Example 9.1

Design a circuit that will linearly transfer any signal with an amplitude of 1.0 volt or less, but will clip all larger signals at an amplitude of 1.0 volt. Assume that supply voltages of +12 volts and −12 volts are available.

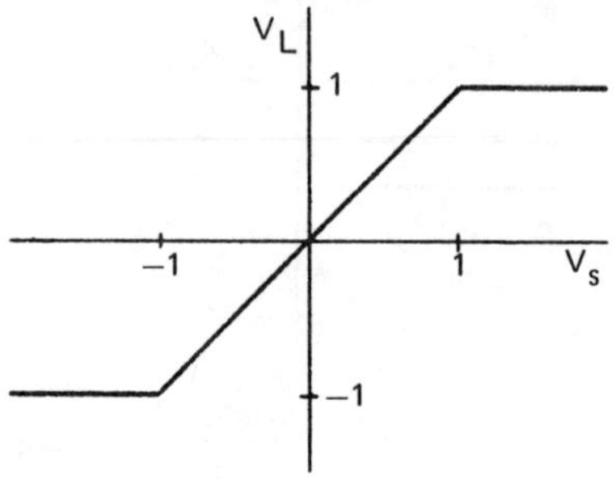

The characteristic required is shown. For V_S greater than 1 volt, an ideal diode must be connected from the output terminal to a 1.0 V source. For the negative voltage situation, an ideal diode must be connected from a −1.0 V source to the output terminal. The one-volt sources are derived from the two 12 volt supply sources. The resulting circuit is shown.

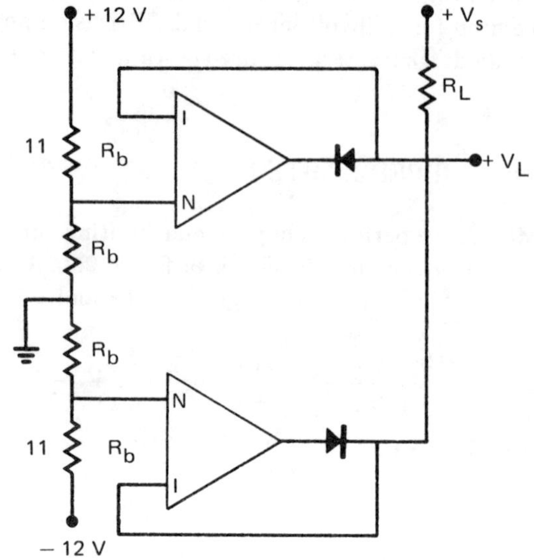

Example 9.2

Obtain the voltage transfer characteristics for the circuit shown.

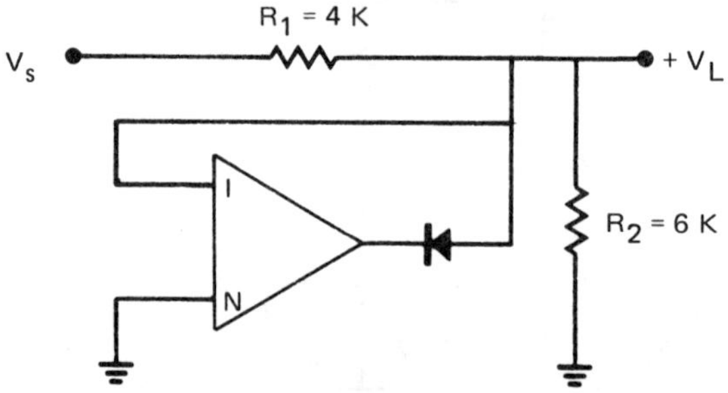

The Thevenin equivalent of the input sources, R_1 and R_2, are needed. R_1 and R_2 form a voltage divider, and the Thevenin equivalent voltage source is

$$\frac{R_2 V_S}{R_1 + R_2} = .6V_S$$

The Thevenin equivalent resistance is the parallel combination of R_1 and R_2.

$$R_{\text{Th}} = \frac{R_1 R_2}{R_1 + R_2}$$

The effective source voltage is $.6V_S$, so the slope is .6. The problem description calls for only negative sensitivity to the source. The resulting characteristic is shown.

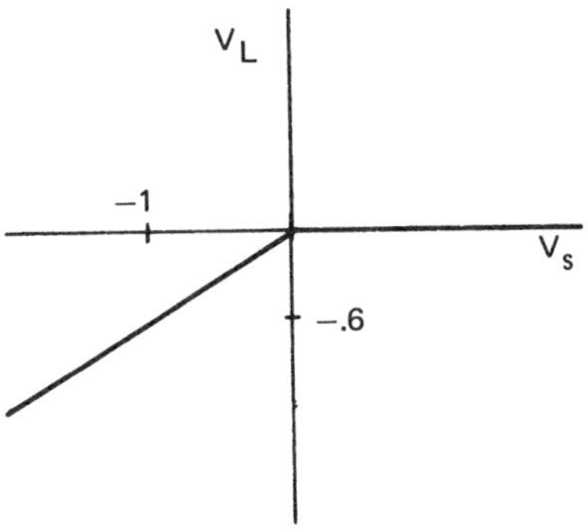

2 VOLTAGE REGULATORS

A. INTRODUCTION

The characteristics shown in figure 9.3(b) are similar to voltage regulator characteristics. The accompanying circuit approximates the equivalent circuit of a Zener diode (e.g., the characteristics of an ordinary diode operated with reverse voltage). A Zener or voltage-regulator diode is manufactured to have a specific breakdown voltage. These devices are used in circuits where the source voltage V_S is kept between limits, and the load current must be able to vary from zero to some maximum value.

Zener diodes are used to control the output voltage within a narrow range of values regardless of changes in the source load voltages.

B. IDEAL ZENER VOLTAGE REGULATOR

An ideal Zener voltage regulator is shown in figure 9.4. (An ideal Zener diode maintains a constant output voltage as long as the Zener current is positive.)

The source resistance R_S which will result in the lowest power loss while maintaining a constant load voltage during changes of V_S and I_L is needed. Design problems also include finding the power rating of the Zener diode, since it absorbs the difference between the maximum and minimum load currents.

From figure 9.4(b), it can be seen that voltage control will not occur unless the ideal diode is forward-biased. This requires that V_L be equal to V_{zo}, so that $V_S - I_L R_S$ must be at least V_{zo} (i.e., $I_S = I_L + I_z$, and I_z must be zero or positive for the ideal diode to be forward biased). To ensure regulation, it is necessary that with the maximum possible load current, the ideal diode passes no negative current. This requires

$$R_S \leq \frac{V_{S,\,\text{min}} - V_{zo}}{I_{L,\text{max}}} \qquad 9.6$$

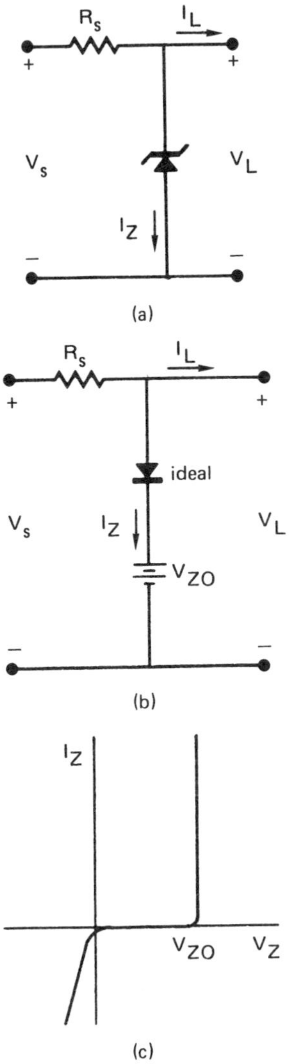

Figure 9.4 (a) Simple Zener diode voltage regulator (b) equivalent circuit with ideal Zener diode, and (c) the voltage-current characteristics of an ideal Zener diode

To minimize the power loss in R_S, the equality is taken. The current through the Zener diode is

$$I_z = \frac{V_S - V_z}{R_S} - I_L \qquad 9.7$$

Equation 9.7 is maximum when V_S is maximum and when I_L is minimum. Then,

$$I_{z,\text{max}} = \frac{V_{S,\,\text{max}} - V_{zo}}{R_S} = \left[\frac{V_{S,\,\text{max}} - V_{zo}}{V_{S,\,\text{min}} - V_{zo}}\right] I_{L,\text{max}} \qquad 9.8$$

The Zener maximum power is $V_{zo} I_{z,\text{max}}$. The source resistance must absorb a maximum power of $I^2_{z,\text{max}} R_S$, and the resistor must be rated accordingly.

Example 9.3

Design a voltage regulator circuit to maintain an output voltage of 5 volts when the source voltage varies between 11 and 14 volts, and the load current varies between 0 and 5 amps.

$$R_S = \frac{11-5}{5} = 1.2 \text{ ohms}$$

$$I_{z,\max} = \left(\frac{14-5}{11-5}\right)5 = 7.5 \text{ amps}$$

$$P_d = 5 \times 7.5 = 37.5 \text{ watts}$$

Therefore, the diode must be rated at 40 watts. Also,

$$P_R = \frac{(14-5)^2}{1.2} = 67.5 \text{ watts}$$

Therefore, the resistor must be rated at 75 watts.

C. PRACTICAL ZENER DIODE REGULATOR

The design used in example 9.3 is not practical for several reasons. First, 5 volts is not a standard Zener voltage. (5.1 is the closest. See page 9-29.) Second, diodes with Zener voltages below about 8 volts are controlled by the Zener effect in which there is considerable resistance. Because of this, the Zener voltage can rise by as much as 20% of the basic value at rated current. In addition, there is a *keep-alive current*, I_{zk}, necessary to maintain the Zener characteristics in the linear region of operation. Therefore, the Zener current must always be maintained at that level or higher.

As might be expected, the Zener diode regulator in a practical circuit may not approximate the ideal case at all. To demonstrate this, consider the equivalent circuit and practical characteristics shown in figure 9.5.

For the circuit of 9.5(a),

$$V_S - V_L = (I_L + I_z)R_S \qquad 9.9$$

$$V_L = V_{zo} + I_z R_z \qquad 9.10$$

Solving simultaneously,

$$V_L = \frac{V_{zo}R_S + V_S R_z}{R_S + R_z} - I_L\left[\frac{R_S R_z}{R_S + R_z}\right] \qquad 9.11$$

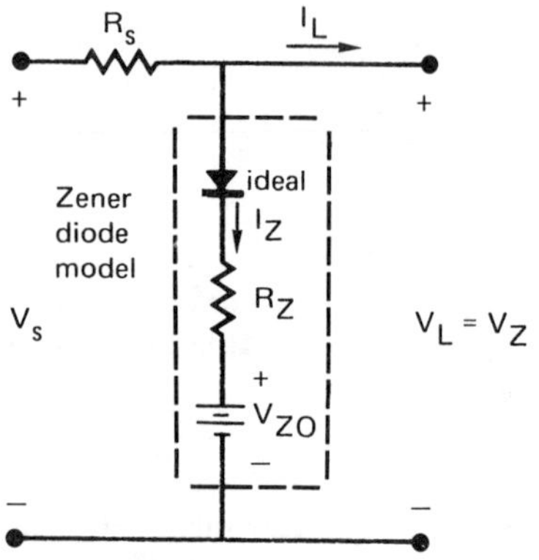

(a)

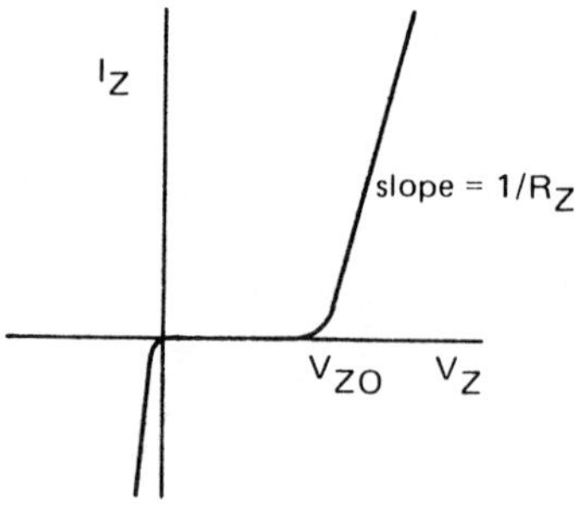

(b)

Figure 9.5 (a) Practical voltage regulator, and (b) characteristics of practical Zener diode

Example 9.4

Determine the regulation for the circuit shown in figure 9.5. Assume $V_{zo} = 5.1$ volts, $I_{zk} = 0.5$ amps, and $R_z = 0.1$ ohms. From example 9.3, $V_{S,\max} = 14$ volts, $V_{S,\min} = 11$ volts, and $I_{L,\max} = 5$ amps.

Because it is necessary to maintain I_z at .5 amps or greater, V_L must be maintained at

$$V_L = V_{zo} + I_{zk}R_z = 5.15 \text{ volts}$$

This minimum voltage can be designed to occur only when the source voltage is at its minimum and the load current is at its maximum. The 5.15 value for V_L will occur when

$$5.15 = \frac{5.1R_S}{0.1+R_S} + \frac{11 \times .1}{0.1+R_S} - 5 \times \left[\frac{0.1 \times R_S}{0.1+R_S}\right]$$

$$R_S = 1.06 \text{ ohms}$$

Substituting into equation 9.11,

$$V_L = 4.66 + 0.086V_S - 0.091I_L$$

For $V_S = 11$ volts, the load voltage reaches its limiting value at $I_L = 5$ amps. For increased I_L, the Zener diode will come out of the breakdown region and cease to regulate. In that case, the load voltage will be simply

$$V_L = V_S - I_L R_S \qquad 9.12$$

The resulting characteristic of V_L versus I_L for the two extremes of source voltage are shown.

The maximum regulation will occur between the maximum source voltage at no load current, and the minimum source voltage at the maximum load current. The minimum load voltage was found to be 5.15 volts. The maximum load voltage occurs at $V_S = 14$ and $I_L = 0$. From equation 9.12, the load voltage is 5.86 volts. Thus, the voltage regulation at maximum is

$$\text{regulation} = \frac{5.86 - 5.15}{5.86} \times 100 = 12.1\%$$

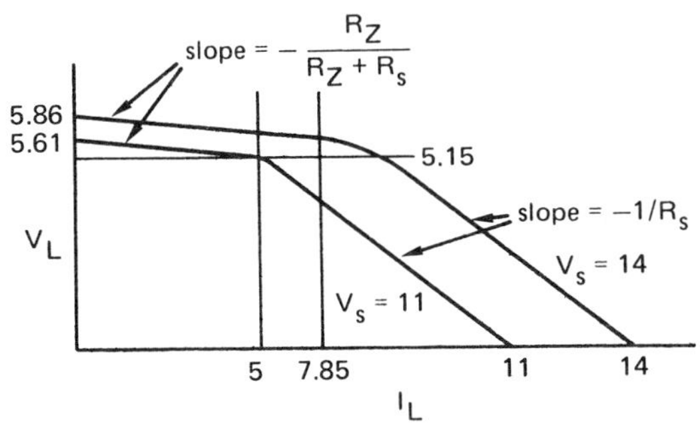

D. AN OPERATIONAL AMPLIFIER VOLTAGE REGULATOR

A Zener diode regulator is useful when current demands are low but not when there are currents as large as were used in example 9.4. It is possible to use a precision diode circuit to improve the voltage regulation. In general, the operational amplifier will have a limited current capacity. The most usual method to reduce the voltage regulation and still supply larger currents is to use the Zener diode as a reference source for the amplifier. This, in turn, drives a *pass transistor* as illustrated in figure 9.6. This circuit has the added advantage that the Zener voltage need not be the same as the desired regulated voltage. The *regulated voltage* can be adjusted by varying the ratio of the resistances R_1 and R_2.

The input to the operational amplifier is $V_z - kV_L$ where

$$k = \frac{R_2}{R_1 + R_2} \qquad 9.13$$

So, the output is $A_v(V_z - kV_L)$. The transistor is modeled by its base-emitter P-N voltage drop, and the current gain β is indicated. The circuit analysis of the output circuit results in

$$A_v(V_z - kV_L) = i_b R_b + V_{be} + \beta R_S i_b + V_L \qquad 9.14$$

$$V_L = (1+\beta) i_b (R_1 + R_2) - I_L (R_1 + R_2) \qquad 9.15$$

$$V_L = \frac{A_v V_z - V_{be} - I_L\left(R_S + \frac{R_b}{\beta+1}\right)}{kA_v + 1 + \frac{k}{R_2}\left(R_S + \frac{R_b}{\beta+1}\right)}$$

$$\approx \frac{V_z}{k} - I_L \frac{\left(R_S + \frac{R_b}{\beta+1}\right)}{kA_v} \qquad 9.16$$

A_v is on the order of 10^5, and the terms other then kA_v in the denominator of equation 9.16 are negligible. The output voltage regulation due to I_L is diminished by a factor of $1/kA_v$ from that of the Zener regulator of figure 9.5.

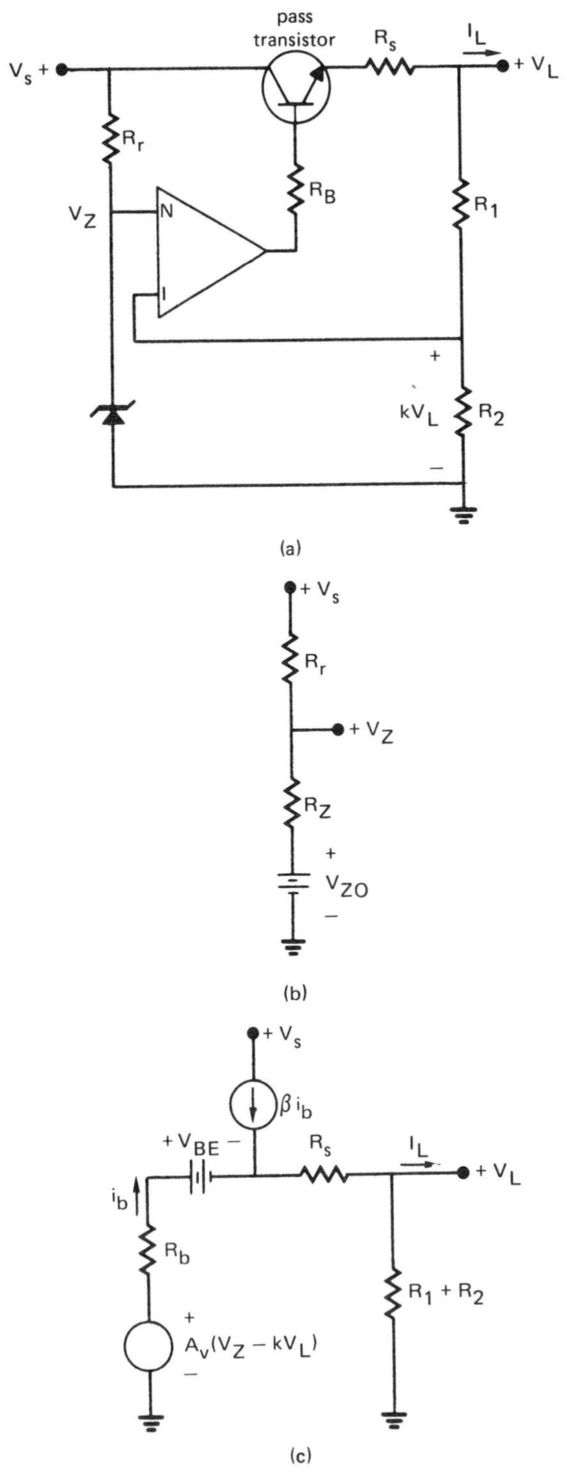

Figure 9.6 (a) Simple operational amplifier voltage regulator circuit, (b) reference voltage circuit, and (c) equivalent output circuit

The variation of V_z with V_S is important. With V_S at its minimum value, R_r is set to allow the minimum keep-alive current to flow.

$$R_r = \frac{V_{S,\min} - V_{zo}}{I_{zk}} - R_z \qquad 9.17$$

Then,

$$V_{z,\text{min}} = V_{zo} + I_{zk}R_z \quad 9.18$$

$$V_{z,\text{max}} = \frac{V_{S,\text{max}}R_z - V_{zo}R_r}{R_r + R_z} \quad 9.19$$

Example 9.5

Design an operational amplifier voltage regulator to meet the specifications of $V_L = 5$ volts, $V_{S,\text{max}} = 14$ volts, $V_{S,\text{min}} = 11$ volts, and $I_{L,\text{max}} = 5$ amps. Determine the worst-case voltage regulation for the circuit.

Begin by choosing a $\frac{1}{4}$ watt Zener diode with a voltage of 4.7 volts at the keep-alive current of 10 mA and a Zener resistance of 35 ohms. From equation 9.18,

$$4.7 = V_{zo} + I_{zk}R_z = V_{zo} + .01(35)$$

So,

$$V_{zo} = 4.35 \text{ volts}$$

At $I_z = 10$ mA and $V_S = 11$ volts, the necessary value for R_r is

$$R_r = \frac{11 - 4.7}{.01} = 630 \text{ ohms}$$

From equations 9.18 and 9.19,

$$V_{z,\text{min}} = 4.7 \text{ volts}$$

$$V_{z,\text{max}} = \frac{14 \times 35}{630 + 35} + \frac{4.35 \times 630}{630 + 35} = 4.86 \text{ volts}$$

Next, choose a current limiting resistance which will permit the transistor some voltage (say one volt from collector to emitter) when the maximum load current and minimum source voltage exist simultaneously.

$$R_S = \frac{V_{S,\text{min}} - V_L - 1}{I_{L,\text{max}}} = \frac{11 - 5 - 1}{5} = 1 \text{ ohm}$$

The transistor has to absorb 4 volts at 5 amps when the source voltage is maximum. Therefore, it must be rated at least 20 watts and 5 amps. Such a transistor will have an input resistance h_{ie} of about 2 ohms and a base-emitter turn-on voltage of about 0.7 volts. (At rated current, the base-emitter voltage is about 1.2 volts, and the current gain is about 15 for a high-current transistor. Thus, the base current is about $I_{\text{rated}}/15$. h_{ie} is $(V_{be} - .6)/I_b = 1.8$ ohms.)

As R_b includes both the transistor input resistance and the operational amplifier output resistance, choose R_b as 50 ohms total.

With an amplifier gain on the order of 10^5, all the terms in equation 9.16 are negligible except V_z/k. So,

$$V_L = \frac{V_z}{k}$$

Setting $V_L = 5$ at the minimum V_S,

$$k = \frac{4.7}{5} = .940$$

Setting $R_1 + R_2 = 10\text{K}\ \Omega$, $R_2 = 9.4\text{K}\ \Omega$, and $R_1 = 600\ \Omega$. This completes the design.

The voltage regulation is

$$\text{regulation} = \frac{V_{z,\text{max}} - V_{z,\text{min}}}{V_{z,\text{min}}} \times 100$$

$$= \frac{4.86 - 4.7}{4.7} \times 100 = 3.4\%$$

This is an improvement over example 9.4. Additional improvement is possible by using either an additional higher voltage Zener diode to supply the reference Zener or a higher voltage reference Zener and a voltage divider.

3 LOGIC CIRCUITS

A. INTRODUCTION

In the previous chapter, the analysis of small signal behavior for amplifying devices assumed that small input signals would not cause radical departures from the operating point. The devices were assumed to operate within their linear regions. Departure from these conditions can lead to saturation or cutoff.

With *saturation*, the input becomes large enough to drive the output current to a limiting value (usually a voltage so low that the output current is limited only by the external resistance of the circuit). *Cutoff* is also a limiting case, but it usually refers to the output current. In the cutoff state, the output appears as an open circuit.

Figure 9.7 shows the saturation and cutoff regions for an NPN transistor. The transistor saturates when the base current is 0.6 mA or more; it is cutoff for a base current of 0.0 mA. The required saturation current to the base is a function of the load-line. When the base-emitter voltage is negative (or less than about 0.6 volts), the transistor is cutoff.

Most logic circuits use these two extreme regions to define the two states needed in binary logic.

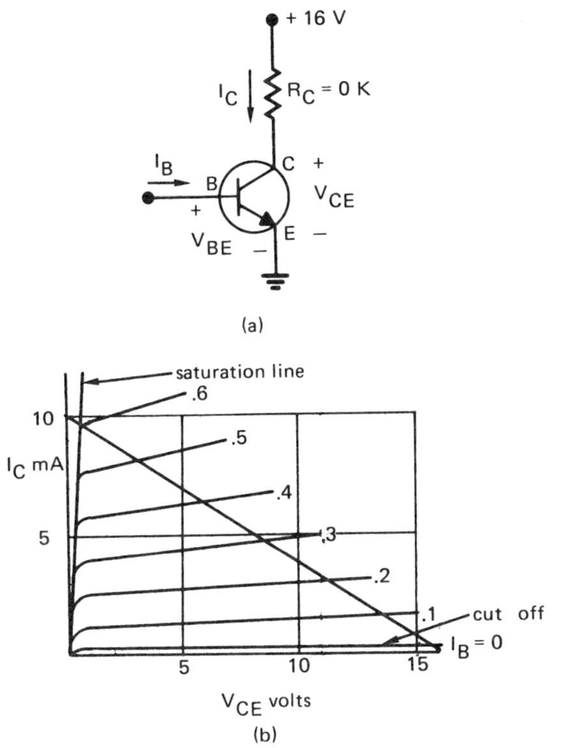

Figure 9.7 Saturation and cutoff characteristics

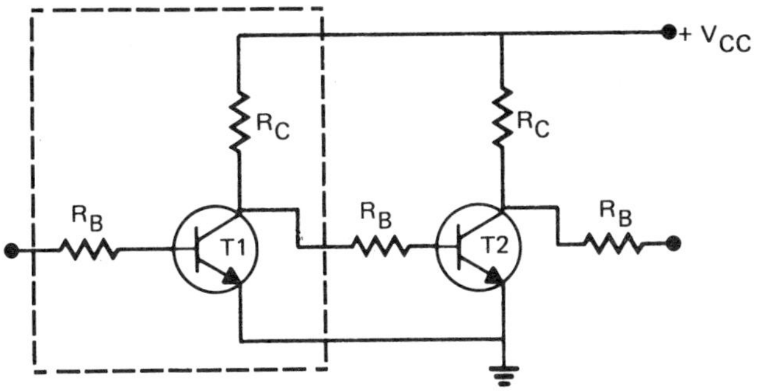

Figure 9.8 NOT gates

B. THE LOGICAL NOT INVERTER CIRCUIT

The common emitter amplifier is the basic *NOT (inverter or complement) circuit,* as shown in figure 9.8. (This is essentially the same as figure 9.7.) If the applied voltage to the base is low (less than 0.6 volts), the transistor will be cutoff, and no collector current will flow. In this condition, the collector voltage will be high. If the base-emitter voltage is high, the transistor will be saturated. The collector-emitter voltage will be low (on the order of 0.2 volts). Such inverter circuits are designed so that the input to one stage is the output of an identical stage.

The analysis of the circuit shown in figure 9.8 is straight-forward. The base current of transistor T2 must be sufficient to saturate T2 when T1 is cutoff (i.e., when V_{CE} is high). When T1 is saturated (i.e., when V_{CE} is less than 0.2 volts), the base voltage of T2 will be less then 0.6 volts, and T2 will be cutoff.

In practical problems, single transistors must be able to drive multiple inputs to subsequent stages. Thus, when a transistor is cutoff, it may have to provide saturation input current to a number of other identical circuits. This is a major design consideration. The ability of a transistor to drive other circuits is called its *fan-out.*

Example 9.6

A transistor inverter has the following parameters: $I_{B,\text{sat}} > 0.5$ mA, $V_{BE,\text{sat}} < 1.5$ volts. $V_{CC} = 5$ volts, and $R_C = 500$ ohms. How many identical inverter circuits can be driven by such an inverter drive if the input base resistance is 1000 ohms?

Using the circuit of figure 9.8, the source current must divide among N identical circuits. All identical circuits must be provided with the minimum base current of 0.5 mA when the base voltage is 1.5 volts. Then, recognizing that $I_C = NI_{B,\text{min}}$ and using Kirchoff's voltage law across R_B and R_C,

$$V_{cc} = NI_{B,\text{min}}R_C + I_{B,\text{min}}R_B + V_{BE,\text{sat}}$$
$$5 = N(.0005 \times 500) + (.0005 \times 1000) + 1.5$$

The fan-out is $N = 12$.

When the inverter is on, the collector-emitter voltage is typically 0.2 volts. Both the collector-base and the emitter-base junctions will be forward-biased. The collector must be able to accept the reverse saturation currents of the load transistors without coming out of saturation. This also limits the number of loads a gate can handle and thereby affects the fan-out.

Logic gates are typically specified in terms of the current they can source (i.e., the current they can supply when the collector voltage is high) and the current they can sink (i.e., the current the collector or output circuit can accept from loads in the low state without being pulled out of saturation).

Within a particular logic family, this may be specified in terms of *unit loads.* A *unit source current* is the input current drawn by a single gate when the input voltage is high. A *sinking unit load* is the current drawn from the input of a single gate when the input voltage is low. The fan-out will be the minimum of these two unit loads that the output of the gate can provide.

Consider figure 9.9. Gate 1 sources a unit load by providing base current to gate 2. With the base of

T1 low (V_{BE} less than 0.6 volts), T1 will be cutoff. Since this appears as an open circuit to the output, the current to the base of T2 will be

$$I_{B,2} = \frac{V_{cc} - V_{BE2}}{R_C + R_B} \qquad 9.20$$

Therefore, I_{B2} is the current being sourced by gate #1.

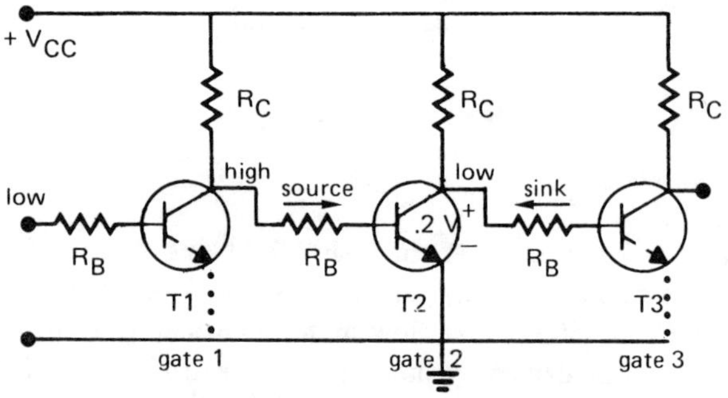

Figure 9.9 Unit loads for fan-out

Gate 2 will be saturated so that the collector-emitter voltage is 0.2 volts. This results in the transistor of gate 3 being cutoff. The base will draw reverse saturation current from the reverse-biased collector-base junction. This unit sink current is over and above the current being drawn from V_{cc}. If I_{CO3} is the unit sink curent, the collector current of T2 is

$$I_{C2} = \frac{V_{cc} - 0.2}{R_C} + I_{CO3} \qquad 9.21$$

For multiple loads, the output circuits are as shown in figure 9.10. Figure 9.10(a) illustrates the sourcing configuration. With $V_{BEF} \approx 0.6$ volts, the current necessary to maintain saturation will require the common emitter characteristic as in figure 9.7 to determine $I_{B,\text{min}}$ for a required I_C. The necessary value of I_C will depend on the number of loads to be sinked, so that the design problem is based on both the sinked and sourced currents. The minimum value of I_B to hold the transistor in saturation will depend on the number of loads being served. The saturation collector current is

$$I_{C,\text{sat}} = \frac{V_{cc} - V_{CE,\text{sat}}}{R_C} + NI_{COE} \qquad 9.22$$

N is the number of load gates, $V_{CE,\text{sat}} \approx 0.2$, and I_{COE} is the reverse collector current with the emitter terminal open. The load line is shown in figure 9.11. $I_{B,\text{min}}$ can be obtained from this figure.

For the sourcing condition of figure 9.10(a), the current through R_C will include the reverse saturation current I_{COE} for the cutoff transistor.

$$\frac{V_{cc} - V_{CE}}{R_C} - I_C = N\left[\frac{V_{CE} - V_{BEF}}{R_B}\right] \qquad 9.23$$

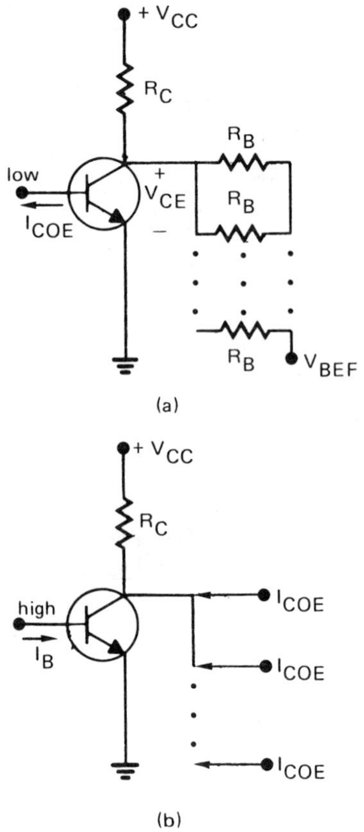

Figure 9.10 Circuits for calculating (a) source loads, (b) sink loads

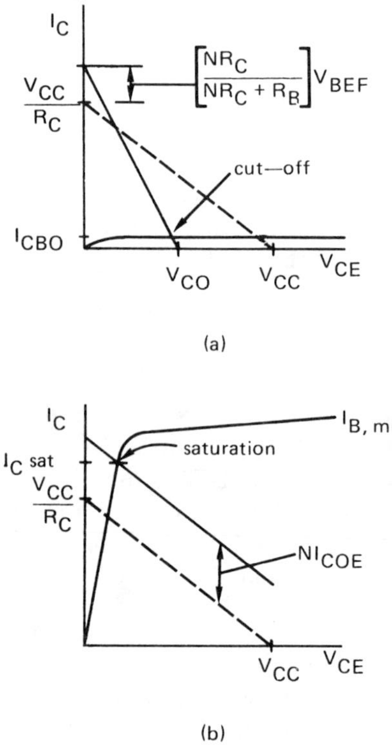

Figure 9.11 Load lines for (a) cutoff, and (b) saturation of an inverter gate with N similar loads

$V_{BEF} \approx 0.6$ volts to hold the load transistors in saturation, and $I_C = I_{COE}$. The load line equation is obtained from equation 9.23.

$$I_C + \left(\frac{1}{R_C} + \frac{N}{R_B}\right)V_{CE} = \frac{V_{cc}}{R_C} + \left(\frac{N}{R_B}\right)V_{BEF} \qquad 9.24$$

For each load input, the base current is

$$I_B = \frac{V_{CE} - V_{BEF}}{R_B} \qquad 9.25$$

The simultaneous solution of equations 9.24 and 9.25 is

$$I_C + \left(N + \frac{R_B}{R_C}\right)I_B = \frac{V_{cc} - V_{BEF}}{R_C} \qquad 9.26$$

The design problem is formulated by setting $I_C = I_{CBO}$ and solving for N.

$$N \leq \frac{R_C I_{C,\text{sat}} + V_{CE,\text{sat}} - V_{cc}}{R_C I_{CBO}} \qquad 9.27$$

$$N + \frac{R_B}{R_C} \leq \frac{V_{cc} - V_{BEF} - R_C I_{CBO}}{I_{B,\text{min}} R_C} \qquad 9.28$$

Example 9.7

A transistor with the characteristics shown has $R_C = 1000$ ohms and $V_{cc} = 5$ volts. Determine the base current necessary for a fan-out of 5.

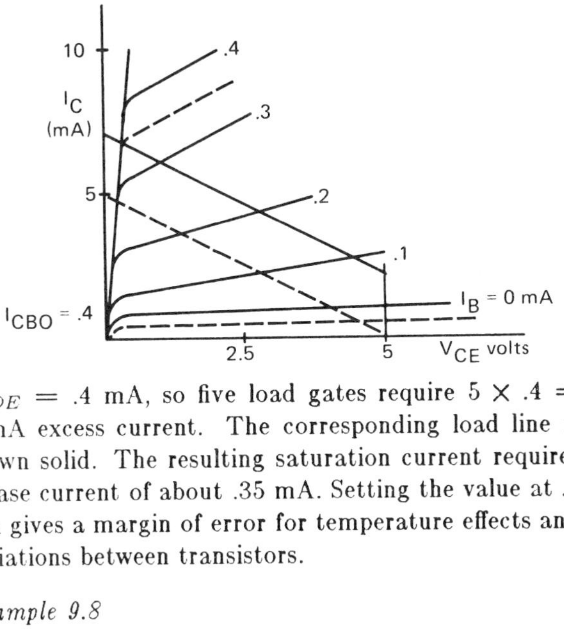

$I_{COE} = .4$ mA, so five load gates require $5 \times .4 = 2$ mA excess current. The corresponding load line is shown solid. The resulting saturation current requires a base current of about .35 mA. Setting the value at .4 mA gives a margin of error for temperature effects and variations between transistors.

Example 9.8

Determine the maximum base resistance R_B which will allow sourcing five gates from the transistor gate of example 9.7.

From equation 9.28 with $V_{BEF} = 0.6$ volts,

$$5 + \frac{R_B}{1000} \leq \frac{5 - .6 - .4}{1000 \times .0004}$$

$$R_{B,\text{max}} = 5000 \text{ ohms}$$

C. GATE DELAYS

Switching between low and high levels does not occur instantaneously. Figure 9.12 shows the relationships between input and output voltages versus time for purposes of identifying the delays.

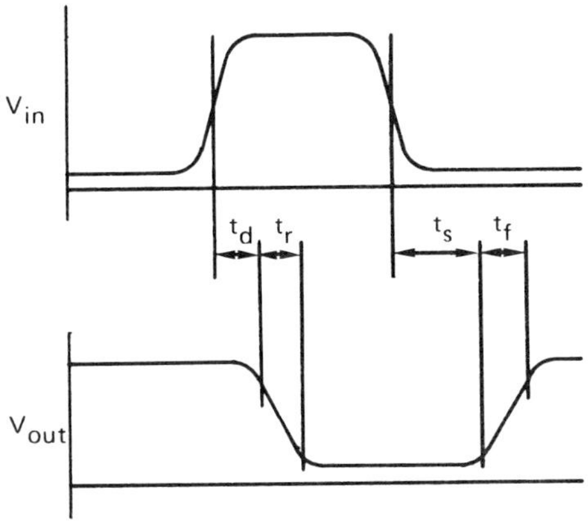

Figure 9.12 Gate delays

When the input voltage rises in an inverter gate, it will take some time for the base-emitter voltage to reach the *turn-on level*. This is called the *delay time*, t_d. It may range from 1 to 10 nanoseconds (ns). Typical values are 2 to 4 ns.

The time for the collector output voltage to fall to the saturation level is called the *rise time*, t_r. It can range from 1 to 100 ns, with a typical value being 50 ns.

When the transistor is driven well into saturation, an excess charge is stored. There will be a time after the base current has dropped before this excess charge is cleared out of the base region. Then, the collector current will drop, allowing the output voltage to rise. The delay is called the *storage time*, t_S. It can range from 5 to 50 ns.

Once the excess charge has been removed from the base, the collector current can begin to fall. The time interval involved here is called the *fall time*, t_f. It can range from 1 to 100 ns, with a typical value of 30 ns.

The minimum time for a transistor to pass a pulse (i.e., the time from saturation to cutoff and back to saturation, or from cutoff to saturation and back to cutoff) is the sum of these delays. Therefore, the maximum frequency at which a transistor can operate is the reciprocal of the sum of the delays.

D. LOGICAL FUNCTIONS

Logical functions can be obtained by combining gates. The basic logical functions are inversion (NOT), logical sum (OR), and logical product (AND). Logical variables and functions have two possible values: true (1) and false (0).

Definition of functions is often done with a *truth table.* The function value is given for each combination of the input variables.

Table 9.1

$C = \overline{A}$ truth table
(NOT)

A	C
0	1
1	0

Table 9.2

$C = A + B$ truth table
(OR)

A	B	C
0	0	0
0	1	1
1	0	1
1	1	1

Table 9.3

$C = A \cdot B$ truth table
(AND)

A	B	C
0	0	0
0	1	0
1	0	0
1	1	1

The most commonly occurring functions in transistor-transistor logic are the negations of AND (NAND) and OR (NOR).

Table 9.4

$C = \overline{A + B}$
(NOR)

A	B	C
0	0	1
0	1	0
1	0	0
1	1	0

Table 9.5

$C = \overline{A \cdot B}$
(NAND)

A	B	C
0	0	1
0	1	1
1	0	1
1	1	0

A basic operation in Boolean algebra is De Morgan's theorem, which can be written in two ways.

$$\overline{A \cdot B} = \overline{A} + \overline{B} \qquad 9.29$$

$$\overline{A + B} = \overline{A} \cdot \overline{B} \qquad 9.30$$

Two useful identities are repeated here.

$$A \cdot \overline{A} = 0 \qquad 9.31$$

$$A + \overline{A} = 1 \qquad 9.32$$

The symbols for logic gates are shown in figure 9.13.

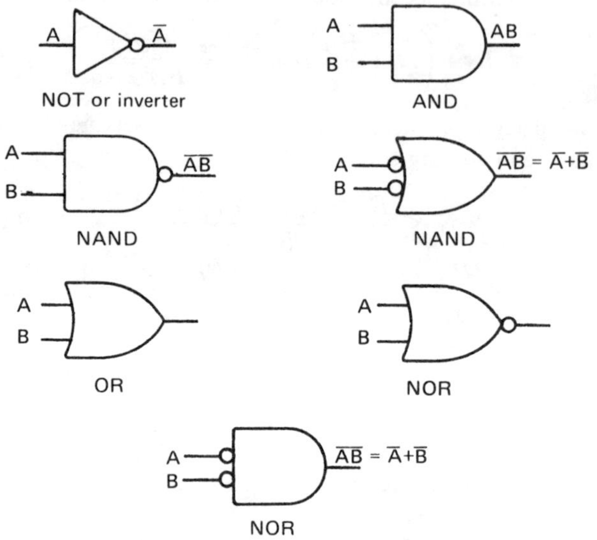

Figure 9.13 Gate symbols

4 TYPES OF LOGIC

A. RESISTOR-TRANSISTOR LOGIC (RTL)

The inverter illustrated in figure 9.8 is an RTL circuit, the basic building block for RTL logic. RTL was popular prior to large-scale integration. The fundamental logical gate is the NOR circuit shown in figure 9.14.

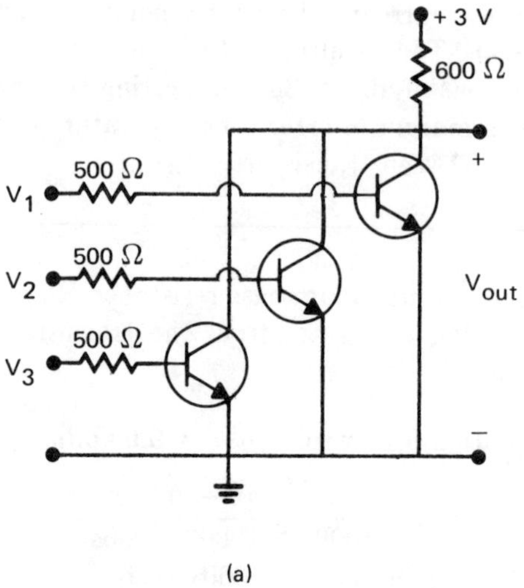

(a)

V_1	V_2	V_3	V_{out}
L	L	L	H
L	L	H	L
L	H	L	L
L	H	H	L
H	L	L	L
H	L	H	L
H	H	L	L
H	H	H	L

H=high voltage (3v)
L=low voltage (.2v)

(b)

Figure 9.14 Basic RTL gate: (a) circuit, (b) voltage truth table

For *positive logic* (where the higher voltage represents a logical 1 and the lower voltage represents a logical 0), the circuit and truth table correspond to a NOR circuit. However, the choice of positive or *negative logic* is often arbitrary.

B. DIODE-TRANSISTOR LOGIC (DTL)

The basic DTL circuit is shown in figure 9.15. The output voltage V_C is high in all cases except when both inputs are low. A desirable feature of this circuit is that when sourcing a load, no current is required from the output. On the other hand, the current necessary to sink the load when the output voltage is low is much more than is required by the RTL circuit. The two diodes in series with the base ensure that the transistor will not be turned on when only one input is low.

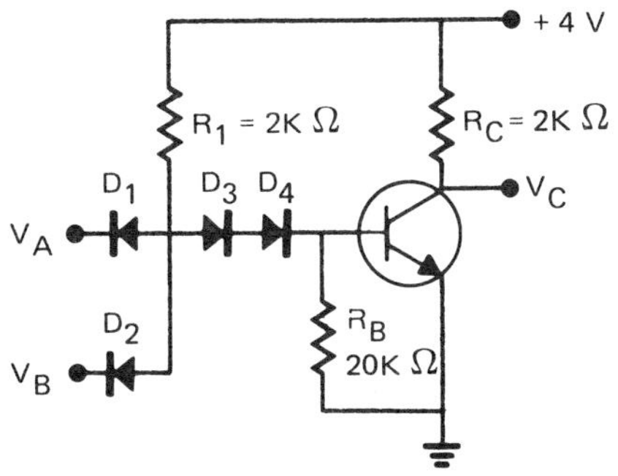

(a)

V_A	V_B	V_C
L	L	H
L	H	H
H	L	H
H	H	L

(b)

Figure 9.15 Basic DTL gate: (a) circuit, (b) voltage truth table

C. TRANSISTOR-TRANSISTOR LOGIC (TTL)

A very fast logic which uses saturation switching (from cutoff to saturation) is the TTL family. The basic two-input circuit is shown in figure 9.16 with an accompanying truth table. For positive logic, the basic gate is the NAND gate.

The basic TTL NAND gate can be considered a modification of the DTL NAND gate, with the input diodes replaced by a multi-emitter transistor's base-emitter junctions. The two series diodes are replaced by the base-collector junction of the multi-emitter input transistor and by the base-emitter junction of T2.

The behavior of the input transistor is unique because bias voltages are not provided. When one or both of the input voltages are low, the base voltage of T2 is pulled down to about 0.8 volts. This voltage attempts to forward bias the string of PN junctions (T1: B-C, T2: B-E, and T3: B-E) to ground. However, the three junctions require $3 \times 0.6 = 1.8$ volts in order to turn on. So, T3 is cutoff and V_C is high.

When both of the input voltages are high, the base of T1 is connected to V_{cc}. This is high enough to bias all three junctions to ground in the forward direction. T2 acts as an emitter-follower, providing the current necessary to hold V_C down to 0.2 volts.

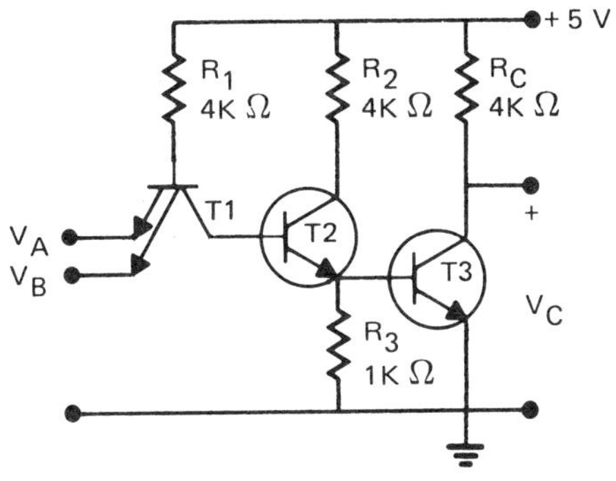

(a)

V_A	V_B	V_C
L	L	H
L	H	H
H	L	H
H	H	L

(b)

Figure 9.16 Basic TTL gate: (a) circuit diagram, (b) voltage truth table

Example 9.9

Determine the power required of the two-input TTL stage shown in figure 9.16 if the fan-out is 5.

When V_C is low (.2 volts), the base-emitter voltage of T3 is .6 volts. The base-emitter voltage of T2 and the base-collector voltage of T1 will also be .6 volts. T2 can be assumed to be saturated since its base current will be 0.8 mA and its saturated current is 2.1 mA. Then, the D.C. equivalent circuit shown can be used. The current through R_C will be 1.2 mA.

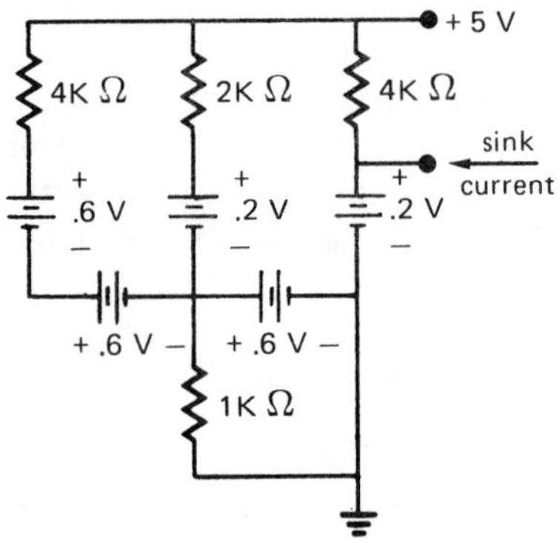

The unit load must be found from the figure shown. The low-output power will be determined after the power for high-output is found.

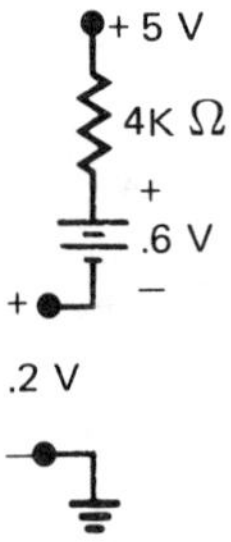

With output high at T3, both T2 and T3 are assumed off. (Even with the emitter of T2 at ground, the voltage at the base of T1 would have to be 1.2 volts to turn on T2, allowing for a .6 volt drop across the two PN junctions.) At least one of the inputs must be low (0.2 volts). With the base-emitter voltage of T1 at 0.6 volts, the voltage across R_1 is 4.2 volts. Then, the unit current is

$$\frac{4.2}{4000} = 1.05\,\text{mA}$$

As the reverse collector saturation current will be on the order of .1μA, the power drain due to collector saturation will be negligible compared to the 1.05 mA through R_1.

The power in the high output state (lost in this stage) is

$$(V_{cc} - V_{\text{in}})I_{R_1} = (5 - .2) \times 1.05 = 5.04\,\text{mW}$$

Returning to the low output state, the unit current from one load will be divided among the sourcing gates which have low outputs. The worst case of sinking is when there is only one sourcing gate (i.e., only one low input). This single gate would receive all of the 1.05 mA unit current.

Thus, for five loads in the low-output state, the maximum sink current would be

$$5 \times 1.05 = 5.25\,\text{mA}$$

The power dissipated for this sinked current is

$$0.2 \times 5.25 = 1.05\,\text{mW}$$

The power directly from V_{cc} to the stage (exclusive of the sinked current) is

$$5 \times (0.8 + 2.1 + 1.2) = 20.5\,\text{mW}$$

The total power dissipated in the stage for low output is

$$20.5 + 1.05 = 21.55\,\text{mW}$$

D. TTL OUTPUT CIRCUITS

A method of improving the fall time of TTL gates is to reduce the collector current in the saturated state. Circuits which accomplish this are shown in figure 9.17. The addition of T4 and D disconnects the collector of T4 from V_{cc} when T4 is on. For T3 to be saturated, T2 also must be saturated, which brings the base of T4 to 0.8 volts. This would be just enough to turn on T4 if the diode D were not present.

However, the presence of diode D requires that the base of T3 be at 1.4 volts to be on. Thus, with T3 saturated, T4 is cutoff, disconnecting T3 from V_{cc} and requiring a collector current only from the loads. When T3 is cutoff, T2 is also cutoff and the 1.6K resistance turns on T4 and so pulls up the collector voltage of T3. This *totem-pole configuration* results in a lower storage time and faster turn-off of the output transistor.[1]

The *open-collector TTL gate* (figure 9.17(b)) is used when two or more outputs are wired together. This results in a so-called *wired OR* or *wired AND* configuration. (Depending on the view-point, it may be called either.) When two or more open collector outputs are connected and tied through an externally added resistor to V_{cc}, all outputs must be high in order to obtain a high output. This is an AND function from the point of view of the outputs.

[1]Totem-pole is also known as *active pull-up*.

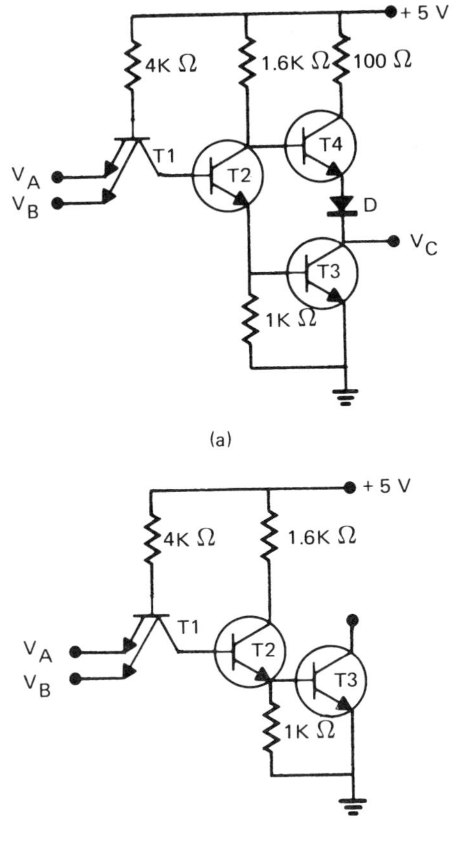

Figure 9.17 Alternative TTL output circuits: (a) totem-pole, (b) open collector

E. EMITTER-COUPLED (ECL) OR CURRENT MODE (CML) LOGIC

The speed limitation on TTL is due to the saturation switching. A faster configuration is based on the emitter-coupled amplifier which does not drive the transistor into saturation. The basic circuit is shown in figure 9.18. Outputs are taken through emitter-followers. This accounts for the difference in input and output voltages in the truth table. ECL can be up to an order of magnitude faster than TTL, but it is much more demanding of power than TTL.

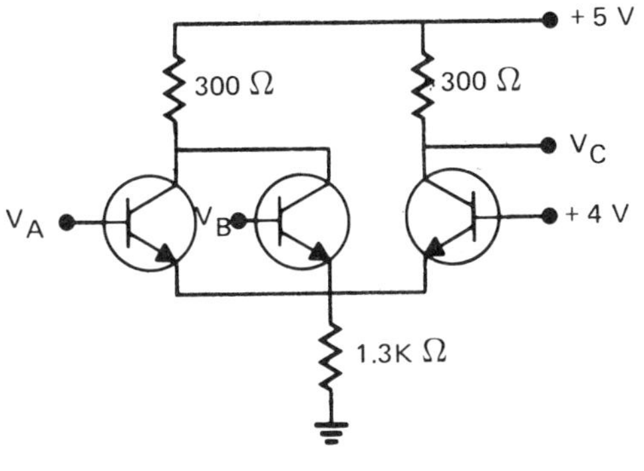

(a)

V_A	V_B	V_C
3.6	3.6	4.2
3.6	4.4	5.0
4.4	3.6	5.0
4.4	4.4	5.0

(b)

A	B	C
0	0	0
0	1	1
1	0	1
1	1	1

(c)

Figure 9.18 ECL or CML basic gate: (a) circuit diagram, (b) actual voltage table, and (c) truth table

F. PMOS LOGIC GATES

PMOS stands for P-channel enhancement mode MOSFET. This logic is used in integrated circuit form, requiring no external resistors or capacitors. The attraction of this logic is its simplicity of manufacture. It requires only one diffusion into N-type substrates, plus three additional steps for oxidation and metalization. PMOS is slower than bipolar junction transistor technology, but much higher gate densities are possible.

A PMOS inverter is shown in figure 9.19. The upper transistor is biased to an on condition. It serves as a somewhat nonlinear resistance. The substrate lead is connected to the most positive voltage available (the ground in figure 9.19) to insure that the P-N junctions between drain and substrate and between source and substrate remain reverse-biased.

Negative logic is typically used in PMOS logic. For negative logic, $-V_{dd}$ corresponds to a logical 1 and zero volts corresponds to a logical 0.

When the gate of T1 is at ground, the output voltage approaches V_{dd}, a negative voltage corresponding to a logical 1. When the gate voltage has a logical 1, (i.e., a negative voltage applied), the source is positive with respect to the gate. T1 will then be turned on, and the output voltage will approach ground, the logical 0 level.

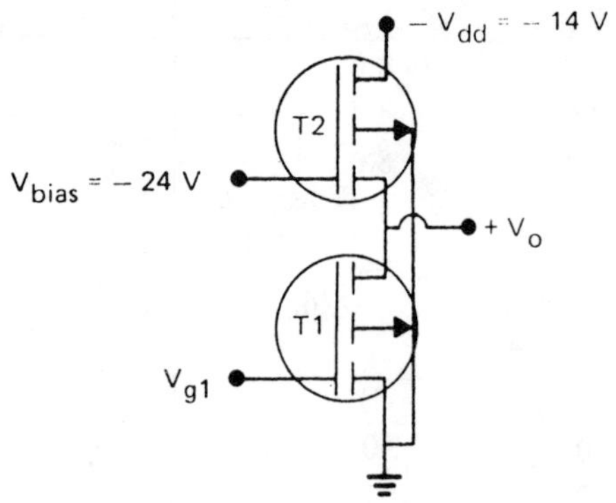

Figure 9.19 PMOS inverter with negative logic

The circuit of figure 9.20 showns a basic PMOS NOR gate. The output is pulled toward ground if either gate V_a or V_b is brought to the 1 voltage (i.e., a negative voltage).

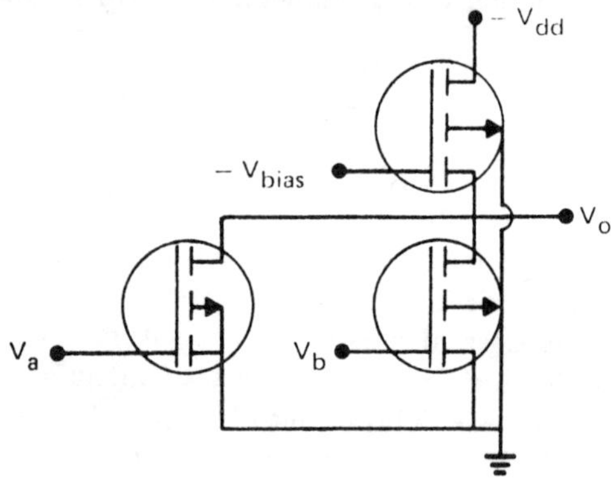

(a)

V_a	V_b	V_o
0	0	$-V_{dd}$
0	$-V_{dd}$	0
$-V_{dd}$	0	0
$-V_{dd}$	$-V_{dd}$	0

(b)

Figure 9.20 (a) PMOS NOR gate with negative logic, and (b) truth table

G. CMOS LOGIC GATES

It is possible (and, in some cases, worthwhile) to employ N-channel enhancement mode MOSFET's in a manner similar to the PMOS. That is called NMOS and is faster than PMOS because electron mobility is about 2.5 times hole mobility. NMOS is more difficult and expensive to manufacture, so it has not been as popular as PMOS.

A combined technology using both N-channel and P-channel enhancement mode MOSFET's has proven to be very useful. It requires much less power than the single-types of logic circuits yet is relatively simple to fabricate. This is called CMOS for complimentary MOSFET's. CMOS requires only a single supply voltage, which allows for additional simplifications.

The basic CMOS inverter is shown in figure 9.21. The upper transistor is P-channel with its substrate lead connected to the most positive voltage available (+10 volts). The lower transistor is N-channel with its substrate lead connected to the most negative voltage available. These connections are made to maintain reverse bias on the substrate P-N junctions.

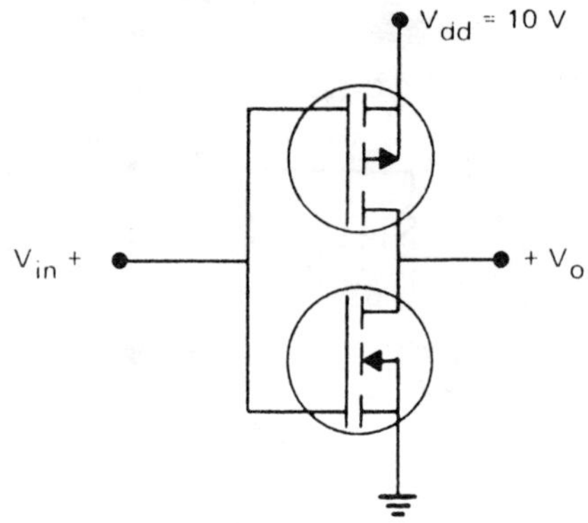

Figure 9.21 CMOS inverter

When the input voltage is high (near +10 volts), the upper transistor has $V_{GS} = 0$ (or at least lower than the turn-on voltage of the upper transistor), so the upper transistor is off. The lower transistor will be turned on by the high V_{GS}, and the output terminal will be grounded through the lower transistor's on-resistance.

When the input voltage is low (near zero volts), the lower transistor is turned off, breaking the path to ground from the output terminal. The upper transistor has a large negative V_{GS} and, being a P-channel device, is turned on. The output terminal is connected to V_{dd} through its internal on-resistance.

The only current which flows is that which is necessary to charge up the gate-channel capacitances of load stages (corresponding to the input). Some current also leaks through the off-transistor. Thus, power consumption is extremely low, but the speed is limited by the long time constants associated with the charging of the off-transistor gates.

Typical CMOS logic gate construction is shown in figure 9.22. For NOR logic (figure 9.22(a)), a positive input at either terminal will turn on one of the lower transistors while turning off one of the upper. Thus, if either or both of the inputs are high, the output is low. Both inputs must be low to turn on the series combination and connect V_{dd} to the output.

In figure 9.22(b), a low voltage at either input terminal turns on at least one of the parallel upper transistors while turning off at least one of the series lower transistors. Thus, a low voltage on either or both of the inputs results in a high output voltage. A high voltage on both inputs is necessary to turn off both upper transistors and connect the output to ground through the two lower transistors.

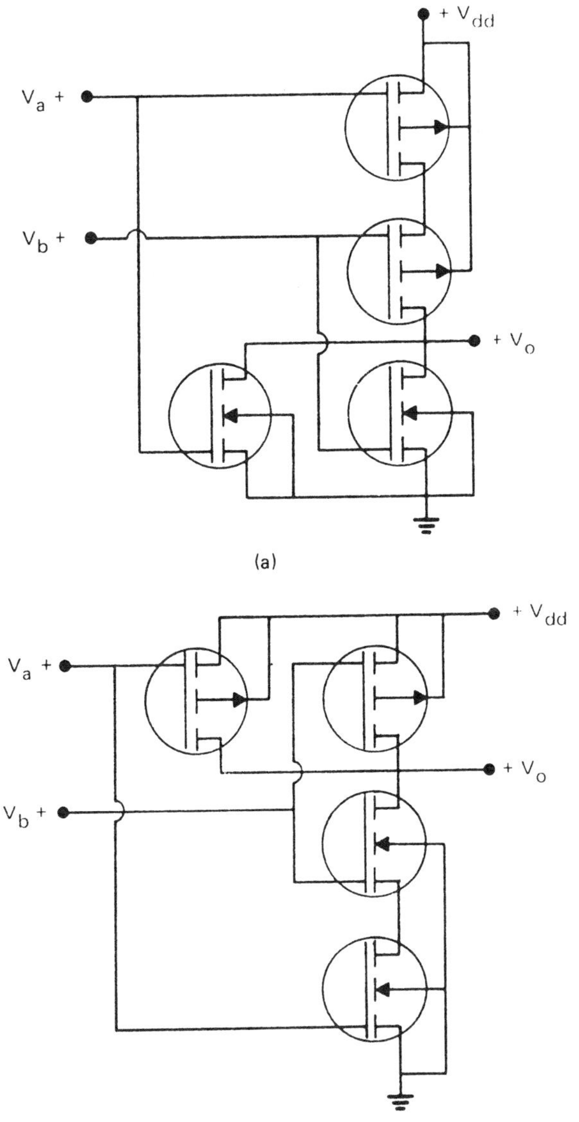

Figure 9.22 Positive logic CMOS (a) NOR gate, and (b) NAND gate

5 ANALOG SWITCHES

A. INTRODUCTION

Electronic analog switches can be controlled by logic inputs (i.e., TTL levels of less than 0.8 volts low and about 3 volts high). They are designed to switch analog signals up to about 75% of the D.C. supply voltage in electronic circuits. Switches are used in sample-and-hold circuits, programmable amplifiers, digital-to-analog converters, and a wide variety of other applications.

Analog switches make use of the on-resistance of FET channels. The channel resistances range from a few ohms for VMOS to 10 or 100 ohms for JFET's and enhancement mode MOSFET's. Because channel resistances vary with current in the channel, it is almost always necessary that the series or load resistance be high in comparison with the nominal channel resistance.

B. CMOS TRANSMISSION SWITCH

The CMOS switch is similar to CMOS logic. The basic switch is shown in figure 9.23. CMOS logic level voltages are expected as the control signal V_c. With V_c low and no signal present, both parallel switch channels are turned off. Either terminal of the channel can be the source (or drain) depending on the polarity of the signal being switched.

Consider the off case first. V_c is virtually zero. This level is applied to the lower transistor in the switching pair. For this transistor to turn on, both terminals must be negative by at least the turn-on voltage of V_{GS} (from 1 to 3 volts). Thus, to maintain an off condition, it is necessary that at least one terminal be connected to a potential more positive than about −1 volt. This requirement is satisfied whenever one terminal is connected to a resistive load.

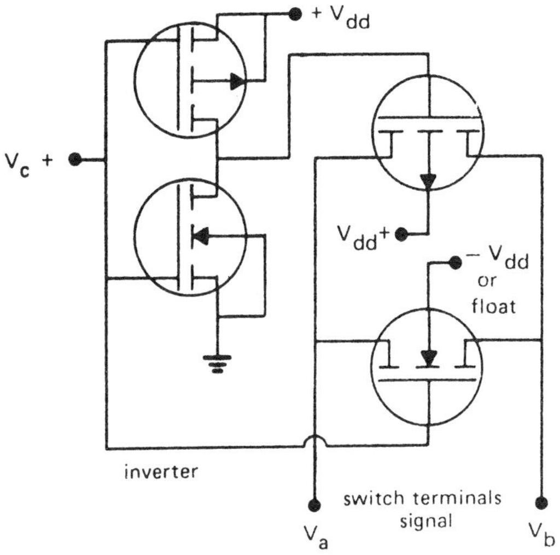

Figure 9.23 CMOS switch

With the lower transistor still in the off condition, consider the upper (P-channel) transistor in the switching pair. The gate is near the V_{dd} level, so the channel cannot be activated unless both terminals are more positive than the supply voltage V_{dd}.

When the control signal is high (i.e., nearly V_{dd}), the lower channel will be turned on for negative signals and for signals up to within a few volts of V_{dd}. The upper channel will turn on when the switching voltages are sufficiently positive to make V_{GS} sufficiently negative to turn on the P-channel. With proper design, the upper channel will compensate for the turning off of the lower channel for positive switching voltages. With proper design of the load impedance, the difference between V_a and V_b will be small when the switch is on.

The limitations on switching are in the switched voltage when the switch is off.

C. JFET ANALOG SWITCHES

Another analog switch device is developed from depletion mode P-channel JFET's. These are on when the gate-source voltage is zero. A positive voltage from gate to channel is required to turn them off. This is the *pinchoff voltage* $V_{DS,\text{off}}$. A major problem with an ordinary JFET is that it cannot have any forward bias on the gate-channel P-N junction. For this reason, JFET's for switching application use a reverse diode in series with the gate lead as indicated in figure 9.24.

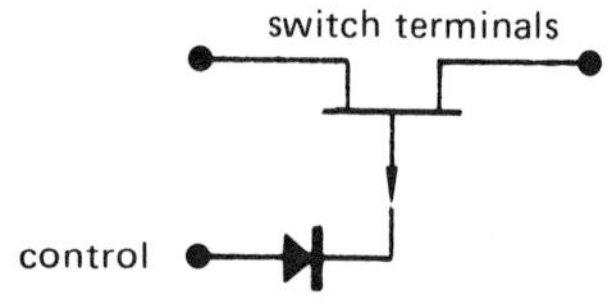

Figure 9.24 JFET analog switch

An advantage of JFET logic over CMOS logic is in a lower on-resistance. Disadvantages are complexity and price. The JFET is restricted to switching voltages which are at least a few volts less than the control voltage, as larger voltages can turn the switch on. Also, a sufficiently large signal while the switch is in the on condition can turn the channel off.

D. VMOS ANALOG SWITCHES

VMOS transistors are available as an N-channel (enhancement mode) switch. It is, therefore, used only for switching of positive voltages (e.g., when the drain is more positive than the source). VMOS transistors will operate with negative voltages only to about $-V_{GS,\text{on}}$. Their advantage is the lower channel resistance, typically a few ohms, in the on condition.

6 POWER SWITCHING

A. INTRODUCTION

Power switching up to approximately 50 amps can be accomplished with power bipolar junction transistors. VMOS transistors can be used up to approximately 10 amps. Switching is typically D.C., but bridges for control of A.C. power also require switching. Hold-off voltages for these devices may be as high as 500 volts with power dissipation up to 200 watts. The power controlled can be many times the device power ratings, as the higher currents occur at the lowest voltage level.

Control of A.C. power is still accomplished, in a large part, by the use of SCR's (silicon controlled rectifiers) and TRIAC's (bi-directional SCR's). These are designed for high (up to 100 volts) hold-off voltages and currents up to 80 amps.

The GTO (Gate Turn Off) is a special SCR which can be turned off from the gate. However, it requires substantial voltage (around 70 volts) to do so. It does not rely on the current falling below an extinguishing value, as is done in SCR's. The turn-on voltage for GTO's, as for SCR's and TRIAC's, is around 2.5 volts.

B. THYRISTOR CONTROL

The SCR has the turn-on characteristic shown in figure 9.25. With no gate current, the characteristic indicates the maximum voltage V_{BO} which can be applied without the SCR breaking down. With increasing gate current, I_G, the breakdown voltage decreases.

At $I_G = I_{GT}$, the SCR behaves essentially like a P-N diode. I_{GT} is a D.C. current. It occurs in conjunction with a gate-cathode (main terminal 1) voltage from 1 to 2.5 volts. The precise value of I_{GT} is specified by the manufacturer, as is V_{GT}.

The minimum value of D.C. current that will turn on the SCR is I_{GT}. This is not sufficient in a pulse condition. In general, I_{GT} must be applied for about 50 microseconds to fully turn on the SCR. For pulse turn-on, the total charge must approximately equal the current-time product ($I_{GT} \times 50\mu$ sec). This current will not flow unless the voltage exceeds V_{GT}.

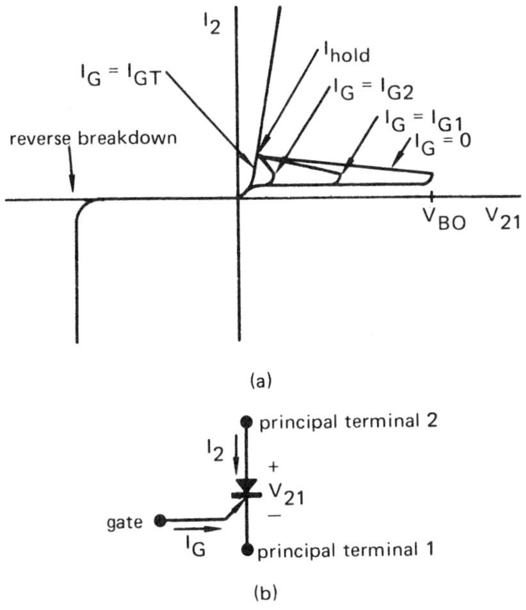

Figure 9.25 (a) SCR and GTO characteristics and (b) SCR symbol

Example 9.10

An SCR has $I_{GT} = 80$ mA. Determine the current necessary to turn on this SCR in 1 μsec. Also determine the voltage on a .1μF capacitor necessary to deliver this charge to the gate in 1 μsec.

The total charge needed in 1 μsec will be $I_{GT} \times 50\,\mu$sec or

$$.08 \times 50 = 4\,\mu C$$

The average current will be

$$\frac{4\mu C}{1\mu\text{sec}} = 4 \text{ amp}$$

The capacitor will discharge from from its initial value to V_{GT} (assumed to be 2.5 volts). Thus, the charge to be delivered will obey

$$C(V_{\text{initial}} - 2.5) = 4\,\mu C$$

With $C = 0.1\,\mu F$, the initial voltage is

$$V_{\text{initial}} = 2.5 + \frac{4\,\mu C}{0.1\,\mu F} = 42.5 \text{ volts}$$

This assumes that the RC time constant will permit the capacitor to discharge in one microsecond.

C. UJT TRIGGER FOR SCR's

A common trigger device for SCR's is the *unijunction transistor*, UJT, sometimes called a *double-base diode.* This device can hold a voltage off until it reaches a critical value. Above the critical value, the UJT releases all stored charge. The charge can be directed to the gate of an SCR as shown in the typical circuit of figure 9.26.

The UJT is structurally similar to the N-channel JFET. The difference is that the P-region is concentrated near base B2. The lead to the P-region is called the *emitter.* While the P-N junction is reverse-biased, the voltage V_1 is simply the *voltage division* of the voltage V_{21}. The voltage division ranges from .5 to .8, with a typical value of .75. The resistance $R_1 + R_2$ varies from $1K\ \Omega$ to $10K\ \Omega$, with a typical value of $5K\ \Omega$.

When the junction voltage $(V_E - V_1)$ exceeds the usual P-N forward voltage of about 0.6 volts, the E-B1 combination becomes a forward-biased diode. The resulting V_{EBI} versus I_f characteristics are shown in figure 9.26(d). I_{hold} is the minimum current necessary to keep the UJT in the triggered or on condition. Below this value, it will revert to the untriggered state as indicated by the arrows.

Note the similarity of the circuit symbol to an N-JFET symbol. The difference is the bent P-lead, which points toward base B1.

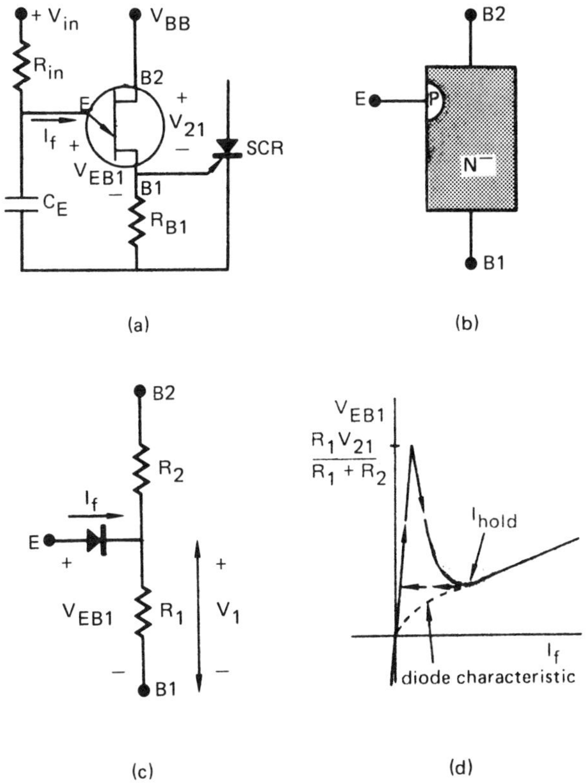

Figure 9.26 (a) UJT-SCR trigger circuit, (b) UJT construction, (c) equivalent UJT circuit prior to triggering, and (d) characteristic

Example 9.11

A UJT circuit as shown in figure 9.26(a) has $R_1 = 4K\,\Omega$, $R_2 = 1K\,\Omega$, $R_{B1} = 1K\,\Omega$, and $V_{BB} = 24$ volts. Determine the triggering voltage.

The UJT will trigger when

$$V_E \geq V_{BB}\left[\frac{R_1 + R_{B1}}{R_1 + R_{B1} + R_2}\right] + .6$$

The .6 volt accounts for the usual diode drop.

$$V_{\text{trigger}} = 24\left[\frac{4+1}{4+1+1}\right] + .6 = 20.6 \text{ volts}$$

Example 9.12

The SCR of example 9.10 had $I_{GT} = 80$ mA and $V_{GT} = 2.5$ volts. Determine the capacitance necessary to trigger the SCR if the UJT trigger voltage is 21.2 volts.

The trigger voltage is 21.2 volts, but 0.6 volts is used for the P-N junction. 2.5 volts are needed to drive the gate of the SCR. This leaves 18.5 volts for charge transfer. The charge needed was found in example 9.10 to be 4 μC, so

$$C_E = \frac{4 \times 10^{-6}}{18.5} = 0.22\,\mu F$$

D. DIAC TRIGGER FOR TRIAC CIRCUITS

A TRIAC is essentially a bi-directional SCR which can be triggered for either polarity of V_{21} (with a corresponding polarity of V_{GT}). The DIAC is a TRIAC with no gate. It is designed to have a controlled *breakover voltage* V_{BO}. The forward and reverse breakdown voltages are nearly equal. Typical breakover voltages range from 25 to 40 volts. Fallback voltages are approximately 9 volts, as indicated in figure 9.27.

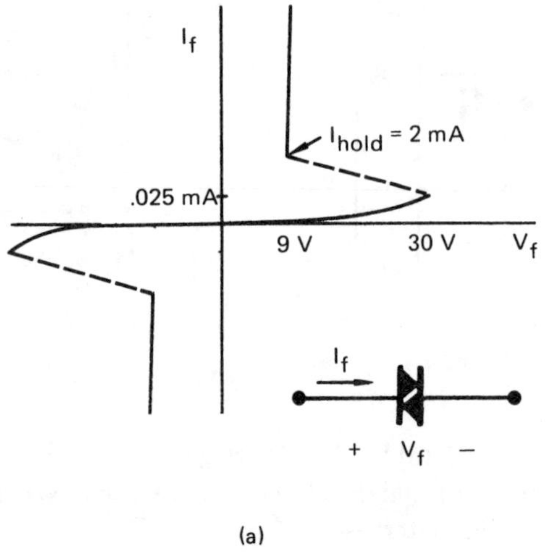

(a)

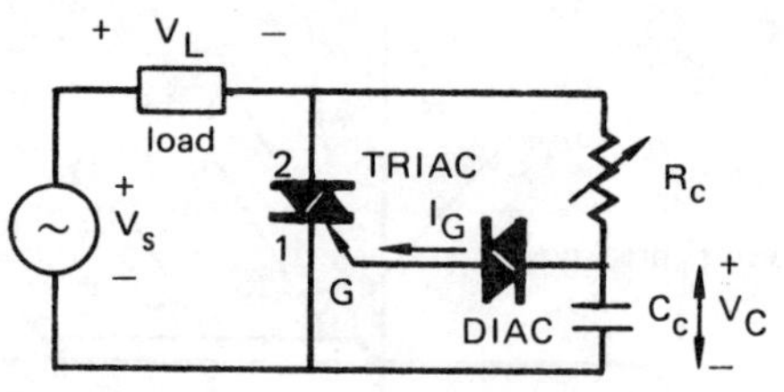

(b)

Figure 9.27 (a) DIAC symbol and characteristic, (b) simple triggering circuit for A.C. power

The firing angle of the TRIAC in figure 9.27(b) is controlled by resistance R_c. The capacitance C_c releases charge through the DIAC to swing the capacitor voltage from the breakover value to the *dynamic on-resistance* region (9 volts in figure 9.27(a)). Once the DIAC has triggered, the capacitor voltage will remain at the dynamic on value until the A.C. source voltage has reversed. The waveforms for the control voltage V_c across the capacitor are shown together with other waveforms in figure 9.28.

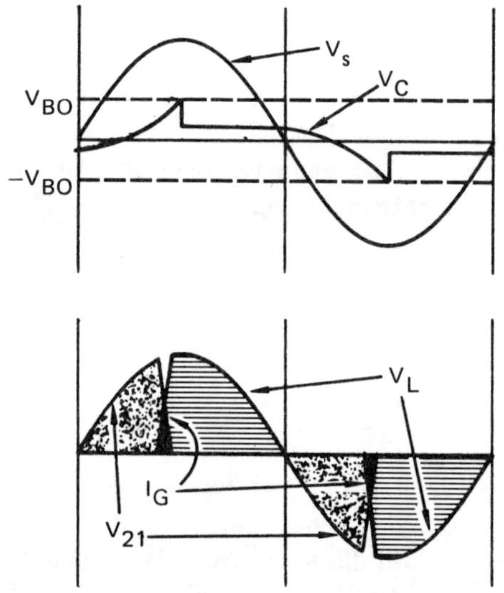

Figure 9.28 DIAC trigger waveforms

Before the TRIAC is triggered, only the current through R_c and C_c flows in the load. As R_c is much greater than the load impedance, it can be assumed that all of V_S appears across R_c and C_c. At the beginning of each half-cycle, the capacitor has the dynamic DIAC voltage across it from the last half-cycle firing. Therefore, it is opposite in polarity from the next firing (V_{BO}) value. The voltage across the capacitor in a steady state phasor analysis tries to follow V_S according to the impedance ratio.

$$V_{c,\text{steady state}} = \frac{V_S}{j\omega RC + 1} \tag{9.33}$$

However, the initial charge on the capacitor must go through a transient to approach the value specified by equation 9.33. Details of this transient are shown in figure 9.28. The firing angle will be where V_c reaches the breakover voltage V_{BO}. The worst case for the firing angle is when the transient has completely died out before the breakover voltage is reached.

When the transient can be ignored (e.g., when it is completed before V_c reaches VBO), the trigger point can be found from the expression of V_c in steady state.

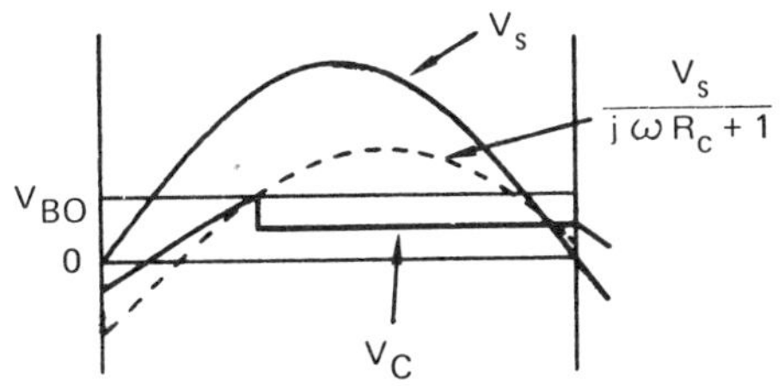

Figure 9.29 Dashed sinusoid showing the steady state curve that the capacitor voltage attempts to follow before the DIAC breaks over

$$V_{c,\text{sinusoidal steady state}} = \frac{V_{S,\,\max}}{\sqrt{\omega^2 R_c^2 C_c^2 + 1}} \times \sin\left[\omega t_f - \arctan\left(\omega R_c C_c\right)\right] = V_{BO} \qquad 9.34$$

Equation 9.34 is a first approximation to actual performance. Complete analysis of the circuit, including initial conditions, yields equation 9.35.

$$V_c(t) = V_{c,\text{sinusoidal steady state}} + \left[V_c(0) + \frac{V_{S,\,\max}\omega R_c C_c}{\omega^2 R_c^2 C_c^2 + 1}\right] e^{-t/R_c C_c} \qquad 9.35$$

$V_c(t)$ must reach V_{BO} at $t = t_f$. Equation 9.35 must be solved by interation beginning with the approximation of equation 9.34. The angles in equations 9.34 and 9.35 are in radians.

Example 9.13

The circuit of figure 9.27(b) uses a DIAC with the characteristics of figure 9.27(a). $R_c = 10K$ and $C_c = 0.1\ \mu F$. V_S has an RMS value of 120 volts, and the freuqency is 60 hz. Determine the firing angle.

Since the frequency is 60 hz, $\omega = 377$. Then,

$$\omega R_c C_c = .377$$

and

$$\omega^2 R_c^2 C_c^2 + 1 = 1.142$$

$$V_{c,\text{sinusoidal steady state}} = \frac{120\sqrt{2}}{\sqrt{1.142}} \sin\left[377 t_f - \arctan(.377)\right] = 30$$

Solving for t_f,

$$377 t_f - .361 = \arcsin(.189) = .190\,\text{radians}$$
$$t_f = .0015\,\text{sec}$$

This is a first approximation to the TRIAC firing time. It is used as a starting value in equation 9.35.

$$30 = \frac{120\sqrt{2}}{\sqrt{1.142}} \sin(377 t_f - .361) + \left[-9 + \frac{120\sqrt{2}(.377)}{1.142}\right] e^{-10^3 t_f}$$

The iteration formula is

$$\sin(377 t_f - .361) = 0.189 - .2961 e^{-10^3 t_f}$$

Substituting $t_f = .0015$ in the right-hand side, solve for t_f from the left-hand side. Repeat until there is no charge in t_f.

$$t_{f1} = 1.28\,\text{msec}$$
$$t_{f2} = 1.24\,\text{msec}$$
$$t_{f3} = 1.23\,\text{msec}$$
$$t_{f4} = 1.23\,\text{msec}$$

The firing angle is

$$377 \times .00123 = .464\,\text{radians} = 26.6^\circ$$

For any sizeable phase shift ($\arctan \omega R_c C_c$), the initial value of the capacitance voltage is usually negligible.

7 NON-SINUSOIDAL WAVEFORM GENERATION

A. INTRODUCTION

Waveform generation can be accomplished with a variety of circuits. These circuits use feedback switching devices and energy storage elements (i.e., capacitors or inductors) to shape waves. A few representative circuits are described here.

B. UJT RELAXATION OSCILLATOR

The unijunction transistor can be used in a simple circuit to generate the non-sinusoidal waveform shown in figure 9.30. The UJT base acts as a voltage divider until the emitter exceeds the voltage divider voltage. Then the UJT's emitter-B2 base junction becomes forward-biased and behaves more or less like an ordinary junction diode. With the UJT described by two resistors (R_1 attached to B1 base and R_2 to B2 base) in the off state, the voltage divider provides a triggering voltage of

$$\frac{V_{bb}R_1}{R_1+R_2+R_{bb}} \qquad 9.36$$

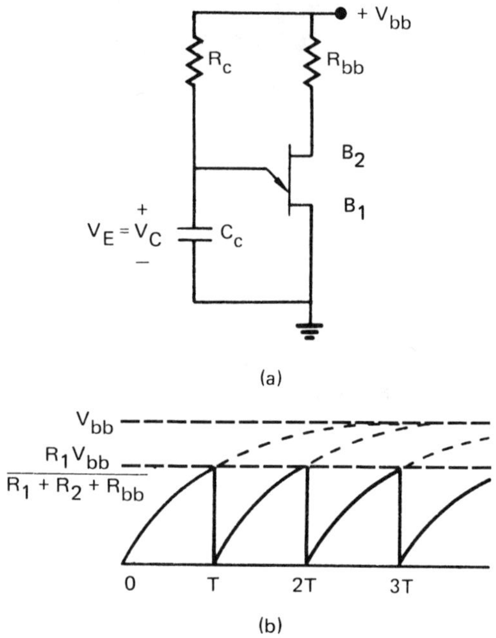

Figure 9.30 (a) UJT relaxation oscillator, (b) oscillation waveform across C_c

The UJT is off until the emitter voltage reaches the value given by equation 9.36, after which time the E-B1 terminals appear as a forward-biased P-N junction. This allows the capacitor to discharge. Once the capacitor is discharged, the UJT turns off, and the capacitor can again begin to charge toward V_{bb} according to equation 9.37.

$$V_C = V_{bb}(1-e^{-t/\tau}) = V_E \qquad 9.37$$

$$\tau = R_cC_c \qquad 9.38$$

Example 9.14

A UJT oscillator with $R_1 = 10K\ \Omega$ and $R_2 = 5K\ \Omega$ is to produce a sawtooth wave with a maximum of 10 volts and a period of 10 msec. A 24 volt D.C. supply is available. Specify the values of the parameters for the circuit of figure 9.30(a) to accomplish this.

The maximum of 10 volts sets the divider ratio.

$$\frac{R_1V_{bb}}{R_1+R_2+R_{bb}} = \frac{(10K)(24)}{10K+5K+R_{bb}}$$

$$R_{bb} = 9\text{K ohms}$$

The time that the voltage on the capacitor reaches 10 volts can be derived from equation 9.37.

$$10 = 24(1-e^{-t/\tau})$$

$$t = \tau\, ln\left[\frac{24}{24-10}\right] = 0.54\tau$$

However, the period, T, was set at 10 msec, so,

$$\tau = \frac{10}{.54} = 19\text{ msec}$$

$$R_cC_c = \tau = 0.019$$

C. NEON BULB RELAXATION OSCILLATOR

A neon bulb behaves much like a UJT when used in a relaxation oscillator. The un-ionized neon bulb acts like an open circuit. It ionizes when the applied voltage is at a sufficient level, V_I. If a neon bulb replaced the UJT in figure 9.30(a), a wave with a peak voltage of V_I would result.

The resistance R_{bb} in the UJT circuit provided a means of adjusting the peak voltage, but this is not available for the neon circuit. The only simple way to provide for peak voltage adjustment in a neon circuit is by a voltage divider on the output as shown in figure 9.31. The output voltage is

$$V_o = \frac{V_CR_1}{R_1+R_2} \qquad 9.39$$

To determine the timing and peak voltage, it is convenient to obtain the Thevenin equivalent circuit as seen from the parallel combination of the capacitance and the bulb.

$$V_{\text{Th}} = V_S\left[\frac{R_1+R_2}{R_1+R_2+R_S}\right] \qquad 9.40$$

$$R_{\text{Th}} = R_S\left[\frac{R_1+R_2}{R_1+R_2+R_S}\right] \qquad 9.41$$

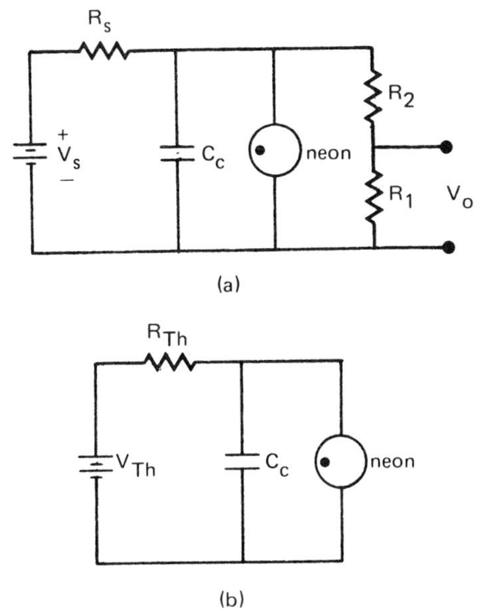

Figure 9.31 (a) Neon relaxation oscillator, (b) Thevenin equivalent circuit

The Thevenin equivalent circuit is shown in figure 9.31(b). In order to achieve oscillatory behavior, V_{Th} must be greater than V_I.

Prior to ionization, the capacitance voltage is

$$V_C = V_{\mathrm{Th}}(1 - e^{-t/\tau}) \qquad 9.42$$

$$\tau = R_{\mathrm{Th}} C_c \qquad 9.43$$

When the capacitance voltage reaches V_I, the neon ionizes and discharges the capacitance. The oscillation period is

$$T = R_{\mathrm{Th}} C_c \, ln\left[\frac{V_{\mathrm{Th}}}{V_{\mathrm{Th}} - V_I}\right] \qquad 9.44$$

The resulting waveform is shown in figure 9.31.

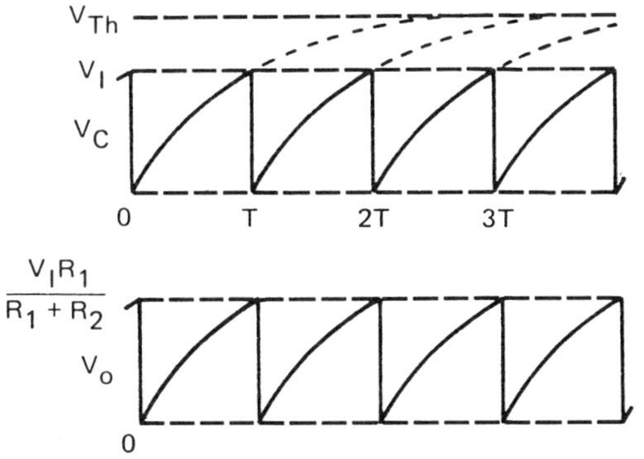

Figure 9.32 Capacitance and output waveforms for neon relaxation oscillator

D. NEGATIVE RESISTANCE OSCILLATOR

A *negative resistance device* is one in which the V-I curve has a negative slope as indicated in figure 9.22(a). (Voltage is limited to single values for any current.) The DIAC has such characteristics, and the UJT with a fixed voltage at base B2 can also be considered to be a negative resistance device. The principal parameters defined in figure 9.33(b) are used to define a piecewise linear model of the device in figure 9.33(c).

In order to obtain oscillatory behavior, it is necessary that the load line of the device-capacitance parallel combination intersect the negative resistance region.

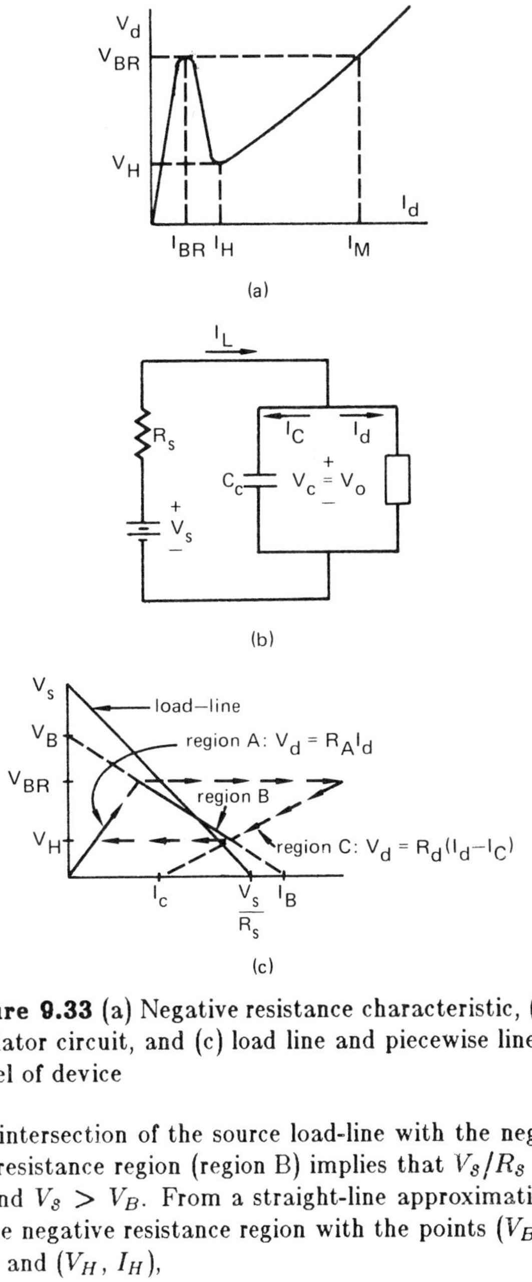

Figure 9.33 (a) Negative resistance characteristic, (b) oscillator circuit, and (c) load line and piecewise linear model of device

The intersection of the source load-line with the negative resistance region (region B) implies that $V_S/R_S < I_B$ and $V_S > V_B$. From a straight-line approximation to the negative resistance region with the points (V_{BR}, I_{BR}) and (V_H, I_H),

$$V_B = \frac{V_{BR} I_H - V_H I_{BR}}{I_H - I_{BR}} < V_S \qquad 9.45$$

$$I_B = \frac{V_{BR} I_H - V_H I_{BR}}{V_{BR} - V_H} > \frac{V_S}{R_S} \qquad 9.46$$

When these conditions are met, the circuit will oscillate. If the source load line intersects the characteristic at a positive slope, the circuit will stabilize at that point and will fail to oscillate.

With the device operating in region A, the source voltage charges the capacitance toward V_S. Since the device and capacitance voltages are identical, the device voltage and current are increasing. The charging current for the capacitance is the difference between the source current, which lies on the load line at the voltage $V_C = V_S$, and the device characteristic at the same voltage.

From figure 9.33(c), the charging current is positive in region A, so the capacitance voltage increases exponentially until V_{BR} is reached. Then the device and capacitance voltage cannot increase further. To remain on the negative resistance portion of the characteristic, the capacitance voltage must decrease. This requires that the capacitance current be negative, which is impossible in the negative resistance region.

The source current must lie on the load line, and the device current must lie on the negative resistance region at V_{BR}. The capacitance voltage cannot change instantaneously, so a jump from positive to negative voltage is physically impossible. Therefore, the capacitance current must remain positive.

The device accomplishes the desired voltage change by increasing its current over the source current. Being constrained to operate on its characteristic, it does this by jumping to the point (V_{BR}, I_M), in the operating region C.

With the device in region C, I_C is negative, and the capacitance will discharge toward the intersection of the straight line representation of region C and the source load line. However, when it reaches the holding current point (V_H, I_H), it faces a similar dilemma of region B operation: to operate in region B, it is necessary to increase the voltage on the capacitance with negative capacitance current which is impossible in this region. Thus, the current jumps back to region A operation.

In cyclic operation, the device begins region A operation with $V_C = V_d = V_H$, as indicated in figure 9.33(c). The device is characterized by the slope of the characteristic

$$R_A = \frac{V_{BR}}{I_{BR}} \qquad 9.47$$

The Kirchoff current law equation of the circuit is

$$\frac{V_S}{R_S} = \frac{V_C}{R_S} + \frac{V_C}{R_A} + C\frac{dV_C}{Cdt} \qquad 9.48$$

The solution to equation 9.48 is

$$V_C = \frac{V_S R_A}{R_S + R_A} + \left[V_H - \frac{V_S R_A}{R_S + R_A}\right]e^{-at} \qquad 9.49$$

$$a = \frac{R_S + R_A}{C_c R_S R_A} \qquad 9.50$$

Region A operation terminates when V_C reaches V_{BR}. This occurs at the time T_A where

$$T_A = \frac{C_c R_S R_A}{R_S + R_A} \times ln\left[\frac{V_S R_A - V_H(R_S + R_A)}{V_S R_A - V_{BR}(R_S + R_A)}\right] \qquad 9.51$$

In region C, the device begins operation with $V_C = V_d = V_{BR}$. In this region, the device is characterized by the straight line connecting points (V_H, I_H) and (V_{BR}, I_M). Then,

$$I_d = G_C V_d + I_C \qquad 9.52$$

$$G_C = \frac{I_M - I_H}{V_{BR} - V_H} \qquad 9.53$$

$$I_C = \frac{V_{BR} I_H - V_H I_M}{V_{BR} - V_H} \qquad 9.54$$

Kirchoff's voltage law equation for region C operation is

$$\frac{V_S}{R_S} - I_C = V_C\left[G_C + \frac{1}{R_S}\right] + C\frac{dV_C}{Cdt} \qquad 9.55$$

$$V_C = \frac{V_S - I_C R_S}{1 + G_C R_S} + \left[V_{BR} - \frac{V_S - I_C R_S}{1 + G_C R_S}\right]e^{-c(t - T_A)} \qquad 9.56$$

$$c = \frac{1 + G_C R_S}{R_S C_c} \qquad 9.57$$

Region C operation terminates when $V_C = V_H$ at time T_C.

$$T_C - T_A = \frac{R_S C_c}{1 + R_S C_c} \times ln\left[\frac{V_S - I_C R_S - V_{BR}(1 + G_C R_S)}{V_S - I_C R_S - V_H(1 + G_C R_S)}\right] \qquad 9.58$$

The waveform for one cycle is shown in figure 9.34. Region A operation is from $t = 0$ to T_A. Region C operation is from T_A to T_C.

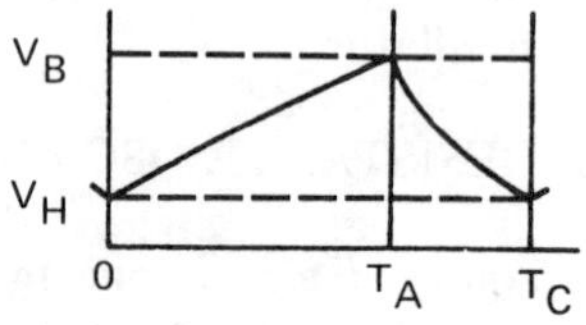

Figure 9.34 Negative resistance oscillator waveform

E. NEGATIVE CONDUCTANCE OSCILLATOR

Negative conductance devices are duals of negative resistance devices. The negative conductance device has single current values for each voltage, in contrast to the negative resistance device which has single values of voltage for each current. A *tunnel diode* is an example of a negative conductance device.

While the negative resistance device used a parallel capacitance to prevent sudden changes in voltage, the negative resistance device uses a series inductance to prevent the current from changing instantaneously. A negative conductance oscillator is shown in figure 9.35(a), with the device characteristic shown in figure 9.35(b).

Oscillatory operation requires the source load line ($V_S = I_t R + V_t$) to intersect the device characteristic in the negative conductance region. This requires $V_S/R_S > I_B$ and $V_S < V_B$.

Of course, $I_t = I_d = I_L$. Kirchhoff's voltage law for the circuit of figure 9.35(a) is

$$V_S = R_S I_L + L\frac{dI_L}{dt} + V_d \qquad 9.59$$

$0 < V_d < V_{BR}$ in region A. Operation is governed by

$$V_d = R_A I_L \qquad 9.60$$

$$R_A = \frac{V_{BR}}{I_{BR}} \qquad 9.61$$

The device enters region A with $I_d = I_H$ at $t = 0$. Then,

$$I_L = \frac{V_S}{R_S + R_A} + \left[I_H - \frac{V_S}{R_S + R_A}\right]e^{-at}$$
$$= \frac{V_d}{R_A} \qquad 9.62$$

$$a = \frac{R_S + R_A}{L} \qquad 9.63$$

Operation in region A stops when $I_S = I_{BR}$ at time $t = T_A$.

$$T_A = \frac{L}{R_S + R_A} \times ln\left[\frac{V_S - (R_S + R_A)I_H}{V_S - (R_S + R_A)I_{BR}}\right] \qquad 9.64$$

The device enters region C ($V_d > V_H$) operation at $t = T_A$ with $I_L = I_{BR}$. Operation is characterized by

$$V_d = V_C + I_L R_C \qquad 9.65$$

$$R_C + \frac{V_M - V_H}{I_{BR} - I_H} \qquad 9.66$$

$$V_C = \frac{V_H I_{BR} - V_M I_H}{I_{BR} - I_H} \qquad 9.67$$

Substituting for V_d in the the KVL equation and solving for the transient,

$$I_L = \frac{V_S - V_C}{R_S + R_C} + I_{BR} - \left[\frac{V_S - V_C}{R_S + R_C}\right]e^{-c(t-T_A)}$$
$$= \frac{1}{R_C}(V_d - V_C) \qquad 9.68$$

Region C operation terminates when $I_L = I_H$. This occurs at $t = T_B$.

$$T_B - T_A = \frac{L}{R_S + R_C} \times ln\left[\frac{I_{BR}(R_S + R_C) + V_C - V_S}{I_H(R_S + R_C) + V_C - V_S}\right] \qquad 9.69$$

The waveform of V_d is shown in figure 9.35(c).

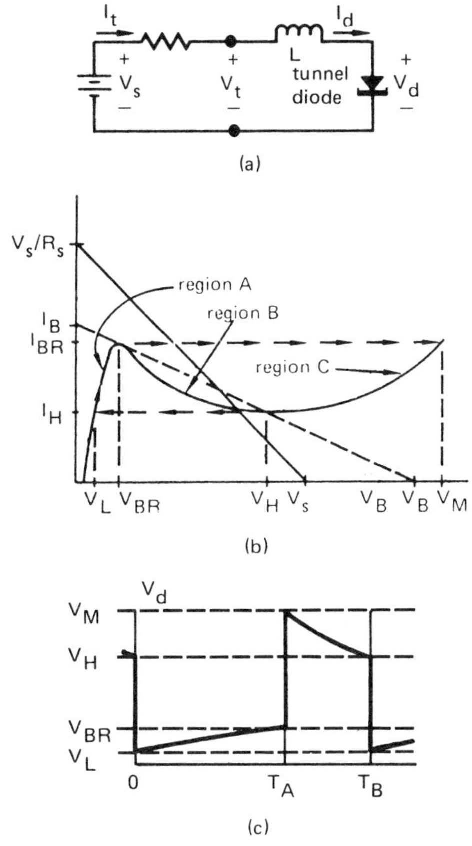

Figure 9.35 (a) Tunnel diode negative conductance oscillator, (b) tunnel diode characteristic and source load-line, (c) tunnel diode voltage waveform

F. OPERATIONAL AMPLIFIER SQUARE-WAVE GENERATOR

An operational amplifier can be used as a switching device for capacitor charging circuits. The amplifier provides the timing for an oscillator, and the output is a rectangular waveform.

A circuit to produce rectangular waves is shown in figure 9.35. The operational amplifier characteristics are limited by the Zener diodes. The operational amplifier is used in its linear zone only during the brief switching interval when it is operating open-loop (i.e., no feedback). It has an essentially infinite slope at that time.

If V_o is $+V_H$, then V_N is kV_H. V_C will charge toward $+V_H$ through diode D_2 and resistance R_2. As long as V_C is less then $kV_H = V_N$, the operational amplifier input voltage V_{in} remains negative. The operational amplifier will remain positively saturated with the output voltage at $+V_H$.

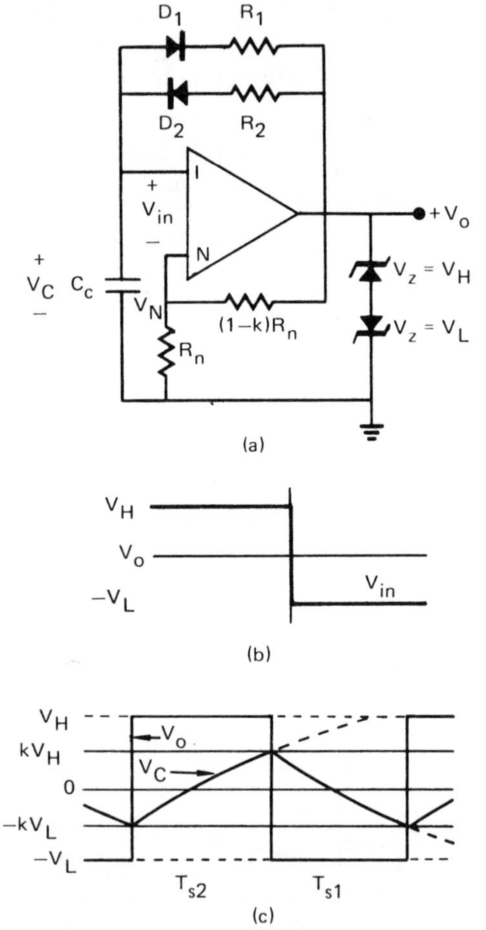

Figure 9.36 (a) Square-wave generator, (b) operational amplifier characteristic with Zener output diodes, (c) capacitor and output voltages

As the capacitance voltage charges past kV_H, the operational amplifier input becomes positive, causing the output to switch to $-V_L$. The voltage at the N input also switches to $-kV_L$, while the capacitance voltage remains at essentially $+kV_H$. The input voltage will become quite positive, holding the output at $-V_L$.

Thus, the capacitance begins charging toward $-V_L$ through the diode D_1 and the resistance R_1, beginning at a value of $+kV_H$. Assuming time starts at $t = 0$, the capacitance charging equation is found from Kirchhoff's voltage law.

$$-V_L = V_C + R_1 C_c \frac{dV_C}{dt} \qquad 9.70$$

$$V_C = -V_L + (kV_H + V_L)e^{-t/R_1 C_c} \qquad 9.71$$

When V_C reaches $-kV_L$, V_{in} will pass through zero and reverse the output to $+V_H$, terminating the negative half-cycle. This occurs when $V_C = -kV_L$. The corresponding time is $t = T_{S1}$.

$$T_{s1} = R_1 C_c \, ln\left[\frac{kV_H + V_L}{V_L(1-k)}\right] \qquad 9.72$$

The capacitance begins the positive half-cycle at a voltage of $-kV_L$. For convenience, time is reset to $t = 0$ at the beginning of this cycle. The solution of the Kirchhoff voltage law equation with the capacitance charging toward V_H through R_2 yields

$$V_C = V_H + (-kV_L - V_H)e^{-t/R_2 C_c} \qquad 9.73$$

When V_C reaches $+kV_H$, the positive half-cycle is terminated. This occurs at a time T_{s2}.

$$T_{s2} = R_2 C_c \, ln\left[\frac{kV_L + V_H}{V_H(1-k)}\right] \qquad 9.74$$

The waveforms of V_o and V_C are shown in figure 9.36(c). A symmetrical wave is obtained by making $R_1 = R_2$ and $V_H = V_L$.

G. ASTABLE MULTIVIBRATOR

The basic circuit of an *astable multivibrator* requires two amplifying elements. These amplifiers can be in the common emitter configuration, or they can have a common emitter resistance. The collector-coupled common emitter configuration is shown in figure 9.37.

Assume transistor T2 is in the saturated condition and T1 is in the cutoff state. For these conditions to hold, V_{C1} must have a voltage greater than −0.4 volts. This, together with the collector saturation voltage (0.2 volts) of T2, does not turn on T1 since the base-emitter voltage is kept less than the turn-on voltage (taken as 0.6 volts). With these conditions, C_1 charges toward $-V_{cc}+$ .2 for the polarities shown. When V_{C1} does reach −.4 volts, it will initiate the regenerative switching of the circuit. The switching process listed here occurs very quickly. The capacitor voltages do not change during switching.

1. V_{C1} reaches -0.4 volts, so V_{BE1} turns on T1.
2. V_{CE1} begines to drop. This drop is coupled to the base of T2 through C_2 causing V_{BE2} to drop.
3. The drop in V_{BE2} causes T2 to come out of saturation and V_{CE2} to rise.
4. The rise in V_{CE2} is coupled to the base of T1 through C_1, driving T1 farther into the on region.
5. V_{CE1} drops farther. This turns off T2 more which turns on T1 more, and the cycle continues.

Begin a more detailed analysis at $t = 0$, the instant V_{BE1} reaches 0.6 volts, turning T1 on and T2 off. The voltage C_1 is then -0.4 volts at $t = 0$. The partial circuits of figure 9.37(b) and (c) are valid from $t = 0$ until the base-emitter voltage of T1 reaches 0.6 volts and initiates switching again. The next switching occurs a half-cycle from $t = 0$, or $T/2$.

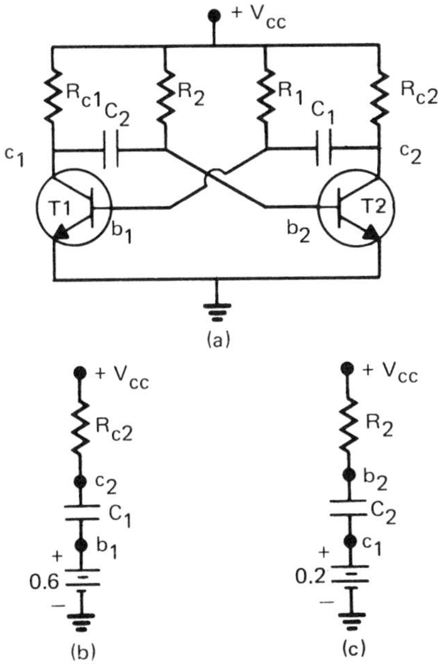

Figure 9.37 Collector coupled astable multivibrator: (a) circuit diagram, (b) C_1 charging circuit, and (c) C_2 charging circuit

Figure 9.36(b) shows the charging path of capacitor C_1. The equation which governs this charging is

$$V_{cc} - .6 = V_{C1} + R_c C \frac{dV_{C1}}{dt} \qquad 9.75$$

Using $V_{C1}(0) = -.4$ volts, the solution is

$$V_{C1} = (V_{cc} - .6) - (V_{cc} - .2)e^{-t/R_c C} \qquad 9.76$$

At $t = T/2$, V_{C1} will have the value that V_{C2} has at $t = 0$.

$$V_{C1}|_{t=t/2} = V_{C2}|_{t=0} = (V_{cc} - .6) - (V_{cc} - .2)e^{-T/2R_c C} \qquad 9.77$$

The behavior of the base-emitter voltage on the cutoff transistor, V_{BE2}, is desired. While T1 is saturated,

$$V_{BE2} = 0.2 - V_{C2} < .6$$

$$V_{C2} > .4$$

As C_2 is charging toward $V_{C2} = -V_{cc} + .2$, it will turn T2 on again at $t = T/2$.

The differential equation for V_{C2} is

$$-V_{cc} + .2 = V_{C2} + R_b C \frac{dV_{C2}}{dt} \qquad 9.78$$

The solution is

$$V_{C2} = -(V_{cc} - .2) + [V_{cc} - .2 + V_{C2}(0)]e^{-t/T_b C} \qquad 9.79$$

At $t = t/2$, V_{C2} will have the same value that V_{C1} had at $t = 0$.

$$V_{C2}|_{t=T/2} = V_{C1}|_{t=0} = -0.4 \qquad 9.80$$

Then,

$$V_{C2}|_{t=t/2} = -.4$$
$$= -(V_{cc} - .2) + [(V_{cc} - .2 + V_{C2}(0)]e^{-T/2R_b C} \qquad 9.81$$

Equations 9.77 and 9.81 contain the same two unknowns: $V_{C2}(0)$ and T. Taking the logarithms of the equations,

$$\frac{T}{2R_c C} = ln\left[\frac{V_{cc} - .2}{V_{cc} - .6 - V_{C2}(0)}\right] \qquad 9.82$$

$$\frac{T}{2R_b C} = ln\left[\frac{V_{cc} - .2 + V_{C2}(0)}{V_{cc} - .6}\right] \qquad 9.83$$

Equations 9.82 and 9.83 can only be solved algebraically when $R_b = R_c$. In other cases, particularly if R_b is much greater than R_c, it can be assumed that $V_{C2}(0) \approx V_{cc} - .2$. Then, solve for the cycle time, which will converge with successive iterations.

Example 9.15

An astable multivibrator, such as the one shown in figure 9.37(a), has $V_{cc} = 16$ volts, $R_c = 1K$ ohms,

$R_b = 5K$ ohms, $C = 2\,nF$. Determine the frequency of oscillation.

Equating equations 9.82 and 9.83 and cancelling the common T and C terms,

$$5\ ln\left[\frac{15.8 + V_{C2}(0)}{15.4}\right] = ln\left[\frac{15.8}{15.4 - V_{C2}(0)}\right]$$

Since $R_c \neq R_b$, trial and error is required. Taking the antilog of both sides,

$$V_{C2}(0) = 15.4 - 15.8e^{-5\,ln([15.8+V_{C2}(0)]/15.4)}$$

Starting with $V_{C2}(0) = 15$ in the right-hand side, $V_{C2}(0)$ on the left is 14.90625. Putting that value into the right-hand side, the left side is 14.89866. A third iteration yields 14.89804, so the value of V_{C2} is taken as 14.90. From equation 9.83,

$$T = 2 \times 5000 \times 2 \times 10^{-9}\ ln\left[\frac{15.8 + 14.90}{15.4}\right]$$
$$= 1.38 \times 10^{-5}$$

The frequency of oscillation is

$$f = \frac{1}{T} = 72.4\,\text{khz}$$

H. A MONOSTABLE MULTIVIBRATOR

A *monostable multivibrator* (also known as a *blocking oscillator*) is one which has one stable state. There is one astable state which is typically controlled by the charging of a capacitor. Such a circuit is shown in figure 9.38. The capacitor C_1 is a *speed-up capacitor* to enhance the regenerative action whenever it begins. It does this by coupling the change at the collector of T2 to the base of T1. The timing capacitor is C_2, which charges through R_{C1} when T2 is on, and through R_2 when T2 is off and T1 is on.

The stable state is when T2 is on, being driven by V_{cc} through R_2. In the quiescent condition, B2 is at 0.6 volts, and V_2 is at about 0.3 volts, as T2 is saturated. Since T1 is off, C_2 is charged to $-V_{cc} + 0.6$ volts, and C_1 is discharged.

When a trigger pulse is applied to B1 for a sufficient time, the following processes occur.

1. T1 turns on and lowers the voltage V_1. Since V_1 is coupled through C_2 (which cannot change voltage instantaneously), T2 begins to turn off.

2. The turn-off of T2 causes the voltage V_2 to increase. This increase is coupled through C_1 to reinforce the turn-on of T1.

3. Steps 1 and 2 are very fast. T1 is turned on and T2 is turned off very quickly. The capacitance conditions do not have time to change significantly.

V_{C2} begins to charge toward $V_{cc} - .3$ since the collector of T1 is at about 0.3 volts. Charging is through R_2 with a time constant R_2C_2. Charging continues until a voltage of about 0.4 volts is reached, whereupon that (together with the 0.3 volts of V_1) will begin to turn T2 on again. This terminates the astable state. The regenerative action quickly turns T1 off and T2 on, returning to the stable state. Before another trigger is accepted, C_2 must be charged to $-V_{cc} + .6$ again.

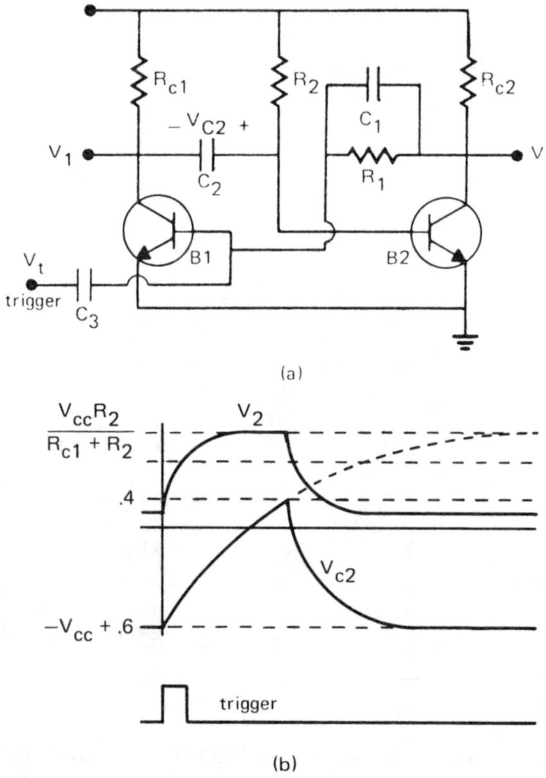

Figure 9.38 (a) Monostable multivibrator circuit, (b) principal waveforms

Appendix A

TABLE OF STANDARD ZENER DIODES with TYPICAL ZENER IMPEDANCES and KEEP-ALIVE CURRENTS

nominal Zener voltage	500 milliwatt		1 watt		5 watt		50 watt	
	Z_z ohms	I_{zk} mA	Z_z ohms	I_{zk} mA	Z_z ohms	I_{zk} mA	Z_z ohms	I_{zk} mA
2.4	30	1.0						
2.7	30	1.0						
3.0	29	1.0						
3.3	28	1.0	10	1.0	3.0	1.0		
3.6	24	1.0	10	1.0	2.5	1.0		
3.9	23	1.0	9.0	1.0	2.0	1.0	0.16	5.0
4.3	22	1.0	9.0	1.0	2.0	1.0	0.16	5.0
4.7	19	1.0	8.0	1.0	2.0	1.0	0.12	5.0
5.1	17	1.0	7.0	1.0	1.5	1.0	0.12	5.0
5.6	11	1.0	5.0	1.0	1.0	1.0	0.12	5.0
6.2	7.0	1.0	2.0	1.0	1.0	1.0	0.14	5.0
6.8	5.0	1.0	1.5	1.0	1.0	1.0	0.16	5.0
7.5	5.5	0.5	1.5	1.0	1.5	1.0	0.24	5.0
8.2	6.5	0.5	4.5	0.5	1.5	1.0	0.4	5.0
8.7					2.0	1.0		
9.1	7.5	0.5	5.0	0.5	2.0	1.0	0.5	5.0
10	8.5	0.25	7.0	0.25	2.0	1.0	0.6	5.0
11	9.5	0.25	8.0	0.25	2.5	1.0	0.8	5.0
12	11.5	0.25	9.0	0.25	2.5	1.0	1.0	5.0
13	13	0.25	10	0.25	2.5	1.0	1.1	5.0
14					2.5	1.0	1.2	5.0
15	16	0.25	14	0.25	2.5	1.0	1.4	5.0
16	17	0.25	16	0.25	2.5	1.0	1.6	5.0
17					2.5	1.0	1.8	5.0
18	21	0.25	20	0.25	2.5	1.0	2.0	5.0
19					3.0	1.0	2.2	5.0
20	25	0.25	22	0.25	3.0	1.0	2.4	5.0
22	29	0.25	23	0.25	3.5	1.0	2.5	5.0
24	33	0.25	25	0.25	3.5	1.0	2.6	5.0
25					4.0	1.0	2.7	5.0
27	41	0.25	35	0.25	5.0	1.0	2.8	5.0
28					6.0	1.0		
30	49	0.25	40	0.25	8.0	1.0	3.0	5.0
33	58	0.25	45	0.25	10	1.0	3.2	5.0
36	70	0.25	50	0.25	11	1.0	3.5	5.0
39	80	0.25	60	0.25	14	1.0	4.0	5.0
43	93	0.25	70	0.25	20	1.0	4.5	5.0
45							4.5	5.0
47	105	0.25	80	0.25	25	1.0	5.0	5.0
50							5.0	5.0
51	125	0.25	95	0.25	27	1.0	5.2	5.0

TABLE OF STANDARD ZENER DIODES
with TYPICAL ZENER IMPEDANCES and KEEP-ALIVE CURRENTS

nominal Zener voltage	500 milliwatt		1 watt		5 watt		50 watt	
	Z_z ohms	I_{zk} mA	Z_z ohms	I_{zk} mA	Z_z ohms	I_{zk} mA	Z_z ohms	I_{zk} mA
56	150	0.25	110	0.25	35	1.0	6.0	5.0
60					40	1.0		
62	185	0.25	125	0.25	42	1.0	7.0	5.0
68	230	0.25	150	0.25	44	1.0	8.0	5.0
75	270	0.25	175	0.25	45	1.0	9.0	5.0
82	330	0.25	200	0.25	65	1.0	11	5.0
87					75	1.0		
91	400	0.25	250	0.25	75	1.0	15	5.0
100			350	0.25	90	1.0	20	5.0
105							25	5.0
110					125	1.0	30	5.0
120			550	0.25	170	1.0	40	5.0
130			700	0.25	190	1.0	50	5.0
140					230	1.0	60	5.0
150			1000	0.25	330	1.0	75	5.0
160			1100	0.25	350	1.0	80	5.0
170					380	1.0		
175							85	5.0
180			1200	0.25	430	1.0	90	5.0
190					450	1.0		
200			1500	0.25	480	1.0	100	5.0

Practice Problems

NONLINEAR WAVESHAPING AND SWITCHING

Warmups

1. Using a precision diode circuit, design a clipping circuit which will pass voltages above −0.25 volts and clip voltages below that value. Assume a stable power supply.

2. Design a pre-reference Zener circuit using an 8.1 volt Zener diode ($V_Z = 8.1$ volts at $I_z = I_{zk} = 8$ mA, with $R_z = 8$ ohms) from which the reference diode of example 9.5 is to be operated. Design the entire reference voltage circuit, and calculate the new voltage regulation for the output circuit of example 9.5.

3. For the transistor circuit shown, determine (a) the necessary base current for this gate to sink 10 identical gates connected to its output, and (b) the minimum size of R_b to allow this gate to source ten identical gates.

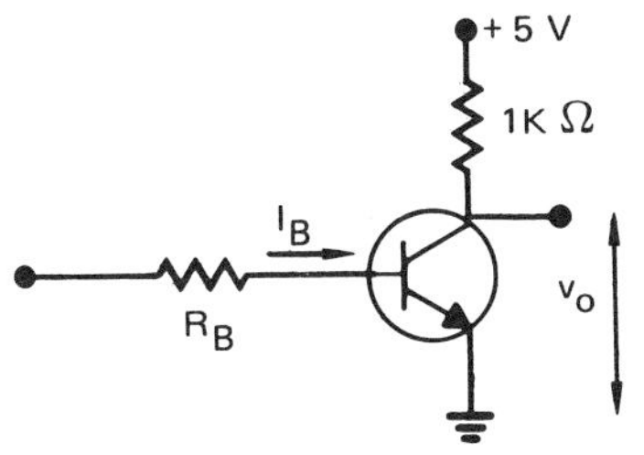

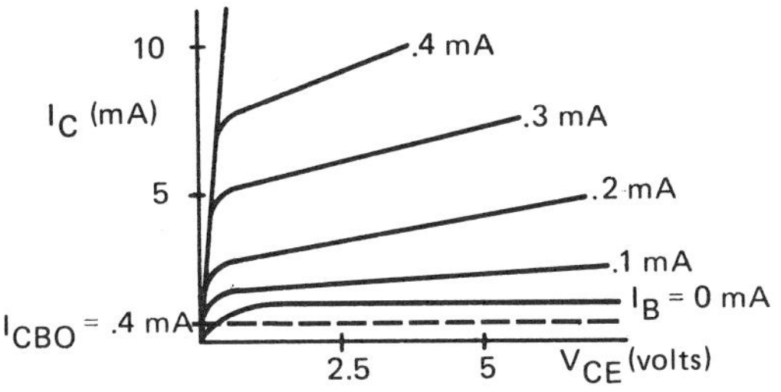

4. A particular logic gate has the following parameters: $t_d = 5$ nsec, $t_r = 50$ nsec, $t_s = 30$ nsec, and $t_f = 30$ nsec. Determine the maximum frequency at which this gate can operate.

5. Use the truth table to show that the circuit of figure 9.15 is a NOR gate for negative logic (i.e., low voltage = TRUE, high voltage = FALSE).

6. A particular SCR has $I_{GT} = 50$ mA. Determine the current pulse amplitude necessary to trigger it in 10 μsec.

7. For the relaxation oscillator of example 9.14, determine the change of peak voltage and period for a 25% variation in V_{bb}.

8. The switch in the circuit shown is alternately opened and closed for 0.5 sec intervals. Determine the voltage across the capacitance, and show the waveform.

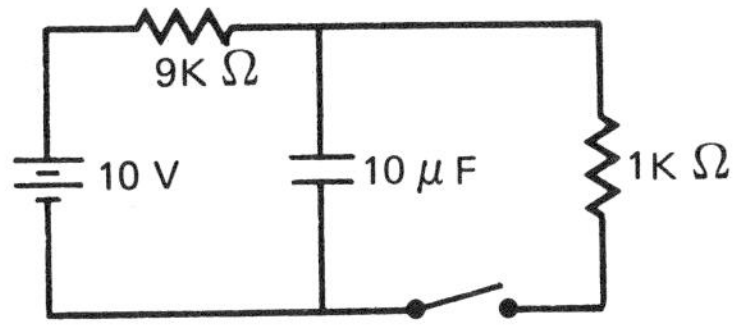

9. Refer to the circuit of figure 9.38. $V_{cc} = 12$ volts, $R_2 = 10K$ ohms, and $C_2 = 0.1\ \mu F$. Determine the duration for which the monostable output V_2 is high.

10. Use an operational amplifier to convert a rectangular wave voltage of ±5 volts and a period of 10 msec to a triangular wave with peak output ±5 volts with the same period.

Concentrates

1. For the circuit shown, determine the transfer characteristic of the output voltage versus the input voltage. The diodes are ideal.

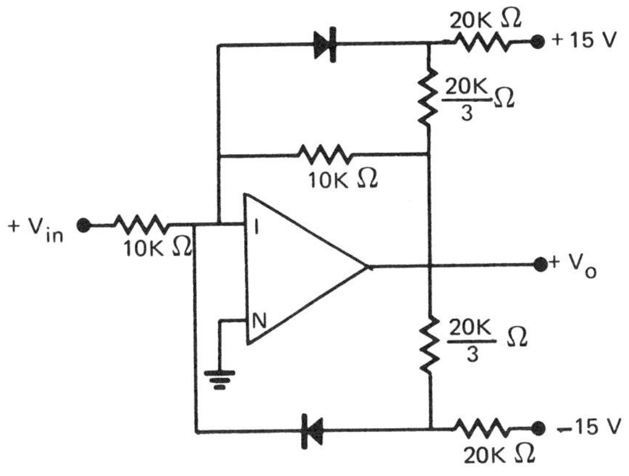

2. Design a voltage regulator which has a maximum supply voltage of 30 volts a minimum of 24 volts, and is to supply a load voltage of 16 volts for currents up to 500 mA. (A 16-volt Zener diode—rated at a current of 155 mA—with $R_z = 4$ ohms is available.) Its maximum current is 530 mA. Determine the maximum regulation of your design, and discuss methods of improving the regulation.

3. For the DTL gate shown, the diode drops may range from 0.6 to 0.8 volts, the base-emitter drop of the transistor will be between .7 and .8 volts, the transistor current gain ranges between 18 and 32, and the saturation

voltage of the transistor is 0.2 volts. The resistance values are nominal, with 20% possible variation.

Determine the maximum reliable fan-out for production of these gates in large quantities.

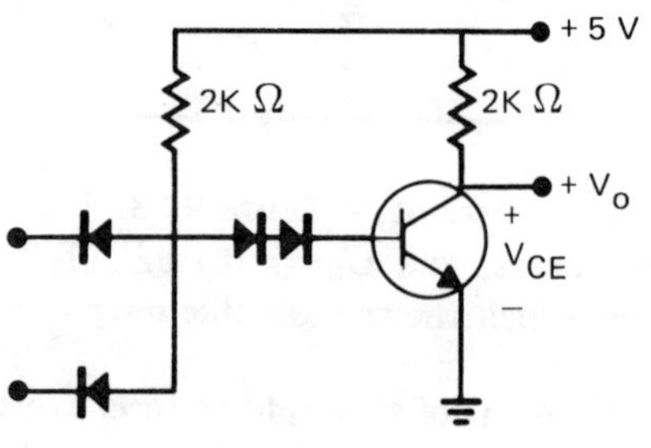

4. The circuit shown below is used to flash a camera lamp. The lamp characteristics also are shown. (a) Determine the time after the switch is closed when the lamp illuminates. (b) Determine the time after the switch is closed when the lamp extinguishes.

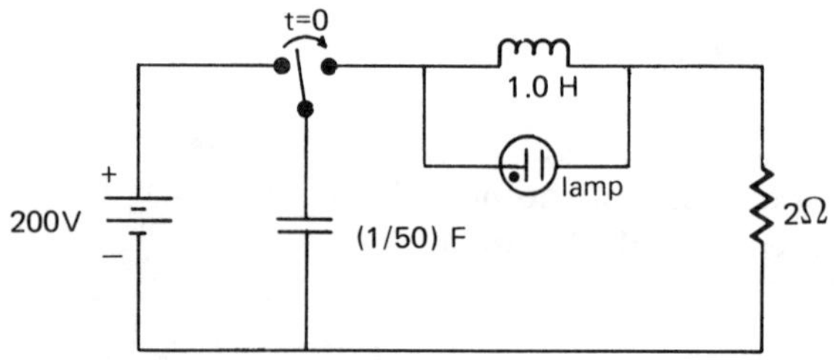

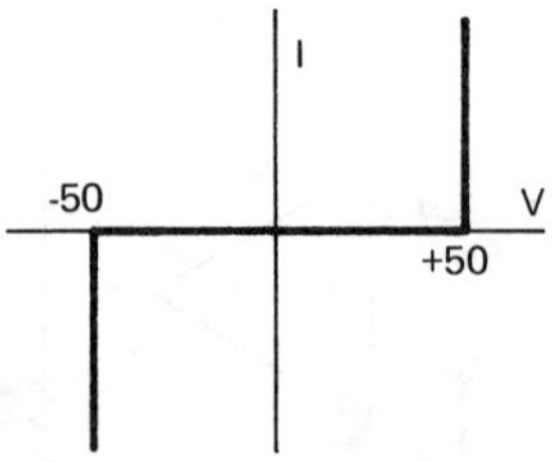

characteristics of camera lamp

5. For the circuit shown, the UJT has $R_1 = 10K$ and $R_2 = 5K$ ohms. Determine the firing angle of the SCR.

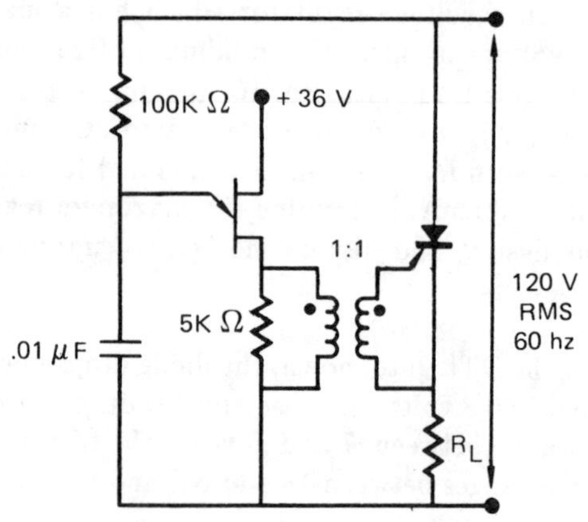

6. A DIAC with characteristics shown is to be used in a negative resistance oscillator circuit shown. Determine a reasonable value of R_s. Select a capacitance to provide a cyclic frequency of 400 hz. Show the resulting waveform.

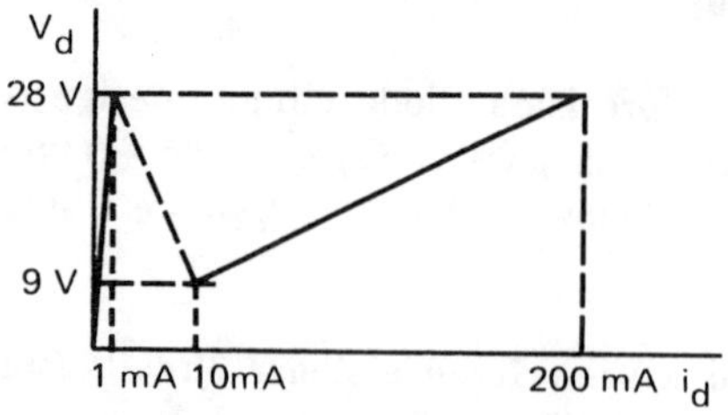

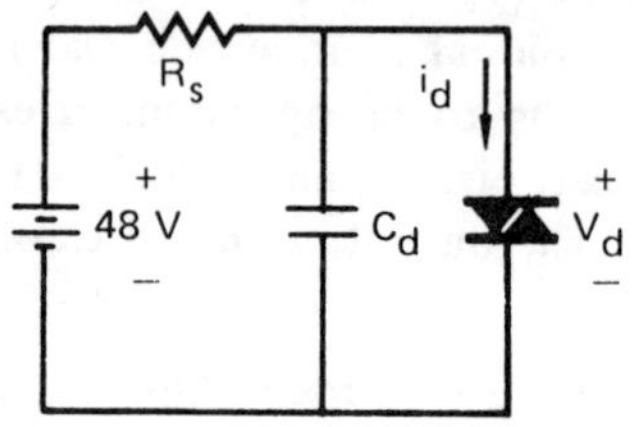

Timed

1. The circuit shown is to provide full-wave rectification of a pure sinusoidal input voltage and to feed that voltage to an average reading voltmeter. The voltmeter has a 10 volt scale and is to indicate the RMS value of the input voltage V_{in} directly (i.e., no scale-change). If the amplifiers saturate at positive and negative 12 volts, what is the maximum voltage which the meter can measure accurately? Determine the values of R_1, R_2, and R_3 necessary for this circuit to operate correctly.

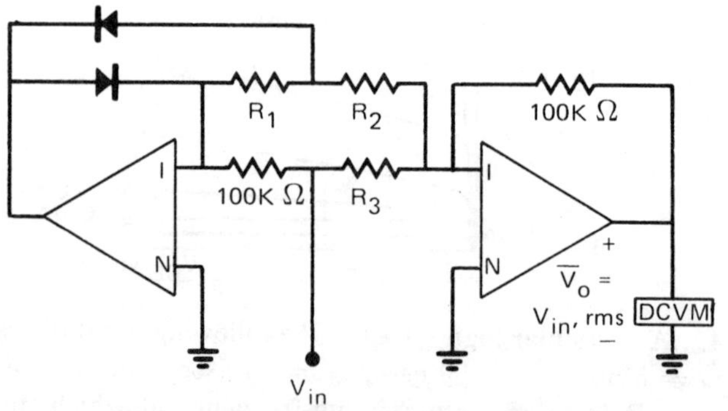

2. Determine the power requirement for the DTL circuit shown for a fan-out of 5.

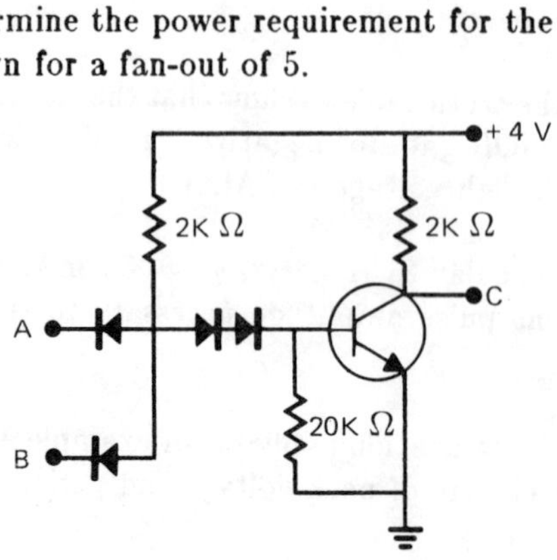

3. For the circuit shown, $R_L = 10$ ohms, $R_c = 100K$ ohms, and $C_c = 0.01$ microfarads. The bilateral breakdown diode (treated as a DIAC) breaks over at ± 4 volts, and the holding voltages are negligible. The power source is 220 volts RMS at 60 hz. (a) Find the conduction angle by calculation. (b) Determine the maximum I_{GT} that the TRIAC can demand. (c) Show the waveforms for the TRIAC gate current and load voltage.

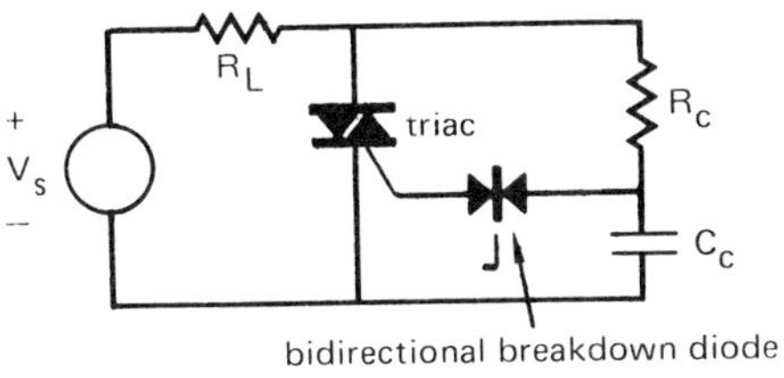

4. The tunnel diode in the circuit has the characteristics given. (a) Find the period of the astable waveform $v_d(t)$. (b) Sketch the voltage waveform of $v_d(t)$. (c) Sketch the current waveform $i_d(t)$.

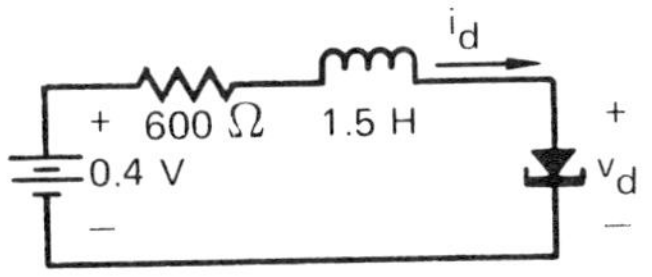

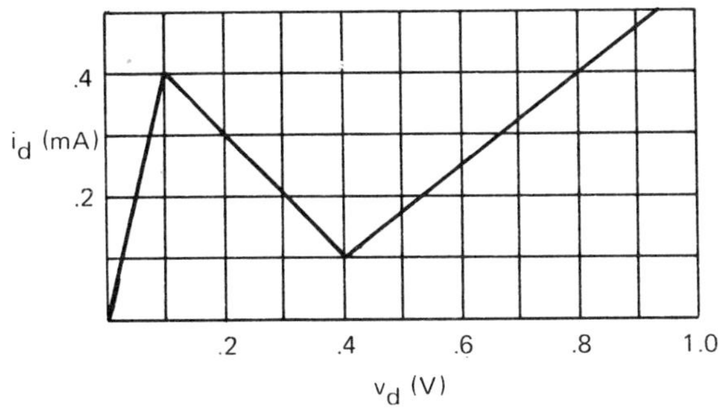

5. Design an astable multivibrator circuit as shown to obtain the waveform shown. The transistors have parametric values of $V_{BE} = 0.7$ volts, $V_{CE,sat} = 0.2$ volts and $\beta_{min} = 25$.

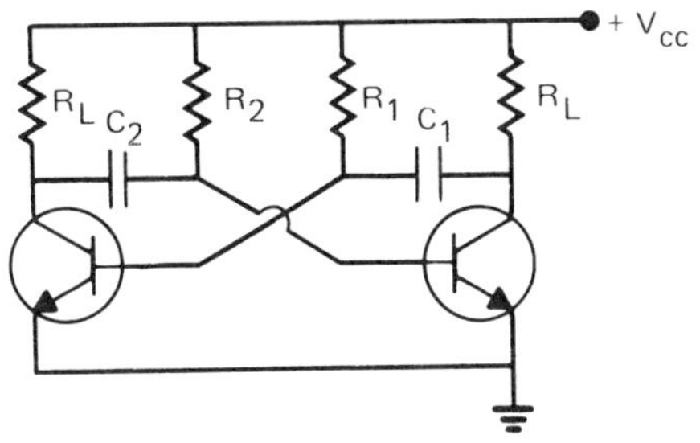

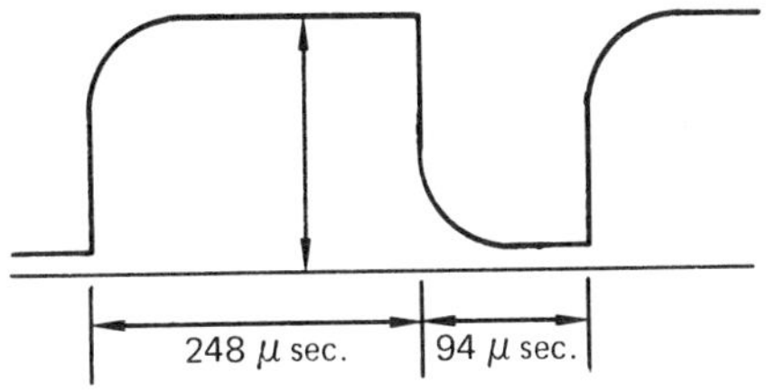

RESERVED FOR FUTURE USE

10 DIGITAL LOGIC

Nomenclature

D	data or delay (flip-flop)
$J-K$	J-K (flip-flop)
m	minterm
M	maxterm
MUX	multiplexer or data-selector
PLA	programmable logic array
POS	product of sums
Q	a state variable
Q^+	the next value of Q
res	residue
ROM	read-only-memory
S	selection variable
SOP	sum of products
$S-R$	set-reset (flip-flop)
T	toggle (flip-flop)
X	a "don't-care" condition

1 FUNDAMENTAL LOGICAL OPERATIONS

A. Symbols, Truth Tables, and Expressions

LOGICAL AND

The symbols and interpretation of logic circuits are introduced with the symbol for an AND gate. Inputs are on the left; output is on the right. The *truth table* shows the input logic values on the left and the corresponding output values in the right-hand column. By convention, a logical "1" means *true*, and a logical "0" means *false*.

The logical expressions utilize the dot "·" to indicate AND and the "+" for OR. In expressions where all variables are represented by single letters, the "·" is commonly omitted.

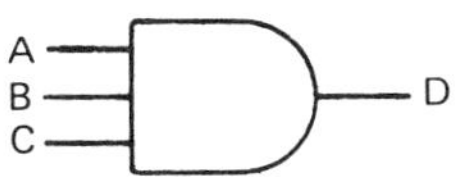

Figure 10.1 Symbol for logic element AND

Table 10.1
Truth table for AND gate

A	B	C	D
0	0	0	0
0	0	1	0
0	1	0	0
0	1	1	0
1	0	0	0
1	0	1	0
1	1	0	0
1	1	1	1

$$D = A \cdot B \cdot C \qquad 10.1$$

$$A \cdot B \cdot C = D \qquad 10.2$$

LOGICAL OR

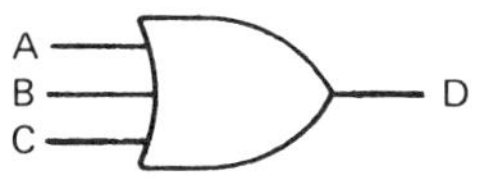

Figure 10.2 OR symbol

Table 10.2
Truth table for OR gate

A	B	C	D
0	0	0	0
0	0	1	1
0	1	0	1
0	1	1	1
1	0	0	1
1	0	1	1
1	1	0	1
1	1	1	1

$$D = A + B + C \qquad 10.3$$

$$A + B + C = D \qquad 10.4$$

LOGICAL NOT

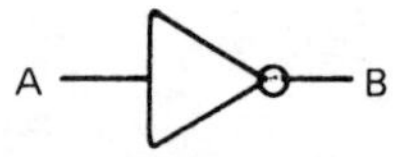

Figure 10.3 NOT symbol

Table 10.3
Truth table for NOT gate

A	B
0	1
1	0

$$B = \overline{A} \qquad 10.5$$

$$\overline{A} = B \qquad 10.6$$

LOGICAL NAND

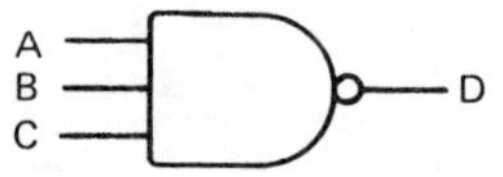

Figure 10.4 NAND symbol

Table 10.4
Truth table for NAND gate

A	B	C	D
0	0	0	1
0	0	1	1
0	1	0	1
0	1	1	1
1	0	0	1
1	0	1	1
1	1	0	1
1	1	1	0

$$D = \overline{A \cdot B \cdot C} \qquad 10.7$$

$$\overline{A \cdot B \cdot C} = D \qquad 10.8$$

LOGICAL NOR

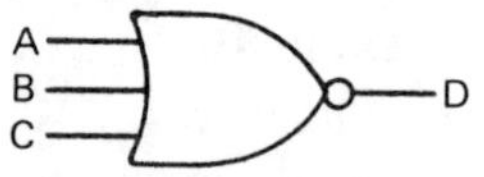

Figure 10.5 NOR symbol

Table 10.5
Truth table for NOR gate

A	B	C	D
0	0	0	1
0	0	1	0
0	1	0	0
0	1	1	0
1	0	0	0
1	0	1	0
1	1	0	0
1	1	1	0

$$D = \overline{A + B + C} \qquad 10.9$$

$$\overline{A + B + C} = D \qquad 10.10$$

EXCLUSIVE OR (XOR)

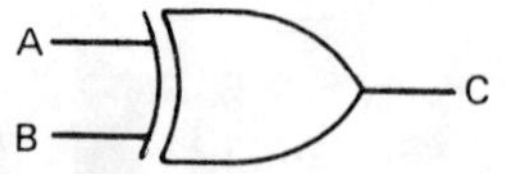

Figure 10.6 XOR symbol

Table 10.6
Truth table for XOR gate

A	B	C
0	0	0
0	1	1
1	0	1
1	1	0

$$C = A \oplus B \qquad 10.11$$

$$A \oplus B = C \qquad 10.12$$

EXCLUSIVE NOR (XNOR)

Figure 10.7 XNOR symbol

Table 10.7
Truth table for XNOR gate

A	B	C
0	0	1
0	1	0
1	0	0
1	1	1

$$C = \overline{A \oplus B} = A \odot B \qquad 10.13$$

$$A \odot B = C \qquad 10.14$$

The XNOR gate is also known as a *coincidence gate* and has a alternate symbol.

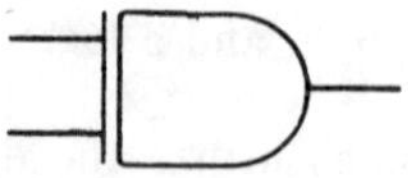

Figure 10.8 Alternate XNOR symbol

B. LOGICAL VARIABLES AND CONSTANTS

A *logical variable* can have one of two possible values: **true** (1) and **false** (0). *Logical constants* have only one possible value: 0 or 1. Two logical variables can take on four possible combinations of values: 00, 01, 11, and 10.

Three logical variables can take on eight possible combinations of values: 000, 001, 010, 011, 100, 101, 110, and 111.

In general, N logical variables can take on a total of 2^N possible values. The position of each variable in the string of 1's and 0's remains the same. For example, when $A = 1$, $B = 0$, and $C = 1$, we can write 101 for the sequence ABC.

C. LOGICAL IDENTITIES

RELATIONSHIPS BETWEEN CONSTANTS

$$0 \cdot 0 = 0 \qquad 10.15$$
$$0 \cdot 1 = 0 \qquad 10.16$$
$$1 \cdot 1 = 1 \qquad 10.17$$
$$0 + 0 = 0 \qquad 10.18$$
$$0 + 1 = 1 \qquad 10.19$$
$$1 + 1 = 1 \qquad 10.20$$
$$\overline{0} = 1 \qquad 10.21$$
$$\overline{1} = 0 \qquad 10.22$$
$$0 \oplus 0 = 0 \qquad 10.23$$
$$0 \oplus 1 = 1 \qquad 10.24$$
$$1 \oplus 0 = 1 \qquad 10.25$$
$$1 \oplus 1 = 0 \qquad 10.26$$

RELATIONSHIPS BETWEEN VARIABLES

$$A + B = B + A \qquad 10.27$$
$$A \cdot B = B \cdot A \qquad 10.28$$
$$A + \overline{A} = 1 \qquad 10.29$$
$$A \cdot \overline{A} = 0 \qquad 10.30$$
$$A + A = A \qquad 10.31$$
$$A \cdot A = A \qquad 10.32$$
$$\overline{\overline{A}} = A \qquad 10.33$$
$$A + (B + C) = A + B + C \qquad 10.34$$
$$A \cdot (B + C) = (A \cdot B) + (A \cdot C) \qquad 10.35$$
$$A + (B \cdot C) = A + B \cdot C \qquad 10.36$$
$$A \oplus B = \overline{A} \cdot B + A \cdot \overline{B} \qquad 10.37$$
$$A \odot B = \overline{A} \cdot \overline{B} + A \cdot B \qquad 10.38$$
$$A \oplus B \oplus C = A \oplus (B \oplus C) = (A \oplus B) \oplus C \qquad 10.39$$

D. LOGICAL THEOREMS

$$A + A \cdot B = A \qquad 10.40$$
$$A + \overline{A} \cdot B = A + B \qquad 10.41$$
$$A \cdot C + \overline{A} \cdot B \cdot C = A \cdot C + B \cdot C \qquad 10.42$$
$$A \cdot B + A \cdot C + \overline{B} \cdot C = A \cdot B + \overline{B} \cdot C \qquad 10.43$$
$$A \cdot B + A \cdot \overline{B} = A \text{ (logical adjacency)} \qquad 10.44$$
$$\overline{A \cdot B \cdot C \cdot D \cdot E \cdot} = \overline{A} + \overline{B} + \overline{C} + \overline{D} + \overline{E} \text{ (De Morgan)} \qquad 10.45$$
$$\overline{A + B + C + D + E} = \overline{A} \cdot \overline{B} \cdot \overline{C} \cdot \overline{D} \cdot \overline{E} \text{ (De Morgan)} \qquad 10.46$$

E. LOGICAL FUNCTIONS

A single logical variable can have two possible values: **true** and **false**.

For two logical variables there are four possible combinations. These can be expressed two different but equivalent ways: as *product terms* and *sum terms*. The sum terms are equivalent to the product terms by De Morgan's theorem, equations 10.45 and 10.46.

All possible functions of two logical variables can be expressed in product terms (called *minterms*) or in sum terms (called *maxterms*).

Table 10.8
Product and Sum Terms

variable combinations A B	product terms (minterms)	sum term ($\overline{\text{maxterms}}$)
0 0	$\overline{A} \cdot \overline{B} = m_{00}$	$\overline{A + B} = \overline{M}_{00}$
0 1	$\overline{A} \cdot B = m_{01}$	$\overline{A + \overline{B}} = \overline{M}_{01}$
1 0	$A \cdot \overline{B} = m_{10}$	$\overline{\overline{A} + B} = \overline{M}_{10}$
1 1	$A \cdot B = m_{11}$	$\overline{\overline{A} + \overline{B}} = \overline{M}_{11}$

Each of the four possible *minterms* given in table 10.8 is true only for the corresponding combination of the two variables. A function which is true for more than one combination is obtained by *ORing* the corresponding minterms. The result is called the *Sum of Products* or *SOP* form of the logical function expression.

Example 10.1

A function of A and B is

$$F_1 = A$$

This is **true** whether B is **true** or **false**. In minterm formulation, all terms which contain A must be included:

$$F_1 = A \cdot B + A \cdot \overline{B}$$

From the logical adjacency theorem, this is $F_1 = A$.

Example 10.2

The XOR or EXOR function of two variables is **true** when one variable is **true** and the other is **false**. This corresponds to

$$F_2 = A \oplus B = \overline{A} \cdot B + A \cdot \overline{B}$$

Example 10.3

The NEXOR OR XNOR function is true when both variables are **false** or when both are **true**.

$$F_3 = A \odot B = \overline{A} \cdot \overline{B} + A \cdot B$$

The minterm formulation of a logical function may not be in its simplest form with respect to the utilization of logic gates. For example, in example 10.1 a blind implementation would require two AND gates and one OR gate, while a straight-through wire would be sufficient.

The *maxterm* formulation of a function is determined somewhat indirectly. To begin with, the *complement* or NOT of the function is formed with a sum of the corresponding maxterms. Then De Morgan's theorem is applied to obtain the function as a Product of Sums (POS) formulation.

Example 10.4

The function is F_1 as in example 10.1. To form the POS or maxterm formulation, first express the NOT of the function as a sum of the corresponding maxterms:

$$\overline{F_1} = \overline{M_{00}} + \overline{M_{01}} = \overline{(A+B)} + \overline{(A+\overline{B})}$$

Then complement the result and apply De Morgan's theorem:

$$F_1 = \overline{\overline{F_1}} = \overline{\overline{M_{00}} + \overline{M_{01}}} = \overline{\overline{M_{00}}} \cdot \overline{\overline{M_{01}}}$$
$$= M_{00} \cdot M_{01}$$
$$F_1 = \overline{\overline{(A+B)} + \overline{(A+\overline{B})}} = \overline{\overline{(A+B)}} \cdot \overline{\overline{(A+\overline{B})}}$$
$$= (A+B) \cdot (A+\overline{B})$$

Functions of three variables are sufficiently complex to justify describing them with a truth table augmented by minterms and maxterms. This is done in example 10.5.

Example 10.5

Given the truth table, obtain the function in both SOP and POS forms (called *canonical forms*).

$XY\ Z$	F_4	m	M
0 0 0	0	–	$X+Y+Z$
0 0 1	1	$\overline{X} \cdot \overline{Y} \cdot Z$	–
0 1 0	1	$\overline{X} \cdot Y \cdot \overline{Z}$	–
0 1 1	0	–	$X+\overline{Y}+\overline{Z}$
1 0 0	1	$X \cdot \overline{Y} \cdot \overline{Z}$	–
1 0 1	0	–	$\overline{X}+Y+\overline{Z}$
1 1 0	0	–	$\overline{X}+\overline{Y}+Z$
1 1 1	1	$X \cdot Y \cdot Z$	–

First, fill in the columns with the corresponding minterms (where F_4 is 1) and maxterms (where F_4 is 0).

Next, combine the minterms and maxterms.

$$SOP: \ F_4 = \overline{X} \cdot \overline{Y} \cdot Z + \overline{X} \cdot Y \cdot \overline{Z}$$
$$+ X \cdot \overline{Y} \cdot \overline{Z} + X \cdot Y \cdot Z$$
$$POS: \ F_4 = (X+Y+Z) \cdot (X+\overline{Y}+\overline{Z})$$
$$\cdot (\overline{X}+Y+\overline{Z}) \cdot (\overline{X}+\overline{Y}+Z)$$

(Note that the truth table entry of maxterms was the actual maxterm instead of the NOTed maxterms in the general table for combinations of two variables.

For the function illustrated in example 10.5, there were the same number of 1's as 0's. This was a coincidence. $A \cdot B \cdot C$ has one 1 and seven 0's, while $A + B + C$ has seven 1's and only one 0.

F. CANONICAL SOP REALIZATION, NAND-NAND

The term *realization* indicates a logic diagram using gates. The SOP expression of example 10.5 indicates a two-level AND-OR realization, requiring three 3-input AND gates. The outputs from those go to a 4-input OR gate as indicated in figure 10.9. There it is assumed that X, Y, and Z (the original variables) and their complements $\overline{X}$, $\overline{Y}$ and $\overline{Z}$ are all available.

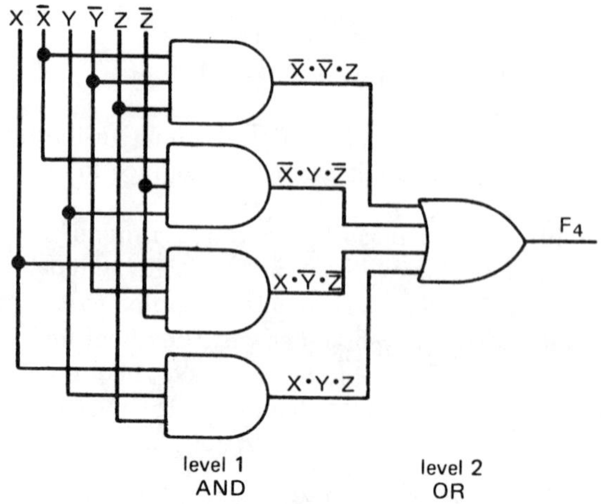

Figure 10.9 Realization of example 10.5

In figure 10.9, the connection points are indicated with a dot; non-connecting crossings of lines do not have the dot. This is standard practice in digital circuits.

There are two levels of logic in figure 10.9. The first is the AND level, and the second is the OR level.

It is possible to achieve the same two-level SOP realization using only NAND gates. This is best seen in a truth table for the NAND gate.

Table 10.9
NAND Gate Truth Table

A	B	$\overline{A \cdot B}$	$\overline{A} + \overline{B}$
0	0	1	1
0	1	1	1
1	0	1	1
1	1	0	0

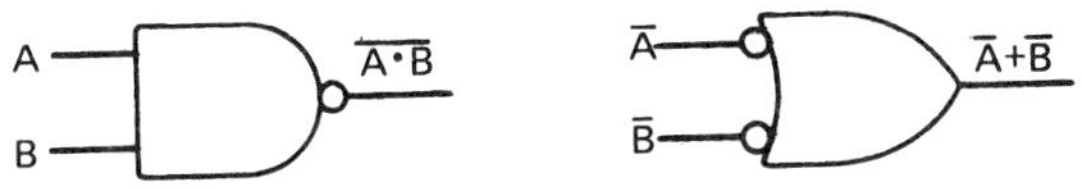

Figure 10.10 Two equivalent symbols for NAND gate

De Morgan's theorem is used to prove the equivalence.

$$\overline{A \cdot B} = \overline{A} + \overline{B} \qquad 10.47$$

Thus, a NAND gate has two representations—an AND gate with a *bubble* on the output, and an OR gate with bubbles on the input as indicated in figure 10.10. The bubbles are (inverting) NOT functions.

The NAND gate at level 1 in figure 10.10 has the inverting bubble on the output, while the NAND gate at level 2 has bubbles on the input. The result is a cancellation of inversions.

A two-level NAND-NAND realization for the function of example 10.5 is shown in figure 10.11. This is cannonical AND-OR realization because it is composed entirely of minterms. It is not a minimum realization since the function can be implemented using fewer gates. (In fact, the function is the XOR of X, Y, and Z. As such it can be realized using two 2-input XOR gates.)

In general, logic functions are minimized for gate logic, resulting in circuit simplification. For minimized functions, it is common for one or more variables to be brought in at the second level (the OR input level). The variables must be inverted (i.e., put through a NOT gate) at this level if they are not also available in the complement.

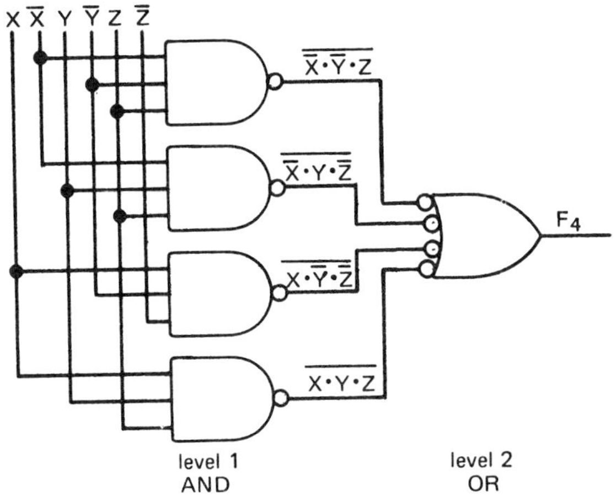

Figure 10.11 NAND-NAND realization of the AND-OR function

G. CANNONICAL POS REALIZATION, NOR-NOR

The product of sums form of a logical function is realized using a first-level OR and second level AND, or an OR-AND realization involving the maxterms of the function. For example, consider the POS form of the function F_4 of example 10.5.

$$F_4 = (X+Y+Z)\cdot(X+\overline{Y}+\overline{Z})\cdot(\overline{X}+Y+\overline{Z})\cdot(\overline{X}+\overline{Y}+Z)$$

The NOR function equivalents from De Morgan's theoem are

$$A \text{ NOR } B = \overline{A+B} = \overline{A} \cdot \overline{B}$$

Thus, the NOR gate can be represented as an OR gate with a bubble on the output, or as an AND gate with bubbles on the input. In the OR-AND realization, as shown in figure 10.12, the first level can be thought of as an OR level, and the second as an AND level. This is apparent since the two bubbles between the output of the OR stage and the input of the AND stage cancel each other.

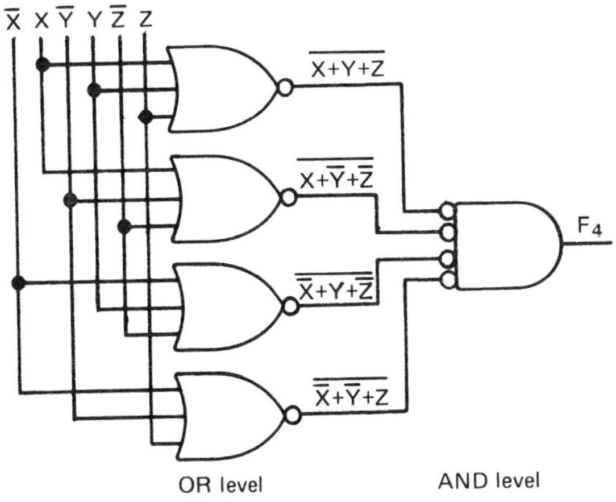

Figure 10.12 OR-AND realization of a POS function using NOR-NOR logic

2 LOGICAL MINIMIZATION TECHNIQUES

The primary method of minimizing logical functions depends on the *theorem of logical adjacency*:

$$A \cdot B + A \cdot \overline{B} = A \qquad 10.48$$

This can be generalized for functions as well as for variables.

$$F \cdot B + F \cdot \overline{B} = F \qquad 10.49$$

$$F \cdot G + F \cdot \overline{G} = F \qquad 10.50$$

F and G are functions, and B is a simple logical variable. Equations 10.49 and 10.50 indicate a method of minimizing minterm expressions (SOP). A simple expansion of maxterm expressions shows that

$$(A+B) \cdot (A+\overline{B}) = A \cdot A + A \cdot \overline{B} + B \cdot A + B \cdot \overline{B}$$
$$= A + A \cdot (B + \overline{B}) + 0 = A \qquad 10.51$$

In general,

$$(F+G) \cdot (F+\overline{G}) = F \qquad 10.52$$

Equation 10.52 is another form of the logical adjacency theorem.

The logical adjacency theorem implies that minimization of SOP and POS functions can be achieved by inspecting the logical adjacency of minterms or maxterms. Adjacency can be determined from inspection of the truth table and by combining minterms (or maxterms) which differ by only one value.

For example, $A \cdot B \cdot C \cdot D = 1111$ differs from $\overline{A} \cdot B \cdot C \cdot D (= 0111)$, from $A \cdot \overline{B} \cdot C \cdot D (= 1011)$, from $A \cdot B \cdot \overline{C} \cdot D (= 1101)$, and from $A \cdot B \cdot C \cdot \overline{D} (= 1110)$, each by only the value of a binary variable. Therefore, any of the latter can be combined with $A \cdot B \cdot C \cdot D$ to eliminate the one variable which differs.

Example 10.6

A function consists of the maxterms $X + \overline{Y} + \overline{Z}$, $X + \overline{Y} + Z$, and $\overline{X} + Y + \overline{Z}$. Simplify the function.

These maxterms can be represented by the logical values of the variables when they are true.

$$X + \overline{Y} + \overline{Z} = 100$$
$$X + \overline{Y} + Z = 101$$
$$\overline{X} + Y + \overline{Z} = 010$$

The first two maxterms are adjacent since they differ in the true value of Z. Thus, they can be combined. The third term is not adjacent to either of the others, and it cannot be simplified by combining. The resulting function is

$$F = (X + \overline{Y}) \cdot (\overline{X} + Y + \overline{Z})$$

Truth tables with multiple input variables are not conducive to visualizing adjacencies. The Karnough map is much easier to work with.

The *Karnaugh map* is a representation of the input variable *space* for a logical function, arranged to visually present the logical adjacencies. There is a one-to-one relationship between the Karnaugh map and the truth table. The binary values associated with the input values in the truth table are converted to decimal values for convenience. These values are then placed in the Karnaugh map to indicate the corresponding location of minterms and maxterms.

decimal value	inputs A	B	output C
0	0	0	0
1	0	1	1
2	1	0	0
3	1	1	0

(a)

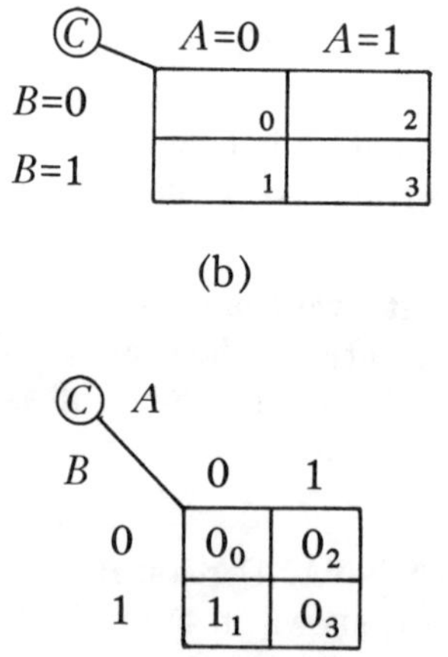

(c)

Figure 10.13 2-input Karnaugh map development

A 2-input Karnough map development appears in figure 10.13. Part (b) shows the Karnaugh map and the correspondence of the map locations to the truth table decimal equivalents of the binary number the input conditions require for each true condition. Part (c) shows the entries from the truth table, with 1's corresponding to the minterms of the function C and 0's to the maxterm values of the function.

decimal value	inputs A B C	output F
0	0 0 0	1
1	0 0 1	1
2	0 1 0	0
3	0 1 1	0
4	1 0 0	0
5	1 0 1	1
6	1 1 0	1
7	1 1 1	0

(a)

Ⓕ C \ AB	00	01	11	10
0	1_0	0_2	1_6	0_4
1	1_1	0_3	0_7	1_5

(b)

Ⓕ BC \ A	0	1
00	1_0	0_4
01	1_1	1_5
11	0_3	0_7
10	0_2	1_6

(c)

Figure 10.14 3-input Karnough map development

The development of two possible three-input Karnough maps is shown in figure 10.14. Part (b) shows one possible arrangement with the first two variables (AB) values across the top and the last (so-called *least-significant* because of its position as a binary bit) on the side. Part (c) shows a different arrangement which contains the same information arranged with the least significant variables down the side.

In Figure 10.14, (b), the decimal term 0 is adjacent to terms 1 and 2, while In (c) the 0 term is adjacent to terms 1 and 4. This seeming inconsistency can be accounted for by visualizing the maps as wrapping around a cylinder. In (b), the right and left columns meet, and adjacencies are obvious, (i.e., term 1 is adjacent to term 5 and term 4mys adjacent to term 0.) In (c), the cylinder axis is in the other direction, so the top and bottom rows are adjacent.

In the 3-input map, the numbering for the dimension containing two variables is not binary, as binary numbering does not insure logical adjacency. The numbering technique is called *unit distance coding,* and adjacent numbers differ by only one bit. This technique of indicating adjacency is used in the Karnaugh map for four inputs shown in figure 10.15. The adjacency is made obvious by repeating the first column on the right, and by repeating the first row on the bottom.

A 5-input Karnaugh map can be made by doubling one dimension on the 4-input map , as in figure 10.16, where adjacencies are somewhat difficult to perceive. For example, that term 17 and term 25 are adjacent could be easily overlooked.

Ⓕ CD \ AB	00	01	11	10	00
00	0	4	12	8	0
01	1	5	13	9	1
11	3	7	15	11	3
10	2	6	14	10	2
repeated 00	0	4	12	8	0

Figure 10.15 A 4-input Karnaugh map

Ⓖ DE \ ABC	000	001	011	010	110	111	101	100
00	0	4	12	8	24	28	20	16
01	1	5	13	9	25	29	21	17
11	3	7	15	11	27	31	23	19
10	2	6	14	10	26	30	22	18

Figure 10.16 Single layer 5-input Karnaugh map

Another representation of the 5-input Karnaugh map conceptually folds the right-hand side of the map of figure 10.16 under the left-hand side to make a three-dimensional map as in figure 10.17. The most significant variable (A) is indicated by dividing each square **diagonally** with $A = 0$ in the upper half and $A = 1$ in the lower half. This representation shows all adjacencies by the extension method of the 4-input map. Terms are adjacent if they are in the same square or if they are in adjacent squares in the same relative half (adjacent squares in the upper half, or adjacent squares in the lower half).

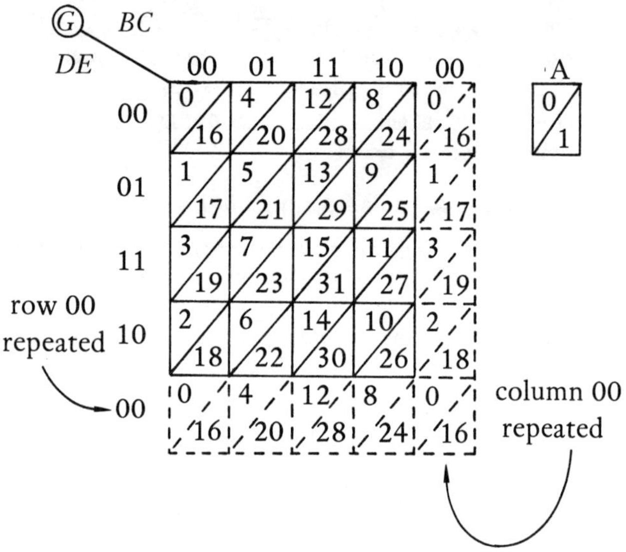

Figure 10.17 Three-dimensional Karnaugh map

A 5-input Karnaugh map is as complex as can be used practically. A 6-input map can be constructed by dividing each square of the 5-dimensional map in half again, but this becomes difficult to perceive, and all adjacencies cannot be shown. For more than 4 inputs, it is useful to achieve minimization by computer.

3 MINIMIZATION OF FUNCTIONS

As SOP and POS expressions are nontrivial only for functions of three or more inputs, minimization methods are needed. These minimization techniques guarantee a minimum number of gates and gate leads only for POS or SOP realizations. There is no guarantee that the reduced functions found are absolute minimizations. They are good starting places for small-scale logic designs. The techniques developed here are very useful in the design of flip-flop logic for sequential circuits.

For canonical forms with three-input variables, there is a total of eight possible terms, part of which correspond to minterms for the SOP realization. The remainder correspond to maxterms for a POS implementation. In an otherwise unrestricted situation, choose the cannonical form which results in the least number of packages. This generally corresponds to the minimum number of gate input leads. Unless otherwise restricted, try minimizing both the POS and the SOP realizations of the function.

The minimization procedure is given here.

Step 1: Construct a Karnaugh map from the truth table or given logic function.

Step 2: Combine adjacent terms into the largest possible groupings.

Step 3: Include a sufficient number of groupings to include each of the original terms in at least one group.

Step 4: Check the resulting function against the original truth table. It must have the same outputs for the same inputs.

This procedure is demonstrated by example 10.7.

Example 10.7

ABC	D
0 0 0	0
0 0 1	0
0 1 0	1
0 1 1	0
1 0 0	1
1 0 1	0
1 1 0	1
1 1 1	0

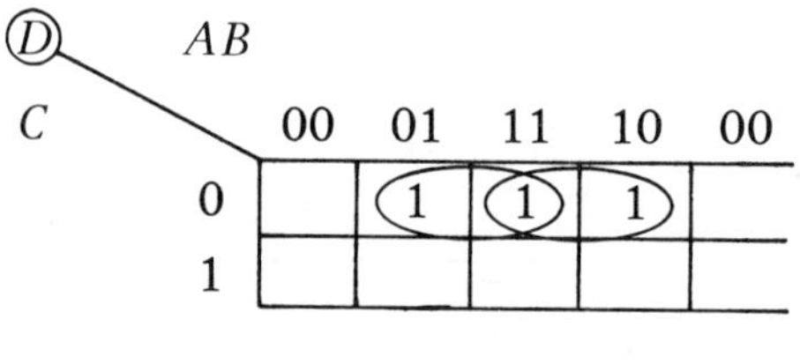

The function **D** (A,B,C) has the truth table shown. Obtain a minimal NAND-NAND or NOR-NOR realization of this function.

The function has only three 1's, so the SOP form is tried first. The minterms are entered in the map. Next, the adjacent terms are circled to indicate that they can be combined. With this function, the largest possible grouping is of two terms.

In general, terms can be grouped in twos, fours and eights (i.e., in powers of 2), as will be demonstrated later. Each grouping has eliminated the one variable that is not common to the group. In the left-hand group $C = 0$ and $B = 1$ is common. This is true when the combination

$\overline{C} \cdot B$ is true.

The right-hand group has $C = 0$ and $A = 1$ in common, which is true when $\overline{C} \cdot A$ is true. Thus, the function written with these two terms is

$$D = \overline{C} \cdot B + \overline{C} \cdot A$$

A	B	C	$B \cdot \overline{C}$	$A \cdot \overline{C}$	$\overline{C} \cdot B + \overline{C} \cdot A$
0	0	0	0	0	0
0	0	1	0	0	0
0	1	0	1	0	1
0	1	1	0	0	0
1	0	0	0	1	1
1	0	1	0	0	0
1	1	0	1	1	1
1	1	1	0	0	0

This can be checked with a truth table. The reduced function has the same outputs as the original function. The realization using NAND-NAND logic, assuming all inputs and their complements are available, is shown.

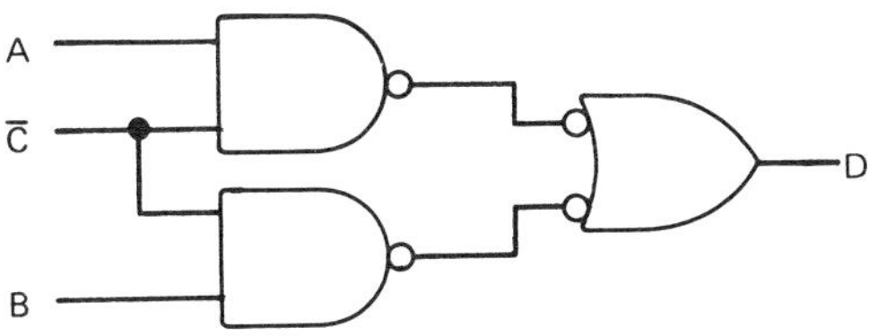

The NAND-NAND function can be further simplified, but it would no longer be a NAND-NAND function. As will be shown, the simplification is the direct result of the POS or maxterm formulation. Because the maxterm formulation is one step more abstract than the minterm formulation, the process will be described here in detail.

To obtain the maxterms, enter the zeros from the truth table as shown. The maxterms are made up of the OR'd complements of the input variable values.

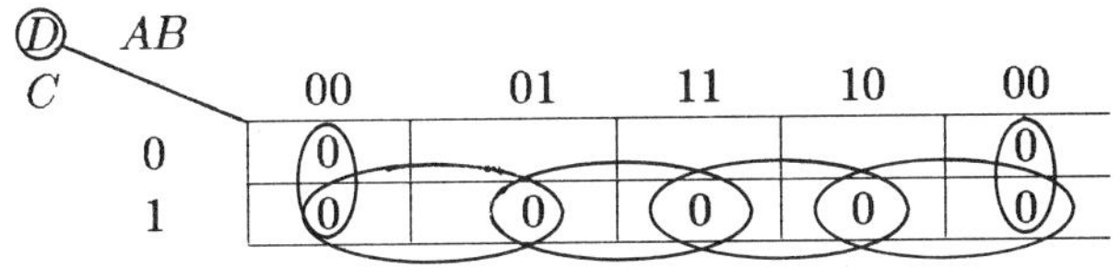

This is explained in terms of a second Karnaugh map. The OR'd complements of the left-hand group contain everything but the group of zeros $M = A + B$.

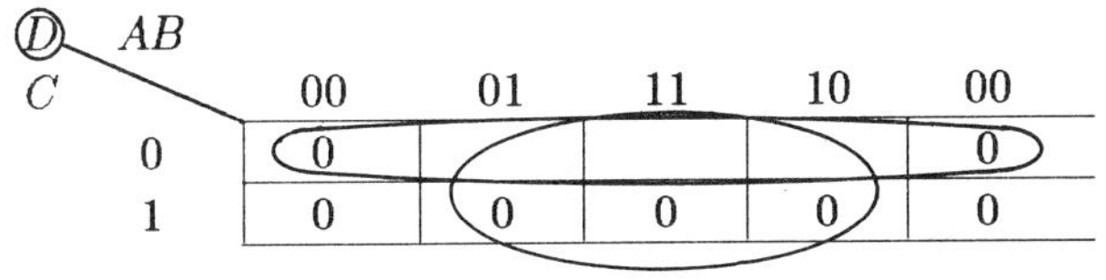

The lower row of zeros can be grouped in four different adjacencies: 001–011, 011–111, 111–101 and 101–001. If arbitrarily grouped in adjacencies, the result is $0X1$, $X11$, $1X1$, and $X01$. X in each case is the location of the variable which was eliminated by the groupings of two terms.

The first and third of these differ only in A, so that they can be grouped to $XX1$. Also, the second and fourth differ only in B, so they can be grouped together as $XX1$. Thus, the lower row of zeros can be grouped together with their common variable, C. The maxterm associated with C is $\overline{C}$. This is interpreted as everything except the row of zeros along the bottom.

The second Karnaugh map, then, contains two OR terms: $A + B$ and $\overline{C}$. The function **D** is the AND of these two OR's.

$$\mathbf{D} = (\overline{C}) \cdot (A + B)$$

This can be realized in NOR-NOR logic. Because the $\overline{C}$ must be introduced at the AND (second) level, it must be in complement form (i.e., NOT $\overline{C}$ or C).

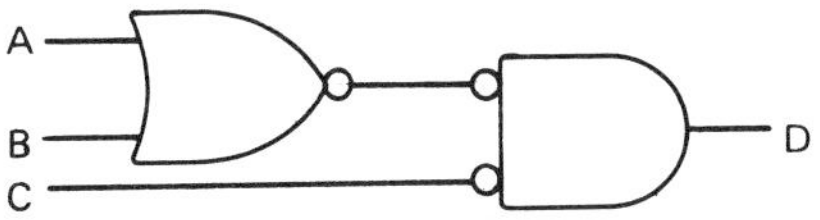

The verification of the truth table is omitted here.

4 MULTIPLE ADJACENCIES AND VEITCH DIAGRAMS

Multiple adjacencies are important in function minimization. They have been briefly introduced in the previous example. The procedure can be formalized by the extended use of the logical adjacency theorem.

$$A \cdot B \cdot \; \overline{C} \cdot \overline{D} + A \cdot B \cdot C \cdot \overline{D} + A \cdot B \cdot \overline{C} \cdot D + A \cdot B \cdot C \cdot D = A \cdot B \qquad 10.53$$

It is useful to introduce another way of viewing the Karnaugh map —in terms of the variables by name rather than by value. First, redefine the rows and columns of a three-input Karnaugh map in terms of the input values which select them. This is shown in figure 10.18(a).

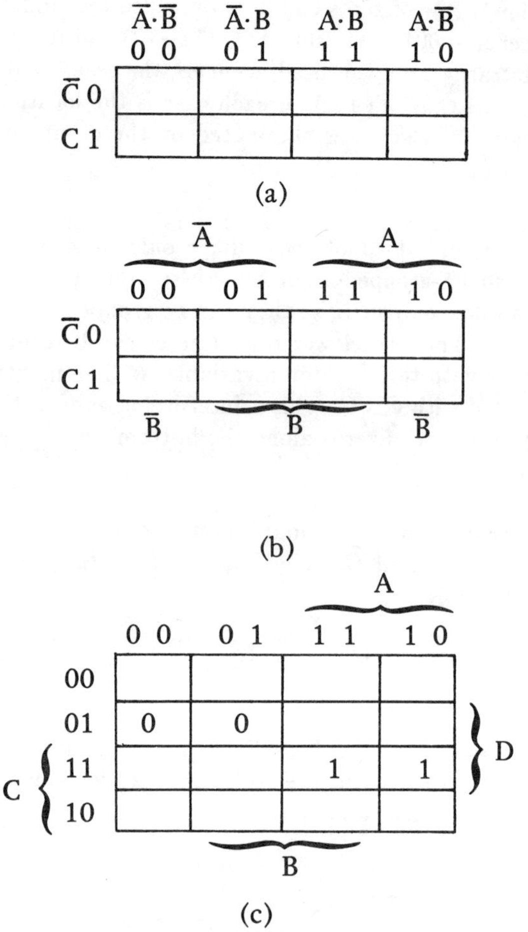

Figure 10.18 (a) 3-inputs by name, (b) Veitch diagram notation, (c) 4-input Veitch diagram.

When $A = 0$, A is false, and the true condition is $\overline{A}$. Thus, wherever A is noted as a 0 on the Karnaugh map, the corresponding value of $\overline{A}$ is true, or $\overline{A} = 1$. Thus, the *Veitch diagram* is simply a Karnaugh map which tells where a variable is true. Where it is not true, the complement is true.

This is an extremely useful notion in interpreting the Karnaugh map for purposes of establishing reduced functions. In the notation indicated in figure 10.18 (b), the location 011 is seen to be $\overline{A}\cdot B\cdot C$ in minterm realization, and $A+\overline{B}+\overline{C}$ in maxterm realization.

A four-input Karnaugh map with Veitch diagram enhancement is shown in figure 10.18 (c). The entry of the maxterms $A+\overline{B}+C+\overline{D}$ and $A+B+C+\overline{D}$ can be seen as the zero's at $\overline{A}\cdot B\cdot \overline{C}\cdot D = 0101$ and at $\overline{A}\cdot \overline{B}\cdot \overline{C}\cdot D = 0001$ as indicated on the diagram. These can be combined to give the maxterm $A+C+\overline{D}$, eliminating the B term.

Likewise, the two one's at positions 1111 and 1011 can be interpreted as $A\cdot B\cdot C\cdot D$ and $A\cdot \overline{B}\cdot C\cdot D$ minterms, respectively. These terms can be combined as $A\cdot C\cdot D$. To do this, note that both terms have A, C, and D in common.

The Veitch diagram is equivalent to the Karnaugh map, but it permits the immediate entry of 1's and 0's from a logical expression, and it further allows the direct writing of a logical expression.

In general, it is recommended, that Karnaugh maps be constructed to include both the minterm information (values) and the Veitch diagram information (names).

As the position of a particular variable in a truth table is random, all locations of combinations of different variables in the K-V (Karnaugh-Veitch) map should be inspected. Figure 10.19 shows the placements of the groups of four terms corresponding to A, $\overline{A}$, B, $\overline{B}$, C, and $\overline{C}$ for three-input functions. (These are shown as minterms, but maxterms replace the 1's and 0's.)

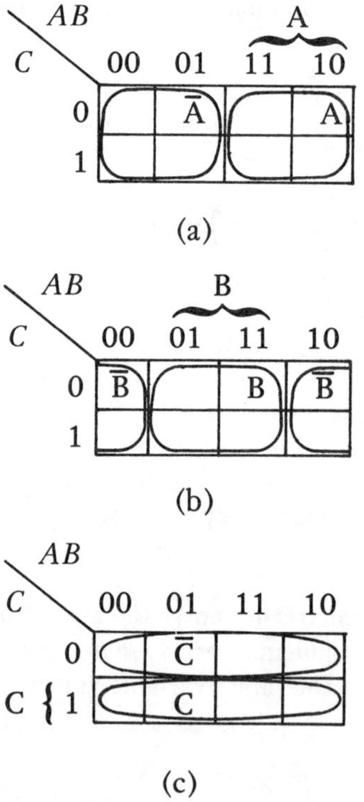

Figure 10.19 (a) A and $\overline{A}$, (b) B and $\overline{B}$, and (c) C and $\overline{C}$.

With four variable inputs, the groupings (also called *encirclements*) are more numerous and, therefore, more complex. The obvious groups of four terms are shown in figure 10.20. For the arbitrary definition of variables to this point, these correspond to (a) $\overline{A}\cdot\overline{B}$, $\overline{A}\cdot B$, $A\cdot B$, and $A\cdot\overline{B}$; (b) $\overline{C}\cdot\overline{D}$, $\overline{C}\cdot D$, $C\cdot D$, and $C\cdot\overline{D}$; and (c) $\overline{A}\cdot\overline{C}$, $\overline{A}\cdot C$, $A\cdot C$, and $A\cdot\overline{C}$. The corresponding maxterms involve the complements of the variables in each case.

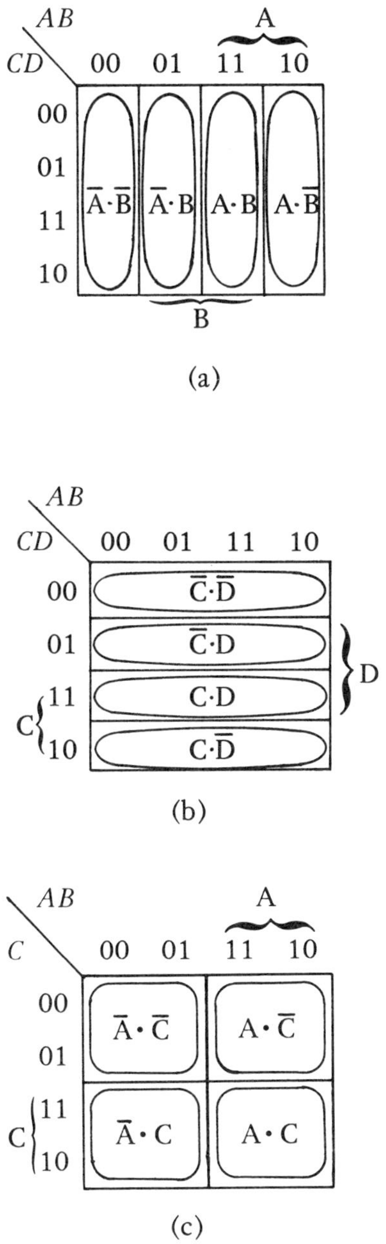

Figure 10.20 (a) functions of four variables dependent only on *A* and *B*, (b) functions dependent only on *C* and *D*, and (c) functions dependent only on *A* and *C*.

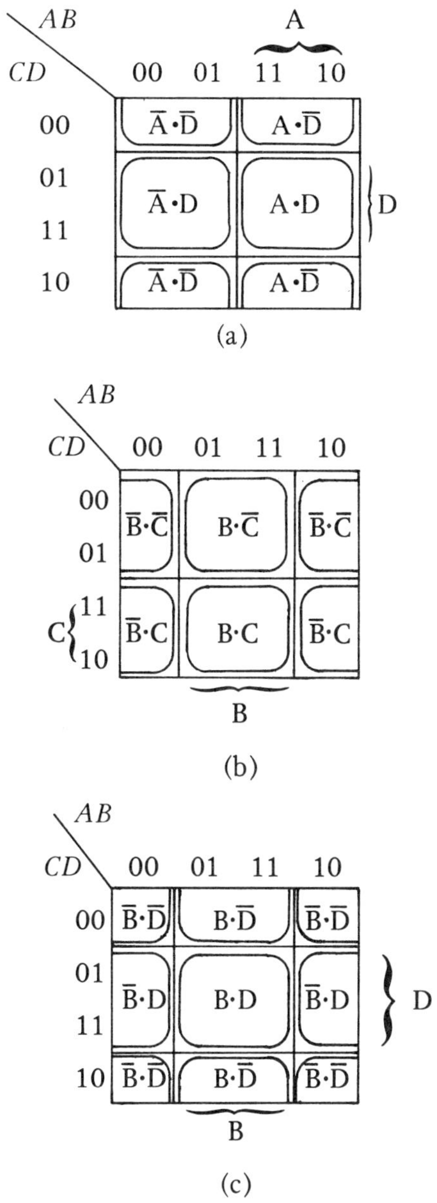

Figure 10.21 (a) functions of four variables dependent only on *A* and *D*, (b) functions dependent only on *B* and *C*, and (c) functions dependent only on *B* and *D*.

The three remaining combinations (shown in figure 10.21) of four terms are more complicated, as they require wrapping a cylinder around the K-V map. Those combinations which eliminate all but *A* and *D* require a cylinder placed so that the top and bottom of the map are joined (a); those removing all but *B* and *C* as variables require a cylinder connecting the left and right edges (b); and those removing all but *B* and *D* as variables require a cylinder which is more like a sphere, in that it has to connect simultaneously the top and bottom and the sides. This joins the four corners (c).

For functions of four variables with eight combinable adjacencies, the results are the values given on the Veitch diagram, or their complements: A, $\overline{A}$, B, $\overline{B}$, C, $\overline{C}$, D, and $\overline{D}$.

For four-input minterm realizations, the minimization involves the recognition of combinable terms in all possible combinations. An interesting case is where there are inputs which cannot occur and are, therefore, called *don't care* inputs.

An example of this is a binary coded decimal seven-segment display. There are four inputs which can vary in decimal values from 0 through 9. The input signals reflect this limitation, and the binary combinations of 1010 (10_{10}) through 1111 (15_{10}) do not occur.

inputs ABCD	a	b	c	d	e	f	g	decimal number
0000	1	1	1	1	1	1	0	0
0001	0	1	1	0	0	0	0	1
0010	1	1	0	1	1	0	1	2
0011	1	1	1	1	0	0	1	3
0100	0	1	1	0	0	1	1	4
0101	1	0	1	1	0	1	1	5
0110	1	0	1	1	1	1	1	6
0111	1	1	1	0	0	0	0	7
1000	1	1	1	1	1	1	1	8
1001	1	1	1	1	0	1	1	9
1010	X	X	X	X	X	X	X	not used
1011	X	X	X	X	X	X	X	not used
1100	X	X	X	X	X	X	X	not used
1101	X	X	X	X	X	X	X	not used
1110	X	X	X	X	X	X	X	not used
1111	X	X	X	X	X	X	X	not used

(a)

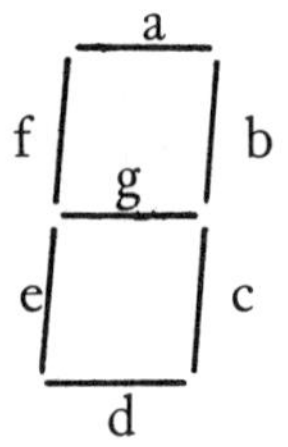

(b)

CD \ AB	00	01	11	10
00	0	4	X_{12}	8
01	1	5	X_{13}	9
11	3	7	X_{15}	X_{11}
10	2	6	X_{14}	X_{10}

(c)

Figure 10.22 Seven-segment display (a)truth table, (b) identification of segments, (c) Karnaugh map with X's to indicate don't-care conditions

The *don't-care* conditions can be used to minimize the logical function by assigning values of true or false as is advantageous. For example, take the segment a from the truth table of figure 10.22. Enter the one's and zero's from there into the Veitch diagram on figure 10.23.

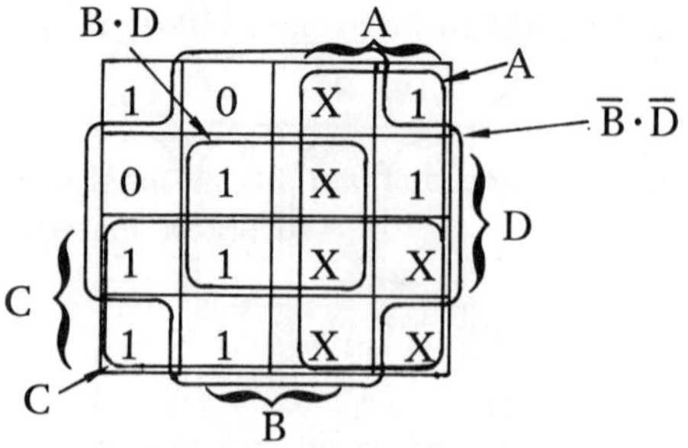

(a)

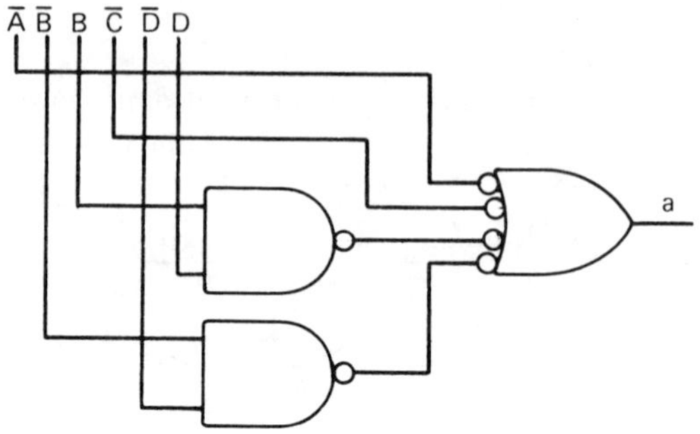

(b)

Figure 10.23 (a) Veitch diagram for segment a, (b) NAND-NAND realization of segment a logical function

The reduced logical expression for a must include all of the one's as minterms. This means that each must be included in at least one chosen group. The minimum function will have the minimum number of reduced minterm products. The unreduced minterms are referred to by number, and they have been enumerated in the Karnaugh map of figure 10.22(c).

If the *don't-care* of minterm 10 is treated as a *one*, the four-corners grouping ($\overline{B} \cdot \overline{D}$) takes care of minterms 0, 2, 8 and 10.

The *don't-care's* of minterms 13 and 15 are treated as one's, to group with minterms 5 and 7 in $B \cdot D$.

When the *don't-care's* of 11 and 14 are added to the list of one's, minterms 2, 3 and 6 can be covered by C. This also covers minterm 7 which had already been covered. There is no reason to avoid multiple coverages of one term if it reduces the complexity.

The only remaining uncovered minterm is 9. By making the one remaining *don't-care*, 12, a *one*, the total region A can be included to cover 9. In doing so, this also covers terms that were previously covered.

The resulting expression for segment a is

$$a = A + \overline{B} \cdot \overline{D} + B \cdot D + C \qquad 10.54$$

This is realized by the NAND-NAND circuit shown in figure 10.23(b). Because A and C are introduced at the second (OR) level, they must be present in their complements.

In this illustration, all of the *don't-care's* were changed to one's. This is not necessary in every case.

An alternative to minterm realization is the *maxterm method,* resulting in *NOR-NOR logic.* For purposes of comparison, the segment a of the seven segment display is used. The Veitch diagram is shown in figure 10.24.

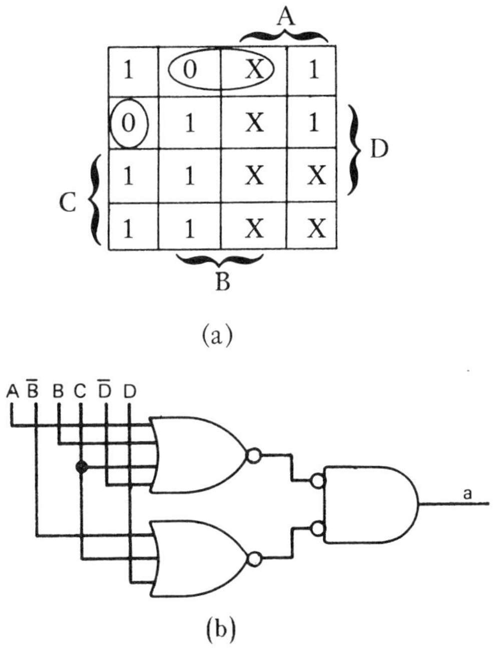

Figure 10.24 POS realization of the a segment (a) Veitch diagram, (b) NOR-NOR realization

As the two zero's cannot be combined, there will be two maxterms. The term at minterm position 1 cannot be combined with the other *zero* or *don't-care.* The maxterm is made up of the complements of the position's minterm

$$A + B + C + \overline{D}$$

The zero at position 4 can be combined with the *don't-care* at position 12. The resulting maxterm is the complement of the combined minterm for the group

$$\overline{B} + C + D$$

Thus the maxterm expression is

$$a = (A + B + C + \overline{D}) \cdot (\overline{B} + C + D) \qquad 10.55$$

Only one *don't-care* was changed to a zero, so the remaining are treated as one's. The NOR-NOR realization of the reduced function is shown.

5 EXCLUXIVE OR AND EXCLUSIVE NOR LOGIC

A. INTRODUCTION

The XOR function is also called a *ring sum,* and it can be considered as an *odd parity function.* For example, *A XOR B* is true only when one of the two variables is true. *A XNOR B* is true when both variables are the same.

When a third variable is included, as in *A XOR B XOR C,* the operation must be done in steps. *A XOR B* is taken first; then *XOR* is applied to that result with C. The order in which this is done is of no importance. *A XOR B XOR C* will be true whenever there is an odd number of trues present

$$A \oplus B \oplus C = (A \cdot B \cdot C) + (\overline{A} \cdot \overline{B} \cdot C) + (\overline{A} \cdot B \cdot \overline{C}) + (A \cdot \overline{B} \cdot \overline{C}) \qquad 10.56$$

It also can be seen from simple permutation theory that

$$\begin{aligned} A \oplus B \oplus C &= \overline{A} \oplus \overline{B} \oplus C = \overline{A} \oplus B \oplus \overline{C} \\ &= A \oplus \overline{B} \oplus \overline{C} \end{aligned} \qquad 10.57$$

The XNOR function is the complement of the XOR function and is an *even parity function.*

$$\begin{aligned} A \odot B \odot C &= \overline{A \oplus B \oplus C} = \overline{A} \oplus \overline{B} \oplus \overline{C} \\ &= \overline{A} \oplus B \oplus C = A \oplus \overline{B} \oplus C \\ &= A \oplus B \oplus \overline{C} \end{aligned} \qquad 10.58$$

In general, an XOR function of any number of variables can have only two values. Either there will be an odd or even number of trues. For any XOR function it is necesary only to count the number of complemented variables in the expression. If this number is even, then the function is true for odd parity. If the number of complemented variables is odd, then the function is true for even parity.

For example, the function $A \oplus B \oplus C \oplus \overline{D}$ is true with even parity when an even number of the variables are true. When $ABCD$ = 0000, 0011, 0101, 1001, 1010, 1100, 0110 or 1111, the function is true.

B. XOR MAPS

The Karnaugh map and Veitch diagram for exclusive-OR functions using all map variables result in a checkerboard pattern as shown in figure 10.25. The alternating one's and zero's result in even parity functions.

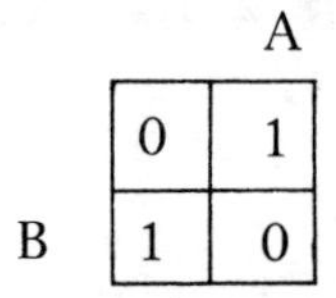

A⊕B

(a)

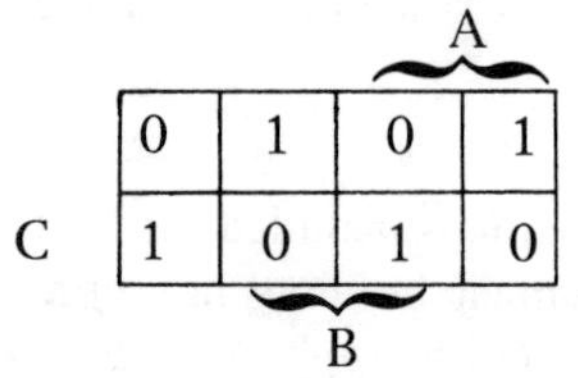

A⊕B⊕C

(b)

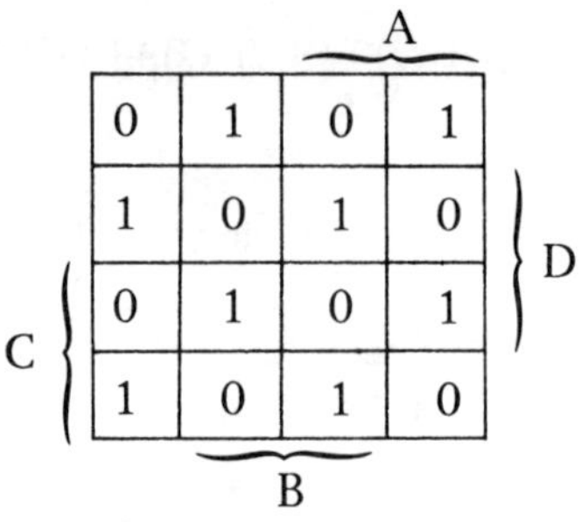

A⊕B⊕C⊕D

(c)

Figure 10.25 (a) XOR patterns for 2-variable, 3-variable, and 4-variable maps

It is useful to recognize XOR functions with fewer than all of the map variables. For example, the XOR functions of two variables on a 3-variable map are shown in figure 10.26.

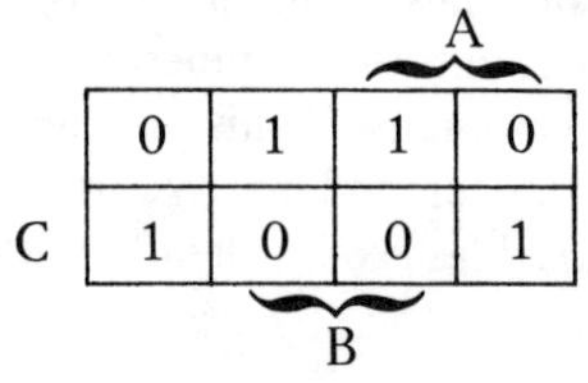

B⊕C

(a)

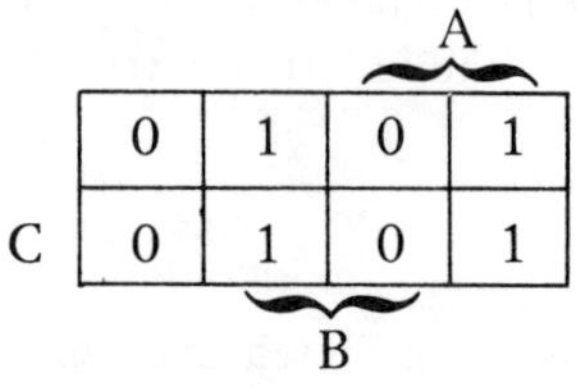

A⊕B

(b)

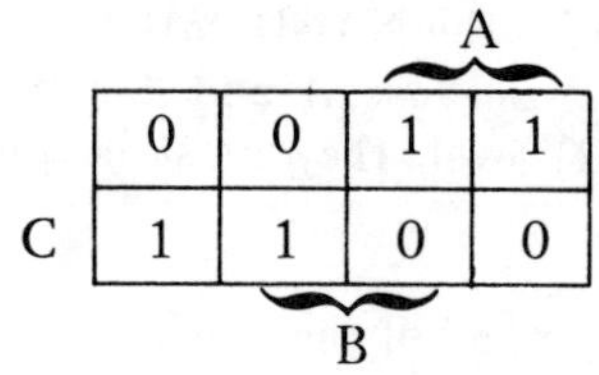

A⊕C

(c)

Figure 10.26 XOR patterns for two variable XOR functions on a 3-variable map

The XOR function of two variables results in groupings of two one's and two zero's. Patterns for a 4-variable map involving XOR of three variables also results in groupings of one's and of zero's pairs, as is shown in figure 10.27.

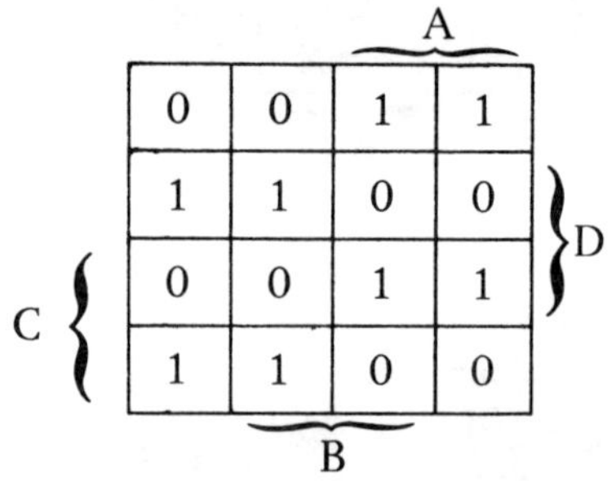

A⊕C⊕D

(a)

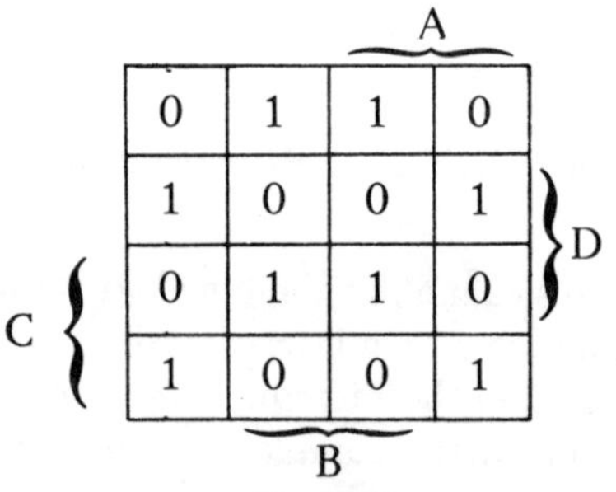

B⊕C⊕D

(b)

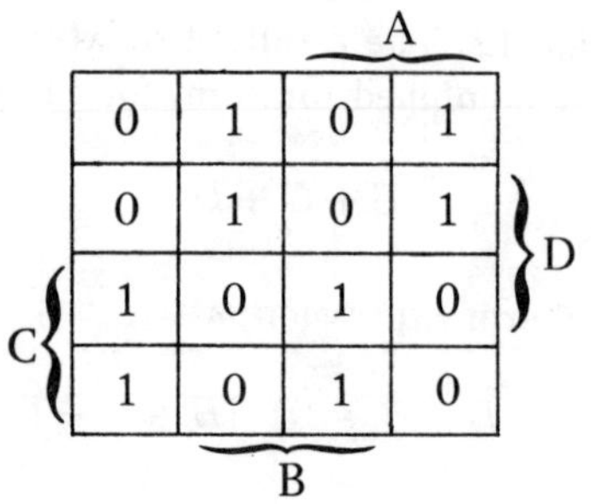

A⊕B⊕C

(c)

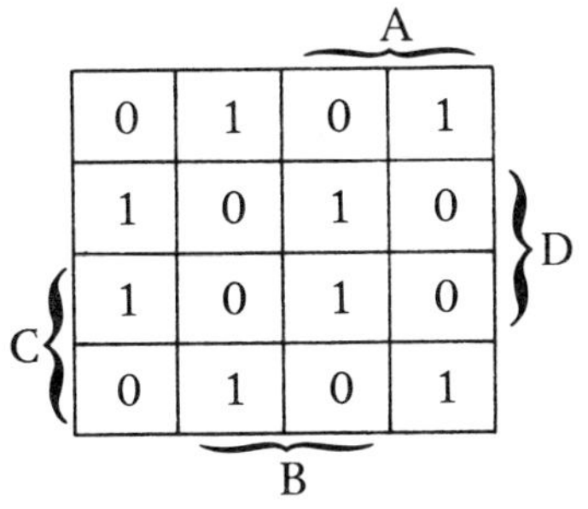

A⊕B⊕D

(d)

Figure 10.27 XOR patterns for functions of three variables on a 4-variable map

More complex, but to some extent more useful, are XOR functions of two variables on a 4-variable map. These result in groupings of four one's and four zero's as shown in figure 10.28.

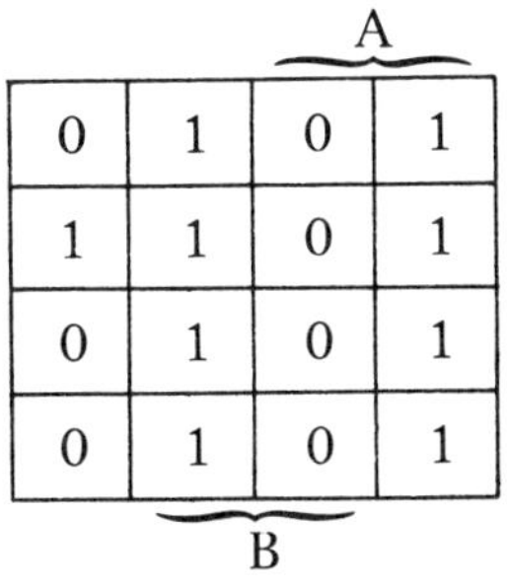

A⊕B

(a)

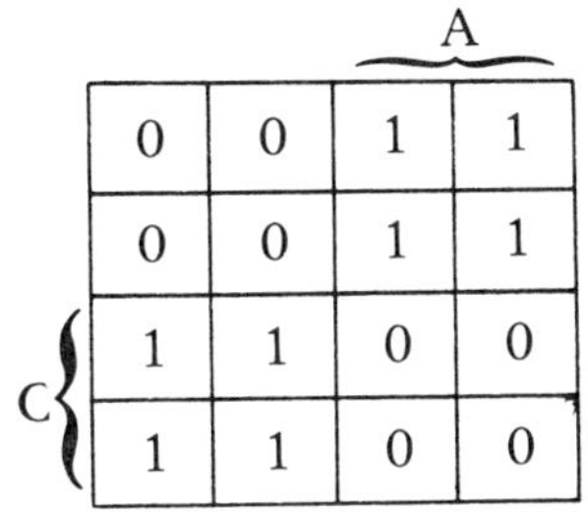

A⊕C

(b)

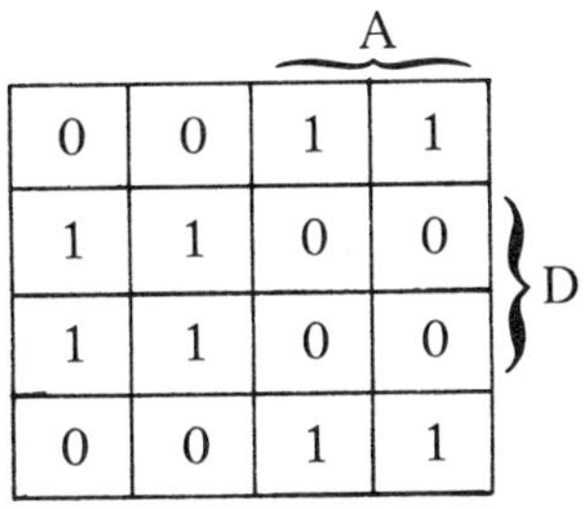

A⊕D

(c)

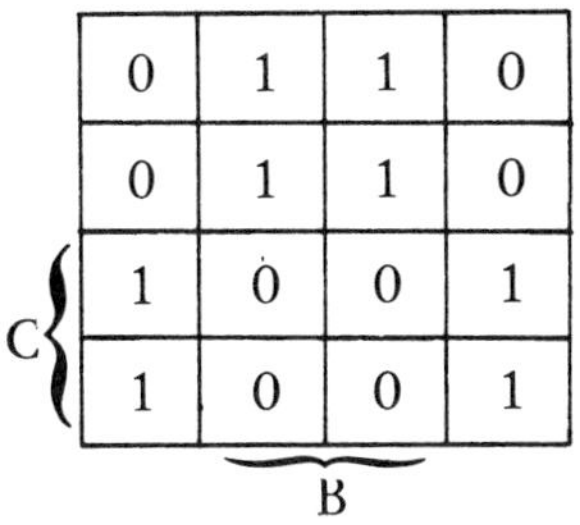

B⊕C

(d)

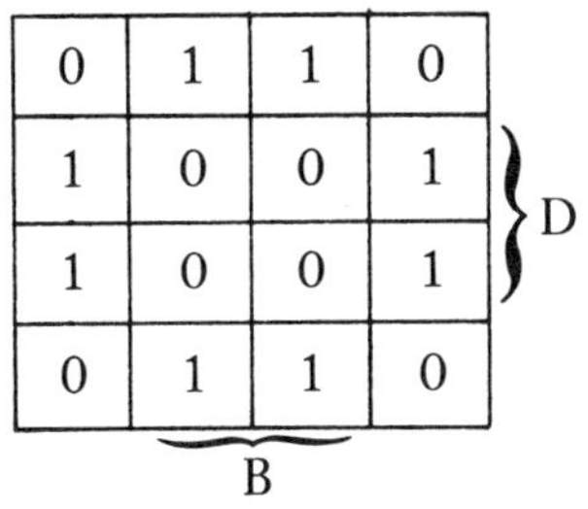

B⊕D

(e)

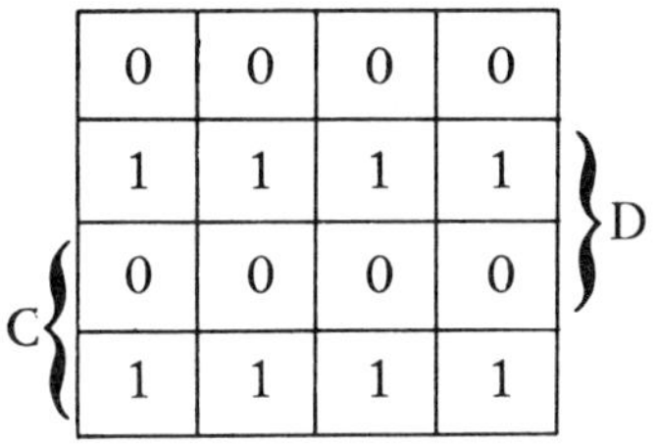

C⊕D

(f)

Figure 10.28 XOR patterns for functions of two variables on a 4-variable map

C. XOR WITH MINTERMS

XOR functions can be included with minterms. Additional logic levels are produced but with the further minimization of gates and gate leads. If the XOR or $\overline{\text{XOR}}$ function involves only two terms, it can be implemented with 2-input XOR or $\overline{\text{XOR}}$ gates. The usefulness of these gates depends on recognizing an XOR function from the map or from the recognition of such a term in the minterm expression (i.e., a term $\overline{A}\cdot B + A\cdot\overline{B}$ recognized as XOR, or $A\cdot B+\overline{A}\cdot\overline{B}$ recognized as XNOR).

An example of this is the segment *a* in a seven-segment display. This has the logical function

$$a = A + (\overline{B}\cdot\overline{C} + B\cdot C) + C$$

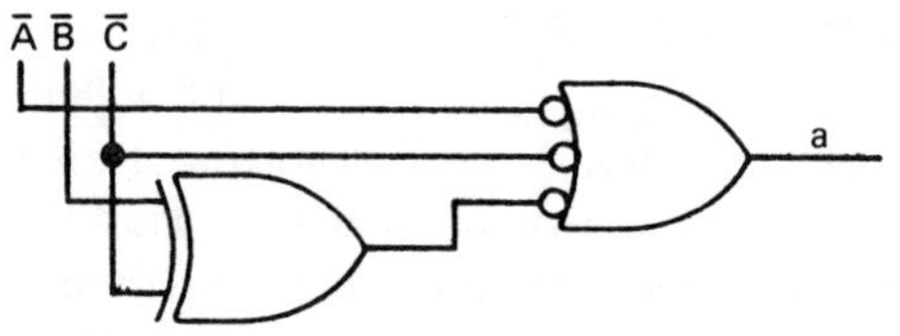

Figure 10.29 Using an XOR gate to further minimize SOP realization

This can be minimized further by using an XOR gate at the first logic level since the function must be complemented at the second logic level. Figure 10.29 illustrates this. Here $\overline{B} \oplus \overline{C}$ is an odd parity function and is, therefore, equivalent to $B \oplus C$. Also, the required function was the even parity $B \oplus C$. At the OR level of NAND-NAND logic, however, it is necesary to invert the inputs.

Example 10.8

The logical function which drives segment d of the seven- segment display of figure 10.22 requires a minterm or SOP function of

$$d = A + \overline{B} \cdot C + \overline{B} \cdot \overline{D} + B \cdot \overline{C} \cdot D + C \cdot \overline{D}$$

Realization of this function requires a very awkward set of gates. An 8-input NAND is required, as there are no 5-input NAND gates available. This function is not an obvious candidate for the use of an XOR realization. However, a careful study of the Karnaugh map, as indicated in figure 10.30, shows that the function can be realized with a 4-input XOR, an AND gate, and an OR gate. The implementation takes advantage of the various odd-parity arrangements, providing the required odd-parity to the complementing input of the OR level.

(d) CD \ AB	00	01	11	10
00	1	0	X	1
01	0	1	X	1
11	1	0	X	X
10	1	1	X	X

=

1	0	1	0
0	1	0	1
1	0	1	0
0	1	0	1

+

1	0	0	1
0	0	0	0
0	0	0	0
1	0	0	1

$$d = \overline{A \oplus B \oplus C \oplus D} + \overline{B}\,\overline{D}$$

(a)

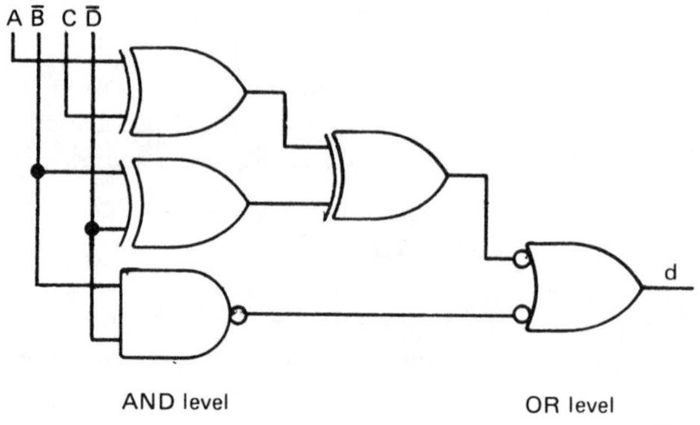

(b)

Figure 10.30 Karnaugh map NAND-NAND implementation (with XOR gates substituted) for segment d

With this type of implementation, there is an additional level of logic added. The extra level slows down the system due to gate delays. However, it requires a smaller number of integrated circuit packages.

D. XOR WITH MAXTERMS

XOR or XNOR gates usually can be substituted for either minterms or maxterms at the expense of increased gating levels. With maxterms it is, as usual, more difficult to visualize the process. For example, consider segment f of the seven-segment display, which has a maxterm representation of

$$f = (A + B + \overline{D}) \cdot (B + \overline{C}) \cdot (\overline{C} + \overline{D}) \qquad 10.59$$

XOR or XNOR terms could be substituted, depending on recognition of basic patterns. Here, the substitution is at the OR level. Careful examination of the Karnaugh map of f, shown in figure 10.31, reveals that the function is the sum of *A XOR B* and *C XNOR D*, except for rows C AND D which must be eliminated. This is accomplished by ORing the two XOR functions and ANDing the result with $\overline{C}$ OR $\overline{D}$. The resulting circuit is not superior to a straight NOR-NOR circuit. However, it requires only 2-input gates. This might be advantageous in some cases.

(f) CD \ AB	00	01	11	10
00	1	1	X	1
10	0	1	X	X
11	0	0	X	X
10	0	1	X	X

$$= \left\{ \begin{array}{|c|c|c|c|} \hline 0&1&0&1\\ \hline 0&1&0&1\\ \hline 0&1&0&1\\ \hline 0&1&0&1\\ \hline \end{array} + \begin{array}{|c|c|c|c|} \hline 1&1&1&1\\ \hline 0&0&0&0\\ \hline 1&1&1&1\\ \hline 0&0&0&0\\ \hline \end{array} \right\} \cdot \begin{array}{|c|c|c|c|} \hline 1&1&1&1\\ \hline 1&1&1&1\\ \hline 0&0&0&0\\ \hline 1&1&1&1\\ \hline \end{array}$$

$$f = (A \oplus B + \overline{C} \oplus D) \cdot (\overline{C} + \overline{D})$$

Figure 10.31 Maxterm (POS) implementation with XOR subsitution of XOR functions at the first logic level for segment f of the seven-segment display.

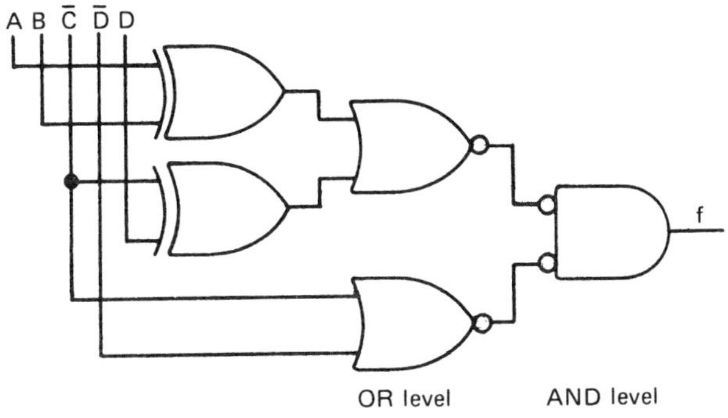

Figure 10.32 XOR gate realization for segment f

6 DATA SELECTORS

A. INTRODUCTION

Data selectors, also called *multiplexers* and abbreviated MUX, are useful for a number of digital functions. These include multiplexing and design of central processor units.

The simplest data selector involves one *select input* and two function inputs, as shown in figure 10.33. The select must be low to allow I_0 to pass through to the output Y; it must be high to allow I_1 to pass through. In effect, the select is acting like a control of a two-position switch. The schematic representation of the 2-input data selector is shown in figure 10.33(c). Data selectors are available commercially in single packages with four or 2-input (*quad*) versions with a common select for all.

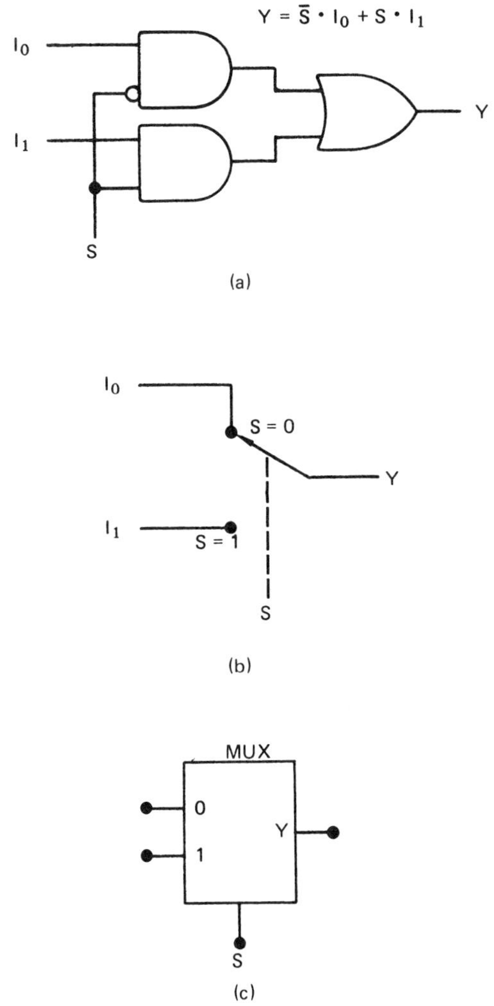

Figure 10.33 (a) 2-input data selector logic diagram, (b) switch analogy, (c) circuit representation

A 4-input data selector has the logic diagram shown in figure 10.34. Note that the input selected corresponds to the line with the decimal number equivalent to the binary number S_1S_0.

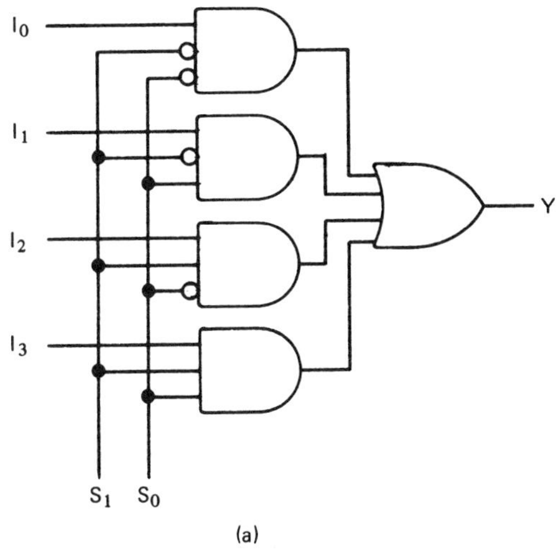

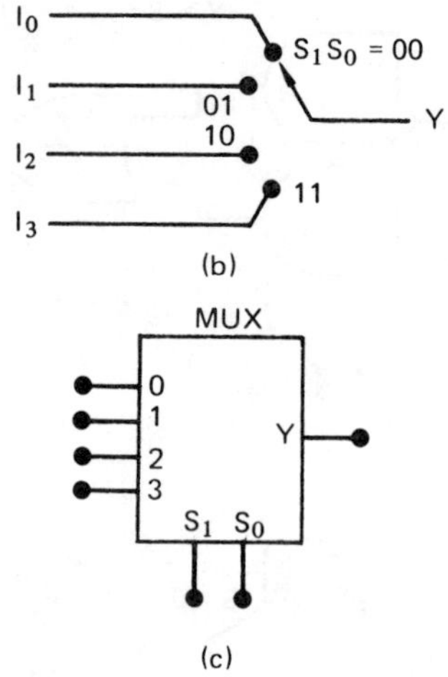

Figure 10.34 (a) 4-input MUX logic diagram, (b) switch analogy, (c) symbol

B. 2-INPUT MUX LOGIC

There are three distinct functions of two variables—AND, OR, and XOR. The AND functions are minterms with four possibilities. The OR functions are maxterms, also with four possibilities. The XOR functions include only two possibilities—the odd-parity and the even-parity functions. All possible functions of two variables are included in this set. (The *trivial functions* of TRUE and FALSE, and the functions which reduce to only one variable are not included.) A two-input MUX is able to produce any of these functions of two variables.

Representatives of the three types of functions making up all nontrivial functions of two variables are shown in the truth table of figure 10.35(a). For these examples, A is used for the select input. The truth table is divided into two parts, one where $A = 0$ and the other where $A = 1$. For each value of A, the functions are compared to B and the corresponding *residue functions* are placed in the reduced truth-table of figure 10.35(b).

For example, in the AND function for $A = 0$, $\overline{A} \cdot B = B$. For $A = 1$, $\overline{A} \cdot B = 0$. These are entered as the residues in the reduced table, and they become the inputs to the 2-input MUX used to generate $\overline{A} \cdot B$ shown in figure 10.35(c).

input		functions		
A	B	$\overline{A} \cdot B$	$A+\overline{B}$	$A \oplus B$
0	0	0	1	0
0	1	1	0	1
1	0	0	1	1
1	1	0	1	0

(a)

select input	residues		
	$\overline{A} \cdot B$	$A+\overline{B}$	$A \oplus B$
$\overline{A}$	B	$\overline{B}$	B
A	0	1	$\overline{B}$

(b)

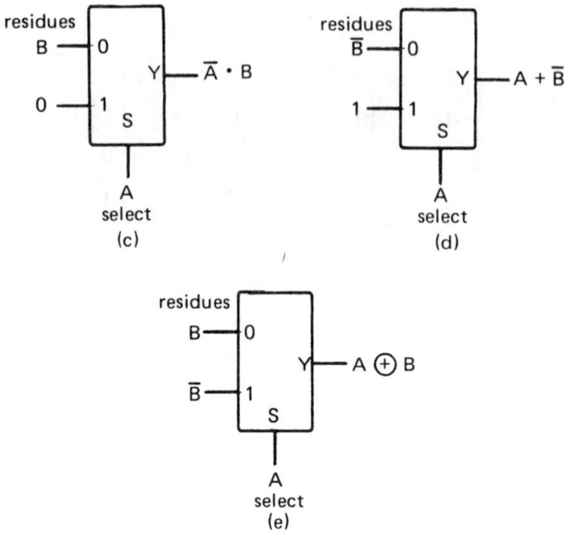

Figure 10.35 2-input MUX logic

The OR function is handled similarly. When $A = 0$, the residue is $\overline{B}$ which is the function to be provided at input 0 of the MUX of figure 10.35 (d). When $A = 1$, the OR residue is 1, and this is the value provided to input 1.

For the XOR function with $A = 0$, the function is B; while for $A = 1$, the function is $\overline{B}$. These are the residues to be applied to inputs 0 and 1 respectively in figure 10.35(e).

These examples show that any function of two variables can be generated with a 2-input MUX.

C. 4-INPUT MUX LOGIC

A 4-input data selector (MUX) requires two select inputs to distinguish between the four data inputs. This permits the realization of single functions with three variables. Although there are fourteen possible nontrivial functions of two variables, there are 206 distinct functions using all 3 variables.

An arbitrary function of three variables can be realized with a 4-input data selector with a procedure similar to that of the previous section: (1) Divide the truth table in half by grouping both possibilities of the right-most variable. (2) Examine the relationship of that variable to the function in each case. (3) Find the residues, which can be 1, 0, the right-most function, or its complement. (4) The residue is the input to the data selector for the input corresponding to the decimal equivalent of the binary value of the two left-most variables.

As an illustration, consider the function given in table 10.10. The truth table is divided into groupings according to the two left-most inputs. The right-most input is compared to the function for that combination of the left-most inputs, which are then used for the two select inputs, S_1 and S_0. The comparison results in the residue indicated, and these become the corresponding inputs to the data selector, as shown in figure 10.36.

Table 10.10 A 3-input truth table with residues

A	B	C	Y	res
0	0	0	0	
0	0	1	0	0
0	1	0	1	
0	1	1	0	$\overline{C}$
1	0	0	0	
1	0	1	1	C
1	1	0	1	
1	1	1	1	1

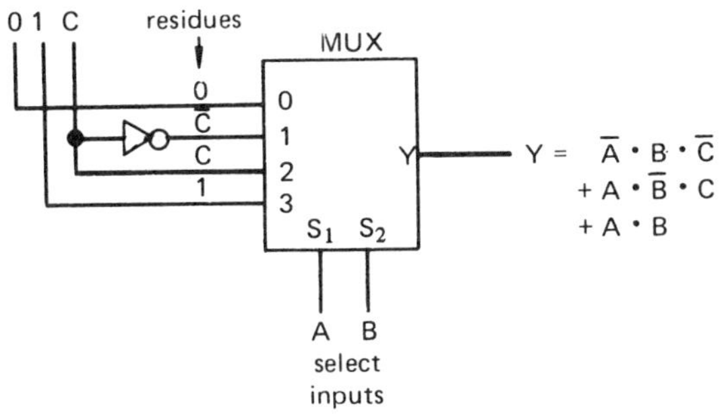

Figure 10.36 4-input data selector realization

This function was chosen to illustrate the four possible residues of one input variable. 4-input data selectors are available in dual packages with common select inputs. Two 3-input functions can be generated for each package.

With additional logic gates, it is possible to realize functions of even more than 3 variables. To do this, it is necesary to extend the concept of residues. The two left-most variables in the truth table are taken as the select inputs; the two right-most variables are used to generate the residues. These are necessarily functions of two variables, which can be generated with 2-input data selectors. As a 4-input data selector can use only four inputs, a quad 2-input package can generate all functions of two variables.

As an illustration, consider the function of four variables given in table 10.11. The truth table is divided into groups by the values of the two left-most variables, which are used for the select inputs.

The 2-variable groups for each value of the left-most pair are further divided to 2-input function generator form. From these latter divisions, the resulting 2-input functions provide the residues for the 4-input data selector with the left-most variables used as select inputs.

Table 10.11
A 3-input truth table with residues

A	B	C	D	Y	D-res
0	0	0	0	1	$\overline{D}$
		0	1	0	
		1	0	0	0
		1	1	0	
0	1	0	0	1	1
		0	1	1	
		1	0	0	D
		1	1	1	
1	0	0	0	1	$\overline{D}$
		0	1	0	
		1	0	1	1
		1	1	1	
1	1	0	0	1	$\overline{D}$
		0	1	0	
		1	0	0	D
		1	1	1	

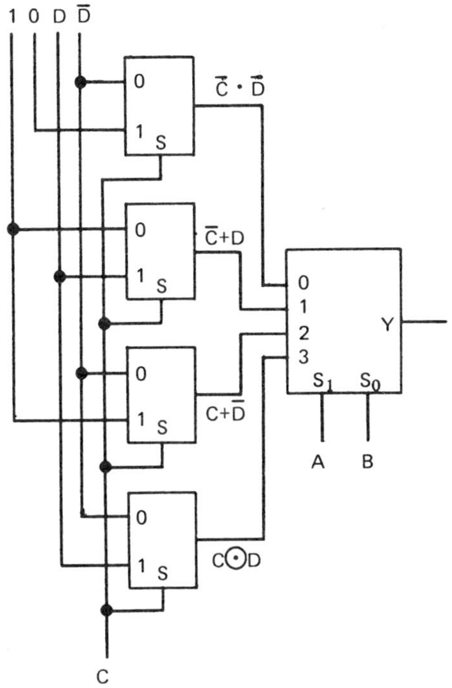

Figure 10.37 2-level logic realization with 4- and 2-input data selectors

While this method does not necesarily produce the minimum realization, it demonstrates a method of using multiplexers to realize functions of more 3-variable functions.

D. 8-INPUT AND HIGHER MUX LOGIC

An 8-input data selector requires three select inputs to account for the addressing of eight distinct inputs. This provides a method for generating functions of 4 variables. The left-most three variables are used for the select, and the right-most variable is used for the generation of residues. The method is the same as with 2- and 4-input data selectors.

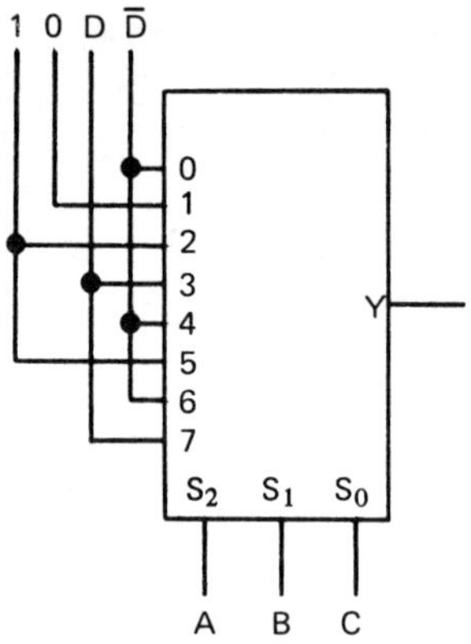

Figure 10.38 8-input data selector use

Consider table 10.11 as an example of 8-input data selector logic. A, B, and C are taken as the select inputs. The D residues are applied to the eight inputs as indicated in figure 10.38.

16-input data selectors are also available. These are capable of generating any function of 5 variables. However, they require about twice the space of the 8:1, dual 4:1, or quad 2:1 data selectors.

Data selector trees, illustrated by figure 10.37, can be used to implement functions of more than five variables, but other techniques (described in the following sections) are more efficient and practical.

7 DECODERS/DEMULTIPLEXERS

A decoder performs the inverse function of a data selector. It selects one of 2^n output lines, and brings that line low. All remaining lines are kept high. The n select inputs determine which of the output lines is brought low. These devices have at least one *enable* input which can be used for demultiplexing.

Decoders can generate all of the (complemented) minterms of the n select inputs.

8 READ-ONLY-MEMORIES AND PROGRAMMABLE LOGIC ARRAYS

Read-only memories (ROM's) can provide multiple functions of the inputs corresponding to the address selection inputs. Typical word lengths for ROM's are eight bits (one *byte*) or multiples of four bits. Thus, they can provide eight distinct functions of 5, 6, 7, or 8 variables. The specific functions are *burned* into the ROM from the function truth table.

In many cases, the ROM solution is wasteful of space, particularly when the functions do not contain a majority of the possible minterms of the input variables. A special device, called *programmable logic array*, has been developed to function as a 14 or 16 input, 8-output Sum-of-Products generator. These are available with provisions for 48 or 96 minterms. Some minimization may be necesary to fit complex multiple functions into the space available.

9 FLIP-FLOPS

The basic elements of sequential logic circuits are *flip-flops*, which are esentially bi-stable multivibrators. Their operation can be explained with either NAND gates or NOR gates connected with positive feedback.

The basic flip-flop is the *S-R (set-reset) flip-flop* which can be constructed from two NAND gates or two NOR gates. The other common flip-flops (T, D, and J-K flip-flops) are constructed from S-R flip-flops and other gates of the same type.

A. THE SET RESET (S-R) FLIP-FLOP

The S-R flip-flop is constructed from either two NAND gates as shown in figure 10.39 or two NOR gates as shown in figure 10.40. The *next condition* of a flip-flop depends on both the present output and the present input. Thus a *next state table* (rather than a truth table) is appropriate. Note that for the NAND realization the inputs are the complements of set and reset.

Table 10.12
Next-state table for S-R flip-flop

present S	present R	present Q	next Q
0	0	0	0
0	1	0	0
1	0	0	1
1	1	0	N
0	0	1	1
0	1	1	0
1	0	1	1
1	1	1	N

N=input not permitted

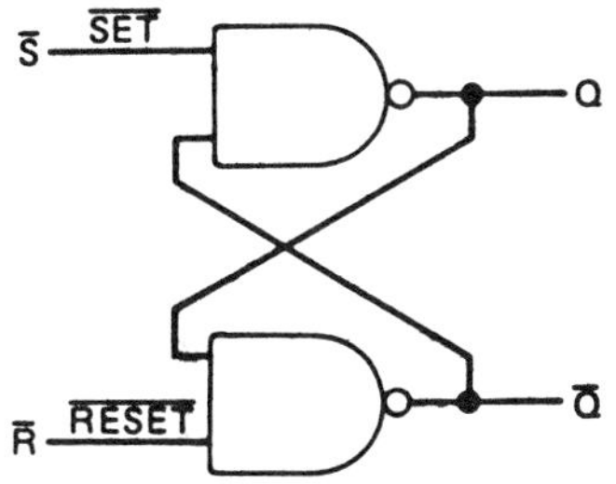

Figure 10.39 NAND gate realization of an S-R flip-flop

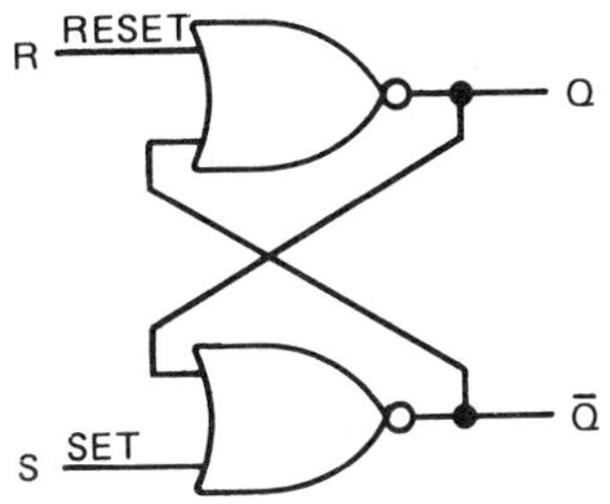

Figure 10.40 NOR gate realization of an S-R flip-flop

B. THE TOGGLE (T) FLIP-FLOP

The toggle flip-flop has only one data input (T) but is clocked so it will change states only when the clock is high. This can be realized with NAND or NOR based S-R flip-flops as shown in figure 10.41.

The necessary configuration is determined from the required inputs for the S-R flip-flop together with the requirements for the T flip-flop. A *transition table* from table 10.12 and the T flip-flop requirements is derived as follows:

Table 10.13
Transition table for T flip-flop

present Q	next Q	T	$\overline{S}$	$\overline{R}$
0	0	0	1	X
0	1	1	0	1
1	0	1	1	0
1	1	0	X	1

X=don't care

It is desired to determine the logic in NAND form for the generation of $\overline{S}$ and $\overline{R}$, the inputs to the NAND S-R flip flop. The inputs are T, the present value of Q, and the complement $\overline{Q}$.

From table 10.13, S is true ($\overline{S}$ is false) when the present Q is false and T is true. Then,

$$S = \overline{Q} \cdot T \qquad 10.60$$

$$\overline{S} = \overline{\overline{Q} \cdot T} \qquad 10.61$$

This is the output of two input NAND gate with inputs $\overline{Q}$ and T. In addition, the clock (**Cl**) must be high for toggling to occur.

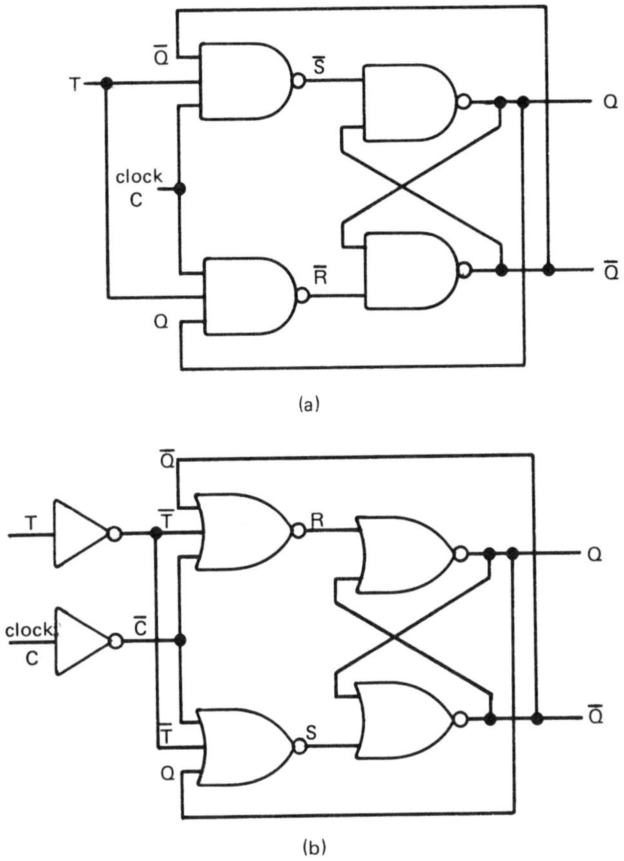

Figure 10.41 Toggle (T) flip-flop configurations: (a) NAND realization, (b) NOR realization

$$S = \overline{Q} \cdot T \cdot \mathbf{Cl} \qquad 10.62$$

Inverting both sides of equation 10.62

$$\overline{S} = \overline{\overline{Q} \cdot T \cdot \mathbf{Cl}} \qquad 10.63$$

The realization of equation 10.63 requires the 3-input NAND gate shown in figure 10.41 (a) with inputs T, $\overline{Q}$ and **Cl**.

Similarly, the transition table shows that R is true ($\overline{R}$ is false) when the present Q is 1 and T is 1. Since the clock must be high,

$$\overline{R} = \overline{Q \cdot T \cdot \mathbf{Cl}} \qquad 10.64$$

This is the output of the lower input NAND gate of figure 10.41(a).

To obtain the NOR gate logic for figure 10.41(b), it is necesary to put equations 10.63 and 10.64 in OR form by applying de Morgan's theorem to the right-hand side. Then, the negation of the NOR gate will provide S and R.

$$\overline{S} = \overline{\overline{Q} \cdot T \cdot \mathbf{Cl}} = \overline{\overline{Q}} + \overline{T} + \overline{\mathbf{Cl}}$$

$$= Q + \overline{T} + \overline{\mathbf{Cl}} \qquad 10.65$$

$$S = \overline{Q + \overline{T} + \mathbf{Cl}} \qquad 10.66$$

This is the output of a 3-input NOR gate such as the lower input gate of figure 10.41(b).

The clock pulse for figure 10.41 can be high no longer than the propagation time through the gates.

C. THE DATA (D) FLIP-FLOP

The D flip-flop is clocked so that the output Q will follow the input when the clock is high, when the clock makes its transition (*edge triggered*) from low to high, or when the clock makes its transition from high to low. The simplest D flip-flop, called the *D-latch*, uses an *enable* (which may be a clock). The transition table, including the S-R input requirements, is shown in table 10.14.

The X's in S and R can be used because they allow the variable to be 0 or 1 as needed. If the X's are 1, the logical expressions for S and R are

$$S = E \cdot D \qquad 10.67$$

$$R = E \cdot \overline{D} \qquad 10.68$$

Table 10.14
D flip-flop transition table

E	D	present Q	next Q	S	R
1	0	0	0	0	X
1	0	1	0	0	1
1	1	0	1	1	0
1	1	1	1	X	0
0	X	X	present	0	0

D. J-K FLIP-FLOPS

The J-K flip-flop has been developed to make use of the two disallowed inputs of the S-R flip-flop. The inputs S, R=1,1 result in a toggled output (i.e., if the present state is $Q = 1$, the next state is $Q = 0$, and vice versa). There are two types of J-K flip-flops, an ordinary J-K and a so-called *master-slave J-K*. The difference is in when the change of output occurs. This will be discussed in the next section.

Table 10.15
J-K flip-flop truth table

J	K	present Q	next Q	S	R
0	0	0	0	0	X
0	0	1	1	X	0
0	1	0	0	0	X
0	1	1	0	0	1
1	0	0	1	1	0
1	0	1	1	X	0
1	1	0	1	1	0
1	1	1	0	0	1

$$S = J \cdot \overline{Q} \qquad 10.69$$

$$R = K \cdot Q \qquad 10.70$$

E. CLOCKED FLIP-FLOPS

Flip-flops can be triggered by either the rising edge or the falling edge of the clock pulse. The master-slave J-K flip-flop consists of two parts, a *master* and a *slave*, as indicated in figure 10.42. For these, the inputs must be stable during the time the clock is high. This sets the first master stage, and the change is propagated to the slave stage on the falling edge of the clock. (There are also master-slave versions of the S-R flip-flop which operate similarly.)

The triangular symbol on the clock input indicates that the flip-flop is triggered at the rising edge of the clock

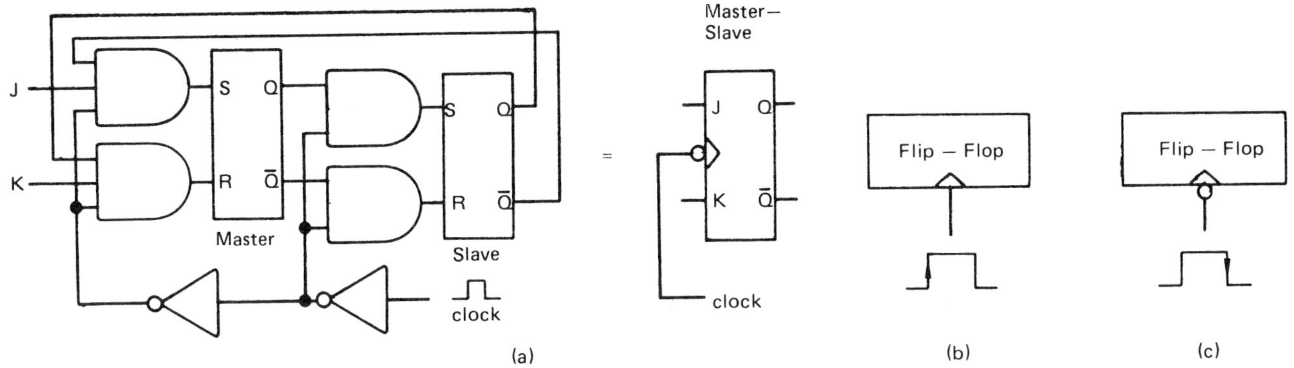

Figure 10.42 (a) master-slave J-K flip-flop, (b) symbol for a rising edge-triggered flip-flop, (c) symbol for a falling edge-triggered flip-flop

input as in figure 10.42 (b). An inversion (bubble) with the triangle indicates that the flip-flop is triggered by the falling edge of the clock pulse.

Ordinary flip-flops (not master-slave units) include D and J-K types which are triggered on either the rising edge or the falling edge of the clock pulse. *T flip-flops* are not commercially available. They must be built from one of the other available types and added logic.

Other inputs usually found on flip-flops are *preset,* which *asynchronously* (immediately) sets the output state to $Q = 1$, and *clear,* which asynchronously sets the output to $Q = 0$. Both provide an over-riding of the logic. They do not generally wait for the proper clock transition.

There are several important *propagation delay times* associated with flip-flops. These are usually different for high-to-low and from low-to-high output transitions. They also differ depending on the origin of the initiator of the change (e.g., clock, preset, or clear). These times determine the maximum frequency at which the flip-flops can operate.

Two additional times are the *set-up time,* which is the minimum time that the inputs must be stable before the triggering clock transition and the *hold time,* which is the minimum time the inputs must remain stable after the triggering clock transition.

10 FLIP-FLOP TRANSITION TABLES

In many design problems involving flip-flops, the *transitions* from the present states to the next output states are needed to design logic to satisfy the problem specifications. The four possible transitions are : 0 to 0, 0 to 1, 1 to 1 and 1 to 0. While 0 to 0 and 1 to 1 may not appear to be transitions, they are necessary in logic considerations. The transitions for the four principal flip-flop types are given in table 10.16.

Table 10.16 shows that a T flip-flop can be constructed from a J-K flip-flop by making $J = K = T$. It can also be constructed from a D flip-flop by making $D = T$ XOR Q, or from an S-R flip-flop by making $S = \overline{Q} \cdot T$ and $R = Q \cdot T$.

11 COUNTERS

Counters are sets of flip-flops. Usually, the flip-flops are all the same type, and they all respond to the same clock edge for triggering. A counter passes through a sequence of *states* in response to one or more inputs. The count is determined by the number of states in the sequence and is limited to 2^n, where n is the number of flip-flops in the counter.

A simple example is the *ripple counter* shown in figure 10.43(a). The input signal is applied to the clock input of the lowest flip-flop. This flip-flop is triggered by the

Table 10.16 Flip-flop transition table

present state Q		next state Q^+	S	R	J	K	T	D
0	⟶	0	0	X	0	X	0	0
0	⟶	1	1	0	1	X	1	1
1	⟶	0	0	1	X	1	1	0
1	⟶	1	X	0	X	0	0	1

$$Q^+ = S + \overline{R} \cdot Q$$
$$Q^+ = J \cdot \overline{Q} + \overline{K} \cdot Q$$
$$Q^+ = T \oplus Q$$
$$Q^+ = D$$

falling edge of the signal. When the lowest flip-flop makes a high-to-low transition, the middle flip-flop is triggered. When the middle flip-flop makes a transition from high-to-low, the upper flip-flop is triggered. Figure 10.43(b) shows the count cycle. The three flip-flops are at their low states at the beginning of the cycle.

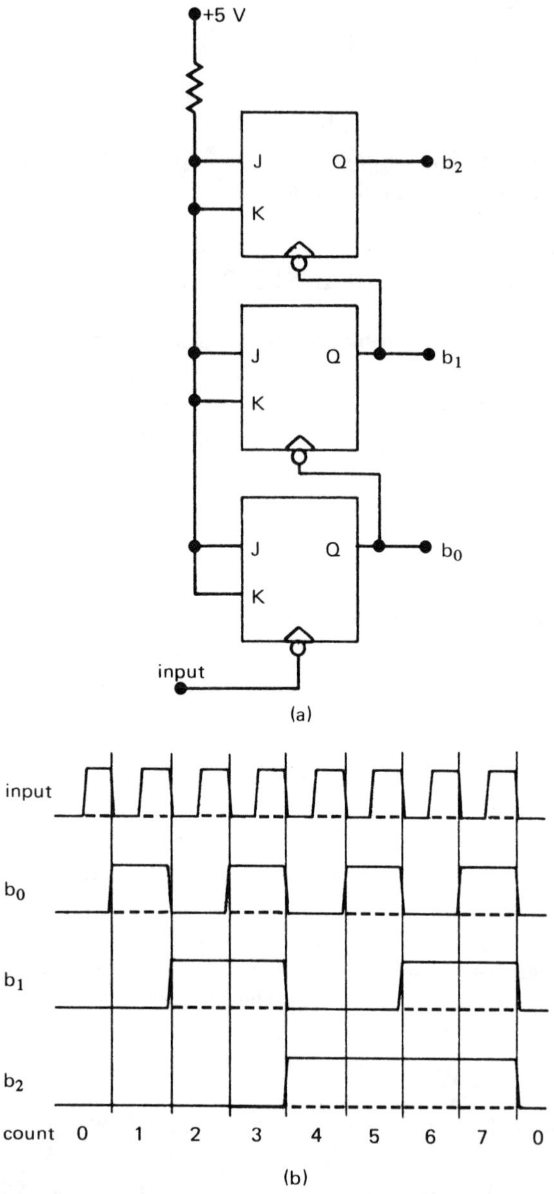

Figure 10.43 (a) A ripple counter (b) timing diagram for one count sequence

The count ripples through the circuit from the input flip-flop, so that delay times become important with additional flip-flops. This type of a counter is called *asynchronous*, in that the outputs of the flip-flops are not actually clocked. This *propagation delay* is a fundamental limitation with this type of counter. Other means (involving common clocking) are required for count lengths of more than a few bits.

A. SYNCHRONOUS BINARY COUNTERS

In order to make a counter (or similar sequential circuit) synchronous, the flip-flops all must make transitions at the same time. A *state table* for the counter can then be constructed where the *state* is the concatenation of the Q outputs of the flip-flops.

For example, the state sequence for the count indicated in figure 10.45(b) is 000, 001, 010, 011, 100, 101, 110, 111, 000, with the positional significance of $Q_2Q_1Q_0$ in that order.

The state table is constructed by listing the state sequence in order as the left-hand column as shown in table 10.17. The next state for each of the present states is shown in the second column. (The superscript + is used to indicate the next state.) The third column contains the set of inputs to the flip-flops which are necessary to obtain the proper transition (see table 10.16).

Table 10.17
State table for a 3-bit binary counter

present state $Q_2\ Q_1\ Q_0$	next state $Q_2^+\ Q_1^+\ Q_0^+$	required T inputs $T_2\ T_1\ T_0$
000	001	001
001	010	011
010	011	001
011	100	111
100	101	001
101	110	011
110	111	001
111	000	111

$$T_2 = Q_0 \cdot Q_1$$
$$T_1 = Q_0$$
$$T_0 = 1$$

The design requires determining the flip-flop inputs as logical functions of the present state values of Q_2, Q_1, and Q_0. This is normally accomplished with Karnaugh maps or Veitch diagrams. For this example, the logic is simple enough to merely inspect the state-table. *T-flip-flops* (which are obtained by tying the J and K imputs together on J-K flip-flops) are assumed. The resulting counter circuit is shown in figure 10.44(a).

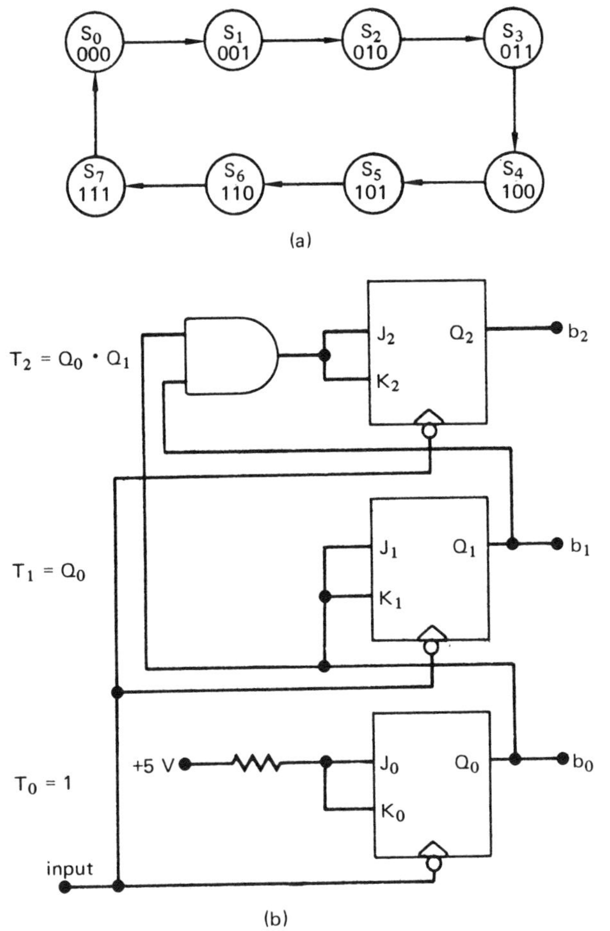

Figure 10.44 3-bit binary counter (a) state diagram, (b) T flip-flop realization

Figure 10.44 (a) shows a *state diagram* for the system. This is a useful device for the design or analysis of sequential logic circuits. The states are indicated by subscripted S's. They do not have to be given the same subscripts as the counts, although they have been in this example.

A major problem with this design is that more than one output can change at the same time. Because there are always differences in the speeds of transitions, there is a time during the transitions in which a false value of the count can occur. For example, in going from S_1 to S_2, both Q_0 and Q_1 are changing. If Q_1 reaches 1 before Q_0 reaches 0, then there is a momentary output of 011 existing. This can cause problems when the outputs are used as logical inputs to some other system.

A better design would have changes occur with only one state variable changing. This is equivalent to making state assignments so that the next state is logically adjacent to the present state. While this is not possible in all designs, it is good practice to try to do so.

B. AN ADJACENT-STATE BINARY COUNTER

Because the binary counter uses all 2^n states, it is possible to assign all next states adjacent to the present state. This is complicated by the fact that the state variables no longer correspond to the correct count. To correct this, it is necessary to use logic to provide the correct counts for each state.

Consider the previous example. Start with state assignments so that the next state is logically adjacent to the present state. These are shown in the second column of table 10.18. The first column has the state it names, which are subscripted to match the correct count.

The third column contains the next state, and the fourth column contains the required outputs in each state. In this example, D flip-flops are used. The required values of inputs for D flip-flops are in the fifth column.

Table 10.18

Adjacent-state binary counter

state	present state $Q_2Q_1Q_0$	next state $Q_2^+Q_1^+Q_0^+$	outputs $b_2b_1b_0$	D inputs $D_2D_1D_0$
S_0	0 0 0	0 0 1	0 0 0	0 0 1
S_1	0 0 1	0 1 1	0 0 1	0 1 1
S_2	0 1 1	0 1 0	0 1 0	0 1 0
S_3	0 1 0	1 1 0	0 1 1	1 1 0
S_4	1 1 0	1 1 1	1 0 0	1 1 1
S_5	1 1 1	1 0 1	1 0 1	1 0 1
S_6	1 0 1	1 0 0	1 1 0	1 0 0
S_7	1 0 0	0 0 0	1 1 1	0 0 0

Once the state table has been established, the logic to obtain flip-flop inputs and the outputs is determined by the use of Karnaugh maps as indicated in figure 10.45(b). Having determined the necessary logic, the circuit is realized in figure 10.45(c). (There is a redundant gate in this circuit diagram. $Q_1 \cdot \overline{Q_0}$, is needed for both D_1 and D_2.) A simplification of the state table is possible when using D flip-flops, because the value of D is always the next state value for the corresponding state variable. Thus, the last column is redundant when using D flip-flops.

Even for this relatively simple circuit, the inclusion of all logic lines from the state variables to the flip-flop inputs is confusing. For that reason, these feedback lines are usually omitted, and the state variables are indicated at the various gate inputs.

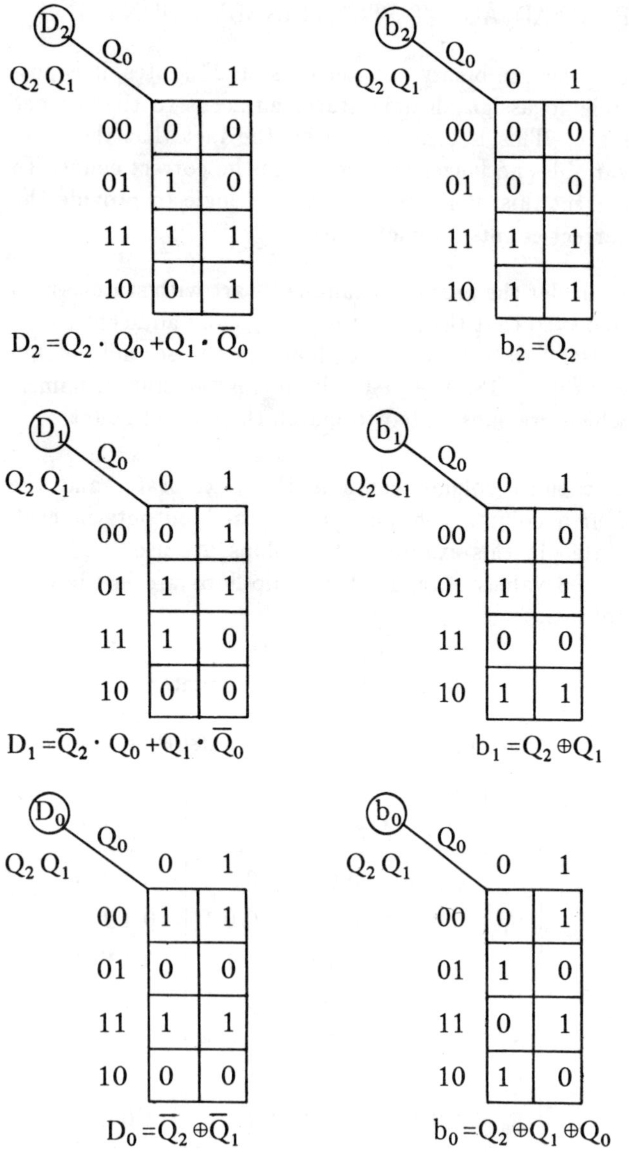

(a)

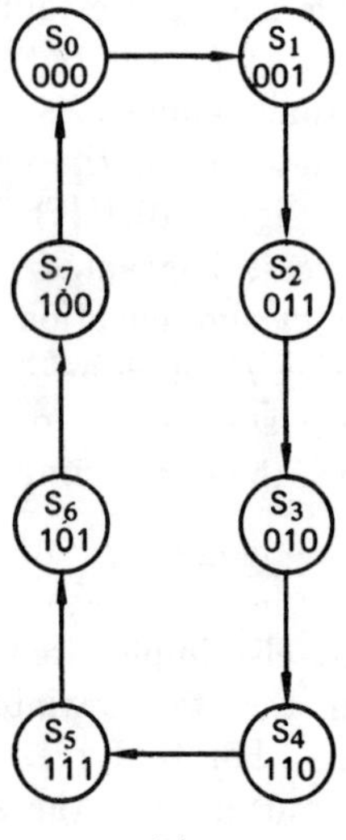

(b)

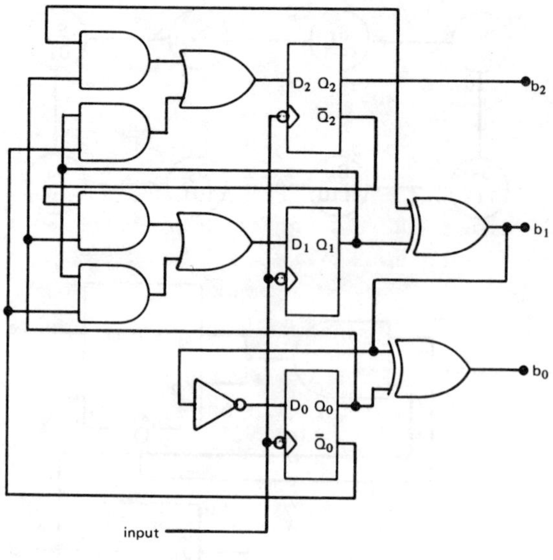

(c)

Figure 10.45 Adjacent-state counter (a) logic, (b) state diagram, (c) circuit

C. COUNTER ANALYSIS

In order to analyze a counter circuit, the most orderly approach is to construct a state table. The logical connections are used to establish the flip-flop inputs from the present state. The usual starting point is at the state with all zero state variables. Each successive state is then entered in sequence.

The following example is intended to illustrate the method. The circuit shown in figure 10.46 uses three different types of flip-flops and two AND gates. The flip-flops are all triggered by the rising edge of the input signal.

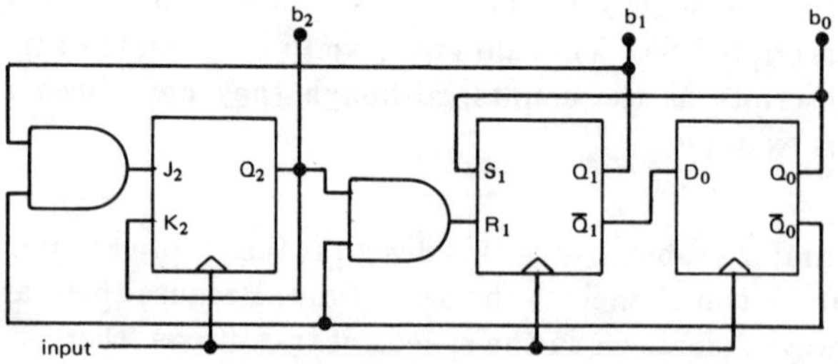

$J_2 = Q_1 \cdot \overline{Q}_0$ $\quad S_1 = Q_0$ $\quad D_0 = \overline{Q}_1$

$K_2 = Q_2$ $\quad R_1 = Q_2 \cdot \overline{Q}_0$

Figure 10.46 Sample counter circuit

From the counter circuit and the flip-flop logical equations given in table 10.16 we can write the next-state variables.

$$Q_2^+ = J_2 \cdot \overline{Q_2} + \overline{K_2} \cdot Q_2 = \overline{Q_2} \cdot Q_1 \cdot \overline{Q_0} + \overline{Q_2} \cdot Q_2 = \overline{Q_2} \cdot Q_1 \cdot \overline{Q_0} \qquad 10.71$$

$$Q_1^+ = S_1 + \overline{R_1} \cdot Q_1 = Q_0 + (\overline{Q_2} + Q_0) \cdot Q_1$$

$$= Q_0 + \overline{Q_2} \cdot Q_1 \qquad 10.72$$

$$Q_0^+ = D_0 = \overline{Q_1} \qquad 10.73$$

These equations are used to generate the second column of the state table. The first four rows in the table complete the count sequence since the next state from S_4 is S_0. Because there are three unused states, it is necessary to check for hang-up states.

Table 10.19
State table for figure 10.46

	present state $Q_2\ Q_1\ Q_0$	next state $Q_2^+\ Q_1^+\ Q_0^+$
S_0	0 0 0	0 0 1
S_1	0 0 1	0 1 1
S_2	0 1 1	0 1 0
S_3	0 1 0	1 1 0
S_4	1 1 0	0 0 0
S_5	1 1 1	0 1 0
S_6	1 0 1	0 1 1
S_7	1 0 0	0 0 1

Hang-up states are two or more unused states which have no next state in the desired sequence. Unless there is an automatic clearing of all flip-flops when a system is turned on, it could come up into one of the unused states and be hung-up there, never getting into the desired sequence. To make sure this doesn't happen, all unused states must have a path into the desired sequence. A check of the states 5, 6, and 7 shows that they all return to the main sequence.

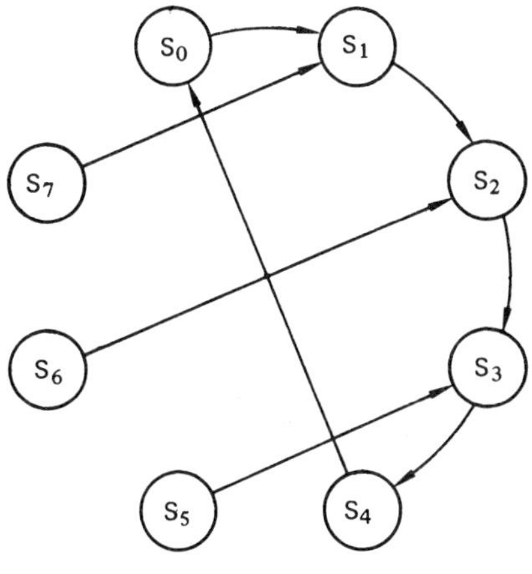

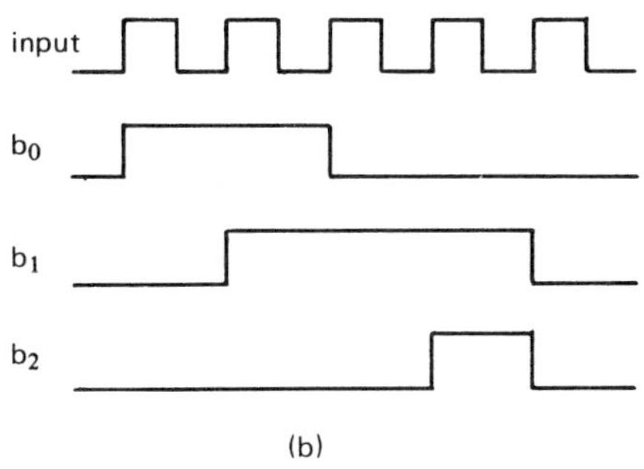

Figure 10.47 (a) state diagram, (b) timing diagram for figure 10.46

The state diagram, with the same designations as in the state table, is shown in figure 10.47(a). The timing diagram in figure 10.47(b) shows the outputs in reference to the input. The resulting sequence is 000, 001, 011, 010, 110, 000. Thus, this is a five state counter.

12 SEQUENCE DETECTION

A *sequence detector* is a sequential circuit which determines when a specific input sequence has occured. For example, a detector might be needed to detect the sequence 1001. The inputs will be monitored using a clock with the same frequency as the monitored data line.

The detector starts in a state S_0 where the previous input was not from a proper sequence (e.g., was a zero following two or more zeros). This will be the *start-up state*. The detector will remain in state S_0 as long as 0's are received and will go to the next state, S_1, as soon as a 1 is received.

The detector will remain in state S_1 as long as 1's are received. It will go to the next state, S_2, when a 0 is received.

S_2 is a state where a 1 has been followed by a 0. S_3 is a state where there has been a 1 followed by two 0's. The detector can go to S_3 only if the next input is a 0. If the next input is a 1, the detector must return to state S_1 (S_1 is the state where a 1 has just been received). The detector can remain in S_2 for only one clock cycle.

S_3 is the state where the sequence 100 has been received. The detector can remain in this state for only one clock cycle. If a zero is the next input, the detector must return to state S_0. If the input is a 1, the correct sequence has been detected, and the detector goes to state S_4 where the correct sequence signal is given. The stay in state S_4 is limited to one clock cycle. If the next input is a 0, the detector makes a transition to state S_2 since a 1 has been followed by a 0. If the next input is a 1, the transition is to state S_1, the state where a 1 has just been received.

This logic is shown in figure 10.48. This state diagram is different from previous examples. There are two paths from each state, and the particular path taken depends on the input. The output is indicated below the horizontal line in each circle.

The procedure used to assign the state variable values for each state is complicated. Three rules can be used to minimize logic costs.

> **Rule 1.** States which have the same next-states for the same inputs should be given adjacent assignments.
>
> **Rule 2.** States which are the next-state of the same state should be given adjacent assignments.
>
> **Rule 3.** States which have the same outputs for a given input should be given adjacent assignments.

In figure 10.48(c), the adjacency rules are entered into a matrix to evaluate the desirability of assigning various states with adjacent variable values. Each entry in the matrix shows the number of times a rule applies. Because there is only one state with an output, rule 3 is of minor importance here. It should be used only to break ties. Giving rules 1 and 2 equal weight, it is obvious that the $S_1 - S_2$ adjacency must be made.

$S_0 - S_1$ is the second priority because of rule 3. Next in priority are the groups $S_1 - S_4$ and $S_0 - S_4$. Then the three groupings of $S_0 - S_2$, $S_0 - S_3$ and $S_1 - S_3$ should be ranked. The final consideration is $S_2 - S_4$ followed by $S_2 - S_3$ and $S_3 - S_4$.

Several combinations are approximately equal, and all groupings are not possible. Because of reset on start-up, it is common to assign 000 to state S_0. From these, one efficient grouping is as shown in table 10.20. The logic for the flip-flop inputs is obtained for J-K flip-flops using table 10.16 relationships.

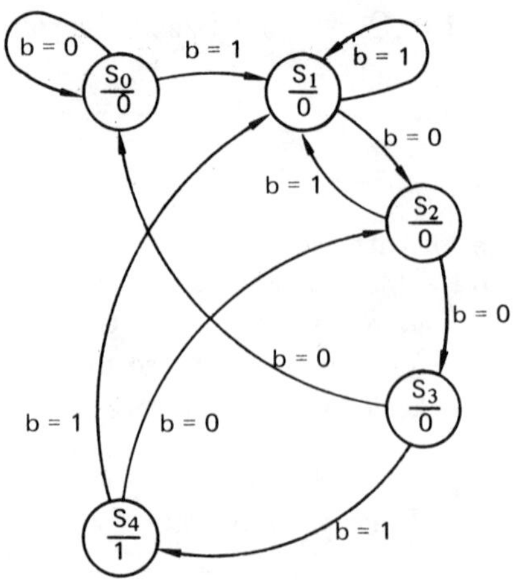

(a)

present state	next state for input b $b=0$	$b=1$	present output
S_0	S_0	S_1	0
S_1	S_2	S_1	0
S_2	S_3	S_1	0
S_3	S_0	S_4	0
S_4	S_2	S_1	1

(b)

	S_4	S_3	S_2	S_1
S_0	1, 2	1, 3	1, 3	1, 2, 3
S_1	1, 1	2, 3	1, 2, 2, 3	
S_2	1	3		
S_3				

(c)

Figure 10.48 (a) State diagram for sequence 1001 detector, (b) state table, (c) adjacency weighting according to state assignment rules

Table 10.20
Stable table for 1001 detector

S	present state $Q_2Q_1Q_0$	next state $b=0$ $Q_2^+Q_1^+Q_0^+$	next state $b=1$ $Q_2^+Q_1^+Q_0^+$	present output Y	J_2	K_2	J_1	K_1	J_0	K_0
S_0	0 0 0	0 0 0	0 0 1	0	0	X	0	X	b	X
S_1	0 0 1	0 1 1	0 0 1	0	0	X	$\overline{b}$	X	X	0
S_2	0 1 1	0 1 0	0 0 1	0	0	X	X	b	X	$\overline{b}$
S_3	0 1 0	0 0 0	1 0 1	0	b	X	X	1	b	X
S_4	1 0 1	0 1 1	0 0 1	1	X	1	$\overline{b}$	X	X	0

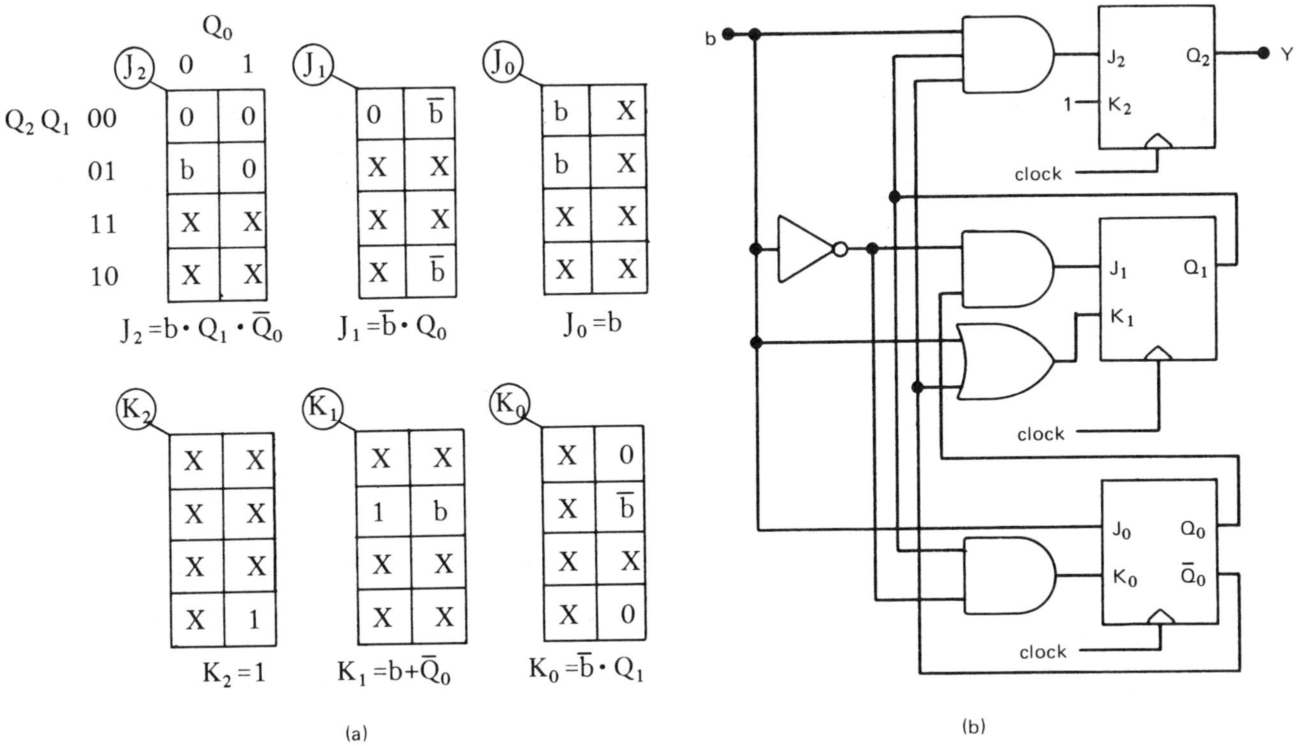

Figure 10.49 (a) J-K logic determination using Karnaugh maps, (b) realization with rising edge-triggered J-K flip-flops

Practice Problems

Digital Logic Problems

Warmups

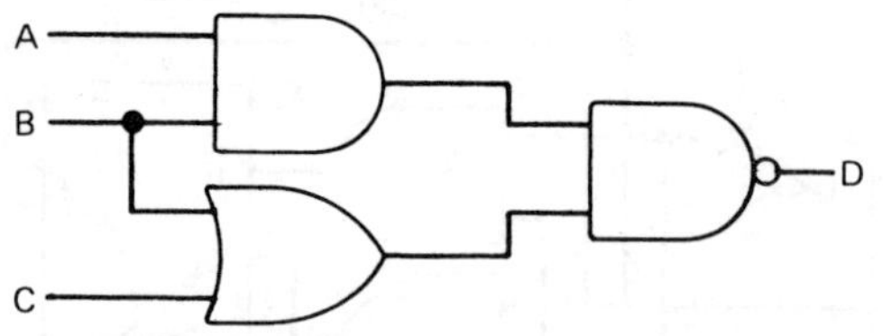

Figure for problems 1, 2, and 3

1. For the circuit shown, obtain the truth table for the function D.

2. Obtain the sum-of-products and the product-of-sums expressions for D.

3. Using identities and theorems, determine the simplest expression for D. Revise your circuit to use a minimum of gates.

4. For the truth table given, obtain the Karnaugh map and Veitch diagram for the outputs D, E and F. An X entry in the truth table indicates a *don't care*, so 0 or 1 can be used for the maximum convenience.

inputs ABC	outputs DEF
0 0 0	0 1 1
0 0 1	0 X 1
0 1 0	X 1 0
0 1 1	0 0 0
1 0 0	1 1 X
1 0 1	1 1 X
1 1 0	0 X X
1 1 1	X 1 X

Truth table for problems 4, 5, and 6

5. For the outputs D, E, and F, obtain the minimum gate realization between the SOP (NAND-NAND) and the POS (NOR-NOR) implementations. (Hint: determine which cannonical form will result in the minimum number of gates after minimization.)

6. Using 4:1 data selectors, determine the residues for each of the functions D, E, and F.

7. In the truth table for a seven-segment display (figure 10.22), notice that segment c requires only one 0, while segment e requires six 0's. From Karnaugh maps for these two segments, determine whether a minterm SOP realization or a maxterm POS realization is preferable.

8. For segment f of the seven-segment display, design an 8:1 data selector realization of the logic. Use the variable D for residues.

9. For the NOR form of the S-R flip-flop (see figure 10.40), consider that the gate with R as an input is faster than the other. Trace the signals when the inputs have been $R = 0$, $S = 1$ for some time when R is suddenly changed to 1. Does it make any difference whether one or the other gate is faster?

10. A sequential circuit is to have the state table given. Using an S-R flip-flop, design a circuit. Q is the present state and output, A and B are inputs, and Q^+ is the next state and output.

Q	A	B	Q^+
0	0	0	0
0	0	1	1
0	1	0	1
0	1	1	0
1	0	0	1
1	0	1	0
1	1	0	0
1	1	1	1

Concentrates

1. You are given the synchronous sequential circuit shown on the following page. Develop the state transition diagram. Show all work.

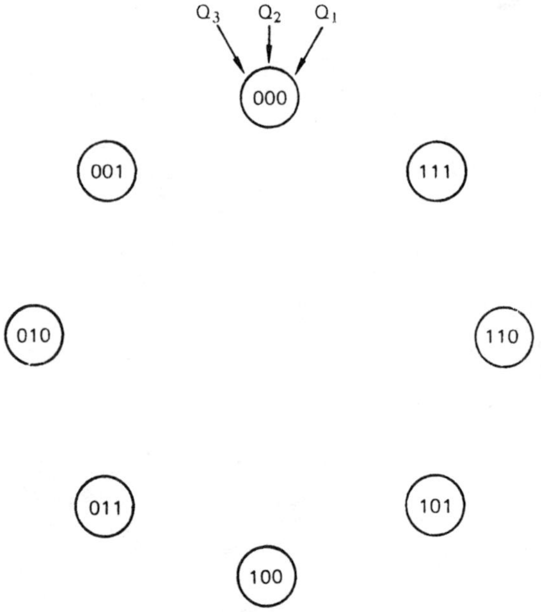

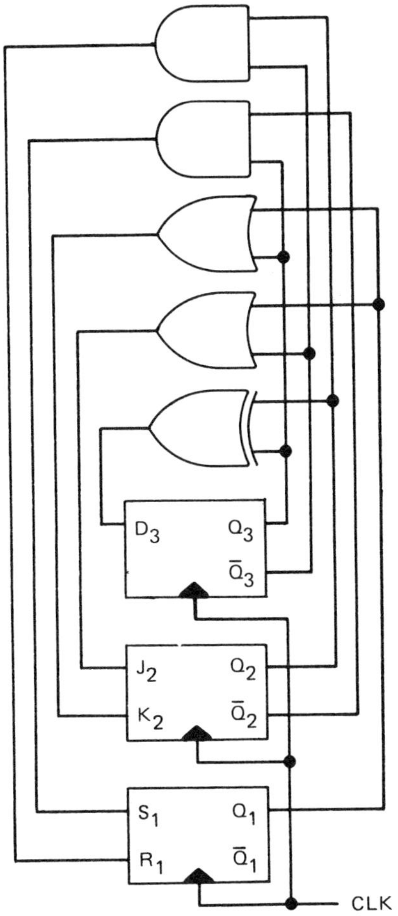

2. A one-bit full-adder cell has inputs A, B, and CI (carry in) and outputs S (sum) and CO (carry out). All of these quantities are one bit. (a) Obtain a NAND-NAND realization of the full-adder cell. (b) Obtain a dual 4:1 data selector realization of the full-adder cell.

3. A shift-register is to be designed using D flip-flops and data selectors. This shift register has four functions: shift left, shift right, broadside load, and do nothing. The register has 6 inputs: L (left input), R (right input), D0, D1, D2, and D3, where the D's are the broadside data bits. The register has four outputs: Q0, Q1, Q2, and Q3. The finished register and its function table are shown.

$C1$	$C0$	$Q3^+$	$Q2^+$	$Q1^+$	$Q0^+$
0	0	$Q3$	$Q2$	$Q1$	$Q0$
0	1	L	$Q3$	$Q2$	$Q0$
1	0	$Q2$	$Q1$	$Q0$	R
1	1	$D3$	$D2$	$D1$	$D0$

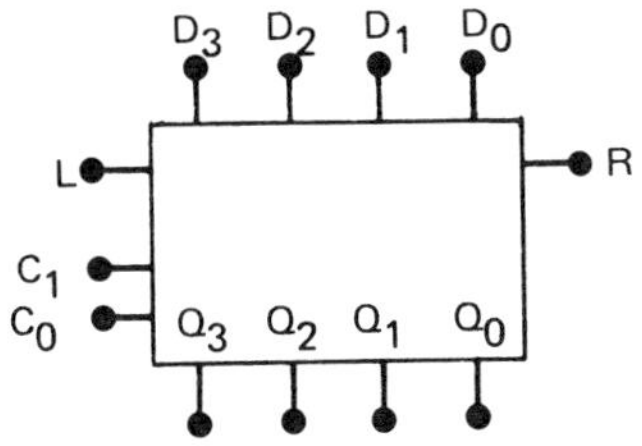

4. For the counter circuit shown, develop the state table and the state transition diagram.

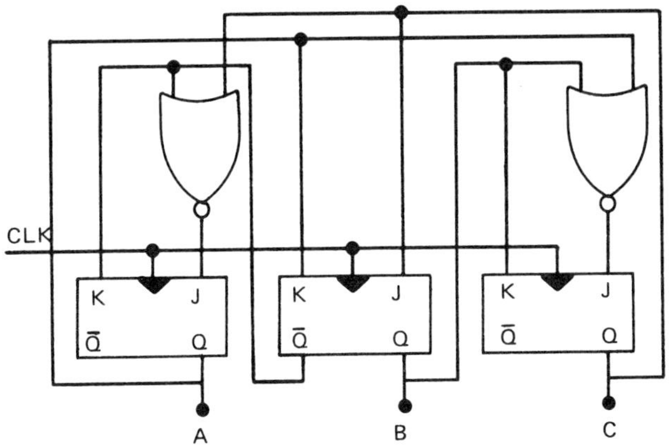

5. Using T flip-flops, design a counter which will produce the following sequence:

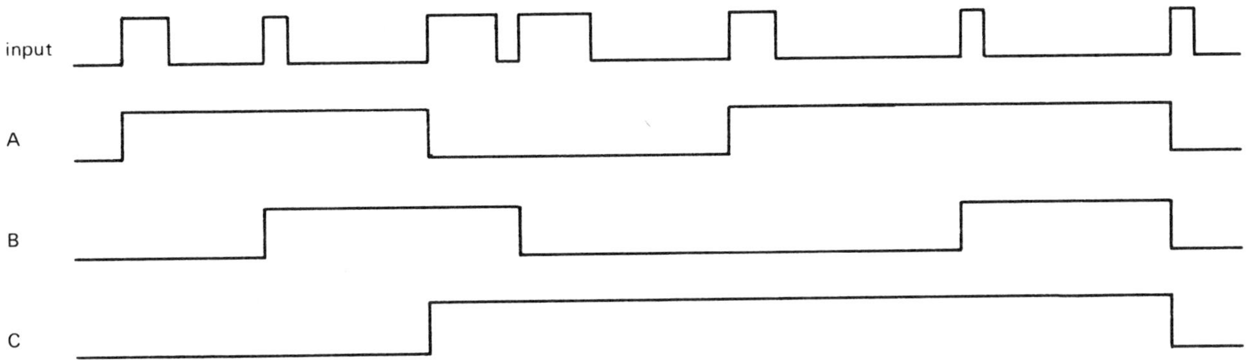

Figure for problem 5.

Timed

1. Given the digital circuit shown, complete the state diagram. Show all transitions and your method of analysis.

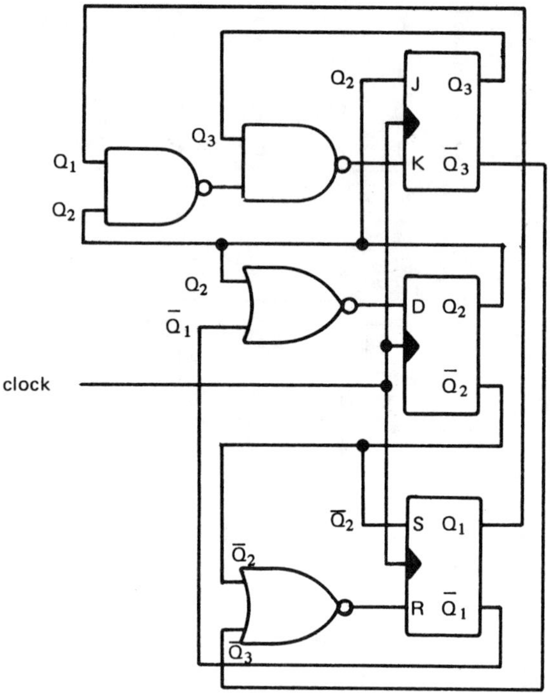

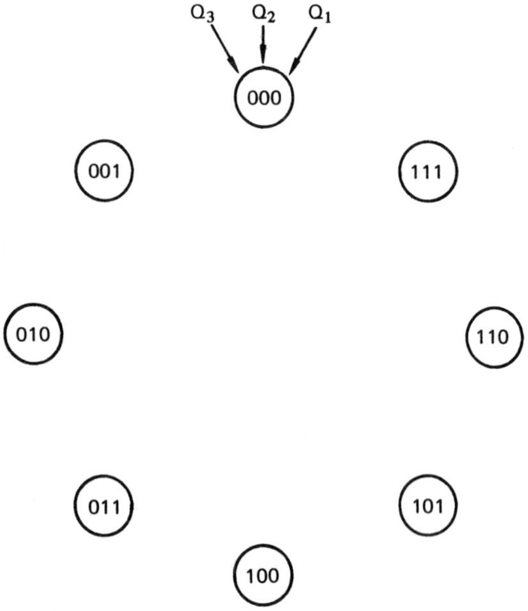

2. A counter is constructed from J-K master-slave flip-flops and NAND gates. Show one complete cycle of the counter. Assume $Q_1 = Q_2 = Q_3 = 0$ initially.

3. Using an 8:1 data selector, obtain the function

$$F = \overline{A} \cdot B \cdot D + A \cdot C + \overline{A} \cdot B \cdot \overline{C} \cdot D + \overline{A} \cdot \overline{B} \cdot \overline{C} \cdot \overline{D}$$

The available inputs are A, $\overline{A}$, B, C, D, 0, and 1.

4. You are given a digital logic function F of four variables. Develop the minimum 2-level NAND realization for F.

W	X	Y	Z	F
0	0	0	0	0
0	0	0	1	0
0	0	1	0	0
0	0	1	1	1
0	1	0	0	0
0	1	0	1	0
0	1	1	0	1
0	1	1	1	1
1	0	0	0	0
1	0	0	1	0
1	0	1	0	0
1	0	1	1	1
1	1	0	0	1
1	1	0	1	0
1	1	1	0	1
1	1	1	1	1

5. You are to design a digital circuit using only 2-input NAND gates and J-K master-slave flip-flops. You may use as many of each as is necessary, but your design should be minimized with respect to packages. (There are four 2-input NAND gates in a package and two J-K flip-flops per package).

The system is to have two inputs: a clock and x. x is a sequence of binary voltages—high=1, low=0. Binary data is presented in sequences of one, two, or three 1's followed by at least one 0. No other sequences will ever occur (i.e., a sequence of four 1's will not occur).

There are to be two outputs, A and B. A is to go high when the first 1 of a sequence occurs and to go low after the second or when the next 0 occurs. B is to go high when the second 1 of a sequence occurs and stay high until the next 0 occurs.

Figure for problem 2

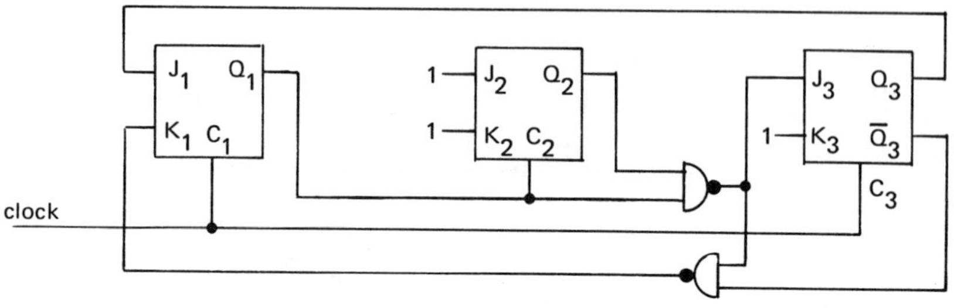

RESERVED FOR FUTURE USE

RESERVED FOR FUTURE USE

11 CONTROL SYSTEMS

Nomenclature

a	a coefficient	-
A	voltage gain	-
b	a coefficient	-
c	controlled variable of time	-
C	Laplace-transformed controlled variable	-
D	mechanical damping coefficient	N-m-sec
e	error variable of time	-
E	Laplace-transformed error variable, or electromotive force	- or V
f	function of time	-
F	function of s	-
G	transfer function of s	-
GH	loop transfer function of s	-
H	feedback transfer function of s	-
I	current	amps
J	mass moment of inertia	N-m-sec^2
K	gain constant	-
K_v	electromotive force constant	V-sec/rad
L	inductance	henrys
p	power	watts
r	reference variable of time	-
R	Laplace-transformed reference variable	-
R	resistance	ohms
s	Laplace transform variable	1/sec
t	time	sec
$u(t)$	unit step function	-
V	voltage	V
x	an arbitrary variable	-

Symbols

ζ	damping ratio	-
θ	angular position	rad
σ	real part of s	nepers
τ	time constant	sec
Ω	mechanical angular velocity	rad/sec
ω	frequency	rad/sec

Subscripts

a	armature
c	compensation or controlled
e	error
f	field
g	generator or forward
h	feedback
m	an integer
n	an integer or natural
p	plant or pole
r	reference
s	source
ss	steady state
t	terminal or tachometer
z	zero

1 GENERAL FEEDBACK THEORY

Linear *control systems* often are represented in block diagram form. Transfer functions typically are given as ratios of polynomials of the complex frequency s, often in factored form. The standard terminology for such a system is shown in figure 11.1(a). C is the controlled variable, R is the reference (input) variable, and E is the error (difference) between R and C for unity feedback.[1]

$$E = R - C \qquad 11.1$$

The forward block indicated by **G** in figure 11.1 performs the transformation

$$C = \mathbf{G}E \qquad 11.2$$

The equations for the output and error in terms of the reference are

$$\frac{C}{R} = \frac{\mathbf{G}}{1+\mathbf{G}} \qquad 11.3$$

$$\frac{E}{R} = \frac{1}{1+\mathbf{G}} \qquad 11.4$$

[1]C, E, and R are Laplace-transformed variables: $C(s) = \mathcal{L}\{c(t)\}$, $E(s) = \mathcal{L}\{c(t)\}$, and $R(s) = \mathcal{L}\{r(t)\}$. In this chapter, lower case variables are functions of t, and upper case variables are functions of s.

In general, **G** is used to represent the *forward transfer function* with unit feedback systems. Systems using other feedback have a transfer function represented by **H** as indicated in figure 11.1 (b). The output and error equations are

$$\frac{C}{R} = \frac{\mathbf{G}}{1 + \mathbf{GH}} \qquad 11.5$$

$$\frac{E}{R} = \frac{1}{1 + \mathbf{GH}} \qquad 11.6$$

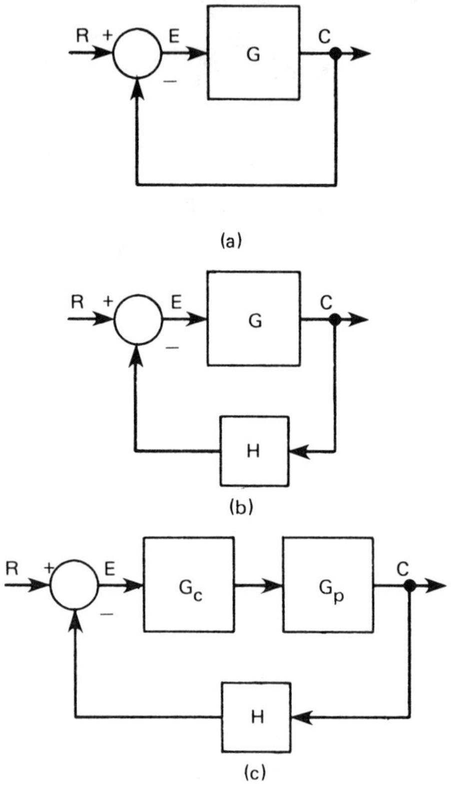

Figure 11.1 Block diagrams: (a) unity feedback, (b) general feedback, and (c) general system with compensation

In the most general case, the forward transfer function will consist of a controlled *plant* and a *compensator* to improve the stability or to reduce the steady-state errors of the system. In this case, the forward transfer function is expressed as a product of the *compensator transfer function* $\mathbf{G}_c$ and the *plant transfer function* $\mathbf{G}_p$. This is indicated in figure 11.1(c).

The terminology for control systems is based on the classical positioning control system, called a *servomechanism.* The *controlled variable* corresponds to a position (typically an angular position). Thus, when a second feedback proportional to the derivative of the output is used, it is called *velocity feedback* or *tachometer feedback.*

Since d/dt corresponds to s in Laplace-transformed variables, a unity feedback augmented with tachometer feedback will have the form

$$\mathbf{H} = 1 + k_t s \qquad 11.7$$

k_t is the *effective tachometer feedback coefficient.* For a true *tachometer*, the terminal voltage is proportional to the angular velocity Ω. (A_t is the ratio of tachometer voltage to the angular velocity in equation 11.8.)

$$V_t = A_t \Omega \qquad 11.8$$

The classical positioning control system is shown in figure 11.2. The output device is a D.C. motor with separate field excitation, and the armature is driven by a power amplifier.

A separately excited D.C. servomotor will have its armature current limited by the back-EMF, which is proportional to the angular velocity of the shaft. The proportionality constant is K_v. It can be found transformed from the steady-state circuit equation

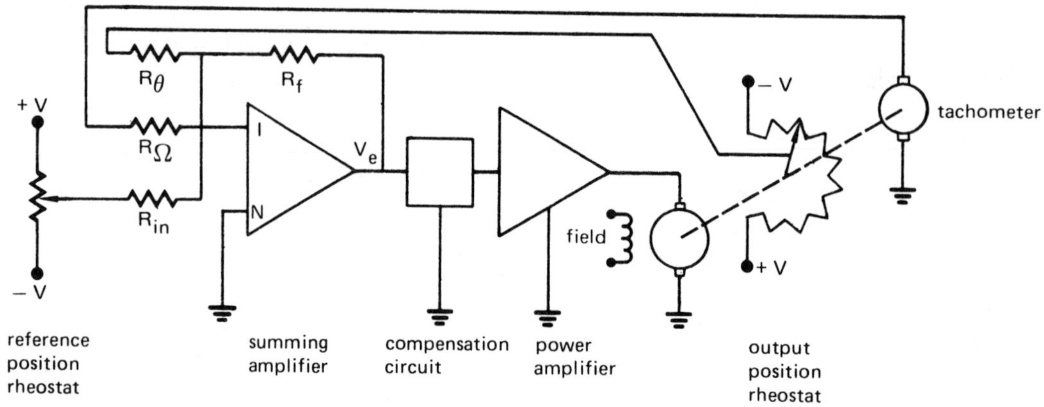

Figure 11.2 D.C. motor positional control system with tachometer feedback

$$V_S = (R_a + R_S)I_a + L_a\frac{dI_a}{dt} + K_v\Omega$$
$$= (R_a + R_S + sL_a)I_a + K_v\Omega \qquad 11.9$$

In steady state, $s = 0$, so the armature voltage V_a is

$$V_a\big|_{\text{steady state}} = I_aR_a + K_v\Omega \qquad 11.10$$

K_v can be calculated when the armature voltage, the armature current, and the shaft velocity are known.

For a *linear system*, the torque applied to the shaft by the armature current is balanced by the friction counter-torque and the change in angular momentum.[2]

$$T = D\Omega + J\frac{d\Omega}{dt}$$
$$= (D + Js)\Omega \qquad 11.11$$

D is the friction coefficient, and J is the mass moment of inertia of the rotating part. The *mechanical time-constant* is J/D.

$$\tau_m = \frac{J}{D} \qquad 11.12$$
$$T = D(1 + \tau_m s)\Omega \qquad 11.13$$

The electromechanical energy conversion relationship is

$$P_{\text{shaft}} = T\Omega = E_gI_a$$
$$= K_v\Omega I_a \qquad 11.14$$

Therefore, equations 11.11 and 11.13 can be written as

$$T = K_vI_a \qquad 11.15$$

Or,

$$I_a = \frac{D}{K_v}(1 + \tau_m s)\Omega \qquad 11.16$$

$s = 0$ in steady state, so D can be determined from speed, armature current, voltage, and armature terminal voltage.

$$D\big|_{\text{steady state}} = \frac{K_vI_a}{\Omega} = \frac{\left(\frac{E_g}{\Omega}\right)I_a}{\Omega}$$
$$= \frac{E_gI_a}{\Omega^2} \qquad 11.17$$

And,

$$E_g\big|_{\text{steady state}} = V_a - I_aR_a \qquad 11.18$$

Substituting equation 11.16 into equation 11.9,

$$\frac{V_S}{\Omega} = [(R_S + R_a) + sL_a](1 + s\tau_m)\frac{D}{K_v} + K_v \qquad 11.19$$

Or,

$$\frac{\Omega}{V_S} = \frac{K_v}{K_v^2 + D(R_a + R_S) + sD[L_a + \tau_m(R_a + R_S)] + s^2\tau_mL_aD} \qquad 11.20$$

[2] D and J are constants in linear systems.

Equation 11.20 normally is written as

$$\frac{\Omega}{V_S} = \frac{K}{1 + \frac{2\zeta}{\omega_n} + \frac{s^2}{\omega_n^2}} \qquad 11.21$$

$$K = \frac{K_v}{K_v^2 + D(R_a + R_S)} \qquad 11.22$$

$$\omega_n^2 = \frac{K_v^2 + D(R_a + R_S)}{DL_a\tau_m} \qquad 11.23$$

$$2\zeta = \frac{\omega_nD[L_a + \tau_m(R_a + R_S)]}{K_v^2 + D(R_a + R_S)} \qquad 11.24$$

The denominator of equation 11.21 has two s roots, p_1 and p_2, called *poles*.

$$s = p_1, p_2\big|_{\zeta > 1} = -\zeta\omega_n \pm \omega_n\sqrt{\zeta^2 - 1} \qquad 11.25$$

$$s = p_1, p_2\big|_{\zeta < 1} = -\zeta\omega_n \pm j\omega_n\sqrt{1 - \zeta^2} \qquad 11.26$$

$$s = p_1, p_2\big|_{\zeta = 1} = -\omega_n \qquad 11.27$$

The position θ, not the velocity, is to be controlled. Since $\Omega = s\theta$, equations 11.25, 11.26, and 11.27 can be rewritten as equations 11.28, 11.29, and 11.30.

$$\frac{\theta}{V_S}\Big|_{\zeta < 1} = \frac{K}{s[1 + (\frac{2\zeta}{\omega_n})s + (\frac{1}{\omega_n^2})s^2]} \qquad 11.28$$

$$\frac{\theta}{V_S}\Big|_{\zeta > 1} = \frac{K}{s[1 + (\frac{s}{p_1})][1 + (\frac{s}{p_2})]} \qquad 11.29$$

$$\frac{\theta}{V_S}\Big|_{\zeta = 1} = \frac{K}{s[1 + (\frac{s}{\omega_n})]^2} \qquad 11.30$$

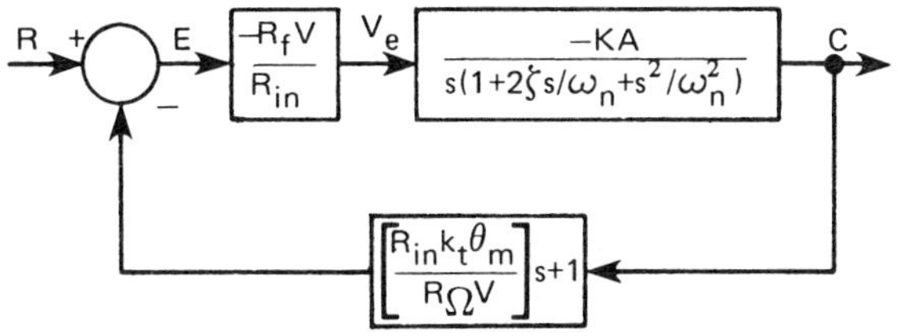

Figure 11.3 Servo block diagram

The position-measuring rheostats have a maximum voltage swing of $-V$ to $+V$, so the angular output and reference can be normalized to these maximum values.

$$V_r = \left(\frac{V}{\theta_m}\right)\theta_r = V\left(\frac{\theta_r}{\theta_m}\right)$$
$$= VR \qquad 11.31$$

$$V_c = -\left(\frac{V}{\theta_m}\right)\theta_c = -V\left(\frac{\theta_c}{\theta_m}\right)$$
$$= -VC \qquad 11.32$$

The operational amplifier used as a summing point adds the inputs weighted by their resistances.

$$V_e = -\left[\left(\frac{R_f V}{R_{\text{in}}\theta_m}\right)\theta_r - \left(\frac{R_f V}{R_\theta \theta_m}\right)\theta_c - \left(\frac{R_f k_t s}{R_\Omega}\right)\theta_c\right]$$

$$= -V\left[\left(\frac{R_f}{R_{\text{in}}}\right)R - \frac{R_f}{R_\theta}\left(1 + \frac{R_\theta k_t \theta_m s}{R_\Omega V}\right)C\right] \qquad 11.33$$

For unity feedback, it is necessary that $R_\theta = R_{\text{in}}$. Then, equation 11.33 would be rewritten as

$$V_e\Big|_{\text{unity feedback}} = \frac{-VR_f}{R_{\text{in}}}\left[R - C\left(1 + \frac{R_{\text{in}} k_t \theta_m s}{VR_\Omega}\right)\right] \qquad 11.34$$

The locations of R, C, and E are shown in figure 11.3.

The gain of the power amplifier must be negative to cancel out the negative effect of the summing amplifier gain. The open-circuit amplifier output voltage V_g is $-A_p V_e$. We can combine the machine's transfer function with the summing amplifier gain.

$$\theta_c = \left(\frac{A_p R_f V}{R_{\text{in}}}\right) E \times \frac{K}{s[1 + (\frac{2\zeta}{\omega_n})s + (\frac{1}{\omega_n^2})s^2]} \qquad 11.35$$

However, $\theta_c = C\theta_m$, so the forward transfer function is

$$G = \frac{C}{E} = \frac{A}{s[1 + (\frac{2\zeta}{\omega_n})s + (\frac{1}{\omega_n^2}))s^2]} \qquad 11.36$$

$$A = \frac{A_p R_f V K}{R_{\text{in}}\theta_m} \qquad 11.37$$

$$H = 1 + ks \qquad 11.38$$

$$k = \frac{R_{\text{in}} k_t \theta_m}{R_\Omega V} \qquad 11.39$$

2 BLOCK DIAGRAM ALGEBRA

The previous section illustrated how an engineering system can be modeled by operational black boxes. Several such boxes can be grouped together to obtain the desired response.

Complex systems of several box diagrams can be simplified by using the equivalent structures shown in table 11.1.

Example 11.1

A complex block system is constructed from five blocks and two summing points as shown.

Simplify the system and determine its overall effect on an input.

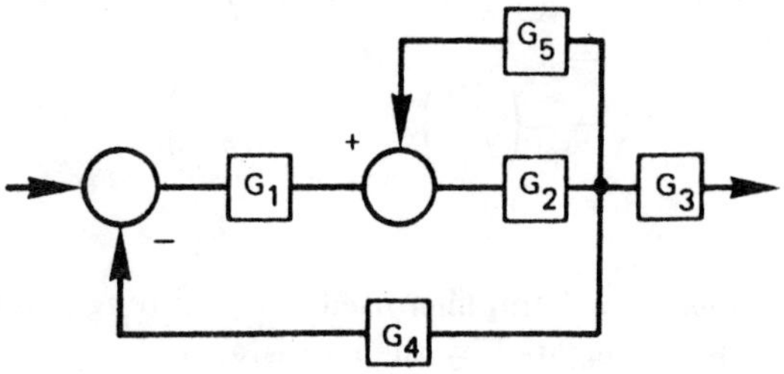

Use case 5 from table 11.1 to move the second summing point back to the first summing point.

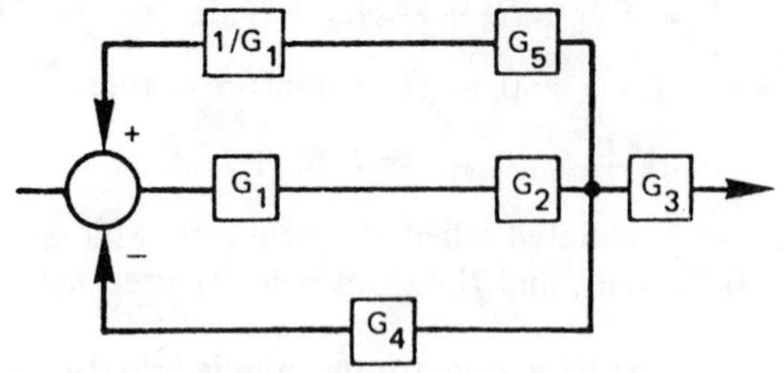

Use case 1 to combine boxes in series.

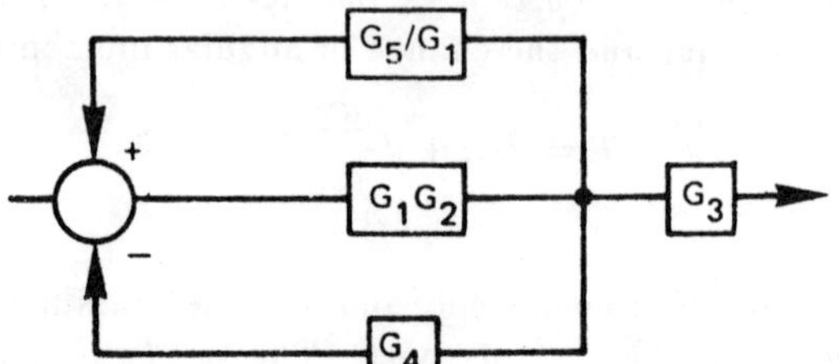

Use case 2 to combine the two feedback loops.

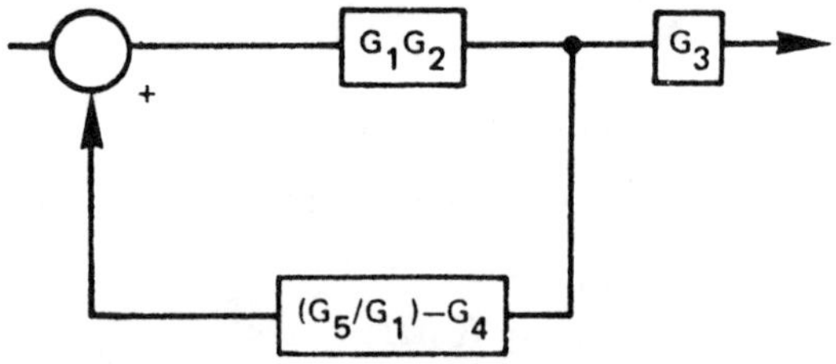

Use case 8 to move the pick-off point outside of the G_3 box.

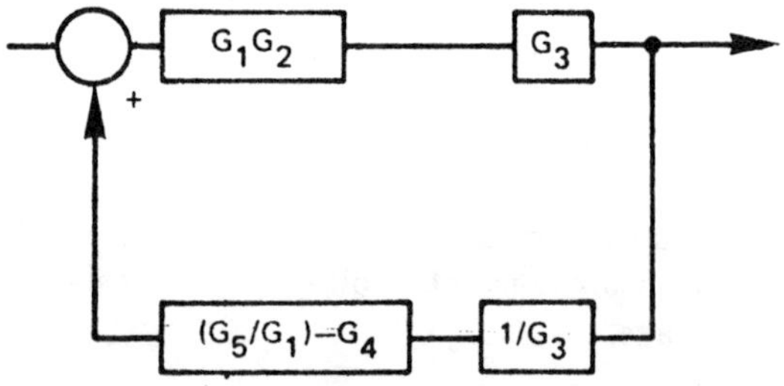

Use case 1 to combine the boxes in series.

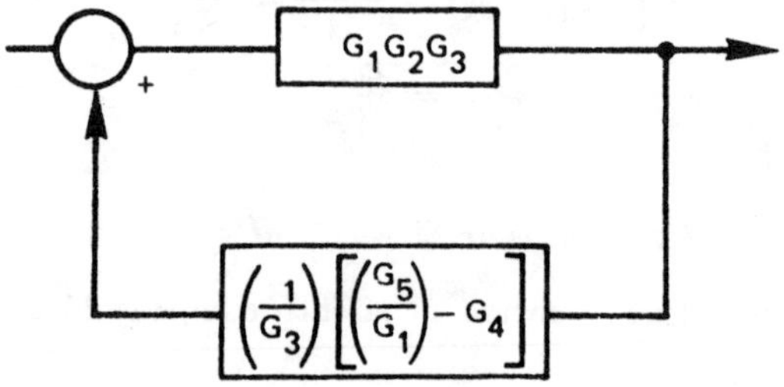

Use case 3 to determine the system gain.

$$\mathbf{G} = \frac{G_1 G_2 G_3}{1 - (G_1 G_2 G_3)(\frac{1}{G_3})[(\frac{G_5}{G_1}) - G_4]}$$

The system gain equation can be simplified.

$$\mathbf{G} = \frac{G_1 G_2 G_3}{1 - G_2 G_5 + G_1 G_2 G_4}$$

Table 11.1 Equivalent Block Diagrams

case	original structure	equivalent structure
1	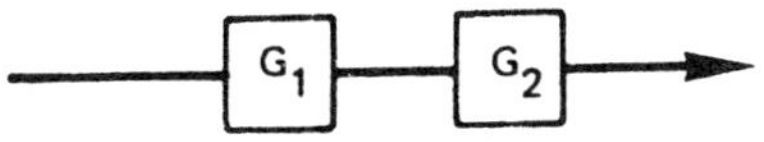	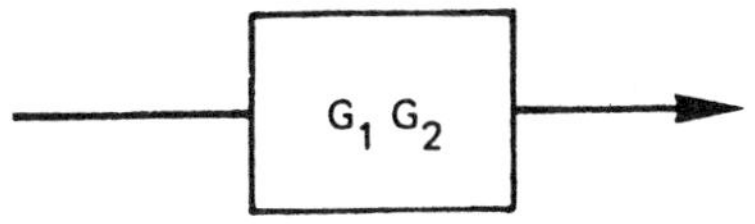
2	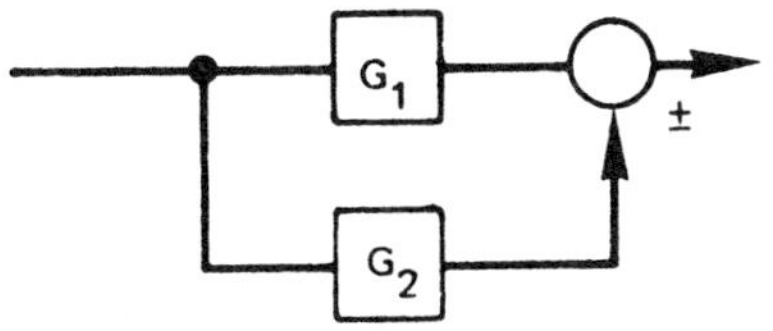	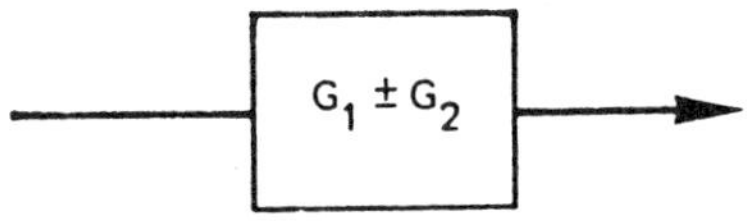
3	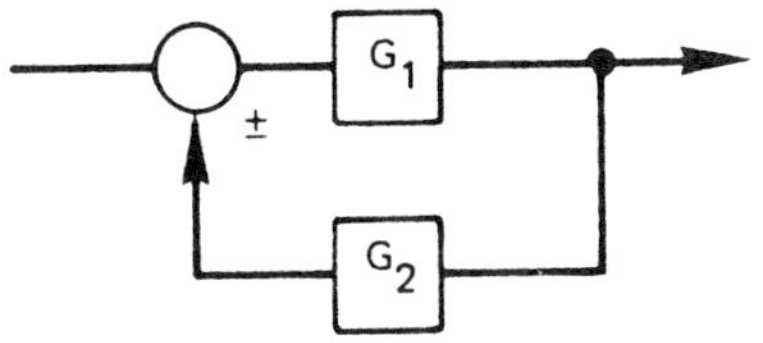	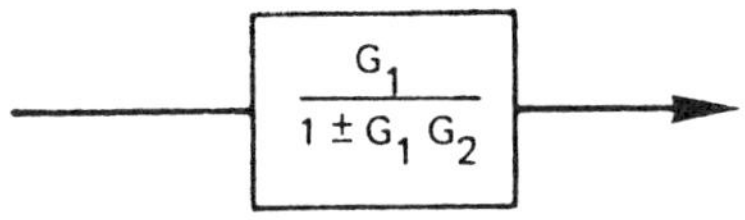
4	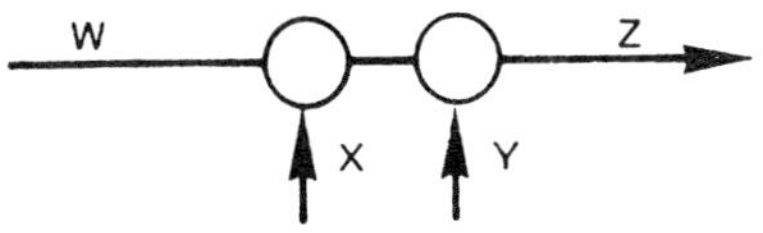	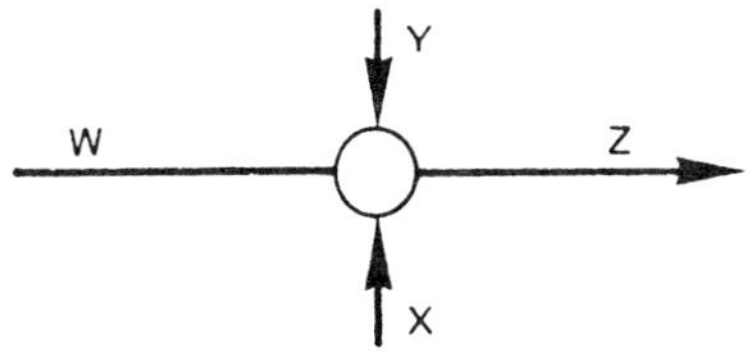
5	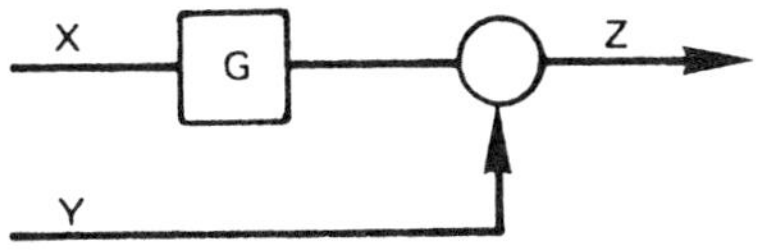	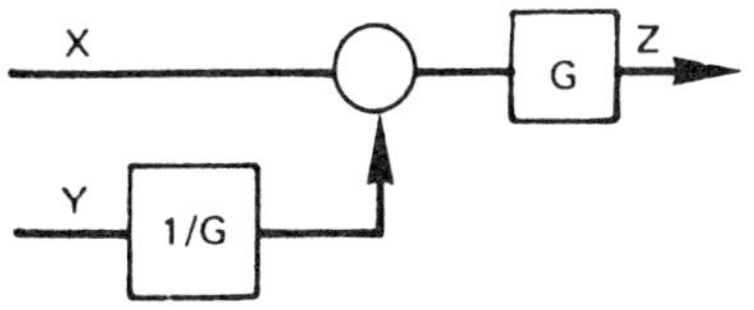
6	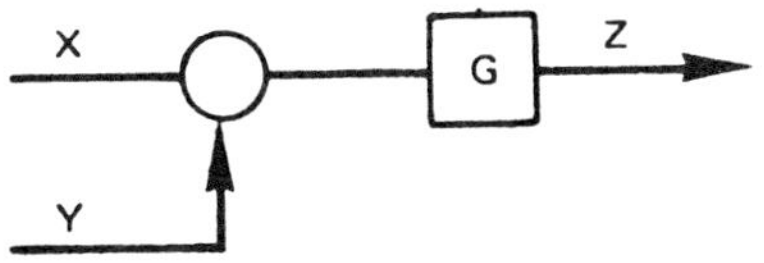	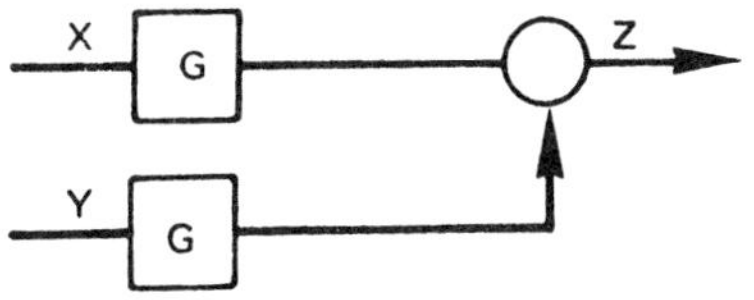
7	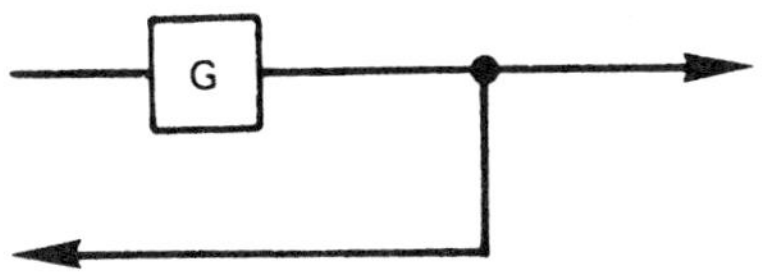	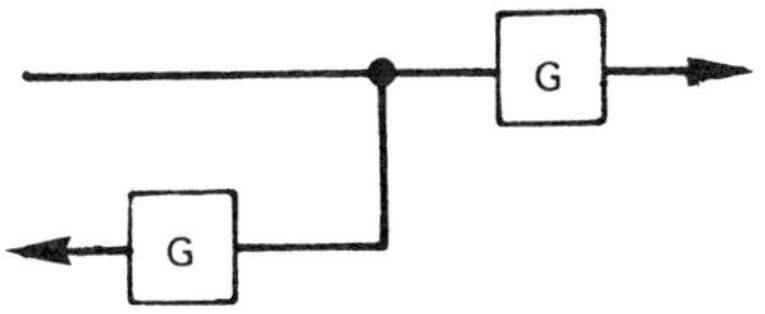
8	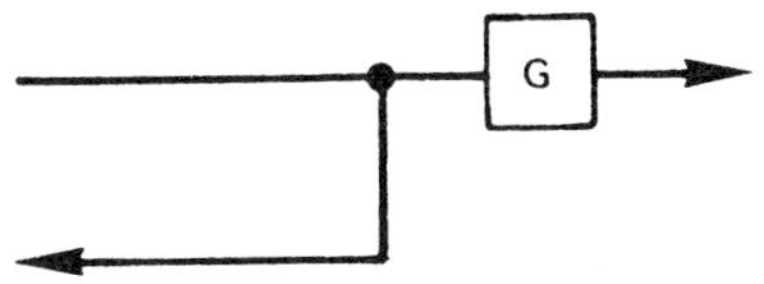	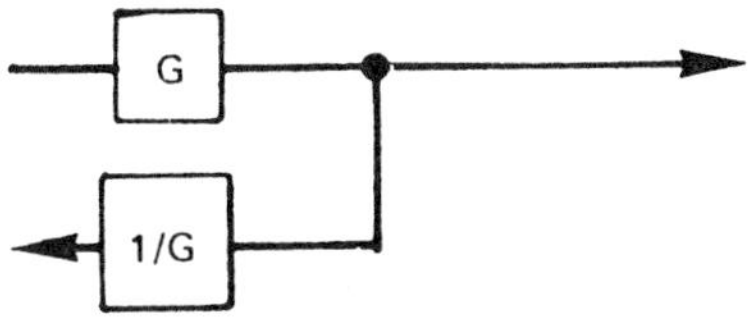

3 BODE DIAGRAMS

A primary design problem with control systems is assuring system stability. If the system is stable, the nature of the response of the system and the accuracy of the system also can be evaluated. A great deal can be determined from the *frequency response* of the system. The most useful form of frequency response data is the *Bode diagram,* where the amplitude (in decibels) and the phase angle (in degrees) are plotted against the logarithm of frequency.

When the open loop transfer function **GH**, is expressed in terms of the transform variable s, frequency response can be determined by replacing s with $j\omega$. The logarithmic nature of the Bode diagram is most useful when the transfer function is expressed in factored form. The transfer function can, in general, be expressed as a ratio of two polynomials.

$$\mathbf{GH}(s) = A\left[\frac{a_n s^n + a_{n-1}s^{n-1} + a_{n-2}s^{n-2} + \cdots + a_1 s^1 + a_0}{b_m s^m + b_{m-1}s^{m-1} + b_{m-2}s^{m-2} + \cdots + b_1 s^1 + b_0}\right] \qquad 11.40$$

This is more conveniently expressed in factored form.

$$\mathbf{GH}(s) = \frac{a_0 A}{b_0}\left[\frac{(1+\frac{s}{z_1})(1+\frac{s}{z_2})\cdots(1+\frac{s}{z_n})}{(1+\frac{s}{p_1})(1+\frac{s}{p_2})\cdots(1+\frac{s}{p_m})}\right] \qquad 11.41$$

The z's are the zeros of **GH**, and the p's are the poles of **GH**. Taking the logarithm of a complex number (s is a complex variable),

$$\log \mathbf{GH} = \log|\mathbf{GH}| + j(\log_e)\angle\mathbf{GH} \qquad 11.42$$

(*log* is understood to be a logarithm to the base 10, and e is the base of a natural logarithm.)

The Bode diagram measures amplitude in decibels.

$$\mathbf{GH}_{db} = 20\log|\mathbf{GH}| \qquad 11.43$$

Consider the fact that $\log(AB) = \log A + \log B$, and $\log(A/B) = \log A - \log B$, expressing $\mathbf{GH}(s)$ in decibels as in equation 11.42, the magnitude part is

$$\mathbf{GH}_{db} = 20\log\left(\frac{a_0 A}{b_0}\right) + 20\log\left|1+\frac{s}{z_1}\right| + 20\log\left|1+\frac{s}{z_2}\right|\cdots - 20\log\left|1+\frac{s}{p_1}\right| - 20\log\left|1+\frac{s}{p_2}\right|\cdots - 20\log\left|1+\frac{s}{p_m}\right| \qquad 11.44$$

Each part of **GH** (in the factored form) can be plotted separately and the results added. However, functions of the form $\mathbf{F} = 20\log\left|1+\frac{s}{s_0}\right|$ have an assymptotic behavior which can be used to advantage. For small values of s, the function approaches 0 *db* ($\log 1 = 0$). For large values of s, the function approaches

$$20\log\left|\frac{s}{s_0}\right| = 20\log|s| - 20\log|s_0|$$

This intersects the low-frequency assymptote ($s = j\omega$, and the low frequency assymptote is 0 *db*) where $\omega = s_o$.

When the amplitude in decibels is plotted against $\log\omega$, the high frequency assymptote ($20\log|s/s_o|$) has a slope of 20 db per unit of $\log\omega$. Since a unit for a logarithmic scale is a change of the argument by a factor of 10 (i.e., $\log 1000 - \log 100 = 3 - 2 = 1$), the logarithmic unit is a *decade* of frequency. Therefore, the slope of **F** is 20 db per decade.

Figure 11.4 shows the assymptotic behavior of functions like **F**, as well as for constants, derivatives, and resonant terms. (Note that when a term is in the denominator of the transfer functions, the slope is negative.)

$$F_a = -20\left|1+\frac{s}{\omega_p}\right| \qquad (a)$$

$$F_b = 20\log|K| \qquad (b)$$

$$F_c = -20\log|s| \qquad (c)$$

$$F_d = -20\log\left|1+\frac{2\varsigma s}{\omega_n}+\frac{s^2}{\omega_n^2}\right| \qquad (d)$$

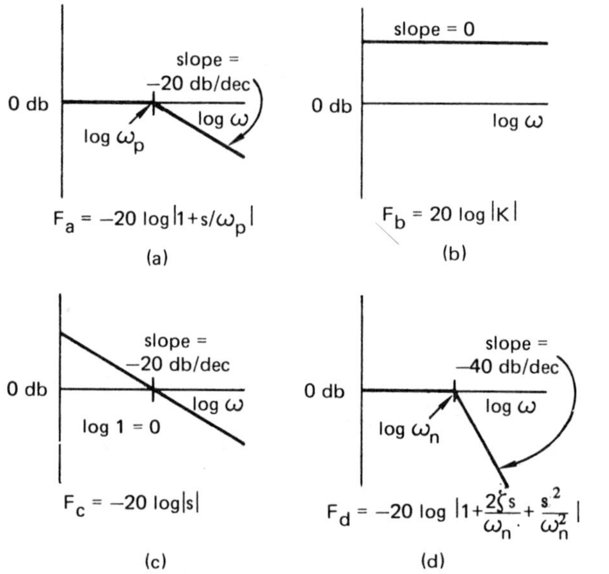

Figure 11.4 Assymptotic behavior plotted on a Bode diagram

The errors involved in the four types of terms plotted assymptotically in figure 11.4 are of considerable interest. For the constant and derivative types plotted in (b) and (c), there is no error. For those representing a single pole or zero, as in (a), the maximum error is 3 db at the frequency of the pole or zero. At frequency one octave (a factor of .5 or 2), the error is approximately 1 db. At two octaves (a factor of 4 or .25) from the pole or zero, the error is approximately .25 db. The errors which arise with a resonant term, as in (d), can be quite

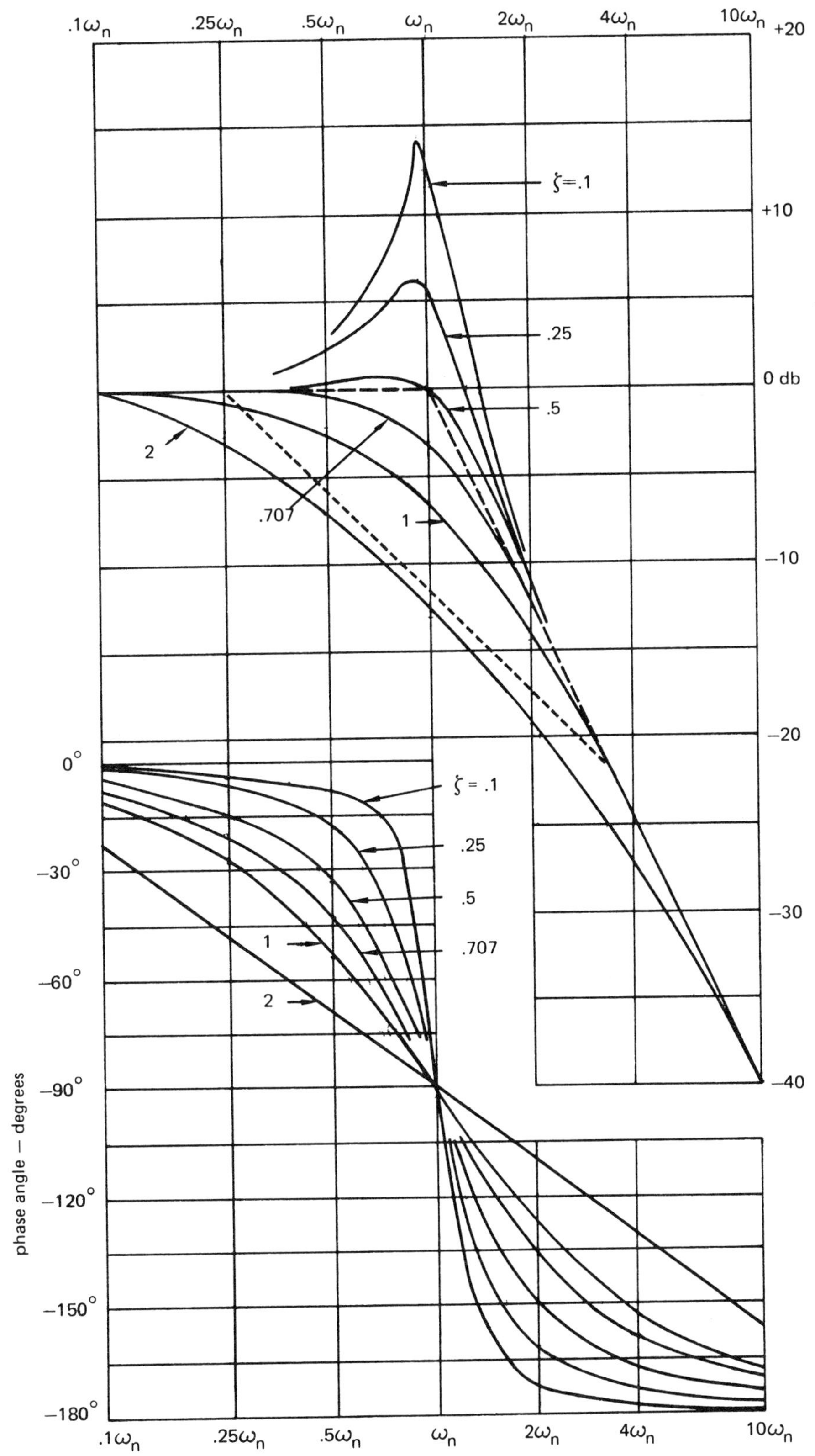

Figure 11.5 Bode diagram for resonant term (as in figure 11.4) showing the error from the assymptotic value for various values of ζ

significant, depending on the value of the damping constant ς. Representative curves are shown in figure 11.5. Note that for $\varsigma = 2$, the term can be broken into two real poles with the dotted assymptote shown.

Using the assymptotes, it is possible to construct the assymptotic Bode diagram by adding the individual contributions. Where the terms involving s are in the form $20 \log |1 + s \times f(s)|$, the effect of a pole or zero does not have to be considered until moving to the positive frequencies beyond the pole or zero. This is because the contribution of a term at low frequencies is zero. The procedure for plotting the assymptotic Bode diagram is given here.

step 1: Find the db value of the constant term. Plot it as a horizontal line (dashed) at that height parallel to the 0 db line.

step 2: For any s^n terms, the slope will be $20 \times n$ db/dec. It will be a positive slope if the numerator term is a zero. It will be a negative slope if the denominator term is a pole. Such a term will cross the 0 db axis at $\omega = 1$. When added to the constant term, the s^n term will intersect the original constant term at $\omega = 1$. This is illustrated in figure 11.6.

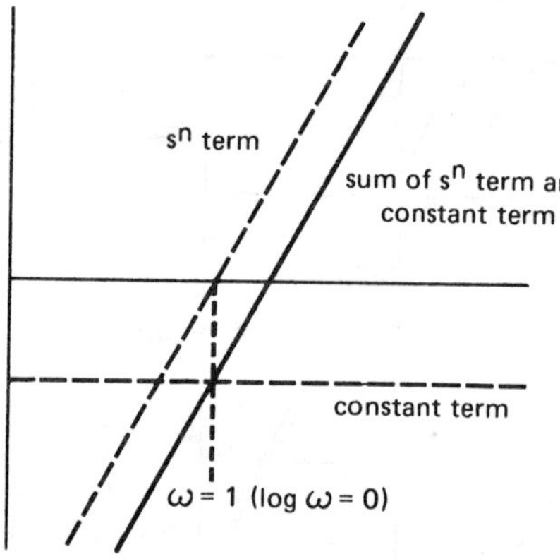

Figure 11.6 Assymptotic addition of constant and s^n terms

The sum assymptote should not be extended to the right of any pole or zero. It should be ended at the first additional pole or zero.

step 3: Upon reaching the next pole or zero, add its slope to the existing slope, beginning at the pole or zero frequency. Simple poles have negative slope, so they subtract 20 db/dec from the slope immediately to the left. Simple zeros add 20 db/dec to the slope immediately to the left. This is illustrated in figure 11.7 for a function $10(1+s/20)/s(1+s/2)$.

step 4: For a resonant term, the slope will change by 40 db/dec from the slope just to the left of the pole or zero. It will be a positive change for a zero and a negative change for a pole.

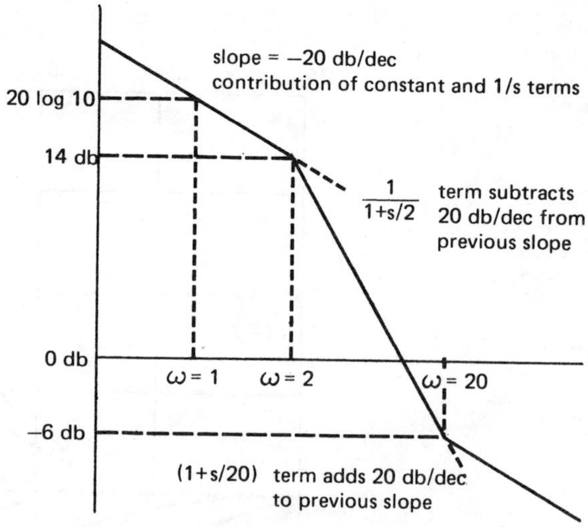

Figure 11.7 Sample assymptotic Bode diagram (not to scale)

The assymptotic Bode plot is typically within 3 db of the actual value unless poles are close together. A decade (factor of 10) is enough to cause an error of no more than 1 db at the pole, but an *octave* (factor of 2) between poles (or between zeros) can cause an error of 4 db at the poles. The adjacency of a pole and a zero tends to diminish the error for both to less then 3 db.

However, there can be significant errors with a resonant term. With ς less than 0.5, the error at the poles natural frequency is not under the assymptotes, but above it as in figure 11.5. The actual curve will be below the assymptotes for a pole and above the assymptotes for a zero for simple poles and zeros or complex poles and zeros with ς greater than .707.

Example 11.2

Find the error at the natural frequency ω_n for a system consisting of a resonant pole pair and a zero one octave below the resonant frequency. Use a damping factor $\varsigma = 0.2$

This can be normalized to the resonant frequency.

$$\mathbf{GH}|_{s = j\omega_n} = K\left[\frac{1+\frac{2s}{\omega_n}}{1+\frac{2(.2)s}{\omega_n}+\frac{s^2}{\omega_n^2}}\right]$$

$$= K\left[\frac{1+j2}{1+j.4-1}\right] = K\left[\frac{1+j2}{j.4}\right]$$

The amplitude is

$$\frac{K\sqrt{5}}{.4} = 5.59K \mapsto (20\log K + 14.95)\text{db}$$

Next, the assymptotic value at the resonant frequency is needed. Up to the zero, the value was 20 log K. The zero adds a slope of 20 db per decade. An octave is a factor of 2 and $\log 2 = .301$, while a decade is a factor of 10 and $\log 10 = 1$. Then, the change in an octave will be

$$20 \times .301 = 6.02\,\text{db}$$

Thus, the asymptotic value at resonance will be

$$20 \log K + 6.02\,\text{db}$$

The error is

$$14.95 - 6.02 = 8.93\,\text{db}$$

This can be verified analytically. Since the curve for $\varsigma = .5$ intersects the assymptote at resonance, and the high-frequency assymptote of the zero is $2s/\omega_n$, it is possible to use the assymptotic zero value and the value of the resonant term with a damping of 0.5.

$$\begin{aligned} \mathbf{GH}|_{\text{assymptotic}} &= 20 \log K + 20 \log 2 - 20 \log 1 \\ &= 20 \log K + 6.02 \end{aligned}$$

The *phase response* of a system is not as easily plotted as the amplitude because the assymptotes are not straight lines. Strictly speaking, they are not assymptotes at all. Rather, they are very predictable curves for simple poles and zeros.

For resonant terms, the phase graph in figure 11.5 is applicable. It contributes negative phase for denominator (poles) terms and positive phase for numerator terms (zeros). For simple poles and zeros, there is a 45° phase shift at the pole or zero frequency. The effect of a pole or zero on phase is much more extensive than the effect on amplitude. The phase effect is felt over several octaves of frequency above and below the pole or zero location. Table 11.2 gives values at octave intervals. Intermediate values can be calculated from equation 11.45.

$$\text{phase angle} = \arctan\left(\frac{\omega}{z}\right) = -\arctan\left(\frac{\omega}{p}\right) \qquad 11.45$$

Phase angles for the entire system must be obtained by summing the contributions of all the zeros and subtracting the contributions of all poles. The phase angle is important because of feedback. The denominator of equations 11.5 and 11.6 is 1+**GH**. This can go to zero for $|\mathbf{GH}| = 1$ when $\angle\mathbf{GH} = 180°$.

For absolute stability of a feedback control system, it is necessary that the gain be less than 1 (0 db) for any frequency at which the phase is +180° or −180°. The usual case in control systems is for the the phase to be negative, so the −180° phase point is of primary interest.

Satisfactory transient and steady-state performance of control systems requires that the phase be no worse than −150° for the frequency at which the gain crosses the 0 db line. The difference between the actual phase and 180° at this point is called the *phase margin*. As *critical damping* is obtained with a phase angle of 45°, the closed loop system may be underdamped if the phase margin is less than 45°. The phase margin and the *gain margin* are shown in figure 11.8. Empirical results indicate that the gain margin should be at least 8 to 10 db.

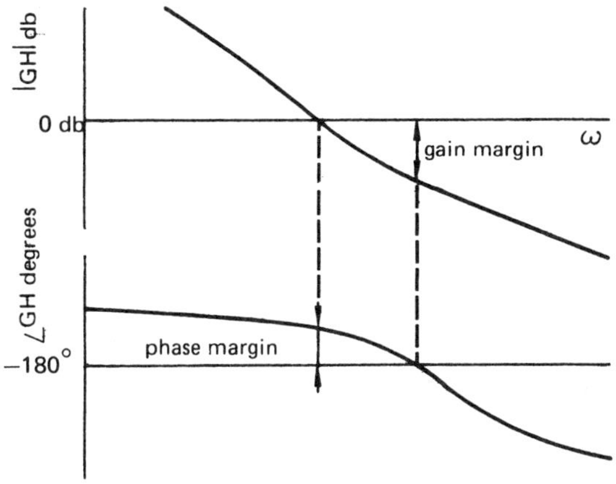

Figure 11.8 Gain and phase margins

4 THE NYQUIST DIAGRAM

There are other graphic methods of representing frequency functions. In the *Nyquist diagram*, the amplitude of GH is plotted against the phase angle in a polar form. In such plots, the points are found by determining both the magnitude and the phase angle at each frequency. Thus, the frequency is a parameter on the curves.

The *absolute stability* of a closed-loop system can be determined from the behavior of the curve near the −1

Table 11.2 Phase angles for simple poles and zeros

octave	−5	−4	−3	−2	−1	0	1	2	3	4	5
ω/ω_{pz}	1/32	1/16	1/8	1/4	1/2	1	2	4	8	16	32
phase °	1.8	3.6	7.1	14.0	26.6	45	63.4	76	82.9	86.4	88.2

point, as indicated in figure 11.9. The system will be absolutely stable if the curve of $|\mathbf{GH}|$ versus $\angle\mathbf{GH}$ does not encircle the -1 point on the Nyquist diagram.

The behavior of a closed-loop system will be determined largely by the nearness of the curve to the -1 point. It is useful to note that the gain and phase margins also are available on the Nyquist diagram. These are indicated in figure 11.9.

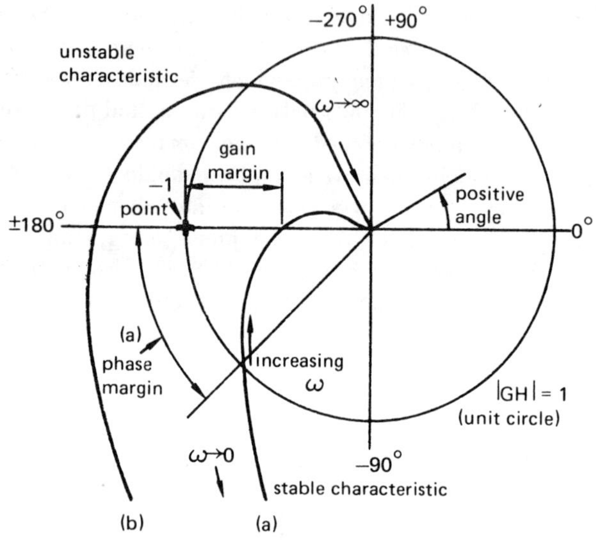

Figure 11.9 Nyquist Diagram: (a) stable function, (b) unstable function

The two curves shown in figure 11.9 are typical of actual control systems with a single integrator. Curve (a) is stable by the Nyquist criterion because it does not encircle the -1 point. However, if the gain of the system is increased sufficiently, it will approach the unstable characteristic form (b) which does encircle the -1 point. Then the system would become unstable by the Nyquist criterion.

In general, the Nyquist diagram is much more difficult to construct than the Bode diagram. It contains no more information than the Bode diagram, so it is not usually chosen as the method to present system information. When the system characteristics are presented in this form, however, it is important to be able to interpret the stability and phase margin.

5 STEADY-STATE ERRORS

A. INTRODUCTION

When the open-loop transfer function is expressed as a function of s, and the input is transformable to a function of s, the error as a function of s can be written as equation 11.46.

$$E(s) = R(s) \times \frac{1}{1 + \mathbf{GH}(s)} \qquad 11.46$$

There are three types of input for which errors are well defined: the *step input* results in a so-called position error; the *ramp* or *velocity input* results in an error in velocity; and the *parabolic input*, corresponds to an error in acceleration.

The type of response is determined by the number of integrators in the open-loop system. Integrators result in denominator terms in **GH** of the form s^n, where n is the number of integrators. A *type one system* has one integrator, a *type two system* has two integrators, and a *type zero system* has no integrators in the open-loop transfer function $\mathbf{GH}(s)$. The corresponding values of n are 1, 2, and 0 respectively.

B. POSITION ERROR

The *position error* is the steady state error when the input $r(t)$ is a step function $R \times u(t)$ of magnitude R. The Laplace transform of the unit step is $1/s$, so $E(s)$ is

$$E(s) = \frac{R}{s} \times \frac{1}{1 + \mathbf{GH}(s)} \qquad 11.47$$

Using the Laplace final value theorem, the steady state positional error is

$$e_{ss}(t) = \lim_{s \to 0} \frac{R}{1 + \mathbf{GH}(s)} \qquad 11.48$$

For the three types of systems under consideration, $\mathbf{GH}(s)$ can be written similarly to equation 11.41.

$$\mathbf{GH}(s) = K\left[\frac{(1+\frac{s}{z_1})(1+\frac{s}{z_2})\cdots(1+\frac{s}{z_k})}{s^n(1+\frac{s}{p_1})(1+\frac{s}{p_2})\cdots(1+\frac{s}{p_m})}\right] \qquad 11.49$$

Taking the limit as s goes to zero, terms with the form $1 + s/s_j$ will become 1. Then,

$$\lim_{s \to 0} \mathbf{GH}(s) = \lim_{s \to 0} \frac{K}{s^n} \qquad 11.50$$

$$e_{ss}(t) = \lim_{s \to 0} \frac{Rs^n}{s^n + K} \qquad 11.51$$

For a type zero system ($n = 0$),

$$e_{ss}(t)|_{\text{type zero step input}} = \frac{R}{1 + K} \qquad 11.52$$

For a type one system ($n = 1$),

$$e_{ss}(t)|_{\text{type one step input}} = \frac{0}{0 + K} = 0 \qquad 11.53$$

For a type two system ($n = 2$),

$$e_{ss}(t)|_{\text{type two step input}} = \frac{0}{0 + K} = 0 \qquad 11.54$$

C. VELOCITY ERROR

The *velocity* is the standard term used for the output function when the input is a *ramp function* characterized by $r(t) = R't$. (R' is the slope of the ramp.) The Laplace transform of $R't$ is R'/s^2, so $E(s)$ is

$$E(s) = \frac{R'}{s^2[1 + \mathbf{GH}(s)]} \quad 11.55$$

Then from the final value theorem, the steady-state error for a ramp input is

$$e_{ss}(t) = \lim_{s\to 0} s \frac{R'}{s^2[1 + \mathbf{GH}(s)]} \quad 11.56$$

By using $\mathbf{GH}(s)$ from equation 11.50,

$$e_{ss}(t) = \lim_{s\to 0} \frac{\frac{R'}{s}}{1 + \frac{K}{s^n}}$$

$$= \lim_{s\to 0} \frac{R' s^{n-1}}{s^n + K} = \lim_{s\to 0} \frac{R' s^{n-1}}{K} \quad 11.57$$

For a type zero system ($n = 0$),

$$e_{ss}(t)|_{\text{type zero ramp input}} = \lim_{s\to 0} \frac{R'}{sK} = \infty \quad 11.58$$

For a type one system ($n = 1$),

$$e_{ss}(t)|_{\text{type one ramp input}} = \lim_{s\to 0} \frac{R'}{K} = \frac{R'}{K} \quad 11.59$$

For a type two system ($n = 2$),

$$e_{ss}(t)|_{\text{type two ramp input}} = \lim_{s\to 0} \frac{R's}{K} = 0 \quad 11.60$$

D. ACCELERATION ERROR

The acceleration error occurs when the input is a constant acceleration term of the form $r(t) = R''t^2$. This has a Laplace transform of $R(s) = 2R''/s^3$, so $E(s)$ is

$$E(s) = \frac{2R''}{s^3}\left[\frac{1}{1 + \mathbf{GH}(s)}\right] \quad 11.61$$

By inserting the limiting value of $\mathbf{GH}(s)$ as s goes to zero from equation 11.50,

$$e_{ss}(t) = \lim_{s\to 0} s \frac{2R''}{s^3\left[1 + \frac{K}{s^n}\right]}$$

$$= \lim_{s\to 0} \frac{2R'' s^{n-2}}{s^n + K} = \lim_{s\to 0} \frac{2R'' s^{n-2}}{K} \quad 11.62$$

For a type zero system ($n = 0$),

$$e_{ss}(t)|_{\text{type zero parabolic input}} = \infty \quad 11.63$$

For a type one system ($n = 1$),

$$e_{ss}(t)|_{\text{type one parbolic input}} = \lim_{s\to 0} \frac{2R'' s^{-1}}{K} = \infty \quad 11.64$$

For a type two system ($n = 2$),

$$e_{ss}(t)|_{\text{type two parabolic input}} = \lim_{s\to 0} \frac{2R'' s^0}{K} = \frac{2R''}{K} \quad 11.65$$

Systems without an integrator in the open loop (*type zero systems*) will have a finite position error and infinite velocity and acceleration errors. An infinite steady-state error indicates that the system cannot track the particular type of function which causes the error. The system will continually lose ground against the function.

A system with one integrator in the open loop will be able to follow a position input with zero steady-state error. That is, it will track the position perfectly, given enough time. A *type one system* will follow a ramp input with a constant error proportional to the velocity input, but it cannot track an acceleration input at all.

A *type two system* can track both position and velocity with no long-run error, but it will have a constant error to an acceleration input.

Systems with more than two integrators in the open loop will have closed loop performance able to track all derivatives up to one less than the number of integrators. However, systems with even two integrations have difficulty damping transients. Systems with more than two integrators are very difficult to stabilize.

6 THE ROUTH-HURWITZ STABILITY CRITERION

A mathematical method for determining the stability of a system from the open-loop transfer function is the *Routh-Hurwitz stability criterion.* This method is based on the concept of multiplying out the closed loop transfer function $\mathbf{G}_c = \mathbf{G}/(1 + \mathbf{GH})$ where $\mathbf{G}$ and $\mathbf{H}$ are expressed as ratios of polynomials in s.

$$\mathbf{G} = \frac{N_g}{D_g} = K\left[\frac{(s + z_{g1})(s + z_{g2})\cdots}{s^n(s + p_{g1})(s + p_{g2})\cdots}\right] \quad 11.66$$

$\mathbf{H}$ is expressed in terms of a numerator polynomial N_h and a denominator polynomial D_h. The rationalization of $\mathbf{G}_c$ is

$$\mathbf{G}_c = \frac{N_g D_h}{D_g D_h + N_g N_h} \quad 11.67$$

$\mathbf{G}_c$ will be absolutely stable if the denominator term has no zeros with positive real parts, since positive real parts will result in positive exponentials which are unstable.

The Routh-Hurwitz criterion starts by expanding the denominator of $\mathbf{G}_c$ in equation 11.67 to

$$D = a_n s^n + a_{n-1} s^{n-1} + a_{n-2} s^{n-2} + \cdots + a_1 s + a_o \quad 11.68$$

When this is done, note whether any of the a_j coefficients have different signs from the rest. If so, the system automatically is unstable since it will have zeros of the denominator (poles) with positive real parts. However, this criterion is not sufficient when there are no sign changes among the coefficients. There still may be poles with positive real parts.

The Routh-Hurwitz criterion is based on the development of an array generated from the coefficients of s in D. The first row is made up of a_n, a_{n-2}, a_{n-4}, etc., and the second row is generated from a_{n-1}, a_{n-3}, a_{n-5}, etc., as shown below:

$$\begin{array}{c|cccc} s^n & a_n & a_{n-2} & a_{n-4} & a_{n-6}\cdots \\ s^{n-1} & a_{n-1} & a_{n-3} & a_{n-5} & a_{n-7}\cdots \end{array}$$

The third row is generated from the first and second rows by taking the determinant of the two left-most coefficients of the first and second rows, along with the coefficients just to the right of the position where the new coefficient is to be placed, and then dividing the determinant by the negated left-most coefficient of the second row.

$$b_{n-1} = \frac{\begin{vmatrix} a_n & a_{n-2} \\ a_{n-1} & a_{n-3} \end{vmatrix}}{-a_{n-1}} \qquad 11.69$$

$$b_{n-3} = \frac{\begin{vmatrix} a_n & a_{n-4} \\ a_{n-1} & a_{n-5} \end{vmatrix}}{-a_{n-1}} \qquad 11.70$$

This results in the rows

$$\begin{array}{c|ccc} s^n & a_n & a_{n-2} & a_{n-4}\cdots \\ s^{n-1} & a_{n-1} & a_{n-3} & a_{n-5}\cdots \\ s^{n-2} & b_{n-1} & b_{n-3} & b_{n-5}\cdots \end{array}$$

The next row is generated from the two above it in a similar manner.

$$c_{n-3} = \frac{\begin{vmatrix} a_{n-1} & a_{n-5} \\ b_{n-1} & b_{n-5} \end{vmatrix}}{-b_{n-1}} \qquad 11.71$$

This process is continued until no more rows with non-zero terms can be generated. Notice the practice of placing decreasing powers of s to the left of the rows. This is based on the theoretical justification of the process, not discussed here. Beginning with the highest power of s at the first row, $n+1$ rows will be generated by the process. The last row will have s^0 to the left.

The Routh-Hurwitz stability criterion states that there are exactly the same number of poles with positive real parts as there are sign changes in going down the first column of the array. If there are no poles with positive real parts (i.e., no sign changes), the system is absolutely stable.

Example 11.3

Evaluate the stability of the system with an open loop transfer function.

$$\mathbf{GH}(s) = \frac{5s^2 + 3s + 2}{3s^4 + 2s^3 + 3s^2 + s}$$

The denominator of $\mathbf{G}_c$ will be

$$\begin{aligned} D &= (3s^4 + 2s^3 + 3s^2 + s) + (5s^2 + 3s + 2) \\ &= 3s^4 + 2s^3 + 8s^2 + 4s + 2 \end{aligned}$$

Then the first two rows of the array are

$$\begin{array}{c|ccc} s^4 & 3 & 8 & 2 \\ s^3 & 2 & 4 & \end{array}$$

So,

$$b_3 = \frac{\begin{vmatrix} 3 & 8 \\ 2 & 4 \end{vmatrix}}{-2} = 2$$

$$b_1 = \frac{\begin{vmatrix} 3 & 2 \\ 2 & 0 \end{vmatrix}}{-2} = 2$$

The third row is

$$\begin{array}{c|cc} s^2 & 2 & 2 \end{array}$$

$$c_3 = \frac{\begin{vmatrix} 2 & 4 \\ 2 & 2 \end{vmatrix}}{-2} = 2$$

$$c_1 = \frac{\begin{vmatrix} 2 & 0 \\ 2 & 0 \end{vmatrix}}{-2} = 0$$

The fourth row is

$$\begin{array}{c|cc} s & 2 & 0 \end{array}$$

$$d_3 = \frac{\begin{vmatrix} 2 & 2 \\ 2 & 0 \end{vmatrix}}{-2} = 2$$

The entire array is

$$\begin{array}{c|ccc} s^4 & 3 & 8 & 2 \\ s^3 & 2 & 4 & \\ s^2 & 2 & 2 & \\ s^1 & 2 & & \\ s^0 & 2 & & \end{array}$$

There are no sign changes in the first column, so the system is stable.

7 ROOT LOCUS

A. INTRODUCTION

In order to evaluate the relative stability and transient performance of a closed loop system, the roots of the denominator of the closed loop transfer function must be determined. From the previous section, the roots of the closed loop system are the zeros of **G** and the poles of **H**.

The *root locus method* deals with the term **GH**, which for $\mathbf{GH} + 1 = 0$ must have a magnitude of 1 and an angle of $180° + 360n°$ (where n is an integer). When these conditions are satisfied, there is a zero-root of the denominator **G**. The root-loci all are possible points where the angle of the function is $180° + 360n°$. This results in curves of constant angles in the complex s-plane where poles and zeros of the transfer function $\mathbf{GH}(s)$ are plotted.

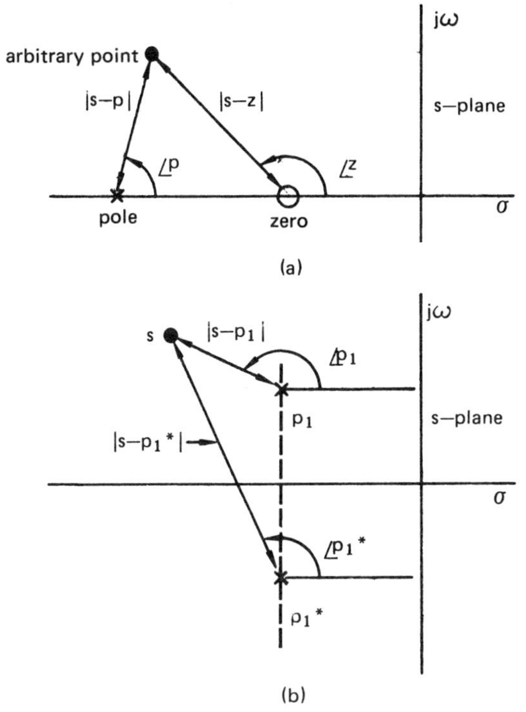

Figure 11.10 Angles and magnitudes in the s-plane

Consider the measurement of an angle at an arbitrary point in the s-plane for a function with one pole and one zero as shown in figure 11.10. Poles are represented by X's and zeros by 0's. Angles are measured counterclockwise from the positive-real direction. The function in figure 11.10(a) is defined by its magnitude and direction.

$$\mathbf{F} = \frac{s - z}{s - p} \qquad 11.72$$

$$|\mathbf{F}| = \frac{|s - z|}{|s - p|} \qquad 11.73$$

$$\angle \mathbf{F} = \angle(s - z) - \angle(s - p) \qquad 11.74$$

Thus, **F** can be evaluated by dividing the distance from the zero to s by the distance from the pole to s. The angle for **F** is the angle $\angle z$ minus the angle $\angle p$.

Figure 11.10(b) shows a pair of complex poles: p_1 and its conjugate p_1^*. The function defined by this pair has magnitude and direction given by equations 11.75 and 11.76.

$$|\mathbf{F}| = \frac{1}{|s - p||s - p_1^*|} \qquad 11.75$$

$$\angle \mathbf{F} = -\angle(s - p_1) - \angle(s - p_1^*) \qquad 11.76$$

B. ROOT LOCUS ON THE REAL AXIS

The root locus on the real axis is independent of any complex poles or zeros. This is because the contribution of angle from a complex conjugate pair is zero on the real axis. Complex zeros and poles always occur in conjugate pairs. The pole or zero in the upper-half-plane will contribute the negative of the angle contributed by the conjugate in the lower half-plane along the real axis.

For any point on the real axis, any zero to the right will contribute $+180°$ of phase angle, and any pole to the right will contribute $-180°$ to the phase angle at that point on the real axis. Thus, a root locus on the real axis will exist whenever there is an odd number of poles and zeros on the real axis to the right.

As the denominator of the closed loop transfer function is of the form $D_g D_h + K N_g N_h$ when the gain K of **GH** is zero, the roots of D are the same as the poles of **GH**. As K goes to infinity, the roots of D will be the roots of $N_g N_h$ (the zeros of the open loop transfer function **GH**). Thus, the roots begin at the poles of **GH** with zero gain and end on the zeros of **GH** with infinite gain.

Example 11.4

A system consisting of an integrator with variable gain has $\mathbf{GH} = K/s$. Determine the root locus for this sytem, and determine the gain necessary to place the closed loop pole at $s = -5$.

The root locus is shown. As there are no poles or zeros to the right of the origin of the s-plane, the positive real axis is not part of the root locus. There is one pole to the right of any point on the negative real axis, an odd number of poles and zeros, so the entire negative real axis is in the root locus. The locus extends from the origin to negative infinity (**GH** is zero at s equal to infinity, so there is a zero at infinity for the locus to terminate on).

To find the value of K for a closed loop pole at $s = -5$, write the equation **GH** $= -1$ and insert the desired value of s to solve for K.

$$GH = \frac{K}{s} = -1$$

$$= \frac{K}{-5}$$

Therefore, $K = 5$.

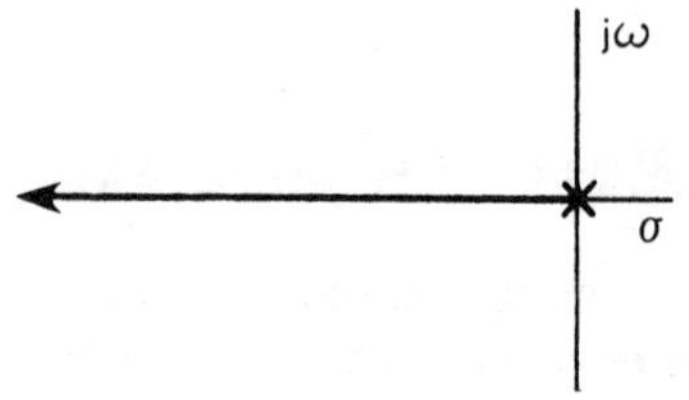

When two poles or two zeros are adjacent on the negative real axis, and there are an even number of poles and zeros to the right of the right-most of the pair, a root locus exists between the pair. That is only part of the root locus diagram because the loci begin on poles and end on zeros. The root loci must leave the negative real axis and become complex. Such root loci do so in conjugate fashion, as roots must occur in conjugate pairs when they have an imaginary part.

The simplest case is an integrator and one time constant of the form

$$\mathbf{GH} = \frac{K}{s(s-p)} \qquad 11.77$$

In this case the roots are defined by

$$s^2 - ps + K = 0 \qquad 11.78$$

The roots are

$$s = \frac{p}{2} \pm \sqrt{\left(\frac{p}{2}\right)^2 - K} \qquad 11.79$$

For $K = (p/2)^2$, there are two roots at $p/2$. Further increase in K will result in complex conjugate roots with a real part of $p/2$. Thus, the root locus will be as in figure 11.11(a). In the general case of only two poles, the root locus will leave the axis at the average of the two roots as indicated in figure 11.11(b).

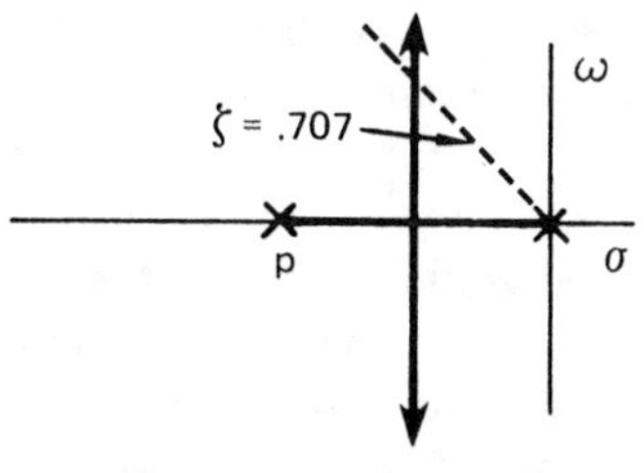

(a)

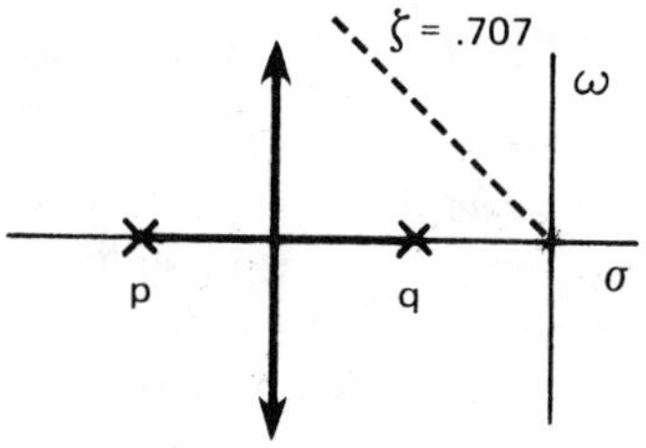

(b)

Figure 11.11 (a) Root loci for **GH** $= K/[s(s-p)]$, (b) root loci for **GH** $= K/[(s-p)(s-q)]$

C. ADJACENT POLES WITH A ZERO

Two adjacent poles with a zero will result in a circular root locus centered on the zero and having a radius which is the geometric mean of the distances from the zero to the poles.

$$\text{radius of root-locus circle} = \sqrt{(z-p_1)(z-p_2)} \qquad 11.80$$

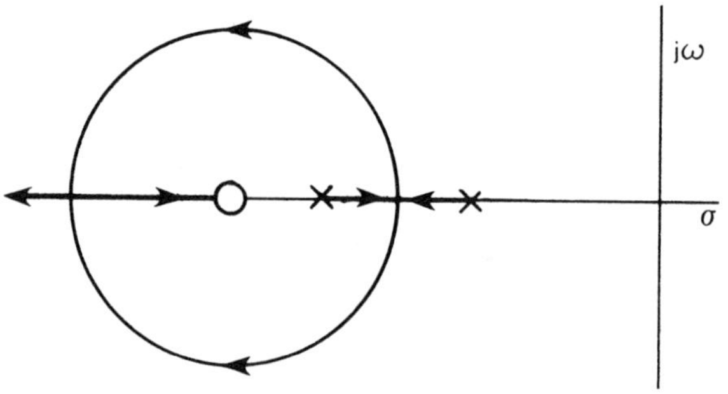

Figure 11.12 Root loci for two adjacent poles on real axis with a zero

D. HIGHER ORDER SYSTEMS

Higher order systems can become unstable if the root locus crosses into the positive half-plane ($\sigma > 0$). Systems with two poles (i.e., a second order system) and all zeros in the left-hand plane unconditionally will be stable. In previous sections, root loci were found on the real axis, parallel to the imaginary axis in the negative half-plane, and in a circle.

There also were a number of root loci going to infinity. The number of loci going to infinity is equal to the number of poles minus the number of zeros. When there is one excess pole, the single locus is on the negative real axis. When there are two excess poles, the loci are at $+90°$ and $-90°$.

These results are general: the number of loci going to infinity is the difference in the number of poles and zeros. The angles at which these loci go to infinity are symmetrical with respect to the negative real axis. The cases for one and two excess poles have been covered. For three excess poles (or excess zeros), the loci are at

$\pm 60°$ and $180°$. For four excess poles, the loci are at $\pm 45°$ and $\pm 135°$.

The root loci do not immediately follow these angles to infinity for third order and higher systems. They approach them as assymptotes. The assymptotes are centered at a point on the real axis which is found from equation 11.81.

$$X_{\text{assymptote}} = \frac{\text{sum of pole values} - \text{sum of zero values}}{\text{number of poles} - \text{number of zeros}} \quad 11.81$$

The angles are given by equation 11.82.

$$\text{angles} = \frac{2i+1}{n_p - n_z} \times 180° \quad \text{for } i = 0, 1 \cdots (n_p - n_z - 1) \quad 11.82$$

In equation, 11.82 n_p is the number of poles, and n_z is the number of zeros.

When there are two adjacent poles on the real axis with an intervening root locus, the root locus must leave the real axis and become complex at some point between the two poles. The point where this happens is called a *break-away point* . The value of this point can be found from equation 11.83 where σ_b is the location of the break-away point.

$$\sum_{i=1}^{n_p} \frac{1}{\sigma_b - p_i} = \sum_{i=1}^{n_z} \frac{1}{\sigma_b - z_i} \quad 11.83$$

The values of p_i and z_i have negative real parts when the poles and zeros are in the left-half of the s-plane, so that $-z_i$ and $-p_i$ typically are positive numbers. The above formula can be solved easily for only simple systems. For more complex systems, an iterative method usually is needed to obtain a good estimate of the break-away point.

When the open loop transfer function contains complex poles, the root locus will leave those poles at an angle (known as the *angle of emergence*) of 180° plus the sum of the angles contributed by all the other poles at the complex pole in question, minus the sum of the angles of the zeros measured to that pole.

$$\begin{aligned}\text{angle of emergence} = {} & 180° \\ & + \sum \text{angles from other poles} \\ & - \sum \text{angles from zeros}\end{aligned} \quad 11.84$$

When the open loop transfer function contains complex zeros, the root locus will converge to those zeros at the *angle of convergence*.

$$\begin{aligned}\text{angle of convergence} = {} & 180° \\ & - \sum \text{angles from poles} \\ & + \sum \text{angles from other zeros}\end{aligned} \quad 11.85$$

8 STATE VARIABLE REPRESENTATION

A. INTRODUCTION

Up to this point, the methods used have been based on Laplace-transformed variables and *frequency domain methods*. A different procedure is the *time domain method*, referred to as the *state-space representation*. The fundamental aspects of this representation and the translation between time domain and frequency domain are covered in this section. The general solutions to state variable problems will not be discussed, as they are more complex.

A simple example of an s-domain transfer function is

$$\frac{C}{R} = \frac{K}{s^2 + a_1 s + a_0} \quad 11.86$$

This can be rewritten as

$$(s^2 + a_1 s + a_0)\frac{C}{K} = R \quad 11.87$$

If this is a Laplace-transformed differential equation, the corresponding differential equation is

$$\frac{d^2 x_1}{d^2} + a_1 \frac{dx_1}{d} + a_0 x_1 = r() \quad 11.88$$

$$x_1 = \frac{c(t)}{K} \quad 11.89$$

The variable change is made to put the results into the standard state variable notation.

Using the prime notation for time derivatives, $dx/dt = x'$ and $d^2x/dt^2 = x''$, so that

$$x_1'' + a_1 x_1' + a_0 x_1 = r(t) \quad 11.90$$

Solving for x_1'',

$$x_1'' = r(t) - a_1 x_1' - a_0 x_1 \quad 11.91$$

This result can be modeled by the use of integrators and amplifiers as indicated in block diagram form in figure 11.13. This constitutes a state variable block diagram for the system described by equation 11.86. This system has no zeros. Therefore, it is readily transformed from the transfer function to the state variable form.

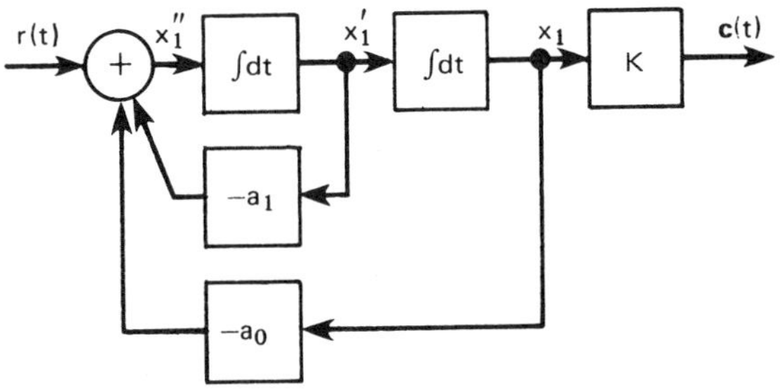

Figure 11.13 State variable block diagram

The state variable approach is to represent a system by a set of first-order differential equations. The variables of these equations are the *state variables.* One method to generate the set of state variable equations from a transfer function or an nth order differential equation in one variable is given here.

step 1: Define the variable x_1 (as in the previous example) as a state variable.

step 2: Define the derivate of x_1 as a second state variable: $x_2 = x_1'$.

step 3: Continue to define new state variables up to, but not including, the highest derivative. In the previous example x_1'' is the highest derivative, and $x_1'' = x_2'$, so no further state variables can be defined.

step 4: Write the set of first-order differential equations. In the previous example, these were

$$x_1' = x_2 \qquad 11.92$$

$$x_2' = r(t) - a_1 x_2 - a_0 x_1 \qquad 11.93$$

Equations 11.92 and 11.93 constitute a set of first-order differential equations in two state-variables. An additional equation is needed to define the output in terms of the state variables. For this example,

$$y(t) = c(t) = K x_1 \qquad 11.94$$

The usual format for state equations and the auxilliary output equation is matrix form. The above equations are written as equations 11.95 and 11.96.

$$\begin{bmatrix} x_1' \\ x_2' \end{bmatrix} = \begin{bmatrix} 0 & 1 \\ -a_0 & -a_1 \end{bmatrix} \begin{bmatrix} x_1 \\ x_2 \end{bmatrix} + \begin{bmatrix} 0 \\ 1 \end{bmatrix} r(t) \qquad 11.95$$

$$y(t) = [K \quad 0] \times \begin{bmatrix} x_1 \\ x_2 \end{bmatrix} + 0 \times r(t) \qquad 11.96$$

In matrix notation, equations 11.95 and 11.96 are written as

$$\mathbf{x}' = \mathbf{A}\mathbf{x} + \mathbf{B}r(t) \qquad 11.97$$

$$y = \mathbf{C}\mathbf{x} + d\,r(t) \qquad 11.98$$

$\mathbf{x}'$ and $\mathbf{x}$ are column matrices, $\mathbf{A}$ is an $n \times n$ square matrix, and $\mathbf{C}$ is a row matrix.

B. TRANSFER FUNCTIONS WITH ZEROS

The previous example did not have any finite zeros. The output variable was a simple function of one state-variable, so that it might appear that a state variable should be directly associated with the output. This is not true except for systems with no finite zeros.

Consider a transfer function of the form given by equation 11.99.

$$\frac{C}{R} = \frac{b_2 s^2 + b_1 s + b_0}{s^3 + a_2 s^2 + a_1 s} = \frac{(b_2 s^2 + b_1 s + b_0)X_1}{s^3 + a_2 s^2 + a_1 s)X_1} \qquad 11.99$$

Making variable changes,

$$R = (s^3 + a_2 s^2 + a_1 s)X_1 \qquad 11.100$$

$$r(t) = x_1''' + a_2 x_1'' + a_1 x_1' \qquad 11.101$$

Therefore,

$$x_1''' = r(t) - a_2 x_1'' - a_1 x_1' \qquad 11.102$$

This defines the feedback part of the state variable block diagram indicated in figure 11.14(a).

It is then necessary to identify the feed forward portion of the diagram as indicated in figure 11.14(b). Combining figure 11.14(a) and (b) produces the state variable block-diagram as shown in figure 11.14(c).

$$C = (b_2 s^2 + b_1 s + b_0)X_1 \qquad 11.103$$

$$y(t) = c(t) = b_2 x_1'' + b_1 x_1' + b_0 x_1 \qquad 11.104$$

The state variables are defined as $x_2 = x_1'$ and $x_3 = x_2'$, so the matrices are

$$\begin{bmatrix} x_1' \\ x_2' \\ x_3' \end{bmatrix} = \begin{bmatrix} 0 & 1 & 0 \\ 0 & 0 & 1 \\ 0 & -a_1 & -a_2 \end{bmatrix} \begin{bmatrix} x_1 \\ x_2 \\ x_3 \end{bmatrix} + \begin{bmatrix} 0 \\ 0 \\ 1 \end{bmatrix} r(t) \qquad 11.105$$

$$y(t) = [b_0 \quad b_1 \quad b_2] \begin{bmatrix} x_1 \\ x_2 \\ x_3 \end{bmatrix} \qquad 11.106$$

Example 11.5

Given the state variable block diagram shown in figure 11.15, find the equivalent open-loop frequency-domain system. Assume unity feedback.

Reverse the procedure presented.

$$\frac{Y(s)}{R(s)} = \frac{C}{R} = \frac{(2s+1)X_1}{(s^3 + s^2 + 3s + 4)X_1}$$

The closed-loop equation is

$$\mathbf{G}_c = \frac{2s+1}{s^3 + s^2 + 3s + 4}$$

However, for unity feedback,

$$\mathbf{G}_c = \frac{\mathbf{G}_o}{1 + \mathbf{G}_o}$$

So,

$$\mathbf{G}_o = \frac{\mathbf{G}_c}{1 - \mathbf{G}_c}$$

$$\mathbf{G}_c = \frac{C}{R}$$

Therefore,

$$\mathbf{G}_o = \frac{C}{R - C} = \frac{(2s+1)X_1}{(s^3 + s^2 + 3s + 4)X_1 - (2s+1)X_1}$$

$$= \frac{2s+1}{s^3 + s^2 + s + 3}$$

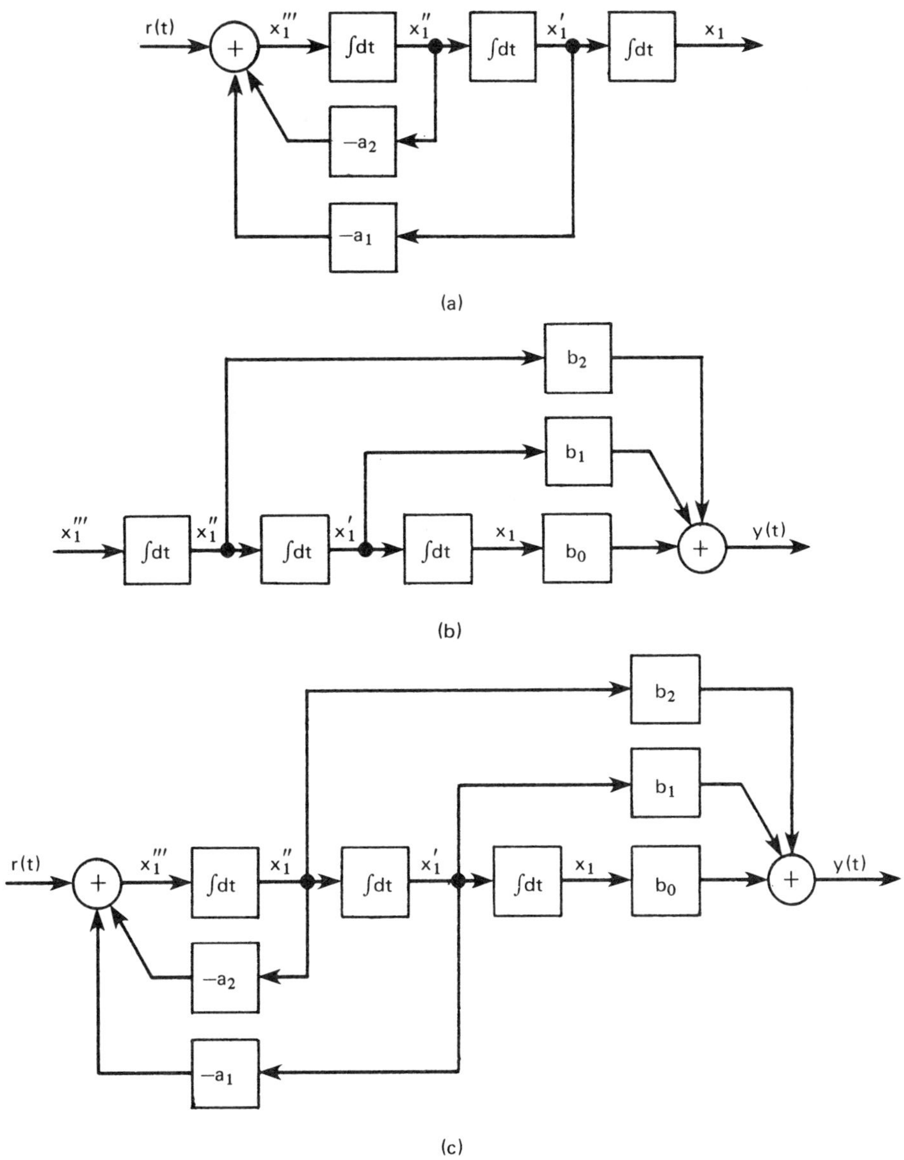

Figure 11.14 State variable block diagram development

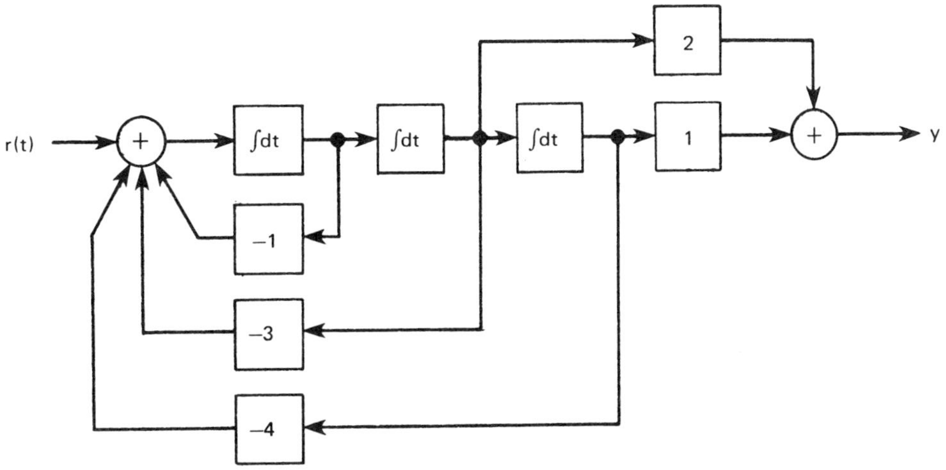

Figure 11.15 for example 11.5

Practice Problems

CONTROL SYSTEMS

Warmups

1. A separately-excited D.C. servomotor (see figure 11.2) has $R_A = 5$ ohms (armature resistance) and $L_A = 0.1$ henrys (armature inductance). The loaded motor runs at 160 rad/sec when the applied armature voltage is 24 volts and the armature current is 1.6 amps. When the armature voltage is removed, the mechanical system has a time constant of 1.8 sec. The output impedance of the power amplifier is 3 ohms (resistive).

(a) Obtain the block diagram of the motor and load. The input to this block should be the power amplifier open circuit voltage (i.e., include the amplifier output impedance within this block). All relations are to be in the Laplace-transformed notation.

(b) The summing amplifier has values $R_f = R_{\text{in}} = R_\theta = 10\text{K}$ ohms. The position rheostats for reference and the output are 500 ohms each. They can be neglected when compared with the input resistances to the summing amplifier. The voltages V are 10 volts. The tachometer output is 0.05Ω, where Ω is the shaft velocity in rad/sec. The tachometer resistance is an adjustable parameter. The position measuring rheostats swing is $-\pi < \theta < \pi$, over which the reference voltage ranges from -10 volts to $+10$ volts. The output rheostat voltage ranges from $+10$ volts to -10 volts. Thus, if the two rheostats are in the same position, they produce equal and opposite voltages. Obtain the relations between $C(s)$ and $R(s)$ and between $E(s)$ and $R(s)$ for this system.

(c) Obtain the system block diagram in the form of figure 11.3. The power amplifier can be considered to be a voltage amplifier. Its gain is an adjustable parameter.

2. A control system has the open loop transfer function

$$\frac{10(s+2)}{s(s+20)}$$

(a) Plot the amplitude assymptotes for this system's Bode diagram. (b) Find the steady-state position error to a unit step input for a closed-loop system. (Assume unity feedback.) (c) Find the steady-state velocity error to a unit ramp input for a closed-loop system.

3. A control system has unity gain and the open loop transfer function

$$\frac{50s}{s^2 + 10s + 400}$$

(a) Construct the assymptotic Bode-amplitude plot. (b) Find the closed-loop steady-state error to a unit step input. (c) Find the closed-loop steady-state error to a unit ramp input.

4. A particular control system has

$$\mathbf{GH}(s) = \frac{5s^2 + 16s + 25}{15s^4 + 225s^3 + 5s^2}$$

Determine the steady-state errors to unit step, ramp, and parabolic inputs.

5. Determine the stability of the following two systems using the Routh-Hurwitz method. If either system is unstable, determine the number of poles in the right-half s-plane.

$$\mathbf{GH}(s) = \frac{40(s+1)}{s^2(s+10)(s+20)} \qquad (a)$$

$$\mathbf{GH}(s) = \frac{s+8}{s(s^2+s+1)} \qquad (b)$$

6. A control system has a single open-loop time constant of 0.25 sec and a variable gain. Show the root locus for this system and determine the gain necessary to produce a closed loop time constant of 0.1 sec.

7. A feedback control system has one open-loop pole and one open-loop zero. The pole is in the left-half of the s-plane. Show all possible root loci for this system.

8. A system with unity feedback has an open-loop transfer function of

$$\frac{K(s+9)}{s(s+3)}$$

Find the values of gain K for which the damping ratio of the closed loop system will be $\varsigma = .707$. Find the closed-loop system response to a unit step for both cases.

9. A unity feedback, open-loop system has

$$\mathbf{G} = \frac{K(s+2)}{s^2(s+10)(s+20)}$$

Determine the assymptotes for large values of K. Estimate the low K region by treating the two poles at the origin and the zero at $s = -2$ as the total system. From these results, sketch the remainder of the root locus. (a) Using the value of s where the high K assymptotes cross the imaginary axis, calculate the value of K for this point. (Note that the imaginary part of K indicates that this point is not actually on the root locus.) (b) Use the Routh-Hurwitz method to determine the gain value for the crossing of the imaginary axis. (c) Determine the frequency corresponding to the gain in part (b).

10. For the system of problem 4, assume unity feedback and obtain the state variable block diagram for the closed-loop system.

Concentrates

1. A unity feedback control system has

$$\mathbf{G} = \frac{20}{s(s+10)}$$

The closed-loop response is considered too slow, and tachometer feedback is suggested to improve the response time. The tachometer feedback is to be applied in parallel with the unity feedback. If the tachometer has a transfer function ks, determine the value of k_T necessary to result in a damping ratio of .707.

2. The control system shown has $\mathbf{H} = 1$, and

$$\mathbf{G}_1 = \frac{1}{s+1}$$
$$\mathbf{G}_2 = \frac{80}{(s+2)(s+10)}$$

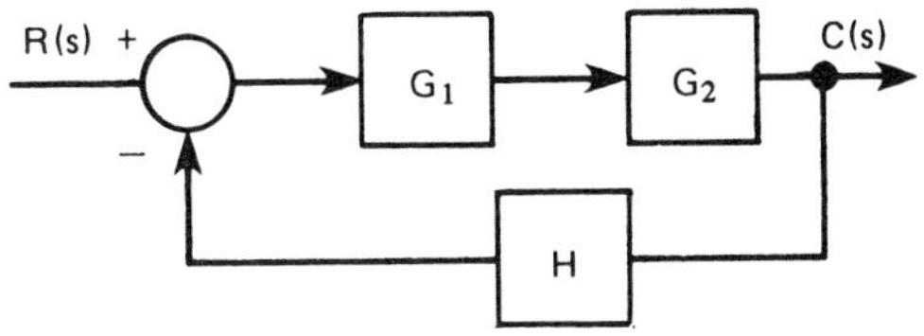

(a) Is this system stable or unstable? (b) Is the system stable if the transfer functions are changed to

$$\mathbf{G}_1 = \frac{1}{s+1}$$
$$\mathbf{G}_2 = \frac{80}{s+2}$$
$$\mathbf{H} = \frac{1}{s+10}$$

How will the time output of this system differ from the time output of (a)?

3. The open-loop transfer function of a feedback system is

$$\mathbf{GH}(s) = \frac{K(s+5)}{s(s+1)}$$

Sketch the root-locus. Find the points for $K = 5$ and $K = 10$. Is this system stable?

4. A control system has an output $c(t) = 20(1 - e^{-5t})$ volts for a step error input of 10 millivolts at $t = 0$. The output is zero before $t = 0$. The feedback is proportional to the output with a value of less than or equal to one times the output. With the input $r(t) = 1$ volt, the output reaches a final value of 10 volts. (a) Determine the feedback ratio. (b) With the feedback ratio set at 0.1, determine the output as a function of time for a step input of one volt.

5. The unity feedback system shown has

$$\mathbf{G}_p = \frac{20}{s(s+2)(s+10)}$$
$$G = G_c G_p$$

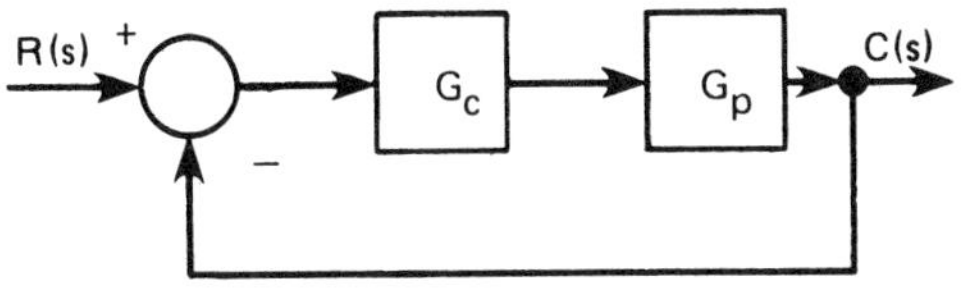

G has the following frequency description: (1) a slope of −20 db/dec below 0.1 rad/sec, (2) a slope of −40 db/dec between 0.1 and 1.5 rad/sec, (3) a slope of −20 db/dec between 1.5 and 18 rad/sec, (4) a slope of −60 db/dec for frequencies above 18 rad/sec and (5) $\mathbf{G} = 1$ a 10 rad/sec. (a) Determine **G**. (b) Determine $\mathbf{G}_c$. (c) Determine the steady state error for $r(t) = (5 + st)u(t)$.

Timed

1. A state-variable simulation of a control system is shown on the next page. Experimental tests made on the closed-loop system show approximate critical damping at a natural frequency of 1 rad/sec. (a) Determine $C(s)/R(s)$ as a ratio of two factored polynomials. (b) The physical system has unity feedback. Obtain the transfer function of the open-loop physical system. (c) Explain which block of the simulation needs to be changed to alter the damping ratio.

2. A unity gain feedback system has the open-loop transfer function

$$\frac{K}{(s+3)(s+6)(s+7)}$$

(a) If the feedback is positive, what range of K will result in a stable system? (b) If the feedback is negative, what range of K will result in a stable system?

3. Consider the control system shown on the next page. (a) Plot the root-locus as a function of K. (b) Discuss the stability. (c) Discuss the steady-state error for a step input.

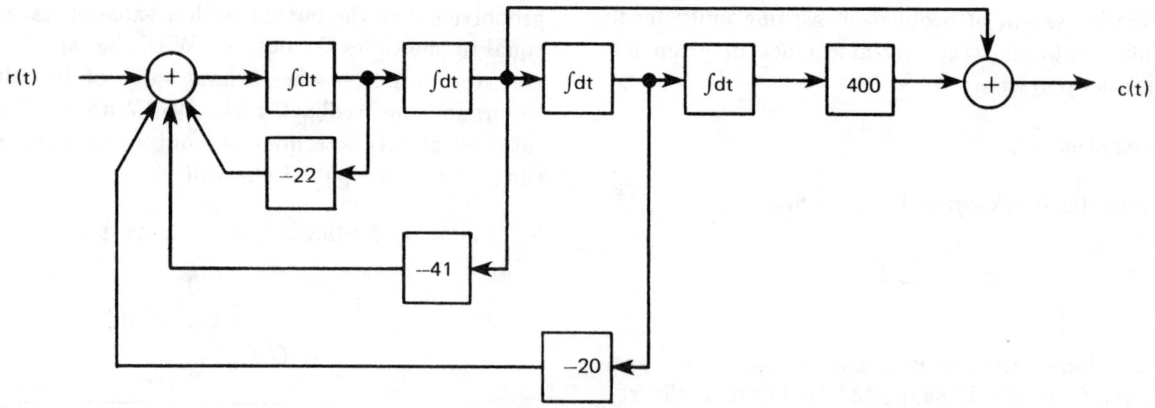

Figure for problem 1

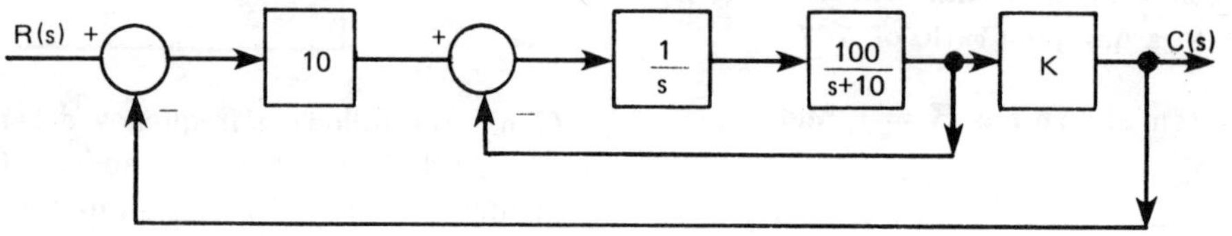

Figure for problem 3

4. A control system has the closed-loop transfer function

$$\frac{K}{s^3 + 2s^2 + 4s + K}$$

For positive K, sketch the root-locus and find the following: (a) any break-away points on the real axis, (b) frequency and gain at any point where instability begins, and (c) the complex frequency and gain where the damping ratio is .707.

5. Two transfer functions are cascaded in the forward path of a unity feedback system. These two transfer functions have the following frequency characteristics:

$G_1(j\omega)$: A break frequency at 10 rad/sec where the assymptotes meet at a gain of 20 db. Below 10 rad/sec, the slope is −20 db/dec. Above 10 rad/sec, the slope is −40 db/dec. The high frequency phase shift is −180°.

$G_2(j\omega)$: A low-frequency assymptote at −20 db, a corner frequency at 1 rad/sec, and a high-frequency slope of −20 db/dec. The high-frequency phase shift is −90°.

(a) Determine $G_1(s)$, $G_2(s)$, and $G(s) = G_1(s)G_2(s)$. (b) Apply Routh's criterion to determine whether the system is stable.

RESERVED FOR FUTURE USE

RESERVED FOR FUTURE USE

12 ILLUMINATION

Nomenclature

CBU	coefficient of beam utilization	-
CCR	ceiling cavity ratio	-
CR	cavity ratio	-
CU	coefficient of utilization	-
E	illumination	fc
E_{rh}	reflected horizontal illumination	fc
E_{rv}	reflected vertical illumination	fc
FCR	floor cavity ratio	-
h_{cc}	ceiling cavity height	ft
Hpf	high power factor	-
H_{fc}	floor cavity height	ft
h_{rc}	room cavity height	ft
Hs	house-side lateral distance	ft
I	luminous intensity	cd
II	initial illumination level	fc
L	length	ft
L	luminance	ft-lamberts
LC_{cc}	ceiling cavity luminance coefficient	-
LC_w	wall luminance coefficient	-
LDD	luminaire dirt depreciation factor	-
LLD	lamp lumen depreciation factor	-
LLF	light loss factor	-
MH	mounting height	ft
MMI	minimum maintained illumination level	fc
ND	part of MH	ft
Npf	normal power factor	-
R	coefficient of reflectance	-
RCR	room cavity ratio	-
RPM	room position multiplier	-
RRC	reflected radiation coefficient	-
Ss	street-side lateral distance	ft
T	coefficient of transmission	-
W	width	ft
WRRC	wall reflected radiation coefficient	-

Symbols

Φ	luminous flux	Lm
α	longitudinal angles	°
β	lateral angles	°

Definitions

Light

Light is an electromagnetic phenomenon dealing with the radiation, refraction, transmission, and absorption of electromagnetic waves. Radiation patterns are similar to those of various antennas. Light radiation is usually a far-field situation, where the power density falls off as $1/\text{distance}^2$.

The language of lighting differs from that of electromagnetic radiation because only the *visible spectrum*, rather than total electromagnetic power, is involved. The human eye is sensitive to wavelengths between 380 and 780 nanometers with maximum sensitivity at 556 nanometers. Light sources are evaluated by their power output in this range.

The unit of light power is the *lumen*, which is weighted power within the visible portion of the spectrum. The amount of light power obtained from the electromagnetic power depends on the spectral content of the electromagnetic wave. For example, a complete conversion of electromagnetic power to light power will result in 680 lumens per watt if the wavelength is exactly 556 nanometers. On the other hand, a wave with uniform power density over the total visible range from 380 to 780 nanometers will result in a light power of about 200 lumens per watt of electric power.

Luminous intensity I

Luminous intensity is the basic standard quantity of light. All other quantities are derived from it. The unit of luminous intensity is the candela (cd). One candela is approximately the intensity of an ordinary wax candle as seen in the horizontal plane.

A *lumen* is a unit of total light flux. The strength of a light source is typically given in candlepower (cp).

A uniform source of one candela radiates 4π lumens of light flux. In general, light sources are quite directional. Candlepower (candelas) is used to describe the apparent strength of a source as viewed from any particular direction. Thus, it is a measure of flux-density on a solid angular basis and has units of candelas, which are equivalent to lumens per steradian. Candlepower is always used to describe a light source and its luminous intensity in some direction and is given in units of candela (cd).

Luminous flux Φ

Luminous flux is defined as one lumen (Lm) when a point source of one candela illuminates one square foot of a sphere having a radius of one foot which is centered on the light source. That is, the flux incident on the one-square-foot area is one lumen.

Illumination E

Illumination is the flux density measured in units of lumens per square foot, or footcandles. Illumination is a point quantity subject to the law of cosines. For a point source of light as shown in figure 12.1, the illumination on a point a distance D from the source, lying on a surface so that the surface normal makes an angle a with the line connecting the point on the surface with the point source is

$$E = I\frac{\cos(a)}{D^2} \qquad 12.1$$

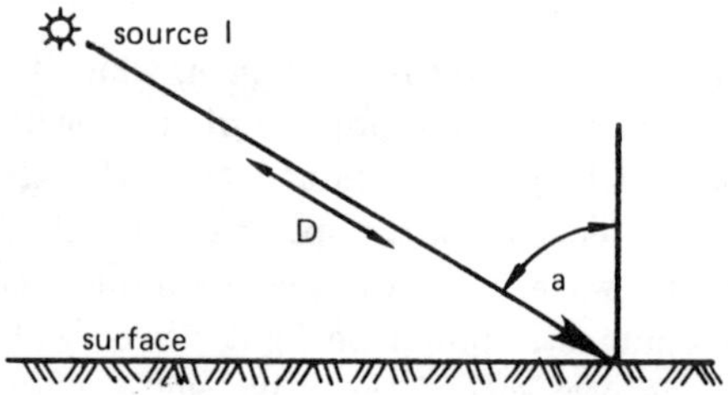

Figure 12.1 Angular dependence of illumination

Luminance or **photometric brightness** L

The apparent luminous intensity of a surface, because it is either emitting light or reflecting light, is called the *luminance* or *photometric brightness*.

Luminance due to *reflected* light is typically calculated from the coefficient of reflectance R and the surface illumination.

$$L(\text{footlamberts}) = E(\text{footcandles}) \times R \qquad 12.2$$

Luminance due to transmitted light through a diffuser is typically calculated from the incident illumination and the coefficient of transmission (T)

$$L(\text{footlamberts}) = E(\text{footcandles}) \times T \qquad 12.3$$

1 LAMPS

General purpose lamps are classified according to their light-producing mechanism as *incandescent* or *vapor discharge*. Vapor discharge includes both mercury discharge and fluorescent lamps. Incandescent lamps provide light described as warm white which is considered the closest to natural light. Although they are more efficient and lower in cost, mercury vapor lamps are not widely used for interior lighting because of a less pleasing color content. Fluorescent lamps, also more efficient than incandescent lamps, are available in a wider range of colors.

A. RELATIVE COSTS

Incandescent lamps typically are much lower in cost than discharge lamps, but they must be replaced more often and have lower efficiency in terms of lumens per watt. A measure of the relative costs of each type of lamp will vary with demand and technological development, so the following information is subject to some variation on a year to year basis.

Taking the cost per 1000 lumen hours (multiplied by the lamp lumen depreciation factor) per watt of electrical input power and using the cost for the incandescent as the unit, approximate costs give the following relative expenses.

Incandescent	=	1
Metal halide	=	0.4
Sodium vapor	=	0.1
Mercury vapor	=	0.1
Fluorescent	=	0.05

There is some variation depending on wattage, so these figures should be used only as a guideline.

B. LAMP CLASSIFICATION

Lamps are classified according to their electrical power dissipation, shape or type, size, and base. The letters used to indicate the shape of a bulb are

A	the common shape used for domestic lighting under 200 watts
BT	bulbous tubular
F	flame shaped
G	globular or round
PAR	parabolic reflector

PS	pear-shaped
R	reflector
S	straight sides
T	tubular

The diameter of a bulb is given in eighths of inches. A lamp designated A-19, for example, indicates the common shaped bulb with a maximum diameter of 19 eighths ($\frac{19}{8}$) or $2\frac{3}{8}$ inches.

Fluorescent tubes have the length given in inches. In this chapter, only the 48″ and 96″ lengths are considered. Metal halide may also be designated BU for base up, or BD for base down. High intensity discharge lamps have different initial lumen ratings depending on whether they are operated horizontally or vertically.

2 BALLAST

Ballast performs several functions. Discharge lamps, presenting negative resistance, require current limitation. To keep the power loss to a minimum, this is done with reactance. The limiting element is usually inductive, so that in *high power factor* (Hpf) *ballasts* some capacitive power factor compensation is used to achieve a power factor of 0.90 or greater. *Normal power factor* (Npf) *ballasts* can have a power factor as low as 0.45, but a value of 0.60 is typical.

Ballasts must provide the proper starting conditions for a discharge lamp. Once the lamp is started, the ballast must provide proper running conditions and, in some cases, protection from overheating. Ballasts are specifically designed for particular voltages and operating conditions.

Ballasts require some power to operate and must be considered in wiring design. Ballasts for two 40 watt fluorescent (F40T12) lamps normally require 12 watts. Normal ballast life of interior units is 12 years.

Tables in the following section give some typical ballast losses. *Auto reg ballasts* are high power factor, and *lag ballasts* are normal power factor.

3 LUMINAIRES

The terms *luminaire* and *fixture* indicate a complete lighting unit including lamps, sockets, ballasts, reflectors, diffusers, and other parts of the unit. Each type of luminaire has its own characteristics which are specified by the manufacturer.

Three important parameters are associated with a luminaire. The *coefficient of utilization* (CU) deals with the fraction of the available light reaching the surface. The *luminaire dirt depreciation factor* (LDD) depends on the environment and frequency of cleaning. The *lamp lumen depreciation factor* (LLD) indicates depreciation from aging.

A. COEFFICIENT OF UTILIZATION (CU)

The *coefficient of utilization* is a measure of the efficiency of a luminaire. It is the ratio of the light reaching the working surface to the light being emitted from the luminaire. The coefficient of utilization takes into account the distribution of light from the luminaire, the mounting height, room conditions such as dimensions and reflectances of surfaces for interior lighting, and, for outdoor lighting, those lumens going beyond the area to be lighted.

Each type of luminaire is unique in its distribution of light. The coefficient of utilization depends on the dimensions of the space to be lighted and the pattern of luminaire placement. For interior lighting, a *room cavity ratio* (RCR) is an important parameter for determination of the coefficient of utilization. As an example, table 12.1 gives the coefficient of utilization for two T-12 fluorescent lamps when the spacing of parallel rows of luminaires does not exceed 1.3 times the mounting height. Each coefficient of utilization table is specific for a given type of luminaire.

B. LUMINAIRE DIRT DEPRECIATION FACTORS (LDD)

The *minimum maintained illumination level* will depend on the cleanness of a luminaire's environment and on how often the luminaire is cleaned. The minimum maintained illumination level typically occurs just before the luminaires are cleaned. Table 12.2 shows characteristic *luminaire dirt depreciation* values for interior lighting.

The types of environments in which luminaires are installed are defined according to their general cleanness.

- **Clean:** commercial offices, light assembly, or inspection
- **Medium:** industrial offices, light machining
- **Dirty:** heavy manufacturing, processing of materials

C. LAMP LUMEN DEPRECIATION FACTOR (LLD)

The *lamp lumen depreciation factor* is a multiplier used with initial lamp lumens to determine the lamp output depreciation due to aging. This factor is required to determine the minimum maintained illumination level or to design for a particular minimum level.

Table 12.1 Coefficients of utilization for two T-12 lamps in a fluorescent luminaire (maximum ratio of spacing to mounting height not to exceed 1.3)

two T-12	REFLECTANCES									
ceiling cavity	80%			50%			10%			0%
Walls	50%	30%	10%	50%	30%	10%	50%	30%	10%	0%
RCR	Coefficients of Utilization									
1	.88	.84	.81	.79	.77	.74	.69	.68	.66	.64
2	.77	.71	.66	.70	.65	.62	.61	.59	.56	.54
3	.68	.61	.56	.61	.56	.52	.54	.51	.48	.46
4	.60	.52	.47	.54	.49	.44	.48	.44	.41	.39
5	.52	.45	.39	.48	.42	.37	.43	.38	.35	.33
6	.47	.39	.34	.43	.37	.32	.38	.34	.30	.28
7	.42	.34	.29	.38	.32	.28	.34	.30	.26	.24
8	.37	.30	.25	.34	.28	.24	.31	.26	.22	.21
9	.33	.26	.21	.31	.25	.21	.31	.23	.19	.18
10	.30	.23	.19	.28	.22	.18	.25	.20	.17	.15

Table 12.2 Luminaire dirt depreciation factors (LDD)

Environment	Enclosed			Open		
Fluorescent →	clean	medium	dirty	clean	medium	dirty
cleaned every year	.88	.83	.77	.94	.90	.84
cleaned every 2 years	.83	.77	.71	.89	.85	.78
cleaned every 3 years	.80	.74	.66	.87	.80	.74
HID						
cleaned every year	.88	.83	.77	.90	.87	.86
cleaned every 2 years	.83	.77	.71	.84	.80	.75
cleaned every 3 years	.80	.74	.66	.79	.74	.68

Table 12.3 Typical characteristics of standard lamps showing lamp lumen depreciation

HIGH INTENSITY DISCHARGE

type	lamp	ballast	rated life (hours)	initial lumens horizontal	initial lumens vertical	light lumen depreciation (LLD)
MERCURY VAPOR (deluxe white, auto reg ballast)						
H38JA	100	20	24,000	4,300	4,400	.79
H39KC	175	39	24,000	8,290	8,500	.98
H37KC	250	40	24,000	12,500	13,000	.85
H33GL	400	59	24,000	22,100	23,000	.86
H36GW	1,000	95	24,000	60,500	63,000	.76
METAL HALIDE (auto reg ballast)						
MH175	175		7,500		14,000	.73
MH250	250		7,500	19,500	20,500	.78
MH400	400	64	15,000	30,000	32,000	.76
MH1000	1,000	80	10,000	95,000	100,000	.79
MH1500	1,500	120	10,000	150,000	155,000	.92

Table 12.3 continued

HIGH PRESSURE SODIUM (lag ballast)

C100S52	100	27	20,000	9,500	9,500	.90
C150S56	150	42	20,000	16,000	16,000	.90
C250S50	250	45	24,000	25,500	25,500	.91
C400S51	400	80	24,000	50,000	50,000	.90
C1000S52	1,000	100	24,000	130,000	130,000	.92

FLUORESCENT (cool white, two lamps per luminaire)

type	wattage per lamp: lamp	wattage per lamp: ballast	rated life (hours)	initial lumens	light lumen depreciation (LLD)
48″ lamps					
F40T12					
rapid start	40.0	6.0	20,000	3,150	.84
F48T12					
slimline	38.5	4.0	9,000	3,000	.80
high output	60.0	12.5	12,000	4,200	.80
super high output	110.0	15.0	12,000	6,900	.79
96″ lamps					
F96T12					
slimline	75.0	12.5	12,000	6,300	.90
high output	100	22.5	12,000	9,000	.86
super high output	215.0	15.0	12,000	15,500	.80

Multiplying factors for lumen output for fluorescent lamp color designations are

cool white:	1.00
white and warm-white:	1.03
daylight:	0.86
warm-white-deluxe and cool white deluxe:	0.72
living white:	0.76
supermarket white:	0.75
merchandising white:	0.78

FILAMENT OR INCANDESCENT LAMPS

type	wattage	rated life (hours)	initial lumens	light lumen depreciation (LLD)
T-19	100	1,000	1,630	.89
T-19	150	1,000	2,650	.89
A-23	200	750	3,900	.89
PS-25	300	750	6,300	.89
PS-35	500	1,000	10,750	.89
T3Q/CL	500	2,000	10,500	.98
T3Q/CL	1,000	2,000	21,000	.98
T3Q/CL	1,500	2,000	33,000	.98

D. LUMINAIRE DETERMINATION

The selection of a particular luminaire is based on loss factors, total quantity of light needed, and the quality of the light. Four of the factors affecting quality of the light are *glare, brightness ratios, diffusion,* and *color.*

•**Glare:** The greater the brightness of the light source, the greater the discomfort, interference with vision, and eye fatigue. The position of the source can produce glare. When the source is in the line of vision, glare can be a factor. Glare is reduced as the source is moved further from the line of vision.

•**Brightness ratios:** Adjacent surfaces with great contrasts in reflectance can produce glare.

•**Diffusion:** Diffusion occurs where light is coming from many directions. It is a function of the size and number of sources contributing to the illumination of a point. Perfectly diffuse light is ideal for many critical visual tasks.

•**Color:** Color has no effect on visual efficiency, but it is important in tasks involving color discrimination. Color has psychological impact, and the various colors of fluorescent lamps are named for their particular applications. (It has been found that sales in women's ready-to-wear shops improve when the most natural light is used, indicating that incandescent lamps should be the choice for such shops.)

The choice of luminaires (and lamps) is influenced by the quality of light needed and by the level of illumination which must be maintained. Minimum levels of maintained illumination for typical usages are given in table 12.4.

The luminaire choice depends, on both the quality of the light and the *minimum maintained illumination* level (MMI).

4 MINIMUM MAINTAINED ILLUMINATION

The *minimum acceptable illumination level* for a specific area is one of the principal design parameters for lighting. This level must be maintained under the expected conditions of lumen depreciation, dirt factor, and coefficient of utilization for the luminaires chosen. Each of these depreciation factors is necessary in determining the number of luminaires required to meet the lighting specifications.

The *minimum maintained illumination* level (MMI) is specified. The *initial illumination level* (II) is calculated from the coefficient of utilization CU, the luminaire dirt depreciation factor LDD, and the lamp lumen depreciation factor LLD.

$$II = \frac{MMI}{CU \times LDD \times LLD} \qquad 12.4$$

Most luminaire tables assume the following reflectances.

ceiling:	80% for commercial
	50% for industrial
wall:	50%
floor:	20%

Table 12.4 Suggested illumination levels

Illumination (fc)	Application
3	for all emergency lighting
10	for stairways, corridors, and hallways with casual use, and for storage areas of bulk items
20	for service area stairways and corridors, elevators, and areas involving non-precision work for short periods of time
30	for stock areas and areas involving casual machining, occasional reading, and rough assembly
50	for general office background and for areas involving sorting, ordinary inspection, sustained non-precision machining, and reading over longer periods of time
100	for areas involving medium difficulty assembly, inspection, and machining
200	for areas involving precision assembly, inspection, and machining
500	for areas such as in an operating room where extremely fine detail work is required

For other surfaces, the following values may be used.

LIGHT COLORS	
white paint:	81%
ivory paint:	79%
cream paint:	74%
cream stone:	69%
MEDIUM COLORS	
buff paint:	63%
light green paint:	63%
light gray paint:	58%
gray stone:	56%
DARK COLORS	
tan paint:	48%
light oak paint:	32%
dark gray paint:	26%
olive green paint:	17%
dark oak paint:	13%
mahogany paint:	8%
natural cement:	25%
red brick:	13%

5 LUMEN METHOD OF ILLUMINATION DESIGN

The process for design of interior lighting based on average illumination of a working surface is interchangeably called the *lumen method* or the *zonal cavity method.*

The lumen method is used when a large area is to be provided with uniform illumination. It considers the total light flux available from the luminaires, the reflection of light from all surfaces, the coefficient of utilization, and the light loss factors, together with the total area to be illuminated. Luminaire spacing is taken into account to insure the uniformity of illumination on the working surface.

A. ZONAL CAVITY RATIOS

The coefficient of utilization takes into account dimensions of the space, reflectivity of the walls, ceiling, and floor surfaces, and light distribution for each specified luminaire. Luminaire spacing specified by the manufacturer has a maximum fraction of the mounting height of the luminaire so as to limit the variation of illumination to one-sixth of the average illumination.

The working surface is assumed to be 30 inches from the floor. The lumen method determines the illumination at that plane by combining the direct component from the luminaires, the component reflected from the ceiling, the component reflected from the walls, and the component reflected from the floor.

The coefficient of utilization is obtained from calculations of three zones in the room: the *room cavity,* which comprises the space between the working surface and the luminaires; the *ceiling cavity,* consisting of the space above the luminaires; and the *floor cavity,* which is the space below the working surface, as illustrated in figure 12.2.

A large room has a high ratio of floor or ceiling area to wall area, so the reflectance of the walls adds little to the average illumination of the work surface. Therefore, one of the basic parameters of the lumen method is the *room cavity ratio* (RCR), the ratio of the room cavity wall area to the base (ceiling or floor) area multiplied by 2.5. A cavity ratio can be expressed as

$$CR = 2.5\left[\frac{\text{wall area of the cavity}}{\text{area of the cavity base}}\right] \qquad 12.5$$

Thus, the cavity ratios for the three zonal cavities corresponding to the room of width W and length L in figure 12.2 are

$$RCR = 5\,h_{rc} \times \frac{L+W}{L \times W} \qquad 12.6$$

$$CCR = 5\,h_{cc} \times \frac{L+W}{L \times W} \qquad 12.7$$

$$FCR = 5\,h_{fc} \times \frac{L+W}{L \times W} \qquad 12.8$$

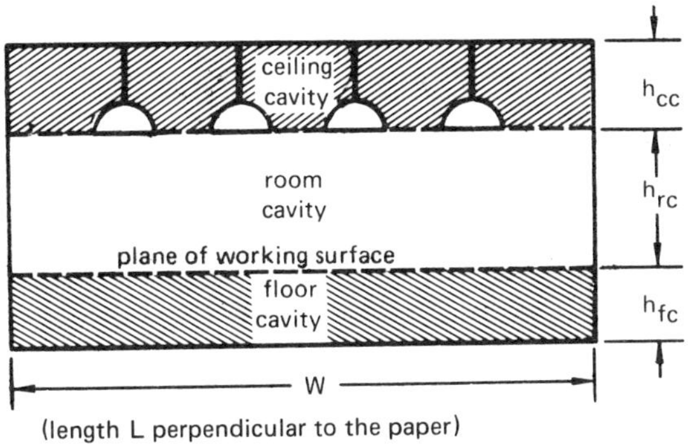

Figure 12.2 Zonal cavities

Example 12.1

A room with a ceiling height of 24 feet and a floor area 40 feet × 60 feet has luminaires suspended 6 feet from the ceiling. Determine the cavity ratios.

Assume the floor cavity height is $30'' = 2.5' = h_{fc}$; the ceiling cavity height is $h_{cc} = 6'$. Thus, the room cavity height is $h_{rc} = 24 - 6 - 2.5 = 15.5'$.

$$\frac{W+L}{W \times L} = \frac{40+60}{40 \times 60} = \frac{1}{24}$$

Then,

$$RCR = 5 \times \frac{15.5}{24} = 3.23$$

$$CCR = 5 \times \frac{6}{24} = 1.25$$

$$FCR = 5 \times \frac{2.5}{24} = 0.52$$

Cavity ratios are used to determine the *effective reflectances* of the ceiling and floor cavities. These reflectances and the room cavity ratios then determine the coefficient of utilization for the source luminaires. In most cases, the effective floor reflectance is taken as 20%, so that only the effective ceiling cavity reflectance must be calculated.

B. EFFECTIVE CEILING REFLECTANCE

To use a manufacturer's table for the coefficient of utilization, it is necessary to determine the effective ceiling reflectance. This is done by using the *ceiling cavity ratio* as calculated from equation 12.7, the base (ceiling) reflectance, and the reflectance of the walls within the ceiling cavity. These values are used with the table 12.7.

Example 12.2

For a 24 feet high room with floor area 40 feet × 60 feet and luminaires suspended 6 feet from the ceiling, the ceiling reflectance is 86%, and the wall reflectance is 60%. $CCR = 1.25$, $RCR = 3.23$, and $FCR = 0.52$. Determine the effective ceiling reflectance using table 12.5.

As none of the values given correspond to entries in table 12.5, an exact solution requires triple interpolation. Linear interpolations are used, so the order in which they are made is inconsequential.

The entries surrounding the given values are

base reflectance	90		80	
wall reflectance	70	50	70	50
cavity ratio = 1.0	80	75	72	67
cavity ratio = 1.5	76	68	68	61

The first interpolation is taken to obtain values for a base reflectance of 86%.

$$(80 - 72) \times \frac{86 - 80}{90 - 80} + 72 = 76.8$$

$$(75 - 67) \times \frac{86 - 80}{90 - 80} + 67 = 71.8$$

$$(76 - 68) \times \frac{86 - 80}{90 - 80} + 68 = 72.8$$

$$(68 - 61) \times \frac{86 - 80}{90 - 80} + 61 = 65.2$$

base reflectance	86	
wall reflectance	70	50
$CR = 1.0$	76.8	71.8
$CR = 1.5$	72.8	65.2

The second interpolation is taken to obtain values for a wall reflectance of 60%.

$$(76.8 - 71.8) \times \frac{60 - 50}{70 - 50} + 71.8 = 74.3$$

$$(72.8 - 65.2) \times \frac{60 - 50}{70 - 50} + 65.2 = 69$$

base reflectance	86
wall reflectance	60
$CR = 1.0$	74.3
$CR = 1.5$	69.0

The final interpolation gives the effective reflectance for $CR = 1.25$.

$$\text{effective reflectance} = (74.3 - 69.0)\left(\frac{1.25 - 1.0}{1.5 - 1.0}\right) + 69$$
$$= 71.65$$

C. DETERMINING THE COEFFICIENT OF UTILIZATION

When the effective ceiling reflectance is known, it is possible to determine the coefficient of utilization from a specific manufacturer's table such as the one shown in table 12.1 for two T-12 fluorescent lamps in a specific luminaire.

Tables for the coefficient of utilization are based on the assumption that the effective floor cavity reflectance is 20%, an assumption which is valid when the floor is a medium reflectance material. When the floor is either dark or light, adjustment to the final design is required. The usual adjustment when the floor covering is dark is to increase the number of luminaires by 1 or 2 per cent. Decrease the number of luminaires by 1 or 2 percent when the floor covering is light.

In table 12.1, the *base reflectance* used is the effective ceiling cavity reflectance, as determined in section B. When the luminaires are recessed into the ceiling, the actual ceiling reflectance is used.

Example 12.3

A room with a 24 feet height and floor area 40 feet × 60 feet having luminaires suspended 6 feet from the ceiling uses two T-12 lamps per luminaire. $RCR = 3.23$, and

Table 12.5 Effective Ceiling Reflectances *(Westinghouse Lighting Handbook,* revised 1978, page 6-9)

Base Refl. %	90						
Wall Refl. %	**90**	**80**	**70**	**50**	**30**	**10**	**0**
CAVITY RATIO 0.2	89	88	88	86	85	84	82
0.4	88	87	86	84	81	79	76
0.6	87	86	84	80	77	74	73
0.8	87	85	82	77	73	69	67
1.0	86	83	80	75	69	64	62
1.5	85	80	76	68	61	55	51
2.0	83	77	72	62	53	47	43
2.5	82	75	68	57	47	40	36
3.0	80	72	64	52	42	34	30
3.5	79	70	61	48	37	31	26
4.0	77	69	58	44	33	25	22
5.0	75	59	53	38	28	20	16
6.0	73	61	49	34	24	16	11
8.0	68	55	42	27	18	12	06
10.0	65	51	36	22	15	09	04

Base Refl. %	80						
Wall Refl. %	**90**	**80**	**70**	**50**	**30**	**10**	**0**
CAVITY RATIO 0.2	79	78	78	77	76	74	72
0.4	79	77	76	74	72	70	68
0.6	78	76	75	71	68	65	63
0.8	78	75	73	69	65	61	57
1.0	77	74	72	67	62	57	55
1.5	75	72	68	61	54	49	46
2.0	74	69	64	56	48	41	38
2.5	73	67	61	51	42	35	32
3.0	72	65	58	47	37	30	27
3.5	71	63	55	43	33	26	24
4.0	70	61	53	40	30	22	20
5.0	68	58	48	35	25	18	14
6.0	66	55	44	31	22	15	10
8.0	62	50	38	25	17	11	05
10.0	59	46	33	21	14	08	03

Base Refl. %	70						
Wall Refl. %	**90**	**80**	**70**	**50**	**30**	**10**	**0**
CAVITY RATIO 0.2	70	69	68	67	66	65	64
0.4	69	68	67	65	63	61	58
0.6	69	67	65	63	59	57	54
0.8	68	66	64	60	56	53	50
1.0	68	65	62	58	53	50	47
1.5	67	62	59	54	46	42	40
2.0	66	60	56	49	40	36	33
2.5	65	60	54	45	36	31	29
3.0	64	58	52	42	32	27	24
3.5	63	57	50	38	29	23	21
4.0	63	55	48	36	26	20	17
5.0	61	52	44	31	22	16	12
6.0	60	51	41	28	19	13	09
8.0	57	46	35	23	15	10	05
10.0	55	43	31	19	12	08	03

Base Refl. %	60						
Wall Refl. %	**90**	**80**	**70**	**50**	**30**	**10**	**0**
CAVITY RATIO 0.2	60	59	59	58	56	55	53
0.4	60	59	59	57	54	52	50
0.6	60	58	57	55	51	50	46
0.8	59	57	56	54	48	46	43
1.0	59	57	55	51	45	43	41
1.5	59	55	52	46	40	37	34
2.0	58	54	50	43	35	31	29
2.5	58	53	47	39	30	25	23
3.0	57	52	46	37	28	23	20
3.5	57	50	44	35	25	20	17
4.0	57	49	42	32	23	18	14
5.0	56	48	40	28	20	14	11
6.0	55	45	37	25	17	11	07
8.0	53	42	33	22	14	08	04
10.0	51	39	29	18	11	07	02

Base Refl. %	50						
Wall Refl. %	**90**	**80**	**70**	**50**	**30**	**10**	**0**
CAVITY RATIO 0.2	50	50	49	48	47	46	44
0.4	50	49	48	47	45	44	42
0.6	50	48	47	45	43	41	38
0.8	50	48	47	44	40	38	36
1.0	50	48	46	43	38	36	34
1.5	50	47	45	40	34	31	26
2.0	50	46	43	37	30	26	24
2.5	50	46	41	35	27	22	21
3.0	50	45	40	32	24	19	17
3.5	50	44	39	30	22	17	15
4.0	50	44	38	28	20	15	12
5.0	50	42	35	25	17	12	09
6.0	50	42	34	23	15	10	06
8.0	49	40	30	19	12	07	03
10.0	47	37	27	17	10	06	02

Base Refl. %	40						
Wall Refl. %	**90**	**80**	**70**	**50**	**30**	**10**	**0**
CAVITY RATIO 0.2	40	40	39	39	38	36	36
0.4	41	40	39	38	36	34	34
0.6	41	40	39	37	34	32	31
0.8	41	40	38	36	33	31	29
1.0	42	30	38	34	32	29	27
1.5	42	39	37	32	28	24	22
2.0	42	39	36	31	25	21	19
2.5	43	39	35	29	23	18	15
3.0	43	39	35	27	21	16	13
3.5	44	39	34	26	20	14	12
4.0	44	38	33	25	18	12	10
5.0	45	38	31	22	15	10	07
6.0	44	37	30	20	13	08	05
8.0	44	35	28	18	11	06	03
10.0	43	34	25	15	08	05	02

Base Refl. %	30						
Wall Refl. %	**90**	**80**	**70**	**50**	**30**	**10**	**0**
CAVITY RATIO 0.2	31	31	30	29	29	28	27
0.4	31	31	30	29	28	26	25
0.6	32	31	30	28	26	25	23
0.8	32	31	30	28	25	23	22
1.0	33	32	30	27	24	22	20
1.5	34	33	30	25	22	18	17
2.0	35	33	29	24	20	16	14
2.5	36	32	29	24	18	14	12
3.0	37	33	29	22	17	12	10
3.5	38	33	29	21	15	10	09
4.0	38	33	28	21	14	09	07
5.0	39	33	28	19	13	08	05
6.0	39	33	27	18	11	06	04
8.0	40	33	26	16	09	04	02
10.0	40	32	24	14	08	03	01

Base Refl. %	20						
Wall Refl. %	**90**	**80**	**70**	**50**	**30**	**10**	**0**
CAVITY RATIO 0.2	21	20	20	20	19	19	17
0.4	22	21	20	20	19	18	16
0.6	23	21	21	19	18	17	15
0.8	24	22	21	19	18	16	14
1.0	25	23	22	19	17	15	13
1.5	26	24	22	18	16	13	11
2.0	28	25	23	18	15	11	09
2.5	29	26	23	18	14	10	08
3.0	30	27	23	17	13	09	07
3.5	32	27	23	17	12	08	05
4.0	33	28	23	17	11	07	07
5.0	35	29	24	16	10	06	04
6.0	36	30	24	16	10	05	02
8.0	37	30	23	15	08	03	01
10.0	37	29	22	13	07	03	01

Base Refl. %	10						
Wall Refl. %	**90**	**80**	**70**	**50**	**30**	**10**	**0**
CAVITY RATIO 0.2	11	11	11	10	10	09	09
0.4	12	11	11	11	10	09	08
0.6	13	13	12	11	10	08	08
0.8	15	14	13	11	10	08	07
1.0	16	14	13	12	10	08	07
1.5	18	16	15	12	10	07	06
2.0	20	18	16	13	09	06	05
2.5	22	20	17	13	09	05	04
3.0	24	21	18	13	09	05	03
3.5	26	22	19	13	09	05	03
4.0	27	23	20	14	09	04	02
5.0	30	25	20	14	08	04	02
6.0	31	26	21	14	08	03	01
8.0	33	27	21	13	07	03	01
10.0	34	28	21	12	07	02	01

Base Refl. %	0						
Wall Refl. %	**90**	**80**	**70**	**50**	**30**	**10**	**0**
CAVITY RATIO 0.2	02	02	02	01	01	00	0
0.4	04	03	03	02	01	00	0
0.6	05	05	04	03	02	01	0
0.8	07	06	05	04	02	01	0
1.0	08	07	06	04	02	01	0
1.5	11	10	08	06	03	01	0
2.0	14	12	10	07	04	01	0
2.5	16	14	12	08	05	02	0
3.0	18	16	13	09	05	02	0
3.5	20	17	15	10	05	02	0
4.0	22	18	15	10	05	02	0
5.0	25	21	17	11	06	02	0
6.0	27	23	18	12	06	02	0
8.0	30	25	20	12	06	02	0
10.0	31	25	20	12	06	02	0

the effective ceiling reflectance is 71.65%. Determine the coefficient of utilization for the room.

Coefficients of utilization for two T-12 lamps per luminaire are given in table 12.1. In this case, interpolations are necessary for the ceiling cavity reflectance and the RCR, and an extrapolation is necessary for the wall reflectance. The interpolation of the ceiling reflectance is done first, followed by the extrapolation on the wall reflectance, and finally by the interpolation of the RCR.

ceiling	from Table 12.1			
cavity	80%		50%	
walls	50%	30%	50%	30%
$RCR = 3$	.68	.61	.61	.56
$RCR = 4$	.60	.52	.54	.49

interpolation		extrapolation	last interpolation
71.65%		71.65%	71.65%
50%	30%	60%	60%
.661	.596	.694	RCR=3.23
.583	.512	.619	.677=CU

D. DETERMINING THE NUMBER OF LUMINAIRES

The coefficient of utilization is applied to a system of luminaires to determine the illumination level at the work plane. The net flux from each luminaire is multiplied by the coefficient of utilization and by the number of luminaires to find the total luminous flux at the work plane. The total flux is then divided by the base area to yield the average illumination at the work plane.

Example 12.4

A room with 24 feet height and floor area 40 feet × 60 feet has a coefficient of utilization of 0.677. The required illumination level is 100 footcandles. Determine the net luminous flux necessary for luminaires with two T-12 lamps.

The total flux at the work surface is

$$100\, fc \times 60' \times 40' = 240{,}000\, fc$$

Total luminous flux required is

$$\text{Flux required} = \frac{240{,}000}{0.677} = 355{,}505\, fc$$

The net flux from a luminaire is determined only after taking into account a number of light loss factors, such as the lamp lumen depreciation factor and luminaire dirt depreciation factors. For fluorescent and other discharge lamps, the initial lumen output is determined after 100 hours of operation, and the lamp lumen depreciation factor is determined after 70% of the life expectancy has elapsed. In addition to these losses, several other factors must be considered.

- Ballast performance contributes a loss of approximately 5%. Thus, unless better information on ballast loss is available, a multiplier of 0.95 is used.

- Voltage fluctuations cause a serious effect in incandescent lamps, with a 3% change in lamp lumens for each 1% change in line voltage. Fluorescent and other discharge lamps are less effected, having only 0.4% change in lumen output for each 1% change in line voltage.

- Lamp outage is a function of the maintenance program. Loss can be minimized by massive replacement at 70% of expected life.

- Ambient temperature is a factor in fluorescent lamps. Flux measurements are made at 70° F, but maximum efficiency occurs at a lower temperature. Efficiency decreases for temperatures above 70° F, so it can be worthwhile to include luminaire cooling, particularly when there are four lamps per luminaire. Some luminaires are designed to serve as cool air returns in ventilating systems, which can increase the lumens/watt up to 20%.

Example 12.5

A room with 24 feet height and floor area 40 feet × 60 feet, using two 40T-12 lamps per luminaire has the following environment: (1) the line voltage is known to drop 5% at the peak loading time of day, (2) burned out lamps are replaced immediately, all lamps will be replaced at the end of 70% of the expected life, and the lamps will burn 12 hours for each start, resulting in a lamp lumen depreciation of 0.84, (3) the environment is clean, and the luminaires are to be cleaned on an annual schedule. Determine the total loss factor.

As these are fluorescent lamps, the voltage fluctuation loss will be (5 × 0.04) or 2%, resulting in a multiplier of 0.98. The LLD factor is 0.84; from table 12.2, LDD is 0.94 (assuming an open luminaire).

Allow for ballast loss, giving an additional multiplier of 0.95. Thus the total light loss factor LLF is

$$LLF = .98 \times .84 \times .94 \times .95 = .735$$

The *minimum maintained illumination* is the product of the number of luminaires, the initial luminous flux of the lamps, the number of lamps per luminaire, the light loss factor, and the coefficient of utilization, divided by the base area of the working surface (i.e., the floor or ceiling area).

$$MMI = \frac{(\#\text{luminaires}) \times \left(\frac{\text{lamps}}{\text{luminaire}}\right) \times (\text{initial flux}) \times LLF \times CU}{\text{area of the working surface}} \qquad 12.9$$

This relationship is rearranged to provide the design equation for determining the number of luminaires required.

$$\#\text{luminaires} = \frac{MMI \times (\text{area of the working surface in sq. ft})}{\left(\frac{\text{lamps}}{\text{luminaire}}\right) \times (\text{initial flux in lumens}) \times LLF \times CU} \qquad 12.10$$

Example 12.6

A room with 24 feet height and floor area 40 feet × 60 feet, using two T-12 lamps per luminaire, has a loss factor of 0.735, a coefficient of utilization of 0.674, and requires a minimum maintained illumination level of 100 footcandles. If 48 inch fluorescent lamps with 3000 initial lumens per lamp are used, determine the number of luminaires required.

Using the previously calculated values,

$$\begin{aligned} MMI &= 100\,fc \\ A &= 2400\,\text{square feet} \\ \text{lamps/luminaires} &= 2 \\ LLF &= .735 \\ CU &= .674 \\ \text{initial lamp lumens} &= 3000 \end{aligned}$$

Then,

$$\begin{aligned} \#\,\text{luminaires} &= \frac{100 \times 2400}{2 \times 3000 \times .735 \times .674} \\ &= 80.7\,\text{luminaires, or } 81 \end{aligned}$$

E. PLACEMENT OF LUMINAIRES

The final step in the lumen method is specifying the placement of the luminaires. The manufacturer's specifications in table 12.1 give the maximum spacing to mounting height, ensuring that the variation is no more than one-sixth of the average. It is customary to place these first and last rows of luminaires no further than one-third of the row spacing from the walls. Spacing from a wall depends on the reflectance of the wall, which usually is not calculated.

Fluorescent luminaires usually are placed in essentially continuous rows, with the long dimension of the luminaire oriented to the direction of the row. The direction of the rows depends on the general direction of vision within the room, being preferably perpendicular to the prevailing direction of vision. This often will be along the shorter dimension of the room.

Example 12.7

A room with 24 feet height and floor area 40 feet × 60 feet, using two 40T-12 lamps per luminaire, requires 81 luminaires. Assuming the luminaires are 52 inches long and are to be placed in rows across the short (40′) dimension of the room, specify a layout for this room.

$$\text{luminaires per row} = \frac{40' \times 12\ \text{inches/foot}}{52''} = 9.23$$

The maximum number of luminaires per row is 9. Thus, with 81 luminaires required, there will be 9 rows of 9 luminaires. With 9 rows, there will be 8 even spaces for the interior of the layout, and two spacings of one-third the interior spacing for the distance from the walls. This results in a total dimension of $8\frac{2}{3}$ times the interior row spacing, which is required to be 60 feet (the long dimension of the room). The interior row spacing is then calculated.

$$\text{interior row spacing} = \frac{\text{room length}}{\#\ \text{rows} - 1 + \frac{2}{3}} = \frac{60'}{8.667} = 6'\ 11''$$

Table 12.1 restricts the maximum spacing to less than 1.3 times the mounting height. The mounting height corresponds to the room cavity height (h_{rc} in figure 12.2) which is 15.5 feet, and the resulting design value of 6 feet 11 inches is well below this value.

6 POINT METHOD OF INTERIOR ILLUMINATION DESIGN

When a particular area requires a higher illumination level than that of the average illumination of the working surface, calculating the level of illumination at a point is necessary, and supplementary local sources must be specified. In such cases, the average illumination has been set by a lumen method design, so that the placement and data on the ambient lighting system are known.

The illumination level at certain points must be calculated to determine whether there is a need for supplementary lighting or for alteration of the ambient lighting system. To obtain the local or point illumination, two components of illumination, the direct and the reflected, must be calculated. The calculations for the reflected component are similar to those for the lumen method, but they exclude the direct component.

The two types of luminaire arrays considered in the calculation of the direct component are an array of point sources and an array of line sources, corresponding to fluorescent fixtures arranged in rows. In both cases, the mathematics are specific to the luminaire, so tables published by manufacturers for each luminaire are used in the calculations.

A. THE REFLECTED COMPONENT

The reflected component is considered first because its calculation is similar to the lumen method. The reflected component is dependent on the position within the room of the point in question. For this reason, the room in figure 12.3 has been divided into a grid system which takes into account the influence of position in the room. The grid uses 11 lines in each direction with the first and last lines corresponding to the walls and the sixth line the centerline. The division is always into 11 grid lines, regardless of the length to width ratio.

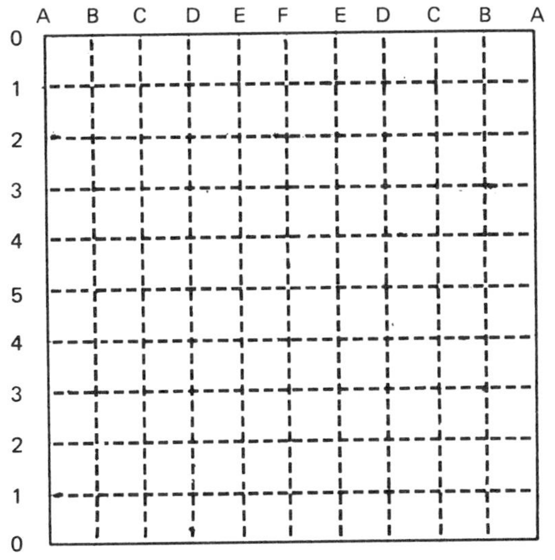

Figure 12.3 Room position grid

The amount of reflected illumination is dependent on room position and the room cavity ratio to take into account the geometric properties. These geometric properties are then obtained from a table to provide a *room position multiplier*, RPM. Table 12.6 gives these values. Because of its symmetry, only one quarter of the grid is necessary in the tabled values. Interpolation between points often is necessary when using the table, particularly for points near the corners and along the walls. Interpolation between room cavity ratios also is required.

Table 12.6 Room position multipliers. (*Westinghouse Lighting Handbook,* revised 1978, page 6-27, used with permission)

	A	B	C	D	E	F
Room Cavity Ratio = 1						
0	.24	.42	.47	.48	.48	.48
1	.42	.74	.81	.83	.84	.84
2	.47	.81	.90	.92	.93	.93
3	.48	.83	.92	.94	.95	.95
4	.48	.84	.93	.95	.96	.97
5	.48	.84	.93	.95	.97	.97
Room Cavity Ratio = 2						
0	.24	.36	.42	.44	.46	.46
1	.36	.51	.60	.63	.66	.68
2	.42	.60	.68	.72	.78	.83
3	.44	.63	.72	.77	.82	.85
4	.46	.66	.78	.82	.85	.86
5	.46	.68	.83	.85	.86	.87
Room Cavity Ratio = 3						
0	.23	.32	.37	.40	.42	.42
1	.32	.40	.48	.51	.53	.57
2	.37	.48	.58	.61	.64	.67
3	.40	.51	.61	.65	.69	.71
4	.42	.53	.64	.69	.73	.75
5	.42	.57	.67	.71	.75	.77
Room Cavity Ratio = 4						
0	.22	.28	.32	.35	.37	.37
1	.28	.33	.40	.42	.44	.48
2	.32	.40	.48	.50	.52	.57
3	.35	.42	.50	.54	.58	.61
4	.37	.44	.52	.58	.62	.64
5	.37	.48	.57	.61	.64	.66
Room Cavity Ratio = 5						
0	.21	.25	.28	.31	.33	.33
1	.25	.29	.33	.36	.38	.42
2	.28	.33	.40	.42	.44	.48
3	.31	.36	.42	.46	.49	.52
4	.33	.38	.44	.49	.52	.54
5	.33	.42	.48	.52	.54	.56
Room Cavity Ratio = 6						
0	.20	.23	.26	.28	.29	.30
1	.23	.26	.29	.31	.33	.36
2	.26	.29	.35	.37	.38	.40
3	.28	.31	.37	.39	.41	.43
4	.29	.33	.38	.41	.43	.45
5	.30	.36	.40	.43	.45	.47
Room Cavity Ratio = 7						
0	.18	.21	.23	.25	.26	.27
1	.21	.23	.26	.28	.29	.30
2	.23	.26	.30	.32	.33	.34
3	.25	.28	.32	.34	.35	.36
4	.26	.29	.33	.35	.37	.37
5	.27	.30	.34	.36	.37	.38
Room Cavity Ratio = 8						
0	.17	.18	.21	.22	.22	.23
1	.18	.20	.23	.25	.26	.26
2	.21	.23	.26	.27	.28	.29
3	.22	.25	.27	.29	.30	.30
4	.22	.26	.28	.30	.31	.32
5	.23	.26	.29	.30	.31	.32
Room Cavity Ratio = 9						
0	.15	.17	.18	.19	.20	.20
1	.17	.18	.20	.21	.22	.23
2	.18	.20	.23	.24	.25	.25
3	.19	.21	.24	.25	.26	.26
4	.20	.22	.25	.26	.26	.27
5	.20	.23	.25	.26	.27	.27
Room Cavity Ratio = 10						
0	.14	.16	.16	.17	.18	.18
1	.16	.17	.18	.19	.19	.20
2	.16	.18	.19	.21	.22	.22
3	.17	.19	.21	.22	.23	.23
4	.18	.19	.22	.23	.23	.24
5	.18	.20	.22	.23	.24	.25

B. REFLECTED RADIATION COEFFICIENT (RRC)

The method for determining the *reflected component* of illumination is the same as that for average ilumination except that the reflected radiation coefficient is substituted for the coefficient of utilization.

The *reflected radiation coefficient* depends on the wall luminance and the effective luminance of the ceiling cavity. These are taken into account by luminance coefficients that are listed by manufacturers in wall luminance coefficient and ceiling cavity luminance coefficient LC_{cc} tables. The reflected radiation coefficient is obtained from equation 12.11.

$$RRC = LC_w + RPM(LC_{cc} - LC_w) \qquad 12.11$$

Values for LC_w and LC_{cc} are obtained from manufacturers' listings, such as table 12.7. Parameters for ceiling and wall reflectances and the room cavity ratio are found on the lower part of the table.

C. REFLECTED ILLUMINATION ON VERTICAL SURFACES

The method of determining the component of reflected illumination on vertical surfaces again uses the same formula (equation 12.6) as the average illumination, with the coefficient of utilization replaced by the *wall reflected radiation coefficient,* WRRC. This is determined by using the *wall direct radiation coefficient,* (WDRC) as found in the left column of table 12.7, the wall luminance coefficient, and the average wall reflectance. Thus, this coefficient is produced from the same information as the reflected radiation coefficient by following the procedure indicated. The wall reflected radiation coefficient is obtained from equation 12.12.

$$WRRC = \frac{\text{wall luminance coefficient (table 12.7)}}{\text{average wall reflectance}} - WDRC \qquad 12.12$$

- Horizontal surface

$$E_{rh} = (\text{number of luminaires}) \times (\text{lamps per luminaire}) \times (\text{lumens per lamp}) \times \frac{LLF \times RRC}{\text{work plane horizontal area}} \qquad 12.13$$

- Vertical surface

$$E_{rv} = FC_{rh} \times \frac{WRRC}{RRC} \qquad 12.14$$

Example 12.8

A room with ceiling height of 24 feet and floor dimensions 40 feet $\times$60 feet has the following parameters: luminaires are hung 6 feet from the ceiling; there are

Table 12.7 Direct illumination components for luminaire height 6 feet above work plane, (*Westinghouse Lighting Handbook*, revised 1978, page 6-28 used with permission)

Direct Illumination Components

β	5	15	25	35	45	55	65	75	5	15	25	35	45	55	65	75
α	**Vertical Surface Illumination Footcandles at a Point On a Plane Parallel to Luminaires**								**Vertical Surface Illumination Footcandles at a Point On a Plane Perpendicular to Luminaires**							
0-10	.9	2.6	3.6	3.9	3.3	1.9	.7	.1	.9	.8	.7	.5	.3	.1	...	...
0-20	1.8	5.0	7.0	7.7	6.6	3.8	1.5	.2	3.6	3.2	2.7	1.9	1.2	.5	.1	...
0-30	2.6	7.2	10.1	11.3	9.8	5.7	2.3	.3	7.7	7.0	5.8	4.3	2.7	1.1	.3	...
0-40	3.2	9.0	12.8	14.5	12.9	7.7	3.2	.5	12.6	11.6	9.7	7.5	4.9	2.1	.6	...
0-50	3.7	10.3	14.9	17.1	15.7	9.6	4.3	.7	17.8	16.6	14.2	11.2	7.7	3.4	1.1	.1
0-60	4.0	11.2	16.3	18.8	17.6	11.3	5.5	1.0	22.6	21.2	18.4	14.7	10.4	5.1	1.9	.2
0-70	4.1	11.6	17.0	19.8	18.9	12.7	6.8	1.4	26.2	24.7	21.8	17.8	13.1	7.2	3.2	.3
0-80	4.1	11.7	17.3	20.2	19.4	13.3	7.4	1.9	28.2	26.7	23.8	19.7	14.9	8.7	4.3	.8
0-90	4.1	11.7	17.3	20.2	19.4	13.4	7.5	2.0	28.6	27.1	24.2	20.1	15.3	9.1	4.7	1.1

F.C. at a Point on Workplane

α	5	15	25	35	45	55	65	75
0-10	10.6	9.5	7.6	5.5	3.3	1.3	.3	...
0-20	20.6	18.5	14.9	10.9	6.6	2.6	.7	...
0-30	29.4	26.5	21.6	16.0	9.8	4.0	1.1	...
0-40	36.5	33.1	27.4	20.6	12.9	5.4	1.5	...
0-50	41.8	38.1	31.9	24.3	15.7	6.7	2.0	.1
0-60	45.2	41.3	34.8	26.8	17.6	7.9	2.6	.2
0-70	46.9	43.0	36.4	28.3	18.9	8.9	3.2	.3
0-80	47.4	43.6	36.9	28.8	19.4	9.3	3.5	.4
0-90	47.5	43.7	37.0	28.8	19.4	9.3	3.5	.4

Category III

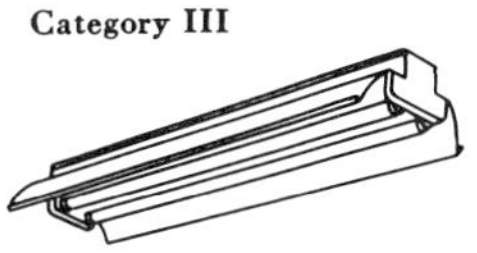

2 T-12 Lamps—Any Loading
For T-10 Lamps—C.U. x 1.02

Luminance Coefficients for 20% Effective Floor Cavity Reflectance

		Reflectances											
Ceiling Cavity		80		50		10		80		50		10	
Walls		50	30	50	30	50	30	50	30	50	30	50	30
WDRC	RCR	Wall Luminance Coefficients						Ceiling Cavity Luminance Coefficients					
.281	1	.246	.140	.220	.126	.190	.109	.230	.209	.135	.124	.025	.023
.266	2	.232	.127	.209	.115	.182	.102	.222	.190	.130	.113	.024	.021
.245	3	.216	.115	.196	.105	.172	.095	.215	.176	.127	.105	.024	.020
.226	4	.202	.102	.183	.097	.161	.088	.209	.164	.124	.099	.023	.019
.212	5	.191	.097	.173	.090	.154	.082	.204	.156	.121	.094	.023	.018
.196	6	.178	.090	.163	.084	.145	.076	.200	.149	.118	.090	.022	.017
.182	7	.168	.083	.153	.078	.136	.071	.194	.144	.115	.087	.022	.017
.170	8	.158	.077	.145	.072	.130	.066	.190	.139	.113	.085	.021	.016
.159	9	.150	.072	.138	.068	.123	.062	.185	.135	.110	.082	.021	.016
.149	10	.141	.068	.130	.064	.116	.059	.180	.131	.107	.080	.020	.016

The direct illumination components are based on data for F40 lamps producing 3100 lumens.

10 rows of two 40T-12 luminaires with spacing of 6′3″ running in the shorter direction, the LLF is .704, the effective ceiling reflectance is 72%, the wall reflectance is 60%, and the RCR is 3.23.

Determine the reflected horizontal component of illumination at a point which is 6 feet from two walls.

Find the room position. Taking the long dimension (60 feet) as the lettered positions, the point is located at grid line B. Along the short dimension (40 feet), 6 feet is half way between 1 and 2 of the numbered grid lines.

Referring to table 12.7, interpolations between RCR's of 3 and 4 and between grid positions B1 and B2 must be made. For $RCR = 3$, the interpolation between the grid positions yields a room position multiplier RPM of .44 and for $RCR = 4$, an RPM of .365. Interpolating between RCRs of 3 and 4 to the actual value of 3.23, the RPM obtained is .423.

With ceiling reflectance of 72%, it is necessary to interpolate table 12.7 (lower) between 80% and 50%. With wall reflectance of 60%, it is necessary to extrapolate from 50% and 30%. Interpolation between RCR's of 3 and 4 is necessary.

	WALL				CEILING			
ceil. refl.	80		50		80		50	
wall refl.	50	30	50	30	50	30	50	30
$RCR = 3$	.216	.115	.196	.105	.215	.176	.127	.105
$RCR = 4$	.202	.102	.183	.097	.209	.164	.124	.099
wall refl.	60		60		60		60	
$RCR = 3$	.267		.242		.235		.138	
$RCR = 4$	.252		.226		.232		.137	
ceil. refl.	72				72			
wall refl.	60				60			
$RCR = 3$	.260				.209			
$RCR = 4$	.245				.207			

$RCR = 3.23$	wall luminance coef. $LC_w = .257$	ceiling luminance coef. $LC_{cc} = .209$

Using equation 12.11, the reflected radiation coefficient is obtained.

$$RRC = .257 + .423(.209 - .257) = .237$$

The reflected illumination is then found using the maintained minimum illumination formula, equation 12.9, substituting RRC for CU.

$$E_{rh} = \frac{90 \times 2 \times 3000 \times .704 \times .237}{2400} = 37.5\, fc$$

D. LINE SOURCE CALCULATION OF DIRECT ILLUMINATION COMPONENT

This method is particularly suited to continuous rows of fluorescent luminaires. It is based on vector integration too complex to include here. The method is based on an angular coordinate system. Imagine a vertical plane through the point where the illumination is to be determined and parallel to the rows of luminaires. Measure the angles from the point in that plane to the ends of the rows, naming these angles α_1 and α_2, as indicated in figure 12.4.

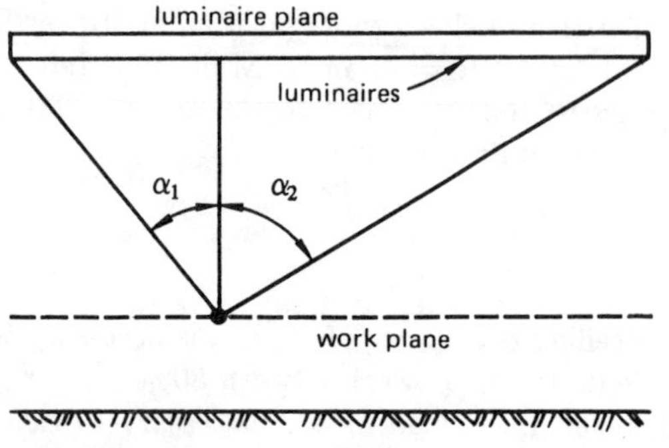

Figure 12.4 Longitudinal angles of a luminaire array, shown in the side view

From the point where the illumination is to be determined, construct another vertical plane perpendicular to the rows of luminaires. Measure the angles from the vertical line through the point to the center line of each row of luminaires. Name these angles by giving them the row subscript together with the variable name β as indicated in figure 12.5. The angles named α are *longitudinal angles,* and those named β are the *lateral angles.* The direct component of illumination is determined directly from a table of direct illumination components for the specific luminaire (see table 12.7).

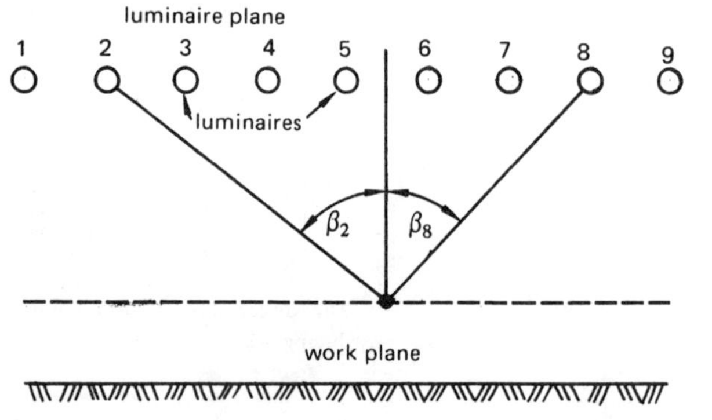

Figure 12.5 Latitudinal angles of a luminaire array, shown in the end view

The tables usually will give the same two values of longitudinal angles α for all the rows with a full set of lateral angles β. A table must be constructed to account for the contributions of each row, as demonstrated in example 12.9. Interpolation of lateral angle values usually is not necessary, but interpolation of the final sums for longitudinal angles is required if the lateral angles differ significantly from the table values.

The values determined from the table are based on a room height (from the working plane to the luminaire plane) of six feet. A correction must be made for other room cavity height : 15.5 feet point of calculation 6 feet from each of 2 walls

row	lateral distance (feet)	lateral angle β (degrees)	near longitudinal angle α_1 20°	bracketing for longitudinal angles 60° α_2	70° α'_2
1	4.125	15.0	18.5 fc	41.3 fc	43.0 fc
2	2.125	8.0	20.6	45.2	46.9
3	8.375	28.0	14.9	34.8	36.4
4	14.625	43.0	6.6	17.6	18.9
5	20.875	53.0	2.6	7.9	8.9
6	27.125	60.2	.7	2.6	3.2
7	33.375	65.0	.7	2.6	3.2
8	39.625	68.0	.7	2.6	3.2
9	45.875	71.0	—	.2	.3
10	52.125	74.0	—	.2	.3
			65.3	155.0	164.3

65° interpolation 159.7

mounting heights—by multiplying by 6 feet and dividing by the room cavity height,

$$\frac{6}{h_{rc}}$$

Example 12.9

For a room with ceiling height of 24 feet, floor dimensions 40 feet ×60 feet, luminaires hung 6 feet from the ceiling, 10 rows of two 40T-12 luminaires (table 12.7) running in the shorter direction, interior spacing of 6′ 3″ and spacing of 1′ 10.5″ from the wall for the edge rows, determine the direct illumination for the point which is 6 feet from two walls.

Latitudinal or lateral angles for the arrangement from the luminaire rows to the point are found to be 14.9°, 7.8°, etc., which are rounded to the nearest degrees. The longitudinal angles are found to be about 20° and 65°. The table contains a column for the 20° angle but requires 60° and 70° columns for interpolation. In choosing values from latitudinal (lateral) columns, the value nearest the given angle is satisfactory, as errors tend to cancel each other.

total uncorrected direct component footcandles

$$159.7 + 65.3 = 225\,fc$$

Using the correction factor $\frac{6}{15.5}$, the direct illumination is

$$225 \times \frac{6}{15.5} = 87 \text{ fc illumination}$$

7 STREET LIGHTING

Outdoor lighting differs from interior lighting due to the absence of reflected light from ceilings and walls. The coefficient of utilization is generally smaller for outdoor than for interior applications. Another difference is in the types of luminaires and lamps typically employed. While interior lighting is dominated by fluorescent fixtures, outdoor luminaires are usually mercury vapor and other discharge lamps, such as sodium and metal halide. They are broadly classified as *roadway* and *flood* lighting.

All the considerations developed for interior lighting are applicable to outdoor lighting. The most significant differences are in determining the coefficient of use for average illumination calculations and the need for iso candela or iso footcandle curves for point calculations.

Figure 12.6 shows the basic dimensions involved in street lighting: mounting height, street-side lateral distance, and house-side lateral distance. An additional dimension is the spacing between luminaires, which is indicated in figure 12.7 for three variations of luminaire arrangements.

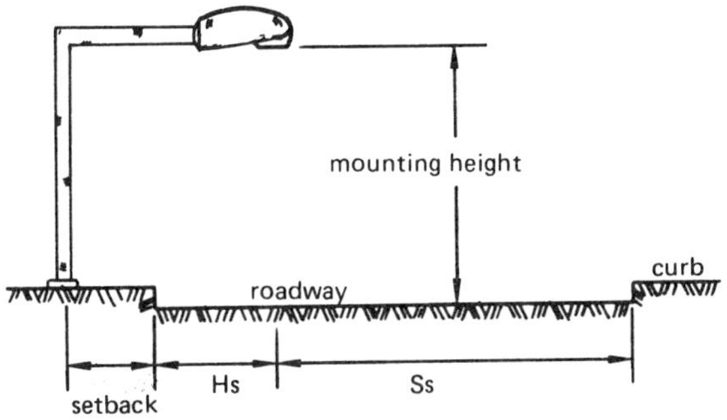

Figure 12.6 Basic dimensions for street lighting

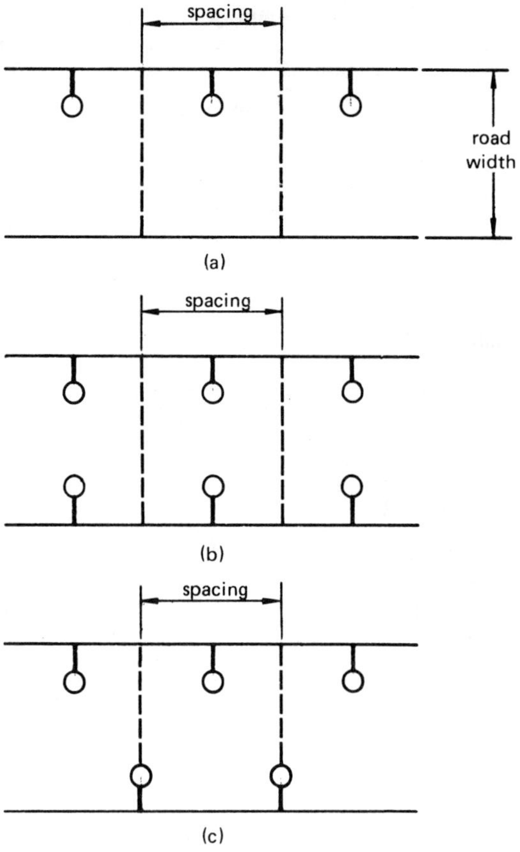

Figure 12.7 Street luminaire spacing and placements: (a) one side, (b) opposite sides, (c) staggered

A. RECOMMENDED LEVELS FOR ROADWAY LIGHTING

Lighting levels depend on the density of motor traffic, parking demand, and the amount of pedestrian traffic in an area. The three area classifications are:

•Commercial: areas where there is high demand for off-street and on-street parking, heavy foot traffic, and periods of peak traffic as in any densely developed business area.

•Intermediate: areas of industrial or moderate business development, including densely developed areas such as apartment complexes, hospitals, public libraries, and recreational centers.

•Residential: single-family and townhouse areas or a mixture of light commercial and residential areas. Areas with less activity are also included in this classification.

The five classifications of roadways are:

•Freeway: all divided highways with full control of access and no crossings at grade.

•Major and Expressway: principal routes for through traffic, including divided arterial highways with partial control of access but having interchanges on grade at major crossroads.

•Collector: roadways which are used for traffic movement in local areas, principally within residential, commercial, or industrial areas.

•Local: roadways providing direct access to residential, commerical, or industrial areas, and carrying no through traffic.

•Alley: Alleys

B. AVERAGE ILLUMINATION DESIGN

For roadside lighting, the coefficient of utilization is determined from two curves for a specific luminaire. Figure 12.8 shows the two curves with the isofootcandles contours.

To develop the concepts of roadway illumination design, it is necessary to know the *mounting height* (MH) and *spacing between luminaires.*

There are two coefficient of utilization curves in figure 12.8, both starting at point zero on the left edge of the graph. This point corresponds to the location of the luminaire. Points on the graph above the zero line correspond to the house side, and points below correspond to the street side. The coefficient of utilization is read on the upper scale for both curves. Note that the lateral distances on the left scale of the graph are normalized to the mounting height. For any situation, both coefficients of utilization are found and then added together to obtain the *effective coefficient of utilization.*

Table 12.8 Recommended illumination levels

ROADWAY	Commercial	Intermediate	Residential
Freeway	0.6 fc	0.6 fc	0.6 fc
Major and Expressway	2.0 fc	1.4 fc	1.0 fc
Collector	1.2 fc	0.9 fc	0.6 fc
Local	0.9 fc	0.6 fc	0.4 fc
Alley	0.6 fc	0.4 fc	0.2 fc

Coefficient of Utilization and Iso Footcandle* Plot
Based on Mounting Height of 30 ft.

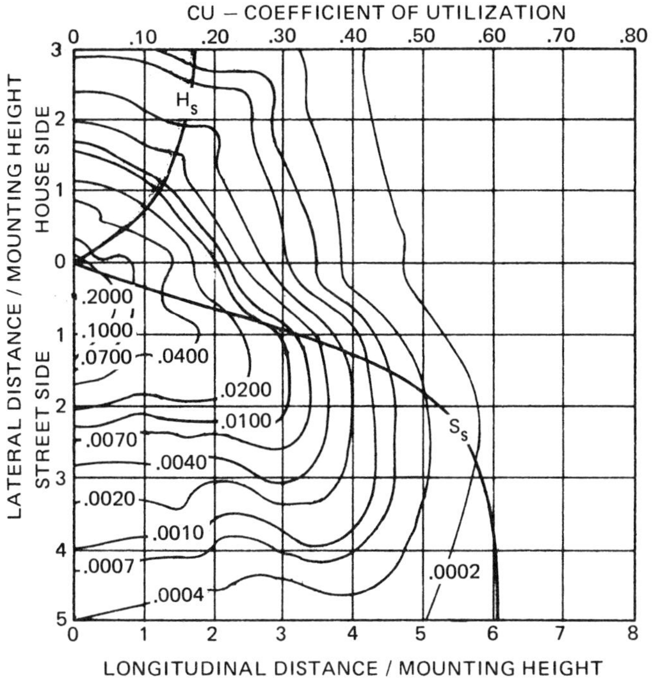

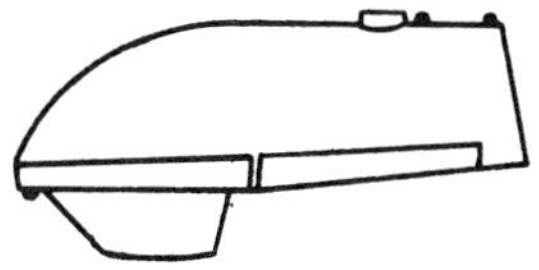

LUMINAIRE AND TEST LAMP DATA

Luminaire: sample

Refractor:

Reflector:

Socket Position:

Other Variables:

Lamp:
LCL: 5¾ inches
Lumen Rating: 25500

Adjustment Factor for Iso Ft-c values other than 30 ft. mounting height.

MTG.HT.	20	21	22	23	24	25	26	27	28	29
FACTOR:	2.25	2.04	1.86	1.70	1.56	1.44	1.33	1.23	1.15	1.07
MTG.HT.	30	31	32	33	34	35	36	37	38	39
FACTOR:	1.00	.94	.88	.83	.78	.73	.69	.66	.62	.59
MTG.HT.	40	41	42	43	44	45	46	47	48	49
FACTOR:	.56	.54	.51	.49	.46	.44	.43	.41	.39	.37

LIGHT FLUX DATA

Distribution by zones: (% of lamp lumens)

Downward Street Side:	62.0%]
Downward House Side:	17.3%
Upward Steet Side:	1.5%
Upward House Side:	1.0%
Total:	81.8%

*NOTE: Iso footcandle, candela, and Iso candela values as printed are per 1000 lamp lumens. For actual values multiply by $\frac{\text{Latest Lamp Rating}}{1000}$

At time of test this was $\frac{25500}{1000} = 25.5$

Candela* Data: (per 1000 lamp lumens)

Max. Cd:	600	at	73°V,	60°H
Nader Cd:	110	at	0°V,	0°H

$$\text{Initial Average Horizontal Ft-c.} = \frac{\text{Lamp Lumens} \times \text{C of U}}{\text{Spacing} \times \text{Width of Street}}$$

$$\text{Spacing} = \frac{\text{Lamp Lumens} \times \text{C of U}}{\text{Initial Ft-c Level} \times \text{Width of Street}}$$

Distribution Calssification: (ANSI D 12.2 1972)

Lateral:	Type IV
Vertical:	Medium
Control:	Semi-cutoff

Figure 12.8 Photometric data for a typical roadside luminaire

Example 12.10

A roadway is 60 feet wide, and the luminaires are suspended 6 feet from one side of the roadway. The luminaire is pictured in figure 12.8, and its mounting height is 30 feet. Determine the coefficient of utilization for this application.

$$\frac{Ss}{MH} = \frac{54}{30} = 1.8$$

From the upper section of figure 12.8, the Ss coefficient of utilization is 0.5.

$$\frac{Hs}{MH} = \frac{6}{30} = .2$$

From the lower section of figure 12.8, the *Hs* coefficient of utilization is 0.04. The total coefficient of utilization is

$$.5 + .04 = .54$$

For roadway luminaires, the luminaire dirt depreciation depends on the following cases of cleaning frequency and environments.

cleaning frequency	ENVIRONMENT clean	medium	dirty
every year	.95	.92	.87
every 3 years	.86	.77	.67
every 6 years	.77	.62	.49

Environmental definitions are given here.

•Clean: Clean pavement, grass, no open loose ground, few adhesive particles in the atmosphere, average or low car and truck traffic.

•Medium: Similar to clean except more adhesives in the atmosphere and heavier than average truck traffic.

•Dirty: Even more adhesives in the atmosphere and heavy truck traffic.

Each luminaire serves an area indicated by the dashed vertical lines in figure 12.7. For the opposite sides layout, there are two luminaires for each area. For the staggered layout there also are two luminaires for each area. The area is (roadway width) × (spacing).

The minimum maintained illumination formula, equation 12.9, is applicable here and is repeated for convenience.

$$MMI = \frac{(\#\text{luminaires}) \times \left(\frac{\text{lamps}}{\text{luminaires}}\right) \times (\text{initial flux}) \times LLF \times CU}{\text{area of the working surface}}$$

Here, the light loss factor (LLF) consists of the product of lamp lumen depreciation factor and the luminaire dirt depreciation factor. The area of the working surface is (roadway width) × (spacing).

Thus, this formula can be rearranged to obtain the spacing.

$$\text{spacing} = \frac{\#\text{ luminaires} \times \frac{\text{lamps}}{\text{luminaire}} \times \text{initial flux} \times LLF \times CU}{MMI \times \text{roadway width}} \qquad 12.15$$

Example 12.11

The roadway of example 12.10 is in a residential area. Assume an OV-25 (figure 12.8) luminaire with a high pressure sodium lamp C250S50, 250 watts. Determine the spacing to give the proper illumination level as recommended in table 12.8, assuming the street is a collector type and the environment is clean. The luminaires are cleaned on a 3-year schedule.

From table 12.8, the recommended illumination level is 0.6 fc. From example 12.10, CU=.54. The C250S50 lamp has initial lumens of 25,500 and an LLD of .91. The luminaire dirt depreciation LDD is .86. Therefore, the light loss factor is

$$\begin{aligned} LLF &= LLD \times LDD \\ &= .91 \times .96 = .78 \end{aligned}$$

Using equation 12.15, the spacing is

$$spacing = \frac{1 \times 1 \times 25,500 \times .78 \times .54}{.6 \times 60} = 298'$$

These simple examples do not take into account a number of possible variations. The effects of other mounting heights and the uniformity of the lighting must also be considered.

C. LUMINAIRE CLASSIFICATIONS

Minimum mounting height and spacing are determined by the initial candlepower of the luminaire lamp(s) and by the luminaire's *light distribution.* Light distribution is classified by *lateral light distribution, vertical light distribution,* and *candlepower distribution,* called *control* in figure 12.8.

The vertical and lateral distribution classifications are made on the basis of an iso-candela diagram on a rectangular coordinate grid which has superimposed a series of longitudinal roadway lines and a series of transverse roadway lines, both in multiples of the mounting height.

1. *Lateral light distributions* have seven classifications: xDpes I through V with additional Type I-4-way and Type II-4-way for intersections. Lateral light distributions are divided into two groups: center of the roadway or side of the roadway. These groups are then subdivided according to the spread (which is half the maximum candlepower) of the isocandela contours.

2. *Vertical light distribution* is split into three groups: short, medium, and long. These classifications depend on the vertical angle of maximum candlepower in relation to mounting height.

3. *Candlepower distribution* is divided into three categories: *cutoff* (first choice for glare control) up to 10%, *semi-cutoff* (second choice for glare control) from 10% to 30%, and *non-cutoff* with no limitation.

D. SPACING AND MOUNTING HEIGHT

The location, spacing, and mounting height of luminaires involves the uniformity ratio between lowest and average illumination level and glare considerations. The minimum mounting heights are shown in table 12.9.

Table 12.9 Minimum mounting heights

maximum luminaire candlepower	cutoff	semi-cutoff	non-cutoff
under 5,000	20	20	20
5,000 - 10,000	20	25	30
10,000 - 15,000	25	30	35
over 15,000	30	35	40

Once a luminaire is selected, the mounting height and spacing can be determined from the above table and the spacing formula (equation 12.15). The following maximum spacing to mounting height ratios must be used to insure that the ratio of minimum illumination to average is not less than $\frac{1}{3}$ ($\frac{1}{6}$ for local residential streets).

vertical distribution	maximum spacing / mounting height
short	4.5
medium	7.5
long	12.0

In addition, the overhang of a street-side mounted luminaire is not to exceed $\frac{1}{4}$ of the mounting height.

The staggered arrangement is preferred. Opposite spacing is used where the width of the street exceeds twice the mounting height. One- side layouts should be used only on very narrow streets. The recommended spacing and distribution types are shown in table 12.10.

At intersections, the illumination level should be the sum of the levels for the intersecting roadways.

Some typical acceptable roadway lighting configurations for two-lane roads are given in table 12.10.

A formula for the minimum mounting height is

$$\text{minimum } MH \text{ (feet)} = \frac{\text{maximum candlepower}}{1000} + ND \qquad 12.16$$

ND depends on the vertical distribution: for short distribution, $ND = 5$ feet; for medium $ND = 10$ feet; and for long distribution $ND = 15$ feet.

Example 12.12

Re-examine the design of example 12.11 and modify it to account for correct mounting height and spacing.

From figure 12.8, the vertical distribution is medium for this luminaire, so that from equation 12.16 with 25,500 initial lumens, the minimum MH is 35.5 feet.

From example 12.10, the roadway is 60 feet wide, and the luminaires are suspended 6 feet from one side of the roadway. With a mounting height of 36 feet, the width of the roadway is $\frac{60}{36} = 1.67$, and is more than $1.5 \times MH$, so that the luminaire placement must be staggered on opposite sides. Here we shall assume a staggered arrangement. The coefficient of utilization for this application then is found as follows:

$$\frac{Ss}{MH} = \frac{54}{36} = 1.5$$

From figure 12.8 (upper), the *Ss* coefficient of utilization is .45.

$$\frac{Hs}{MH} = \frac{6}{36} = .17$$

From figure 12.8 (lower), the *Hs* coefficient of utiilization is .01.

The total CU is

$$.45 + .01 = .46$$

Table 12.10 Recommended spacing and distribution types

roadway width	luminaire location	type of spacing	lateral distribution type
up to $1.5 \times MH$	side of roadway	staggered or one side	II, III or II-4-way for intersections
beyond $1.5 \times MH$	side of roadway	staggered or opposite	III or IV
up to $2 \times MH$	center of roadway		I with I-4-way or V for intersections

Table 12.11 Some acceptable lighting configurations for 2-lane roads, with $LLF = 0.7$

Road Type: Two-lane, major rural, 10-foot shoulders
Road Width: 24 feet (two 12-foot lanes)
Desired Illumination Level: 0.9 average horizontal footcandles
Desired Uniformity Ratio: 3:1

Light Distribution			Lamp Data Mercury		Pole Data		Arrangement	Spacing (feet)	Illumination* (footcandles)		Uniformity Ratio
Type	Vertical	Control	Designation	Light Output (lumens)	Mounting Height (feet)	Overhang (feet)			Average	Minimum	
II	S	C.O.	H39KB 175 W Clear	7,700	30	4	One Side	92	0.9	0.36	2.5:1
II	M	Semi C.O.	H37KB 250 W Clear	12,100	30	4	One Side	135	0.9	0.49	1.8:1
II	M	C.O.	H33CD 400 W Clear	21,000	30	4	One Side	100	2.13	0.71	3.0:1
II	M	C.O.	H33CD 400 W Clear	21,000	30	4	Staggered	150	1.41	0.47	3.0:1
III	L	Semi C.O.	H35NA 700 W Clear	39,000	35	4	One Side	200	1.49	0.51	2.9:1
III	L	Semi C.O.	H35NA 700 W Clear	39,000	35	4	Staggered	210	1.47	0.49	3.0:1

* Based on 70 percent maintenance factor.

The recommended illumination level is 0.6 fc., and the LLF remains at .78. From equation 12.15,

$$\text{spacing} = \frac{1 \times 1 \times 25{,}500 \times .78 \times .45}{.6 \times 60} = 248'$$

The $\frac{\text{spacing}}{MH}$ is $\frac{248}{36} = 6.9$, which is just acceptable for this luminaire to maintain the ratio of minimum illumination to average at $\frac{1}{3}$ or more. If this ratio were higher, the mounting height would have to be increased for a non-residential application.

E. POINT CALCULATIONS

It is necessary to check for illumination uniformity by making point calculations, which involve the use of the iso-footcandle contours. The points of minimum illumination can be found by overlaying the lighted zone for one luminaire with its transverse boundaries. At some point on this boundary, the point of minimum illumination will occur, and it is to be found by common sense and the luminaire pattern chosen. For instance, it will be at the opposite curb for one-sided patterns and at the middle of the street for staggered or opposite-sides patterns.

Then, it will be necessary to read the iso-footcandle contours and make the adjustments indicated in figure 12.8. The minimum illumination level will be twice the calculated value for one-side and staggered arrangements, and four times the value for opposite-sides arrangements.

8 FLOODLIGHTING

Much of what has been covered up to this point also applies to floodlighting. Some changes are necessary, particularly in the coefficient of utilization, although this is not applicable unless extensive calculations are made. The basic formula for floodlighting is

$$\#\text{floodlights} = \frac{\text{illumination level} \times \text{square feet}}{\text{beam lumens} \times (\text{lamps/floodlight}) \times LLF \times CBU} \qquad 12.17$$

Beam lumens are the total lumens contained within the beam spread. The beam spread is the angle between the two directions in which the candlepower is 10% of the maximum candlepower (which is typically at the center of the beam). This information is available as catalog and photometric data.

CBU is the *coefficient of beam utilization*, which is the ratio of the lumens striking the surface to the beam lumens.

Beam spreads are categorized by seven types.

- **type 1:** beam spread is 10° to 18°
- **type 2:** beam spread is 18° to 29°
- **type 3:** beam spread is 29° to 46°
- **type 4:** beam spread is 46° to 70°
- **type 5:** beam spread is 70° to 100°
- **type 6:** beam spread is 100° to 130°
- **type 7:** beam spread is 130° or greater

Floodlights not having a perfectly conical beam shape are classified by the minimum and maximum angles, such as a type 5 × 6.

Design of floodlighting is best accomplished mechanically by pictorial drafting. This is perhaps more of an art than a science where the illumination is to be at a certain level, and consists of a trial and error method of laying out circular and/or elliptical areas within the beam spreads of the various floodlights.

Because the candlepower at the beam edge is only 10% of that at the middle, it is necessary to have some overlap of beams at the lighted surface. The overlap will account for the coefficient of beam utilization, which should lie between .6 and .9. If the coefficient of beam utilization is less than .6, the plan may be more expensive than necessary. If greater than .9, the resulting lighting pattern tends to be spotty.

9 WIRING DESIGN AND THE NATIONAL ELECTRIC CODE

Safety of persons and property is a primary concern with electrical wiring. As a dynamic and highly concentrated form of energy, electricity constitutes a hazard in the vicinity of its use. The *National Electric Code,*[1] hereafter referred to as the *Code*, recommends safeguards to protect against these hazards. The *Code* has been adopted as the basis of many state and local ordinances regulating electrical installations.

The principal concern in this section is the design of wiring systems, including conductor specifications and circuit protection. More complete treatment of *Code* requirements is found in other reference books.[2,3]

[1]National Fire Protection Association, *National Electric Code* 1978, 470 Atlantic Avenue, Boston, MA 02210.

[2]*Guide to the National Electric Code*, T. Harman and C. Allen, Prentice-Hall Inc., Englewood Cliffs, NJ 07632, 1978.

[3]*National Electric Code Reference Book*, J.D. Garland, Prentice-Hall Inc., Englewood Cliffs, NJ 07632, second edition, 1978.

The *Code* does not deal with the distribution system of a utility company, but begins with the *service* provided by the utility. Services can range from the common $\frac{115}{230}$ V single phase service typically provided to residences, to 13.8 KV three-phase service provided to large industrial users. All portions beyond the "last pole" of the utility company are covered by the *Code* .

The wiring design can be divided into three parts: the *service entrance, feeders,* and *branch circuits.*

A. SERVICE

The *service entrance* is composed of the overhead *service drop* or the underground *lateral conductors, raceways, service disconnect, main overcurrent protection, grounding electrode conductor,* the *main bonding jumper,* and any protective housings. It extends from the service drop or the lateral conductors to the service disconnect. Service requirements are covered primarily in *Code* article 230, but other articles are pertinent to the service entrance. Further references to the *Code* articles are given in brackets [ ].

Of particular interest to electrical designers are sections [230 E] covering service entrance conductors, and [230 J] covering service equipment-overcurrent protection.

[230-41] deals with size and rating of the conductors, stating that conductors shall be of sufficient size to carry the loads in accordance with article [200], with ampacity determined from tables [310-16] through [310-19] and all applicable notes to these tables. Article [220] discusses the determination of loads but refers to [430-22], [430-24], [430-25], and [430-26] for motor loads and to article [440] for air conditioning and refrigeration. Also of interest are [440.5] and [440 D] dealing with ampacity and branch circuit conductors, respectively.

In [230 J] on overcurrent protection, section [230-90] requires that service entrance overcurrent protection be no greater than the ampacity of the conductors. Exception No. 1 refers to articles [430-52], [430-62], and [430-63] on motor circuits, and exception No. 2 permits protection at the next higher fuse or circuit breaker rating, in turn referring to sections [240-3], exception No. 1, and [240-6] dealing with the standard ratings for fuses and inverse-time circuit breakers. Fuses in grounded neutrals are prohibited in [230-90(b)], which permits circuit breakers in grounded neutrals only when all ungrounded lines are broken with the same circuit breaker.

Other applicable references are service raceways [chapter 9: tables 3, 4, and 5], service disconnects [230-79], the grounding electrode conductor [230-94], and the main bonding jumper [250-79(c)].

B. FEEDERS AND BRANCH CIRCUITS

Feeders are used in large installations where the service is remote from parts of the load. A feeder is defined in [100] as "all conductors between the service equipment and the final branch overcurrent device." Small installations, such as residences, often have no feeders but have branch circuits radiating from the service equipment.

Feeders are covered generally in article [215] - Feeders, article [220] - Branch-Circuit and Feeder Calculations (part B), article [225] - Outside Branch Circuits and Feeders, and in article [430] - Motor Circuits and Controllers.

Branch circuits are defined as "the circuit conductors between the final overcurrent device protecting the circuit and the outlets." Branch circuits are covered in articles [210] - Branch Circuits, [220] - Branch and Feeder Calculations (part A), [225] - Outside Branch Circuits and Feeders, and [430] on motors.

C. CONDUCTORS FOR GENERAL WIRING

Ampacity (the current-carrying capacity of electrical conductors expressed in amperes [100]) ratings for conductors are specified for various types of insulations, uses, and conditions in [310].Conductors which are integral parts of equipment are not included.

In general, the minimum wire size for copper conductors is AWG[4] #14. For aluminum or copper clad aluminum, it is AWG #12 [310-5]. Important exceptions are for fixture wire [410-23] and flexible cords [400-12] which may be AWG # 18. Other exceptions are given in [310-5].

Conductor types are established by the insulation used, so the insulation type, together with the metal (copper or aluminum) used, is descriptive of the conductor. Table [310-13] lists the conductor types for nominal voltages up to 600 volts. (Higher voltage conductor types are covered in sections [310-30] through [310-61] but are not considered here.) The meanings of the letters in the insulation and conductor types are

A - asbestos
C - cotton
H - heat resistant
L - lead covered
MI - mineral insulated
R - rubber
RU - latex rubber
S - silicone rubber
T - thermoplastic
V - varnished cambric
W - moisture resistant

Table [310-13] gives the maximum operating temperature and the approved applications for each certified insulation type. Note that wet locations, including underground installations and raceways or conduits subject to moisture condensation, require a W in the insulation type.

Tables [310-16] through [310-19] give the allowable ampacities for various types. Table [310-16] is the most applicable in branch circuit design. This table and the footnotes to tables [310-16] through [310-19] are very important in the design of branch circuits and are equally applicable to feeders and services below a voltage of 600 volts nominal.

[110-3(b)] requires that "listed or labeled equipment shall be used or installed in accordance with any instructions included in the listing or labeling." This requirement incorporates the listings of Underwriters Laboratories (UL) and other laboratories into the provisions of the Code.[5] Its effect on the selection of conductors for branch circuits is significant.

A basic UL rule requires that terminations rated at 100 amps or less use conductors rated at 60° C and that terminations rated at more than 100 amps use conductors rated at 75° C[5]. This includes fuses and circuit breakers which set the current rating of a circuit. Because the maximum circuit rating for branch circuits is 50 amps [210-23], branch circuit conductors must be rated at 60° C. Some of the advantages of using higher temperature conductors (75° C and 90° C) are lost by this rule.

D. CABLES AND RACEWAYS

For the direct application of tables [310-16] and [310-18], a cable or raceway can contain no more than three conductors. Additional conductors require an ampacity derating according to the requirements of [note 8] following table [310-19]. According to [note 10(a)], the neutral of a normally balanced 3-phase circuit is not counted as an extra conductor. [Note 10(b)] and [note 10(c)] do require that the neutral be counted as an extra conductor in cases of unbalanced loads and electric discharge lighting loads.

[4]Wire sizes are designated by American Wire Gage (AWG) numbers, or in thousands of circular mils of cross-sectional area (MCM). AWG numbers begin with the largest at 0000, which has a cross-sectional area of 212,000 circular mils, followed by 000, 00, 0, 1, 2 etc., with increasing numbers corresponding to smaller cross-section area and lower ampacity. Larger conductors begin at 250 MCM and increase in increments of 50,000 circular mils of cross-sectional area.

[5]*A Guide to Basic Electrical Design*, J. F. McPartland, Electrical Cosntruction and Maintenance, July 1978, part 3, p 72 ff.

[300-20] requires that all phases, the neutral conductor, and the associated grounding conductor for a single circuit be grouped together in metallic enclosures and raceways, including conduits and tubing.

[300-3] permits the grouping of any number of separate circuits of nominal voltage less than 600 within the same raceway or cable.

[Chapter 9] gives the number of conductors permitted in a conduit or tubing. [Table 1] lists the percentage of the interior that can be filled with conductor (including insulation); [table 2] is for fixture wire; [table 3] shows the number of conductors for conduits or tubing of trade sizes using general purpose wiring. [Tables 2 and 3] are based on identical conductors within the conduit or tubing. When sizes or types are mixed, [table 1] is used to determine the minimum size of conduit or tubing.

Associated information, which is useful in calculations, is given in [table 4] on trade sizes of conduit and tubing and in [tables 5, 6 and 7] listing dimensions of approved conductors.

E. OVERCURRENT PROTECTION [240]

Overcurrent protection prevents overheating of conductors and equipment, and is covered in article [240]. [240-2] gives references for specific equipment. Of particular interest are items [210-20] on branch circuits, and [430 parts C, D, E, and H] and table [430-152] on motors, motor circuits, and controllers. Diagram [430-1] gives a directory through the article.

The basic code rule is that overcurrent devices must be rated no higher than the ampacity of the conductors after all derating factors have been included. Exceptions to this rule are given in [240-3]. Circuit breakers CB and fuses are the devices used for overcurrent protection. Standard ratings are given in [240-6].

F. BRANCH CIRCUIT DESIGN - ELECTRIC DISCHARGE LIGHTING

Article [100] defines a multi-wire branch circuit as consisting of "two or more ungrounded conductors having a potential difference between them, and an identified grounded conductor having equal potential difference between it and each ungrounded conductor of the circuit and which is connected to the neutral conductor of the system." This permits 120/240 volt single-phase and three-phase, 4-wire branch circuits, such as 120/208 wye connected circuits, to be classified as multi-wire branch circuits. [210-4] requires that these circuits supply only line-to-neutral loads except when all ungrounded conductors of the circuit are opened simultaneously by the branch circuit overcurrent protection device [exception No. 2 to 210-4]. This, in effect, prohibits the use of fuses for overcurrent protection and requires the use of multi-pole circuit breakers.

[210-6(a)] limits the maximum voltage to ground for branch circuits supplying lampholders, fixtures, or standard receptacles rated at 15 amperes to 150 volts. [Exception No. 1] permits up to 300 volts to ground for industrial installations supplying only lighting fixtures with mogul-base screw-shell lampholders or other approved types of lampholders, provided that the fixtures are not less than 8 feet above the floor. [Exception No. 2] permits up to 300 volts to ground for branch circuits in stores, health-care facilities, office buildings, schools, or public and commercial areas of other buildings such as hotels or transportation terminals which supply only ballasts for electric discharge lamp fixtures which are permanently installed. For screw-shell lampholders, the fixtures must be no less than 8 feet above the floor.

[210-6(b)] permits voltages up to 500 volts between conductors for branch circuits supplying only the ballasts of electric-discharge lamps which are permanently mounted in fixtures at least 22 feet above the ground in outdoor applications and at least 18 feet above the floor in other structures such as tunnels.

[210-6(c)] limits the voltage between conductors to 150 volts on branch circuits supplying screw-shell lamp holders, receptacles or appliances in dwelling units, hotel guest rooms and similar occupancies, and in any use where medium base screw-shell lampholders are supplied.

The above restrictions virtually eliminate the possibility of the use of multi-wire branch circuits for lighting circuits in residences and similar occupancies but permit their use in industrial and commercial occupancies.

[210-10] permits the use of tapped conductors for A.C. circuits having two or more ungrounded conductors and a grounded neutral conductor. [210-19(c)] requires that such tapped conductors have an ampacity sufficient for the load, but that AWG #14 is the minimum tap conductor size. For circuits rated at less than 40 amps, the minimum ampacity of tap conductors is 15 amps; and for circuits rated at 40 or 50 amps, the minimum ampacity of tap conductors is 20 amperes. Such taps, as used for individual lampholders or fixtures, are restricted further to extend no more than 18 inches beyond the lampholder or fixture.

[210-20] requires that branch circuit protection be provided by overcurrent devices with ratings or settings

not exceeding those specified by [240-3]. This section refers to tables [310-16 through 310-19] and the following [notes] for lighting circuit conductors with the exception of tap conductors in [210-19(c)] discussed above and as provided for in [210-21(a)] for lampholders. [210-21(a)] requires that lampholders connected to branch circuits having circuit ratings in excess of 20 amperes be of a heavy-duty type with ratings of 660 watts if of the admedium type and at least 750 watts if of other types.

[210-22] requires that the total load of a branch circuit not exceed the rating of the branch circuit, which is determined from the rating or setting of the overcurrent device. Further requirements restrict the load to 80% of the rating of the branch circuit for lighting loads. In addition, [210-22(b)] requires that circuits supplying lighting units with ballasts, transformers, or autotransformers have the load computed using the actual current rather than the power and voltage. This includes all discharge lighting systems. [Exception No. 2] to [210-22(c)] permits only one derating for lighting circuit conductor ampacity to satisfy the 80% figure in this article and the requirements of [note 8 to tables 310-16 through 310-19]. If the derating of that note exceeds that of this article, the severest restriction applies (i.e., if note 8 is required, a derating factor of 70% would prevail over the 80% factor of this article).

[210-23] specifies the permissible load for various rated branch circuits. In particular, 25, 30, 40, and 50 amp circuits require heavy duty lampholders in fixed lighting units. They are permitted only in occupancies other than dwellings. [Table 210-24] summarizes the branch circuit requirements and is useful in wiring design.

[220-2] deals with the computation of branch circuit loads. (b) and [table 220-2(b)] describe the minimum unit load for listed occupancies on the basis of watts per square foot of floor area. These unit loads are minimum values and typically will be less than those determined from a lighting design procedure as outlined earlier in this chapter. In any particular case, the actual wattage per square foot must be no less than that given in table [220-2(b)].

Article [410-18] requires that all fixed metal lighting fixtures be grounded. [250-9] permits equipment grounding conductors to be wires and metal raceways including rigid conduit and EMT. Wire sizes for grounding conductors are given in table [250-95].

Example 12.13

Example 12.8 is a lighting design calling for 90 luminaires with two 48T-12 fluorescent bulbs per luminaire. Consider the branch circuit design for that example. A 120/208 volt four-wire three-phase service is available.

The room involved has 2400 square feet of floor area, and the light power, at 40 watts per bulb and 180 bulbs, is nominally 7200 watts, or 3 watts per square foot. This is not necessarily a listed occupancy of table [200-2(b)] but would qualify for commercial-industrial buildings or schools, among others.

However, the lighting power for fluorescent and other electric discharge lamps cannot be calculated from the wattage of the lamps, as there are inductive components involved in the ballasting. [210-22(b)] requires that the load be calculated from the actual current rather than from the wattage for electric discharge lighting units. Garland[6] gives a "rule of thumb" of adding 25% to the wattage to account for the ballast effects. This is the same as assuming a power factor of 0.8. Thus, for 40 watt fluorescent bulbs, the volt-amperes per bulb would be VI = 40 watts ×1.25 = 50 volt-amperes, which is to be used to determine the current load.

Then, the load per bulb is 50 VA/120 volts=.417 amps (using a 120 volt line-to-neutral circuit). It is sensible to divide the load into three equal parts, assigning one part to each phase. Thus, each phase would supply a current of

$$30 \text{ luminaires} \times 2\frac{\text{bulbs}}{\text{luminaire}} \times .417\frac{\text{amps}}{\text{bulb}} = 25 \text{ amps}$$

A circuit rated at more than 20 amps [210-20(a)] requires lampholders to be of a heavy-duty type, meaning admedium or mogul. Most fluorescent bulbs are supplied with medium bipin bases (table 2.3 shows a 40W 60T-12 lamp with a mogul bipin base), and circuit ratings are to be restricted to 20 amp ratings for most fluorescent applications. Garland[6] interprets [410-14] to permit only mogul based screw-shell lampholders being used on circuits rated up to 50 amps, and restricting fluorescents of all types to circuits rated at 20 amps or 15 amps. (If that were a correct interpretation, there would be little reason for manufacturers to produce mogul-based fluorescent bulbs.)

Here we divide each phase into two branches, each having a load of 12.5 amps. [210-22(c)] requires a minimum of 80% derating for lighting circuits. [Note 8 to tables 310-16 through 310-19] can cause a further derating if more than three current-carrying conductors are routed through the same raceway or cable.

The luminaire layout indicates that with ten rows and with the two branch circuits per phase it would be

[6] *National Electrical Code Reference Book*, J.D. Garland, Prentice-Hall, Englewood Cliffs, NJ 2d ed. page 29.

sensible to use two multi-conductor branch circuits with each serving five rows of 9 luminaires each. Then, each cable or raceway would contain the three phase conductors and a neutral. [Note 10(c) to tables 310-16 through 310-19] requires that the neutral here be counted as a current carrying conductor.

[220-22] (which actually applies to feeders) requires that the portion of the neutral load supplying electric discharge lighting units cannot be reduced in ampacity from that portion of the phase currents. This in effect requires that the neutral have the same ampacity as the ungrounded conductors. Here, there is an 80% derating from [note 8 of tables 310-16 through 310-19]. The total derating will be 80% for the ampacity of the conductors, as [210-22(c)] only requires 80% derating when [note 8 of tables 310-16 through 310-19] has not been in force. Then, the required ampacity is

$$12.5 \text{ amps} \times 1.25 = 15.625 \text{ amps}$$

$$\frac{12.5}{.8} = 15.625 \text{ amps}$$

As the overcurrent protection device cannot operate continuously at more than 80% of its rating [210-23], such devices here must be rated at

$$\frac{15.625}{0.8} = 19.53 \text{ amps}$$

This capacity is not standard, so [240-3, exception no. 1] permits the next larger size, which is 20 amps. (Note that [210-3] restricts branch circuits serving more than one load to ratings of 15, 20, 30, 40, and 50 amperes).

The branch ampacity is over 15 amps, so the footnote (†) to table [310-16], which limits the overcurrent protection device for AWG #14 wire to 15 amps, eliminates the possibility of using 90 °C conductors with their higher ampacity to achieve a smaller wire size. Thus, a 60 °C AWG #12 conductor is specified for the branch conductors. Table 3A of [chapter 9] allows up to seven AWG #12 conductors of the 90° temperature class in $\frac{1}{2}$ inch electrical metal tubing.

If the luminaires are of the type which permit the circuit conductors to be run through the luminaires, tap conductors will not be needed, as the branch conductors will be within 18 inches of the fixture, allowing fixture wire of AWG #18[240-4]. This wire should have insulation to protect it in environments up to 90 °C, as [410-68] permits units to reach such temperatures. [Table 402-3] gives information on fixture wire, and TFN is a suitable choice. If the maximum operating temperature is known, a lower rated fixture wire could be specified.

If the luminaires are to be supplied by individual taps from overhead branch circuits, table [210-24] requires AWG #14 conductors to be used. Again a 90 °C conductor should be specified in the absence of luminaire temperature limits.

The design specifications are:

Overcurrent protection: 20 amps (6 required)

Branch conductors: AWG #12, 60 °C type (type T is adequate)

(a) Fixture wire: AWG #13, 90 °C type (type TFN is adquate)

(b) Tap conductors: AWG #14, 90 °C type (type THHN is adequate)

$\frac{1}{2}$ inch EMT (Electrical metal tubing) which will be used as the grounding conductor.

(a) or (b) above is to be used depending on the luminaire. Two 4-wire branch circuits rated at 20 amps each are required.

Practice Problems: ILLUMINATION

Warmups

1. A room with floor area 40 feet × 60 feet and a 24 foot ceiling height has luminaires mounted flush with the ceiling. Determine the cavity ratios.

2. A room with floor area 40 feet × 60 feet and a 24 foot ceiling height has luminaires suspended 12 feet from the ceiling. Determine the cavity ratios.

3. For the room of problem 2, the ceiling reflectance is 90% and the wall reflectance is 70%. Determine the effective ceiling reflectance.

4. For the room of problem 3, use two T-12 lamps per luminaire, and determine the coefficient of utilization.

5. For the room of problem 4, the required illumination level is 100 footcandles. Determine the net luminous flux necessary for luminaires with two T-12 lamps.

6. The room of problem 5 has the following environment: (1) the line voltage is known to drop 5% at the peak loading time of day, (2) high quality material has been used in the luminaires; (3) all burned out lamps are replaced immediately, (4) all lamps will be replaced at the end of 70% of the expected life, and the lamps will burn 12 hours for each start, resulting in a lamp lumen depreciation of 0.84. The luminaire is as in table 12.1, the environment is clean, and the luminaires are to be cleaned on an annual schedule. Determine the total loss factor.

7. The room of problem 6 requires a minimum maintained illumination level of 100 footcandles. If two 48 foot T-12 lamps, with 3000 initial lumens per lamp, are used per luminaire, determine the number of luminaires required.

8. For the room of problem 7, assume the luminaires are 52 inches long, and are to be placed in rows across the short (40 feet) dimension of the room. Provide a layout for this room.

9. For the room of problem 8, determine the reflected horizontal component of illumination at a point which is six feet from two walls.

10. Determine the direct illumination at the point of calculation of problem 9.

Concentrates

1. A display room is 56 feet × 28 feet with a 14 foot ceiling and has display counters oriented along the length of the room. The ceiling reflectance is 90%, and the wall reflectance is 50%. Design a lighting system using warm white deluxe lamps in the 2 T-12 luminaire of table 12.1. The maintained minimum illumination is to be 90 footcandles 30 inches from the floor. The environment is clean, and the cleaning schedule is annual. Make a sketch of the resulting design.

2. The floor area of an industrial plant is 96 feet × 180 feet and the ceiling height is 29 feet. The long axis of the machines is parallel to the long dimension of the building. The illumination proposal is to use 8 foot (100 inches) luminaires with four 4 foot lamps per luminaire, and to finish any rows with 4 foot (50 inches) luminaires with two 4 foot lamps per luminaire. The basic design is

minimum maintained illumination: 100 fc
initial lumens per lamp: 3100 Lm
coefficient of utilization: 0.75
maintenance factor (LLF): 0.62
Mounting height: 18 feet
8 foot luminaire wattage (including ballast): 285 W
4 foot luminaire wattage (including ballast): 190 W

(a) Determine the number of 8 foot and 4 foot luminaires required, (b) sketch the layout, giving spacings, and (c) determine the operating cost if energy is 1.5 cents per kilowatt hour and the factory is operated 17 hours a day for 50 weeks a year.

3. Design the lighting for a 40 foot wide roadway in a commercial area where the roadway is a major carrier, the environment is medium, and the luminaires are cleaned every three years. Use the luminaire from figure 12.8.

Timed

1. Two proposed lighting installations are being considered to maintain an illumination level of 30 footcandles in a warehouse with 20,000 square feet of floor area. Assume an electric energy cost of 75 mils and 3200 hours of lighting per year. Assume a coefficient of utilization of 78% and a maintenance factor of 72%. All costs involved in installation are prorated into the fixture costs. Replacement costs include lamps, installation, and cleaning. The two proposed systems are

Luminaire A: Incandescent, 150 W, 115 V, 2600 Lm initally, 750 hours life, fixture cost=$ 75, replacement cost=$4.35.

Luminaire B: Mercury vapor, 175 W, 115V, 7700 inital Lm, 24000 hours life, fixture=$162.50, replacement cost=$10, ballast loss 25 W.

Compare the two systems with respect to installation cost and annual cost of operation.

2. A parking lot of dimensions 420 feet × 360 feet is to use 40 foot poles with a maximum of four luminaires per pole. The minimum maintained illumination level is to be 6.25 footcandles (average). 1000 W, 65,000 lumen lamps are to be used. Assume an LLD of 74%, an LDD of 48%, and a 25 W/lamp ballast loss. Design for a coefficient of beam utilization of 65%.

(a) Determine the layout, giving a sketch and the recommended luminaire type (i.e., beam angle), (b) specify the wiring, assuming 220 V, three-phase voltage (wire size and distribution), and (c) specify the protective breakers (voltage, current, and estimated surge current assuming proportionality to average current from 1.75 to 2).

RESERVED FOR FUTURE USE

13 ENGINEERING ECONOMIC ANALYSIS

Nomenclature

A	annual amount or annuity	\$
B	present worth of all benefits	\$
BV_j	book value at the end of the jth year	\$
C	cost, or present worth of all costs	\$
d	declining balance depreciation rate	decimal
D_j	depreciation in year j	\$
$D.R.$	present worth of after-tax depreciation recovery	\$
e	natural logarithm base (2.718)	–
$EUAC$	equivalent uniform annual cost	\$
f	federal income tax rate	decimal
F	future amount or future worth	\$
G	uniform gradient amount	\$
i	effective rate per period (usually per year)	decimal
k	number of compounding periods per year	–
n	number of compounding periods, or life of asset	–
P	present worth or present value	\$
P_t	present worth after taxes	\$
ROR	rate of return	decimal
ROI	return on investment	\$
r	nominal rate per year (rate per annum)	decimal
s	state income tax rate	decimal
S_n	expected salvage value in year n	\$
t	composite tax rate, or time	decimal,-
z	a factor equal to $\frac{(1+i)}{(1-d)}$	decimal
ϕ	nominal rate per period	decimal

1 EQUIVALENCE

Industrial decision makers using engineering economics are concerned with the timing of a project's cash flows as well as with the total profitability of that project. In this situation, a method is required to compare projects involving receipts and disbursements occurring at different times.

By way of illustration, consider \$100 placed in a bank account which pays 5% effective annual interest at the end of each year. After the first year, the account will have grown to \$105. After the second year, the account will have grown to \$110.25.

Assume that you will have no need for money during the next two years and that any money received would immediately go into your 5% bank account. Then, which of the following options would be more desirable?

option a: \$100 now
option b: \$105 to be delivered in one year
option c: \$110.25 to be delivered in two years

In light of the previous illustration, none of the options is superior under the assumptions given. If the first option is chosen, you will immediately place \$100 into a 5% account, and in two years the account will have grown to \$110.25. In fact, the account will contain \$110.25 at the end of two years regardless of the option chosen. Therefore, these alternatives are said to be *equivalent.*

2 CASH FLOW DIAGRAMS

Although they are not always necessary in simple problems (and they are often unwieldly in very complex problems), *cash flow diagrams* may be drawn to help visualize and simplify problems having diverse receipts and disbursements.

The conventions below are used to standardize cash flow diagrams.

- The horizontal (time) axis is marked off in equal increments, one per period, up to the duration or horizon of the project.

- All disbursements and receipts (cash flows) are assumed to take place at the end of the year in which they occur. This is known as the *year-end convention.* The exception to the year-end convention is any initial cost (purchase cost) which occurs at $t = 0$.

- Two or more transfers in the same year are placed end-to-end, and these may be combined.

- Expenses incurred before $t = 0$ are called *sunk costs.* Sunk costs are not relevant to the problem.

Example 13.1

A mechanical device will cost $20,000 when purchased. Maintenance will cost $1,000 each year. The device will generate revenues of $5,000 each year for 5 years after which the salvage value is expected to be $7,000. Draw and simplify the cash flow diagram.

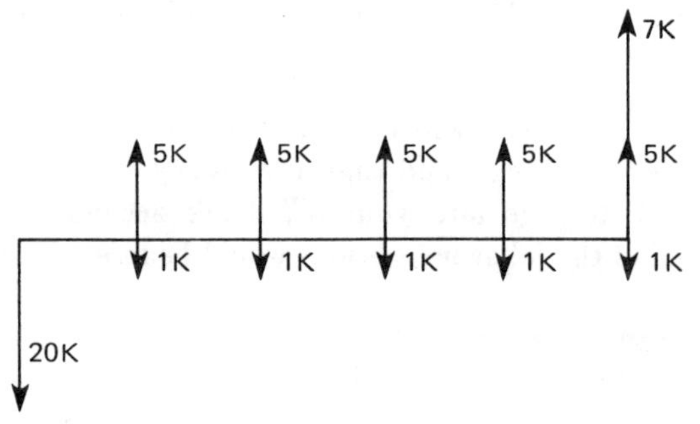

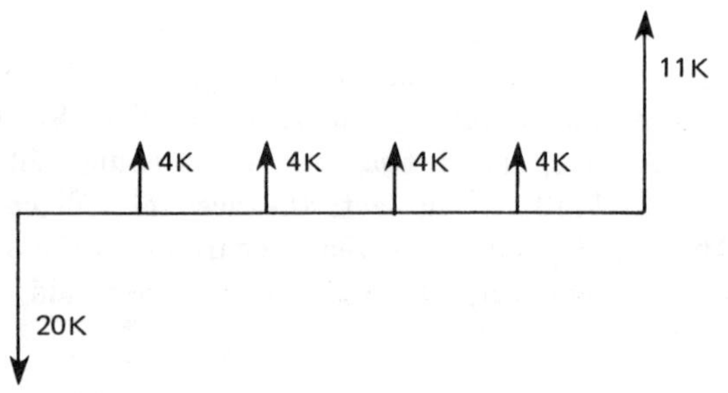

3 TYPICAL PROBLEM FORMAT

With the exception of some investment and rate of return problems, the typical problem involving engineering economics will have the following characteristics:

- An interest rate will be given.

- Two or more alternatives will be competing for funding.

- Each alternative will have its own cash flows.

- It is necessary to select the best alternative.

Example 13.2

Investment **A** costs $10,000 today and pays back $11,500 two years from now. Investment **B** costs $8,000 today and pays back $4,500 each year for two years. If an interest rate of 5% is used, which alternative is superior?

The solution to this example is not difficult, but it will be postponed until methods of calculating equivalence have been covered.

4 CALCULATING EQUIVALENCE

It was previously illustrated that $100 now is equivalent at 5% to $105 in one year. The equivalence of any present amount, P, at $t = 0$ to any future amount, F, at $t = n$ is called the *future worth* and can be calculated from equation 13.1.

$$F = P(1+i)^n \qquad 13.1$$

The factor $(1+i)^n$ is known as the *compound amount factor* and has been tabulated at the end of this chapter for various combinations of i and n. Rather than actually writing the formula for the compound amount factor, the convention is to use the standard functional notation $(F/P, i\%, n)$. Thus,

$$F = P(F/P, i\%, n) \qquad 13.2$$

Similarly, the equivalence of any future amount to any present amount is called the *present worth* and can be calculated from

$$P = F(1+i)^{-n} = F(P/F, i\%, n) \qquad 13.3$$

The factor $(1+i)^{-n}$ is known as the *present worth factor,* with functional notation $(P/F, i\%, n)$. Tabulated values are also given for this factor at the end of this chapter.

Example 13.3

How much should you put into a 10% savings account in order to have $10,000 in 5 years?

This problem could also be stated: What is the equivalent present worth of $10,000 5 years from now if money is worth 10%?

$$P = F(1+i)^{-n} = 10,000(1+.10)^{-5} = 6,209$$

The factor .6209 would usually be obtained from the tables.

A cash flow which repeats regularly each year is known as an *annual amount.* Although the equivalent value for each of the n annual amounts could be calculated and then summed, it is much easier to use one of the *uniform series factors*, as illustrated in example 13.4 below.

Example 13.4

Maintenance costs for a machine are \$250 each year. What is the present worth of these maintenance costs over a 12 year period if the interest rate is 8%?

Notice that

$$(P/A, 8\%, 12) = (P/F, 8\%, 1) + (P/F, 8\%, 2) + \cdots + (P/F, 8\%, 12)$$

Then

$$P = A(P/A, i\%, n) = -250(7.5361) = -1,884$$

A common complication involves a uniformly increasing cash flow. Such an increasing cash flow should be handled with the *uniform gradient factor*, $(P/G, i\%, n)$. The uniform gradient factor finds the present worth of a uniformly increasing cash flow which starts in year 2 (not year 1) as shown in example 13.5.

Example 13.5

Maintenance on an old machine is \$100 this year but is expected to increase by \$25 each year thereafter. What is the present worth of 5 years of maintenance? Use an interest rate of 10%.

In this problem, the cash flow must be broken down into parts. Notice that the 5-year gradient factor is used even though there are only 4 non-zero gradient cash flows.

$$P = A(P/A, 10\%, 5) + G(P/G, 10\%, 5)$$
$$= -100(3.7908) - 25(6.8618) = -551$$

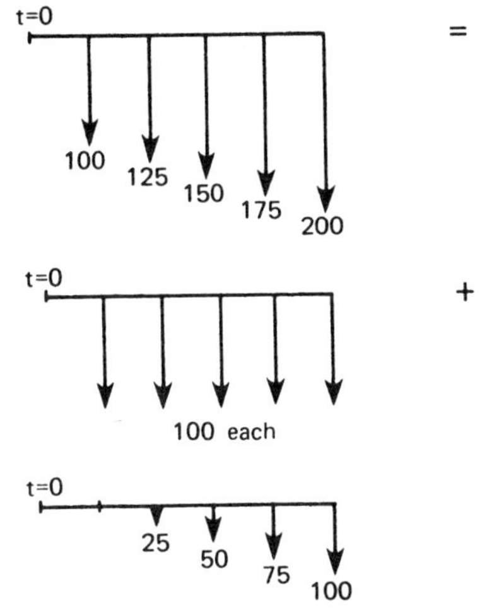

Various combinations of the compounding and discounting factors are possible. For instance, the annual cash flow that would be equivalent to a uniform gradient may be found from

$$A = G(P/G, i\%, n)(A/P, i\%, n) \qquad 13.4$$

Table 13.1 Discount Factors for Discrete Compounding

Factor Name	*Converts*	*Symbol*	*Formula*
Single Payment Compound Amount	P to F	$(F/P, i\%, n)$	$(1+i)^n$
Present Worth	F to P	$(P/F, i\%, n)$	$(1+i)^{-n}$
Uniform Series Sinking Fund	F to A	$(A/F, i\%, n)$	$\frac{i}{(1+i)^n-1}$
Capital Recovery	P to A	$(A/P, i\%, n)$	$\frac{i(1+i)^n}{(1+i)^n-1}$
Compound Amount	A to F	$(F/A, i\%, n)$	$\frac{(1+i)^n-1}{i}$
Equal Series Present Worth	A to P	$(P/A, i\%, n)$	$\frac{(1+i)^n-1}{i(1+i)^n}$
Uniform Gradient	G to P	$(P/G, i\%, n)$	$\frac{(1+i)^n-1}{i^2(1+i)^n} - \frac{n}{i(1+i)^n}$

Formulas for all of the compounding and discounting factors are contained in table 13.1. Normally, it will not be necesary to calculate factors from the formulas. The tables at the end of this chapter are adequate for solving most problems.

5 THE MEANING OF 'PRESENT WORTH' AND 'i'

It is clear that \$100 invested in a 5% bank account will allow you to remove \$105 one year from now. If this investment is made, you will clearly receive a *return on investment* (ROI) of \$5. The cash flow diagram and the present worth of the two transactions are

$$P = -100 + 105(P/F, 5\%, 1)$$
$$= -100 + 105(.9524) = 0$$

Figure 13.1

Notice that the present worth is zero even though you did receive a 5% return on your investment.

However, if you are offered \$120 for the use of \$100 over a one-year period, the cash flow diagram and present worth (at 5%) would be

$$P = -100 + 120(P/F, 5\%, 1)$$
$$= -100 + 120(.9524) = 14.29$$

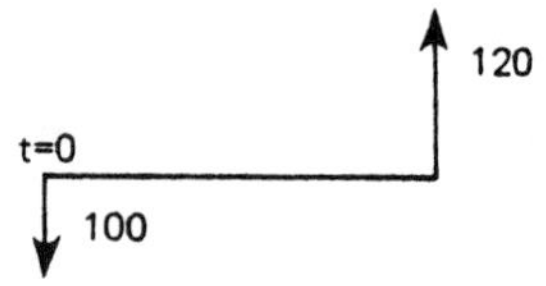

Figure 13.2

Therefore, it appears that the present worth of an alternative is equal to the equivalent value at $t = 0$ of the increase in return above that which you would be able to earn in an investment offering i% per period. In the above case, \$14.29 is the present worth of (\$20−\$5), the difference in the two *ROI's*.

Alternatively, the actual earned interest rate, called *rate of return (ROR)*, can be defined as the rate which makes the present worth of the alternative zero.

The *present worth* is also the amount that you would have to be given to dissuade you from making an investment, since placing the initial investment amount along with the present worth into a bank account earning i% will yield the same eventual *ROI*. Relating this to the previous paragraphs, you could be dissuaded against investing \$100 in an alternative which would return \$120 in one year by a $t = 0$ payment of \$14.29. Clearly, (\$100+\$14.29) invested at $t = 0$ will also yield \$120 in one year at 5%.

The selection of the interest rate is difficult in engineering economics problems. Usually it is taken as the average rate of return that an individual or business organization has realized in past investments. Fortunately, an interest rate is usually given. A company may not know what effective interest rate to use in an economic analysis. In such a case, the company can establish a minimum acceptable return on its investment. This *minimum attractive rate of return (MARR)* should be used as the effective interest rate i in economic analyses.

It should be obvious that alternatives with negative present worths are undesirable, and that alternatives with positive present worths are desirable because they increase the average earning power of invested capital.

6 CHOICE BETWEEN ALTERNATIVES

A variety of methods exists for selecting a superior alternative from among a group of proposals. Each method has its own merits and applications.

The *Present Worth Method* has already been implied. When two or more alternatives are capable of performing the same functions, the superior alternative will have the largest present worth. This method is suitable for ranking the desirability of alternatives. The present worth method is restricted to evaluating alternatives that are mutually exclusive and which have the same lives.

Returning to example 13.2, the present worth of each alternative should be found in order to determine which alternative is superior.

Example 13.2, continued

$$P(\mathbf{A}) = -10,000 + 11,500(P/F, 5\%, 2) = 431$$
$$P(\mathbf{B}) = -8,000 + 4,500(P/A, 5\%, 2) = 367$$

Alternative **A** is superior and should be chosen.

The present worth of a project with an infinite life is known as the *capitalized cost*. Capitalized cost is the amount of money at $t = 0$ needed to perpetually support the project on the earned interest only. Capitalized cost is a positive number when expenses exceed income.

$$\frac{\text{Capitalized}}{\text{Cost}} = \frac{\text{Initial}}{\text{Cost}} + \frac{\text{Annual Costs}}{i} \qquad 13.5$$

If disbursements occur irregularly instead of annually, the capitalized cost must be calculated from the equivalent uniform annual cost instead of the actual annual expenses.

$$\frac{\text{Capitalized}}{\text{Cost}} = \frac{\text{Initial}}{\text{Cost}} + \frac{EUAC}{i} \qquad 13.6$$

In comparing two alternatives, each of which is infinitely lived, the superior alternative will have the lowest capitalized cost.

Alternatives which accomplish the same purpose but which have unequal lives must be compared by the *Annual Cost Method.* The annual cost method assumes that each alternative will be replaced by an identical twin at the end of its useful life (infinite renewal). This method, which may also be used to rank alternatives according to their desirability, is also called the *Annual Return Method* and *Capital Recovery Method.*

Restrictions are that the alternatives must be mutually exclusive and infinitely renewed up to the duration of the longest-lived alternative. The calculated annual cost is known as the *Equivalent Uniform Annual Cost, EUAC.* Cost is a positive number when expenses exceed income.

Example 13.6

Which of the following alternatives is superior over a 30 year period if the interest rate is 7%?

	A	**B**
type	brick	wood
life	30 years	10 years
cost	$1,800	$450
maintenance	$5/year	$20/year

$$EUAC(A) = 1,800(A/P,7\%,30) + 5 = 150$$
$$EUAC(B) = 450(A/P,7\%,10) + 20 = 84$$

Alternative **B** is superior since its annual cost of operation is the lowest. It is assumed that three wood facilities, each with a life of 10 years and a cost of $450, will be built to span the 30 year period.

The *Benefit-Cost Ratio Method* is often used in municipal project evaluations where benefits and costs accrue to different segments of the community. With this method, the present worth of all benefits (regardless of the beneficiary) is divided by the present worth of all costs. The project is considered acceptable if the ratio exceeds *one.*

When the benefit-cost ratio method is used, disbursements by the initiators or sponsors are *costs.* Disbursements by the users of the project are known as *disbenefits.* It is often difficult to determine whether a cash flow is a cost or a disbenefit (whether to place it in the numerator or denominator of the benefit-cost ratio calculation).

Regardless of where the cash flow is placed, an acceptable project will always have a benefit-cost ratio greater than one, although the actual numerical result will depend on the placement. For this reason, the benefit-cost ratio method should not be used to rank competing projects.

The benefit-cost ratio method may be used to rank alternative proposals only if an *incremental analysis* is used. First, determine that the ratio is greater than one for each alternative. Then, calculate the ratio of benefits to costs

$$\frac{(B_2 - B_1)}{(C_2 - C_1)}$$

for each possible pair of alternatives. If the ratio exceeds one, alternative 2 is superior to alternative 1. Otherwise, alternative 1 is superior.

Perhaps no method of analysis is less understood than the *Rate of Return* (ROR) *Method.* As was stated previously, the ROR is the interest rate that would yield identical profits if all money were invested at that rate. The present worth of any such investment is zero.

The ROR is defined as the interest rate that will discount all cash flows to a total present worth equal to the initial required investment. This definition is used to determine the ROR of an alternative. The ROR should not be used to rank or compare alternatives unless an incremental analysis is used. The advantage of the ROR method is that no knowledge of an interest rate is required.

To find the ROR of an alternative, proceed as follows:

step 1: Set up the problem as if to calculate the present worth.

step 2: Arbitrarily select a reasonable value for i. Calculate the present worth.

step 3: Choose another value of i (not too close to the original value) and again solve for the present worth.

step 4: Interpolate or extrapolate the value of i which gives a zero present worth.

step 5: For increased accuracy, repeat steps (2) and (3) with two more values that straddle the value found in step (4).

A common, although incorrect, method of calculating the ROR involves dividing the annual receipts or returns by the initial investment. However this technique ignores such items as salvage, depreciation, taxes, and the time value of money. This technique also fails when the annual returns vary.

Example 13.7

What is the return on invested capital if $1,000 is invested now with $500 being returned in year 4 and $1,000 being returned in year 8?

First, set up the problem as a present worth calculation.

$$P = -1,000 + 500(P/F, i\%, 4) + 1,000(P/F, i\%, 8)$$

Arbitrarily select $i = 5\%$. The present worth is then found to be $88.15. Next take a higher value of i to reduce the present worth. If $i = 10\%$, the present worth is −$192. The ROR is found from simple interpolation to be approximately 6.6%.

7 TREATMENT OF SALVAGE VALUE IN REPLACEMENT STUDIES

An investigation into the retirement of an existing process or piece of equipment is known as a *replacement study*. Replacement studies are similar in most respects to other alternative comparison problems: an interest rate is given, two alternatives exist, and one of the previously mentioned methods of comparing alternatives is used to choose the superior alternative.

In replacement studies, the existing process or piece of equipment is known as the *defender*. The new process or piece of equipment being considered for purchase is known as the *challenger*.

Because most defenders still have some market value when they are retired, the problem of what to do with the salvage arises. It seems logical to use the salvage value of the defender to reduce the initial purchase cost of the challenger. This is consistent with what would actually happen if the defender were to be retired.

By convention, however, the salvage value is subtracted from the defender's present value. This does not seem logical, but it is done to keep all costs and benefits related to the defender with the defender. In this case, the salvage value is treated as an opportunity cost which would be incurred if the defender is not retired.

If the defender and the challenger have the same lives and a present worth study is used to choose the superior alternative, the placement of the salvage value will have no effect on the net difference between present worths for the challenger and defender. Although the values of the two present worths will be different depending on the placement, the difference in present worths will be the same.

If the defender and the challenger have different lives, an annual cost comparison must be made. Since the salvage value would be 'spread over' a different number of years depending on its placement, it is important to abide by the conventions listed in this section.

There are a number of ways to handle salvage value. The best way is to think of the EUAC of the defender as the cost of keeping the defender from now until next year. In addition to the usual operating and maintenance costs, that cost would include an opportunity interest cost incurred by not selling the defender and also a drop in the salvage value if the defender is kept for one additional year. Specifically,

$$\begin{aligned}\text{EUAC(defender)} = &\text{ maintenance costs}\\ &+ i(\text{current salvage value})\\ &+(\text{current salvage}-\text{next year's salvage})\end{aligned} \qquad 13.7$$

It is important in retirement studies not to double count the salvage value. That is, it would be incorrect to add the salvage value to the defender and at the same time subtract it from the challenger.

8 BASIC INCOME TAX CONSIDERATIONS

Assume that an organization pays $f\%$ of it profits to the federal government as income taxes. If the organizaiton also pays a state income tax of $s\%$, and if state taxes paid are recognized by the federal government as expenses, then the composite tax rate is

$$t = s + f - sf \qquad 13.8$$

The basic principles used to incorporate taxation into economic analyses are listed below.

a. Initial purchase cost is unaffected by income taxes.

b. Salvage value is unaffected by income taxes.

c. Deductible expenses, such as operating costs, maintenance costs, and interest payments, are reduced by $t\%$ (e.g., multiplied by the quantity $(1-t)$).

d. Revenues are reduced by $t\%$ (e.g., multiplied by the quantity $(1-t)$).

e. Depreciation is multiplied by t and added to the appropriate year's cash flow, increasing that year's present worth.

Income taxes and depreciation have no bearing on municipal or governmental projects since municipalities, states, and the U.S. Government pay no taxes.

Example 13.8

A corporation which pays 53% of its revenue in income taxes invests \$10,000 in a project which will result in \$3,000 annual revenue for 8 years. If the annual expenses are \$700, salvage after 8 years is \$500, and 9% interest is used, what is the after-tax present worth? Disregard depreciation.

$$\begin{aligned} P_t &= -10,000 + 3000(P/A, 9\%, 8)(1 - .53) \\ &\quad - 700(P/A, 9\%, 8)(1 - .53) \\ &\quad + 500(P/F, 9\%, 8) \\ &= -3,766 \end{aligned}$$

It is interesting that the alternative evaluated in example 13.8 is undesirable if income taxes are considered but is desirable if income taxes are omitted.

9 DEPRECIATION

Although depreciation calculations may be considered independently in examination questions, it is important to recognize that depreciation has no effect on engineering economic calculations unless income taxes are also considered.

Generally, tax regulations do not allow the cost of equipment[1] to be treated as a deductible expense in the year of purchase. Rather, portions of the cost may be allocated to each of the years of the item's economic life (which may be different from the actual useful life). Each year, the book value (which is initially equal to the purchase price) is reduced by the depreciation in that year. Theoretically, the book value of an item will equal the market value at any time within the economic life of that item.

Since tax regulations allow the depreciation in any year to be handled as if it were an actual operating expense, and since operating expenses are deductible from the income base prior to taxation, the after-tax profits will be increased. If D is the depreciation, the net result to the after-tax cash flow will be the addition of tD.

The present worth of all depreciation over the economic life of the item is called the *depreciation recovery.* Although originally established to do so, depreciation recovery can never fully replace an item at the end of its life.

Depreciation is often confused with amortization and depletion. While depreciation spreads the cost of a fixed asset over a number of years, *amortization* spreads the cost of an intangible asset (e.g., a patent) over some basis such as time or expected units of production.

Depletion is another artificial deductible operating expense designed to compensate mining organizations for decreasing mineral reserves. Since original and remaining quantities of minerals are seldom known accurately, the *depletion allowance* is calculated as a fixed percentage of the organization's gross income. These percentages are usually in the 10%-20% range and apply to such mineral deposits as oil, natural gas, coal, uranium, and most metal ores.

There are four common methods of calculating depreciation. The book value of an asset depreciated with the *Straight Line* (SL) *Method* (also known as the *Fixed Percentage Method*) decreases linearly from the initial purchase at $t = 0$ to the estimated salvage at $t = n$. The depreciated amount is the same each year. The quantity $(C - S_n)$ in equation 13.9 is known as the *depreciation base.*

$$D_j = \frac{(C - S_n)}{n} \qquad 13.9$$

Double Declining Balance[2] (DDB) depreciation is independent of salvage value. Furthermore, the book value never stops decreasing, although the depreciation decreases in magnitude. Usually, any remaining book value is written off in the last year of the asset's estimated life. Unlike any of the other depreciation methods, DDB depends on accumulated depreciation.

$$D_j = \frac{2(C - \sum_{i=1}^{j-1} D_i)}{n} \qquad 13.10$$

In *Sum-of-the-Year's-Digits* (SOYD) depreciation, the digits from 1 to n inclusive, are summed. The total, T, can also be calculated from

$$T = \frac{1}{2}n(n+1) \qquad 13.11$$

[1]The IRS tax regulations allow depreciation on almost all forms of *property* except land. The following types of property are distinguished: *real* (e.g., buildings used for business), *residential* (e.g., buildings used as rental property), and *personal* (e.g., equipment used for business). Personal property does *not* include items for personal use, despite its name. *Tangible* personal property is distinguished from *intangible property* (e.g., goodwill, copyrights, patents, trademarks, franchises, and agreements not to compete).

[2]Double declining balance depreciation is a particular form of *declining balance depreciation*, as defined by the IRS tax regulations. Declining balance depreciation also includes 125% declining balance and 150% declining balance depreciations which can be calculated by substituting 1.25 and 1.50 respectively for the 2 in equation 13.10.

The depreciation can be found from

$$D_j = \frac{(C - S_n)(n - j + 1)}{T} \qquad 13.12$$

The *Sinking Fund Method* is seldom used in industry because the initial depreciation is low. The formula for sinking fund depreciation is

$$D_j = (C - S_n)(A/F, i\%, n) \qquad 13.13$$

The above discussion gives the impression that any form of depreciation may be chosen regardless of the nature and circumstances of the purchase. In reality, the IRS tax regulations place restrictions on the higher-rate ("accelerated") methods such as DDB and SOYD. Furthermore, the *Economic Recovery Act of 1981* substantially changed the laws relating to personal and corporate income taxes.

Property placed into service in 1981 or after must use the *Accelerated Cost Recovery System (ACRS)*. Other methods (straight-line, declining balance, etc.) cannot be used except in special cases.

Property placed into service in 1980 or before must continue to be depreciated according to the method originally chosen (e.g., straight-line, declining balance, or sum-of-years-digits). *ACRS* cannot be used.

Under *ACRS*, the cost recovery amount in the jth year of an asset's cost recovery period is calculated by multiplying the initial cost by a factor found from a table 13.2.

$$D_j = (\text{initial cost})(\text{factor}) \qquad 13.14$$

The initial cost used is not reduced by the asset's salvage value for either the regular or alternate *ACRS* calculations. The factor found from the table depends on the asset's cost recovery period.

Table 13.2
Table of Recovery Factors

	recovery period		
year	3 yrs	5 yrs	10 yrs
1	.25	.15	.08
2	.38	.22	.14
3	.37	.21	.12
4		.21	.10
5		.21	.10
6			.10
7			.09
8			.09
9			.09
10			.09

The *ACRS* groups tangible property into four categories depending on the cost recovery period. Tangible property is divided into *real property* and *personal property*. Real property is real estate (i.e., land and its structures). Land is never eligible for cost recovery. Personal property consists of all assets (other than real assets) used for the production of business income. Personal property is not for personal use.

A. 3-YEAR PERSONAL PROPERTY

The cost of cars, light trucks, special tools, all R&D equipment, and other assets with short lives should be recovered over 3 years. The full amount applies regardless of when the asset is placed into service during the first year.

If desired, the cost of the asset can be recovered over a longer period. Straight-line depreciation can be used with either a 5- or a 12-year life. This is known as the *alternate ACRS calculation*. No other lives or depreciation methods are allowed. If this alternate method is used, only one-half of the cost recovery amount can be claimed in the first year. This is known as the *half-year* rule. The remaining one-half can be claimed in the year following the end of the recovery period.

B. 5-YEAR PERSONAL PROPERTY

Machinery, single-purpose agricultural structures, and equipment not recovered over 3 years should be recovered over 5 years. The full percentage applies regardless of when the asset is placed into service during the first year.

The alternate method uses 12- and 25-year periods with straight-line calculations. No other lives or depreciation methods can be used. As with 3-year property, the half-year rule applies to amounts calculated with this alternate method.

C. 10-YEAR PERSONAL AND REAL PROPERTY

Selected heavy equipment and machinery (e.g., rail cars), theme park structures, and some public utility property should be recovered over 10 years. The full percentage applies regardless of when the asset is placed into service during the first year.

If desired, the cost of the asset can be recovered over 35 or 45 years using straight-line calculations for the alternate method. No other lives or depreciation methods can be used. The half-year rule applies to amounts calculated for the first year.

Table 13.3
Recovery Factors for 15-Year Property

	Month Placed in Service											
Year	1	2	3	4	5	6	7	8	9	10	11	12
1	.12	.11	.10	.09	.08	.07	.06	.05	.04	.03	.02	.01
2	.10	.10	.11	.11	.11	.11	.11	.11	.11	.11	.11	.12
3	.09	.09	.09	.09	.10	.10	.10	.10	.10	.10	.10	.10
4	.08	.08	.08	.08	.08	.08	.09	.09	.09	.09	.09	.09
5	.07	.07	.07	.07	.07	.07	.08	.08	.08	.08	.08	.08
6	.06	.06	.06	.06	.07	.07	.07	.07	.07	.07	.07	.07
7	.06	.06	.06	.06	.06	.06	.06	.06	.06	.06	.06	.06
8	.06	.06	.06	.06	.06	.06	.06	.06	.06	.06	.06	.06
9	.06	.06	.06	.06	.05	.06	.05	.05	.05	.06	.06	.06
10	.05	.06	.05	.06	.05	.05	.05	.05	.05	.05	.06	.05
11	.05	.05	.05	.05	.05	.05	.05	.05	.05	.05	.05	.05
12	.05	.05	.05	.05	.05	.05	.05	.05	.05	.05	.05	.05
13	.05	.05	.05	.05	.05	.05	.05	.05	.05	.05	.05	.05
14	.05	.05	.05	.05	.05	.05	.05	.05	.05	.05	.05	.05
15	.05	.05	.05	.05	.05	.05	.05	.05	.05	.05	.05	.05
16	–	–	.01	.01	.02	.02	.03	.03	.04	.04	.04	.05

D. 15-YEAR REAL PROPERTY

Real property, including residential real estate, should be recovered over 15 years. Unlike the previous asset categories, however, the factors used depend on the month in which the asset was placed into service.

If desired, the asset can be recovered over 35 or 45 years using straight-line calculations. The half-year rule limiting first-year deductions does not apply to 15-year assets.

* * *

Three other depreciation methods should be mentioned, not because they are currently accepted or in widespread use, but because they have seen recent use on the licensing examinations.

The *sinking-fund plus interest on first cost* depreciation method, like the following two methods, is an attempt to include the *opportunity interest cost* on the purchase price with the depreciation. That is, the purchasing company not only incurs an annual loss due to the drop in book value, but it also loses the interest on the purchase price. The formula for this method is

$$D_j = (C - S_n)(A/F, i\%, n) + (C)(i) \qquad 13.15$$

The *straight-line plus interest on first cost* method is similar. Its formula is

$$D_j = (\frac{1}{n})(C - S_n) + (C)(i) \qquad 13.16$$

The *straight-line plus average interest method* assumes that the opportunity interest cost should be based on the book value only, not on the full purchase price. Since the book value changes each year, an average value is used. The depreciation formula is

$$D_j = \left(\frac{1}{n}\right)(C - S_n) + \frac{1}{2}(i)(C - S_n)\left(\frac{n+1}{n}\right) \qquad 13.17$$

These three depreciation methods are not to be used in the usual manner (e.g., in conjunction with the income tax rate). These methods are attempts to calculate a more accurate annual cost of an alternative. Sometimes they work, and sometimes they give misleading answers. Their use cannot be recommended. They are included in this chapter only for the sake of completeness.

Example 13.9

An asset is purchased for $9,000. Its estimated economic life is 10 years, after which it will be sold for $1,000. Find the depreciation in the first three years using SL, DDB, SOYD, and Sinking Fund at 6%.

SL: $D = \frac{(9{,}000-1{,}000)}{10}$ =800 each year

DDB: $D_1 = \frac{2(9{,}000)}{10}$ =1,800 in year 1

$D_2 = \frac{2(9{,}000-1{,}800)}{10}$ =1,440 in year 2

$D_3 = \frac{2(9{,}000-3{,}240)}{10}$ =1,152 in year 3

SOYD: $T = \frac{1}{2}(10)(11) = 55$

$D_1 = (\frac{10}{55})(9,000 - 1,000)$ =1455 in year 1

$D_2 = (\frac{9}{55})(8,000)$ =1309 in year 2

$D_3 = (\frac{8}{55})(8,000)$ =1164 in year 3

Sinking Fund: $D = (9,000 - 1,000)(A/F, 6\%, 10)$ =607 each year

Example 13.10

For the asset described in example 13.9, calculate the book value during the first three years if SOYD depreciation is used.

The book value at the beginning of year 1 is $9,000. Then,

$$BV_1 = 9,000 - 1,455 = 7,545$$
$$BV_2 = 7,545 - 1,309 = 6,236$$
$$BV_3 = 6,236 - 1,164 = 5,072$$

Example 13.11

For the asset described in example 13.9, calculate the after-tax depreciation recovery with SL and SOYD depreciation methods. Use 6% interest with 48% income taxes.

SL: $D.R. = .48(800)(P/A, 6\%, 10) = 2,826$

SOYD: The depreciation series can be thought of as a constant 1,454 term with a negative 145 gradient.

$$\begin{aligned} D.R. &= .48(1,454)(P/A, 6\%, 10) \\ &\quad - .48(145)(P/G, 6\%, 10) \\ &= 3,076 \end{aligned}$$

Finding book values, depreciation, and depreciation recovery is particularly difficult with DDB depreciation, since all previous years' quantities seem to be required. It appears that the depreciation in the 6th year cannot be calculated unless the values of depreciation for the first five years are calculated first. Questions asking for depreciation or book value in the middle or at the end of an asset's economic life may be solved from the following equations:

$$d = \frac{2}{n} \quad 13.18$$

$$z = \frac{(1+i)}{(1-d)} \quad 13.19$$

$$(P/EG) = \frac{z^n - 1}{z^n(z-1)} \quad 13.20$$

Then, assuming that the remaining book value at $t = n$ is written off in one lump sum, the present worth of the depreciation recovery is

$$D.R. = t\left[\frac{(d)(C)}{(1-d)}(P/EG) + (1-d)^n(C)(P/F, i\%, n)\right] \quad 13.21$$

$$D_j = (d)(C)(1-d)^{j-1} \quad 13.22$$

$$BV_j = C(1-d)^j \quad 13.23$$

Example 13.12

What is the after-tax present worth of the asset described in example 13.8 if SL, SOYD, and DDB depreciation methods are used?

The after-tax present worth, neglecting depreciation, was previously found to be –3766.

Using SL, the depreciation recovery is

$$\begin{aligned} D.R. &= (.53)\frac{(10,000 - 500)}{8}(P/A, 9\%, 8) \\ &= 3,483 \end{aligned}$$

Using SOYD, the depreciation recovery is calculated as follows:

$$T = \frac{1}{2}(8)(9) = 36$$

$$\text{Depreciation base} = (10,000 - 500) = 9,500$$

$$D_1 = \frac{8}{36}(9,500) = 2111$$

$$\begin{aligned} G = \text{gradient} &= \frac{1}{36}(9,500) \\ &= 264 \end{aligned}$$

$$\begin{aligned} D.R. &= (.53)[2,111(P/A, 9\%, 8) - 264(P/G, 9\%, 8)] \\ &= 3,829 \end{aligned}$$

Using DDB, the depreciation recovery is calculated as follows:

$$d = \frac{2}{8} = .25$$

$$z = \frac{1.09}{.75} = 1.453$$

$$(P/EG) = \frac{(1.453)^8 - 1}{(1.453)^8(.453)} = 2.096$$

$$D.R. = .53\left[\frac{(.25)(10,000)}{.75}(2.096) + (.75)^8(10,000)(P/F, 9\%, 8)\right]$$

$$= 3,969$$

The after-tax present worths including depreciation recovery are:

SL: $P_t = -3,766 + 3,483 = -283$
SOYD: $P_t = -3,766 + 3,829 = 63$
DDB: $P_t = -3,766 + 3,969 = 203$

10 ADVANCED INCOME TAX CONSIDERATIONS

There are a number of specialized techniques that are needed infrequently These techniques are related more to the accounting profession than to the engineering profession. Nevertheless, it is occasionally necessary to use these techniques.

A. IMMEDIATE WRITE-OFF

The cost of small assets and part of the cost of large assets can be written off entirely in the year of purchase. The limits on the amounts that can be written off each year are given in the following table.

Table 13.4
Limits on Immediate Write-Offs

year	amount
1982	$5,000
1983	5,000
1984	7,500
1985	7,500
1986 and on	10,000

B. TAX CREDITS

Investment tax credits for the purchase of assets are allowed in the year of purchase. These credits are a fraction of the assets' cost.

$$\text{tax credit} = \begin{pmatrix}\text{initial}\\ \text{cost}\end{pmatrix}\begin{pmatrix}\text{decimal}\\ \text{amount}\end{pmatrix} \qquad 13.24$$

The decimal amounts are taken from the accompanying table. Only assets with lives of 3-years or greater qualify for tax credits. The decimal amounts for 1 and 2 years assume the asset disposed of prior to the 3 years.

Table 13.5
Tax Credit Amounts

life (years)	decimal amount
1	.02
2	.04
3	.06
4 or more	.10

C. GAIN ON THE SALE OF A DEPRECIATED ASSET

If an asset is sold for more than its current book value, the difference between selling price and book value is taxable income. The gain is taxed at capital gains rates. Excluded from this preferential treatment is non-residential real property depreciated under regular *ACRS* provisions. However, non-residential real property depreciated under the straight-line alternate method qualifies for the capital gains rate.

D. CAPITAL GAINS AND LOSSES

A *gain* is defined as the difference between selling and purchase prices of a capital asset. The gain is called a *regular gain* if the item sold has been kept less than one year. The gain is called a *capital gain* if the item sold has been kept for longer than one year. Capital gains are taxed at the taxpayer's usual rate, but 60% of the gain is excluded from taxation. That is, only 40% of the gain is taxed. The maximum capital gains tax is 20% for individuals in the maximum (50%) tax bracket.

Regular (as defined above) *losses* are fully deductible in the year of their occurrence. The IRS tax regulations should be consulted to determine the treatment of *capital losses*.

11 RATE AND PERIOD CHANGES

All of the foregoing calculations were based on compounding once a year at an *effective interest rate, i.* However, some problems specify compounding more frequently than annually. In such cases, a *nominal interest rate, r,* will be given. The nominal rate does not include the effect of compounding and is not the same as the effective rate, *i*. A nominal rate may be used to calculate the effective rate by using equation 13.25 or 13.26.

$$i = (1 + \frac{r}{k})^k - 1 \qquad 13.25$$

$$= (1 + \phi)^k - 1 \qquad 13.26$$

A problem may also specify an effective rate per period, ϕ, (e.g., per month). However, that will be a simple problem since compounding for n periods at an effective rate per period is not affected by The definition or length of the period.

The following rules may be used to determine which interest rate is given in a problem:

- Unless specifically qualified in the problem, the interest rate given is an annual rate.
- If the compounding is annually, the rate given is the effective rate. If compounding is other than annually, the rate given is the nominal rate.
- If the type of compounding is not specified, assume annual compounding.

In the case of continuous compounding, the appropriate discount factors may be calculated from the formulas in table 13.6.

Table 13.6 Discount Factors for Continuous Compounding

(F/P)	e^{rn}
(P/F)	e^{-rn}
(A/F)	$(e^r - 1)/(e^{rn} - 1)$
(F/A)	$(e^{rn} - 1)/(e^r - 1)$
(A/P)	$(e^r - 1)/(1 - e^{-rn})$
(P/A)	$(1 - e^{-rn})/(e^r - 1)$

Example 13.13

A savings and loan offers $5\frac{1}{4}\%$ compounded daily. What is the annual effective rate?

method 1: $r = .0525, k = 365$

$$i = (1 + \tfrac{.0525}{365})^{365} = .0539$$

method 2: Assume daily compounding is the same as continuous compounding.

$$i = (F/P) - 1$$

$$= e^{.0525} - 1 = .0539$$

12 PROBABILISTIC PROBLEMS

Thus far, all of the cash flows included in the examples have been known exactly. If the cash flows are not known exactly but are given by some implicit or explicit probability distribution, the problem is *probabilistic.*

Probabilistic problems typically possess the following characteristics:

- There is a chance of extreme loss that must be minimized.
- There are multiple alternatives that must be chosen from. Each alternative gives a different degree of protection against the loss or failure.
- The outcome is independent of the alternative chosen. Thus, as illustrated in example 13.14, the size of the dam that is chosen for construction will not alter the rainfall in successive year. However, it will alter the effects on the downstream watershed areas.

Probabilistic problems are typically solved using annual costs and expected values. An *expected value* is similar to an 'average value' since it is calculated as the mean of the given probability distribution. If cost 1 has a probability of occurrence p_1, cost 2 has a probability of occurrence of p_2, and so on, the expected value is

$$E(\text{cost}) = p_1(\text{cost } 1) + p_2(\text{cost } 2) + \cdots \qquad 13.27$$

Example 13.14

Flood damage in any year is given according to the table below. What is the present worth of flood damage for a 10-year period? Use 6%.

Damage	Probability
0	.75
\$10,000	.20
\$20,000	.04
\$30,000	.01

The expected value of flood damage is

$$E(\text{damage}) = (0)(.75) + (10{,}000)(.20) + (20{,}000)(.04) + (30{,}000)(.01)$$

$$= \$3,100$$

$$\text{present worth} = 3,100(P/A, 6\%, 10)$$

$$= 22,816$$

Probabilities in probabilistic problems may be given to you in the problem (as in the example above) or you may have to obtain them from some named probability distribution. In either case, the probabilities are known explicitly and such problems are known as *explicit probability problems.*

Example 13.15

A dam is being considered on a river which periodically overflows and causes $600,000 damage. The damage is essentially the same each time the river causes flooding. The project horizon is 40 years. A 10% interest rate is being used.

Three different designs are available, each with different costs and storage capacities.

design alternative	cost	maximum capacity
A	500,000	1 unit
B	625,000	1.5 units
C	900,000	2.0 units

The U.S. Weather Service has provided a statistical analysis of annual rainfall in the area draining into the river.

units annual rainfall	probability
0	.10
.1-.5	.60
.6-1.0	.15
1.1-1.5	.10
1.6-2.0	.04
2.0 or more	.01

Which design alternative would you choose assuming the dam is essentially empty at the start of each rainfall season?

The sum of the construction cost and the expected damage needs to be minimized. If alternative **A** is chosen, it will have a capacity of 1 unit. Its capacity will be exceeded (causing $600,000 damage) when the annual rainfall exceeds 1 unit. Therefore, the annual cost of **A** is

$$\begin{aligned}\text{EUAC}(\mathbf{A}) &= 500{,}000(A/P, 10\%, 40)\\ &\quad + 600{,}000(.10 + .04 + .01)\\ &= 141{,}150\end{aligned}$$

Similarly,

$$\begin{aligned}\text{EUAC}(\mathbf{B}) &= 625{,}000(A/P, 10\%, 40)\\ &\quad + 600{,}000(.04 + .01)\\ &= 93{,}940\\ \text{EUAC}(\mathbf{C}) &= 900{,}000(A/P, 10\%, 40)\\ &\quad + 600{,}000(.01)\\ &= 98{,}070\end{aligned}$$

Alternative **B** should be chosen.

In other problems, a probability distribution will not be given even though some parameter (such as the life of an alternative) is not known with certainty. Such problems are known as *implicit probability problems* since they require a reasonable assumption about the probability distribution.

Implicit probability problems typically involve items whose *maximum lives* are known. The key to such problems is in recognizing that a maximum life is not the same as a fixed life.

Reasonable assumptions can be made about the form of probability distributions in implicit probability problems.

One such reasonable assumption is that of a *rectangular distribution.* A rectangular distribution is one which is assumed to give an equal probability of failure in each year. Such an assumption is illustrated in example 13.16.

Example 13.16

A bridge is needed for 20 years. Failure of the bridge at any time will require a 50% reinvestment. Evaluate the two design alternatives below using 6% interest.

design alternative	initial cost	maximum life	annual costs	salvage at t=20
A	15,000	9 years	1,200	0
B	22,000	43 years	1,000	0

It will be assumed that each alternative has an annual probability of failure that is inversely proportional to its maximum life. Then, for alternative **A**, the probability of failure in any year is $(\frac{1}{9})$. Similarly, the annual failure probability for alternative **B** is $(\frac{1}{43})$.

$$\begin{aligned}\text{EUAC}(\mathbf{A}) &= 15{,}000(A/P, 6\%, 20)\\ &\quad + 15{,}000(.5)(\frac{1}{9}) + 1{,}200\\ &= 3{,}341\\ \text{EUAC}(\mathbf{B}) &= 22{,}000(A/P, 6\%, 20)\\ &\quad + 22{,}000(.5)(\frac{1}{43}) + 1{,}000\\ &= 3{,}174\end{aligned}$$

Alternative **B** should be chosen.

13 ESTIMATING ECONOMIC LIFE

As assets grow older, their operating and maintenance costs typcially increase each year. Eventually, the cost to keep an asset in operation becomes prohibitive, and the asset is retired of replaced. However, it is not always obvious when an asset should be retired or replaced.

As the asset's maintenance is increasing each year, the amortized cost of its initial purchase is decreasing. It is the sum of these two costs that should be evaluated to determine the point at which the asset should be returned or replaced. Since an asset's initial purchase price is likely to be high, the amortized cost will be the controlling factor in those years when the maintenance costs are low. Therefore, the EUAC of the asset will decrease in the initial part of its life.

However, as the asset grows older, the change in its amortized cost decreases while maintenance increases. Eventually the sum of the two costs reaches a mininum and then starts to increase. The age of the asset at the minimum cost point is known as the *economic life* of the asset.

The determination of an asset's economic life is illustrated by example 13.17.

Example 13.17

A bus in a municipal transit system has the characteristics listed below. When should the city replace its buses if money can be borrowed at 8%?

Initial cost: $120,000

year	maintenance cost	salvage value
1	35,000	60,000
2	38,000	55,000
3	43,000	45,000
4	50,000	25,000
5	65,000	15,000

If the bus is kept for 1 year and then sold, the annual cost will be

$$\begin{aligned}\text{EUAC}(1) &= 120{,}000(A/P,8\%,1) + 35{,}000(A/F,8\%,1) \\ &\quad - 60{,}000(A/F,8\%,1) \\ &= 104{,}600\end{aligned}$$

If the bus is kept for 2 years and then sold, the annual cost will be

$$\begin{aligned}\text{EUAC}(2) &= [120{,}000 + 35{,}000(P/F,8\%,1)](A/P,8\%,2) \\ &\quad + (38{,}000 - 55{,}000)(A/F,8\%,2) \\ &= 77{,}300\end{aligned}$$

If the bus is kept for 3 years and then sold, the annual cost will be

$$\begin{aligned}\text{EUAC}(3) &= [120{,}000 + 35{,}000(P/F,8\%,1) \\ &\quad + 38{,}000(P/F,8\%,2)](A/P,8\%,3) \\ &\quad + (43{,}000 - 45{,}000)(A/F,8\%,3) \\ &= 71{,}200\end{aligned}$$

This process is continued until EUAC begins to increase. In this example, EUAC(4) is 71,700. Therefore, the bus should be retired after 3 years.

14 BASIC COST ACCOUNTING

Cost accounting is the system which determines the cost of manufactured products. Cost accounting is called *job cost accounting* if costs are accumulated by part number or contract. It is called *process cost accounting* if costs are accumulated by departments or manufacturing processes.

Three types of costs (direct material, direct labor, and all indirect costs) make up the total manufacturing cost of a product.

Direct material costs are the costs of all materials that go into the product, priced at the original purchase cost.

Indirect material and labor costs are generally limited to costs incurred in the factory, excluding costs incurred in the office area. Examples of indirect materials are cleaning fluids, assembly lubricants, and temporary routing tags. Examples of indirect labor are stock-picking, inspection, expediting, and supervision labor.

Here are some important points concerning basic cost accounting:

- The sum of direct material and direct labor costs is known as the *prime cost.*
- Indirect costs may be called *indirect manufacturing expenses* (IME).
- Indirect costs may also include the overhead sector of the company (e.g., secretaries, engineers, and corporate administration). In this case, the indirect cost is usually called *burden* or *overhead.* Burden may also include the EUAC of non-regular costs which must be spread evenly over several years.

- The cost of a product is usually known in advance from previous manufacturing runs or by estimation. Any deviation from this known cost is called a *variance*. Variance may be broken down into *labor variance* and *material variance*.

- Indirect cost per item is not easily measured. The method of allocating indirect costs to a product is as follows:

  Step 1: Estimate the total expected indirect (and overhead) costs for the upcoming year.

  Step 2: Decide on some convenient vehicle for allocating the overhead to production. Usually, this vehicle is either the number of units expected to be produced or the number of direct hours expected to be worked in the upcoming year.

  Step 3: Estimate the quantity or size of the overhead vehicle.

  Step 4: Divide expected overhead costs by the expected overhead vehicle to obtain the unit overhead.

  Step 5: Regardless of the true size of the overhead vehicle during the upcoming year, one unit of overhead cost is allocated per product.

- Although estimates of production for the next year are always somewhat inaccurate, the cost of the product is assumed to be independent of forecasting errors. Any difference between true cost and calculated cost goes into a variance account.

- *Burden (overhead) variance* will be caused by errors in forecasting both the actual overhead for the upcoming year and the vehicle size. In the former case, the variance is called *burden budget variance*; in the latter, it is called *burden capacity variance*.

Example 13.18

A small company expects to produce 8,000 items in the coming year. The current material cost is \$4.54 each. Sixteen minutes of direct labor are required per unit. Workers are paid \$7.50 per hour. 2,133 direct labor hours are forecast for the product. Miscellaneous overhead costs are estimated at \$45,000.

Find the expected direct material cost, the direct labor cost, the prime cost, the burden as a function of production and direct labor, and the total cost.

- The direct material cost was given as \$4.54.

- The direct labor cost is $(\frac{16}{60})(\$7.50) = \$2.00$.

- The prime cost is $\$4.54 + \$2.00 = \$6.54$.

- If the burden vehicle is production, the burden rate is $(\$\frac{45,000}{8,000}) = \$5.63$ per item, making the total cost $\$4.54 + \$2.00 + \$5.63 = \$12.17$.

- If the burden vehicle is direct labor hours, the burden rate is $(\frac{45,000}{2,133}) = \$21.10$ per hour, making the total cost $\$4.54 + \$2.00 + \frac{16}{60}(\$21.10) = \$12.17$.

Example 13.19

The actual performance of the company in example 13.18 is given by the following figures:

actual production: 7,560

actual overhead costs: \$47,000

What are the burden budget variance and the burden capacity variance?

The burden capacity variance is

$$\$45,000 - 7,560(\$5.63) = \$2,437$$

The burden budget variance is

$$\$47,000 - \$45,000 = \$2,000$$

The overall burden variance is

$$\$47,000 - 7,560(\$5.63) = \$4,437$$

15 BREAK-EVEN ANALYSIS

Break-even analysis is a method of determining when costs exactly equal revenue. If the manufactured quantity is less than the break-even quantity, a loss is incurred. If the manufactured quantity is greater than the break-even quantity, a profit is incurred.

Consider the following special variables:

f	a fixed cost which does not vary with production
a	an incremental cost which is the cost to produce one additional item. It may also be called the *marginal cost* or *differential cost*.
Q	the quantity sold

p the incremental revenue
R the total revenue
C the total cost

Assuming no change in the inventory, the *break-even point* can be found from $C = R$, where

$$C = f + aQ \qquad 13.28$$

$$R = pQ \qquad 13.29$$

An alternate form of the break-even problem is to find the number of units per period for which two alternatives have the same total costs. Fixed costs are to be spread over a period longer than one year. One of the alternatives will have a lower cost if production is less than the break-even point. The other will have a lower cost for production greater than the break-even point.

Example 13.20

Two plans are available for a company to obtain automobiles for its salesmen. How many miles must the cars be driven each year for the two plans to have the same costs? Use an interest rate of 10%.

Plan A Lease the cars and pay $.15 per mile

Plan B Purchase the cars for $5,000. Each car has an economic life of three years, after which it can be sold for $1,200. Gas and oil cost $.04 per mile. Insurance is $500 per year.

Let x be the number of miles driven per year. Then, the EUAC for both alternatives is:

$$\text{EUAC}(A) = .15x$$

$$\text{EUAC}(B) = .04x + 500 + 5{,}000(A/P, 10\%, 3) - 1{,}200(A/F, 10\%, 3) = .04x + 2{,}148$$

Setting EUAC(A) and EUAC(B) equal and solving for x yields 19,527 miles per year as the break-even point.

16 HANDLING INFLATION

It is important to perform economic studies in terms of *constant value dollars.* One method of converting all cash flows to constant value dollars is to divide the flows by some annual *economic indicator* or price index. Such indicators would normally be given to you as part of a problem.

If indicators are not available, this method can still be used by assuming that inflation is relatively constant at a decimal rate e per year. Then, all cash flows can be converted to '$t = 0$' dollars by dividing by $(1+e)^n$ where n is the year of the cash flow.

Example 13.21

What is the uninflated present worth of $2,000 in two years if the average inflation rate is 6% and i is 10%?

$$P = \frac{\$2{,}000}{(1.10)^2(1.06)^2} = \$1{,}471.07$$

An alternative is to replace i with a value corrected for inflation. This corrected value, i', is

$$i' = i + e + ie \qquad 13.30$$

This method has the advantage of simplifying the calculations. However, pre-calculated factors may not be available for the non-integer values of i'. Therefore, table 13.1 will have to be used to calculate the factors.

Example 13.22

Repeat example 13.21 using i'.

$$i' = .10 + .06 + (.10)(.06) = .166$$

$$P = \frac{\$2{,}000}{(1.166)^2} = \$1{,}471.07$$

17 LEARNING CURVES

The more products that are made, the more efficient the operation becomes due to experience gained. Therefore, direct labor costs decrease. Usually, a *learning curve* is specified by the decrease in cost each time the quantity produced doubles. If there is a 20% decrease per doubling, the curve is said to be an 80% learning curve.

Consider the following special variables:

T_1 time or cost for the first item
T_n time or cost for the nth item
n total number of items produced
b learning curve constant

Table 13.7

Learning Curve Constants

learning curve	b
75%	.415
80%	.322
85%	.234
90%	.152
95%	.074

Then, the time to produce the nth item is given by

$$T_n = T_1(n)^{-b} \qquad 13.31$$

The total time to produce units from quantity n_1 to n_2 inclusive is

$$\int_{n_1}^{n_2} T_n dn \approx \frac{T_1}{(1-b)}\left[\left(n_2+\frac{1}{2}\right)^{1-b} - \left(n_1-\frac{1}{2}\right)^{1-b}\right] \qquad 13.32$$

The average time per unit over the production from n_1 to n_2 is the above total time from equation 13.32 divided by the quantity produced, $(n_2 - n_1 + 1)$.

It is important to remember that learning curve reductions apply only to direct labor costs. They are not applied to indirect labor or direct material costs.

Example 13.23

A 70% learning curve is used with an item whose first production time was 1.47 hours. How long will it take to produce the 11th item? How long will it take to produce the 11th through 27th items?

First, find b.

$$\frac{T_2}{T_1} = .7 = (2)^{-b} \quad \text{or } b = .515$$

Then, $T_{11} = 1.47(11)^{-.515} = .428$ hours.

The time to produce the 11th item through 27th item is approximately

$$T = \frac{1.47}{1-.515}[(27.5)^{1-.515} - (10.5)^{1-.515}]$$
$$= 5.643 \text{ hours}$$

18 ECONOMIC ORDER QUANTITY

The *economic order quantity* (EOQ) is the order quantity which minimizes the inventory costs per unit time. Although there are many different EOQ models, the simplest is based on the following assumptions:

- Reordering is instantaneous. The time between order placement and receipt is zero.
- Shortages are not allowed.
- Demand for the inventory item is deterministic (i.e., is not a random variable).
- Demand is constant with respect to time.
- An order is placed when the on-hand quantity is zero.

The following special variables are used:

a	the constant depletion rate $\left(\frac{\text{items}}{\text{unit time}}\right)$
h	the inventory storage cost $\left(\frac{\$}{\text{item-unit time}}\right)$
H	the total inventory storage cost between orders (\$)
K	the fixed cost of placing an order (\$)
Q_0	the order quantity

If the original quantity on hand is Q_0, the stock will be depleted at

$$t^* = \frac{Q_0}{a}$$

The total inventory storage cost between t_0 and t^* is

$$H = \frac{1}{2}h\frac{Q_o^2}{a} \qquad 13.33$$

The total inventory and ordering cost per unit time is

$$C_t = \frac{aK}{Q_0} + \frac{1}{2}hQ_0 \qquad 13.34$$

C_t can be minimized with respect to Q_0. The EOQ and time between orders are:

$$Q_0^* = \sqrt{2\frac{ak}{h}} \qquad 13.35$$

$$t^* = \frac{Q_0^*}{a} \qquad 13.36$$

19 CONSUMER LOANS

Many consumer loans cannot be handled by the equivalence formulas presented up to this point. Many different arrangements can be made between lender and borrower. Four of the most common consumer loan arrangements are presented below. Refer to a real estate or investment analysis book for more complex loans.

A. SIMPLE INTEREST

Interest due does not compound with a *simple interest* loan. The interest due is merely proportional to the length of time the principal is outstanding. Because of this, simple interest loans are seldom made for long periods (e.g., longer than one year).

Example 13.24

A \$12,000 simple interest loan is taken out at 16% per annum. The loan matures in one year with no intermediate payments. How much will be due at the end of the year?

Amount due=(1+.16)($12,000)=$13,920

For loans less than one year, it is commonly assumed that a year consists of 12 months of 30 days each.

Example 13.25

$4000 is borrowed for 75 days at 16% per annum simple interest. How much will be due at the end of 75 days?

$$\text{Amount due} = \$4000 + (.16)\left(\frac{75}{360}\right)(4000) = \$4,133$$

B. LOANS WITH CONSTANT AMOUNT PAID TOWARDS PRINCIPAL

With this loan type, the payment is not the same each period. The amount paid towards the principal is constant, but the interest varies from period to period. The following special symbols are used.

BAL_j	balance after the jth payment
LV	total value loaned (cost minus down payment)
j	payment or period number
N	total number of payments to pay off the loan
PI_j	jth interest payment
PP_j	jth principal payment
PT_j	jth total payment
ϕ	effective rate per period (r/k)

The equations which govern this type of loan are

$$BAL_j = LV - (j)(PP) \quad 13.37$$

$$PI_j = \phi(BAL_j) \quad 13.38$$

$$PT_j = PP + PI_j \quad 13.39$$

C. DIRECT REDUCTION LOANS

This is the typical 'interest paid on unpaid balance' loan. The amount of the periodic payment is constant, but the amounts paid towards the principal and interest both vary.

The same symbols are used with this type of loan as are listed above.

$$N = -\frac{ln[\frac{-\phi(LV)}{PT} + 1]}{ln(1+\phi)} \quad 13.40$$

$$BAL_{j-1} = PT\left[\frac{1-(1+\phi)^{j-1-N}}{\phi}\right] \quad 13.41$$

$$PI_j = \phi(BAL_j) \quad 13.42$$

$$PP_j = PT - PI_j \quad 13.43$$

$$BAL_j = BAL_{j-1} - PP_j \quad 13.44$$

Example 13.26

A $45,000 loan is financed at 9.25% per annum. The monthly payment is $385. What are the amounts paid toward interest and principal in the 14th period? What is the remaining principal balance after the 14th payment has been made?

The effective rate per month is

$$\phi = \frac{r}{k} = \frac{.0925}{12}$$

$$= .007708$$

$$N = -\frac{ln\left[\frac{-(.007708)(45,000)}{385} + 1\right]}{ln(1+.007708)} = \$301$$

$$BAL_{13} = 385\left[\frac{1-(1+.007708)^{14-1-301}}{.007708}\right]$$

$$= \$44,476.39$$

$$PI_{14} = (.007708)(\$44,476.39) = \$342.82$$

$$PP_{14} = \$385 - \$342.82 = \$42.18$$

$$BAL_{14} = \$44,476.39 - \$42.18 = \$44,434.20$$

Equation 13.40 calculates the number of payments necessary to pay off a loan. This equation can be solved with effort for the total periodic payment (PT) or the initial value of the loan (LV). It is easier, however, to use the $(A/P, i\%, n)$ factor to find the payment and loan value.

$$PT = (LV)(A/P, i\%, n)$$

If the loan is repaid in yearly installments, then i is the effective annual rate. If the loan is paid off monthly, then i should be replaced by the effective rate per month (ϕ from equation 13.26). For monthly payments, n is the number of months in the payback period.

D. DIRECT REDUCTION LOAN WITH BALLOON PAYMENT

This type of loan has a constant periodic payment, but the duration of the loan is insufficient to completely pay back the principal. Therefore, all remaining unpaid principal must be paid back in a lump sum when the loan matures. This large payment is known as a *balloon payment.*

Equations 13.40 through 13.44 can also be used with this type of loan. The remaining balance after the last payment is the balloon payment. This balloon payment must be repaid along with the last regular payment calculated.

Practice Problems

ENGINEERING ECONOMICS

Warm-ups

1. How much will be accumulated at 6% if $1000 is invested for 10 years?

2. What is the present worth at 6% of $2000 which becomes available in 4 years?

3. How much will it take to accumulate $2000 in 20 years at 6%?

4. What year-end annual amount over 7 years at 6% is equivalent to $500 invested now?

5. $50 is invested at the end of each year for 10 years. What will be the accumulated amount at the end of 10 years at 6%?

6. How much should you deposit at 6%at the start of each year for 10 years in order to empty the fund by drawing out $200 at the end of each year for 10 years?

7. You need $2000 on the date of your last deposit. How much should be deposited at the start of each year for 5 years at 6%?

8. How much will be accumulated at 6% in 10 years if you deposit 3 payments of $100 every other year for 4 years, with the first payment occuring at $t = 0$?

9. $500 is compounded monthly at 6% annual rate. How much will you have in 5 years?

10. What is the rate of return of an $80 investment that pays back $120 in 7 years?

Concentrates

1. A new machine will cost $17,000 and will have a value of $14,000 in 5 years. Special tooling will cost $5000, and it will have a resale value of $2500 after 5 years. Maintenance will be $200 per year. What will be the average cost of ownership during the next 5 years if interest is 6%?

2. An old highway bridge may be strengthened at a cost of $9000, or it may be replaced for $40,000. The present salvage value of the old bridge is $13,000. It is estimated that the reinforced bridge will last for 20 years with an annual cost of $500 and will have a salvage value of $10,000 at the end of 20 years. The estimated salvage of the new bridge after 25 years is $15,000. The maintenance for the new bridge will be $100 annually. Which is the best alternative at 8%?

3. A firm expects to receive $32,000 each year for 15 years from the sale of a product. It will require an initial investment of $150,000. Expenses will run $7530 per year. Salvage is zero, and straight-line depreciation is used. The tax rate is 48%. What is the after-tax rate of return?

4. A public works project has initial costs of $1,000,000, benefits of $1,500,000, and disbenefits of $300,000. (a) What is the benefit-cost ratio? (b) What is the excess of benefits over costs?

5. A speculator in land pays $14,000 for property that he expects to hold for 10 years. $1000 is spent in renovation, and a monthly rent of $75 is collected from the tenants. Taxes are $150 per year, and maintenance costs are $250. What must be the sales price in 10 years to realize a 10% rate of return? Use the year-end convention.

6. What is the effective interest rate for a payment plan of 30 equal payments of $89.30 per month when a lump sum of $2000 would have been an outright purchase?

7. An apartment complex is purchased for $500,000. What is the depreciation in each of the first 3 years if the salvage value is $100,000 in 25 years? Use (a) straight line, (b) sum-of-the-year's digits, and (c) double declining balance.

8. Equipment, which is expected to be sold after 10 years for $2000, is purchased for $12,000. The estimated maintenance is $1000 the first year but is expected to increase $200 each year thereafter. Using 10%, find (a) the present worth and (b) the annual cost.

9. One of 5 grades of pipe with maximum lives (in years) and costs (in dollars) of (9, 1500), (14, 1600), (30, 1750), (52, 1900), and (86, 2100) is to be chosen for a 20-year project. A failure of the pipe at any time during the project will result in a cost equal to 35% of the original cost. Annual costs are 4% of the initial cost, and the pipes are not recoverable. At 6%, which pipe is superior? HINT: Assume a waveform distribution.

10. A grain combine with 20 year life can remove 7 pounds of rocks from its harvest per hour. Any rocks left in its output will cause $25,000 damage in subsequent processes. Several investments are available to

increase its removal capacity. At 10%, what should be done?

rock rate (lb/hr)	probability of exceeding rock rate	required investment to meet rock rate
7	.15	no cost
8	.10	$15,000
9	.07	$20,000
10	.03	$30,000

Timed (1 hour allowed for each)

1. A structure costing $10,000 has the operating costs and salvage values given below. (a) What is the economic life of the structure? (b) Assuming that the structure has been owned and operated for 4 years, what is the cost of owning the structure for exactly 1 more year? Use 20%.

Year 1: maintenance $2000, salvage $8000
Year 2: maintenance $3000, salvage $7000
Year 3: maintenance $4000, salvage $6000
Year 4: maintenance $5000, salvage $5000
Year 5: maintenance $6000, salvage $4000

2. For personal use of 15,000 miles per year, a man purchases a car for $5000. It costs him $200 per year for insurance and $150 per year for maintenance. He gets 15 mpg, and gasoline costs $.60 per gallon. The resale value after 5 years is $1000. Because of unexpected business driving (5000 miles extra per year), his insurance is increased to $300 per year and maintenance to $200. Salvage is reduced to $500. Use 10%. (a) The man's company offers $.10 per mile reimbursement. Is that adequate? (b) How many miles must be driven per year at $.10 per mile to justify the company's buying the car for the man's use? The cost would be $5000, but insurance, maintenance, and salvage would be $250, $200, and $800 respectively.

STANDARD CASH FLOW FACTORS

MULTIPLY — *BY* — *TO OBTAIN*

F (P/F,i%,n) P

P (F/P,i%,n) F

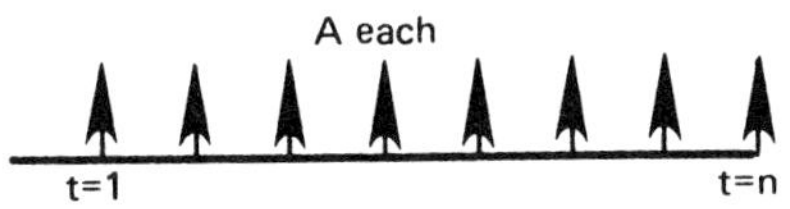

A (P/A,i%,n) P

P (A/P,i%,n) A

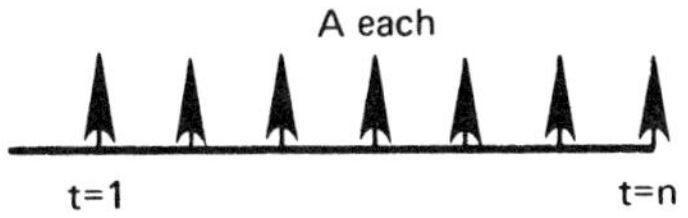

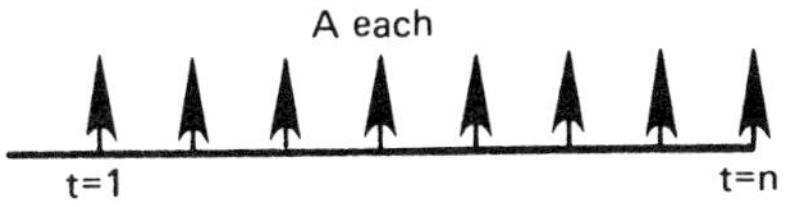

A (F/A,i%,n) F

F (A/F,i%,n) A

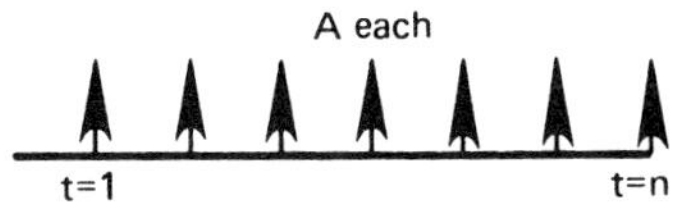

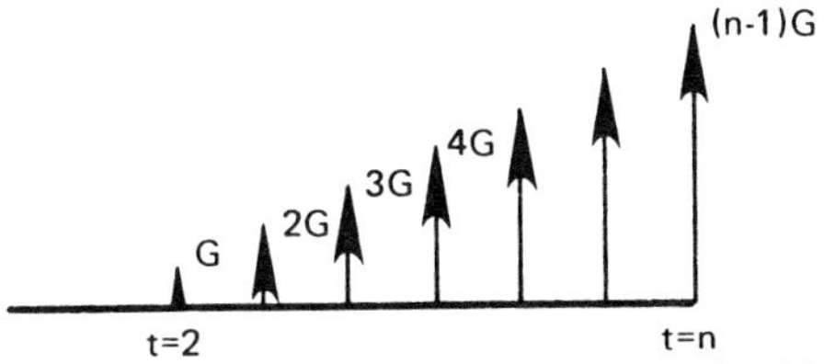

G (P/G,i%,n) P

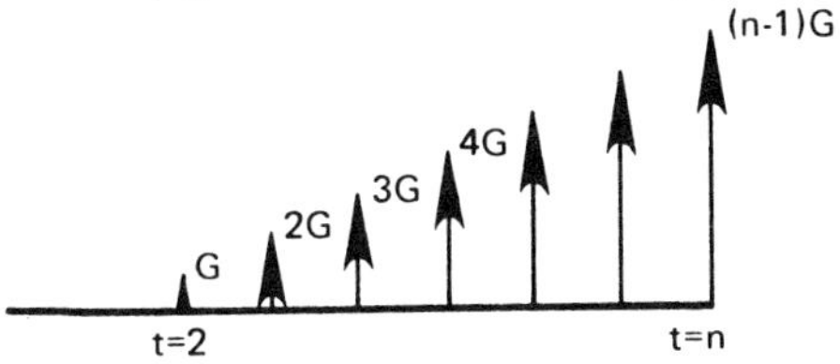

G (A/G,i%,n) A

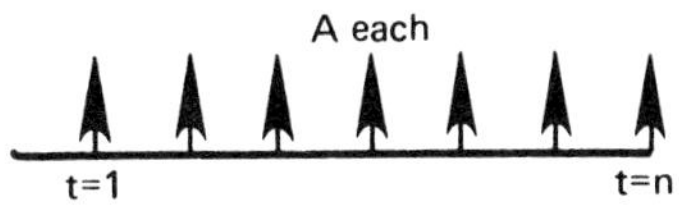

I = 0.50 %

N	(P/F)	(P/A)	(P/G)	(F/P)	(F/A)	(A/P)	(A/F)	(A/G)	N
1	.9950	0.9950	−0.0000	1.0050	1.0000	1.0050	1.0000	−0.0000	1
2	.9901	1.9851	0.9901	1.0100	2.0050	0.5038	0.4988	0.4988	2
3	.9851	2.9702	2.9604	1.0151	3.0150	0.3367	0.3317	0.9967	3
4	.9802	3.9505	5.9011	1.0202	4.0301	0.2531	0.2481	1.4938	4
5	.9754	4.9259	9.8026	1.0253	5.0503	0.2030	0.1980	1.9900	5
6	.9705	5.8964	14.6552	1.0304	6.0755	0.1696	0.1646	2.4855	6
7	.9657	6.8621	20.4493	1.0355	7.1059	0.1457	0.1407	2.9801	7
8	.9609	7.8230	27.1755	1.0407	8.1414	0.1278	0.1228	3.4738	8
9	.9561	8.7791	34.8244	1.0459	9.1821	0.1139	0.1089	3.9668	9
10	.9513	9.7304	43.3865	1.0511	10.2280	0.1028	0.0978	4.4589	10
11	.9466	10.6770	52.8526	1.0564	11.2792	0.0937	0.0887	4.9501	11
12	.9419	11.6189	63.2136	1.0617	12.3356	0.0861	0.0811	5.4406	12
13	.9372	12.5562	74.4602	1.0670	13.3972	0.0796	0.0746	5.9302	13
14	.9326	13.4887	86.5835	1.0723	14.4642	0.0741	0.0691	6.4190	14
15	.9279	14.4166	99.5743	1.0777	15.5365	0.0694	0.0644	6.9069	15
16	.9233	15.3399	113.4238	1.0831	16.6142	0.0652	0.0602	7.3940	16
17	.9187	16.2586	128.1231	1.0885	17.6973	0.0615	0.0565	7.8803	17
18	.9141	17.1728	143.6634	1.0939	18.7858	0.0582	0.0532	8.3658	18
19	.9096	18.0824	160.0360	1.0994	19.8797	0.0553	0.0503	8.8504	19
20	.9051	18.9874	177.2322	1.1049	20.9791	0.0527	0.0477	9.3342	20
21	.9006	19.8880	195.2434	1.1104	22.0840	0.0503	0.0453	9.8172	21
22	.8961	20.7841	214.0611	1.1160	23.1944	0.0481	0.0431	10.2993	22
23	.8916	21.6757	233.6768	1.1216	24.3104	0.0461	0.0411	10.7806	23
24	.8872	22.5629	254.0820	1.1272	25.4320	0.0443	0.0393	11.2611	24
25	.8828	23.4456	275.2686	1.1328	26.5591	0.0427	0.0377	11.7407	25
26	.8784	24.3240	297.2281	1.1385	27.6919	0.0411	0.0361	12.2195	26
27	.8740	25.1980	319.9523	1.1442	28.8304	0.0397	0.0347	12.6975	27
28	.8697	26.0677	343.4332	1.1499	29.9745	0.0384	0.0334	13.1747	28
29	.8653	26.9330	367.6625	1.1556	31.1244	0.0371	0.0321	13.6510	29
30	.8610	27.7941	392.6324	1.1614	32.2800	0.0360	0.0310	14.1265	30
31	.8567	28.6508	418.3348	1.1672	33.4414	0.0349	0.0299	14.6012	31
32	.8525	29.5033	444.7618	1.1730	34.6086	0.0339	0.0289	15.0750	32
33	.8482	30.3515	471.9055	1.1789	35.7817	0.0329	0.0279	15.5480	33
34	.8440	31.1955	499.7583	1.1848	36.9606	0.0321	0.0271	16.0202	34
35	.8398	32.0354	528.3123	1.1907	38.1454	0.0312	0.0262	16.4915	35
36	.8356	32.8710	557.5598	1.1967	39.3361	0.0304	0.0254	16.9621	36
37	.8315	33.7025	587.4934	1.2027	40.5328	0.0297	0.0247	17.4317	37
38	.8274	34.5299	618.1054	1.2087	41.7354	0.0290	0.0240	17.9006	38
39	.8232	35.3531	649.3883	1.2147	42.9441	0.0283	0.0233	18.3686	39
40	.8191	36.1722	681.3347	1.2208	44.1588	0.0276	0.0226	18.8359	40
41	.8151	36.9873	713.9372	1.2269	45.3796	0.0270	0.0220	19.3022	41
42	.8110	37.7983	747.1886	1.2330	46.6065	0.0265	0.0215	19.7678	42
43	.8070	38.6053	781.0815	1.2392	47.8396	0.0259	0.0209	20.2325	43
44	.8030	39.4082	815.6087	1.2454	49.0788	0.0254	0.0204	20.6964	44
45	.7990	40.2072	850.7631	1.2516	50.3242	0.0249	0.0199	21.1595	45
46	.7950	41.0022	886.5376	1.2579	51.5758	0.0244	0.0194	21.6217	46
47	.7910	41.7932	922.9252	1.2642	52.8337	0.0239	0.0189	22.0831	47
48	.7871	42.5803	959.9188	1.2705	54.0978	0.0235	0.0185	22.5437	48
49	.7832	43.3635	997.5116	1.2768	55.3683	0.0231	0.0181	23.0035	49
50	.7793	44.1428	1035.6966	1.2832	56.6452	0.0227	0.0177	23.4624	50
51	.7754	44.9182	1074.4670	1.2896	57.9284	0.0223	0.0173	23.9205	51
52	.7716	45.6897	1113.8162	1.2961	59.2180	0.0219	0.0169	24.3778	52
53	.7677	46.4575	1153.7372	1.3026	60.5141	0.0215	0.0165	24.8343	53
54	.7639	47.2214	1194.2236	1.3091	61.8167	0.0212	0.0162	25.2899	54
55	.7601	47.9814	1235.2686	1.3156	63.1258	0.0208	0.0158	25.7447	55
60	.7414	51.7256	1448.6458	1.3489	69.7700	0.0193	0.0143	28.0064	60
65	.7231	55.3775	1675.0272	1.3829	76.5821	0.0181	0.0131	30.2475	65
70	.7053	58.9394	1913.6427	1.4178	83.5661	0.0170	0.0120	32.4680	70
75	.6879	62.4136	2163.7525	1.4536	90.7265	0.0160	0.0110	34.6679	75
80	.6710	65.8023	2424.6455	1.4903	98.0677	0.0152	0.0102	36.8474	80
85	.6545	69.1075	2695.6389	1.5280	105.5943	0.0145	0.0095	39.0065	85
90	.6383	72.3313	2976.0769	1.5666	113.3109	0.0138	0.0088	41.1451	90
95	.6226	75.4757	3265.3298	1.6061	121.2224	0.0132	0.0082	43.2633	95
100	.6073	78.5426	3562.7934	1.6467	129.3337	0.0127	0.0077	45.3613	100

I = 0.75 %

N	(P/F)	(P/A)	(P/G)	(F/P)	(F/A)	(A/P)	(A/F)	(A/G)	N
1	.9926	0.9926	-0.0000	1.0075	1.0000	1.0075	1.0000	-0.0000	1
2	.9852	1.9777	0.9852	1.0151	2.0075	0.5056	0.4981	0.4981	2
3	.9778	2.9556	2.9408	1.0227	3.0226	0.3383	0.3308	0.9950	3
4	.9706	3.9261	5.8525	1.0303	4.0452	0.2547	0.2472	1.4907	4
5	.9633	4.8894	9.7058	1.0381	5.0756	0.2045	0.1970	1.9851	5
6	.9562	5.8456	14.4866	1.0459	6.1136	0.1711	0.1636	2.4782	6
7	.9490	6.7946	20.1808	1.0537	7.1595	0.1472	0.1397	2.9701	7
8	.9420	7.7366	26.7747	1.0616	8.2132	0.1293	0.1218	3.4608	8
9	.9350	8.6716	34.2544	1.0696	9.2748	0.1153	0.1078	3.9502	9
10	.9280	9.5996	42.6064	1.0776	10.3443	0.1042	0.0967	4.4384	10
11	.9211	10.5207	51.8174	1.0857	11.4219	0.0951	0.0876	4.9253	11
12	.9142	11.4349	61.8740	1.0938	12.5076	0.0875	0.0800	5.4110	12
13	.9074	12.3423	72.7632	1.1020	13.6014	0.0810	0.0735	5.8954	13
14	.9007	13.2430	84.4720	1.1103	14.7034	0.0755	0.0680	6.3786	14
15	.8940	14.1370	96.9876	1.1186	15.8137	0.0707	0.0632	6.8606	15
16	.8873	15.0243	110.2973	1.1270	16.9323	0.0666	0.0591	7.3413	16
17	.8807	15.9050	124.3887	1.1354	18.0593	0.0629	0.0554	7.8207	17
18	.8742	16.7792	139.2494	1.1440	19.1947	0.0596	0.0521	8.2989	18
19	.8676	17.6468	154.8671	1.1525	20.3387	0.0567	0.0492	8.7759	19
20	.8612	18.5080	171.2297	1.1612	21.4912	0.0540	0.0465	9.2516	20
21	.8548	19.3628	188.3253	1.1699	22.6524	0.0516	0.0441	9.7261	21
22	.8484	20.2112	206.1420	1.1787	23.8223	0.0495	0.0420	10.1994	22
23	.8421	21.0533	224.6682	1.1875	25.0010	0.0475	0.0400	10.6714	23
24	.8358	21.8891	243.8923	1.1964	26.1885	0.0457	0.0382	11.1422	24
25	.8296	22.7188	263.8029	1.2054	27.3849	0.0440	0.0365	11.6117	25
26	.8234	23.5422	284.3888	1.2144	28.5903	0.0425	0.0350	12.0800	26
27	.8173	24.3595	305.6387	1.2235	29.8047	0.0411	0.0336	12.5470	27
28	.8112	25.1707	327.5416	1.2327	31.0282	0.0397	0.0322	13.0128	28
29	.8052	25.9759	350.0867	1.2420	32.2609	0.0385	0.0310	13.4774	29
30	.7992	26.7751	373.2631	1.2513	33.5029	0.0373	0.0298	13.9407	30
31	.7932	27.5683	397.0602	1.2607	34.7542	0.0363	0.0288	14.4028	31
32	.7873	28.3557	421.4675	1.2701	36.0148	0.0353	0.0278	14.8636	32
33	.7815	29.1371	446.4746	1.2796	37.2849	0.0343	0.0268	15.3232	33
34	.7757	29.9128	472.0712	1.2892	38.5646	0.0334	0.0259	15.7816	34
35	.7699	30.6827	498.2471	1.2989	39.8538	0.0326	0.0251	16.2387	35
36	.7641	31.4468	524.9924	1.3086	41.1527	0.0318	0.0243	16.6946	36
37	.7585	32.2053	552.2969	1.3185	42.4614	0.0311	0.0236	17.1493	37
38	.7528	32.9581	580.1511	1.3283	43.7798	0.0303	0.0228	17.6027	38
39	.7472	33.7053	608.5451	1.3383	45.1082	0.0297	0.0222	18.0549	39
40	.7416	34.4469	637.4693	1.3483	46.4465	0.0290	0.0215	18.5058	40
41	.7361	35.1831	666.9144	1.3585	47.7948	0.0284	0.0209	18.9556	41
42	.7306	35.9137	696.8709	1.3686	49.1533	0.0278	0.0203	19.4040	42
43	.7252	36.6389	727.3297	1.3789	50.5219	0.0273	0.0198	19.8513	43
44	.7198	37.3587	758.2815	1.3893	51.9009	0.0268	0.0193	20.2973	44
45	.7145	38.0732	789.7173	1.3997	53.2901	0.0263	0.0188	20.7421	45
46	.7091	38.7823	821.6283	1.4102	54.6898	0.0258	0.0183	21.1856	46
47	.7039	39.4862	854.0056	1.4207	56.1000	0.0253	0.0178	21.6280	47
48	.6986	40.1848	886.8404	1.4314	57.5207	0.0249	0.0174	22.0691	48
49	.6934	40.8782	920.1243	1.4421	58.9521	0.0245	0.0170	22.5089	49
50	.6883	41.5664	953.8486	1.4530	60.3943	0.0241	0.0166	22.9476	50
51	.6831	42.2496	988.0050	1.4639	61.8472	0.0237	0.0162	23.3850	51
52	.6780	42.9276	1022.5852	1.4748	63.3111	0.0233	0.0158	23.8211	52
53	.6730	43.6006	1057.5810	1.4859	64.7859	0.0229	0.0154	24.2561	53
54	.6680	44.2686	1092.9842	1.4970	66.2718	0.0226	0.0151	24.6898	54
55	.6630	44.9316	1128.7869	1.5083	67.7688	0.0223	0.0148	25.1223	55
60	.6387	48.1734	1313.5189	1.5657	75.4241	0.0208	0.0133	27.2665	60
65	.6153	51.2963	1507.0910	1.6253	83.3709	0.0195	0.0120	29.3801	65
70	.5927	54.3046	1708.6065	1.6872	91.6201	0.0184	0.0109	31.4634	70
75	.5710	57.2027	1917.2225	1.7514	100.1833	0.0175	0.0100	33.5163	75
80	.5500	59.9944	2132.1472	1.8180	109.0725	0.0167	0.0092	35.5391	80
85	.5299	62.6838	2352.6375	1.8873	118.3001	0.0160	0.0085	37.5318	85
90	.5104	65.2746	2577.9961	1.9591	127.8790	0.0153	0.0078	39.4946	90
95	.4917	67.7704	2807.5694	2.0337	137.8225	0.0148	0.0073	41.4277	95
100	.4737	70.1746	3040.7453	2.1111	148.1445	0.0143	0.0068	43.3311	100

I = 1.00 %

N	(P/F)	(P/A)	(P/G)	(F/P)	(F/A)	(A/P)	(A/F)	(A/G)	N
1	.9901	0.9901	-0.0000	1.0100	1.0000	1.0100	1.0000	-0.0000	1
2	.9803	1.9704	0.9803	1.0201	2.0100	0.5075	0.4975	0.4975	2
3	.9706	2.9410	2.9215	1.0303	3.0301	0.3400	0.3300	0.9934	3
4	.9610	3.9020	5.8044	1.0406	4.0604	0.2563	0.2463	1.4876	4
5	.9515	4.8534	9.6103	1.0510	5.1010	0.2060	0.1960	1.9801	5
6	.9420	5.7955	14.3205	1.0615	6.1520	0.1725	0.1625	2.4710	6
7	.9327	6.7282	19.9168	1.0721	7.2135	0.1486	0.1386	2.9602	7
8	.9235	7.6517	26.3812	1.0829	8.2857	0.1307	0.1207	3.4478	8
9	.9143	8.5660	33.6959	1.0937	9.3685	0.1167	0.1067	3.9337	9
10	.9053	9.4713	41.8435	1.1046	10.4622	0.1056	0.0956	4.4179	10
11	.8963	10.3676	50.8067	1.1157	11.5668	0.0965	0.0865	4.9005	11
12	.8874	11.2551	60.5687	1.1268	12.6825	0.0888	0.0788	5.3815	12
13	.8787	12.1337	71.1126	1.1381	13.8093	0.0824	0.0724	5.8607	13
14	.8700	13.0037	82.4221	1.1495	14.9474	0.0769	0.0669	6.3384	14
15	.8613	13.8651	94.4810	1.1610	16.0969	0.0721	0.0621	6.8143	15
16	.8528	14.7179	107.2734	1.1726	17.2579	0.0679	0.0579	7.2886	16
17	.8444	15.5623	120.7834	1.1843	18.4304	0.0643	0.0543	7.7613	17
18	.8360	16.3983	134.9957	1.1961	19.6147	0.0610	0.0510	8.2323	18
19	.8277	17.2260	149.8950	1.2081	20.8109	0.0581	0.0481	8.7017	19
20	.8195	18.0456	165.4664	1.2202	22.0190	0.0554	0.0454	9.1694	20
21	.8114	18.8570	181.6950	1.2324	23.2392	0.0530	0.0430	9.6354	21
22	.8034	19.6604	198.5663	1.2447	24.4716	0.0509	0.0409	10.0998	22
23	.7954	20.4558	216.0660	1.2572	25.7163	0.0489	0.0389	10.5626	23
24	.7876	21.2434	234.1800	1.2697	26.9735	0.0471	0.0371	11.0237	24
25	.7798	22.0232	252.8945	1.2824	28.2432	0.0454	0.0354	11.4831	25
26	.7720	22.7952	272.1957	1.2953	29.5256	0.0439	0.0339	11.9409	26
27	.7644	23.5596	292.0702	1.3082	30.8209	0.0424	0.0324	12.3971	27
28	.7568	24.3164	312.5047	1.3213	32.1291	0.0411	0.0311	12.8516	28
29	.7493	25.0658	333.4863	1.3345	33.4504	0.0399	0.0299	13.3044	29
30	.7419	25.8077	355.0021	1.3478	34.7849	0.0387	0.0287	13.7557	30
31	.7346	26.5423	377.0394	1.3613	36.1327	0.0377	0.0277	14.2052	31
32	.7273	27.2696	399.5858	1.3749	37.4941	0.0367	0.0267	14.6532	32
33	.7201	27.9897	422.6291	1.3887	38.8690	0.0357	0.0257	15.0995	33
34	.7130	28.7027	446.1572	1.4026	40.2577	0.0348	0.0248	15.5441	34
35	.7059	29.4086	470.1583	1.4166	41.6603	0.0340	0.0240	15.9871	35
36	.6989	30.1075	494.6207	1.4308	43.0769	0.0332	0.0232	16.4285	36
37	.6920	30.7995	519.5329	1.4451	44.5076	0.0325	0.0225	16.8682	37
38	.6852	31.4847	544.8835	1.4595	45.9527	0.0318	0.0218	17.3063	38
39	.6784	32.1630	570.6616	1.4741	47.4123	0.0311	0.0211	17.7428	39
40	.6717	32.8347	596.8561	1.4889	48.8864	0.0305	0.0205	18.1776	40
41	.6650	33.4997	623.4562	1.5038	50.3752	0.0299	0.0199	18.6108	41
42	.6584	34.1581	650.4514	1.5188	51.8790	0.0293	0.0193	19.0424	42
43	.6519	34.8100	677.8312	1.5340	53.3978	0.0287	0.0187	19.4723	43
44	.6454	35.4555	705.5853	1.5493	54.9318	0.0282	0.0182	19.9006	44
45	.6391	36.0945	733.7037	1.5648	56.4811	0.0277	0.0177	20.3273	45
46	.6327	36.7272	762.1765	1.5805	58.0459	0.0272	0.0172	20.7524	46
47	.6265	37.3537	790.9938	1.5963	59.6263	0.0268	0.0168	21.1758	47
48	.6203	37.9740	820.1460	1.6122	61.2226	0.0263	0.0163	21.5976	48
49	.6141	38.5881	849.6237	1.6283	62.8348	0.0259	0.0159	22.0178	49
50	.6080	39.1961	879.4176	1.6446	64.4632	0.0255	0.0155	22.4363	50
51	.6020	39.7981	909.5186	1.6611	66.1078	0.0251	0.0151	22.8533	51
52	.5961	40.3942	939.9175	1.6777	67.7689	0.0248	0.0148	23.2686	52
53	.5902	40.9844	970.6057	1.6945	69.4466	0.0244	0.0144	23.6823	53
54	.5843	41.5687	1001.5743	1.7114	71.1410	0.0241	0.0141	24.0945	54
55	.5785	42.1472	1032.8148	1.7285	72.8525	0.0237	0.0137	24.5049	55
60	.5504	44.9550	1192.8061	1.8167	81.6697	0.0222	0.0122	26.5333	60
65	.5237	47.6266	1358.3903	1.9094	90.9366	0.0210	0.0110	28.5217	65
70	.4983	50.1685	1528.6474	2.0068	100.6763	0.0199	0.0099	30.4703	70
75	.4741	52.5871	1702.7340	2.1091	110.9128	0.0190	0.0090	32.3793	75
80	.4511	54.8882	1879.8771	2.2167	121.6715	0.0182	0.0082	34.2492	80
85	.4292	57.0777	2059.3701	2.3298	132.9790	0.0175	0.0075	36.0801	85
90	.4084	59.1609	2240.5675	2.4486	144.8633	0.0169	0.0069	37.8724	90
95	.3886	61.1430	2422.8811	2.5735	157.3538	0.0164	0.0064	39.6265	95
100	.3697	63.0289	2605.7758	2.7048	170.4814	0.0159	0.0059	41.3426	100

I = 1.50 %

N	(P/F)	(P/A)	(P/G)	(F/P)	(F/A)	(A/P)	(A/F)	(A/G)	N
1	.9852	0.9852	-0.0000	1.0150	1.0000	1.0150	1.0000	-0.0000	1
2	.9707	1.9559	0.9707	1.0302	2.0150	0.5113	0.4963	0.4963	2
3	.9563	2.9122	2.8833	1.0457	3.0452	0.3434	0.3284	0.9901	3
4	.9422	3.8544	5.7098	1.0614	4.0909	0.2594	0.2444	1.4814	4
5	.9283	4.7826	9.4229	1.0773	5.1523	0.2091	0.1941	1.9702	5
6	.9145	5.6972	13.9956	1.0934	6.2296	0.1755	0.1605	2.4566	6
7	.9010	6.5982	19.4018	1.1098	7.3230	0.1516	0.1366	2.9405	7
8	.8877	7.4859	25.6157	1.1265	8.4328	0.1336	0.1186	3.4219	8
9	.8746	8.3605	32.6125	1.1434	9.5593	0.1196	0.1046	3.9008	9
10	.8617	9.2222	40.3675	1.1605	10.7027	0.1084	0.0934	4.3772	10
11	.8489	10.0711	48.8568	1.1779	11.8633	0.0993	0.0843	4.8512	11
12	.8364	10.9075	58.0571	1.1956	13.0412	0.0917	0.0767	5.3227	12
13	.8240	11.7315	67.9454	1.2136	14.2368	0.0852	0.0702	5.7917	13
14	.8118	12.5434	78.4994	1.2318	15.4504	0.0797	0.0647	6.2582	14
15	.7999	13.3432	89.6974	1.2502	16.6821	0.0749	0.0599	6.7223	15
16	.7880	14.1313	101.5178	1.2690	17.9324	0.0708	0.0558	7.1839	16
17	.7764	14.9076	113.9400	1.2880	19.2014	0.0671	0.0521	7.6431	17
18	.7649	15.6726	126.9435	1.3073	20.4894	0.0638	0.0488	8.0997	18
19	.7536	16.4262	140.5084	1.3270	21.7967	0.0609	0.0459	8.5539	19
20	.7425	17.1686	154.6154	1.3469	23.1237	0.0582	0.0432	9.0057	20
21	.7315	17.9001	169.2453	1.3671	24.4705	0.0559	0.0409	9.4550	21
22	.7207	18.6208	184.3798	1.3876	25.8376	0.0537	0.0387	9.9018	22
23	.7100	19.3309	200.0006	1.4084	27.2251	0.0517	0.0367	10.3462	23
24	.6995	20.0304	216.0901	1.4295	28.6335	0.0499	0.0349	10.7881	24
25	.6892	20.7196	232.6310	1.4509	30.0630	0.0483	0.0333	11.2276	25
26	.6790	21.3986	249.6065	1.4727	31.5140	0.0467	0.0317	11.6646	26
27	.6690	22.0676	267.0002	1.4948	32.9867	0.0453	0.0303	12.0992	27
28	.6591	22.7267	284.7958	1.5172	34.4815	0.0440	0.0290	12.5313	28
29	.6494	23.3761	302.9779	1.5400	35.9987	0.0428	0.0278	12.9610	29
30	.6398	24.0158	321.5310	1.5631	37.5387	0.0416	0.0266	13.3883	30
31	.6303	24.6461	340.4402	1.5865	39.1018	0.0406	0.0256	13.8131	31
32	.6210	25.2671	359.6910	1.6103	40.6883	0.0396	0.0246	14.2355	32
33	.6118	25.8790	379.2691	1.6345	42.2986	0.0386	0.0236	14.6555	33
34	.6028	26.4817	399.1607	1.6590	43.9331	0.0378	0.0228	15.0731	34
35	.5939	27.0756	419.3521	1.6839	45.5921	0.0369	0.0219	15.4882	35
36	.5851	27.6607	439.8303	1.7091	47.2760	0.0362	0.0212	15.9009	36
37	.5764	28.2371	460.5822	1.7348	48.9851	0.0354	0.0204	16.3112	37
38	.5679	28.8051	481.5954	1.7608	50.7199	0.0347	0.0197	16.7191	38
39	.5595	29.3646	502.8576	1.7872	52.4807	0.0341	0.0191	17.1246	39
40	.5513	29.9158	524.3568	1.8140	54.2679	0.0334	0.0184	17.5277	40
41	.5431	30.4590	546.0814	1.8412	56.0819	0.0328	0.0178	17.9284	41
42	.5351	30.9941	568.0201	1.8688	57.9231	0.0323	0.0173	18.3267	42
43	.5272	31.5212	590.1617	1.8969	59.7920	0.0317	0.0167	18.7227	43
44	.5194	32.0406	612.4955	1.9253	61.6889	0.0312	0.0162	19.1162	44
45	.5117	32.5523	635.0110	1.9542	63.6142	0.0307	0.0157	19.5074	45
46	.5042	33.0565	657.6979	1.9835	65.5684	0.0303	0.0153	19.8962	46
47	.4967	33.5532	680.5462	2.0133	67.5519	0.0298	0.0148	20.2826	47
48	.4894	34.0426	703.5462	2.0435	69.5652	0.0294	0.0144	20.6667	48
49	.4821	34.5247	726.6884	2.0741	71.6087	0.0290	0.0140	21.0484	49
50	.4750	34.9997	749.9636	2.1052	73.6828	0.0286	0.0136	21.4277	50
51	.4680	35.4677	773.3629	2.1368	75.7881	0.0282	0.0132	21.8047	51
52	.4611	35.9287	796.8774	2.1689	77.9249	0.0278	0.0128	22.1794	52
53	.4543	36.3830	820.4986	2.2014	80.0938	0.0275	0.0125	22.5517	53
54	.4475	36.8305	844.2184	2.2344	82.2952	0.0272	0.0122	22.9217	54
55	.4409	37.2715	868.0285	2.2679	84.5296	0.0268	0.0118	23.2894	55
60	.4093	39.3803	988.1674	2.4432	96.2147	0.0254	0.0104	25.0930	60
65	.3799	41.3378	1109.4752	2.6320	108.8028	0.0242	0.0092	26.8393	65
70	.3527	43.1549	1231.1658	2.8355	122.3638	0.0232	0.0082	28.5290	70
75	.3274	44.8416	1352.5600	3.0546	136.9728	0.0223	0.0073	30.1631	75
80	.3039	46.4073	1473.0741	3.2907	152.7109	0.0215	0.0065	31.7423	80
85	.2821	47.8607	1592.2095	3.5450	169.6652	0.0209	0.0059	33.2676	85
90	.2619	49.2099	1709.5439	3.8189	187.9299	0.0203	0.0053	34.7399	90
95	.2431	50.4622	1824.7224	4.1141	207.6061	0.0198	0.0048	36.1602	95
100	.2256	51.6247	1937.4506	4.4320	228.8030	0.0194	0.0044	37.5295	100

I = 2.00 %

N	(P/F)	(P/A)	(P/G)	(F/P)	(F/A)	(A/P)	(A/F)	(A/G)	N
1	.9804	0.9804	-0.0000	1.0200	1.0000	1.0200	1.0000	-0.0000	1
2	.9612	1.9416	0.9612	1.0404	2.0200	0.5150	0.4950	0.4950	2
3	.9423	2.8839	2.8458	1.0612	3.0604	0.3468	0.3268	0.9868	3
4	.9238	3.8077	5.6173	1.0824	4.1216	0.2626	0.2426	1.4752	4
5	.9057	4.7135	9.2403	1.1041	5.2040	0.2122	0.1922	1.9604	5
6	.8880	5.6014	13.6801	1.1262	6.3081	0.1785	0.1585	2.4423	6
7	.8706	6.4720	18.9035	1.1487	7.4343	0.1545	0.1345	2.9208	7
8	.8535	7.3255	24.8779	1.1717	8.5830	0.1365	0.1165	3.3961	8
9	.8368	8.1622	31.5720	1.1951	9.7546	0.1225	0.1025	3.8681	9
10	.8203	8.9826	38.9551	1.2190	10.9497	0.1113	0.0913	4.3367	10
11	.8043	9.7868	46.9977	1.2434	12.1687	0.1022	0.0822	4.8021	11
12	.7885	10.5753	55.6712	1.2682	13.4121	0.0946	0.0746	5.2642	12
13	.7730	11.3484	64.9475	1.2936	14.6803	0.0881	0.0681	5.7231	13
14	.7579	12.1062	74.7999	1.3195	15.9739	0.0826	0.0626	6.1786	14
15	.7430	12.8493	85.2021	1.3459	17.2934	0.0778	0.0578	6.6309	15
16	.7284	13.5777	96.1288	1.3728	18.6393	0.0737	0.0537	7.0799	16
17	.7142	14.2919	107.5554	1.4002	20.0121	0.0700	0.0500	7.5256	17
18	.7002	14.9920	119.4581	1.4282	21.4123	0.0667	0.0467	7.9681	18
19	.6864	15.6785	131.8139	1.4568	22.8406	0.0638	0.0438	8.4073	19
20	.6730	16.3514	144.6003	1.4859	24.2974	0.0612	0.0412	8.8433	20
21	.6598	17.0112	157.7959	1.5157	25.7833	0.0588	0.0388	9.2760	21
22	.6468	17.6580	171.3795	1.5460	27.2990	0.0566	0.0366	9.7055	22
23	.6342	18.2922	185.3309	1.5769	28.8450	0.0547	0.0347	10.1317	23
24	.6217	18.9139	199.6305	1.6084	30.4219	0.0529	0.0329	10.5547	24
25	.6095	19.5235	214.2592	1.6406	32.0303	0.0512	0.0312	10.9745	25
26	.5976	20.1210	229.1987	1.6734	33.6709	0.0497	0.0297	11.3910	26
27	.5859	20.7069	244.4311	1.7069	35.3443	0.0483	0.0283	11.8043	27
28	.5744	21.2813	259.9392	1.7410	37.0512	0.0470	0.0270	12.2145	28
29	.5631	21.8444	275.7064	1.7758	38.7922	0.0458	0.0258	12.6214	29
30	.5521	22.3965	291.7164	1.8114	40.5681	0.0446	0.0246	13.0251	30
31	.5412	22.9377	307.9538	1.8476	42.3794	0.0436	0.0236	13.4257	31
32	.5306	23.4683	324.4035	1.8845	44.2270	0.0426	0.0226	13.8230	32
33	.5202	23.9886	341.0508	1.9222	46.1116	0.0417	0.0217	14.2172	33
34	.5100	24.4986	357.8817	1.9607	48.0338	0.0408	0.0208	14.6083	34
35	.5000	24.9986	374.8826	1.9999	49.9945	0.0400	0.0200	14.9961	35
36	.4902	25.4888	392.0405	2.0399	51.9944	0.0392	0.0192	15.3809	36
37	.4806	25.9695	409.3424	2.0807	54.0343	0.0385	0.0185	15.7625	37
38	.4712	26.4406	426.7764	2.1223	56.1149	0.0378	0.0178	16.1409	38
39	.4619	26.9026	444.3304	2.1647	58.2372	0.0372	0.0172	16.5163	39
40	.4529	27.3555	461.9931	2.2080	60.4020	0.0366	0.0166	16.8885	40
41	.4440	27.7995	479.7535	2.2522	62.6100	0.0360	0.0160	17.2576	41
42	.4353	28.2348	497.6010	2.2972	64.8622	0.0354	0.0154	17.6237	42
43	.4268	28.6616	515.5253	2.3432	67.1595	0.0349	0.0149	17.9866	43
44	.4184	29.0800	533.5165	2.3901	69.5027	0.0344	0.0144	18.3465	44
45	.4102	29.4902	551.5652	2.4379	71.8927	0.0339	0.0139	18.7034	45
46	.4022	29.8923	569.6621	2.4866	74.3306	0.0335	0.0135	19.0571	46
47	.3943	30.2866	587.7985	2.5363	76.8172	0.0330	0.0130	19.4079	47
48	.3865	30.6731	605.9657	2.5871	79.3535	0.0326	0.0126	19.7556	48
49	.3790	31.0521	624.1557	2.6388	81.9406	0.0322	0.0122	20.1003	49
50	.3715	31.4236	642.3606	2.6916	84.5794	0.0318	0.0118	20.4420	50
51	.3642	31.7878	660.5727	2.7454	87.2710	0.0315	0.0115	20.7807	51
52	.3571	32.1449	678.7849	2.8003	90.0164	0.0311	0.0111	21.1164	52
53	.3501	32.4950	696.9900	2.8563	92.8167	0.0308	0.0108	21.4491	53
54	.3432	32.8383	715.1815	2.9135	95.6731	0.0305	0.0105	21.7789	54
55	.3365	33.1748	733.3527	2.9717	98.5865	0.0301	0.0101	22.1057	55
60	.3048	34.7609	823.6975	3.2810	114.0515	0.0288	0.0088	23.6961	60
65	.2761	36.1975	912.7085	3.6225	131.1262	0.0276	0.0076	25.2147	65
70	.2500	37.4986	999.8343	3.9996	149.9779	0.0267	0.0067	26.6632	70
75	.2265	38.6771	1084.6393	4.4158	170.7918	0.0259	0.0059	28.0434	75
80	.2051	39.7445	1166.7868	4.8754	193.7720	0.0252	0.0052	29.3572	80
85	.1858	40.7113	1246.0241	5.3829	219.1439	0.0246	0.0046	30.6064	85
90	.1683	41.5869	1322.1701	5.9431	247.1567	0.0240	0.0040	31.7929	90
95	.1524	42.3800	1395.1033	6.5617	278.0850	0.0236	0.0036	32.9189	95
100	.1380	43.0984	1464.7527	7.2446	312.2323	0.0232	0.0032	33.9863	100

I = 3.00 %

N	(P/F)	(P/A)	(P/G)	(F/P)	(F/A)	(A/P)	(A/F)	(A/G)	N
1	.9709	0.9709	-0.0000	1.0300	1.0000	1.0300	1.0000	-0.0000	1
2	.9426	1.9135	0.9426	1.0609	2.0300	0.5226	0.4926	0.4926	2
3	.9151	2.8286	2.7729	1.0927	3.0909	0.3535	0.3235	0.9803	3
4	.8885	3.7171	5.4383	1.1255	4.1836	0.2690	0.2390	1.4631	4
5	.8626	4.5797	8.8888	1.1593	5.3091	0.2184	0.1884	1.9409	5
6	.8375	5.4172	13.0762	1.1941	6.4684	0.1846	0.1546	2.4138	6
7	.8131	6.2303	17.9547	1.2299	7.6625	0.1605	0.1305	2.8819	7
8	.7894	7.0197	23.4806	1.2668	8.8923	0.1425	0.1125	3.3450	8
9	.7664	7.7861	29.6119	1.3048	10.1591	0.1284	0.0984	3.8032	9
10	.7441	8.5302	36.3088	1.3439	11.4639	0.1172	0.0872	4.2565	10
11	.7224	9.2526	43.5330	1.3842	12.8078	0.1081	0.0781	4.7049	11
12	.7014	9.9540	51.2482	1.4258	14.1920	0.1005	0.0705	5.1485	12
13	.6810	10.6350	59.4196	1.4685	15.6178	0.0940	0.0640	5.5872	13
14	.6611	11.2961	68.0141	1.5126	17.0863	0.0885	0.0585	6.0210	14
15	.6419	11.9379	77.0002	1.5580	18.5989	0.0838	0.0538	6.4500	15
16	.6232	12.5611	86.3477	1.6047	20.1569	0.0796	0.0496	6.8742	16
17	.6050	13.1661	96.0280	1.6528	21.7616	0.0760	0.0460	7.2936	17
18	.5874	13.7535	106.0137	1.7024	23.4144	0.0727	0.0427	7.7081	18
19	.5703	14.3238	116.2788	1.7535	25.1169	0.0698	0.0398	8.1179	19
20	.5537	14.8775	126.7987	1.8061	26.8704	0.0672	0.0372	8.5229	20
21	.5375	15.4150	137.5496	1.8603	28.6765	0.0649	0.0349	8.9231	21
22	.5219	15.9369	148.5094	1.9161	30.5368	0.0627	0.0327	9.3186	22
23	.5067	16.4436	159.6566	1.9736	32.4529	0.0608	0.0308	9.7093	23
24	.4919	16.9355	170.9711	2.0328	34.4265	0.0590	0.0290	10.0954	24
25	.4776	17.4131	182.4336	2.0938	36.4593	0.0574	0.0274	10.4768	25
26	.4637	17.8768	194.0260	2.1566	38.5530	0.0559	0.0259	10.8535	26
27	.4502	18.3270	205.7309	2.2213	40.7096	0.0546	0.0246	11.2255	27
28	.4371	18.7641	217.5320	2.2879	42.9309	0.0533	0.0233	11.5930	28
29	.4243	19.1885	229.4137	2.3566	45.2189	0.0521	0.0221	11.9558	29
30	.4120	19.6004	241.3613	2.4273	47.5754	0.0510	0.0210	12.3141	30
31	.4000	20.0004	253.3609	2.5001	50.0027	0.0500	0.0200	12.6678	31
32	.3883	20.3888	265.3993	2.5751	52.5028	0.0490	0.0190	13.0169	32
33	.3770	20.7658	277.4642	2.6523	55.0778	0.0482	0.0182	13.3616	33
34	.3660	21.1318	289.5437	2.7319	57.7302	0.0473	0.0173	13.7018	34
35	.3554	21.4872	301.6267	2.8139	60.4621	0.0465	0.0165	14.0375	35
36	.3450	21.8323	313.7028	2.8983	63.2759	0.0458	0.0158	14.3688	36
37	.3350	22.1672	325.7622	2.9852	66.1742	0.0451	0.0151	14.6957	37
38	.3252	22.4925	337.7956	3.0748	69.1594	0.0445	0.0145	15.0182	38
39	.3158	22.8082	349.7942	3.1670	72.2342	0.0438	0.0138	15.3363	39
40	.3066	23.1148	361.7499	3.2620	75.4013	0.0433	0.0133	15.6502	40
41	.2976	23.4124	373.6551	3.3599	78.6633	0.0427	0.0127	15.9597	41
42	.2890	23.7014	385.5024	3.4607	82.0232	0.0422	0.0122	16.2650	42
43	.2805	23.9819	397.2852	3.5645	85.4839	0.0417	0.0117	16.5660	43
44	.2724	24.2543	408.9972	3.6715	89.0484	0.0412	0.0112	16.8629	44
45	.2644	24.5187	420.6325	3.7816	92.7199	0.0408	0.0108	17.1556	45
46	.2567	24.7754	432.1856	3.8950	96.5015	0.0404	0.0104	17.4441	46
47	.2493	25.0247	443.6515	4.0119	100.3965	0.0400	0.0100	17.7285	47
48	.2420	25.2667	455.0255	4.1323	104.4084	0.0396	0.0096	18.0089	48
49	.2350	25.5017	466.3031	4.2562	108.5406	0.0392	0.0092	18.2852	49
50	.2281	25.7298	477.4803	4.3839	112.7969	0.0389	0.0089	18.5575	50
51	.2215	25.9512	488.5535	4.5154	117.1808	0.0385	0.0085	18.8258	51
52	.2150	26.1662	499.5191	4.6509	121.6962	0.0382	0.0082	19.0902	52
53	.2088	26.3750	510.3742	4.7904	126.3471	0.0379	0.0079	19.3507	53
54	.2027	26.5777	521.1157	4.9341	131.1375	0.0376	0.0076	19.6073	54
55	.1968	26.7744	531.7411	5.0821	136.0716	0.0373	0.0073	19.8600	55
60	.1697	27.6756	583.0526	5.8916	163.0534	0.0361	0.0061	21.0674	60
65	.1464	28.4529	631.2010	6.8300	194.3328	0.0351	0.0051	22.1841	65
70	.1263	29.1234	676.0869	7.9178	230.5941	0.0343	0.0043	23.2145	70
75	.1089	29.7018	717.6978	9.1789	272.6309	0.0337	0.0037	24.1634	75
80	.0940	30.2008	756.0865	10.6409	321.3630	0.0331	0.0031	25.0353	80
85	.0811	30.6312	791.3529	12.3357	377.8570	0.0326	0.0026	25.8349	85
90	.0699	31.0024	823.6302	14.3005	443.3489	0.0323	0.0023	26.5667	90
95	.0603	31.3227	853.0742	16.5782	519.2720	0.0319	0.0019	27.2351	95
100	.0520	31.5989	879.8540	19.2186	607.2877	0.0316	0.0016	27.8444	100

I = 4.00 %

N	(P/F)	(P/A)	(P/G)	(F/P)	(F/A)	(A/P)	(A/F)	(A/G)	N
1	.9615	0.9615	-0.0000	1.0400	1.0000	1.0400	1.0000	-0.0000	1
2	.9246	1.8861	0.9246	1.0816	2.0400	0.5302	0.4902	0.4902	2
3	.8890	2.7751	2.7025	1.1249	3.1216	0.3603	0.3203	0.9739	3
4	.8548	3.6299	5.2670	1.1699	4.2465	0.2755	0.2355	1.4510	4
5	.8219	4.4518	8.5547	1.2167	5.4163	0.2246	0.1846	1.9216	5
6	.7903	5.2421	12.5062	1.2653	6.6330	0.1908	0.1508	2.3857	6
7	.7599	6.0021	17.0657	1.3159	7.8983	0.1666	0.1266	2.8433	7
8	.7307	6.7327	22.1806	1.3686	9.2142	0.1485	0.1085	3.2944	8
9	.7026	7.4353	27.8013	1.4233	10.5828	0.1345	0.0945	3.7391	9
10	.6756	8.1109	33.8814	1.4802	12.0061	0.1233	0.0833	4.1773	10
11	.6496	8.7605	40.3772	1.5395	13.4864	0.1141	0.0741	4.6090	11
12	.6246	9.3851	47.2477	1.6010	15.0258	0.1066	0.0666	5.0343	12
13	.6006	9.9856	54.4546	1.6651	16.6268	0.1001	0.0601	5.4533	13
14	.5775	10.5631	61.9618	1.7317	18.2919	0.0947	0.0547	5.8659	14
15	.5553	11.1184	69.7355	1.8009	20.0236	0.0899	0.0499	6.2721	15
16	.5339	11.6523	77.7441	1.8730	21.8245	0.0858	0.0458	6.6720	16
17	.5134	12.1657	85.9581	1.9479	23.6975	0.0822	0.0422	7.0656	17
18	.4936	12.6593	94.3498	2.0258	25.6454	0.0790	0.0390	7.4530	18
19	.4746	13.1339	102.8933	2.1068	27.6712	0.0761	0.0361	7.8342	19
20	.4564	13.5903	111.5647	2.1911	29.7781	0.0736	0.0336	8.2091	20
21	.4388	14.0292	120.3414	2.2788	31.9692	0.0713	0.0313	8.5779	21
22	.4220	14.4511	129.2024	2.3699	34.2480	0.0692	0.0292	8.9407	22
23	.4057	14.8568	138.1284	2.4647	36.6179	0.0673	0.0273	9.2973	23
24	.3901	15.2470	147.1012	2.5633	39.0826	0.0656	0.0256	9.6479	24
25	.3751	15.6221	156.1040	2.6658	41.6459	0.0640	0.0240	9.9925	25
26	.3607	15.9828	165.1212	2.7725	44.3117	0.0626	0.0226	10.3312	26
27	.3468	16.3296	174.1385	2.8834	47.0842	0.0612	0.0212	10.6640	27
28	.3335	16.6631	183.1424	2.9987	49.9676	0.0600	0.0200	10.9909	28
29	.3207	16.9837	192.1206	3.1187	52.9663	0.0589	0.0189	11.3120	29
30	.3083	17.2920	201.0618	3.2434	56.0849	0.0578	0.0178	11.6274	30
31	.2965	17.5885	209.9556	3.3731	59.3283	0.0569	0.0169	11.9371	31
32	.2851	17.8736	218.7924	3.5081	62.7015	0.0559	0.0159	12.2411	32
33	.2741	18.1476	227.5634	3.6484	66.2095	0.0551	0.0151	12.5396	33
34	.2636	18.4112	236.2607	3.7943	69.8579	0.0543	0.0143	12.8324	34
35	.2534	18.6646	244.8768	3.9461	73.6522	0.0536	0.0136	13.1198	35
36	.2437	18.9083	253.4052	4.1039	77.5983	0.0529	0.0129	13.4018	36
37	.2343	19.1426	261.8399	4.2681	81.7022	0.0522	0.0122	13.6784	37
38	.2253	19.3679	270.1754	4.4388	85.9703	0.0516	0.0116	13.9497	38
39	.2166	19.5845	278.4070	4.6164	90.4091	0.0511	0.0111	14.2157	39
40	.2083	19.7928	286.5303	4.8010	95.0255	0.0505	0.0105	14.4765	40
41	.2003	19.9931	294.5414	4.9931	99.8265	0.0500	0.0100	14.7322	41
42	.1926	20.1856	302.4370	5.1928	104.8196	0.0495	0.0095	14.9828	42
43	.1852	20.3708	310.2141	5.4005	110.0124	0.0491	0.0091	15.2284	43
44	.1780	20.5488	317.8700	5.6165	115.4129	0.0487	0.0087	15.4690	44
45	.1712	20.7200	325.4028	5.8412	121.0294	0.0483	0.0083	15.7047	45
46	.1646	20.8847	332.8104	6.0748	126.8706	0.0479	0.0079	15.9356	46
47	.1583	21.0429	340.0914	6.3178	132.9454	0.0475	0.0075	16.1618	47
48	.1522	21.1951	347.2446	6.5705	139.2632	0.0472	0.0072	16.3832	48
49	.1463	21.3415	354.2689	6.8333	145.8337	0.0469	0.0069	16.6000	49
50	.1407	21.4822	361.1638	7.1067	152.6671	0.0466	0.0066	16.8122	50
51	.1353	21.6175	367.9289	7.3910	159.7738	0.0463	0.0063	17.0200	51
52	.1301	21.7476	374.5638	7.6866	167.1647	0.0460	0.0060	17.2232	52
53	.1251	21.8727	381.0686	7.9941	174.8513	0.0457	0.0057	17.4221	53
54	.1203	21.9930	387.4436	8.3138	182.8454	0.0455	0.0055	17.6167	54
55	.1157	22.1086	393.6890	8.6464	191.1592	0.0452	0.0052	17.8070	55
60	.0951	22.6235	422.9966	10.5196	237.9907	0.0442	0.0042	18.6972	60
65	.0781	23.0467	449.2014	12.7987	294.9684	0.0434	0.0034	19.4909	65
70	.0642	23.3945	472.4789	15.5716	364.2905	0.0427	0.0027	20.1961	70
75	.0528	23.6804	493.0408	18.9453	448.6314	0.0422	0.0022	20.8206	75
80	.0434	23.9154	511.1161	23.0498	551.2450	0.0418	0.0018	21.3718	80
85	.0357	24.1085	526.9384	28.0436	676.0901	0.0415	0.0015	21.8569	85
90	.0293	24.2673	540.7369	34.1193	827.9833	0.0412	0.0012	22.2826	90
95	.0241	24.3978	552.7307	41.5114	1012.7846	0.0410	0.0010	22.6550	95
100	.0198	24.5050	563.1249	50.5049	1237.6237	0.0408	0.0008	22.9800	100

I = 5.00 %

N	(P/F)	(P/A)	(P/G)	(F/P)	(F/A)	(A/P)	(A/F)	(A/G)	N
1	.9524	0.9524	-0.0000	1.0500	1.0000	1.0500	1.0000	-0.0000	1
2	.9070	1.8594	0.9070	1.1025	2.0500	0.5378	0.4878	0.4878	2
3	.8638	2.7232	2.6347	1.1576	3.1525	0.3672	0.3172	0.9675	3
4	.8227	3.5460	5.1028	1.2155	4.3101	0.2820	0.2320	1.4391	4
5	.7835	4.3295	8.2369	1.2763	5.5256	0.2310	0.1810	1.9025	5
6	.7462	5.0757	11.9680	1.3401	6.8019	0.1970	0.1470	2.3579	6
7	.7107	5.7864	16.2321	1.4071	8.1420	0.1728	0.1228	2.8052	7
8	.6768	6.4632	20.9700	1.4775	9.5491	0.1547	0.1047	3.2445	8
9	.6446	7.1078	26.1268	1.5513	11.0266	0.1407	0.0907	3.6758	9
10	.6139	7.7217	31.6520	1.6289	12.5779	0.1295	0.0795	4.0991	10
11	.5847	8.3064	37.4988	1.7103	14.2068	0.1204	0.0704	4.5144	11
12	.5568	8.8633	43.6241	1.7959	15.9171	0.1128	0.0628	4.9219	12
13	.5303	9.3936	49.9879	1.8856	17.7130	0.1065	0.0565	5.3215	13
14	.5051	9.8986	56.5538	1.9799	19.5986	0.1010	0.0510	5.7133	14
15	.4810	10.3797	63.2880	2.0789	21.5786	0.0963	0.0463	6.0973	15
16	.4581	10.8378	70.1597	2.1829	23.6575	0.0923	0.0423	6.4736	16
17	.4363	11.2741	77.1405	2.2920	25.8404	0.0887	0.0387	6.8423	17
18	.4155	11.6896	84.2043	2.4066	28.1324	0.0855	0.0355	7.2034	18
19	.3957	12.0853	91.3275	2.5270	30.5390	0.0827	0.0327	7.5569	19
20	.3769	12.4622	98.4884	2.6533	33.0660	0.0802	0.0302	7.9030	20
21	.3589	12.8212	105.6673	2.7860	35.7193	0.0780	0.0280	8.2416	21
22	.3418	13.1630	112.8461	2.9253	38.5052	0.0760	0.0260	8.5730	22
23	.3256	13.4886	120.0087	3.0715	41.4305	0.0741	0.0241	8.8971	23
24	.3101	13.7986	127.1402	3.2251	44.5020	0.0725	0.0225	9.2140	24
25	.2953	14.0939	134.2275	3.3864	47.7271	0.0710	0.0210	9.5238	25
26	.2812	14.3752	141.2585	3.5557	51.1135	0.0696	0.0196	9.8266	26
27	.2678	14.6430	148.2226	3.7335	54.6691	0.0683	0.0183	10.1224	27
28	.2551	14.8981	155.1101	3.9201	58.4026	0.0671	0.0171	10.4114	28
29	.2429	15.1411	161.9126	4.1161	62.3227	0.0660	0.0160	10.6936	29
30	.2314	15.3725	168.6226	4.3219	66.4388	0.0651	0.0151	10.9691	30
31	.2204	15.5928	175.2333	4.5380	70.7608	0.0641	0.0141	11.2381	31
32	.2099	15.8027	181.7392	4.7649	75.2988	0.0633	0.0133	11.5005	32
33	.1999	16.0025	188.1351	5.0032	80.0638	0.0625	0.0125	11.7566	33
34	.1904	16.1929	194.4168	5.2533	85.0670	0.0618	0.0118	12.0063	34
35	.1813	16.3742	200.5807	5.5160	90.3203	0.0611	0.0111	12.2498	35
36	.1727	16.5469	206.6237	5.7918	95.8363	0.0604	0.0104	12.4872	36
37	.1644	16.7113	212.5434	6.0814	101.6281	0.0598	0.0098	12.7186	37
38	.1566	16.8679	218.3378	6.3855	107.7095	0.0593	0.0093	12.9440	38
39	.1491	17.0170	224.0054	6.7048	114.0950	0.0588	0.0088	13.1636	39
40	.1420	17.1591	229.5452	7.0400	120.7998	0.0583	0.0083	13.3775	40
41	.1353	17.2944	234.9564	7.3920	127.8398	0.0578	0.0078	13.5857	41
42	.1288	17.4232	240.2389	7.7616	135.2318	0.0574	0.0074	13.7884	42
43	.1227	17.5459	245.3925	8.1497	142.9933	0.0570	0.0070	13.9857	43
44	.1169	17.6628	250.4175	8.5572	151.1430	0.0566	0.0066	14.1777	44
45	.1113	17.7741	255.3145	8.9850	159.7002	0.0563	0.0063	14.3644	45
46	.1060	17.8801	260.0844	9.4343	168.6852	0.0559	0.0059	14.5461	46
47	.1009	17.9810	264.7281	9.9060	178.1194	0.0556	0.0056	14.7226	47
48	.0961	18.0772	269.2467	10.4013	188.0254	0.0553	0.0053	14.8943	48
49	.0916	18.1687	273.6418	10.9213	198.4267	0.0550	0.0050	15.0611	49
50	.0872	18.2559	277.9148	11.4674	209.3480	0.0548	0.0048	15.2233	50
51	.0831	18.3390	282.0673	12.0408	220.8154	0.0545	0.0045	15.3808	51
52	.0791	18.4181	286.1013	12.6428	232.8562	0.0543	0.0043	15.5337	52
53	.0753	18.4934	290.0184	13.2749	245.4990	0.0541	0.0041	15.6823	53
54	.0717	18.5651	293.8208	13.9387	258.7739	0.0539	0.0039	15.8265	54
55	.0683	18.6335	297.5104	14.6356	272.7126	0.0537	0.0037	15.9664	55
60	.0535	18.9293	314.3432	18.6792	353.5837	0.0528	0.0028	16.6062	60
65	.0419	19.1611	328.6910	23.8399	456.7980	0.0522	0.0022	17.1541	65
70	.0329	19.3427	340.8409	30.4264	588.5285	0.0517	0.0017	17.6212	70
75	.0258	19.4850	351.0721	38.8327	756.6537	0.0513	0.0013	18.0176	75
80	.0202	19.5965	359.6460	49.5614	971.2288	0.0510	0.0010	18.3526	80
85	.0158	19.6838	366.8007	63.2544	1245.0871	0.0508	0.0008	18.6346	85
90	.0124	19.7523	372.7488	80.7304	1594.6073	0.0506	0.0006	18.8712	90
95	.0097	19.8059	377.6774	103.0347	2040.6935	0.0505	0.0005	19.0689	95
100	.0076	19.8479	381.7492	131.5013	2610.0252	0.0504	0.0004	19.2337	100

I = 6.00 %

N	(P/F)	(P/A)	(P/G)	(F/P)	(F/A)	(A/P)	(A/F)	(A/G)	N
1	.9434	0.9434	-0.0000	1.0600	1.0000	1.0600	1.0000	-0.0000	1
2	.8900	1.8334	0.8900	1.1236	2.0600	0.5454	0.4854	0.4854	2
3	.8396	2.6730	2.5692	1.1910	3.1836	0.3741	0.3141	0.9612	3
4	.7921	3.4651	4.9455	1.2625	4.3746	0.2886	0.2286	1.4272	4
5	.7473	4.2124	7.9345	1.3382	5.6371	0.2374	0.1774	1.8836	5
6	.7050	4.9173	11.4594	1.4185	6.9753	0.2034	0.1434	2.3304	6
7	.6651	5.5824	15.4497	1.5036	8.3938	0.1791	0.1191	2.7676	7
8	.6274	6.2098	19.8416	1.5938	9.8975	0.1610	0.1010	3.1952	8
9	.5919	6.8017	24.5768	1.6895	11.4913	0.1470	0.0870	3.6133	9
10	.5584	7.3601	29.6023	1.7908	13.1808	0.1359	0.0759	4.0220	10
11	.5268	7.8869	34.8702	1.8983	14.9716	0.1268	0.0668	4.4213	11
12	.4970	8.3838	40.3369	2.0122	16.8699	0.1193	0.0593	4.8113	12
13	.4688	8.8527	45.9629	2.1329	18.8821	0.1130	0.0530	5.1920	13
14	.4423	9.2950	51.7128	2.2609	21.0151	0.1076	0.0476	5.5635	14
15	.4173	9.7122	57.5546	2.3966	23.2760	0.1030	0.0430	5.9260	15
16	.3936	10.1059	63.4592	2.5404	25.6725	0.0990	0.0390	6.2794	16
17	.3714	10.4773	69.4011	2.6928	28.2129	0.0954	0.0354	6.6240	17
18	.3503	10.8276	75.3569	2.8543	30.9057	0.0924	0.0324	6.9597	18
19	.3305	11.1581	81.3062	3.0256	33.7600	0.0896	0.0296	7.2867	19
20	.3118	11.4699	87.2304	3.2071	36.7856	0.0872	0.0272	7.6051	20
21	.2942	11.7641	93.1136	3.3996	39.9927	0.0850	0.0250	7.9151	21
22	.2775	12.0416	98.9412	3.6035	43.3923	0.0830	0.0230	8.2166	22
23	.2618	12.3034	104.7007	3.8197	46.9958	0.0813	0.0213	8.5099	23
24	.2470	12.5504	110.3812	4.0489	50.8156	0.0797	0.0197	8.7951	24
25	.2330	12.7834	115.9732	4.2919	54.8645	0.0782	0.0182	9.0722	25
26	.2198	13.0032	121.4684	4.5494	59.1564	0.0769	0.0169	9.3414	26
27	.2074	13.2105	126.8600	4.8223	63.7058	0.0757	0.0157	9.6029	27
28	.1956	13.4062	132.1420	5.1117	68.5281	0.0746	0.0146	9.8568	28
29	.1846	13.5907	137.3096	5.4184	73.6398	0.0736	0.0136	10.1032	29
30	.1741	13.7648	142.3588	5.7435	79.0582	0.0726	0.0126	10.3422	30
31	.1643	13.9291	147.2864	6.0881	84.8017	0.0718	0.0118	10.5740	31
32	.1550	14.0840	152.0901	6.4534	90.8898	0.0710	0.0110	10.7988	32
33	.1462	14.2302	156.7681	6.8406	97.3432	0.0703	0.0103	11.0166	33
34	.1379	14.3681	161.3192	7.2510	104.1838	0.0696	0.0096	11.2276	34
35	.1301	14.4982	165.7427	7.6861	111.4348	0.0690	0.0090	11.4319	35
36	.1227	14.6210	170.0387	8.1473	119.1209	0.0684	0.0084	11.6298	36
37	.1158	14.7368	174.2072	8.6361	127.2681	0.0679	0.0079	11.8213	37
38	.1092	14.8460	178.2490	9.1543	135.9042	0.0674	0.0074	12.0065	38
39	.1031	14.9491	182.1652	9.7035	145.0585	0.0669	0.0069	12.1857	39
40	.0972	15.0463	185.9568	10.2857	154.7620	0.0665	0.0065	12.3590	40
41	.0917	15.1380	189.6256	10.9029	165.0477	0.0661	0.0061	12.5264	41
42	.0865	15.2245	193.1732	11.5570	175.9505	0.0657	0.0057	12.6883	42
43	.0816	15.3062	196.6017	12.2505	187.5076	0.0653	0.0053	12.8446	43
44	.0770	15.3832	199.9130	12.9855	199.7580	0.0650	0.0050	12.9956	44
45	.0727	15.4558	203.1096	13.7646	212.7435	0.0647	0.0047	13.1413	45
46	.0685	15.5244	206.1938	14.5905	226.5081	0.0644	0.0044	13.2819	46
47	.0647	15.5890	209.1681	15.4659	241.0986	0.0641	0.0041	13.4177	47
48	.0610	15.6500	212.0351	16.3939	256.5645	0.0639	0.0039	13.5485	48
49	.0575	15.7076	214.7972	17.3775	272.9584	0.0637	0.0037	13.6748	49
50	.0543	15.7619	217.4574	18.4202	290.3359	0.0634	0.0034	13.7964	50
51	.0512	15.8131	220.0181	19.5254	308.7561	0.0632	0.0032	13.9137	51
52	.0483	15.8614	222.4823	20.6969	328.2814	0.0630	0.0030	14.0267	52
53	.0456	15.9070	224.8525	21.9387	348.9783	0.0629	0.0029	14.1355	53
54	.0430	15.9500	227.1316	23.2550	370.9170	0.0627	0.0027	14.2402	54
55	.0406	15.9905	229.3222	24.6503	394.1720	0.0625	0.0025	14.3411	55
60	.0303	16.1614	239.0428	32.9877	533.1282	0.0619	0.0019	14.7909	60
65	.0227	16.2891	246.9450	44.1450	719.0829	0.0614	0.0014	15.1601	65
70	.0169	16.3845	253.3271	59.0759	967.9322	0.0610	0.0010	15.4613	70
75	.0126	16.4558	258.4527	79.0569	1300.9487	0.0608	0.0008	15.7058	75
80	.0095	16.5091	262.5493	105.7960	1746.5999	0.0606	0.0006	15.9033	80
85	.0071	16.5489	265.8096	141.5789	2342.9817	0.0604	0.0004	16.0620	85
90	.0053	16.5787	268.3946	189.4645	3141.0752	0.0603	0.0003	16.1891	90
95	.0039	16.6009	270.4375	253.5463	4209.1042	0.0602	0.0002	16.2905	95
100	.0029	16.6175	272.0471	339.3021	5638.3681	0.0602	0.0002	16.3711	100

I = 7.00 %

N	(P/F)	(P/A)	(P/G)	(F/P)	(F/A)	(A/P)	(A/F)	(A/G)	N
1	.9346	0.9346	-0.0000	1.0700	1.0000	1.0700	1.0000	-0.0000	1
2	.8734	1.8080	0.8734	1.1449	2.0700	0.5531	0.4831	0.4831	2
3	.8163	2.6243	2.5060	1.2250	3.2149	0.3811	0.3111	0.9549	3
4	.7629	3.3872	4.7947	1.3108	4.4399	0.2952	0.2252	1.4155	4
5	.7130	4.1002	7.6467	1.4026	5.7507	0.2439	0.1739	1.8650	5
6	.6663	4.7665	10.9784	1.5007	7.1533	0.2098	0.1398	2.3032	6
7	.6227	5.3893	14.7149	1.6058	8.6540	0.1856	0.1156	2.7304	7
8	.5820	5.9713	18.7889	1.7182	10.2598	0.1675	0.0975	3.1465	8
9	.5439	6.5152	23.1404	1.8385	11.9780	0.1535	0.0835	3.5517	9
10	.5083	7.0236	27.7156	1.9672	13.8164	0.1424	0.0724	3.9461	10
11	.4751	7.4987	32.4665	2.1049	15.7836	0.1334	0.0634	4.3296	11
12	.4440	7.9427	37.3506	2.2522	17.8885	0.1259	0.0559	4.7025	12
13	.4150	8.3577	42.3302	2.4098	20.1406	0.1197	0.0497	5.0648	13
14	.3878	8.7455	47.3718	2.5785	22.5505	0.1143	0.0443	5.4167	14
15	.3624	9.1079	52.4461	2.7590	25.1290	0.1098	0.0398	5.7583	15
16	.3387	9.4466	57.5271	2.9522	27.8881	0.1059	0.0359	6.0897	16
17	.3166	9.7632	62.5923	3.1588	30.8402	0.1024	0.0324	6.4110	17
18	.2959	10.0591	67.6219	3.3799	33.9990	0.0994	0.0294	6.7225	18
19	.2765	10.3356	72.5991	3.6165	37.3790	0.0968	0.0268	7.0242	19
20	.2584	10.5940	77.5091	3.8697	40.9955	0.0944	0.0244	7.3163	20
21	.2415	10.8355	82.3393	4.1406	44.8652	0.0923	0.0223	7.5990	21
22	.2257	11.0612	87.0793	4.4304	49.0057	0.0904	0.0204	7.8725	22
23	.2109	11.2722	91.7201	4.7405	53.4361	0.0887	0.0187	8.1369	23
24	.1971	11.4693	96.2545	5.0724	58.1767	0.0872	0.0172	8.3923	24
25	.1842	11.6536	100.6765	5.4274	63.2490	0.0858	0.0158	8.6391	25
26	.1722	11.8258	104.9814	5.8074	68.6765	0.0846	0.0146	8.8773	26
27	.1609	11.9867	109.1656	6.2139	74.4838	0.0834	0.0134	9.1072	27
28	.1504	12.1371	113.2264	6.6488	80.6977	0.0824	0.0124	9.3289	28
29	.1406	12.2777	117.1622	7.1143	87.3465	0.0814	0.0114	9.5427	29
30	.1314	12.4090	120.9718	7.6123	94.4608	0.0806	0.0106	9.7487	30
31	.1228	12.5318	124.6550	8.1451	102.0730	0.0798	0.0098	9.9471	31
32	.1147	12.6466	128.2120	8.7153	110.2182	0.0791	0.0091	10.1381	32
33	.1072	12.7538	131.6435	9.3253	118.9334	0.0784	0.0084	10.3219	33
34	.1002	12.8540	134.9507	9.9781	128.2588	0.0778	0.0078	10.4987	34
35	.0937	12.9477	138.1353	10.6766	138.2369	0.0772	0.0072	10.6687	35
36	.0875	13.0352	141.1990	11.4239	148.9135	0.0767	0.0067	10.8321	36
37	.0818	13.1170	144.1441	12.2236	160.3374	0.0762	0.0062	10.9891	37
38	.0765	13.1935	146.9730	13.0793	172.5610	0.0758	0.0058	11.1398	38
39	.0715	13.2649	149.6883	13.9948	185.6403	0.0754	0.0054	11.2845	39
40	.0668	13.3317	152.2928	14.9745	199.6351	0.0750	0.0050	11.4233	40
41	.0624	13.3941	154.7892	16.0227	214.6096	0.0747	0.0047	11.5565	41
42	.0583	13.4524	157.1807	17.1443	230.6322	0.0743	0.0043	11.6842	42
43	.0545	13.5070	159.4702	18.3444	247.7765	0.0740	0.0040	11.8065	43
44	.0509	13.5579	161.6609	19.6285	266.1209	0.0738	0.0038	11.9237	44
45	.0476	13.6055	163.7559	21.0025	285.7493	0.0735	0.0035	12.0360	45
46	.0445	13.6500	165.7584	22.4726	306.7518	0.0733	0.0033	12.1435	46
47	.0416	13.6916	167.6714	24.0457	329.2244	0.0730	0.0030	12.2463	47
48	.0389	13.7305	169.4981	25.7289	353.2701	0.0728	0.0028	12.3447	48
49	.0363	13.7668	171.2417	27.5299	378.9990	0.0726	0.0026	12.4387	49
50	.0339	13.8007	172.9051	29.4570	406.5289	0.0725	0.0025	12.5287	50
51	.0317	13.8325	174.4915	31.5190	435.9860	0.0723	0.0023	12.6146	51
52	.0297	13.8621	176.0037	33.7253	467.5050	0.0721	0.0021	12.6967	52
53	.0277	13.8898	177.4447	36.0861	501.2303	0.0720	0.0020	12.7751	53
54	.0259	13.9157	178.8173	38.6122	537.3164	0.0719	0.0019	12.8500	54
55	.0242	13.9399	180.1243	41.3150	575.9286	0.0717	0.0017	12.9215	55
60	.0173	14.0392	185.7677	57.9464	813.5204	0.0712	0.0012	13.2321	60
65	.0123	14.1099	190.1452	81.2729	1146.7552	0.0709	0.0009	13.4760	65
70	.0088	14.1604	193.5185	113.9894	1614.1342	0.0706	0.0006	13.6662	70
75	.0063	14.1964	196.1035	159.8760	2269.6574	0.0704	0.0004	13.8136	75
80	.0045	14.2220	198.0748	224.2344	3189.0627	0.0703	0.0003	13.9273	80
85	.0032	14.2403	199.5717	314.5003	4478.5761	0.0702	0.0002	14.0146	85
90	.0023	14.2533	200.7042	441.1030	6287.1854	0.0702	0.0002	14.0812	90
95	.0016	14.2626	201.5581	618.6697	8823.8535	0.0701	0.0001	14.1319	95
100	.0012	14.2693	202.2001	867.7163	12381.6618	0.0701	0.0001	14.1703	100

I = 8.00 %

N	(P/F)	(P/A)	(P/G)	(F/P)	(F/A)	(A/P)	(A/F)	(A/G)	N
1	.9259	0.9259	-0.0000	1.0800	1.0000	1.0800	1.0000	-0.0000	1
2	.8573	1.7833	0.8573	1.1664	2.0800	0.5608	0.4808	0.4808	2
3	.7938	2.5771	2.4450	1.2597	3.2464	0.3880	0.3080	0.9487	3
4	.7350	3.3121	4.6501	1.3605	4.5061	0.3019	0.2219	1.4040	4
5	.6806	3.9927	7.3724	1.4693	5.8666	0.2505	0.1705	1.8465	5
6	.6302	4.6229	10.5233	1.5869	7.3359	0.2163	0.1363	2.2763	6
7	.5835	5.2064	14.0242	1.7138	8.9228	0.1921	0.1121	2.6937	7
8	.5403	5.7466	17.8061	1.8509	10.6366	0.1740	0.0940	3.0985	8
9	.5002	6.2469	21.8081	1.9990	12.4876	0.1601	0.0801	3.4910	9
10	.4632	6.7101	25.9768	2.1589	14.4866	0.1490	0.0690	3.8713	10
11	.4289	7.1390	30.2657	2.3316	16.6455	0.1401	0.0601	4.2395	11
12	.3971	7.5361	34.6339	2.5182	18.9771	0.1327	0.0527	4.5957	12
13	.3677	7.9038	39.0463	2.7196	21.4953	0.1265	0.0465	4.9402	13
14	.3405	8.2442	43.4723	2.9372	24.2149	0.1213	0.0413	5.2731	14
15	.3152	8.5595	47.8857	3.1722	27.1521	0.1168	0.0368	5.5945	15
16	.2919	8.8514	52.2640	3.4259	30.3243	0.1130	0.0330	5.9046	16
17	.2703	9.1216	56.5883	3.7000	33.7502	0.1096	0.0296	6.2037	17
18	.2502	9.3719	60.8426	3.9960	37.4502	0.1067	0.0267	6.4920	18
19	.2317	9.6036	65.0134	4.3157	41.4463	0.1041	0.0241	6.7697	19
20	.2145	9.8181	69.0898	4.6610	45.7620	0.1019	0.0219	7.0369	20
21	.1987	10.0168	73.0629	5.0338	50.4229	0.0998	0.0198	7.2940	21
22	.1839	10.2007	76.9257	5.4365	55.4568	0.0980	0.0180	7.5412	22
23	.1703	10.3711	80.6726	5.8715	60.8933	0.0964	0.0164	7.7786	23
24	.1577	10.5288	84.2997	6.3412	66.7648	0.0950	0.0150	8.0066	24
25	.1460	10.6748	87.8041	6.8485	73.1059	0.0937	0.0137	8.2254	25
26	.1352	10.8100	91.1842	7.3964	79.9544	0.0925	0.0125	8.4352	26
27	.1252	10.9352	94.4390	7.9881	87.3508	0.0914	0.0114	8.6363	27
28	.1159	11.0511	97.5687	8.6271	95.3388	0.0905	0.0105	8.8289	28
29	.1073	11.1584	100.5738	9.3173	103.9659	0.0896	0.0096	9.0133	29
30	.0994	11.2578	103.4558	10.0627	113.2832	0.0888	0.0088	9.1897	30
31	.0920	11.3498	106.2163	10.8677	123.3459	0.0881	0.0081	9.3584	31
32	.0852	11.4350	108.8575	11.7371	134.2135	0.0875	0.0075	9.5197	32
33	.0789	11.5139	111.3819	12.6760	145.9506	0.0869	0.0069	9.6737	33
34	.0730	11.5869	113.7924	13.6901	158.6267	0.0863	0.0063	9.8208	34
35	.0676	11.6546	116.0920	14.7853	172.3168	0.0858	0.0058	9.9611	35
36	.0626	11.7172	118.2839	15.9682	187.1021	0.0853	0.0053	10.0949	36
37	.0580	11.7752	120.3713	17.2456	203.0703	0.0849	0.0049	10.2225	37
38	.0537	11.8289	122.3579	18.6253	220.3159	0.0845	0.0045	10.3440	38
39	.0497	11.8786	124.2470	20.1153	238.9412	0.0842	0.0042	10.4597	39
40	.0460	11.9246	126.0422	21.7245	259.0565	0.0839	0.0039	10.5699	40
41	.0426	11.9672	127.7470	23.4625	280.7810	0.0836	0.0036	10.6747	41
42	.0395	12.0067	129.3651	25.3395	304.2435	0.0833	0.0033	10.7744	42
43	.0365	12.0432	130.8998	27.3666	329.5830	0.0830	0.0030	10.8692	43
44	.0338	12.0771	132.3547	29.5560	356.9496	0.0828	0.0028	10.9592	44
45	.0313	12.1084	133.7331	31.9204	386.5056	0.0826	0.0026	11.0447	45
46	.0290	12.1374	135.0384	34.4741	418.4261	0.0824	0.0024	11.1258	46
47	.0269	12.1643	136.2739	37.2320	452.9002	0.0822	0.0022	11.2028	47
48	.0249	12.1891	137.4428	40.2106	490.1322	0.0820	0.0020	11.2758	48
49	.0230	12.2122	138.5480	43.4274	530.3427	0.0819	0.0019	11.3451	49
50	.0213	12.2335	139.5928	46.9016	573.7702	0.0817	0.0017	11.4107	50
51	.0197	12.2532	140.5799	50.6537	620.6718	0.0816	0.0016	11.4729	51
52	.0183	12.2715	141.5121	54.7060	671.3255	0.0815	0.0015	11.5318	52
53	.0169	12.2884	142.3923	59.0825	726.0316	0.0814	0.0014	11.5875	53
54	.0157	12.3041	143.2229	63.8091	785.1141	0.0813	0.0013	11.6403	54
55	.0145	12.3186	144.0065	68.9139	848.9232	0.0812	0.0012	11.6902	55
60	.0099	12.3766	147.3000	101.2571	1253.2133	0.0808	0.0008	11.9015	60
65	.0067	12.4160	149.7387	148.7798	1847.2481	0.0805	0.0005	12.0602	65
70	.0046	12.4428	151.5326	218.6064	2720.0801	0.0804	0.0004	12.1783	70
75	.0031	12.4611	152.8448	321.2045	4002.5566	0.0802	0.0002	12.2658	75
80	.0021	12.4735	153.8001	471.9548	5886.9354	0.0802	0.0002	12.3301	80
85	.0014	12.4820	154.4925	693.4565	8655.7061	0.0801	0.0001	12.3772	85
90	.0010	12.4877	154.9925	1018.9151	12723.9386	0.0801	0.0001	12.4116	90
95	.0007	12.4917	155.3524	1497.1205	18701.5069	0.0801	0.0001	12.4365	95
100	.0005	12.4943	155.6107	2199.7613	27484.5157	0.0800	0.0000	12.4545	100

I = 9.00 %

N	(P/F)	(P/A)	(P/G)	(F/P)	(F/A)	(A/P)	(A/F)	(A/G)	N
1	.9174	0.9174	-0.0000	1.0900	1.0000	1.0900	1.0000	-0.0000	1
2	.8417	1.7591	0.8417	1.1881	2.0900	0.5685	0.4785	0.4785	2
3	.7722	2.5313	2.3860	1.2950	3.2781	0.3951	0.3051	0.9426	3
4	.7084	3.2397	4.5113	1.4116	4.5731	0.3087	0.2187	1.3925	4
5	.6499	3.8897	7.1110	1.5386	5.9847	0.2571	0.1671	1.8282	5
6	.5963	4.4859	10.0924	1.6771	7.5233	0.2229	0.1329	2.2498	6
7	.5470	5.0330	13.3746	1.8280	9.2004	0.1987	0.1087	2.6574	7
8	.5019	5.5348	16.8877	1.9926	11.0285	0.1807	0.0907	3.0512	8
9	.4604	5.9952	20.5711	2.1719	13.0210	0.1668	0.0768	3.4312	9
10	.4224	6.4177	24.3728	2.3674	15.1929	0.1558	0.0658	3.7978	10
11	.3875	6.8052	28.2481	2.5804	17.5603	0.1469	0.0569	4.1510	11
12	.3555	7.1607	32.1590	2.8127	20.1407	0.1397	0.0497	4.4910	12
13	.3262	7.4869	36.0731	3.0658	22.9534	0.1336	0.0436	4.8182	13
14	.2992	7.7862	39.9633	3.3417	26.0192	0.1284	0.0384	5.1326	14
15	.2745	8.0607	43.8069	3.6425	29.3609	0.1241	0.0341	5.4346	15
16	.2519	8.3126	47.5849	3.9703	33.0034	0.1203	0.0303	5.7245	16
17	.2311	8.5436	51.2821	4.3276	36.9737	0.1170	0.0270	6.0024	17
18	.2120	8.7556	54.8860	4.7171	41.3013	0.1142	0.0242	6.2687	18
19	.1945	8.9501	58.3868	5.1417	46.0185	0.1117	0.0217	6.5236	19
20	.1784	9.1285	61.7770	5.6044	51.1601	0.1095	0.0195	6.7674	20
21	.1637	9.2922	65.0509	6.1088	56.7645	0.1076	0.0176	7.0006	21
22	.1502	9.4424	68.2048	6.6586	62.8733	0.1059	0.0159	7.2232	22
23	.1378	9.5802	71.2359	7.2579	69.5319	0.1044	0.0144	7.4357	23
24	.1264	9.7066	74.1433	7.9111	76.7898	0.1030	0.0130	7.6384	24
25	.1160	9.8226	76.9265	8.6231	84.7009	0.1018	0.0118	7.8316	25
26	.1064	9.9290	79.5863	9.3992	93.3240	0.1007	0.0107	8.0156	26
27	.0976	10.0266	82.1241	10.2451	102.7231	0.0997	0.0097	8.1906	27
28	.0895	10.1161	84.5419	11.1671	112.9682	0.0989	0.0089	8.3571	28
29	.0822	10.1983	86.8422	12.1722	124.1354	0.0981	0.0081	8.5154	29
30	.0754	10.2737	89.0280	13.2677	136.3075	0.0973	0.0073	8.6657	30
31	.0691	10.3428	91.1024	14.4618	149.5752	0.0967	0.0067	8.8083	31
32	.0634	10.4062	93.0690	15.7633	164.0370	0.0961	0.0061	8.9436	32
33	.0582	10.4644	94.9314	17.1820	179.8003	0.0956	0.0056	9.0718	33
34	.0534	10.5178	96.6935	18.7284	196.9823	0.0951	0.0051	9.1933	34
35	.0490	10.5668	98.3590	20.4140	215.7108	0.0946	0.0046	9.3083	35
36	.0449	10.6118	99.9319	22.2512	236.1247	0.0942	0.0042	9.4171	36
37	.0412	10.6530	101.4162	24.2538	258.3759	0.0939	0.0039	9.5200	37
38	.0378	10.6908	102.8158	26.4367	282.6298	0.0935	0.0035	9.6172	38
39	.0347	10.7255	104.1345	28.8160	309.0665	0.0932	0.0032	9.7090	39
40	.0318	10.7574	105.3762	31.4094	337.8824	0.0930	0.0030	9.7957	40
41	.0292	10.7866	106.5445	34.2363	369.2919	0.0927	0.0027	9.8775	41
42	.0268	10.8134	107.6432	37.3175	403.5281	0.0925	0.0025	9.9546	42
43	.0246	10.8380	108.6758	40.6761	440.8457	0.0923	0.0023	10.0273	43
44	.0226	10.8605	109.6456	44.3370	481.5218	0.0921	0.0021	10.0958	44
45	.0207	10.8812	110.5561	48.3273	525.8587	0.0919	0.0019	10.1603	45
46	.0190	10.9002	111.4103	52.6767	574.1860	0.0917	0.0017	10.2210	46
47	.0174	10.9176	112.2115	57.4176	626.8628	0.0916	0.0016	10.2780	47
48	.0160	10.9336	112.9625	62.5852	684.2804	0.0915	0.0015	10.3317	48
49	.0147	10.9482	113.6661	68.2179	746.8656	0.0913	0.0013	10.3821	49
50	.0134	10.9617	114.3251	74.3575	815.0836	0.0912	0.0012	10.4295	50
51	.0123	10.9740	114.9420	81.0497	889.4411	0.0911	0.0011	10.4740	51
52	.0113	10.9853	115.5193	88.3442	970.4908	0.0910	0.0010	10.5158	52
53	.0104	10.9957	116.0593	96.2951	1058.8349	0.0909	0.0009	10.5549	53
54	.0095	11.0053	116.5642	104.9617	1155.1301	0.0909	0.0009	10.5917	54
55	.0087	11.0140	117.0362	114.4083	1260.0918	0.0908	0.0008	10.6261	55
60	.0057	11.0480	118.9683	176.0313	1944.7921	0.0905	0.0005	10.7683	60
65	.0037	11.0701	120.3344	270.8460	2998.2885	0.0903	0.0003	10.8702	65
70	.0024	11.0844	121.2942	416.7301	4619.2232	0.0902	0.0002	10.9427	70
75	.0016	11.0938	121.9646	641.1909	7113.2321	0.0901	0.0001	10.9940	75
80	.0010	11.0998	122.4306	986.5517	10950.5741	0.0901	0.0001	11.0299	80
85	.0007	11.1038	122.7533	1517.9320	16854.8003	0.0901	0.0001	11.0551	85
90	.0004	11.1064	122.9758	2335.5266	25939.1842	0.0900	0.0000	11.0726	90
95	.0003	11.1080	123.1287	3593.4971	39916.6350	0.0900	0.0000	11.0847	95
100	.0002	11.1091	123.2335	5529.0408	61422.6755	0.0900	0.0000	11.0930	100

I = 10.00 %

N	(P/F)	(P/A)	(P/G)	(F/P)	(F/A)	(A/P)	(A/F)	(A/G)	N
1	.9091	0.9091	−0.0000	1.1000	1.0000	1.1000	1.0000	−0.0000	1
2	.8264	1.7355	0.8264	1.2100	2.1000	0.5762	0.4762	0.4762	2
3	.7513	2.4869	2.3291	1.3310	3.3100	0.4021	0.3021	0.9366	3
4	.6830	3.1699	4.3781	1.4641	4.6410	0.3155	0.2155	1.3812	4
5	.6209	3.7908	6.8618	1.6105	6.1051	0.2638	0.1638	1.8101	5
6	.5645	4.3553	9.6842	1.7716	7.7156	0.2296	0.1296	2.2236	6
7	.5132	4.8684	12.7631	1.9487	9.4872	0.2054	0.1054	2.6216	7
8	.4665	5.3349	16.0287	2.1436	11.4359	0.1874	0.0874	3.0045	8
9	.4241	5.7590	19.4215	2.3579	13.5795	0.1736	0.0736	3.3724	9
10	.3855	6.1446	22.8913	2.5937	15.9374	0.1627	0.0627	3.7255	10
11	.3505	6.4951	26.3963	2.8531	18.5312	0.1540	0.0540	4.0641	11
12	.3186	6.8137	29.9012	3.1384	21.3843	0.1468	0.0468	4.3884	12
13	.2897	7.1034	33.3772	3.4523	24.5227	0.1408	0.0408	4.6988	13
14	.2633	7.3667	36.8005	3.7975	27.9750	0.1357	0.0357	4.9955	14
15	.2394	7.6061	40.1520	4.1772	31.7725	0.1315	0.0315	5.2789	15
16	.2176	7.8237	43.4164	4.5950	35.9497	0.1278	0.0278	5.5493	16
17	.1978	8.0216	46.5819	5.0545	40.5447	0.1247	0.0247	5.8071	17
18	.1799	8.2014	49.6395	5.5599	45.5992	0.1219	0.0219	6.0526	18
19	.1635	8.3649	52.5827	6.1159	51.1591	0.1195	0.0195	6.2861	19
20	.1486	8.5136	55.4069	6.7275	57.2750	0.1175	0.0175	6.5081	20
21	.1351	8.6487	58.1095	7.4002	64.0025	0.1156	0.0156	6.7189	21
22	.1228	8.7715	60.6893	8.1403	71.4027	0.1140	0.0140	6.9189	22
23	.1117	8.8832	63.1462	8.9543	79.5430	0.1126	0.0126	7.1085	23
24	.1015	8.9847	65.4813	9.8497	88.4973	0.1113	0.0113	7.2881	24
25	.0923	9.0770	67.6964	10.8347	98.3471	0.1102	0.0102	7.4580	25
26	.0839	9.1609	69.7940	11.9182	109.1818	0.1092	0.0092	7.6186	26
27	.0763	9.2372	71.7773	13.1100	121.0999	0.1083	0.0083	7.7704	27
28	.0693	9.3066	73.6495	14.4210	134.2099	0.1075	0.0075	7.9137	28
29	.0630	9.3696	75.4146	15.8631	148.6309	0.1067	0.0067	8.0489	29
30	.0573	9.4269	77.0766	17.4494	164.4940	0.1061	0.0061	8.1762	30
31	.0521	9.4790	78.6395	19.1943	181.9434	0.1055	0.0055	8.2962	31
32	.0474	9.5264	80.1078	21.1138	201.1378	0.1050	0.0050	8.4091	32
33	.0431	9.5694	81.4856	23.2252	222.2515	0.1045	0.0045	8.5152	33
34	.0391	9.6086	82.7773	25.5477	245.4767	0.1041	0.0041	8.6149	34
35	.0356	9.6442	83.9872	28.1024	271.0244	0.1037	0.0037	8.7086	35
36	.0323	9.6765	85.1194	30.9127	299.1268	0.1033	0.0033	8.7965	36
37	.0294	9.7059	86.1781	34.0039	330.0395	0.1030	0.0030	8.8789	37
38	.0267	9.7327	87.1673	37.4043	364.0434	0.1027	0.0027	8.9562	38
39	.0243	9.7570	88.0908	41.1448	401.4478	0.1025	0.0025	9.0285	39
40	.0221	9.7791	88.9525	45.2593	442.5926	0.1023	0.0023	9.0962	40
41	.0201	9.7991	89.7560	49.7852	487.8518	0.1020	0.0020	9.1596	41
42	.0183	9.8174	90.5047	54.7637	537.6370	0.1019	0.0019	9.2188	42
43	.0166	9.8340	91.2019	60.2401	592.4007	0.1017	0.0017	9.2741	43
44	.0151	9.8491	91.8508	66.2641	652.6408	0.1015	0.0015	9.3258	44
45	.0137	9.8628	92.4544	72.8905	718.9048	0.1014	0.0014	9.3740	45
46	.0125	9.8753	93.0157	80.1795	791.7953	0.1013	0.0013	9.4190	46
47	.0113	9.8866	93.5372	88.1975	871.9749	0.1011	0.0011	9.4610	47
48	.0103	9.8969	94.0217	97.0172	960.1723	0.1010	0.0010	9.5001	48
49	.0094	9.9063	94.4715	106.7190	1057.1896	0.1009	0.0009	9.5365	49
50	.0085	9.9148	94.8889	117.3909	1163.9085	0.1009	0.0009	9.5704	50
51	.0077	9.9226	95.2761	129.1299	1281.2994	0.1008	0.0008	9.6020	51
52	.0070	9.9296	95.6351	142.0429	1410.4293	0.1007	0.0007	9.6313	52
53	.0064	9.9360	95.9679	156.2472	1552.4723	0.1006	0.0006	9.6586	53
54	.0058	9.9418	96.2763	171.8719	1708.7195	0.1006	0.0006	9.6840	54
55	.0053	9.9471	96.5619	189.0591	1880.5914	0.1005	0.0005	9.7075	55
60	.0033	9.9672	97.7010	304.4816	3034.8164	0.1003	0.0003	9.8023	60
65	.0020	9.9796	98.4705	490.3707	4893.7073	0.1002	0.0002	9.8672	65
70	.0013	9.9873	98.9870	789.7470	7887.4696	0.1001	0.0001	9.9113	70
75	.0008	9.9921	99.3317	1271.8954	12708.9537	0.1001	0.0001	9.9410	75
80	.0005	9.9951	99.5606	2048.4002	20474.0021	0.1000	0.0000	9.9609	80
85	.0003	9.9970	99.7120	3298.9690	32979.6903	0.1000	0.0000	9.9742	85
90	.0002	9.9981	99.8118	5313.0226	53120.2261	0.1000	0.0000	9.9831	90
95	.0001	9.9988	99.8773	8556.6760	85556.7605	0.1000	0.0000	9.9889	95
100	.0001	9.9993	99.9202	13780.6123	137796.1234	0.1000	0.0000	9.9927	100

I = 12.00 %

N	(P/F)	(P/A)	(P/G)	(F/P)	(F/A)	(A/P)	(A/F)	(A/G)	N
1	.8929	0.8929	-0.0000	1.1200	1.0000	1.1200	1.0000	-0.0000	1
2	.7972	1.6901	0.7972	1.2544	2.1200	0.5917	0.4717	0.4717	2
3	.7118	2.4018	2.2208	1.4049	3.3744	0.4163	0.2963	0.9246	3
4	.6355	3.0373	4.1273	1.5735	4.7793	0.3292	0.2092	1.3589	4
5	.5674	3.6048	6.3970	1.7623	6.3528	0.2774	0.1574	1.7746	5
6	.5066	4.1114	8.9302	1.9738	8.1152	0.2432	0.1232	2.1720	6
7	.4523	4.5638	11.6443	2.2107	10.0890	0.2191	0.0991	2.5515	7
8	.4039	4.9676	14.4714	2.4760	12.2997	0.2013	0.0813	2.9131	8
9	.3606	5.3282	17.3563	2.7731	14.7757	0.1877	0.0677	3.2574	9
10	.3220	5.6502	20.2541	3.1058	17.5487	0.1770	0.0570	3.5847	10
11	.2875	5.9377	23.1288	3.4785	20.6546	0.1684	0.0484	3.8953	11
12	.2567	6.1944	25.9523	3.8960	24.1331	0.1614	0.0414	4.1897	12
13	.2292	6.4235	28.7024	4.3635	28.0291	0.1557	0.0357	4.4683	13
14	.2046	6.6282	31.3624	4.8871	32.3926	0.1509	0.0309	4.7317	14
15	.1827	6.8109	33.9202	5.4736	37.2797	0.1468	0.0268	4.9803	15
16	.1631	6.9740	36.3670	6.1304	42.7533	0.1434	0.0234	5.2147	16
17	.1456	7.1196	38.6973	6.8660	48.8837	0.1405	0.0205	5.4353	17
18	.1300	7.2497	40.9080	7.6900	55.7497	0.1379	0.0179	5.6427	18
19	.1161	7.3658	42.9979	8.6128	63.4397	0.1358	0.0158	5.8375	19
20	.1037	7.4694	44.9676	9.6463	72.0524	0.1339	0.0139	6.0202	20
21	.0926	7.5620	46.8188	10.8038	81.6987	0.1322	0.0122	6.1913	21
22	.0826	7.6446	48.5543	12.1003	92.5026	0.1308	0.0108	6.3514	22
23	.0738	7.7184	50.1776	13.5523	104.6029	0.1296	0.0096	6.5010	23
24	.0659	7.7843	51.6929	15.1786	118.1552	0.1285	0.0085	6.6406	24
25	.0588	7.8431	53.1046	17.0001	133.3339	0.1275	0.0075	6.7708	25
26	.0525	7.8957	54.4177	19.0401	150.3339	0.1267	0.0067	6.8921	26
27	.0469	7.9426	55.6369	21.3249	169.3740	0.1259	0.0059	7.0049	27
28	.0419	7.9844	56.7674	23.8839	190.6989	0.1252	0.0052	7.1098	28
29	.0374	8.0218	57.8141	26.7499	214.5828	0.1247	0.0047	7.2071	29
30	.0334	8.0552	58.7821	29.9599	241.3327	0.1241	0.0041	7.2974	30
31	.0298	8.0850	59.6761	33.5551	271.2926	0.1237	0.0037	7.3811	31
32	.0266	8.1116	60.5010	37.5817	304.8477	0.1233	0.0033	7.4586	32
33	.0238	8.1354	61.2612	42.0915	342.4294	0.1229	0.0029	7.5302	33
34	.0212	8.1566	61.9612	47.1425	384.5210	0.1226	0.0026	7.5965	34
35	.0189	8.1755	62.6052	52.7996	431.6635	0.1223	0.0023	7.6577	35
36	.0169	8.1924	63.1970	59.1356	484.4631	0.1221	0.0021	7.7141	36
37	.0151	8.2075	63.7406	66.2318	543.5987	0.1218	0.0018	7.7661	37
38	.0135	8.2210	64.2394	74.1797	609.8305	0.1216	0.0016	7.8141	38
39	.0120	8.2330	64.6967	83.0812	684.0102	0.1215	0.0015	7.8582	39
40	.0107	8.2438	65.1159	93.0510	767.0914	0.1213	0.0013	7.8988	40
41	.0096	8.2534	65.4997	104.2171	860.1424	0.1212	0.0012	7.9361	41
42	.0086	8.2619	65.8509	116.7231	964.3595	0.1210	0.0010	7.9704	42
43	.0076	8.2696	66.1722	130.7299	1081.0826	0.1209	0.0009	8.0019	43
44	.0068	8.2764	66.4659	146.4175	1211.8125	0.1208	0.0008	8.0308	44
45	.0061	8.2825	66.7342	163.9876	1358.2300	0.1207	0.0007	8.0572	45
46	.0054	8.2880	66.9792	183.6661	1522.2176	0.1207	0.0007	8.0815	46
47	.0049	8.2928	67.2028	205.7061	1705.8838	0.1206	0.0006	8.1037	47
48	.0043	8.2972	67.4068	230.3908	1911.5898	0.1205	0.0005	8.1241	48
49	.0039	8.3010	67.5929	258.0377	2141.9806	0.1205	0.0005	8.1427	49
50	.0035	8.3045	67.7624	289.0022	2400.0182	0.1204	0.0004	8.1597	50
51	.0031	8.3076	67.9169	323.6825	2689.0204	0.1204	0.0004	8.1753	51
52	.0028	8.3103	68.0576	362.5243	3012.7029	0.1203	0.0003	8.1895	52
53	.0025	8.3128	68.1856	406.0273	3375.2272	0.1203	0.0003	8.2025	53
54	.0022	8.3150	68.3022	454.7505	3781.2545	0.1203	0.0003	8.2143	54
55	.0020	8.3170	68.4082	509.3206	4236.0050	0.1202	0.0002	8.2251	55
60	.0011	8.3240	68.8100	897.5969	7471.6411	0.1201	0.0001	8.2664	60
65	.0006	8.3281	69.0581	1581.8725	13173.9374	0.1201	0.0001	8.2922	65
70	.0004	8.3303	69.2103	2787.7998	23223.3319	0.1200	0.0000	8.3082	70
75	.0002	8.3316	69.3031	4913.0558	40933.7987	0.1200	0.0000	8.3181	75
80	.0001	8.3324	69.3594	8658.4831	72145.6925	0.1200	0.0000	8.3241	80
85	.0001	8.3328	69.3935	15259.2057	127151.7140	0.1200	0.0000	8.3278	85
90	.0000	8.3330	69.4140	26891.9342	224091.1185	0.1200	0.0000	8.3300	90
95	.0000	8.3332	69.4263	47392.7766	394931.4719	0.1200	0.0000	8.3313	95
100	.0000	8.3332	69.4336	83522.2657	696010.5477	0.1200	0.0000	8.3321	100

I = 15.00 %

N	(P/F)	(P/A)	(P/G)	(F/P)	(F/A)	(A/P)	(A/F)	(A/G)	N
1	.8696	0.8696	-0.0000	1.1500	1.0000	1.1500	1.0000	-0.0000	1
2	.7561	1.6257	0.7561	1.3225	2.1500	0.6151	0.4651	0.4651	2
3	.6575	2.2832	2.0712	1.5209	3.4725	0.4380	0.2880	0.9071	3
4	.5718	2.8550	3.7864	1.7490	4.9934	0.3503	0.2003	1.3263	4
5	.4972	3.3522	5.7751	2.0114	6.7424	0.2983	0.1483	1.7228	5
6	.4323	3.7845	7.9368	2.3131	8.7537	0.2642	0.1142	2.0972	6
7	.3759	4.1604	10.1924	2.6600	11.0668	0.2404	0.0904	2.4498	7
8	.3269	4.4873	12.4807	3.0590	13.7268	0.2229	0.0729	2.7813	8
9	.2843	4.7716	14.7548	3.5179	16.7858	0.2096	0.0596	3.0922	9
10	.2472	5.0188	16.9795	4.0456	20.3037	0.1993	0.0493	3.3832	10
11	.2149	5.2337	19.1289	4.6524	24.3493	0.1911	0.0411	3.6549	11
12	.1869	5.4206	21.1849	5.3503	29.0017	0.1845	0.0345	3.9082	12
13	.1625	5.5831	23.1352	6.1528	34.3519	0.1791	0.0291	4.1438	13
14	.1413	5.7245	24.9725	7.0757	40.5047	0.1747	0.0247	4.3624	14
15	.1229	5.8474	26.6930	8.1371	47.5804	0.1710	0.0210	4.5650	15
16	.1069	5.9542	28.2960	9.3576	55.7175	0.1679	0.0179	4.7522	16
17	.0929	6.0472	29.7828	10.7613	65.0751	0.1654	0.0154	4.9251	17
18	.0808	6.1280	31.1565	12.3755	75.8364	0.1632	0.0132	5.0843	18
19	.0703	6.1982	32.4213	14.2318	88.2118	0.1613	0.0113	5.2307	19
20	.0611	6.2593	33.5822	16.3665	102.4436	0.1598	0.0098	5.3651	20
21	.0531	6.3125	34.6448	18.8215	118.8101	0.1584	0.0084	5.4883	21
22	.0462	6.3587	35.6150	21.6447	137.6316	0.1573	0.0073	5.6010	22
23	.0402	6.3988	36.4988	24.8915	159.2764	0.1563	0.0063	5.7040	23
24	.0349	6.4338	37.3023	28.6252	184.1678	0.1554	0.0054	5.7979	24
25	.0304	6.4641	38.0314	32.9190	212.7930	0.1547	0.0047	5.8834	25
26	.0264	6.4906	38.6918	37.8568	245.7120	0.1541	0.0041	5.9612	26
27	.0230	6.5135	39.2890	43.5353	283.5688	0.1535	0.0035	6.0319	27
28	.0200	6.5335	39.8283	50.0656	327.1041	0.1531	0.0031	6.0960	28
29	.0174	6.5509	40.3146	57.5755	377.1697	0.1527	0.0027	6.1541	29
30	.0151	6.5660	40.7526	66.2118	434.7451	0.1523	0.0023	6.2066	30
31	.0131	6.5791	41.1466	76.1435	500.9569	0.1520	0.0020	6.2541	31
32	.0114	6.5905	41.5006	87.5651	577.1005	0.1517	0.0017	6.2970	32
33	.0099	6.6005	41.8184	100.6998	664.6655	0.1515	0.0015	6.3357	33
34	.0086	6.6091	42.1033	115.8048	765.3654	0.1513	0.0013	6.3705	34
35	.0075	6.6166	42.3586	133.1755	881.1702	0.1511	0.0011	6.4019	35
36	.0065	6.6231	42.5872	153.1519	1014.3457	0.1510	0.0010	6.4301	36
37	.0057	6.6288	42.7916	176.1246	1167.4975	0.1509	0.0009	6.4554	37
38	.0049	6.6338	42.9743	202.5433	1343.6222	0.1507	0.0007	6.4781	38
39	.0043	6.6380	43.1374	232.9248	1546.1655	0.1506	0.0006	6.4985	39
40	.0037	6.6418	43.2830	267.8635	1779.0903	0.1506	0.0006	6.5168	40
41	.0032	6.6450	43.4128	308.0431	2046.9539	0.1505	0.0005	6.5331	41
42	.0028	6.6478	43.5286	354.2495	2354.9969	0.1504	0.0004	6.5478	42
43	.0025	6.6503	43.6317	407.3870	2709.2465	0.1504	0.0004	6.5609	43
44	.0021	6.6524	43.7235	468.4950	3116.6334	0.1503	0.0003	6.5725	44
45	.0019	6.6543	43.8051	538.7693	3585.1285	0.1503	0.0003	6.5830	45
46	.0016	6.6559	43.8778	619.5847	4123.8977	0.1502	0.0002	6.5923	46
47	.0014	6.6573	43.9423	712.5224	4743.4824	0.1502	0.0002	6.6006	47
48	.0012	6.6585	43.9997	819.4007	5456.0047	0.1502	0.0002	6.6080	48
49	.0011	6.6596	44.0506	942.3108	6275.4055	0.1502	0.0002	6.6146	49
50	.0009	6.6605	44.0958	1083.6574	7217.7163	0.1501	0.0001	6.6205	50
51	.0008	6.6613	44.1360	1246.2061	8301.3737	0.1501	0.0001	6.6257	51
52	.0007	6.6620	44.1715	1433.1370	9547.5798	0.1501	0.0001	6.6304	52
53	.0006	6.6626	44.2031	1648.1075	10980.7167	0.1501	0.0001	6.6345	53
54	.0005	6.6631	44.2311	1895.3236	12628.8243	0.1501	0.0001	6.6382	54
55	.0005	6.6636	44.2558	2179.6222	14524.1479	0.1501	0.0001	6.6414	55
60	.0002	6.6651	44.3431	4383.9987	29219.9916	0.1500	0.0000	6.6530	60
65	.0001	6.6659	44.3903	8817.7874	58778.5826	0.1500	0.0000	6.6593	65
70	.0001	6.6663	44.4156	17735.7200	118231.4669	0.1500	0.0000	6.6627	70
75	.0000	6.6665	44.4292	35672.8680	237812.4532	0.1500	0.0000	6.6646	75
80	.0000	6.6666	44.4364	71750.8794	478332.5293	0.1500	0.0000	6.6656	80
85	.0000	6.6666	44.4402	144316.6470	962104.3133	0.1500	0.0000	6.6661	85
90	.0000	6.6666	44.4422	290272.3252	1935142.1680	0.1500	0.0000	6.6664	90
95	.0000	6.6667	44.4433	583841.3276	3892268.8509	0.1500	0.0000	6.6665	95
100	.0000	6.6667	44.4438	1174313.4507	7828749.6713	0.1500	0.0000	6.6666	100

I = 20.00 %

N	(P/F)	(P/A)	(P/G)	(F/P)	(F/A)	(A/P)	(A/F)	(A/G)	N
1	.8333	0.8333	-0.0000	1.2000	1.0000	1.2000	1.0000	-0.0000	1
2	.6944	1.5278	0.6944	1.4400	2.2000	0.6545	0.4545	0.4545	2
3	.5787	2.1065	1.8519	1.7280	3.6400	0.4747	0.2747	0.8791	3
4	.4823	2.5887	3.2986	2.0736	5.3680	0.3863	0.1863	1.2742	4
5	.4019	2.9906	4.9061	2.4883	7.4416	0.3344	0.1344	1.6405	5
6	.3349	3.3255	6.5806	2.9860	9.9299	0.3007	0.1007	1.9788	6
7	.2791	3.6046	8.2551	3.5832	12.9159	0.2774	0.0774	2.2902	7
8	.2326	3.8372	9.8831	4.2998	16.4991	0.2606	0.0606	2.5756	8
9	.1938	4.0310	11.4335	5.1598	20.7989	0.2481	0.0481	2.8364	9
10	.1615	4.1925	12.8871	6.1917	25.9587	0.2385	0.0385	3.0739	10
11	.1346	4.3271	14.2330	7.4301	32.1504	0.2311	0.0311	3.2893	11
12	.1122	4.4392	15.4667	8.9161	39.5805	0.2253	0.0253	3.4841	12
13	.0935	4.5327	16.5883	10.6993	48.4966	0.2206	0.0206	3.6597	13
14	.0779	4.6106	17.6008	12.8392	59.1959	0.2169	0.0169	3.8175	14
15	.0649	4.6755	18.5095	15.4070	72.0351	0.2139	0.0139	3.9588	15
16	.0541	4.7296	19.3208	18.4884	87.4421	0.2114	0.0114	4.0851	16
17	.0451	4.7746	20.0419	22.1861	105.9306	0.2094	0.0094	4.1976	17
18	.0376	4.8122	20.6805	26.6233	128.1167	0.2078	0.0078	4.2975	18
19	.0313	4.8435	21.2439	31.9480	154.7400	0.2065	0.0065	4.3861	19
20	.0261	4.8696	21.7395	38.3376	186.6880	0.2054	0.0054	4.4643	20
21	.0217	4.8913	22.1742	46.0051	225.0256	0.2044	0.0044	4.5334	21
22	.0181	4.9094	22.5546	55.2061	271.0307	0.2037	0.0037	4.5941	22
23	.0151	4.9245	22.8867	66.2474	326.2369	0.2031	0.0031	4.6475	23
24	.0126	4.9371	23.1760	79.4968	392.4842	0.2025	0.0025	4.6943	24
25	.0105	4.9476	23.4276	95.3962	471.9811	0.2021	0.0021	4.7352	25
26	.0087	4.9563	23.6460	114.4755	567.3773	0.2018	0.0018	4.7709	26
27	.0073	4.9636	23.8353	137.3706	681.8528	0.2015	0.0015	4.8020	27
28	.0061	4.9697	23.9991	164.8447	819.2233	0.2012	0.0012	4.8291	28
29	.0051	4.9747	24.1406	197.8136	984.0680	0.2010	0.0010	4.8527	29
30	.0042	4.9789	24.2628	237.3763	1181.8816	0.2008	0.0008	4.8731	30
31	.0035	4.9824	24.3681	284.8516	1419.2579	0.2007	0.0007	4.8908	31
32	.0029	4.9854	24.4588	341.8219	1704.1095	0.2006	0.0006	4.9061	32
33	.0024	4.9878	24.5368	410.1863	2045.9314	0.2005	0.0005	4.9194	33
34	.0020	4.9898	24.6038	492.2235	2456.1176	0.2004	0.0004	4.9308	34
35	.0017	4.9915	24.6614	590.6682	2948.3411	0.2003	0.0003	4.9406	35
36	.0014	4.9929	24.7108	708.8019	3539.0094	0.2003	0.0003	4.9491	36
37	.0012	4.9941	24.7531	850.5622	4247.8112	0.2002	0.0002	4.9564	37
38	.0010	4.9951	24.7894	1020.6747	5098.3735	0.2002	0.0002	4.9627	38
39	.0008	4.9959	24.8204	1224.8096	6119.0482	0.2002	0.0002	4.9681	39
40	.0007	4.9966	24.8469	1469.7716	7343.8578	0.2001	0.0001	4.9728	40
41	.0006	4.9972	24.8696	1763.7259	8813.6294	0.2001	0.0001	4.9767	41
42	.0005	4.9976	24.8890	2116.4711	10577.3553	0.2001	0.0001	4.9801	42
43	.0004	4.9980	24.9055	2539.7653	12693.8263	0.2001	0.0001	4.9831	43
44	.0003	4.9984	24.9196	3047.7183	15233.5916	0.2001	0.0001	4.9856	44
45	.0003	4.9986	24.9316	3657.2620	18281.3099	0.2001	0.0001	4.9877	45
46	.0002	4.9989	24.9419	4388.7144	21938.5719	0.2000	0.0000	4.9895	46
47	.0002	4.9991	24.9506	5266.4573	26327.2863	0.2000	0.0000	4.9911	47
48	.0002	4.9992	24.9581	6319.7487	31593.7436	0.2000	0.0000	4.9924	48
49	.0001	4.9993	24.9644	7583.6985	37913.4923	0.2000	0.0000	4.9935	49
50	.0001	4.9995	24.9698	9100.4382	45497.1908	0.2000	0.0000	4.9945	50
51	.0001	4.9995	24.9744	10920.5258	54597.6289	0.2000	0.0000	4.9953	51
52	.0001	4.9996	24.9783	13104.6309	65518.1547	0.2000	0.0000	4.9960	52
53	.0001	4.9997	24.9816	15725.5571	78622.7856	0.2000	0.0000	4.9966	53
54	.0001	4.9997	24.9844	18870.6685	94348.3427	0.2000	0.0000	4.9971	54
55	.0000	4.9998	24.9868	22644.8023	113219.0113	0.2000	0.0000	4.9976	55
60	.0000	4.9999	24.9942	56347.5144	281732.5718	0.2000	0.0000	4.9989	60
65	.0000	5.0000	24.9975	140210.6469	701048.2346	0.2000	0.0000	4.9995	65
70	.0000	5.0000	24.9989	348888.9569	1744439.7847	0.2000	0.0000	4.9998	70
75	.0000	5.0000	24.9995	868147.3693	4340731.8466	0.2000	0.0000	4.9999	75

I = 25.00 %

N	(P/F)	(P/A)	(P/G)	(F/P)	(F/A)	(A/P)	(A/F)	(A/G)	N
1	.8000	0.8000	0.0	1.2500	1.0000	1.2500	1.0000	0.0	1
2	.6400	1.4400	0.6400	1.5625	2.2500	0.6944	0.4444	0.4444	2
3	.5120	1.9520	1.6640	1.9531	3.8125	0.5123	0.2623	0.8525	3
4	.4096	2.3616	2.8928	2.4414	5.7656	0.4234	0.1734	1.2249	4
5	.3277	2.6893	4.2035	3.0518	8.2070	0.3718	0.1218	1.5631	5
6	.2621	2.9514	5.5142	3.8147	11.2588	0.3388	0.0888	1.8683	6
7	.2097	3.1611	6.7725	4.7684	15.0735	0.3163	0.0663	2.1424	7
8	.1678	3.3289	7.9469	5.9605	19.8419	0.3004	0.0504	2.3872	8
9	.1342	3.4631	9.0207	7.4506	25.8023	0.2888	0.0388	2.6048	9
10	.1074	3.5705	9.9870	9.3132	33.2529	0.2801	0.0301	2.7971	10
11	.0859	3.6564	10.8460	11.6415	42.5661	0.2735	0.0235	2.9663	11
12	.0687	3.7251	11.6020	14.5519	54.2077	0.2684	0.0184	3.1145	12
13	.0550	3.7801	12.2617	18.1899	68.7596	0.2645	0.0145	3.2437	13
14	.0440	3.8241	12.8334	22.7374	86.9495	0.2615	0.0115	3.3559	14
15	.0352	3.8593	13.3260	28.4217	109.6868	0.2591	0.0091	3.4530	15
16	.0281	3.8874	13.7482	35.5271	138.1085	0.2572	0.0072	3.5366	16
17	.0225	3.9099	14.1085	44.4089	173.6357	0.2558	0.0058	3.6084	17
18	.0180	3.9279	14.4147	55.5112	218.0446	0.2546	0.0046	3.6698	18
19	.0144	3.9424	14.6741	69.3889	273.5558	0.2537	0.0037	3.7222	19
20	.0115	3.9539	14.8932	86.7362	342.9447	0.2529	0.0029	3.7667	20
21	.0092	3.9631	15.0777	108.4202	429.6809	0.2523	0.0023	3.8045	21
22	.0074	3.9705	15.2326	135.5253	538.1011	0.2519	0.0019	3.8365	22
23	.0059	3.9764	15.3625	169.4066	673.6264	0.2515	0.0015	3.8634	23
24	.0047	3.9811	15.4711	211.7582	843.0329	0.2512	0.0012	3.8861	24
25	.0038	3.9849	15.5618	264.6978	1054.7912	0.2509	0.0009	3.9052	25
26	.0030	3.9879	15.6373	330.8722	1319.4890	0.2508	0.0008	3.9212	26
27	.0024	3.9903	15.7002	413.5903	1650.3612	0.2506	0.0006	3.9346	27
28	.0019	3.9923	15.7524	516.9879	2063.9515	0.2505	0.0005	3.9457	28
29	.0015	3.9938	15.7957	646.2349	2580.9394	0.2504	0.0004	3.9551	29
30	.0012	3.9950	15.8316	807.7936	3227.1743	0.2503	0.0003	3.9628	30
31	.0010	3.9960	15.8614	1009.7420	4034.9678	0.2502	0.0002	3.9693	31
32	.0008	3.9968	15.8859	1262.1774	5044.7098	0.2502	0.0002	3.9746	32
33	.0006	3.9975	15.9062	1577.7218	6306.8872	0.2502	0.0002	3.9791	33
34	.0005	3.9980	15.9229	1972.1523	7884.6091	0.2501	0.0001	3.9828	34
35	.0004	3.9984	15.9367	2465.1903	9856.7613	0.2501	0.0001	3.9858	35
36	.0003	3.9987	15.9481	3081.4879	12321.9516	0.2501	0.0001	3.9883	36
37	.0003	3.9990	15.9574	3851.8599	15403.4396	0.2501	0.0001	3.9904	37
38	.0002	3.9992	15.9651	4814.8249	19255.2994	0.2501	0.0001	3.9921	38
39	.0002	3.9993	15.9714	6018.5311	24070.1243	0.2500	0.0000	3.9935	39
40	.0001	3.9995	15.9766	7523.1638	30088.6554	0.2500	0.0000	3.9947	40
41	.0001	3.9996	15.9809	9403.9548	37611.8192	0.2500	0.0000	3.9956	41
42	.0001	3.9997	15.9843	11754.9435	47015.7740	0.2500	0.0000	3.9964	42
43	.0001	3.9997	15.9872	14693.6794	58770.7175	0.2500	0.0000	3.9971	43
44	.0001	3.9998	15.9895	18367.0992	73464.3969	0.2500	0.0000	3.9976	44
45	.0000	3.9998	15.9915	22958.8740	91831.4962	0.2500	0.0000	3.9980	45
46	.0000	3.9999	15.9930	28698.5925	114790.3702	0.2500	0.0000	3.9984	46
47	.0000	3.9999	15.9943	35873.2407	143488.9627	0.2500	0.0000	3.9987	47
48	.0000	3.9999	15.9954	44841.5509	179362.2034	0.2500	0.0000	3.9989	48
49	.0000	3.9999	15.9962	56051.9386	224203.7543	0.2500	0.0000	3.9991	49
50	.0000	3.9999	15.9969	70064.9232	280255.6929	0.2500	0.0000	3.9993	50
51	.0000	4.0000	15.9975	87581.1540	350320.6161	0.2500	0.0000	3.9994	51
52	.0000	4.0000	15.9980	109476.4425	437901.7701	0.2500	0.0000	3.9995	52
53	.0000	4.0000	15.9983	136845.5532	547378.2126	0.2500	0.0000	3.9996	53
54	.0000	4.0000	15.9986	171056.9414	684223.7658	0.2500	0.0000	3.9997	54
55	.0000	4.0000	15.9989	213821.1768	855280.7072	0.2500	0.0000	3.9997	55
60	.0000	4.0000	15.9996	652530.4468	2610117.7872	0.2500	0.0000	3.9999	60

Reserved for Future Use

Reserved for Future Use

14 ELECTRICAL ENGINEERING

Sample Examination

EXAMINATION IN THE PRINCIPLES

AND

PRACTICE OF ENGINEERING

PART 1

Morning Section

THIS IS AN OPEN BOOK EXAMINATION

You may use textbooks, handbooks, bound reference materials, any battery operated silent calculator, or slide rule. However, no writing tablets, unbound tables, or unbound notes are permitted in the examination room. Sufficient paper for scratchwork is provided in each solution pamphlet. You are not permitted to exchange reference materials or aids during the examination.

You will have four hours in which to work this test. Your score will be determined by the number of problems through four (4) that you solve correctly. Each correct solution counts ten points. The maximum possible score for this part of the examination is 40 points. Partial credit for partially correct solutions will be given.

Work only four of the problems according to your proctor's instructions. Do not submit solutions or partial solutions for more than four problems. Only the first four solutions listed will be scored.

State and justify all assumptions.

After checking your solutions and making sure that you have followed the instructions printed above, turn in all examination materials to your proctor.

You are advised to use your time effectively.

1. Given the power system shown, find the output and power factor of generator G_1.

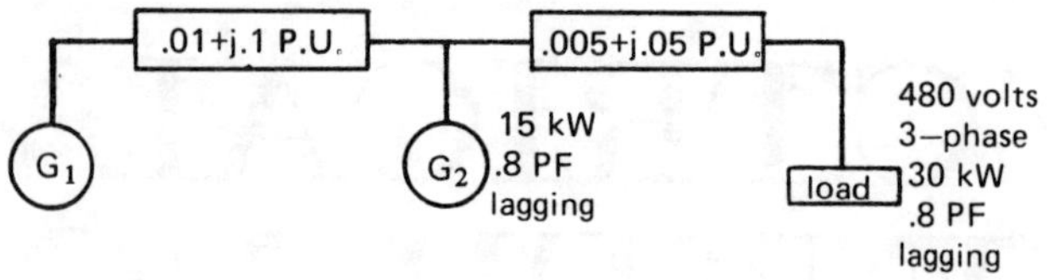

2. Use an 8:1 multiplexer with permissible inputs of A, $\bar{A}$, 0, and 1 to design the function

$$f = \bar{A}BD + AC + \bar{A}B\bar{C}D + \overline{ABCD}$$

3. The tunnel diode in the circuit shown has the characteristics given. (a) Find the period of astable waveform $v_d(t)$. (b) Sketch the voltage waveform $v_d(t)$. (c) Sketch the current waveform $i_d(t)$.

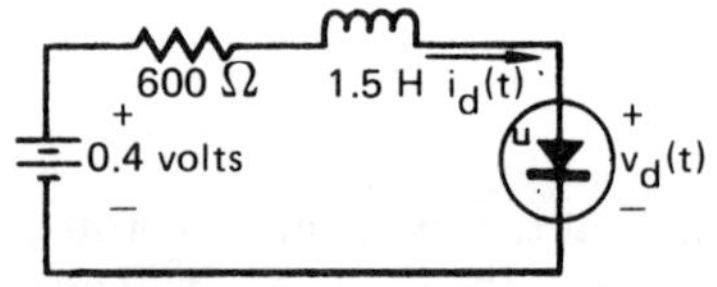

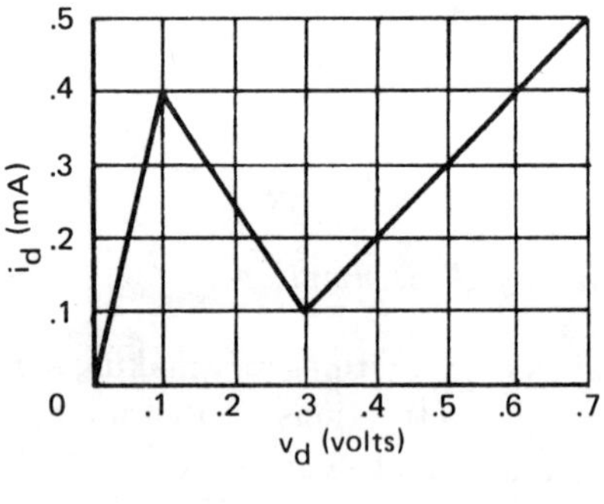

4. A 3-winding power transformer has the following ratings: primary – 115 KV, 40 KVA; secondary – 13.8 KV, 20 KVA; tertiary – 13.8 KV, 20 KVA. The following impedance measurements are taken:

Z_{primary} | secondary shorted, tertiary open $= (5 + j50)\,K\Omega$

$Z_{\text{secondary}}$ | primary shorted, tertiary open $= 80 + j720\,\Omega$

Z_{tertiary} | primary shorted, secondary open $= 80 + j720\,\Omega$

Determine the three winding impedances.

5. In the system shown, $f(t)$ is band limited between ω_1 and ω_2. (ω_{c1} and ω_{c2} are both much greater than ω_2.) Describe the spectral content of $f_1(t)$ and $f_2(t)$ for (a) $\omega_{c2} = \omega_{c1}$ and (b) $\omega_{c2} = K\omega_{c1}$ where $.99 \leq K \leq 1.01$.

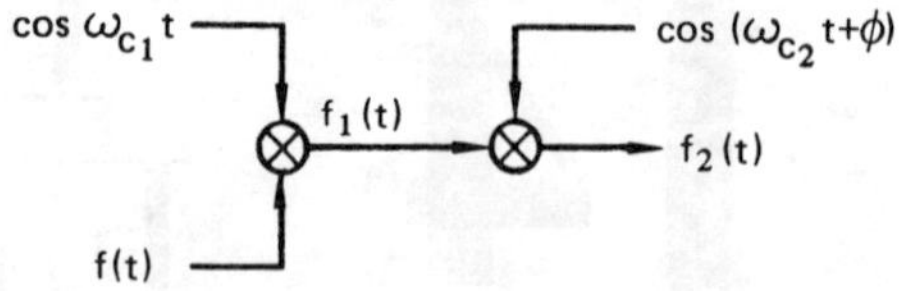

6. In the following circuit, the meter is a d'Arsonval movement with the scale calibrated to indicate RMS voltage values of pure sinusoids. Assume ideal diodes are used. (a) Calculate the RMS value of $V_1 + V_2$. (b) Sketch the waveform of the voltage form X to Y. (c) Calculate the voltage which will be indicated on the meter.

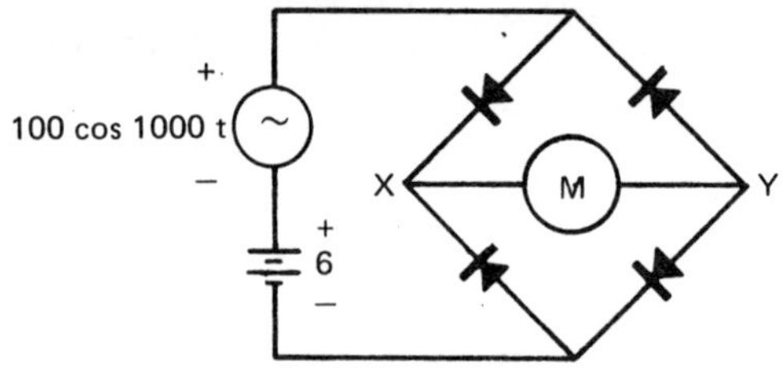

7. A MOSFET amplifier has $g_m = 5 \times 10^{-3}$ mhos and $r_d = 20\,K\Omega$. Neglect capacitor effects and determine (a) the voltage gain and (b) the input impedance.

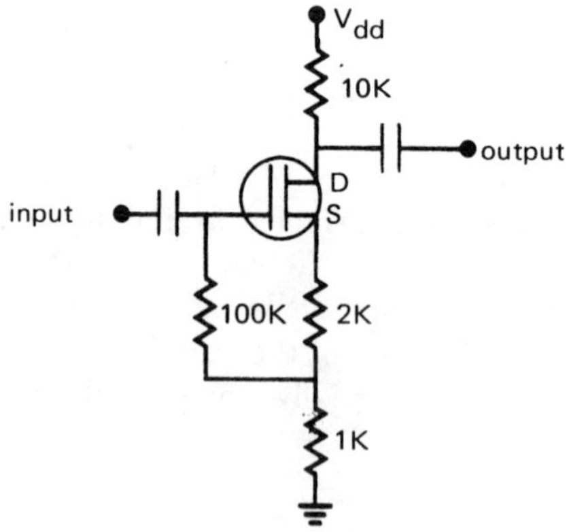

8. For the control circuit shown, (a) plot the root-locus as a function of K, (b) discuss the stability, and (c) discuss the response C for a unit step R.

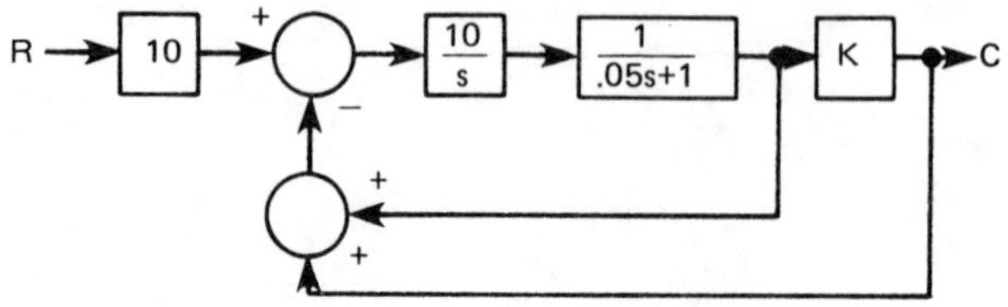

9. A 13.8 KV generator supplies two loads: a balanced 30 kW wye-connected load with 0.8 power factor lagging, and a delta-connected load with $Z_1 = 10K + j10K$ ohms, $Z_2 = 5K + j15K$ ohms, and $Z_3 = j20K$ ohms. Determine (a) the line currents and (b) the generator power output.

10. For the counter shown, draw the complete state diagram and write the transition table.

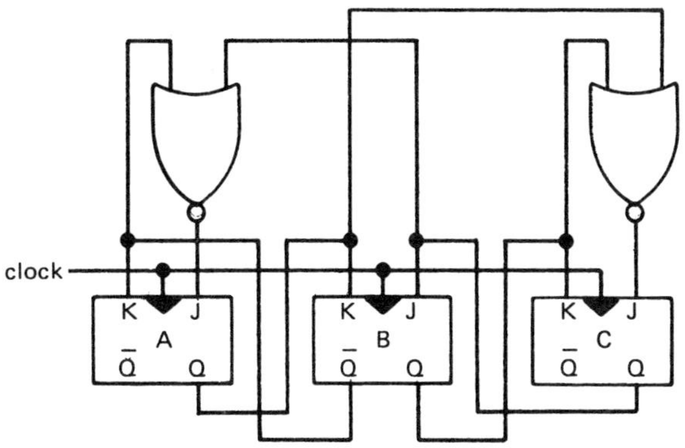

EXAMINATION IN THE PRINCIPLES

AND

PRACTICE OF ENGINEERING

PART 2

Afternoon Section

THIS IS AN OPEN BOOK EXAMINATION

You may use textbooks, handbooks, bound reference materials, any battery operated silent calculator, or slide rule. However, no writing tablets, unbound tables, or unbound notes are permitted in the examination room. Sufficient paper for scratchwork is provided in each solution pamphlet. You are not permitted to exchange reference materials or aids during the examination.

You will have four hours in which to work this test. Your score will be determined by the number of problems through four (4) that you solve correctly. Each correct solution counts ten points. The maximum possible score for this part of the examination is 40 points. Partial credit for partially correct solutions will be given.

Work only four of the problems according to your proctor's instructions. Do not submit solutions or partial solutions for more than four problems. Only the first four solutions listed will be scored.

State and justify all assumptions.

After checking your solutions and making sure that you have followed the instructions printed above, turn in all examination materials to your proctor.

You are advised to use your time effectively.

11. The operational amplifier in the circuit shown can be considered ideal.

$$R = 5K(1 + \delta) \text{ where } -1 \le \delta \le +1$$

(a) Plot e_o versus δ. (b) Specify the type of resistance element for R and the limits of change so that the e_o versus δ plot is linear.

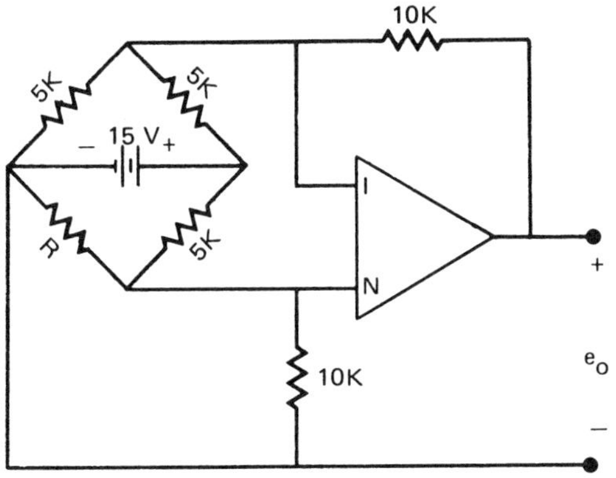

12. A coaxial cable has $z_o = 50$ ohms. Determine the length (in wavelengths) of cable necessary to produce (a) a reactance of −25 ohms, and (b) a reactance of +50 ohms. The stubs may be either open or shorted.

13. A shunt excited D.C. motor is rated 1150 rpm, 10 hp, 80 amps, 125 volts, and 40° temperature rise. The motor drives a load for which the torque is to be constant regardless of speed. Draw and identify the circuit elements for (a) a motor speed of 1300 rpm and (b) a motor speed of 800 rpm.

14. In the circuit shown, the bilateral breakdown diode (DIAC) triggers at ±4 volts. Neglecting holding voltages, determine the conduction angle of the TRIAC.

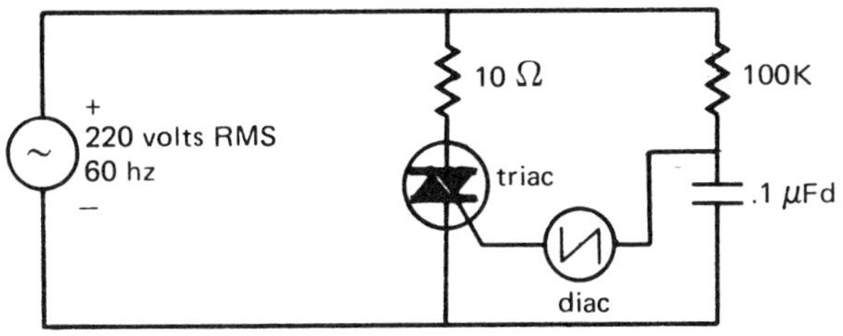

15. The device shown in the circuit has the following V-I characteristics:

v_d	0	±.5	±1.0	±1.5	volts
i_d	0	.0054	.033	.079	amps

Determine the spectral content of v_d.

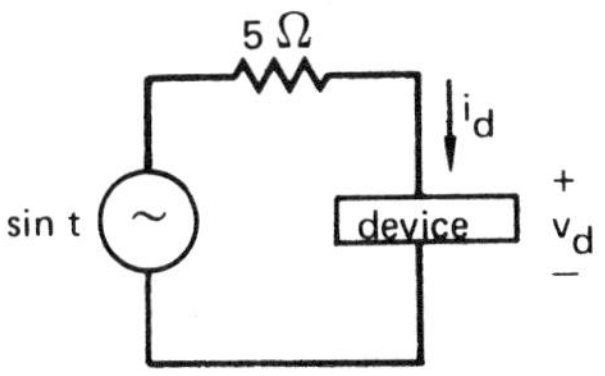

16. Given the digital circuit shown, complete the state diagram showing transitions. Assume an initial state of 000. Show your method of analysis.

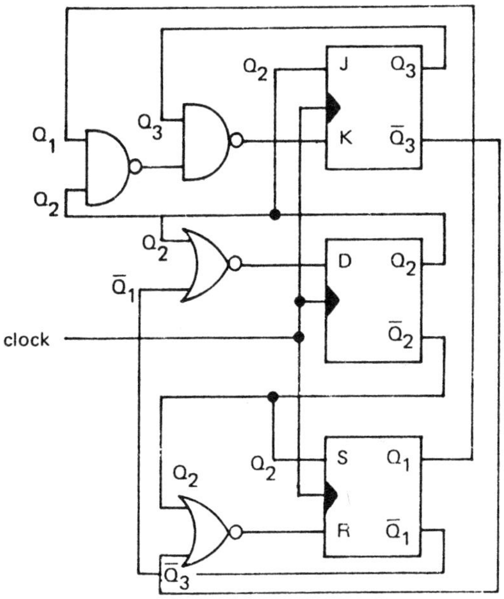

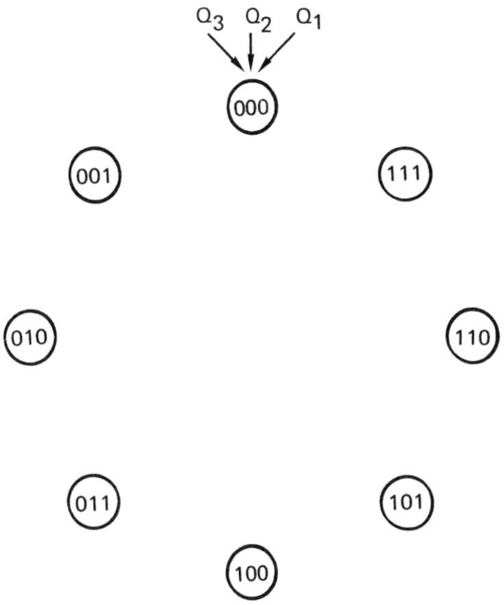

17. The transistor in the circuit shown is germanium with $\beta = 35$. Determine the values for R_1 and R_2. With those values, determine the voltage gain, the input impedance, and the output impedance at mid-frequency.

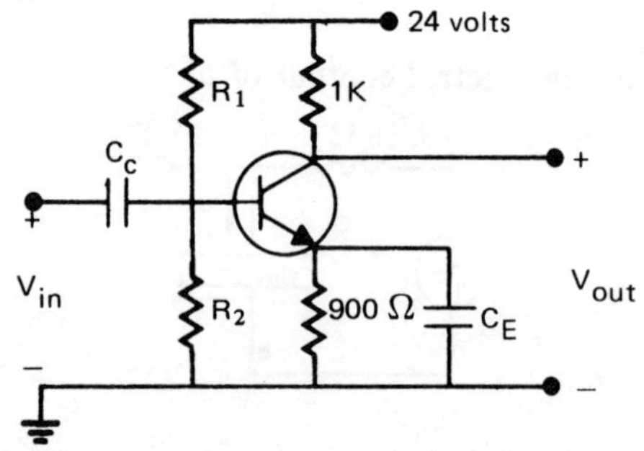

18. Design a Zener diode regulator to supply a variable 15 to 130 ohm load with a constant 15 volts D.C. The supply is 20 volts ± 5% with 3-ohms internal resistance. Sketch the complete circuit and specify all components.

19. A three-phase fault occurs at the output of a delta-wye, 115 KV – 13.8 KV, 30/40/50 MVA transformer with $z = .01 + j.08$ **P.U.** on the 50 MVA base. The feeder to the transformer has $R = .1$ and $X = .1$ **P.U.** with a 240 amp rating. What is the fault current?

20. A state-variable simulation of a control system is shown. From experimental tests on the closed-loop system, it has been shown that the damping ratio ξ is about 0.75, and the natural frequency ω is about 10 radians/sec. (a) Determine $C(s)/R(s)$ as a ratio of two factored polynomials and give the poles and zeros of the closed-loop system. (b) Determine the plant open loop transfer function **G**(s) as a ratio of two polynomials. (c) To reduce the damping ratio, which block would be changed? (Limit your choice to the blocks labeled −75, −1000, and −6000.) Assume that the real poles and zeros, as well as the natural frequency, are held constant. Indicate the approximate range of gain change permitted.

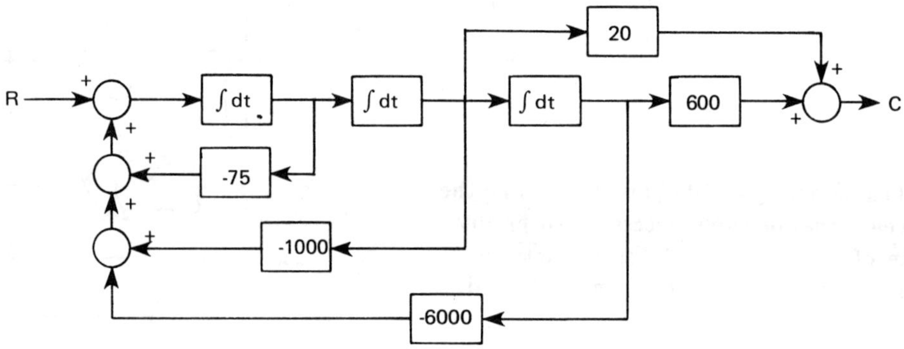

Figure for problem 20.

RESERVED FOR FUTURE USE

RESERVED FOR FUTURE USE

INDEX

ABCD parameters 5-7, 6-8, 6-9, 6-12
absolute stability 11-9
A.C. steady state impedances 2-19
Accelerated Cost Recovery System 13-8
accelerating relays 7-9
acceleration error 11-11
accuracy 3-8
ACRS 13-8
active circuits 2-7
active filter 8-4,8-6
addition of complex numbers 1-2
adjacent poles 11-14
admittance matrix 2-15
air-gap mechanical power 7-19
air-gap torque 7-19
alley 12-16
alternate ACRS calculation 13-8
ambient temperature 12-10
American Wire Gage 12-22, 6-20
amortization 13-7
ampacity 12-21
amplifier noise 8-30
analog switches 9-17
analytic geometry 1-5
AND gate 10-1
angle of convergence 11-15
angle of emergence 11-15
angular momentum 11-3
angular velocity 11-2
annual amount 13-2
Annual Cost Method 13-5
Annual Return Method 13-5
apparent power 3-14
armature 7-17, 7-18
armature reaction 7-2
assymptotic Bode diagram 11-8
astable multivibrator 9-26
asymmetrical faults 5-20
asynchronously 10-23
attenuation constant 6-11
auto reg ballast 12-3
auto-transformer 5-15, 7-16
auto-transformer controller 7-16
avalanche breakdown 8-8
avalanche current 2-7
avalanche diode 8-31
avalanche effect 8-8
avalanche voltage 2-7
average illumination design 12-16
average power 3-14
average value 1-6, 3-1, 3-6
balanced delta-
delta connection 5-4
balanced wye-delta connection 5-4, 5-5
balanced wye-wye connection 5-3
ballast 12-3
ballast performance 12-10
balloon payment 13-18
band-pass filter 8-6
bandwidth 4-23, 8-31
bandwidth limiting of noise 8-30
base 8-17
base reflectance 12-8
base-spreading resistance 8-21
branch circuits 12-21
beam lumens 12-20
Benefit-Cost Ratio Method 13-5
bi lateral circuit 2-7
bias load-line 8-14
binary counter 10-25
bipolar 8-17
bipolar junction transistor 8-17, 8-36
bipolar junction transistor biasing 8-18
bipolar junction transistor stability 8-19
bisection method 4-23
BJT 8-17, 8-36
block diagram algebra 11-4
blocked rotor test 7-14
blocking oscillator 9-28
Bode diagram 11-8
branch circuit 12-21
branch circuit design 12-23

break-even anaylsis 13-15
break-even point 13-16
breakaway point 11-15
breakover voltage 9-20
brightness ratios 12-6
bubble 10-5
bulk base resistance 8-21
burden 13-14
burden budget variance 13-15
burden capacity variance 13-15
burden variance 13-15
bypass capacitors 8-27
cables and raceways 12-22
candela 12-1
candlepower 12-2
candlepower distribution 12-18
canonical forms 10-4
canonical POS realization 10-5
canonical SOP realization 10-4
capacitance 2-2, 2-19, 4-16
capacitance of a pair of parallel solid wires 6-4
capacitive compensation 6-13
capacitive first-order circuit 4-6
capital gain 13-11
capital loss 13-11
Capital Recovery Method 13-5
capitalized cost 13-4
cascaded amplifiers 8-22, 8-28
cascading of networks 2-15
cascode amplifier 8-22, 8-28
cash flow diagrams 13-1
cavity ratios 12-7
ceiling cavity 12-7
ceiling cavity ratio 12-8
chain parameters 2-15, 5-7
challenger 13-6
channel-substrate capacitance 8-17
characteristic impedance 6-11
circuit breakers 5-10
circuit reduction techniques 4-1
circular measure 2-2
circular mils 2-2
cleaning frequency 12-16
clipper 9-3
clipper-limiter circuit 9-3
clocked flip-flops 10-22
CML 9-15
CMOS inverter 9-16
CMOS logic gates 9-16
CMOS transmission switch 9-17
coefficient of beam utilization 12-20
coefficient of utilization 12-3, 12-8
coincidence gate 10-2
cold junction 3-10
collector 8-17, 12-16
color 12-6
combined form 3-6
common base amplifier 8-23
common base characteristics 8-17
common base hybrid parameters 8-23
common base y-parameters 2-20
common collector amplifier 8-22
common drain FET amplifiers 8-24
common emitter amplifier 8-22
common emitter characteristics 8-17
common mode 8-24
common mode gain 8-22
common mode rejection ratio 8-24
common source FET amplifiers 8-23
commutator brushes 7-2
compensating reactance 6-14
compensating winding 7-5
compensator 11-2
compensator transfer function 11-2
complement 10-4
complex conjugate 1-2
complex numbers 1-1, 3-12
complex plane 1-1, 3-12
complex poles 4-22
compound amount factor 13-2
compound D.C. motors 7-10
compound generators 7-5
compound transistors 8-22
compounding 7-3
conductance 2-2
conductive part 3-15
conductor types 12-22
conductors for general wiring 12-21
conjugate 11-13
conjugate pair 11-13
constant function 3-2
constant value dollars 13-16
constants of three-phase synchronous machines 5-19
consumer loans 13-17
continuous compounding 13-12
control systems 11-1
controlled sources 2-7, 2-16
controlled variable 11-1, 11-2
controller 7-9
copper 2-10
copper conductors 12-22
copper-wire table 6-20
corner frequency 8-3, 8-27
corrected current 3-16
cost 13-5
cost accounting 13-14
counter analysis 10-26
counters 10-23
coupling capacitors 8-27
coupling coefficient 2-4

critical damping 11-9
critically damped 1-7, 4-12, 4-14
cross-talk 8-30
cumulative compounding 7-10
current direction 5-2
current feedback biasing 8-16
current gain 2-9
current limiting resistors 7-9
current mode logic 9-15
current regulator diode 8-39
current source 4-3
current source rules 4-2
cutoff 9-8
cylindrical rotor 7-17
D flip flop 10-22
D-latch 10-22
D.C. ammeters 3-9
D.C. component 5-17
D.C. generators 7-2
D.C. machines 7-2
D.C. motors 7-7
D.C. series motors 7-11
D.C. shunt motors 7-7
D.C. steady state impedance analysis 2-19
D.C. voltmeter 3-8
D-MOSFET 8-37
dark current 2-8
Darlington circuit 8-22
d'Arsonval movement 3-8
data (D) flip-flop 10-22
data selector trees 10-20
data selectors 10-17
DDB depreciation 13-10
decade 11-8
decibel 4-21
decimal (base 10) system 1-1
declining balance depreciation 13-7
decoders 10-20
defender 13-6
delay time 9-11
delta configuration 5-3
delta-wye 4-4
delta-wye transformation 6-8
De Morgan's theorem 10-4
demultiplexers 10-20
dependent node 2-15, 2-16
dependent sources 2-5, 2-7
depletion 13-7
depletion allowance 13-7
depletion mode pinchoff 8-10
depletion mode turn-off 8-10
depletion mode MOSFET 8-11, 8-37
depletion mode metal-oxide-
 semiconductor field effect transistors 8-11, 8-37
depletion region 8-7, 8-8
depletion region capacitance 8-9
depreciation 13-7
depreciation recovery 13-7
determinants 1-12
DIAC trigger 9-20
dielectric constant 2-2, 6-21
differential amplifier 8-24
differential compounding 7-10
differential cost 13-15
differential equations 1-6
differential gain 8-22
differential operation 8-26
differentiation 1-5
diffusion 12-6
diffusion capacitance 8-9
digital meters 3-11
diode characteristics 2-9
diode circuits 9-1
diode-transistor logic 9-13
direct material costs 13-14
direct reduction loans 13-18
disbenefits 13-5
discount factors for discrete compounding 13-3
distribution transformer 5-11
distribution transformer impedances 5-11
division of complex numbers 1-2
don't care conditions 10-12
don't care inputs 10-11
doping level 8-8
double-base diode 8-10, 9-19
Double Declining Balance 13-7
drain 8-10
drain current 8-10
drain-source resistance 8-17
DTL 9-13
dynamic on-resistance 9-20
E-MOSFET 8-38
ECL 9-15
economic indicator 13-16
economic life 13-14
economic order quantity 12-17
Economic Recovery Act of 1981 13-8
edge triggered 10-22
effective coefficient of utilization 12-16
effective interest rate 13-11
effective reflectances 12-8
effective tachometer feedback coefficient 11-2
effective value 3-1, 3-4
effective value phasor notation 3-12
electrical degrees 7-18
electrodynamometer movement 3-9
electronic analog meters 3-11
emitter 8-17
emitter-coupled logic 9-15
emitter follower 8-22

enable input 10-20
encirclements 10-10
energy stored in a capacitor 2-2, 4-6
energy stored in an inductor 2-3, 4-7
enhancement mode 8-12
enhancement mode metal-oxide-semiconductor field effect transistors 8-3, 8-14
enhancement mode MOSFET 8-3, 8-14
environment 2-16
EOQ 13-17
equivalence 13-1
equivalent current source 4-2
equivalent uniform annual cost 13-5
equivalent voltage source 4-2
error 11-1, 11-10
EUAC 13-5
Euler's equation 1-2
even parity function 10-13
exclusive NOR (XNOR) 10-2
exclusive NOR logic 10-13
exclusive OR (XOR)10-2
exclusive OR logic 10-13
expected value 13-12
explicit probability problems 13-13
exponentiation 1-3
external conductance 6-2
external inductance 6-3
fall time 9-11
fan-out 9-9
far-field situation 12-2
fault calculations 5-17
feedback theory 11-1
feeder circuits 12-21
FET amplifiers 8-23
FET biasing 8-13
field effect transistors 8-9
15-year real property 13-9
filter 8-5
filter effective impedance 8-28
final value theorem 1-11, 11-10
first-order differential equation 1-6
first-order solutions 4-7
first-order switching transients 4-7
first-order systems 4-17
first-order transient analysis 4-6
5-year personal property 13-8
fixed part 3-12
Fixed Percentage Method 13-7
fixture 12-3
flat speed-torque 7-11
flicker 5-14
flip-flops 10-20
floodlighting 12-15, 12-20
floor cavity 12-7
flux-density 12-2
flux per pole 7-12
footcandles 12-2
footlamberts 12-2
forced air-cooled transformers 5-16
forced oil-cooled transformers 5-16
forced response 4-7
forward-bias condition 8-8
forward block 11-1
forward transfer function 11-2
forward voltage drop 2-6, 9-1
four-wire system 5-1
Fourier analysis 1-8
Fourier series 1-8, 3-1, 3-6
freeway 12-16
frequency domain 4-20, 11-15
frequency of the rotor current 7-13
frequency response analysis 4-20
frequency spectra 3-6
future worth 13-2
g-parameters 2-14
gain 2-8, 13-11
gain-bandwidth product 8-4, 8-29
gain margin 11-9
gain on the sale of a depreciated asset 13-11
gap length 7-18
gate 8-10
gate-channel capacitance 8-17
gate delays 9-11
gate turn-off 9-18
gate turn-off silicon controlled rectifier 8-40
general form 1-5
generated voltage 7-2
generation level 5-10
generators, D.C. 7-2
geometric mean 6-7
geometric mean radius 6-4
germanium diodes 9-1
glare 12-6
grid form 5-11
ground impedance 5-21
grounding electrode conductor 12-21
GTO 8-40, 9-18
h-parameters 2-14
half-power frequencies 4-22
half-power points 4-22
half-wave rectified sinusoid 3-5
half-wave sinusoids 3-3
half-year rule 13-8
Hall effect 2-10
Hall voltage 2-10
hang-up states 10-27
harmonics 8-30
heating value 3-1
high frequency cutoff 8-28
high frequency effects on FET 8-28

high frequency limitations 8-28
high frequency transmission lines 6-12
high-pass filter 8-5, 8-6
high power factor 12-3
hold time 10-23
Hurwitz stability criterion 11-11
hot junction 3-10
hybrid equivalent circuit 2-8
hybrid parameters 2-8, 8-22
Hybrid pi model 8-28
hyperbolic functions 6-12
hysteretic behavior 8-3
ideal diode 8-9
ideal P-N diode equation 8-8
ideal transformer 2-4
ideal Zener voltage regulator 9-5
illumination 12-2
illumination levels 12-16
imaginary part 1-1, 3-12
immediate write-off 13-11
impedance 3-14
impedance analysis 2-18
impedance forms 2-19
impedance locus circle 6-13
impedance per unit length 6-2
implicit probability problems 13-13
in-phase component 3-16
incandescent 12-2
income tax 13-6, 13-11
incremental analysis 13-5
incremental resistance 8-8
incremental transconductance 8-14
independent sources 2-5
indirect manufacturing expenses 13-14
indirect material and labor costs 13-14
inductance 2-3, 2-19, 4-17
induction motor starting 5-14
induction motor tests 7-14
induction motors 7-12
inductive first-order circuit 4-6
infinite bus 5-11
inflation 13-16
initial conditions 4-14
initial illumination level 12-6
initial value theorem 1-11
input impedance 6-13, 8-22
input port 2-13
input terminals 2-13
integrals 1-6
integrator 8-2, 8-5
intercept form 1-5
interference noise 8-30
internal inductance 6-2
interpoles 7-2
inversion layer 8-15
inverting operational amplifier 8-3
inverting terminal 8-29
iterative method of circuit reduction 4-3
J-K flip-flops 10-22
JFET 8-9, 8-37
JFET characteristics 8-11
job cost accounting 13-14
Johnson noise 8-30
junction capacitances 8-9
junction FET 8-9, 8-37
junction field effect transistor 8-9, 8-37
junction transistors 2-8
Karnaugh map 10-6
keep-alive current 9-6
Kirchhoff's current law 4-2
Kirchhoff's voltage law 4-2
Kramer's rule 1-13
labor variance 13-15
lag 3-13
lag ballast 12-3
lagging 3-16
lamp lumen depreciation factor 12-3
lamp outage 12-10
lamps 12-2
Laplace transform analysis 1-9
Laplace transform pairs (table) 1-10
Laplace transforms 4-16
large signal current gain 8-18
lateral conductors 12-21
lateral light distribution 12-18
lead-lag (or lag-lead) operational amplifier filter 8-5
leading 3-13, 3-16
leakage reactance 5-16
learning curves 13-16
least-significant bit 10-7
light 12-1
light-current relays 7-9
light distribution 12-18
lighting levels 12-16
limiter 9-3
limiting resistors 7-9
line-to-line conductance 6-2
line-to-line fault 5-26
line-to-line voltage 5-3
line-to-neutral voltage 5-3
linear amplifiers 8-2
linear circuit elements 2-1
linear differential equations 1-6
linear simultaneous equations 1-13
linear source models 2-4
linear system 11-3
load studies 5-10
loans 13-17
local 12-16

local maximum 1-5
local minimum 1-5
logarithms 1-3
logic circuits 9-8
logical adjacency 10-6
logical AND 10-1
logical constants 10-2
logical functions 9-11
logical minimization 10-6
logical NAND 10-2
logical NOR 10-2
logical NOT 10-2
logical NOT inverter circuit 9-9
logical variable 10-2
long division 1-4
long shunt 7-10
long shunt connection 7-5
long transmission lines 6-10
loop current analysis 2-17
Lorentz force equation 2-10
low frequency behavior 8-27
low frequency noise 8-30
low-pass filter 8-5
low-pass Sallen and Key circuit 8-6
lumen 12-1, 12-2
lumen method 12-7
lumens per square foot 12-2
luminaire 12-3
luminaire classifications 12-18
luminaire dirt depreciation 12-3
luminaire dirt depreciation factor 12-3
luminance 12-2
luminous flux 12-2
luminous intensity 12-1
magnetic conversion factors 7-26
magnetic core 2-4
magnetization curve 7-2
magnetizing current 5-16
magnetomotive force 2-4
main bonding jumper 12-21
main overcurrent protection 12-21
major roadway and expressway 12-16
marginal cost 13-15
MARR 13-4
master 10-22
master-slave J-K 10-22
matched condition 6-14
material variance 13-15
matrixes 1-12
maxima 1-5
maximum lives 13-13
Maximum Power Transfer Theorem 2-12, 8-22
maximum torque 7-13
maxterm 10-4
maxterm method 10-13
mechanical degrees 7-18
mechanical time-constant 11-3
medium length transmission lines 6-9
mhos 2-8
microphones 2-5
mid-frequency behavior 8-21
Miller effective capacitance 8-28
Miller effective impedance 8-28
mils 6-2
minima 1-5
minimization of functions 10-8
minimum acceptable illumination level 12-6
minimum attractive rate of return 13-4
minimum maintained illumination 12-6
minimum maintained illumination level 12-6
minimum maintained level 12-3
minimum wire size 12-22
minor 1-12
minority carrier conduction 8-8
monostable multivibrator 9-28
MOSFET 8-11
motor tests 7-14
motors, D.C. 7-7
mounting height 12-16,12-18
moving iron instruments 3-10
multiple adjacencies 10-9
multiplexers 10-17
multiplication of complex numbers 1-2
mutual flux 2-4
mutual inductance 2-4
MUX 10-17
N-type 8-7
N-type carrier 8-17
NAND gate 10-2
NAND gate truth table 10-5
NAND-NAND realization 10-5
nanometers 12-1
National Electric Code 12-21
natural response 4-7
negative conductance devices 9-25
negative conductance oscillator 9-25
negative logic 9-13
negative resistance device 9-23
negative resistance oscillator 9-23
negative sequence 5-20
negative sequence impedances 5-20
negative sequence reactance 5-20
neon bulb relaxation oscillator 9-22
neutral node 5-2
NEXOR 10-4
next condition 10-20
next state table 10-20, 10-21
no-load test 7-14
no-load voltage 2-6
node merging 2-17

node voltage equations 2-15
noise 8-30
noise bandwidth 8-31
noise figure 8-32, 8-34
noise immunity 8-3
noise power 8-32
noise power to an input resistance 8-33
noise resistance 8-34
noise source equivalent circuit 8-31
noise temperature 8-34
nominal interest rate 13-11
non-cutoff 12-18
non-inverting operational amplifier 8-3
non-inverting terminal 8-29
non-sinusoidal waveform generation 9-21
NOR gate 10-2
normal power factor 12-3
normalized load impedance 6-13, 6-14
normalized magnetization curve 7-2
Norton circuit 4-1
Norton equivalent current 2-11
Norton's theorem 2-11
NOT gate 10-2
number of luminaires 12-9
number of poles 7-17, 7-18
Nyquist diagram 11-9
octave 11-8
odd parity function 10-13
off-condition 8-10
Ohm's law 2-2
one-line diagrams 5-10
op-amp 8-2
open circuit 2-6
open circuit armature 7-2
open circuit armature voltage 7-2
open circuit test 5-5
open circuit voltage 2-6
open circuit voltage polarity 4-2
open-collector TTL gate 9-14
open delta transformer 5-15
operating point 8-13
operating point slope 7-6
operational amplifier 8-2, 8-29
operational amplifier square-wave generator 9-25
operational amplifier voltage regulator 9-7
opportunity interest cost 13-9
out-of-phase 3-16
output capacitance 8-17
output impedance 8-22
output port 2-13
output terminals 2-13
over-excited case 7-21
overcurrent protection 12-22
overdamped 1-7, 4-10, 4-14
overhead 13-14
overhead variance 13-15
overload contact 7-9
P-N diode 8-35
P-N junction 2-8
P-N junctions 8-7, 8-8
P-type 8-7
P-type carrier 8-17
parabolic input 11-10
parallel voltage sources 4-2
parallel LC circuit 4-15
partial fraction expansion 1-3, 1-9
partial reductions 4-2
partial short circuit tests 5-9
pass transistor 9-7
peak detection 3-10
pedestrian traffic 12-16
pen-recorders 3-8
per-unit length 6-2
per-unit system 5-8
periodic functions 1-8
periodic signal 3-1
permanent magnet movement 3-8
permeability 2-3
permeability of nonmagnetic materials 6-3
permittivity 2-2
permittivity of free space 2-2
personal property 13-7, 13-8
phase 4-21
phase constant 6-11
phase currents 5-4
phase inductance 6-7
phase margin 11-9
phase response 11-9
phase sequence 5-3, 7-12
phase sequence abc 5-2, 5-3
phase sequence acb 5-2, 5-3
phase voltage 5-3
phasor analysis 1-8
phasor form 1-1
phasor power 3-13
phasor representation 2-19
photodiode 2-5, 2-8
photometric brightness 12-2
phototransistor 2-8
pi approximation 6-9
pi-tee 4-4
piecewise linear models 2-6
pinchoff voltage 8-10, 8-11, 9-18
placement of luminaires 12-10
plant 11-2
plant transfer function 11-2
PMOS logic 9-15
PMOS logic gates 9-15
point calculations 12-20
point method 12-11

point of inflection 1-5
polar forms 3-12
poles 4-22, 11-3, 11-13
POS 10-4
position error 11-10
positional number systems 1-1
positioning control system 11-2
positive feedback 8-2
positive logic 9-13
positive sequence 5-20
positive sequence impedances 5-20
power 3-13
power and energy conversions 7-27
power dissipated by a resistor 2-2
power factor 3-13
power factor angle 3-13, 7-20
power factor correction 3-16, 5-13, 7-21
power gain 8-31
power loss per phase 7-13
power sources 2-5
power switching 9-18
power (three-phase) 5-5
power transformers 5-5
power transmission lines 6-4
power triangle 3-13
practical operational amplifiers 8-29
practical Zener diode regulator 9-6
precision diode 9-1, 9-3
prefixes 1-14
present worth 13-2, 13-4
present worth factor 13-2
Present Worth Method 13-4
prime cost 13-14
probabilistic problems 13-12
process cost acccounting 13-14
Product Of Sums formulation 10-4
programmable logic array 10-20
propagation constant 6-10
propagation delay 10-24
propagation delay times 10-23
property 13-7
pulse transients 4-6, 4-9
pure sinusoid 3-2
quadratic equations 1-2
quadrature 5-13
quadrature plane 7-2
quadrature reactances 5-18
R-C integrator 8-5
raceways 12-21
radix point 1-1
ramp function 11-10
ramp input 11-10
rate of return 13-4
Rate of Return Method 13-5
reactance 3-14
reactive part 3-15
reactive power 3-14
read-only memories 10-20
real numbers 1-1
real part 1-1, 3-13
real property 13-8
realization 10-4
recovery factors 13-8, 13-9
rectangular distribution 13-13
rectangular forms 3-12
rectification detection 3-10
rectified sinusoid 3-3
rectifier diodes 3-3
rectifier meter 3-10
redundant impedances 4-5
reference bus 5-21
reference (input) variable 11-1
reference phasor 3-12
reflected component 12-11, 12-12
reflected light 12-2
reflected radiation coefficient 12-12
regular gain 13-11
reflection coefficient 6-11
regular losses 13-11
regulated voltage 9-7
regulation 2-6, 6-8
replacement study 13-6
residential property 13-7
residue functions 10-18
resistance 2-1, 2-19
resistance per unit length 6-2
resistive part 3-15
resitivity 2-2
resistor-transistor logic 9-12
resonant terms 11-9
return on investment 13-4
reverse-bias condition 8-8
reverse breakdown 8-8
reverse current 2-7
reverse saturation current 8-8
reverse voltage 2-7
reverse voltage break down value 9-1
rheostats 11-3
ring sum 10-13
ripple counter 10-23
rise time 9-11
RMS value 3-1, 3-4, 3-6
roadway lighting 12-15, 12-16
roadways 12-16
ROI 13-4
ROM 10-20
room cavity 12-7
room cavity ratio 12-3, 12-7
room position grid 12-11
room position mulitplier 12-11

root locus 11-12
root locus method 11-13
roots 1-3, 11-13
ROR 13-4, 13-5
rotor-flux 7-17
Routh-Hurwitz stability criterion 11-11
RTL 9-12
running power-factor 5-17
s-domain transfer function 11-15
s-plane 11-13
S-R flip flops 10-20
salient pole machines 7-19
salient poles 7-17
Sallen and Key band-pass circuit 8-7
Sallen and Key circuit 8-6
saturation 9-8
saturation curve 7-2
SCR 8-40
second-order differential equations 1-7
second-order response 4-21
second-order systems 4-19
second-order transient analysis 4-10
Seebeck coefficient 2-10
Seebeck effect 2-10
select input 10-17
self-biasing circuit 8-13
self-excited D.C. generators 7-3
self-excited generation 7-3
self-excited generators 7-3
self-inductance 2-4
semi-cutoff 12-18
semiconductor junction 8-7
sensitivity 3-8
separately-excited D.C. generators 7-3
separately-excited D.C. motors 7-7
separately-excited generation 7-3
separately-excited motors 7-7
sequence detector 10-27
sequence impedances 5-20
sequence networks 5-22
series current sources 4-3
series LC circuits 4-14
series motors 7-11
service 12-21
service disconnect 12-21
service entrance 12-21
servo block diagram 11-3
set reset (S-R) flip-flop 10-20
set-up time 10-23
shaft power 7-2
short circuit tests 5-6
short shunt 7-10
short shunt connection 7-5
short transmission lines 6-8
shot noise 8-30
shunt capacitance 3-16
shunt conductance 6-2
shunt motors 7-7
signal spectra 3-6
silicon controlled rectifier 8-40, 9-18
silicon diodes 9-1
simple interest 13-17
simple poles 4-22
simultaneous piecewise equations 1-13
single-line analysis 5-4
single-phase distribution transformers 5-11
single-phase faults to ground 5-25
single-pole operational amplifier 8-5
Sinking Fund Method 13-8
sinking fund plus interest on first cost 13-9
sinking unit load 9-9
sinusoidal analysis 4-20
sinusoidal steady state 4-8
sinusoidal voltage 3-4
skin depth 6-2
skin effect 6-2
SL 13-7
slave 10-22
slip 7-12
slip at maximum torque 7-13
slope-intercept form 1-5
small-signal amplifiers 8-21
small-signal FET parameters 8-17
small-signal parameters 8-20
Smith chart 6-13, 6-22
SOP 10-4
source 8-10
source conversion 4-1
source follower 8-24
spacing 12-18
spacing between luminaires 12-16, 12-18
spectral inputs 4-20
spectral power density 8-30
speed-up capacitor 9-28
squirrel-cage rotor winding 7-20
stability 8-19, 11-9, 11-10
stability studies 5-10
start-up state 10-27
starting induction motors 7-15
starting resistors 7-9
starting torque 7-11, 7-13
state diagram 10-25
state-space representation 11-15
state table 10-24
state variables 11-15, 11-16
stator reference point 7-18
steady state 4-8
steady state errors 11-10
steady state value 4-7
step input 11-10

step response 4-9
step response of second-order systems 4-13
storage time 9-11
straight line 1-5
Straight Line Method 13-7
straight line plus average interest 13-9
straight line plus interest on first cost 13-9
street lighting 12-15
subtraction of complex numbers 1-2
subtransient reactances 5-18
subtransient time interval 5-18
subtransmission systems 5-11
suggested illumination levels 12-6
Sum-of-the-Year's-Digits 13-7
sunk costs 13-2
superposition theorem 2-12
susceptance 3-14
susceptive part 3-15
swamping resistor 3-8, 3-9
switching transients 4-6
symbols for one-line diagrams 5-11
symmetrical componenets 5-20, 5-22, 6-8
symmetrical three-phase faults 5-22
synchronous binary counters 10-4
synchronous capacitors 7-21
synchronous generators 5-16, 7-19, 7-21
synchronous machines 7-17
synchronous motors 5-16, 7-20
synchronous reactance 7-18
synchronous speed 7-12, 7-17
T flip-flops 10-21, 10-23, 10-24
tachometer 11-2
tachometer feedback 11-2
tangible 13-7
tangible property 13-7
tax credits 13-11
tee approximation 6-9
tee-pi 4-4
temperature coefficient of resistivity 2-2
temperature effect 6-2
10-year personal property 13-8
theorem of logical adjacency 10-6
thermal limit switch 7-9
thermal noise 8-30
thermocouple instruments 3-10
thermocouples 2-5, 2-10
Thevenin circuit 4-1
Thevenin equivalent impedance 2-11
Thevenin equivalent voltage source 2-11
Thevenin's theorem 2-11
third-harmonic 5-14
third-order system 4-23
three-phase circuits 5-1
three-phase induction motor 7-12
three-phase transformer connections 5-14
three-phase transformers (per-phase) 5-11
three-phase transmission lines 6-6
three-phase voltage generators 5-2
three-terminal device 2-7
three-winding transformers 5-9
three-wire systems 5-1
3-year personal property 13-8
thyristor control 9-18
time constant 4-7
time domain 11-15
time variable part 3-12
toggle (T) flip-flop 10-21
torque 7-2
total inductance per unit length 6-3
total solution 4-8
totem-pole configuration 9-14
traction motors 7-11
transconductance 8-17, 8-21
transconductance curve 8-13
transducer 2-5, 2-8
transfer function 4-20, 11-1
transfer functions with zeros 11-16
transformed rotor resistance 7-14
transformer parameter measurements 5-5
transient analysis 1-9, 2-19
transient component 5-17
transient problems 4-6
transistor-transistor logic 9-13
transition capacitance 8-9
transistion table 10-21
transmission lines 5-10, 6-1
transresistance 8-21
traveling waves 6-11
TRIAC 8-40, 9-18
triangular wave 3-5
triangular waveform 3-3
trigonometric identities 1-4
trivial functions 10-18
truth table 9-12, 10-1
TTL 9-13
TTL output circuits 9-14
tunnel diode 8-39, 9-25
turn-on level 9-11
two-point form 1-5
two-pole high-pass filter 8-6
two-pole low-pass filter 8-6
two-reactance theory 5-18
two terminal networks 2-11
two-winding transformer model 5-5
type one system 11-10, 11-11
type two system 11-10, 11-11
type zero system 11-10, 11-11
types of logic 9-12
UJT relaxation oscillator 9-22
UJT trigger 9-19

under-excited case 7-22
underdamped 1-8, 4-12, 4-14
undetermined coefficients 1-3
uniform gradient factor 13-3
uniform pulse train 3-2, 3-4
uniform series factors 13-3
unijunction transistor 8-10, 9-19
unilateral devices 2-7
unit distance coding 10-7
unit load 9-9
unit source current 9-9
unit step 4-10
unity feedback 11-1
vapor discharge 12-2
variable frequencies 4-20
variance 13-15
vee curve 7-23
vee transformer 5-15
Veitch diagram 10-9, 10-10
velocity 11-10
velocity error 11-10
velocity feedback 11-2
velocity input 11-10
vertical enhancement mode MOSFET 8-16
vertical light distribution 12-18
visible spectrum 12-1
VMOS 8-16
voltage barrier 8-7
voltage control 5-15
voltage divider 4-6
voltage drop 5-2
voltage feedback biasing 8-15
voltage fluctuation 12-10
voltage follower 8-4
voltage gain 8-22, 8-31
voltage regulation 2-6, 5-12, 6-8
voltage regulator diodes 8-8
voltage regulators 9-5
voltage source rules 4-2
voltage sources 4-2
voltage-squared input spectrum 8-30
voltage stability factor 8-19
voltage standing wave ratio 6-13
VSWR 6-13
wall direct radiation coefficient 12-12
wall reflected radiation coefficient 12-12
wattmeters 3-10
wax candle 12-1
white noise 8-30
winding factor 7-12
wire sizes 12-22 , 6-20
wired AND 9-14
wired OR 9-14
wound rotor induction motors 7-17
wye connection 5-2
wye-delta 4-4
wye-wye balanced circuit 5-4
XNOR gate 10-2
XOR gate 10-2
XOR maps 10-13
XOR with maxterms 10-16
XOR with minterms 10-15
y-parameter 2-8, 2-14
year-end convention 13-2
z-parameter 2-8
Zener 2-7
Zener breakdown 8-8
Zener diode 2-7, 8-8, 8-35, 9-5
Zener effect 8-8
zero-frequency term 3-1
zero sequence 5-20
zero sequence impedances 5-21, 5-23
zeros 1-3, 11-13
zonal cavity method 12-7
zonal cavity ratios 12-7

Send more information

Please send me descriptions and prices of all available E-I-T and P.E. review books.

____
____
____
____

I just took the exam!

I just took the [ ] **E-I-T** [ ] **P. E.** examination in the state of______________. This is what you should know about that examination.

This is what you need in the next edition of this book.

Contributed by (optional): ______________________________
____

We disagree

I think there is an error on page_____. Here is the way I think it should be.

[ ] Please tell me if I am correct.

Contributed by (optional): ______________________________
____

Affix

Postage

PROFESSIONAL PUBLICATIONS, INC.

Post Office Box 199

San Carlos, CA 94070

Affix

Postage

PROFESSIONAL PUBLICATIONS

Post Office Box 199

San Carlos, CA 94070

Affix

Postage

PROFESSIONAL PUBLICATIONS

Post Office Box 199

San Carlos, CA 94070